STUDENT SOLUTIONS MANUAL FOR

CALCULUS
THIRD EDITION

By Berkey/Blanchard

Judy Coomes
William Paterson College

Tohien Hoang

Dennis Kletzing
Stetson University

Gloria Langer
University of Colorado

Michael Motto
Ball State University

David Wagner
University of Houston-University Park

SAUNDERS COLLEGE PUBLISHING
Harcourt Brace Jovanovich College Publishers
Fort Worth • Philadelphia

Boston • New York • Chicago • Orlando • San Francisco • Atlanta
Dallas • London • Toronto • Austin • San Antonio

Coomes/Hoang/Kletzing/Langer/Motto/Wagner: Student Solutions Manual to accompany
Berkey/Blanchard's <u>CALCULUS</u>, 3/E

IBSN 0-03-049507-5

34 021 98765432

Preface

This Student Solutions Manual is a supplement for the third edition of Berkey/Blanchard's _Calculus_. It contains complete, detailed solutions for all the odd-numbered section and chapter review exercises. Answers to most of the odd-numbered problems are also given at the back of your text.

Keep in mind that a _solution_ to a problem is different from an _answer_. A _solution_ begins with what is given in the problem, uses what you have learned, and presents a step-by-step argument with reasons showing why the _answer_ is a necessary consequence. Many problems can be solved in more than one way. You solution may be correct even if it is not exactly like the one presented. Compare the ways of solving the problem to see which is easier (or more elegant).

Suggestions for using this manual: Always _try to solve the problem first_ before looking at the solution presented. Then compare your solution with the one here, checking organization and computations. Remember it is not just the answer that is important; it is _how_ and _why_ it must be the answer. Fill in extra steps and computational details. Write your solutions as a well-organized sequence of steps with reasons. Develop the habit of checking your answer to see that it solves the original problem. Always ask "Is this answer reasonable?" Practice, practice, practice each type of problem until you feed confident you have mastered it. Learn the definitions and formulas in the Summary Outlines at the end of each chapter. The _easiest_ way to do this is by working lots of exercises using them. It is not particularly useful just to memorize them.

Great care has been taken to avoid errors. If you have corrections or suggestions for this Student Solutions Manual, please send your correspondence to: Mathematics Editor, Saunders College Publishing, The Public Ledger Building Suite 560, 620 Chestnut Street, Philadelphia, PA 19106-3477.

Good luck with your studies and best wishes for your mathematical future.

Judy Coomes
(William Paterson College)

Tohien Hoang

Dennis Kletzing
(Stetson University)

Gloria Langer
(University of Colorado)

Michael Motto
(Ball State University)

David Wagner
(University of Houston - University Park)

Table of Contents

Solutions to Chapter 1   1

Solutions to Chapter 2   38

Solutions to Chapter 3   59

Solutions to Chapter 4   100

Solutions to Chapter 5   182

Solutions to Chapter 6   204

Solutions to Chapter 7   263

Solutions to Chapter 8   310

Solutions to Chapter 9   336

Solutions to Chapter 10   358

Solutions to Chapter 11   385

Solutions to Chapter 12   397

Solutions to Chapter 13   422

Solutions to Chapter 14   461

Solutions to Chapter 15   475

Solutions to Chapter 16   504

Solutions to Chapter 17   523

Solutions to Chapter 18   547

Solutions to Chapter 19   584

Solutions to Chapter 20   621

Solutions to Chapter 21   642

Chapter 1

Review of Precalculus Concepts

1.1 The Real Number System

1. Rational. The ratio of two integers is rational, by definition. See text, p. 3.

3. Irrational. Although there is a pattern to the digits, there is no fixed repeating pattern of digits.

5. Rational. The digit pattern 163 repeats.

7. $\sqrt{256} = 16$ which is an integer, and hence rational.

9. Irrational. The sum of rational number and an irrational number is always irrational. Proof: If $a = \frac{p_1}{q_1}$, with p_1 and q_1 integers, and b is irrational, and $c = a + b$, then if c were rational, say, $c = \frac{p_2}{q_2}$, (p_2, q_2 integers) then $b = c - a = \frac{p_2 q_1 - p_1 q_2}{q_1 q_2}$. Hence c would have to be rational. Since we assumed that c is irrational, b must then be irrational.

11. $[-2, 5] \cup (-1, 6) = [-2, 6)$

13. $[-2, 5] \cap (-\infty, 0) = [-2, 0)$

15. $(-1, 6) \cap (-\infty, 0) = (-1, 0)$

17. a True. Since the circumference $C = 2\pi r$, where r is the radius, C will be irrational whenever r is rational.

   b True. If A is the area, then $A = \pi r^2$. If r is rational, then so is r^2. Hence A must be irrational in this case.

   c True. $C = 2\pi r \Leftrightarrow r = \frac{C}{2\pi}$ Hence $A = \pi r^2 = \pi \times (\frac{C}{2\pi})^2 = \frac{C^2}{4\pi}$. If C is rational, then so is $\frac{C^2}{4}$. Hence A must be irrational in this case.

19. $x + 4 \leq 3x \Leftrightarrow 4 \leq 2x \Leftrightarrow x \geq 2$

21. $6x - 6 \leq 8x + 8 \Leftrightarrow 6x \leq 8x + 14 \Leftrightarrow -2x \leq 14 \Leftrightarrow x \geq -7$

23. The quantity $(x-3)(x+1)$ changes sign only where $(x-3)(x+1) = 0$, that is, at $x = 3$ or $x = -1$. If $x < -1$ then $x - 3 < 0$ and $x + 1 < 0$, hence the product of these factors is positive. If $-1 < x < 3$ then the factor $x + 1$ is positive while the other factor is negative. Hence in this case the product is negative. Similarly, when $x > 3$ both factors are positive, so the product is positive.

   Hence the solution to $(x-3)(x+1) < 0$ is $-1 < x < 3$.

25. $x(x+6) > -8 \Leftrightarrow x^2 + 6x + 8 > 0 \Leftrightarrow (x+4)(x+2) > 0$ The quantity $(x+4)(x+2)$ changes sign only where $(x+4)(x+2) = 0$, that is, at $x = -4$ or $x = -2$. If $x < -4$ then $x + 4 < 0$ and $x + 2 < 0$, hence the product of these two factors is positive. If $-4 < x < -2$, then $x + 4$ is positive and $x + 2$ is negative. Hence in this case the product is negative. If $x > -2$, then both factors are positive and the product is positive.

Hence the solution to $x(x+6) > -8$ is $\{x \mid x < -4 \text{ or } x > -2\}$.

27. Factor $x^3 + x^2 - 2x$ as $x(x^2 + x - 2) = x(x+2)(x-1)$. This quantity is positive if it is not zero and the number of negative factors is even. This is true if $-2 < x < 0$ or $x > 1$.

29. Factor $x^4 - 9x^2$ as $x^2(x-3)(x+3)$. Since $x^2 \geq 0$, $x^2(x-3)(x+3)$ is negative when the other two factors, $(x-3)$ and $(x+3)$, have opposite signs, and $x \neq 0$. Hence $x^4 - 9x^2 < 0 \Leftrightarrow -3 < x < 0$ or $0 < x < 3$, that is, $0 < |x| < 3$.

31. $|5x - 2| = 0 \Leftrightarrow 5x - 2 = 0 \Leftrightarrow x = \frac{2}{5}$.

33. $2|3x - 1| = 22 \Leftrightarrow |3x - 1| = 11$. The last equality holds when $3x - 1 = 11$ or $3x - 1 = -11$. Solving these two equations gives the solutions $x = 4$ and $x = -\frac{10}{3}$.

35. $(|x| + 6)^2 = 49 \Leftrightarrow |x| + 6 = 7$ or $|x| + 6 = -7$. The second equation has no solutions. The first equation is equivalent to $|x| = 1$. Hence the solutions are $x = 1$ and $x = -1$.

37. $x + 3|x| = 8 \Leftrightarrow |3x| = 8 - x$. The last equation holds when $3x = 8 - x$ or $-3x = 8 - x$. Solving these two equations gives the solutions $x = 2$ and $x = -4$.

39. $|x - 4| = |x - 7|$ when either $x - 4 = x - 7$ or $x - 4 = -(x - 7)$.

$x - 4 = x - 7$ has no solution.

$x - 4 = -(x - 7) \Leftrightarrow 2x = 11 \Leftrightarrow x = \frac{11}{2}$.

41. $|x - 3| \leq 2 \Leftrightarrow -2 \leq (x - 3) \leq 2 \Leftrightarrow 1 \leq x \leq 5$

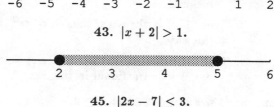

41. $|x - 3| \leq 2$.

43. $|x + 2| > 1$ when $x + 2 < -1$ or $x + 2 > 1$. Solving these two inequalities gives the solution set $\{x \mid x < -3 \text{ or } x > -1\}$.

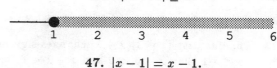

43. $|x + 2| > 1$.

45. $|2x - 7| \leq 3 \Leftrightarrow -3 \leq 2x - 7 \leq 3 \Leftrightarrow 4 \leq 2x \leq 10 \Leftrightarrow 2 \leq x \leq 5$

45. $|2x - 7| \leq 3$.

47. $|x - 1| = x - 1 \Leftrightarrow x - 1 \geq 0$.

47. $|x - 1| = x - 1$.

49. This inequality is true when $(8 - 3x) \leq -5$ or $8 - 3x \geq 5$. Solving these two inequalities give the solution $x \geq \frac{13}{3}$ or $x \leq 1$.

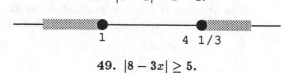

49. $|8 - 3x| \geq 5$.

2

51. $|x+3| + |x-2| < 7 \Leftrightarrow |x+3| < 7 - |x-2|$. Since the right hand side of the inequality is always positive, we can square both sides of the inequality to obtain:

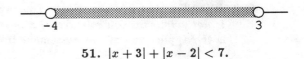

51. $|x+3| + |x-2| < 7.$

$$(x+3)^2 < 49 - 14|x-2| + (x-2)^2$$
$$\Leftrightarrow \quad 14|x-2| < 44 - 10x$$
$$\Leftrightarrow \quad 10x - 44 < 14(x-2) < 44 - 10x$$
$$\Leftrightarrow \quad -4 < x < 3.$$

53. The distance between x and 1 is the same as the distance between x and -3. Hence $x = -1$.

55. The inequality describes all the numbers x that lie more than 2 and less than 5 units from 4. Hence, $-1 < x < 2$ or $6 < x < 9$.

57. $|x+2| = 2|x-12|$

59. Exactly one of the following holds: $y - x$ is positive, $y - x$ is negative, or $x = y$.

$y - x$ is positive if and only if $x < y$.

$y - x$ is negative if and only if $x - y$ is positive if and only if $x > y$.

Hence, exactly one of the following holds: $x = y$, $x < y$ or $x > y$.

61. False. If $a = -1$ and $b = -2$, then $a^2 < b^2$, but $b < a$.

63. Let $a = x + 6$ and let $b = 7 > 0$. Then $(x+6)^2 < 49$ is the same as $a^2 < b^2$. By exercise 62, $-7 < x + 6 < 7$. Solving for x gives us $-13 < x < 1$.

65. **a.** Actually 'proof by contradiction' is more properly called 'proof by contrapositive'. The contrapositive of the implication $p \Rightarrow q$ is the implication $(\neg q) \Rightarrow (\neg p)$. Here $\neg p$ is the symbol for the negation of p. A logical implication is always logically equivalent to its contrapositive. Thus to prove $p \Rightarrow q$ by 'contradiction', we assume $\neg q$ and prove that this implies $\neg p$. Here p represents the hypotheses of the theorem, and q represents the conclusion. Constructing 'proofs by contradiction' in this manner, using careful reasoning, helps avoid mistakes where a 'contradiction' simply results from poor reasoning or a mathematical error. For the theorem at hand the hypothesis is the (true) statement that 2 is a prime number. The conclusion is the statement that $\sqrt{2}$ is not rational. The statement of the theorem is then *"If 2 is prime, then $\sqrt{2}$ is not rational."* However it sounds a little silly to say *"If 2 is prime,"* because 2 is always prime. The theorem sounds much better if we pose it as: *"If x is prime, then $\sqrt{x}$ is not rational."* This is also a much better theorem. The contrapositive of this, is then: *"If $\sqrt{x}$ is rational, then x is not prime."* In order to prove this, we assume some important results from the theory of the natural numbers, particularly the theorem that every natural number has a unique prime factorization.

b. Alternatively, we could simply say that $\sqrt{x} = \frac{p}{q}$, where p and q are natural numbers. This is the definition of what it means for a number to be rational. Let $p = p_1^{j_1} \cdots p_n^{j_n}$ and $q = q_1^{k_1} \cdots q_m^{k_m}$ be the unique prime factorizations of p and q, respectively.

c. Multiplying by q and squaring both sides, we have $xq^2 = p^2$. Hence $xq_1^{2k_1} \cdots q_m^{2k_m} = p_1^{2j_1} \cdots p_n^{2j_n}$.

d.,e. Since the right side of this equation is the unique prime factorization for the left hand side, and since this prime factorization has an even number of factors, while the left hand side is an even number of prime factors, times x, it must be that x has an even number of prime factors.

f. This implies that x is not prime. Thus we have proved that *If the square root of x is rational, then x is not prime*, or, equivalently, *If x is prime, then the square root of x is not rational*. The exercise is the application of this statement to the case $x = 2$.

67. Adding the two inequalities $-|x| \le x \le |x|$ and $-|y| \le y \le |y|$, we obtain $-(|x|+|y|) \le x+y \le |x|+|y|$. Then Theorem 2 gives the result: $|x+y| \le |x|+|y|$.

69. If $a < b$ and $ab > 0$, then $\frac{1}{a} > \frac{1}{b}$

   Proof: If $a < b$ and $ab > 0$, then both $(b-a)$ and ab are positive. Hence, $\frac{1}{a} - \frac{1}{b} = \frac{1}{ab}(b-a)$ is positive. By definition, $\frac{1}{a} - \frac{1}{b} > 0$

71. Yes, simply apply the theorem in the above solution for #65 to the case $x = 3$.

73. We will first prove: $|x| < a \Leftrightarrow -a < x < a$

   Case 1: $x \ge 0$. In this case $|x| = x$, so $|x| < a \Leftrightarrow x < a$.

   Case 2: $x < 0$. In this case $|x| = -x$, so $|x| < a \Leftrightarrow -x < a \Leftrightarrow -a < x$

   Hence, $|x| < a \Leftrightarrow -a < x < a$.

   Now, we will prove that the inequality $|x| > a$ holds if either $x > a$ or $x < -a$.

   If $x \ge 0$, then $|x| > a \Leftrightarrow x > a$.

   If $x < 0$, then $|x| > a \Leftrightarrow -x > a \Leftrightarrow x < -a$.

1.2   The Coordinate Plane, Distance, and Circles

1a. $\sqrt{(0-2)^2 + [2-(-1)]^2} = \sqrt{2^2 + 3^2} = \sqrt{13}$

1b. $\sqrt{(1-3)^2 + (3-1)^2} = \sqrt{2^2 + 2^2} = \sqrt{8} = 2\sqrt{2}$

1c. $\sqrt{(1-0)^2 + (-9-2)^2} = \sqrt{1^2 + 11^2} = \sqrt{122}$

1d. $\sqrt{(6-1)^2 + [6-(-3)]^2} = \sqrt{5^2 + 9^2} = \sqrt{106}$

1e. $\sqrt{(-1-1)^2 + (-1-1)^2} = \sqrt{2^2 + 2^2} = \sqrt{8} = 2\sqrt{2}$

1f. $\sqrt{[1-(-2)]^2 + [2-(-2)]^2} = \sqrt{3^2 + 4^2} = \sqrt{25} = 5$

3. Let (x, y) be a point which lies a distance of 4 units from both $(-2, 3)$ and $(2, 3)$. Then x and y must satisfy the two equations:

$$\sqrt{[x-(-2)]^2 + (y-3)^2} = 4$$
$$\sqrt{[(x-2)^2 + (y-3)^2} = 4$$

Squaring both sides of both equations gives us the equations:

$$(x+2)^2 + (y-3)^2 = 16 \tag{1.1}$$
$$(x-2)^2 + (y-3)^2 = 16 \tag{1.2}$$

Note that $(x+2)^2 = 16 - (y-3)^2 = (x-2)^2$. Hence, $(x+2) = \pm(x-2)$. Solving for x, we have $x = 0$. As equations 1.1 and 1.2 are the same when $x = 0$, we need only solve the following equation:

$$2^2 + (y-3)^2 = 16 \Leftrightarrow (y-3)^2 = 12 \Leftrightarrow y = 3 \pm 2\sqrt{3}.$$

Hence, the two points are $(0, 3 - 2\sqrt{3})$ and $(0, 3 + 2\sqrt{3})$.

5. **A triangle is isosceles when any two sides have the same length.**

Case 1: The distance from $(a, 0)$ to $(-1, 0)$ is the same as the distance from $(a, 0)$ to $(2, 3)$:

$$\sqrt{(-1 - a)^2 + 0^2} = \sqrt{(2 - a)^2 + 3^2}$$
$$\Leftrightarrow \quad (a + 1)^2 = (a - 2)^2 + 9$$
$$\Leftrightarrow \quad a^2 + 2a + 1 = a^2 - 4a + 13$$
$$\Leftrightarrow \quad 6a = 12$$
$$\Leftrightarrow \quad a = 2$$

Case 2: The distance from $(a, 0)$ to $(-1, 0)$ is the same as the distance from $(-1, 0)$ to $(2, 3)$:

$$\sqrt{(-1 - a)^2 + 0^2} = \sqrt{[2 - (-1)]^2 + (3 - 0)^2}$$
$$\Leftrightarrow \quad (a + 1)^2 = 18$$
$$\Leftrightarrow \quad a = -1 \pm 3\sqrt{2}$$

Case 3: The distance from $(a, 0)$ to $(2, 3)$ is the same as the distance from $(-1, 0)$ to $(2, 3)$:

$$\sqrt{(2 - a)^2 + 3^2} = \sqrt{18}$$
$$\Leftrightarrow \quad (a - 2)^2 + 9 = 18$$
$$\Leftrightarrow \quad a = 2 \pm 3$$

Note that $a = -1$ is not a solution, for then $(a, 0) = (-1, 0)$. From the three cases we see that there are four solutions for a: $2, -1 \pm 3\sqrt{2}, 5$.

7. **Let (x, y) be any point lying equidistant from the points $(-1, -1)$ and $(3, 1)$. This means that:**

$$\sqrt{(x + 1)^2 + (y + 1)^2} = \sqrt{(x - 3)^2 + (y - 1)^2}$$
$$\Leftrightarrow \quad x^2 + 2x + 1 + y^2 + 2y + 1 = x^2 - 6x + 9 + y^2 - 2y + 1$$
$$\Leftrightarrow \quad 4y = -8x + 8$$
$$\Leftrightarrow \quad y = -2x + 2$$

9. **The distance from (x_1, y_1) to $(\frac{x_1 + x_2}{2}, \frac{y_1 + y_2}{2})$ is $\sqrt{(x_1 - x_2)^2 + (y_1 - y_2)^2}/2$.**
The distance from $(\frac{x_1 + x_2}{2}, \frac{y_1 + y_2}{2})$ to (x_2, y_2) is the same.

11. **The distance from (x, y) to $(-3, 1)$ is $\sqrt{(x + 3)^2 + (y - 1)^2}$. Thus the set of all points (x, y) that lie more that $\sqrt{2}$ units from $(-3, 1)$ is described by the inequality:**

$$\sqrt{(x + 3)^2 + (y - 1)^2} > \sqrt{2} \qquad (1.3)$$
$$\text{or} \quad (x + 3)^2 + (y - 1)^2 > 2 \qquad (1.4)$$

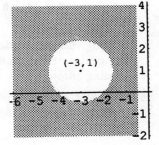

11. $(x + 3)^2 + (y - 1)^2 > 2$.

13. Again, this set is most simply described by two inequalities. The distance from (x, y) to the x-axis is $|x|$. We need both of these quantities to be less than 2. This is expressed by the inequalities $|x| \leq 2$, $|y| \leq 2$.

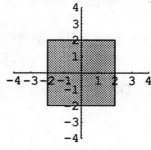

13. $|x| \leq 2$, $|y| \leq 2$

15. The equation for a circle with center (h, k) and radius r is $(x - h)^2 + (y - k)^2 = r^2$. Here $h = k = 0$ and $r = 3$. So the equation for the circle is $x^2 + y^2 = 9$.

17. $(x + 6)^2 + (y + 4)^2 = 25$.

19. There are two points whose distance to the points $(-1, -4)$ and $(4, 1)$, is 5. To solve for the coordinates (h, k) of these points, we write equations for the distance from the centers to the given points:

$$(h + 1)^2 + (k + 4)^2 = 25,$$

$$(h - 4)^2 + (k - 1)^2 = 25.$$

Subtracting the second equation from the first, we obtain:

$$10h + 10k = 0.$$

Thus both of the centers are on the line $k = -h$. Substituting this in the distance equations, we obtain

$$(h + 1)^2 + (-h + 4)^2 = 25.$$

or

$$h^2 - 3h - 4 = 0.$$

Solving this quadratic equation, we obtain $h = 4$ or $h = -1$. Since the centers lie on the line $k = -h$, the centers must have the coordinates $(4, -4)$, and $(-1, 1)$. Thus the equations of the two circles passing through $(-1, -4)$ and $(4, 1)$ are:

$$(x - 4)^2 + (y + 4)^2 = 25,$$

$$(x + 1)^2 + (y - 1)^2 = 25.$$

21. Complete squares:

$$x^2 + y^2 + 4x + 2y - 11 = 0$$
$$\Leftrightarrow \quad x^2 + 4x + 4 - 4 + y^2 + 2y + 1 - 1 - 11 = 0$$
$$\Leftrightarrow \quad (x + 2)^2 + (y + 1)^2 = 16.$$

Thus the radius is 4 and the center is $(-2, -1)$.

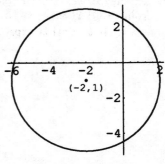

21. $(x + 2)^2 + (y + 1)^2 = 16.$

23. Complete squares:

$$x^2 + y^2 - 2x - 6y + 3 = 0$$
$$\Leftrightarrow \quad x^2 - 2x + 1 - 1 + y^2 - 6y + 9 - 9 + 3 = 0$$
$$\Leftrightarrow \quad (x - 1)^2 + (y - 3)^2 = 1 + 9 - 3 = 7.$$

Thus the radius is $\sqrt{7}$ and the center is $(1, 3)$.

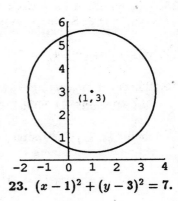

23. $(x - 1)^2 + (y - 3)^2 = 7.$

25. Complete squares:

$$x^2 + y^2 - 2by + b^2 - a^2 = 0$$
$$\Leftrightarrow \quad x^2 + (y - b)^2 = a^2.$$

Thus the radius is a and the center is $(0, b)$. The figure shows the case where $b^2 > a^2$ and $b > 0$.

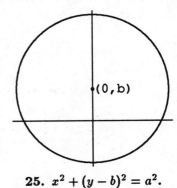

25. $x^2 + (y - b)^2 = a^2.$

27. We solve the first equation for y to get $y = x - 5$. Substituting for y in the equation for the circle gives:

$$x^2 - 8x + (x - 5)^2 - 4(x - 5) + 11 = 0$$
$$\Leftrightarrow \quad x^2 - 11x + 28 = 0$$
$$\Leftrightarrow \quad (x - 7)(x - 4) = 0$$
$$\Leftrightarrow \quad x = 7 \text{ and } x = 4$$

Substituting these values for x in the equation $y = x - 5$ gives us the two points of intersection $(7, 2)$ and $(4, -1)$.

29. Let (x, y) be any point whose distance from the point $(-2, 1)$ is twice the distance from the point $(4, -2)$. Then (x, y) satisfies the equation:

$$\sqrt{(x+2)^2 + (y-1)^2} = 2\sqrt{(x-4)^2 + (y+2)^2}$$
$$\Leftrightarrow \quad (x+2)^2 + (y-1)^2 = 4\left((x-4)^2 + (y+2)^2\right)$$
$$\Leftrightarrow \quad x^2 + 4x + 4 + y^2 - 2y + 1 = 4\left(x^2 - 8x + 16 + y^2 + 4y + 4\right)$$
$$\Leftrightarrow \quad 3x^2 - 36x + 3y^2 + 18y + 75 = 0$$
$$\Leftrightarrow \quad x^2 - 12x + y^2 + 6y + 25 = 0$$
$$\Leftrightarrow \quad (x-6)^2 - 36 + (y+3)^2 - 9 + 25 = 0$$
$$\Leftrightarrow \quad (x-6)^2 + (y+3)^2 = 20$$

This is the equation of the circle with center $(6, -3)$ and radius $2\sqrt{5}$.

31. All circles containing the two points $(1, 2)$ and $(-1, 2)$ have centers on the set of points, L_1, which are equidistant from the two points. The equation for L_1 is given by $x = 0$. Let $(0, k)$ be the center of any circle containing the point $(1, 2)$. The radius, r, is given by $r^2 = 1 + (k-2)^2$. Hence, the family of circles containing the two given points is given by the equation:

$$x^2 + (y-k)^2 = 1 + (k-2)^2,$$

where k is any real number.

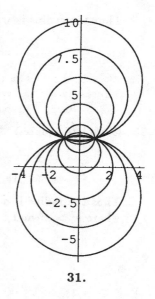

31.

33. Since $x^2 - x - 6 = (x-3)(x+2)$, and since the product of two real numbers is negative if and only if one number is positive and the other is negative, we have that $x^2 - x - 6 \leq 0 \Leftrightarrow -2 \leq x \leq 3$

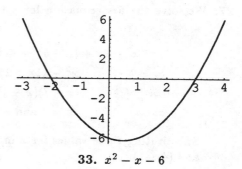

33. $x^2 - x - 6$

8

35. $2x + 3 = -x^2 + 4x + 7 \Leftrightarrow x = 1 \pm \sqrt{5}$. These are the only points where the inequality $2x + 3 > -x^2 + 4x + 7$ can change direction. Using this fact together with the graphs of $2x + 3$ and $-x^2 + 4x + 7$, we see that $2x + 3 > -x^2 + 4x + 7$ if and only if ($x < 1 - \sqrt{5}$ or $x > 1 + \sqrt{5}$).

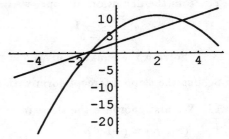

35. $y = 2x + 3$, $y = -x^2 + 4x + 7$

37. Solving for y using the quadratic formula (or using the *Mathematica* "Solve" function), we have $y = \left(x \pm \sqrt{30 - 23 * x^2} \right) / 6$. Graphing both of these solutions together we obtain the ellipse at right.

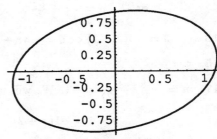

37. $y = \left(x \pm \sqrt{30 - 23 * x^2} \right) / 6$

39. Solving for y using the quadratic formula (or using the *Mathematica* "Solve" function), we have $y = -1 - x \pm \sqrt{328}$. Graphing both of these solutions together we obtain the two straight lines (surprise!) at right.

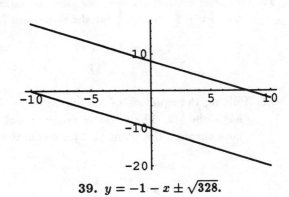

39. $y = -1 - x \pm \sqrt{328}$.

1.3   Linear Equations

1. a. $m = \frac{2 - 1}{-1 - 3} = -\frac{1}{4}$

   b. $m = \frac{-1 - (-2)}{-1 - 6} = -\frac{1}{7}$

   c. $m = \frac{a - b}{b - a} = -1$

   d. $m = \frac{a - 1}{a - 1} = 1$

3. False. Every slope determines a family of parallel lines.

5. True.

7. From the definition of slope, we have the following equation:

$$2 = \frac{-5 - 1}{1 - b}$$

Solving the equation gives $b = 4$.

9. Use the slope-intercept form with $m = -2$ and $b = 5$ to obtain the equation $y = -2x + 5$.

11. We first compute the slope $m = \frac{12 - 6}{4 - (-1)} = \frac{6}{5}$. Now we use the point-slope form of the equation with $(x_1, y_1) = (-1, 6)$:

$$y - 6 = \frac{6}{5}(x + 1)$$
$$\text{or} \quad y = \frac{6}{5}x + \frac{36}{5}$$

13. $y - 3 = -3(x - 1)$ or $y = -3x + 6$.

15. $x = -3$

17. $m = \frac{8 - 4}{-6 - (-2)} = -1$. We will use the point-slope form of the equation with the point $(-2, 4)$:

$$y - 4 = -1(x + 2)$$
$$\text{or} \quad y = -x + 2$$

19. To determine the slope, we put the equation of the line $2x - 6y + 5 = 0$ in slope-intercept form, $y = \frac{1}{3}x + \frac{5}{6}$. Using $\frac{1}{3}$ for the slope and $(1, 4)$ for the point, the equation of the line is:

$$y - 4 = \frac{1}{3}(x - 1)$$
$$\text{or} \quad y = \frac{1}{3}x + \frac{11}{3}$$

21. Putting the equation of the line $3x + y = 7$ in slope-intercept form, $y = -3x + 7$, we find that this line has slope -3. The negative reciprocal of -3 is $\frac{1}{3}$. Hence, the perpendicular line has slope $\frac{1}{3}$. Since it goes through the point $(1, 3)$, its equation is:

$$y - 3 = \frac{1}{3}(x - 1)$$
$$\text{or} \quad y = \frac{1}{3}x + \frac{8}{3}$$

23. False. This is only true if $(0, 0)$ satisfies the linear equation.

25. $y = -x + 7$. The slope is -1, the y-intercept is 7, and the x-intercept is 7.

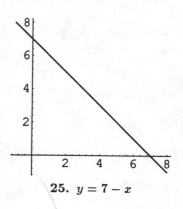

25. $y = 7 - x$

10

27. Let $y = 0$ and solve the equation $x + 3 = 0$ to obtain the x-intercept of -3. Put the equation into the slope-intercept form of $y = -x - 3$ to obtain the slope of -1 and the y-intercept of -3.

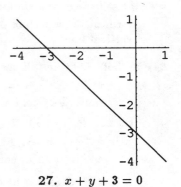

27. $x + y + 3 = 0$

29. There is no x-intercept. The slope is 0 and the y-intercept is 5.

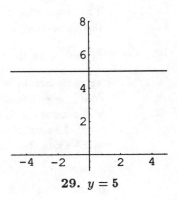

29. $y = 5$

31. There is neither a slope nor a y-intercept. The x-intercept is 4.

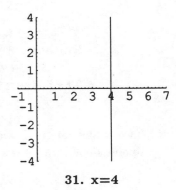

31. x=4

33. Solving the first equation for y, we have $y = 3x - 1$. Substitute this into the second equation:

$$x + 3x - 1 - 3 = 0$$

Solving this equation x gives us $x = 1$ and $y = 3x - 1 = 2$. The point of intersection is $(1, 2)$.

35. Solving the first equation for x gives us $x = 3y - 3$. Substituting this into the second equation gives us $2(3y - 3) - 3y + 6 = 0$. Solving for y, we have $y = 0$ and $x = 3y - 3 = -3$. The two lines intersect at $(-3, 0)$.

37. Both lines have a slope of 2. Since the two lines are parallel and do not coincide, there are no points of intersection.

39. If $b \neq 0$, then the equation $ax + by + c = 0$ is the same as $y = -\frac{a}{b}x - \frac{c}{b}$, and this is the slope-intercept form of the equation of a line. If $b = 0$, the equation becomes $x = -\frac{c}{a}$, and this is the equation of a vertical line.

41. a. Let P, Q and R respectively be the points $(1,3)$, $(-2,0)$ and $(4,6)$. The slope of the line PQ is $m_1 = \frac{0-3}{-2-1} = 1$. The slope of the line QR is $m_2 = \frac{6-0}{4+2} = 1$. Since PQ and QR have the same slope and a point in common, P, Q, and R must lie on the same line.

b. Let P, Q and R be the points $(2,-7)$, $(-2,-3)$, and $(-1,-4)$. The slope of the line PQ is $\frac{-3+7}{-2-2} = -1$. The slope of the line QR is $\frac{-4+3}{-1+2} = -1$. Since PQ and QR have the same slope and a point in common, P, Q, and R must lie on the same line.

c. Let P, Q and R be the points $(5,15)$, $(0,3)$ and $(-2,-7)$. The slope of the line PQ is $\frac{3-15}{0-5} = \frac{12}{5}$. The slope of the line QR is $\frac{-7-3}{-2-0} = 5$. Since the slopes of the two line segments are different, the three points do not lie on a common line.

43. Let P, Q and R be the points $(-2,2)$, $(4,4)$ and $(0,a)$.

a. A triangle has three vertices. The right angle could occur at any of these angles. The angle at $(0,a)$ could be either above or below the line segment PQ. Thus there are two possible values of a that will produce a right angle at this point. A line perpendicular to PQ through $(4,4)$ will determine a third value of a, and a line perpendicular to PQ through $(-2,2)$ will determine a fourth value of a. Thus there are 4 possible values of a.

b. The slope of the line PQ is $m_1 = \frac{4-2}{4+2} = \frac{1}{3}$. The slope of the line QR is $m_2 = \frac{a-4}{0-4}$. The slope of the line PR is $m_3 = \frac{a-2}{0+2}$.

- The lines PQ and QR are perpendicular when $m_2 = -\frac{1}{m_1}$, that is when $\frac{4-a}{4} = -3$, or $4 - a = -12$. Thus $a = 16$.
- The line PR is perpendicular to PQ if $m_3 = -\frac{1}{m_1}$, or $\frac{a-2}{2} = -3$. This is equivalent to $a - 2 = -6$, or $a = -4$.
- The line PR is perpendicular to QR if $m_3 = -\frac{1}{m_1}$, or $\frac{2}{a-2} = -3$. This is equivalent to $(a-2)(a-4) = 8$, or $a^2 - 6a = 0$. Thus $a = 0$ or $a = 6$.

45. Two values for (P, d) are $(20, 12)$ and $(60, 6)$. The slope of the line is $m = \frac{6-12}{60-20} = -\frac{3}{20}$. Using the point-slope form of the equation of a line, we have:

$$d - 12 = -\frac{3}{20}(P - 20)$$
$$\text{or} \qquad d = -\frac{3}{20}P + 15$$

When $d = 0$, we have $P = \frac{20}{3} \times 15 = 100$. Hence, the demand is 0 when the product is priced at \$100.

47. Let F be the temperature in degrees Fahrenheit, and let C be the temperature in degrees Celsius. The freezing point of water is 0 degrees Celsius and 32 degrees Fahrenheit. The boiling point of water is 100 degrees Celsius and 212 degrees Fahrenheit. Hence two values for (C, F) are $(0, 32)$ and $(100, 212)$. The F-intercept is 32 and the slope is $m = \frac{212-32}{100-0} = \frac{9}{5}$. The linear equation is:

$$F = \frac{9}{5}C + 32$$

49. a. Two values for (t, T) are $(0, 22)$ and $(10, 42)$. The T-intercept is 22 and the slope is $m = \frac{42 - 22}{10 - 0} = 2$. Hence the linear equation is $T = 2t + 22$. Note here that $(20, 62)$ also satisfies this equation.

   b. When $t = 25$, $T = 2 \times 25 + 22 = 72°C$. When $t = 35$, $T = 2 \times 35 + 22 = 92°C$.

51. Since the rod is 100cm long at temperature $0°C$, we see that $l_0 = 100$. Substituting this into the equation $l = l_0(1 + at)$, gives us $l = 100(1 + at)$. Since the rod is 100.2 cm long at temperature $50°C$, we have $100(1 + 50a) = 100.2$. Thus, $a = .00004$.

53. We complete the squares for both equations to put them in the standard form for the equation of a circle:

$$(x - 1)^2 + (y - 2)^2 = 4$$
$$x^2 + (y + 1)^2 = 1$$

The two centers are $(1, 2)$ and $(0, -1)$. The y-intercept is -1, and the slope is $m = \frac{-1 - 2}{0 - 1} = 3$. Hence the equation of the line through the two centers is $y = 3x - 1$

55. a. $m = \frac{1.5 - 1.4}{2.25 - 1.96} = \frac{10}{29}$

   b. The equation for l is $y - 1.5 = .345(x - 2.25)$ or $y = .345x + .724$. When $x = 2$, $y = .345 + .724 = 1.414$. Hence the point is $(2, 1.414)$.

   c. An estimate for $\sqrt{2}$ is 1.41.

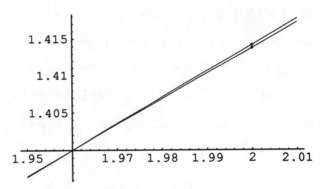

   d. The secant line is *very* close to the curve

1.4 Functions

1.

$f(x)$	$f(-2)$	$f(0)$	$f(4)$	$f(5)$	$\lvert f(3) \rvert$	$f(f(0))$
$1 - 3x^2$	-11	1	-47	-74	26	-2
$\dfrac{1}{x + 2}$	undefined	$\frac{1}{2}$	$\frac{1}{6}$	$\frac{1}{7}$	$\frac{1}{5}$	$\frac{2}{5}$
$\dfrac{(x - 3)^2}{x^2 + 1}$	5	9	$\frac{1}{17}$	$\frac{4}{26}$	0	$\frac{36}{82}$
$\sqrt{x + 4}$	$\sqrt{2}$	2	$2\sqrt{2}$	3	$\sqrt{7}$	$\sqrt{6}$
$\dfrac{1}{\sqrt{16 - x^2}}$	$\frac{1}{2\sqrt{3}}$	$\frac{1}{4}$	undefined	undefined	$\frac{1}{\sqrt{7}}$	$\frac{4}{\sqrt{255}}$
$\begin{cases} 1 - x, & x < -3 \\ x - 1, & x > 1 \end{cases}$	undefined	undefined	3	4	2	undefined

3. $y = \frac{1}{x}$ is a function of x, $x \neq 0$.

5. y is not a function of x. For $x = 0$, y may either be $\sqrt{8}$ or $-\sqrt{8}$

7. y is a function of x.

9. The domain is the set of all real numbers $(-\infty, \infty)$.

11. We must have $x + 4 \geq 0$, or $x \geq -4$. Hence the domain is $[-4, \infty)$.

13. $[0, \infty)$

15. We must have $16 - s^2 \geq 0$. Hence the domain is $[-4, 4]$.

17. The domain is given to be $(-\infty, 0) \cup (0, \infty)$.

19. Both conditions $x - 2 \neq 0$ and $x(x+2) = x^2 + 2x \geq 0$ must be satisfied. The second condition is satisfied when both x and $x + 2$ have the same sign, that is, when $x \leq -2$ and when $x \geq 0$. The domain is

$$(-\infty, -2] \cup [0, 2) \cup (2, \infty).$$

21. $x - 2 \geq 0$. The domain is $[2, \infty)$.

23. Since $s^2 + 1 \geq 0$ for all real numbers s, the domain is $(-\infty, \infty)$.

25. $\{x : x \neq -1\}$. Since the exponent is negative, we $g(x)$ is not defined when $x \neq -1$.

27. False. If $h(x) = x^2 + 1$, then $f(x)$ is defined for all real numbers x.

29. We solve the equation, $3x^2 - 2y - 8 = 0$, for y: $y = \frac{3}{2}x^2 - 4$.

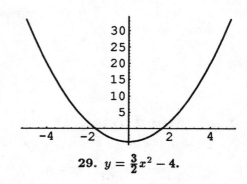

29. $y = \frac{3}{2}x^2 - 4$.

31. We first solve the equation $y - 3x^2 + x - \frac{1}{2} = 0$ for y:

$$y = 3x^2 - x + \frac{1}{2}$$
$$= 3\left(x^2 - \frac{1}{3}x + \frac{1}{36}\right) - \frac{1}{12} + \frac{1}{2}$$
$$= 3\left(x - \frac{1}{6}\right)^2 + \frac{5}{12}$$

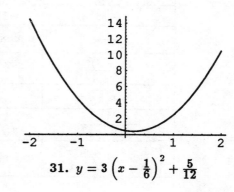

31. $y = 3\left(x - \frac{1}{6}\right)^2 + \frac{5}{12}$

14

33. Solve the equation, $4x^2 + y - 24x + 34 = 0$, for y:

$$y = -4x^2 + 24x - 34$$
$$= -4(x^2 - 6x + 9) + 36 - 34$$
$$= -4(x - 3)^2 + 2$$

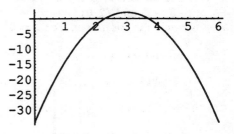

33. $y = -4(x - 3)^2 + 2$

35. Solve the equation, $y^2 + x - 2 = 0$, for x: $x = -y^2 + 2$.

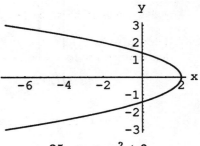

35. $x = -y^2 + 2$.

37. Solve the equation, $3x + 2y^2 - 3 = 0$ for x: $x = -\frac{2}{3}y^2 + 1$.

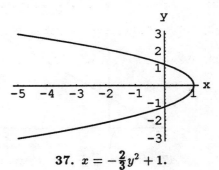

37. $x = -\frac{2}{3}y^2 + 1$.

39. Solve the equation, $x - 2y^2 + 4y - 2 = 0$: for x:

$$x = 2y^2 - 4y + 2$$
$$= 2(y^2 - 2y + 1) - 2 + 2$$
$$= 2(y - 1)^2$$

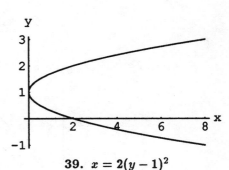

39. $x = 2(y - 1)^2$

41. Solve $x - y - 2 = 0$ for y to obtain $y = x - 2$. Now substitute for y in the equation $y = 4 - x^2$ and solve for x:

$$x - 2 = 4 - x^2 \Leftrightarrow x^2 + x - 6 = 0 \Leftrightarrow (x + 3)(x - 2) = 0 \Leftrightarrow x = -3, 2$$

When $x = -3$, $y = -3 - 2 = -5$. When $x = 2$, $y = 2 - 2 = 0$. Hence, the points of intersection are $(-3, -5)$ and $(2, 0)$.

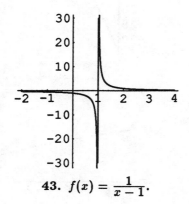

43. $f(x) = \dfrac{1}{x - 1}$.

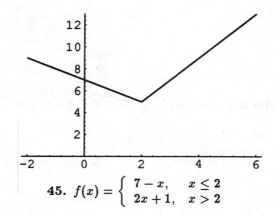

45. $f(x) = \begin{cases} 7 - x, & x \le 2 \\ 2x + 1, & x > 2 \end{cases}$

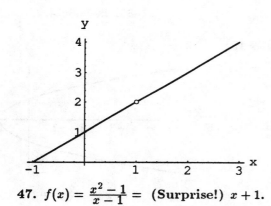

47. $f(x) = \dfrac{x^2 - 1}{x - 1} =$ **(Surprise!)** $x + 1$.

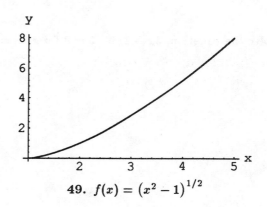

49. $f(x) = (x^2 - 1)^{1/2}$

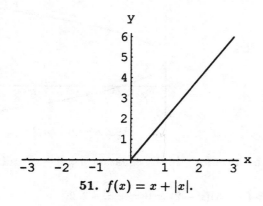

51. $f(x) = x + |x|$.

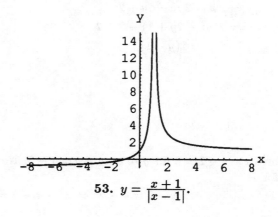

53. $y = \dfrac{x + 1}{|x - 1|}$.

16

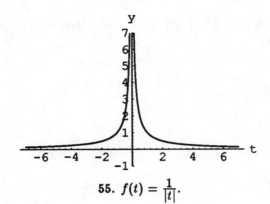

55. $f(t) = \frac{1}{|t|}$.

57. **a.** Since $f(x) = x^2$, $f(-x) = (-x)^2 = x^2 = f(x)$ for all real numbers x. Hence, $f(x)$ is even.

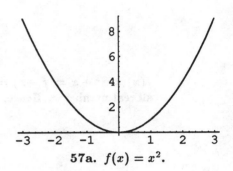

57a. $f(x) = x^2$.

b. Since $f(x) = 2 - x^2$, $f(-x) = 2 - (-x)^2 = 2 - x^2 = f(x)$ for all real numbers x. Hence, $f(x)$ is an even function.

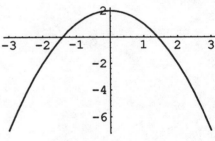

57b. $f(x) = 2 - x^2$.

c. $f(x) = x^3 \Rightarrow f(-x) = (-x)^3 = -x^3 = -f(x)$ for all real numbers x. Hence $f(x)$ is an odd function.

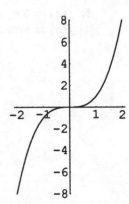

57c. $f(x) = x^3$.

17

d. $f(x) = 1 - x^3 \Rightarrow f(-x) = 1 - (-x)^3 = 1 + x^3$. Hence, $f(x)$ is neither odd nor even.

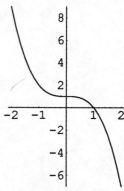

57d. $f(x) = 1 - x^3$.

e. $f(x) = x^3 + x \Rightarrow f(-x) = (-x)^3 + (-x) = -(x^3 + x) = -f(x)$ for all real number x. Hence, $f(x)$ is an odd function.

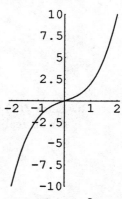

57e. $f(x) = x^3 + x$.

f. $f(x) = 2x^4 + x^2 \Rightarrow f(-x) = 2(-x)^4 + (-x)^2 = 2x^4 + x^2$. Hence, $f(x)$ is even.

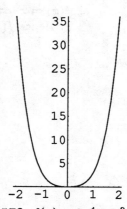

57f. $f(x) = 2x^4 + x^2$.

18

g. $f(x) = |x| + 2 \Rightarrow f(-x) = |-x| + 2 = |x| + 2 = f(x)$
for all real numbers x. Hence $f(x)$ is an even function.

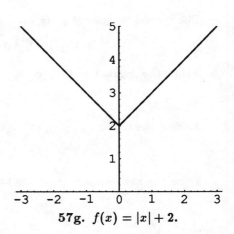

57g. $f(x) = |x| + 2.$

h. $f(x) = x^2 + x \Rightarrow f(-x) = (-x)^2 + (-x) = x^2 - x$. Hence, $f(x)$ is neither odd nor even.

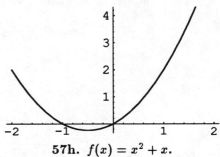

57h. $f(x) = x^2 + x.$

59. $f \circ g(x) = f(x^3) = 3x^3 + 1$

61. $g \circ f(x) = g(3x + 1) = (3x + 1)^3$

63. $h(f(x)) = h(3x + 1) = \sqrt{3x + 1}$

65. $f(g(h(x))) = f(g(\sqrt{x})) = f((\sqrt{x})^3) = f(x^{\frac{3}{2}}) = 3x^{\frac{3}{2}} + 1$

67. a. Since the domain of $f(x) = x^2$ is $\{x| -1 \le x \le 3\}$ the domain of $f \circ g$ is restricted by

$$-1 \le g(x) \le 3 \Leftrightarrow -1 \le (2x + 6) \le 3 \Leftrightarrow -\frac{7}{2} \le x \le -\frac{3}{2}$$

Hence the domain of $f \circ g$ is $\{x| -\frac{7}{2} \le x \le -\frac{3}{2}\}$

b. The range of $f \circ g$ is the same as the range of f. Since $f(x) = x^2 \ge 0$ and $-1 \le x \le 0$ or $0 \le x \le 3$ the range of f and $f \circ g$ is $[0, 9]$.

c. The domain of $g \circ f$ is the same as the domain for f which is $\{x| -1 \le x \le 3\}$.

d. $g \circ f(x) = g(f(x)) = 2f(x) + 6$ From part b. we know that $0 \le f(x) \le 9$. Hence $6 \le 2f(x) + 6 \le 24$. The range of $g \circ f$ is $[6, 24]$.

69. $f(x) = \dfrac{1}{3 - x^2}$ and $g(x) = \sqrt{x^2 - 1}$. To find the domain of $f \circ g(x)$, we must consider the restrictions

$$x^2 - 1 \ge 0 \qquad\qquad 3 - (g(x))^2 \ne 0$$

From the first inequality we see that:

$$x^2 - 1 \ge 0 \Leftrightarrow x^2 \ge 1 \Leftrightarrow |x| \ge 1.$$

The second inequality gives us:

$$3 - (g(x))^2 \neq 0 \Leftrightarrow (g(x))^2 \neq 3 \Leftrightarrow x^2 - 1 \neq 3 \Leftrightarrow x^2 \neq 4 \Leftrightarrow x \neq \pm 2$$

Hence the domain of $f \circ g$ is

$$(-\infty, -2) \cup (-2, -1] \cup [1, 2) \cup (2, \infty)$$

To find the domain of $g \circ f(x)$, we must consider the restrictions:

$$3 - x^2 \neq 0 \qquad\qquad (f(x))^2 - 1 \geq 0$$

From the first inequality we see that:

$$3 - x^2 \neq 0 \Leftrightarrow x^2 \neq 3 \Leftrightarrow |x| \neq \sqrt{3}$$

From the second inequality, we see that:

$$(f(x))^2 \geq 1 \Leftrightarrow \left(\frac{1}{3 - x^2}\right)^2 \geq 1 \Leftrightarrow (3 - x^2)^2 \leq 1 \Leftrightarrow -1 \leq 3 - x^2 \leq 1$$

$$\Leftrightarrow -4 \leq -x^2 \leq -2 \Leftrightarrow 2 \leq x^2 \leq 4 \Leftrightarrow \sqrt{2} \leq |x| \leq 2$$

Hence the domain of $g \circ f$ is

$$[-2, -\sqrt{3}) \cup (-\sqrt{3}, -\sqrt{2}] \cup [\sqrt{2}, \sqrt{3}) \cup (\sqrt{3}, 2]$$

71. $f(x) = x(x^2 + 1)^{-\frac{1}{2}} - \frac{2\sqrt{x^2 + 1}}{x^3} + \frac{3}{x^3\sqrt{x^2 + 1}}$ Note that $x^2 + 1$ is always positive, so it will never cause any problems. The only restriction is that $x \neq 0$. Hence, the domain is $(-\infty, 0) \cup (0, +infty)$. Since $x \neq 0$, the range of f contains the number 0 if and only if the range of $x^3\sqrt{x^2 + 1}f$ contains 0, that is if and only if:

$$x^4 - 2(x^2 + 1) + 3 = 0 \Leftrightarrow x^4 - 2x^2 + 1 = 0 \Leftrightarrow (x^2 - 1)^2 = 0 \Leftrightarrow x = \pm 1$$

$$f(1) = f(-1) = \frac{1}{\sqrt{2}} - 2\sqrt{2} + \frac{3}{\sqrt{2}} = \frac{\sqrt{2}}{2} - 2\sqrt{2} + \frac{3\sqrt{2}}{2} = 0$$

Hence, the range of f does contain 0.

73. True. Every polynomial is the quotient of the polynomial and the constant polynomial $g(x) = 1$.

75. Since the truck depreciates 30% per year, a truck costing $20,000$ has a value $V(t) = \$20,000(.7)^t$ after t years. The cost of the insurance policy is 8% of V. Hence,

$$C(t) = .08(\$20,000)(.7)^t$$

77. a. $p - 30$ is the increase in the daily rental fee over $30. This causes $20(p - 30)$ fewer jackhammers to be rented per year. The number of jackhammers rented per year is therefore $500 - 20(p - 30)$. Hence the yearly revenue is $R = p[500 - 20(p - 30)]$.

b. Since $p \geq 30$, $R = 0$ when:

$$500 - 20(p - 30) = 0 \Leftrightarrow p - 30 = 25 \Leftrightarrow p = 55.$$

The revenues will be 0 when the price reaches $55.00.

79. With $c(n) = \sqrt{1 + 0.2n}$, and $n(t) = 10,000 + 50t^{\frac{2}{3}}$, we compute

$$c \circ n(t) \;=\; \sqrt{1 + 0.2n(t)} \;=\; \sqrt{1 + 0.2(10,000 + 50t^{\frac{2}{3}})} \;=\; \sqrt{2001 + 10t^{\frac{2}{3}}}.$$

81. Let V be the volume and l be the length of one of the edges of a cube. Then

$$V = f(l) = l^3$$

83. $\qquad \gamma = a(Z - b)^2 \qquad Z \geq 0$

This is the right half of a parabola with a vertex at $(b, 0)$ which opens upward.

85.

$$\begin{aligned}
f(x) &= Ax^2 + Bx + C\\
&= A[x^2 + \tfrac{B}{A}x + (\tfrac{B}{2A})^2] - A(\tfrac{B}{2A})^2 + C\\
&= A[x - (-\tfrac{B}{2A})]^2 + C - \tfrac{B^2}{4A}\\
&= A(x - D)^2 + E,
\end{aligned}$$

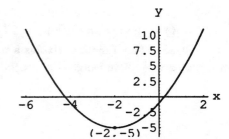

where $D = -\frac{B}{2A}$ and $E = C - \frac{B^2}{4A}$. The figure at right show the case where $A = 1$, $B = 4$, and $C = -1$. In this case $D = -2$ and $E = -5$. Thus the vertex of the parabola is at $(-2, -5)$.

85. $f(x) = x^2 + 4x - 1 = (x + 2)^2 - 5$

87. a. $g(x) = \sqrt{x + 4}$. The range is $[0, \infty)$.

b. $f(t) = 1 + t^2$, $t \geq 0$. The range is $[1, \infty)$.

c. $f(x) = \frac{1}{|x|}$ for $x \neq 0$. The range is $(0, \infty)$.

d. $f(x) = \sqrt{6 - |x + 2|}$. Since the quantity within the square root sign must always be non-negative, we have $0 \leq 6 - |x + 2| \leq 6$. Hence, the range of f is $[0, \sqrt{6}]$.

e. $h(s) = (s^2 + 1)^{\frac{3}{2}}$. $s^2 + 1 \geq 1 \Leftrightarrow h(s) \geq 1$. Hence the range is $[1, \infty)$.

f. $g(x) = (x + 1)^{-\frac{2}{3}}$. Since the exponent is negative, we must have $x + 1 \neq 0$. Since $(x + 1)^2 > 0$, we also have $g(x) = (x + 1)^{-\frac{2}{3}} > 0$. Hence the range is $(0, \infty)$.

g. $f(x) = \sqrt{2x + 1}$. $2x + 1 \geq 0 \Leftrightarrow f(x) \geq 0$. Hence the range is $[0, \infty)$.

h. $f(x) = (x + 2)^{\frac{2}{3}} = [(x + 2)^{\frac{1}{3}}]^2 \geq 0$. Hence, the range is $[0, \infty)$.

89. $y = a|b(x + c)| + d$. Since the graph opens downward, a must be negative; thus y attains its maximum value when $(x + c) = 0$. Since the highest point on the graph occurs at $(-2, 2)$, we must have $c = 2$ and $d = 2$. Thus $y = a|b||x + 2| + 2$. Since $(2, 0)$ is on the graph, we have $0 = 4a|b| + 2$. Hence $a|b| = -\frac{1}{2}$, and

$$y = -\frac{1}{2}|x + 2| + 2 = -\frac{1}{2}f(x + 2) + 2$$

91. $y = \frac{a}{b(x + c)} + d$. First note that 1 is not in the domain of the graph, so that $c = -1$. The range does not contain 0, so that $d = 0$. The funtion is $y = \frac{a}{b(x - 1)}$. We see from the graph that $(0, -3)$ is on the graph of the function. Therefore, $-3 = \frac{a}{b(0 - 1)} \Leftrightarrow \frac{a}{b} = 3$. Hence the function is

$$y = \frac{3}{x - 1}.$$

93. a. $f(x) = 2x + 1 \qquad g(x) = \sqrt{x}$
$(f+g)(x) = 2x + \sqrt{x} + 1$. The domain is $[0, \infty)$. Since the function is increasing on its domain, the minimum value occurs at $x = 0$. Hence the range is $[1, \infty)$.

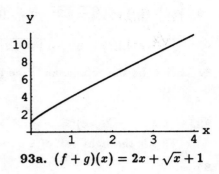

93a. $(f+g)(x) = 2x + \sqrt{x} + 1$

b. $(f-g)(x) = 2x - \sqrt{x} + 1 = 2\left(\sqrt{x} - \frac{1}{4}\right) + \frac{7}{8}$. The domain is $[0, \infty)$. This function attains a minimum value in its domain, at $x = \frac{1}{4}$. The range is $[\frac{7}{8}, \infty)$.

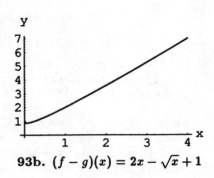

93b. $(f-g)(x) = 2x - \sqrt{x} + 1$

c. $(fg)(x) = (2x+1)\sqrt{x}$. The domain is $[0, \infty)$. The function takes on the value 0 at $x = 0$. At all other values of x the function is positive. It is continuous and increases without bound as $x \to \infty$. Hence the range is $[0, \infty)$.

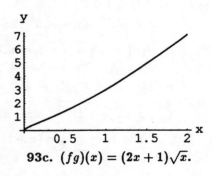

93c. $(fg)(x) = (2x+1)\sqrt{x}$.

d. $\left(\frac{f}{g}\right)(x) = 2\sqrt{x} + \frac{1}{\sqrt{x}}$. The domain is $(0, \infty)$ and the range is $[2\sqrt{2}, \infty)$.

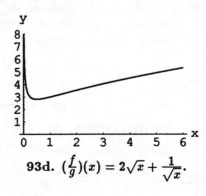

93d. $\left(\frac{f}{g}\right)(x) = 2\sqrt{x} + \frac{1}{\sqrt{x}}$.

22

e. $\left(\frac{g}{f}\right)(x) = \frac{\sqrt{x}}{2x+1}$. The domain is $[0,\infty)$ and the range is $[\frac{\sqrt{2}}{4}, \infty)$

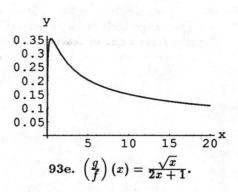

93e. $\left(\frac{g}{f}\right)(x) = \frac{\sqrt{x}}{2x+1}$.

f. $(f \circ g)(x) = f(\sqrt{x}) = 2\sqrt{x} + 1$. The domain is $[0,\infty)$ and the range is $[0,\infty)$.

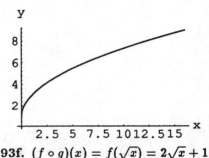

93f. $(f \circ g)(x) = f(\sqrt{x}) = 2\sqrt{x} + 1$.

g. $(g \circ f)(x) = g(2x+1) = \sqrt{x+1}$. The domain is $[-1,\infty)$ and the range is $[0,\infty)$.

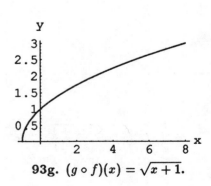

93g. $(g \circ f)(x) = \sqrt{x+1}$.

95. $f(x) = \sqrt{x+1}$, $g(x) = x^2 - 2x + 1$.

a. $(f+g)(x) = x^2 - 2x + \sqrt{x+1} + 1$. The domain is $[-1, \infty)$. The range includes at least the set $[\sqrt{2}, \infty)$. However the minimum value of this function is difficult to determine.

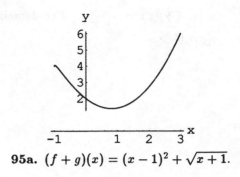

95a. $(f+g)(x) = (x-1)^2 + \sqrt{x+1}$.

b. $(f-g)(x) = \sqrt{x+1} - x^2 + 2x - 1$. The domain is $[-1, \infty)$. The range includes at least the set $(-\infty, \sqrt{2}]$. However the maximum value of this function is difficult to determine.

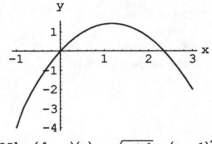

95b. $(f-g)(x) = \sqrt{x+1} - (x-1)^2$.

c. $(fg)(x) = \sqrt{x+1}(x-1)^2$. The domain is $[-1, \infty)$. The range is $[0, \infty)$.

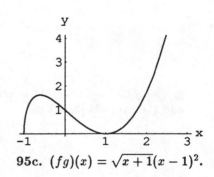

95c. $(fg)(x) = \sqrt{x+1}(x-1)^2$.

d. $(f/g)(x) = \dfrac{\sqrt{x+1}}{(x-1)^2}$. The domain is $\{x : x \neq 1 \text{ and } -1 \leq x < \infty\}$. The range is $[0, \infty)$.

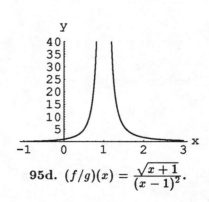

95d. $(f/g)(x) = \dfrac{\sqrt{x+1}}{(x-1)^2}$.

e. $(g/f)(x) = \dfrac{(x-1)^2}{\sqrt{x+1}}$. The domain is $(-1, \infty)$ and the range is $[0, \infty)$.

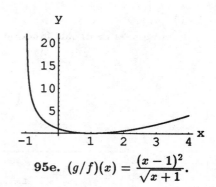

95e. $(g/f)(x) = \dfrac{(x-1)^2}{\sqrt{x+1}}$.

f. $(f \circ g)(x) = \sqrt{x^2 - 2x + 2}$. The domain is $(-\infty, \infty)$ and the range is $[1, \infty)$.

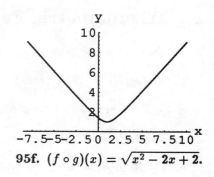

95f. $(f \circ g)(x) = \sqrt{x^2 - 2x + 2}$.

g. $(g \circ f)(x) = x + 2 - 2\sqrt{x+1}$. The domain is $[-1, \infty)$, and the range is $[0, \infty)$.

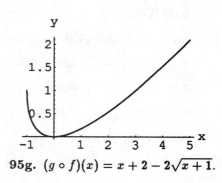

95g. $(g \circ f)(x) = x + 2 - 2\sqrt{x+1}$.

97. From the graph at right, it is apparent that the interpolating polynomial $p(x) = -\dfrac{1}{520}x^6 + \dfrac{9}{130}x^4 - \dfrac{59}{104}x^2 + 1$ does not faithfully represent the rational function $f(x) = \dfrac{1}{1+x^2}$.

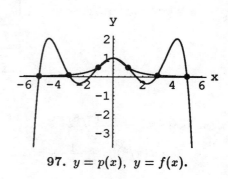

97. $y = p(x)$, $y = f(x)$.

25

99.

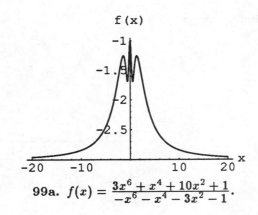

f(x)

99a. $f(x) = \dfrac{3x^6 + x^4 + 10x^2 + 1}{-x^6 - x^4 - 3x^2 - 1}.$

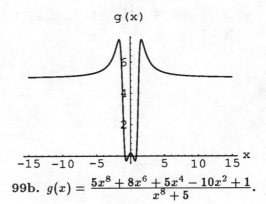

g(x)

99b. $g(x) = \dfrac{5x^8 + 8x^6 + 5x^4 - 10x^2 + 1}{x^8 + 5}.$

1.5 Trigonometric Functions

1. a. 30 degrees $= (30 \text{ degrees}) * \dfrac{2\pi \text{ radians}}{360 \text{ degrees}} = \dfrac{\pi}{6}$ radians.

   b. 75 degrees $= (75 degrees) * \dfrac{2\pi \text{ radians}}{360 \text{ degrees}} = \dfrac{5\pi}{12}$ radians.

   c. -15 degrees $= (-15 \text{ degrees}) * \dfrac{2\pi \text{ radians}}{360 \text{ degrees}} = -\dfrac{\pi}{12}$ radians.

   d. 315 degrees $= (315 \text{ degrees}) * \dfrac{2\pi \text{ radians}}{360 \text{ degrees}} = \dfrac{7\pi}{4}$ radians.

   e. $x + 30$ degrees $= (x + 30 \text{ degrees}) * \dfrac{2\pi \text{ radians}}{360 \text{ degrees}} = \left(\dfrac{x}{180} + \dfrac{1}{6}\right)\pi$ radians.

   f. 2910 degrees $= (2910 \text{ degrees}) * \dfrac{2\pi \text{ radians}}{360 \text{ degrees}} = 16\dfrac{1}{6}\pi$ radians.

3. If $\sin(\theta) = 0$ and $\cos(\theta) = 1$, then θ must be $\pi/2$. Then $\tan(\theta) = 0$ and $\sec(\theta) = 1$. Similarly, if $\sin(\theta) = \dfrac{\sqrt{2}}{2}$ and $\tan(\theta) = 1$, then $\cos(\theta) = \sin(\theta) = \dfrac{\sqrt{2}}{2}$ and $\sec(\theta) = \sqrt{2}$. In addition, θ must be $\pi/4$. If $\sin(\theta) = \dfrac{3}{5}$ and $\sec(\theta) = \dfrac{5}{4}$, then $\cos(\theta) = \dfrac{1}{\sec(\theta)} = \dfrac{4}{5}$, and $\tan(\theta) = \dfrac{\sin(\theta)}{\cos(\theta)} = \dfrac{3}{4}$. In this case θ is $\arctan(\dfrac{3}{4}) \approx .6435$ radians. If $\tan(\theta) = \dfrac{5}{12}$ and $\sec(\theta) = \dfrac{13}{12}$, then $\cos(\theta) = \dfrac{1}{\sec(\theta)} = \dfrac{12}{13}$, and $\sin(\theta) = \tan(\theta)\cos(\theta) = \dfrac{5}{12}$.

θ	$\sin(\theta)$	$\cos(\theta)$	$\tan(\theta)$	$\sec(\theta)$
0	0	1	0	1
$\dfrac{\pi}{4}$	$\dfrac{\sqrt{2}}{2}$	$\dfrac{\sqrt{2}}{2}$	1	$\sqrt{2}$
???	$\dfrac{3}{5}$	$\dfrac{4}{5}$	$\dfrac{3}{4}$	$\dfrac{5}{4}$
????	$\dfrac{5}{13}$	$\dfrac{12}{13}$	$\dfrac{5}{12}$	$\dfrac{13}{12}$

1.5.3

5. a. $\tan(\pi/6) = \dfrac{\sin(\pi/6)}{\cos(\pi/6)} = \dfrac{1/2}{\sqrt{3}/2} = \dfrac{1}{\sqrt{3}}.$ $\cot(\pi/6) = \dfrac{1}{\tan(\pi/6)} = \sqrt{3}.$ $\sec(\pi/6) = \dfrac{1}{\cos(\pi/6)} = \dfrac{2\sqrt{3}}{3}.$
   $\csc(\pi/6) = \dfrac{1}{\sin(\pi/6)} = 2.$

26

b. $\tan(5\pi/2) = \tan(\pi/2)$ and is undefined. $\sec(5\pi/2) = \dfrac{1}{\cos(\pi/2)}$ and is undefined. $\cot(5\pi/2) = \dfrac{\cos(\pi/2)}{\sin(\pi/2)} = 0$. $\csc(5\pi/2) = \dfrac{1}{\sin(\pi/2)} = 1$.

c. Since $-\dfrac{13\pi}{3} = -(4\frac{1}{3})\pi$, $\sin(x) = \sin(-\pi/3) = -\dfrac{\sqrt{3}}{2}$, and $\cos(x) = \cos(-\pi/3) = \frac{1}{2}$. Thus $\tan(x) = -\sqrt{3}$, $\cot(x) = \dfrac{1}{\tan(x)} = -\dfrac{\sqrt{3}}{3}$, $\sec(x) = 2$ and $\csc(x) = -\dfrac{2}{\sqrt{3}}$.

d. $\sin(9\pi) = \sin(\pi) = 0$, and $\cos(9\pi) = \cos(\pi) = -1$. Then $\tan(9\pi) = \tan(\pi) = 0$, $\cot(9\pi) = \dfrac{1}{\tan(9\pi)}$ is undefined, $\sec(9\pi) = \sec(\pi) = -1$, and $\csc(9\pi)$ is undefined.

e. $\sin(-9\pi/4) = \sin(-\pi/4) = -\sin(\pi/4) = -\dfrac{\sqrt{2}}{2}$. $\cos(-9\pi/4) = \cos(-\pi/4) = \cos(\pi/4) = -\dfrac{\sqrt{2}}{2}$. Thus $\tan(-9\pi/4) = -1$, $\cot(-9\pi/4) = \dfrac{1}{\tan(-9\pi/4)} = -1$, $\sec(-9\pi/4) = \sqrt{2}$, and $\csc(-9\pi/4) = -\sqrt{2}$.

f. $\sin(7\pi/6) = \sin(\pi - 7\pi/6) = \sin(-\pi/6) = -\sin(\pi/6) = -\frac{1}{2}$. Also $\cos(7\pi/6) = \cos(7\pi/6 - 2\pi) = \cos(-5\pi/6) = \cos(5\pi/6) = -\dfrac{\sqrt{3}}{2}$. Thus $\tan(7\pi/6) = -\dfrac{1}{\sqrt{3}}$, $\cot(7\pi/6) = \dfrac{1}{\tan(7\pi/6)} = -\sqrt{3}$, $\sec(7\pi/6) = \dfrac{2}{\sqrt{3}}$, and $\csc(7\pi/6) = -2$.

7. In $-2\pi \le x < 2\pi$,

    a. $\cos(x) = 0$ if and only if $-3\pi/2,\ -\pi/2,\ \pi/2,\ 3\pi/2$.

    b. $\sin(x) = \cos(x)$ if and only if $x = -7\pi/4,\ -3\pi/4,\ \pi/4$, or $5\pi/4$.

    c. $\tan(x) = -1$ if and only if $x = -5\pi/4,\ -\pi/4,\ 3\pi/4$, or $7\pi/4$.

    d. $\sin(3x) = 0$ if and only if $3x = n\pi$ for some integer n. Then $x = n\pi/3$. The corresponding values in $-2\pi \le x < 2\pi$ are $x = -2\pi,\ -5\pi/3,\ -4\pi/3,\ -\pi,\ -2\pi/3,\ -\pi/3,\ 0,\ \pi/3,\ 2\pi/3,\ \pi,\ 4\pi/3,\ 5\pi/3,\ 2\pi$.

    e. $\sin(2x) = \cos(x)$ is equivalent to the equation $2\sin(x)\cos(x) = \cos(x)$. This equation factors as $\cos(x)(2\sin(x) - 1) = 0$. The solutions are $\cos(x) = 0$, $\rightarrow x = -3\pi/2,\ -\pi/2,\ \pi/2,\ 3\pi/2$, and $\sin(x) = 1/2$, $\rightarrow x = -11\pi/6,\ -7\pi/6,\ \pi/6,\ 5\pi/6$.

    f. $3\cos^2(x) - \sin^2(x) = 4\cos^2(x) - 1 = 0$ if and only if $\cos(x) = \pm 1/2$. The solutions are $x = -5\pi/3,\ -4\pi/3,\ -2\pi/3,\ -\pi/3,\ \pi/3,\ 2\pi/3,\ 4\pi/3, 5\pi/3$.

    g. $\cos^2(x) + \sin(x) = 0 \Leftrightarrow 1 - \sin^2(x) + \sin(x) = 0$. This is a quadratic equation in $y = \sin(x)$. Using the quadratic formula we find that the solutions of this equation are $y = 1/2 \pm \sqrt{5}/2$. The value $y = 1/2 + \sqrt{5}/2$ cannot be a value of $\sin(x)$ since it is greater than 1. The other value is ≈ -0.618034. Computing $\arcsin(-0.618034)$ we obtain $-.666239$ radians. Other solutions in $-2\pi < x < 2\pi$ are $x = -.666239 + 2\pi$ radians, $x = \pi - (-.666239)$ radians, and $x = .666239 - \pi$ radians.

9. $\sin(t) = \sqrt{3}\cos(t) \Leftrightarrow \tan(t) = \sqrt{3}$. In $0 \le t < 2\pi$, $\cos(t) < 0 \Leftrightarrow \pi/2 < t < 3\pi/2$. The solution to these two conditions is $t = 4\pi/3$.

11. In $0 \le t < 2\pi$, $\sin(t) > 0 \Leftrightarrow 0 < t < \pi$, and $\tan(t) > 0 \Leftrightarrow 0 < t < \pi/2$, or $\pi < t < 3\pi/2$. The intersection of these two solutions sets is $0 < t < \pi/2$.

13. Since $\sec(t) = 1/\cos(t)$, these two inequalities are mutually exclusive. Note that if $\cos(t) = 0$, then $\sec(t)$ is undefined.

15. $\sin(5\pi/12) = \sin(2\pi/3 - \pi/4)$

$$= \sin(2\pi/3)\cos(\pi/4) - \sin(\pi/4)\cos(2\pi/3)$$

$$= \frac{\sqrt{3}}{2}\frac{\sqrt{2}}{2} - \frac{\sqrt{2}}{2}\frac{-1}{2}$$

$$= \frac{\sqrt{6}}{4} + \frac{\sqrt{2}}{4}$$

$$= \frac{\sqrt{6} + \sqrt{2}}{4}.$$

17. Using the half-angle formula, we obtain $\cos(\pi/12) = \cos\left(\dfrac{\pi/6}{2}\right) = \sqrt{\dfrac{1 + \cos(\pi/6)}{2}}$, Evaluating this expression, we obtain $\sqrt{\dfrac{2 + \sqrt{3}}{4}}$. Alternatively, using methods like those used above for #15, we obtain $\cos(\pi/12) = \dfrac{\sqrt{6} + \sqrt{2}}{4}$. Can you reconcile these two answers?

19. $\sec(11\pi/12) = \dfrac{1}{\cos(11\pi/12)} = -\dfrac{1}{\cos(\pi/12)} = -\dfrac{4}{\sqrt{6} + \sqrt{2}}$. See #17.

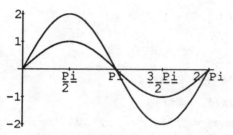

21. $\sin(x),\ 2\sin(x)$.

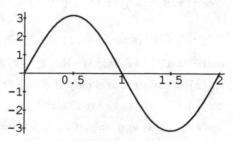

23. $\pi\sin(\pi x)$.

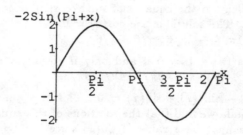

25. $-2\sin(\pi + x)$.

27. The functions $\sin(x)$ and $\cos(x)$ are defined for any real number x. Both functions have range $-1 \le x \le 1$. The functions $\tan(x)$ and $\sec(x)$ are defined whenever $\cos(x) \ne 0$, that is, for $x \ne \pi/2 + n\pi$, $n = 0, \pm1, \pm2, \ldots$. The functions $\cot(x)$ and $\csc(x)$ are defined whenever $\sin(x) \ne 0$, that is, for $x \ne n\pi$, $n = 0, \pm1, \pm2, \ldots$ The functions $\tan(x)$ and $\cot(x)$ both have range $(-\infty, \infty)$. The functions $\sec(x)$ and $\csc(x)$ both have ranges $(-\infty, -1] \cup [1, \infty)$.

29. Since $\cos(-x) = \cos(x)$, cos is an even function.

31. Here $y = f(x) = \dfrac{\sin(x)}{x}$, and $f(-x) = \dfrac{\sin(-x)}{-x} = \dfrac{\sin(x)}{x} = f(x)$. Therefore f is an even function.

33. Theorem: The product of two odd functions is always even (verify this for yourself!). Thus $f(x) = \sin^2(x)$ is an even function.

35. Let (x_1, y_1) and (x_2, y_2) be two points on this line. Then $m = \frac{y_2 - y_1}{x_2 - x_1}$. If we form a right triangle with vertices at (x_1, y_1), (x_2, y_2), and (x_2, y_1), then we compute that $\tan(\theta) = \frac{y_2 - y_1}{x_2 - x_1} = m$.

37. To prove: $\cos(\theta + \phi) = \cos(\theta)\cos(\phi) - \sin(\theta)\sin(\phi)$, for θ and ϕ between 0 and $\pi/2$.

  a. Refer to figure 5.13 in the text. $P_1 = (1, 0)$, $P_2 = (\cos(\phi), \sin(\phi))$, $P_3 = (\cos(\phi + \theta), \sin(\phi + \theta))$, $P_4 = (\cos(-\theta), \sin(-\theta))$.

  b. Since the triangles OP_1P_3 and OP_4P_2 both have two sides which are radii of the unit circle, they are congruent if the central angles $\angle P_1OP_3$ and $\angle P_4OP_2$ are the same. But both angles equal $\theta + \phi$. Thus the two triangles are congruent. Therefore the distance from P_3 to P_1 is the same as the distance from P_2 to P_4.

  c. The square of the distance from $P_1 = (1, 0)$ to $P_3 = (\cos(\phi + \theta), \sin(\phi + \theta))$ is $(\cos(\phi + \theta) - 1)^2 + (\sin(\phi + \theta))^2$. The square of the distance from $P_4 = (\cos(-\theta), \sin(-\theta))$ to $P_2 = (\cos(\phi), \sin(\phi))$ is $(\cos(\theta) - \cos(\phi))^2 + (\sin(\phi) - \sin(-\theta))^2$. By part b, these two distances must be equal.

  d. Expanding the squares, we obtain
  $$\cos^2(\theta + \phi) - 2\cos(\theta + \phi) + 1 + \sin^2(\theta + \phi) =$$
  $$\cos^2(\phi) - 2\cos(\phi)\cos(\theta) + \cos^2(\theta) + \sin^2(\phi) + 2\sin(\phi)\sin(\theta) + \sin^2(\theta).$$

  Successively applying the identity $\sin^2(\alpha) + \cos^2(\alpha) = 1$ with $\alpha = \theta + \phi$, ϕ, and θ, we obtain:
  $$2 - 2\cos(\theta + \phi) = 2 - 2\cos(\theta)\cos(\phi) + 2\sin(\theta)\sin(\phi)$$

  e. Thus
  $$\cos(\theta + \phi) = \cos(\theta)\cos(\phi) - \sin(\theta)\sin(\phi)$$

39. By definition, $\tan(x + T) = \frac{\sin(x + T)}{\cos(x + T)}$. This, in turn, equals $\frac{\sin(x)\cos(T) + \sin(T)\cos(x)}{\cos(x)\cos(T) - \sin(x)\sin(T)}$. Dividing both numerator and denominator by $\cos(x)\cos(T)$, this becomes $\frac{\tan(x) + \tan(T)}{1 - \tan(x)\tan(T)}$ (another formula to remember!). Thus $\tan(x + T) = \tan(x)$, if

$$\frac{\tan(x) + \tan(T)}{1 - \tan(x)\tan(T)} = \tan(x)$$
$$\frac{\tan(x) + \tan(T) - \tan(x) + \tan(x)^2\tan(T)}{1 - \tan(x)\tan(T)} = 0$$
$$\tan(T)(1 + \tan(x)^2) = \tan(T)\sec(x)^2 = 0$$

Thus we conclude that $\tan(x + T) = \tan(x)$, if and only if $\tan(T) = 0$. The solution set of this equation is $\{T \mid T = n\pi, \ n = 0, \pm 1, \pm 2, \pm 3, \cdots\}$. The minimum positive such value of T is π. Thus tan has period π.

Since $\cot(x) = \frac{1}{\tan(x)}$, the period of cot is the same as the period of tan.

41. The pendulum travels a distance equal to the change in the angle, measured in radians, times the radius. The total change in the angle over one cycle is $2\pi/3$. Thus the distance travelled is $4\pi/3m$.

43. The distance from B to C is $(50m) \times \tan(\pi/3) = 50m\sqrt{3}$.

45. Since the altitude is the length of the side of a right triangle, and this side is the side opposite an angle measuring $\pi/4$ radians, and the horizontal distance is 300 meters, and this is the length of the adjacent side, the altitude A equals $\tan(\pi/4) * (300 \text{ meters}) = 300 \text{ meters}$

47. The ladder would form the hypotenuse of a $30° - 60° - 90°$ triangle, with a 3 meter side adjacent to the $30°$ ($\pi/6$) angle. Thus the length of the ladder must be $(3m)/\cos(\pi/6) = (3m)2/\sqrt{3} = 2\sqrt{3}m$.

49. Examining the graph at right, we see that the solution lies somewhere between 0.7 and 0.8. An approximate solution, accurate to 6 digits, is $x = 0.739085$.

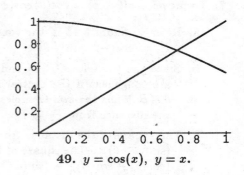

49. $y = \cos(x)$, $y = x$.

51. Since $y \to \infty$ as $x \to \infty$ on the line $y = x$, the points of intersection of this line with the graph of $y = \tan(x)$ will approach the vertical asymptotes of this graph. These asymptotes occur at $x = \dfrac{(2n+1)\pi}{2}$. Thus the solutions tend to these values of x as $x \to \infty$.

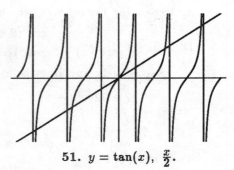

51. $y = \tan(x)$, $\dfrac{x}{2}$.

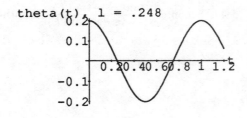

53.a $l = .248$, $T = 1s$.

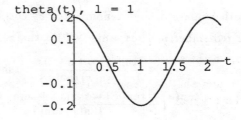

53.b $l = 1$, $T = 2s$.

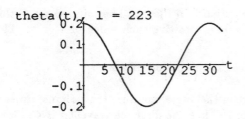

53.c $l = 223$, $T = 30s$

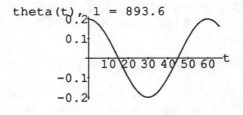

53.d $l = 893.6$, $T = 1$ min. .

Review Exercises–Chapter 1

1. If $A = [2,7)$ and $B = (3,9)$, then $A \cup B = [2,9)$, and $A \cap B = (3,7)$.

3. $2x - 7 \geq 9 \Leftrightarrow 2x \geq 16 \Leftrightarrow x \geq 8$.

5. $A = \{x | x \geq 4\}, B = \{x | -3 \leq x \leq 3\}, C = \{-4, -3, -2, -1, 0, 1, 2, 3, 4\}$

   a. $A \cap B = [-3, -2] \cup [2, 3]$
   b. $A \cup B = (-\infty, \infty)$
   c. $A \cap C = \{-4, -3, -2, 2, 3, 4\}$
   d. $B \cap C = \{-3, -2, -1, 0, 1, 2, 3\}$
   e. $A \cup (B \cap C) = (-\infty, -2] \cup \{-1, 0, 1\} \cup [2, \infty)$
   f. $A \cap (B \cup C) = [-3, -2] \cup [2, 3] \cup \{-4, 4\}$

7. $|x - 7| \leq 12 \Leftrightarrow -12 \leq (x - 7) \leq 12 \Leftrightarrow -5 \leq x \leq 19$.

9. $A \cup B = A$ and $A \cap B = B$

11. $-3 \leq |2x + 1| \leq 11 \Leftrightarrow 0 \leq |2x + 1| \leq 11 \Leftrightarrow -11 \leq 2x + 1 \leq 11 \Leftrightarrow -6 \leq x \leq 5$

13. $|2x^2 + 1| \leq 9 \Leftrightarrow 2x^2 + 1 \leq 9 \Leftrightarrow x^2 \leq 4 \Leftrightarrow -2 \leq x \leq 2$.

15. $\cos x < \frac{\sqrt{3}}{2}$, $0 \leq x \leq 2\pi \Leftrightarrow \frac{\pi}{6} < x < 2\pi - \frac{\pi}{6} = \frac{11\pi}{6}$.

17. $\left(\sin(x) - \frac{1}{2}\right)\left(\cos(x) + \frac{1}{2}\right) = 0 \Leftrightarrow \sin(x) = \frac{1}{2}$ or $\cos(x) = -\frac{1}{2}$. This occurs at $x = \frac{\pi}{6}$, $\frac{2\pi}{3}$, $\frac{5\pi}{6}$ and at $x = \frac{4\pi}{3}$. These are the only points where the quantity $\left(\sin(x) - \frac{1}{2}\right)\left(\cos(x) + \frac{1}{2}\right)$ can change sign. Evaluating $\left(\sin(x) - \frac{1}{2}\right)\left(\cos(x) + \frac{1}{2}\right)$ at points in the intervals $(0, \pi/6)$, $(\pi/6, 2\pi/3$, $(2\pi/3, 5\pi/6)$, $(5\pi/6, 4\pi/3)$, and $(4\pi/3, 2\pi)$, we find that this quantity is positive if and only if $\frac{\pi}{6} < x < \frac{2\pi}{3}$, or $\frac{5\pi}{6} < x < \frac{4\pi}{3}$.

19. $|x - 2| = |x + 2| \Leftrightarrow x - 2 = x + 2$ or $x - 2 = -x - 2$. The first equation has no solutions. Hence $x = 0$ (from the second equation).

21. Consider the line connecting $(0, 0)$ and $(5, 0)$ to be the base, b, of the triangle. Then $b = 5$. Since the third vertex is at $(2, 3)$, the height of the triangle is $h = 3$. Hence the area is

$$A = \frac{1}{2}bh = \frac{15}{2}.$$

23. Since the diameter is 10, the radius is $r = 5$. Since the center is at $(2, 4)$, the equation of the circle is $(x - 2)^2 + (y - 4)^2 = 25$.

25. $x^2 - 4x + y^2 = 0 \Leftrightarrow (x - 2)^2 + y^2 = 4$. Hence the center is at $(2, 0)$ and the radius is 2.

27. $x^2 - 2x + y^2 + 2y = 14 \Leftrightarrow (x^2 - 2x + 1) + (y^2 + 2y + 1) = 14 + 2 \Leftrightarrow (x - 1)^2 + (y + 1)^2 = 16$. Hence the center is $(1, -1)$ and the radius is 4.

29. $x - 3y + 4 = 0 \Leftrightarrow y = \frac{1}{3}x + \frac{4}{3}$. Note that when $y = 0$, $x = -4$. Hence the slope is $\frac{1}{3}$, the y-intercept is $\frac{4}{3}$ and the x-intercept is -4.

31. $7x - 7y + 21 = 0 \Leftrightarrow y = x + 3$. The slope is 1, the y-intercept is 3 and the x-intercept is -3.

33. $3x = 6 \Leftrightarrow x = 2$. The line is vertical and has neither a slope nor a y-intercept. The x-intercept is 2.

35. $2x - 4y = 14 \Leftrightarrow y = \frac{1}{2}x - \frac{7}{2}$. The slope of the parallel line is $\frac{1}{2}$. Since the line passes through the point $(2, -4)$, the equation is given by:

$$y + 4 = \frac{1}{2}(x - 2) \Leftrightarrow y = \frac{1}{2}x - 5.$$

37. $3x - 6y = 8 \Leftrightarrow y = \frac{1}{2}x - \frac{4}{3}$. The slope of this line is $\frac{1}{2}$. The slope of the perpendicular line is the negative reciprocal of $\frac{1}{2}$, which is -2. Since the line passes through the origin, its equation is $y = -2x$.

39. The line $ax + by + c = 0$ crosses the x-axis when $y = 0$:
$ax + 0y + c = 0 \Leftrightarrow x = -\frac{c}{a}$. Hence, the line crosses the x-axis at the point $(-\frac{c}{a}, 0)$.

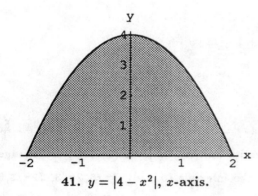

41. $y = |4 - x^2|$, x-axis.

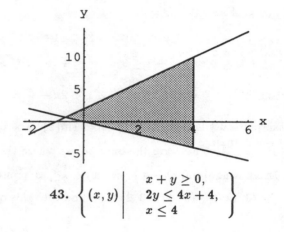

43. $\left\{ (x, y) \middle| \begin{array}{l} x + y \geq 0, \\ 2y \leq 4x + 4, \\ x \leq 4 \end{array} \right\}$

45.

θ	$\sin(\theta)$	$\cos(\theta)$	$\tan(\theta)$	$\cot(\theta)$	$\sec(\theta)$	$\csc(\theta)$
$\frac{\pi}{3}$	$\frac{\sqrt{3}}{2}$	$\frac{1}{2}$	$\sqrt{3}$	$\frac{1}{\sqrt{3}}$	2	$\frac{2}{\sqrt{3}}$
$\frac{-3\pi}{4}$	$-\frac{\sqrt{2}}{2}$	$-\frac{\sqrt{2}}{2}$	1	1	$-\sqrt{2}$	$-\sqrt{2}$
$\frac{7\pi}{3}$	$\frac{\sqrt{3}}{2}$	$\frac{1}{2}$	$\sqrt{3}$	$\frac{1}{\sqrt{3}}$	2	$\frac{2}{\sqrt{3}}$
$\frac{9\pi}{4}$	$\frac{\sqrt{2}}{2}$	$\frac{\sqrt{2}}{2}$	1	1	$\sqrt{2}$	$\sqrt{2}$
$\frac{\pi}{12}$	$\frac{\sqrt{6} - \sqrt{2}}{4}$	$\frac{\sqrt{6} + \sqrt{2}}{4}$	$2 - \sqrt{3}$	$2 + \sqrt{3}$	$\sqrt{6} - \sqrt{2}$	$\sqrt{6} + \sqrt{2}$
$\frac{5\pi}{12}$	$\frac{\sqrt{6} + \sqrt{2}}{4}$	$\frac{\sqrt{6} - \sqrt{2}}{4}$	$2 + \sqrt{3}$	$2 - \sqrt{3}$	$\sqrt{6} + \sqrt{2}$	$\sqrt{6} - \sqrt{2}$

47. Let $x = -6.214\overline{214}$. Then $1000x = -6214.214\overline{214}$.
   $999x = (1000 - 1)x = -6214.214\overline{214} + 6.214\overline{214} = -6208$.
   Hence $-6.214\overline{214} = -\frac{6208}{999}$

49. $x + 3y - 6 = 0$ and $x - y = 2$. solve the second equation for x, we have $x = y + 2$. substitute this into the first equation:
$y + 2 + 3y - 6 = 0 \leftrightarrow 4y = 4 \leftrightarrow y = 1$. also $x = y + 2 = 3$. the point of intersection is $(3, 1)$.

51. let k be the temperature in kelvin and c be the temperature in cesius. two values for (k, c) are given to be $(373, 100)$ and $(273, 0)$. the slope of the line connecting the two points is $\frac{100 - 0}{373 - 273} = 1$. using the point $(273, 0)$, the equation of the line is:
$c = k - 273$.

53. first we find the equation of the line through $(4, 1)$ which is perpendicular to $x + y = 2$, that is $y = -x + 2$. the slope of the given line is -1, so the slope of the perpendicular line is $+1$. the equation of the perpendicular line through $(4, 1)$ is $y - 1 = (x - 4)$ or $y = x - 3$. the point of intersection of the two lines is the desired point. $-x + 2 = y = x - 3$. hence, $x = \frac{5}{2}$ and $y = -\frac{1}{2}$. the point on the line $x + y = 2$ nearest to $(4, 1)$ is $(\frac{5}{2}, -\frac{1}{2})$.

55. no, the equation is that of a circle. the value of y is not uniquely determined for every value of x.

57. yes. y is a constant function of x.

59. we must have $x^2 - 1 \geq 0$. the domain is $(-\infty, 1] \cup [1, \infty)$.

61. since the domain of the sin function is the set of real numbers, there are no restrictions on the domain. the domain is the set of all real numbers x.

63. we must have $1 - \sin^2 x > 0$. this inequality holds for all real numbers $x \neq \frac{\pi}{2} + \pi k$ for some integer k. hence the domain is $\{x | x \neq \frac{\pi}{2} + \pi k, \ k \ any \ integer\}$.

65. since $f(x) = x^3 \sin x$, $f(-x) = (-x)^3 \sin(-x) = (-x^3)(-\sin x) = x^3 \sin x = f(x)$. hence f is an even funtion.

67. $f(x) = (x - 1)(x + 1)$
$f(-x) = (-x - 1)(-x + 1) = (x + 1)(x - 1) = f(x)$. hence f is an even function.

69. Completing squares:
$$y = 2x^2 - 2x + \frac{7}{2}$$
$$= 2(x^2 - x + \frac{1}{4}) - \frac{1}{2} + \frac{7}{2}$$
$$= 2(x - \frac{1}{2})^2 + 3.$$

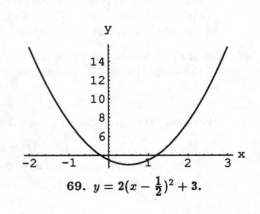

69. $y = 2(x - \frac{1}{2})^2 + 3$.

71. Completing squares:

$$y^2 + 2x - y + \frac{3}{4} = 0$$
$$\Leftrightarrow y^2 - y + \frac{1}{4} = -2x - \frac{1}{2}$$
$$\Leftrightarrow (y - \frac{1}{2})^2 = -2(x + \frac{1}{4})$$
$$\Leftrightarrow x = -\frac{1}{2}(y - \frac{1}{2})^2 - \frac{1}{4}.$$

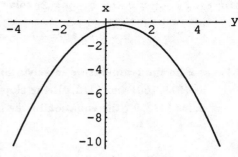

71. $x = -\frac{1}{2}(y - \frac{1}{2})^2 - \frac{1}{4}$.

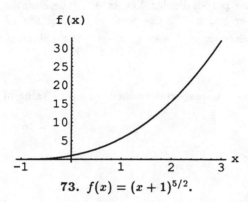

73. $f(x) = (x + 1)^{5/2}$.

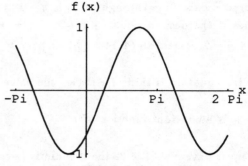

75. $f(x) = \sin(x - \pi/4)$.

77. a. if $\cos(\theta) < 0$ then $\pi < \theta < 2\pi$. then $\sin(\theta) = \sqrt{3}/2 \to \theta = \frac{4\pi}{3}, \frac{5\pi}{3}$.

   b. if $\sec(\theta) = \sqrt{2}$ then $\theta = \frac{\pi}{4}, \frac{7\pi}{4}$. but $\tan(\pi/4) = 1$, while $\tan(7\pi4) = -1$. hence $\theta = \frac{7\pi}{4}$.

In exercise 79, we use $f(x) = x^3 - x$, $g(x) = \frac{1}{1 - x}$, and $h(x) = \sin x$.

79. $g(h(x)) = g(\sin x) = \frac{1}{1 - \sin x}$.

81. $h(g(x)) = h(\frac{1}{1 - x}) = \sin(\frac{1}{1 - x})$

83. $g(f(h(x))) = g(f(\sin x)) = g(\sin^3 x - \sin x) = \frac{1}{1 - (\sin^3 x - \sin x)} = \frac{1}{1 - \sin^3 x + \sin x}$

85. let f and g be odd functions and let $h = fg$. since f and g are odd, we know that
$f(-x) = -f(x)$ and $g(-x) = -g(x)$.
$h(-x) = f(-x)g(-x) = [-f(x)][-g(x)] = f(x)g(x) = h(x)$. Hence, h is an even function.

87. The tangent line to the circle $(x - 6)^2 + (y - 4)^2 = 25$ goes through the point $(3, 8)$. Let m be the slope of the tangent line. Then the equation is $(y - 8) = m(x - 3)$ or

$$y = m(x - 3) + 8.$$

We substitute this value of y into the equation of the circle and solve for x:

$$(x - 6)^2 + [m(x - 3) + 4]^2 = 25$$
$$\Leftrightarrow \quad x^2 - 12x + 36 + m^2x^2 - 6m^2x + 9m^2 + 8mx - 24m + 16 = 25$$
$$\Leftrightarrow \quad (1 + m^2)x^2 + (-12 - 6m^2 + 8m)x + 9m^2 - 24m + 27 = 0$$

34

We recall that the solution to the quadratic equation $ax^2 + bx + c = 0$ is $x = \frac{-b \pm \sqrt{b^2 - 4ac}}{2a}$. The number of solutions for x depends on the value of $D = b^2 - 4ac$. If $D > 0$ there are 2 solutions, if $D = 0$, there is 1 solution, and if $D < 0$ there are no solutions. In our case, we are looking for a tangent line to a circle. There must be only one point of intersection and therefore we must have $D = 0$. In our case, $a = (1 + m^2)$, $b = -12 - 6m^2 + 8m$, and $c = 9m^2 - 24m + 27$. We now solve the equation $D = 0$:

$$(-12 - 6m^2 + 8m)^2 - 4(1 + m^2)(9m^2 - 24m + 27) = 0$$
$$\Leftrightarrow \quad (-6 - 3m^2 + 4m)^2 - (1 + m^2)(9m^2 - 24m + 27) = 0$$
$$\Leftrightarrow \quad 16m^2 - 48m - 24m^3 + 36 + 36m^2 + 9m^4 - 9m^4 + 24m^3 - 36m^2 + 24m - 27 = 0$$
$$\Leftrightarrow \quad 16m^2 - 24m + 9 = 0$$
$$\Leftrightarrow \quad (4m - 3) = 0$$
$$\Leftrightarrow \quad m = \tfrac{3}{4}$$

Hence the equation for the tangent line is

$$y = \frac{3}{4}(x - 3) + 8$$

89. $f(x) = 4 - x^2$, $-\infty < x < \infty$ and $g(x) = \sin x$, $0 \leq x \leq 2\pi$. Since the domain of f is unrestricted, the domain of $f \circ g$ is the same as the domain of g. Hence the domain of $f \circ g$ is $[0, 2\pi]$. Now, $f \circ g(x) = f(\sin x) = 4 - \sin^2 x$.

$$-1 \leq \sin x \leq 1 \quad \Rightarrow \quad 0 \leq \sin^2 x \leq 1 \quad \Rightarrow \quad 3 \leq 4 - \sin^2 x \leq 4.$$

Hence the range of $f \circ g$ is $[3, 4]$.

91. Let P, Q and R be the points $(4, 3)$, $(9, 13)$ and $(20, 15)$. We will consider the base, b, of the triangle to be the line segment, PR. We will compute the height, h, of the triangle by finding the line through Q which is perpendicular to PR. The slope of the line PR is $m = \frac{15 - 3}{20 - 4} = \frac{3}{4}$. The equation of the line PR is $y - 3 = \frac{3}{4}(x - 4)$ or

$$y = \frac{3}{4}x$$

The slope of the perpendicular line is $-\frac{1}{m} = -\frac{4}{3}$. This line goes through $Q = (9, 13)$ and its equation is $y - 13 = -\frac{4}{3}(x - 9)$ or

$$y = -\frac{4}{3}x + 25$$

To find the intersection of the two lines, sustitute $y = \frac{3}{4}x$ into the last equation:

$$\frac{3}{4}x = -\frac{4}{3}x + 25 \quad \Leftrightarrow \quad 9x = -16x + 12 \times 25 \quad \Leftrightarrow \quad x = 12$$

$y = \frac{3}{4}x = 9$. The height of the triangle is the distance from $(12, 9)$ to $Q = (9, 13)$. Hence, $h = \sqrt{(13 - 9)^2 + (9 - 12)^2} = \sqrt{25} = 5$. The base of the triangle is the distance from $P = (4, 3)$ to $R = (20, 15)$ which is $b = \sqrt{(15 - 3)^2 + (20 - 4)^2} = \sqrt{400} = 20$. Hence the area of the triangle is

$$\frac{1}{2}bh = \frac{1}{2} \times 20 \times 5 = 50$$

93.
$$R = \begin{cases} 500x & \text{if } x \leq 5 \\ (500 - 10(x - 5))x & \text{if } 5 < x \leq 25 \\ 300x & \text{if } x > 25 \end{cases}$$

Readiness Test

1. Completing squares,

$$x^2 - 2x + y^2 + 3y = 4$$

$$\Leftrightarrow \quad (x-1)^2 - 1 + \left(y + \frac{3}{2}\right)^2 - \frac{9}{4} = 4$$

$$\Leftrightarrow \quad (x-1)^2 + \left(y + \frac{3}{2}\right)^2 = \frac{29}{4}$$

Thus the center is $(1, -3/2)$ and the radius is $\sqrt{29}/2$.

3. The slope of the given line is $-\frac{3}{2}$. The point–slope form of the parallel line through $(-1, 2)$ is

$$y - 2 = -\frac{3}{2}(x + 1).$$

The slope–intercept form of this equation is

$$y = -\frac{3}{2}x + \frac{1}{2}.$$

5. $|x + 3| > 3 \Leftrightarrow -3 > x + 3$ or $x + 3 > 3$. This is equivalent to $x < -6$ or $x > 0$. The solution set is $(-\infty, -6) \cup (0, \infty)$.

7. Completing squares, $2x^2 - 4x + 3 = 2(x-1)^2 + 1$. Thus the vertex of the parabola is at $(1, 1)$.

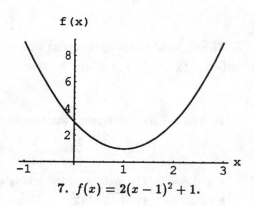

7. $f(x) = 2(x - 1)^2 + 1$.

9. $x^2 - 2x - 3 > 0 \Leftrightarrow (x-1)^2 > 4 \Leftrightarrow |x - 1| > 2$. Thus $\{x \mid x^2 - 2x - 3 > 0\} = (-infty, -1) \cup (3, \infty)$.

11. $(|x| - 3)^2 = 9 \Leftrightarrow |x| - 3 = \pm 3$. This is true if and only if $|x| = 3 \pm 3 \Leftrightarrow x = \pm 3 \pm 3 = -6,\ 0,\ 6$.

13. $\sin(\theta) < 0 \Leftrightarrow (2n-1)\pi < \theta < 2n\pi$ for some integer n. For such values of θ, $\cos(\theta) = \frac{\sqrt{3}}{2} \Leftrightarrow \theta = 2n\pi - \frac{\pi}{6}$.

15. The simplest way to do this is to let the opposite side have length 2, and let the hypotenuse have length 5. Then the adjacent side has length $\sqrt{21}$.

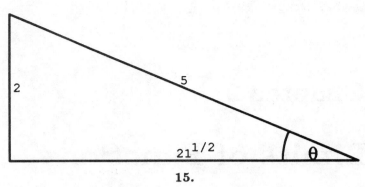

15.

17. The domain of $f(x) = \frac{1}{1 + \sin(x)}$ is $\{x \mid \sin(x) \neq -1\}$. This set is $\left\{x \mid x \neq \frac{3\pi}{2} + 2n\pi\right\}$.

19. The desired set is the perpendicular bisector of the line segment between $(-2, 1)$ and $(0, 0)$. **This line** segment has slope $-\frac{1}{2}$ and midpoint $(-1, 1/2)$. Thus the desired set is $\{(x, y) \mid y - 1/2 = 2(x + 1)\}$.

21. a. $x^4 - 1$ is a polynomial of degree 4.

     b. $\frac{x}{x^2 + 1}$ is a rational function which is not a polynomial.

23. The domain of $f(x) = (x - 2)^{-2/3}$ is $\{x \mid x \neq 2\}$. $f(2)$ is not defined due to the **negative exponent.**

25. $Y(x) = 400 - 2x$ pounds.

Chapter 2

Limits of Functions

2.1   Tangents, Areas, and Limits

1. Here $x_0 = 2$. Hence the slope of the tangent line is

$$m = \lim_{h \to 0} \frac{f(2+h) - f(2)}{h} = \lim_{h \to 0} \frac{[3(2+h) - 2] - 4}{h} = \lim_{h \to 0} 3 = 3.$$

3. Here $x_0 = 3$. Hence the slope of the tangent line is

$$m = \lim_{h \to 0} \frac{f(3+h) - f(3)}{h} = \lim_{h \to 0} \frac{2(3+h)^2 - 18}{h} = \lim_{h \to 0} \frac{12h + 2h^2}{h} = \lim_{h \to 0} (12 + 2h) = 12.$$

5. Here $x_0 = 1$. Hence the slope of the tangent line is

$$m = \lim_{h \to 0} \frac{f(1+h) - f(1)}{h} = \lim_{h \to 0} \frac{[2(1+h)^2 + 3] - 5}{h} = \lim_{h \to 0} \frac{4h + 2h^2}{h} = \lim_{h \to 0} (4 + 2h) = 4.$$

7. Here $x_0 = -2$. Hence the slope of the tangent line is

$$m = \lim_{h \to 0} \frac{f(-2+h) - f(-2)}{h} = \lim_{h \to 0} \frac{[3(-2+h)^2 + 4(-2+h) + 2] - 6}{h}$$

$$= \lim_{h \to 0} \frac{-8h + 3h^2}{h} = \lim_{h \to 0} (-8 + 3h) = -8.$$

9. Here $x_0 = 2$. Hence the slope of the tangent line is

$$m = \lim_{h \to 0} \frac{f(2+h) - f(2)}{h} = \lim_{h \to 0} \frac{[(2+h)^3 + 3] - 11}{h}$$

$$= \lim_{h \to 0} \frac{12h + 6h^2 + h^3}{h} = \lim_{h \to 0} (12 + 6h + h^2) = 12.$$

11. Here $x_0 = -2$. Hence the slope of the tangent line is

$$m = \lim_{h \to 0} \frac{f(-2+h) - f(-2)}{h} = \lim_{h \to 0} \frac{(-2+h)^4 - 16}{h}$$

$$= \lim_{h \to 0} \frac{-32h + 24h^2 - 8h^3 + h^4}{h} = \lim_{h \to 0} (-32 + 24h - 8h^2 = h^3) = -32.$$

13. Here $x_0 = 1$. Hence the slope of the tangent line is

$$m = \lim_{h \to 0} \frac{f(1+h) - f(1)}{h} = \lim_{h \to 0} \frac{[a(1+h)^3 + b(1+h)^2 + c(1+h) + d] - [a+b+c+d]}{h}$$

$$= \lim_{h \to 0} \frac{3ah + 3ah^2 + ah^3 + 2bh + bh^2 + ch}{h} = \lim_{h \to 0} (3a + 3ah + ah^2 + 2b + bh + c) = 3a + 2b + c.$$

15. Here $x_0 = -2$. Hence the slope of the tangent line is

$$m = \lim_{h \to 0} \frac{f(-2+h) - f(-2)}{h} = \lim_{h \to 0} \frac{\frac{1}{h+1} - 1}{h} = \lim_{h \to 0} \frac{1 - (h+1)}{h(h+1)} = \lim_{h \to 0} \frac{-h}{h(h+1)} = \lim_{h \to 0} \frac{-1}{h+1} = -1.$$

17. Here $x_0 = 2$. Hence the slope of the tangent line is

$$m = \lim_{h \to 0} \frac{f(2+h) - f(2)}{h} = \lim_{h \to 0} \frac{\frac{4}{(2+h)^2} - 1}{h} = \lim_{h \to 0} \frac{4 - (2+h)^2}{h(2+h)^2}$$

$$= \lim_{h \to 0} \frac{-4h + h^2}{h(2+h)^2} = \lim_{h \to 0} \frac{-4 + h}{(2+h)^2} = \frac{-4}{4} = -1.$$

19. Here $x_0 = -1$. Therefore the slope of the tangent line is

$$m = \lim_{h \to 0} \frac{f(-1+h) - f(-1)}{h} = \lim_{h \to 0} \frac{3(-1+h)^2 - 3}{h} = \lim_{h \to 0} \frac{-6h + 3h^2}{h} = \lim_{h \to 0} (-6 + 3h) = -6.$$

Hence the equation of the tangent line at $(-1, 3)$ is $y - 3 = -6(x + 1)$, or $y = -6x - 3$.

21. Here $x_0 = 1$. Therefore the slope of the tangent line is

$$m = \lim_{h \to 0} \frac{f(1+h) - f(1)}{h} = \lim_{h \to 0} \frac{a(1+h)^2 - a}{h} = \lim_{h \to 0} \frac{2ah + ah^2}{h} = \lim_{h \to 0} (2a + ah) = 2a.$$

Hence the equation of the tangent line at $(1, a)$ is $y - a = 2a(x - 1)$, or $y = 2ax - a$.

23. Here $x_0 = 1$. Therefore the slope of the tangent line is

$$m = \lim_{h \to 0} \frac{f(1+h) - f(1)}{h} = \lim_{h \to 0} \frac{[2(1+h)^3 + (1+h)] - 3}{h}$$

$$= \lim_{h \to 0} \frac{7h + 6h^2 + 2h^3}{h} = \lim_{h \to 0} (7 + 6h + 2h^2) = 7.$$

Since $f(1) = 3$, the equation of the tangent line at $(1, 3)$ is $y - 3 = 7(x - 1)$, or $y = 7x - 4$.

25. False. The tangent line is horizontal at a point x_0 if the slope $m = 0$. For example, if $f(x) = x^2$, then the slope of the tangent line at the point $(0, 0)$ is

$$m = \lim_{h \to 0} \frac{f(h) - f(0)}{h} = \lim_{h \to 0} \frac{h^2}{h} = \lim_{h \to 0} h = 0,$$

and hence the equation of the tangent line is $y = 0$, which is horizontal.

27. (a) In general, we find that

$$\frac{f(2+h) - f(2)}{h} = \frac{[(2+h)^3 - 3(2+h)] - 2}{h} = \frac{9h + 6h^2 + h^3}{h} = 9 + 6h + h^2.$$

For example, if $h = 0.1000$,

$$\frac{f(2+h) - f(2)}{h} = 9 + 6(0.1) + (0.1)^2 = 9.61.$$

Table 1.3 summarizes the slopes of the secant lines for the indicated values of h.

(b) The slope of the tangent line at $x_0 = 2$ is

$$m = \lim_{h \to 0} \frac{f(2+h) - f(2)}{h} = \lim_{h \to 0}(9 + 6h + h^2) = 9.$$

The values in Table 1.3 show that the slopes of the secant lines become closer to the limiting value as $h \to 0$.

Table 1.3

x_0	h	$\dfrac{f(x_0 + h) - f(x_0)}{h}$
2	0.1000	9.61
2	0.0100	9.0601
2	0.0010	9.006001
2	0.0001	9.00060001
2	−0.0001	8.99940001
2	−0.0010	8.994001
2	−0.0100	8.9401
2	−0.1000	8.41

Exercise 27

Exercise 29

29. (a) Estimating the slopes at $x = -3/2$, -1, and $1/2$, we obtain

$$m_{-3/2} = -7/(-4) = 7/4, \quad m_{-1} = -1, \quad m_{1/2} = 1/(-5) = -1/5.$$

(b) In general, the slope of the tangent line at $x = x_0$ is

$$m = \lim_{h \to 0} \frac{f(x_0 + h) - f(x_0)}{h} = \lim_{h \to 0} \frac{[(x_0 + h)^3 + (x_0 + h)^2 - 2(x_0 + h)] - [x_0^3 + x_0^2 - 2x_0]}{h}$$

$$= \lim_{h \to 0} \frac{3x_0^2 h + 3x_0 h^2 + h^3 + 2x_0 h + h^2 - 2h}{h} = \lim_{h \to 0}(3x_0^2 + 3x_0 h + h^2 + 2x_0 + h - 2) = 3x_0^2 + 2x_0 - 2.$$

Hence the slopes of the tangent lines at $x_0 = -3/2$, -1, and $1/2$ are:

$$x_0 = -3/2 \implies m = 3(-3/2)^2 + 2(-3/2) - 2 = 7/4$$
$$x_0 = -1 \implies m = 3(-1)^2 + 2(-1) - 2 = -1$$
$$x_0 = 1/2 \implies m = 3(1/2)^2 + 2(1/2) - 2 = -1/4.$$

31. (a) In general, we find that

$$\frac{f(h) - f(0)}{h} = \frac{\sin h - \sin 0}{h} = \frac{\sin h}{h}.$$

For example, if $h = 0.1000$,

$$\frac{f(h) - f(0)}{h} = \frac{\sin 0.1}{0.1} = 0.99833417.$$

Table 1.3, below, summarizes the slopes of the secant lines for the indicated values of h and shows that the slope of the tangent line at $x_0 = 0$ is approximately 1.

(b) The slope of the tangent line at $x_0 = 0$ is

$$m = \lim_{h \to 0} \frac{f(h) - f(0)}{h} = \lim_{h \to 0} \frac{\sin h}{h}.$$

At present, we have no method for evauating this expression.

Table 1.3

x_0	h	$\dfrac{f(x_0 + h) - f(x_0)}{h}$
0	0.1000	0.99833417
0	0.0100	0.99998333
0	0.0010	0.99999983
0	0.0001	1
0	−0.0001	1
0	−0.0010	0.99999983
0	−0.0100	0.99998333
0	−0.1000	0.99833417

Exercise 31

Table 1.3

x_0	h	$\dfrac{f(x_0 + h) - f(x_0)}{h}$
0	0.1000	1
0	0.0100	1
0	0.0010	1
0	0.0001	1
0	−0.0001	−1
0	−0.0010	−1
0	−0.0100	−1
0	−0.1000	−1

Exercise 39

33. The slope of the tangent line at $x = a$ is

$$m = \lim_{h \to 0} \frac{f(a + h) - f(a)}{h} = \lim_{h \to 0} \frac{[(a + h)^2 + 6(a + h) + 1] - [a^2 + 6a + 1]}{h}$$

$$= \lim_{h \to 0} \frac{2ah + h^2 + 6h}{h} = \lim_{h \to 0} (2a + h + 6) = 2a + 6.$$

In particular, the slope of the tangent line is $0 \iff m = 2a + 6 = 0 \iff a = -3$.

35. In general, the slope of the tangent line at $x = a$ is

$$m = \frac{f(a + h) - f(a)}{h} = \lim_{h \to 0} \frac{[(a + h)^2 - 3(a + h) + 1] - [a^2 - 3a + 1]}{h}$$

$$= \lim_{h \to 0} \frac{2ah + h^2 - 3h}{h} = \lim_{h \to 0} (2a + h - 3) = 2a - 3.$$

Hence, the slope of the tangent line at the point $(a, f(a))$ is equal to the y-coordinate $\iff m = 2a - 3$ $= f(a) = a^2 - 3a + 1 \iff a^2 - 5a + 4 = (a - 4)(a - 1) = 0 \iff a = 1, 4$. Thus, there are two points at which the slope of the tangent line is equal to the y-coordinate: $(1, f(1)) = (1, -1)$ and $(4, f(4)) = (4, 5)$.

37. y-intercept $(0,5) \Longrightarrow f(0) = c = 5$; graph contains the point $(1,2) \Longrightarrow f(1) = a+b+c = 2$; slope of tangent line at $x = 2$ is $3 \Longrightarrow 2a(2) + b = 3$. Hence $a + b = -3$, $4a + b = 3$ and therefore $a = 2$, $b = -5$, $c = 5$.

39. In general, we find that

$$\frac{f(0+h) - f(0)}{h} = \frac{|h|}{h} = \begin{cases} +1, & \text{if } h > 0 \\ -1, & \text{if } h < 0. \end{cases}$$

Table 1.3, above, summarizes the slopes of the secant lines for the indicated values of h and shows that the absolute value function does not have a tangent line at the point $(0,0)$.

41. (a) The table below summarizes the values of h, m_h^+, and m_h^- for $h = 1$, 0.1, 0.01, 0.001, and 0.0001.

h	m_h^+	m_h^-
1	5	3
0.1	4.1	3.9
0.01	4.01	3.99
0.001	4.001	3.999
0.0001	4.0001	3.9999

(b) The graphs of $y = f(x)$, $y = m_h^+(x-3) + 4$, and $y = m_h^-(x-3) + 4$ with range $[-1,6] \times [0,10]$ and **Xscl** = **Yscl** = 1 are shown above on the right.

(c) Slope of tangent line at $a = 3$ is 4.

43. (a) The table below summarizes the values of h, m_h^+, and m_h^- for $h = 1$, 0.1, 0.01, 0.001, and 0.0001.

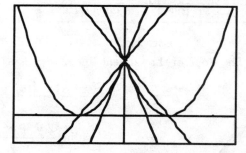

h	m_h^+	m_h^-
1	-1	1
0.1	-1.9	1.9
0.01	-1.99	1.99
0.001	-1.999	1.999
0.0001	-1.9999	1.9999

(b) The graphs of $y = f(x)$, $y = m_h^+ x + 1$, and $y = m_h^- x + 1$ with range $[-2.5, 2.5] \times [-1/2, 2]$ and **Xscl** = **Yscl** = 1 are shown above on the right.

(c) The tangent line at $a = 0$ does not exist.

2.2   Limits of Functions

1. $\lim_{x \to a} f(x)$ exists and equals $f(a)$.

3. $\lim_{x \to a} f(x)$ does not exist.

5. $\lim_{x \to a} f(x)$ exists but does not equal $f(a)$.

7. $\lim_{x \to 0} f(x) = 2$

9. $\lim_{x \to 0} f(x) = 1 = 0$

11. $\lim_{x \to 2} (3 + 7x) = 17$

13. $\lim_{x \to 0} \dfrac{(3 + x)^2 - 9}{x} = \lim_{x \to 0} \dfrac{x(x + 6)}{x} = \lim_{x \to 0} (x + 6) = 6$

15. $\lim_{h \to 0} \dfrac{h^2 - 1}{h - 1} = \dfrac{-1}{-1} = 1$

17. $\lim_{x \to -2} \dfrac{x^2 - x - 6}{x + 2} = \lim_{x \to -2} \dfrac{(x - 3)(x + 2)}{x + 2} = \lim_{x \to -2} (x - 3) = -5$

19. $\lim_{x \to 0} \dfrac{1 - \cos^2 x}{\sin x \cos x} = \lim_{x \to 0} \dfrac{\sin^2 x}{\sin x \cos x} = \lim_{x \to 0} \dfrac{\sin x}{\cos x} = \dfrac{0}{1} = 0$

21. $\lim_{h \to 0} \dfrac{\sin 2h}{\sin h} = \lim_{h \to 0} \dfrac{2 \sin h \cos h}{\sin h} = \lim_{h \to 0} (2 \cos h) = 2$

23. $\lim_{x \to \pi/2} \sin 2x \csc x = \lim_{x \to \pi/2} \dfrac{\sin 2x}{\sin x} = \dfrac{0}{1} = 0$

25. $\lim_{x \to -1} \dfrac{x^2 - 2x - 3}{x + 1} = \lim_{x \to -1} \dfrac{(x - 3)(x + 1)}{x + 1} = \lim_{x \to -1} (x - 3) = -4$

27. $\lim_{x \to \pi/2} \dfrac{\sec x \cos x}{x} = \lim_{x \to \pi/2} \dfrac{1}{x} = \dfrac{2}{\pi}$

29. $\lim_{x \to -1} \dfrac{x^3 + 3x^2 - x - 3}{x^2 - 1} = \lim_{x \to -1} \dfrac{x^2(x + 3) - (x + 3)}{x^2 - 1} = \lim_{x \to -1} \dfrac{(x + 3)(x^2 - 1)}{x^2 - 1} = \lim_{x \to -1} (x + 3) = 2$

31. $\lim_{x \to -3} \dfrac{x^3 - 7x + 6}{x^2 + 2x - 3} = \lim_{x \to -3} \dfrac{(x - 1)(x + 3)(x - 2)}{(x + 3)(x - 1)} = \lim_{x \to -3} (x - 2) = -5$

33. $\lim_{x \to 2} \dfrac{x^{13/4} - 2x^{9/4}}{x^{5/4} - 2x^{1/4}} = \lim_{x \to 2} \dfrac{x^{9/4}(x - 2)}{x^{1/4}(x - 2)} = \lim_{x \to 2} x^2 = 4$

35. $\lim_{h \to 0} \dfrac{3 - \sqrt{9 + h}}{h} = \lim_{h \to 0} \dfrac{3 - \sqrt{9 + h}}{h} \cdot \dfrac{3 + \sqrt{9 + h}}{3 + \sqrt{9 + h}} = \lim_{h \to 0} \dfrac{-h}{h(3 + \sqrt{9 + h})} = \lim_{h \to 0} \dfrac{-1}{3 + \sqrt{9 + h}} = -\dfrac{1}{6}$

37. $\lim_{x \to 1} \dfrac{(1/x) - 1}{x - 1} = \lim_{x \to 1} \dfrac{1 - x}{x(x - 1)} = \lim_{x \to 1} \dfrac{-1}{x} = -1$

39. $\lim_{x \to 0} f(x) = 2$

41. $\lim_{x \to 0} f(x)$ does not exist since, if x is close to zero and positive, $f(x)$ is close to 8, but if x is close to zero and negative, $f(x)$ is close to 4.

43

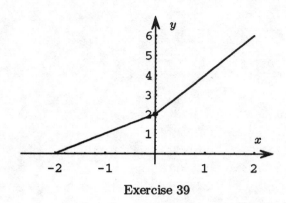

Exercise 39

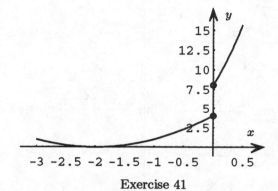

Exercise 41

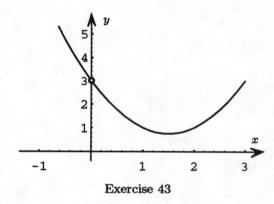

Exercise 43

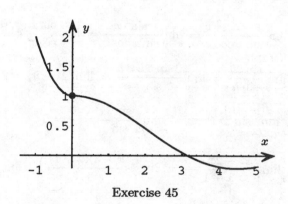

Exercise 45

43. First observe that if $x \neq 0$, $\dfrac{(x-1)^3 + 1}{x} = \dfrac{x^3 - 3x^2 + 3x}{x} = x^2 - 3x + 3$. Hence $\lim\limits_{x \to 0} f(x) = 3$.

45. $\lim\limits_{x \to 0} f(x) = 1$

47. $\lim\limits_{x \to 0} \dfrac{x^2}{1 - \cos x} = 2$; see the table of values on the following page.

49. $\lim\limits_{x \to 0} \dfrac{x - \sin x}{x^3} = \dfrac{1}{6}$; see the table of values on the following page.

51. $\lim\limits_{x \to 0} \dfrac{1 - \cos x^2}{x^4} = \dfrac{1}{2}$; see the table of values on the following page.

53. $c = \pm\sqrt{2}/2$; for, if x is close to zero and negative, $f(x)$ is close to $1/2$, while if x is close to zero and is positive, $f(x)$ is close to c^2. Hence, in order for $\lim_{x \to 0} f(x)$ to exist, $c^2 = 1/2$, or $c = \pm\sqrt{1/2} = \pm\sqrt{2}/2$.

55. False; the function whose graph appears in Exercise 10 has the property that $f(0)$ does not exist but $\lim_{x \to 0} f(x) = 0$.

x	$\dfrac{x^2}{1-\cos x}$	$\dfrac{x-\sin x}{x^3}$	$\dfrac{1-\cos x^2}{x^4}$
1.000	2.17534265	0.15852902	0.45969769
0.500	2.04219271	0.16459569	0.49740125
0.100	2.00166750	0.16658335	0.49999583
0.050	2.00041672	0.16664583	0.49999974
0.010	-1.9999333	0.16666583	0.50000000
0.005	2.00000417	0.16666646	0.50000004
-0.005	2.00000417	0.16666646	0.50000004
-0.010	2.00001667	0.16666583	0.50000000
-0.050	2.00041672	0.16664583	0.49999974
-0.100	2.00166750	0.16658335	0.49999583
-0.500	2.04219271	0.16459569	0.49740125
-1.000	2.17534265	0.15852902	0.45969769

57. Graph with the range: $[-6,-2] \times [-15, 5]$, **Xscl** $= 1$, **Yscl** $= 1$. $\lim_{x \to -4} f(x) = -11$.

h	1	0.1	0.01	0.001	0.0001	0.00001
$f(a+h)$	-9	-10.8	-10.98	-10.998	-10.9998	-10.99998
$f(a-h)$	-13	-11.2	-11.02	-11.002	-11.0002	-11.00002

59. Graph with the range: $[-1.5, 1.5] \times [0, 1/2]$, **Xscl** $= 0.5$, **Yscl** $= 0.25$. $\lim_{x \to 0} f(x) = 1/3$.

h	1	0.1	0.01	0.001	0.0001	0.00001
$f(a+h)$	0.280490	0.332778	0.333328	0.333333	0.333333	0.333333
$f(a-h)$	0.280490	0.332778	0.333328	0.333333	0.333333	0.333333

61. $3/2$.

63. 16.4852814.

2.3   The Formal Definition of Limit

1. (a) $|(2x+5)-9| < \varepsilon \iff |2x-4| < \varepsilon \iff |x-4| < \varepsilon/2$. Thus, in general, set $\delta = \varepsilon/2$, although any smaller value will do.

   (b) If $\varepsilon = 2$, set $\delta = \varepsilon/2 = 1$
   If $\varepsilon = 0.4$, set $\delta = \varepsilon/2 = 0.2$
   If $\varepsilon = 0.05$, set $\delta = \varepsilon/2 = 0.025$

3. (a) $|(x^2+3)-3| < \varepsilon \iff |x|^2 < \varepsilon \iff |x| < \sqrt{\varepsilon}$. Thus, in general, set $\delta = \sqrt{\varepsilon}$, although any smaller value will do.

   (b) If $\varepsilon = 2$, set $\delta = \sqrt{2}$
   If $\varepsilon = 1$, set $\delta = 1$
   If $\varepsilon = 0.3$, set $\delta = \sqrt{0.3}$

5. First observe that if $x \neq 1$,

$$\frac{x^2-1}{x-1} = \frac{(x-1)(x+1)}{x-1} = x+1.$$

45

Now, $|(x+1)-2| < \varepsilon \iff |x-1| < \varepsilon$. Hence, for $x \ne 1$,

$$0 < |x-1| < \varepsilon \implies \left|\frac{x^2-1}{x-1} - 2\right| = |x-1| < \varepsilon.$$

Thus, in general, let $\delta = \varepsilon$, although any smaller value will do. In particular, if $\varepsilon = 2$, set $\delta = 2$; if $\varepsilon = 0.8$, set $\delta = 0.8$; and if $\varepsilon = 0.05$, set $\delta = 0.05$.

7. Let $\varepsilon > 0$ be given. Then

$$|(x+3)-6| < \varepsilon \iff |x-3| < \varepsilon.$$

Thus, if we set $\delta = \varepsilon$, it follows that

$$0 < |x-3| < \delta \implies |(x+3)-6| < \varepsilon$$

and hence $\lim_{x\to 3}(x+3) = 6$.

9. Let $\varepsilon > 0$ be given. Then

$$|(7-3x)+5| = |-3x+12| = 3|x-4| < \varepsilon \iff |x-4| < \varepsilon/3.$$

Thus, if we set $\delta = \varepsilon/3$, it follows that

$$0 < |x-4| < \delta \implies |(7-3x)+5| < \varepsilon$$

and hence $\lim_{x\to 4}(7-3x) = -5$.

11. Let ε be given. Then, if $x \ne 3$,

$$\left|\frac{x^2-9}{x-3} - 6\right| = \left|\frac{(x-3)(x+3)}{x-3} - 6\right| = |x-3|.$$

Thus, if we set $\delta = \varepsilon$, it follows that

$$0 < |x-3| < \delta \implies \left|\frac{x^2-9}{x-3} - 6\right| < \varepsilon$$

and hence $\lim_{x\to 3}\frac{x^2-9}{x-3} = 6$.

13. Let $\varepsilon > 0$ be given. Then, if x is sufficiently close to 3, say within one unit, $|x+3| < 7$ and hence

$$|x^2-9| = |(x-3)(x+3)| = |x-3|\,|x+3| < 7|x-3| < \varepsilon \iff |x-3| < \varepsilon/7.$$

Thus, if we set δ equal to the smaller of the numbers 1 and $\varepsilon/7$, then

$$0 < |x-3| < \delta \implies |x^2-9| < 7(\varepsilon/7) = \varepsilon$$

and hence $\lim_{x\to 3} x^2 = 9$.

15. Let $\varepsilon > 0$ be given. Then

$$|(x^2-2x+4)-3| = |x^2-2x+1| = |x-1|^2 < \varepsilon \iff |x-1| < \sqrt{\varepsilon}.$$

Thus, if we set $\delta = \sqrt{\varepsilon}$, it follows that

$$0 < |x-1| < \delta \implies |(x^2-2x+4)-3| < \varepsilon$$

and hence $\lim_{x\to 1}(x^2-2x+4) = 3$.

46

17. Let $\varepsilon > 0$ be given. Then, if x is sufficiently close to 4, say within one unit, $|\sqrt{x} + 2| > \sqrt{3} + 2$ and hence

$$\left|\sqrt{x} - 2\right| = \left|\frac{(\sqrt{x} - 2)(\sqrt{x} + 2)}{\sqrt{x} + 2}\right| = \frac{|x - 4|}{|\sqrt{x} + 2|} < \frac{|x - 4|}{\sqrt{3} + 2} < \varepsilon \iff |x - 4| < \varepsilon(\sqrt{3} + 2).$$

Thus, if we set δ equal to the smaller of the numbers 1 and $\varepsilon(\sqrt{3} + 2)$, it follows that

$$0 < |x - 4| < \delta \implies \left|\sqrt{x} - 2\right| < \frac{|x - 4|}{\sqrt{3} + 2} < \frac{\varepsilon(\sqrt{3} + 2)}{\sqrt{3} + 2} = \varepsilon$$

and hence $\lim\limits_{x \to 4} \sqrt{x} = 2$.

19. Let $\varepsilon > 0$ be given. Then, if x is sufficiently close to 2, say within one unit, $|x| < 3$ and hence

$$\left|(x^2 - 2x + 2) - 2\right| = |x|\,|x - 2| < 3|x - 2| < \varepsilon \iff |x - 2| < \varepsilon/3.$$

Thus, if we let δ be the smaller of the numbers 1 and $\varepsilon/3$, it follows that

$$0 < |x - 2| < \delta \implies \left|(x^2 - 2x + 2) - 2\right| = |x|\,|x - 2| < 3(\varepsilon/3) = \varepsilon$$

and hence $\lim\limits_{x \to 2}(x^2 - 2x + 2) = 2$.

21. Let $\varepsilon > 0$ and set $\delta = \varepsilon/2$. Using the inequality $|\sin x| \le |x|$ from Example 5, it follows that

$$0 < |x| < \delta \implies |\sin 2x| \le |2x| = 2|x| < 2(\varepsilon/2) = \varepsilon$$

and hence $\lim\limits_{x \to 0} \sin 2x = 0$.

23. Let $\varepsilon > 0$ be given and set $\delta = \sqrt{\varepsilon}$. Then, since $|\sin x| \le |x|$,

$$0 < |x| < \delta \implies |x \sin x - 0| = |x|\,|\sin x| < |x|^2 < \left(\sqrt{\varepsilon}\right)^2 = \varepsilon$$

and hence $\lim\limits_{x \to 0} x \sin x = 0$.

25. Let $\varepsilon > 0$ be given and let $\delta = \varepsilon/3$. Then

$$-\delta < x < 0 \implies |f(x) - 2| = 2|x| < 2\delta < 2(\varepsilon/2) < \varepsilon$$
$$0 < x < \delta \implies |f(x) - 2| = 3|x| < 3\delta = \varepsilon.$$

Therefore $0 < |x| < \delta \implies |f(x) - 2| < \varepsilon$ and hence $\lim\limits_{x \to 0} f(x) = 2$.

Remark: In this case there are two choices for δ, namely $\varepsilon/2$ and $\varepsilon/3$, depending on whether x is positive or negative. We choose the smaller of the two values, $\varepsilon/3$, to insure that it works for both cases.

27. $\lim_{x \to 1/2}(2x^2 - x - 1) = 1$.

ε	0.5	0.25	0.1	0.05
δ	0.25	0.125	0.0625	0.3125

29. $\lim_{x \to 1/10}(1/x) = 10$.

ε	0.5	0.25	0.1	0.05
δ	.003906	0.001953	0.000977	0.000488

31. Yes, it is possible to find a δ that works for all values of a. At $a = 0$ a value of $\delta = 0.0078$ works. If you now draw the graph of $f(x) = \sin x$ using the range $[a - \delta, a + \delta] \times [f(a) - 0.01, f(a) + 0.01]$ with a any real number, you will see that $f(x)$ stays between the lines $y = f(x) \pm 0.01$. It is not possible to do this with $f(x) = 1/x$. For example, if $a = 1$ then $\delta = 0.0078$ will work, but this value of δ will not work at $a = 0.5$. To see this, draw the graph of $f(x) = 1/x$ on the range $[0.4922, 0.5078] \times [1.99, 2.01]$ and observe that the graph of $f(x)$ is not contained in this graphing window.

2.4   Properties of Limits

1. $\lim\limits_{x \to 3}(3x - 7) = 3 \lim\limits_{x \to 3} x - \lim\limits_{x \to 3} 7 = 3(3) - 7 = 2$

3. $\lim\limits_{x \to 2} \dfrac{x^2 + 3x}{x - 3} = \dfrac{\lim\limits_{x \to 2}(x^2 + 3x)}{\lim\limits_{x \to 2}(x - 3)} = \dfrac{\lim\limits_{x \to 2} x^2 + 3 \lim\limits_{x \to 2} x}{\lim\limits_{x \to 2} x - \lim\limits_{x \to 2} 3} = \dfrac{2^2 + 3(2)}{2 - 3} = -10$

5. $\lim\limits_{x \to 4} \sqrt{x}\,(1 - x^2) = \left(\lim\limits_{x \to 4} \sqrt{x}\right)\left(\lim\limits_{x \to 4}(1 - x^2)\right) = \sqrt{4}\,(1 - 4^2) = -30$

7. $\lim\limits_{x \to 1}\left(3x^9 - \sqrt[3]{x}\right) = 3 \lim\limits_{x \to 1} x^9 - \lim\limits_{x \to 1} \sqrt[3]{x} = 3(1)^9 - \sqrt[3]{1} = 2$

9. $\lim\limits_{x \to 3} \dfrac{x^2 - 2x - 3}{x - 3} = \lim\limits_{x \to 3} \dfrac{(x - 3)(x + 1)}{x - 3} = \lim\limits_{x \to 3}(x + 1) = 4$

11. $\lim\limits_{x \to 4} \dfrac{x^{3/2} + 2\sqrt{x}}{x^{5/2} + \sqrt{x}} = \dfrac{\lim\limits_{x \to 4} x^{3/2} + 2 \lim\limits_{x \to 4} \sqrt{x}}{\lim\limits_{x \to 4} x^{5/2} + \lim\limits_{x \to 4} \sqrt{x}} = \dfrac{4^{3/2} + 2\sqrt{4}}{4^{5/2} + \sqrt{4}} = \dfrac{6}{17}$

13. $\lim\limits_{x \to 0} \dfrac{\tan x}{\sin 2x} = \lim\limits_{x \to 0} \dfrac{\sin x}{\cos x} \dfrac{1}{2 \sin x \cos x} = \dfrac{1}{2} \lim\limits_{x \to 0} \dfrac{1}{\cos^2 x} = \dfrac{1}{2}$

15. $\lim\limits_{x \to -8} \dfrac{x^{2/3} - x}{x^{5/3}} = \dfrac{(-8)^{2/3} - (-8)}{(-8)^{5/3}} = -\dfrac{3}{8}$

17. $\lim\limits_{x \to 4} \dfrac{x - 4}{\sqrt{x} - 2} = \lim\limits_{x \to 4} \dfrac{(\sqrt{x} - 2)(\sqrt{x} + 2)}{\sqrt{x} - 2} = \lim\limits_{x \to 4}(\sqrt{x} + 2) = \sqrt{4} + 2 = 4$

19. $\lim\limits_{x \to -1}\left(x^{7/3} - 2x^{2/3}\right)^2 = \left[(-1)^{7/3} - 2(-1)^{2/3}\right]^2 = 9$

21. $\lim\limits_{x \to a} 3 \cdot f(x) \cdot g(x) = \left(\lim\limits_{x \to a} 3\right)\left(\lim\limits_{x \to a} f(x)\right)\left(\lim\limits_{x \to a} g(x)\right) = 3(2)(-3) = -18$

23. $\lim\limits_{x \to a} \dfrac{6f(x) - 4[g(x)]^2}{g(x) - 4f(x)} = \dfrac{6 \lim\limits_{x \to a} f(x) - 4\left[\lim\limits_{x \to a} g(x)\right]^2}{\lim\limits_{x \to a} g(x) - 4 \lim\limits_{x \to a} f(x)} = \dfrac{6(2) - 4(-3)^2}{(-3) - 4(2)} = \dfrac{24}{11}$

25. $\lim\limits_{x \to 0} \dfrac{\sin x}{2x} = \dfrac{1}{2} \lim\limits_{x \to 0} \dfrac{\sin x}{x} = \dfrac{1}{2}(1) = \dfrac{1}{2}$

27. $\lim\limits_{x \to 0} \dfrac{\tan x}{4x} = \dfrac{1}{4} \lim\limits_{x \to 0}\left(\dfrac{\sin x}{x} \dfrac{1}{\cos x}\right) = \dfrac{1}{4}(1)(1) = \dfrac{1}{4}$

29. $\lim\limits_{x \to 0} x^2 \cot x = \lim\limits_{x \to 0}\left(\dfrac{x}{\sin x}\right)(x \cos x) = (1)(0)(1) = 0$

48

31. $\lim\limits_{x\to 0}\dfrac{\sin x}{\sqrt[5]{x}} = \lim\limits_{x\to 0}\left(\dfrac{\sin x}{x}\right)x^{4/5} = (1)(0) = 0$

33. $\lim\limits_{x\to 0} f(x) = 1$ since $1 = \lim\limits_{x\to 0}(1 - x^2) \le \lim\limits_{x\to 0} f(x) \le \lim\limits_{x\to 0}(1 + x^2) = 1$

35. $\lim\limits_{x\to 0} f(x) = 1$ since $1 = \lim\limits_{x\to 0}(1 - x^4) \le \lim\limits_{x\to 0} f(x) \le \lim\limits_{x\to 0}\sec x = \lim\limits_{x\to 0}\dfrac{1}{\cos x} = 1$

37. $\lim\limits_{x\to a}[f(x)g(x)h(x)] = \left(\lim\limits_{x\to a}[f(x)g(x)]\right)\left(\lim\limits_{x\to a} h(x)\right) = \left(\lim\limits_{x\to a} f(x)\right)\left(\lim\limits_{x\to a} g(x)\right)\left(\lim\limits_{x\to a} h(x)\right)$

39. (a) If $n = 1$, $\lim\limits_{x\to a}[f(x)]^n = \lim\limits_{x\to a} f(x) = L = L^n$.

    (b) Let $m > 1$ and assume that $\lim\limits_{x\to a}[f(x)]^m = L^m$. Then

$$\begin{aligned}
\lim_{x\to a}[f(x)]^{m+1} &= \lim_{x\to a}[f(x)]^m f(x) \\[2mm]
&= \left(\lim_{x\to a}[f(x)]^m\right)\left(\lim_{x\to a} f(x)\right) \qquad \text{since the limit of a product} \\
&\hphantom{=\left(\lim_{x\to a}[f(x)]^m\right)\left(\lim_{x\to a} f(x)\right)} \text{is the product of the limits} \\[2mm]
&= (L^m)(L) \qquad\qquad\qquad\qquad \text{by the induction hypothesis} \\[2mm]
&= L^{m+1}.
\end{aligned}$$

    (c) It follows by mathematical induction that $\lim\limits_{x\to a}[f(x)]^n = L^n$ for all positive integers n.

41. Let $f(x) = x^2$, $a = 1$, $b = -1$. Then $\lim\limits_{x\to 1} x^2 = \lim\limits_{x\to -1} x^2 = 1$.

43. Let $f(x) = |x|/x$ and $g(x) = -|x|/x$, $x \ne 0$. Then $f(x)g(x) = -1$ for all $x \ne 0$ and hence $\lim\limits_{x\to 0} f(x)g(x)$
    $= -1$. But neither $\lim\limits_{x\to 0} f(x)$ nor $\lim\limits_{x\to 0} g(x)$ exist.

45. $\lim\limits_{x\to a}(x^2 - 2x - 5) = a^2 - 2a - 5 = 10 \implies a^2 - 2a - 15 = (a - 5)(a + 3) = 0 \implies a = -3, 5$.

47. Let a be any number and let $\varepsilon > 0$. Set $\delta = \varepsilon$. Then

$$0 < |x - a| < \delta \quad \implies \quad |x - a| < \varepsilon$$

and hence $\lim\limits_{x\to a} x = a$.

49. $\lim\limits_{x\to 0}\dfrac{1 - \cos x}{x^{1/3}} = \lim\limits_{x\to 0}\left(\dfrac{1 - \cos x}{x} \cdot x^{2/3}\right) = \left(\lim\limits_{x\to 0}\dfrac{1 - \cos x}{x}\right)\left(\lim\limits_{x\to 0} x^{2/3}\right) = 0 \cdot 0 = 0$.

51. Let $-\pi/2 < x < 0$. Then $0 < -x < \pi/2$ and hence, by inequality (8),

$$\frac{1}{\cos(-x)} \ge \frac{\sin(-x)}{-x} \ge \cos(-x).$$

But $\sin(-x) = -\sin x$ and $\cos(-x) = \cos x$. Therefore

$$\frac{1}{\cos x} \ge \frac{-\sin x}{-x} = \frac{\sin x}{x} \ge \cos x.$$

Hence inequality (8) is true for $-\pi/2 < x < 0$.

53. The graph of these functions with the range $[-1,1] \times [0,2]$ is shown below. The graphs show that $\lim_{x \to 0} (\sin x / x) = 1$.

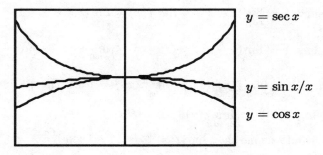

$y = \sec x$

$y = \sin x / x$

$y = \cos x$

2.5 One-Sided Limits

1. (a) $\displaystyle\lim_{x \to 1^-} f(x) = 2$

   (b) $\displaystyle\lim_{x \to 1} f(x) = 2$

   (c) $\displaystyle\lim_{x \to 0} f(x) = 0$

3. (a) $\displaystyle\lim_{x \to -2} f(x) = 0$

   (b) $\displaystyle\lim_{x \to 2^-} f(x) = 2$

   (c) $\displaystyle\lim_{x \to 2^+} f(x) = -1$

5. (a) All numbers except $a = 1$.

   (b) All numbers.

   (c) All numbers except $a = 1$.

   (d) All numbers except $a = 1$.

7. $\displaystyle\lim_{x \to 2^+} \sqrt{x - 2} = 0$

9. $\displaystyle\lim_{x \to 0^+} (x^{5/2} - 5x^{3/2}) = 0$

11. $\displaystyle\lim_{x \to 2^-} \frac{x^2 - 4x + 4}{x - 2} = \lim_{x \to 2^-} \frac{(x - 2)^2}{x - 2} = \lim_{x \to 2^-} (x - 2) = 0$

13. $\displaystyle\lim_{x \to 0^+} \frac{\sqrt{x} + x^{4/3}}{6 - x^{3/4}} = 0$

15. $\displaystyle\lim_{x \to 3^-} [\![1 + x]\!] = 3$

17. $\displaystyle\lim_{x \to 3^-} x[\![x]\!] = 6$

19. $\displaystyle\lim_{x \to 4^+} \frac{\sqrt{x - 4}}{x + 2} = 0$

21. $\displaystyle\lim_{x \to 3^+} \frac{[\![x - 7]\!]}{[\![x + 4]\!]} = -\frac{4}{7}$

23. $\lim\limits_{x \to 0^+} [\![2 - x^2]\!] = 1$

25. $\lim\limits_{x \to 0} [\![2 - x^2]\!] = 1$

27. (a) $\lim\limits_{x \to 0^-} f(x) = \lim\limits_{x \to 0^-} \cos x = 1$

    (b) $\lim\limits_{x \to 0^+} f(x) = \lim\limits_{x \to 0^+} (1 - x) = 1$

    (c) Yes; $\lim\limits_{x \to 0} f(x) = 1$

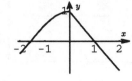

29. (a) $\lim\limits_{x \to 0^-} f(x) = \lim\limits_{x \to 0^-} [\![x + 3]\!] = 2$

    (b) $\lim\limits_{x \to 0^+} f(x) = \lim\limits_{x \to 0^+} \dfrac{(x + 1)^3 - 1}{x}$
    $= \lim\limits_{x \to 0^+} (x^2 + 3x + 3) = 3$

    (c) No; the one-sided limits are not equal.

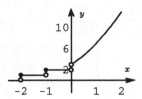

31. Since
$$\lim\limits_{x \to 3^-} f(x) = \lim\limits_{x \to 3^-} (x + 2) = 5 \quad \text{and} \quad \lim\limits_{x \to 3^+} f(x) = \lim\limits_{x \to 3^+} (2x - 1) = 5$$
are equal, $\lim\limits_{x \to 3} f(x)$ exists and is equal to 5.

33. (a) The graph is shown to the right.

    (b) No; $\lim\limits_{x \to -1^-} f(x) = \lim\limits_{x \to -1^-} 2 = 2$ and $\lim\limits_{x \to -1^+} f(x)$
    $= \lim\limits_{x \to -1^+} (-x) = 1$. Therefore $\lim\limits_{x \to -1} f(x)$ does not
    exist since the one-sided limits are not equal.

    (c) Yes; $\lim\limits_{x \to 1^-} f(x) = \lim\limits_{x \to 1^-} (-x) = -1$ and $\lim\limits_{x \to 1^+} f(x)$
    $= \lim\limits_{x \to 1^+} (-x^2) = -1$. Therefore $\lim\limits_{x \to 1} f(x)$ exists and
    is equal to -1.

    (d) See (b) and (c).

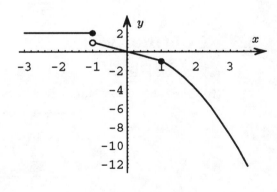

35. The two one-sided limits must be equal:
$$\lim\limits_{x \to -2^-} f(x) = \lim\limits_{x \to -2^-} (3 - x^2) = -1 = \lim\limits_{x \to -2^+} f(x) = \lim\limits_{x \to -2^+} (ax + b) = -2a + b$$

and
$$\lim\limits_{x \to 2^-} f(x) = \lim\limits_{x \to 2^-} (ax + b) = 2a + b = \lim\limits_{x \to 2^+} f(x) = \lim\limits_{x \to 2^+} (x^2/2) = 2.$$

Hence $-2a + b = -1$ and $2a + b = 2 \implies a = 3/4, b = 1/2$.

37. The graph below on the left is that of $f(x) = \sin x$ and $g(x) = (1 - \cos x)/x$ with the range $[0, \pi/2] \times [-1/2, 1]$. It shows that $0 < g(x) < \sin x$. Therefore $\lim_{x \to 0^+} g(x) = 0$. The graph on the right is that of the same functions with the range $[-\pi/2, 0] \times [-1, 1/2]$. This graph shows that $\sin x < g(x) < 0$ and hence $\lim_{x \to 0^-} g(x) = 0$.

51

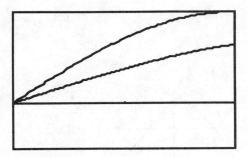
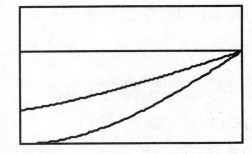

39. The graph shown below with range $[-3, 3] \times [-3, 9]$ indicates that $\lim_{x \to -2-} f(x) = \lim_{x \to -2+} f(x) = 2$ and $\lim_{x \to 2-} f(x) = 6$, $\lim_{x \to 2+} f(x) = 2$.

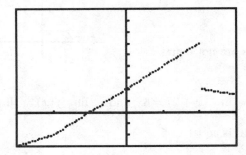

2.6   Continuity

1. (a) Discontinuous at $x = 1, 2, 4, 5, 6$.

   (b) The table below summarizes continuity or discontinuity on each interval.

(0, 1)	(1, 2)	(2, 3)	(3, 4)	(4, 5)	(5, 6)	(6, 7)
cont	cont	cont	cont	cont	cont	cont
[0, 1)	[1, 2)	[2, 3)	[3, 4)	[4, 5)	[5, 6)	[6, 7)
cont	cont	cont	cont	disc	disc	disc
(0, 1]	(1, 2]	(2, 3]	(3, 4]	(4, 5]	(5, 6]	(6, 7]
disc	disc	cont	cont	disc	disc	cont
[0, 1]	[1, 2]	[2, 3]	[3, 4]	[4, 5]	[5, 6]	[6, 7]
disc	disc	cont	cont	disc	disc	disc

3. $((k - 1/2)\pi, (k + 1/2)\pi)$, k an integer

5. $(k\pi, (k + 1)\pi)$, k an integer

7. $(-\infty, -1), (-1, 2), (2, +\infty)$

9. $(-\infty, 0), (0, +\infty)$

11. $(-\infty, 2], (2, +\infty)$

13. $(-\infty, -1], (-1, +\infty)$

15. $(k\pi, (k+1)\pi)$, k an integer

17. $(-7, +\infty)$

19. $(-\infty, +\infty)$

21. $((k-1/2)\pi, (k+1/2)\pi)$, k an integer

23. $\ldots, (-3\pi/2, -\pi/2), (-\pi/2, -1), (-1, \pi/2), (\pi/2, 2), (2, 3\pi/2), (3\pi/2, 5\pi/2), \ldots$

25. Since
$$\frac{x^2 - 1}{x - 1} = \frac{(x-1)(x+1)}{x-1} = x + 1 \quad \text{for } x \neq 1,$$
set
$$f(1) = \lim_{x \to 1} \frac{x^2 - 1}{x - 1} = \lim_{x \to 1} (x + 1) = 2$$
to remove the discontinuity at $a = 1$.

27. Since
$$\frac{\cos^2 x - 1}{\sin x} = \frac{-\sin^2 x}{\sin x} = -\sin x \quad \text{for } \sin x \neq 0,$$
set
$$f(0) = \lim_{x \to 0} \frac{\cos^2 x - 1}{\sin x} = \lim_{x \to 0} (-\sin x) = 0$$
to remove the discontinuity at $a = 0$.

29. $\lim\limits_{x \to 2^-} f(x) = \lim\limits_{x \to 2^-} x^k = 2^k$ and $\lim\limits_{x \to 2^+} f(x) = \lim\limits_{x \to 2^+} (10 - x) = 8 \Longrightarrow 2^k = 8 \Longrightarrow k = 3.$

31. $\lim\limits_{x \to 2^-} f(x) = \lim\limits_{x \to 2^-} (x - k)(x + k) = (2 - k)(2 + k)$ and $\lim\limits_{x \to 2^+} f(x) = \lim\limits_{x \to 2^+} (kx + 5) = 2k + 5 \Longrightarrow 2k + 5$
$= (2 - k)(2 + k) = 4 - k^2 \Longrightarrow k^2 + 2k + 1 = (k + 1)^2 = 0 \Longrightarrow k = -1.$

33. $\lim\limits_{x \to 8} \left(1 + \sqrt[3]{x}\right)^5 = \left(\lim\limits_{x \to 8} \left(1 + \sqrt[3]{x}\right)\right)^5 = 3^5 = 243$

35. $\lim\limits_{x \to \pi/2} \cos(\pi + x) = \cos\left(\lim\limits_{x \to \pi/2} (\pi + x)\right) = \cos(3\pi/2) = 0.$

37. $\lim\limits_{x \to 1} \left(3x^9 - \sqrt[3]{x}\right)^6 = \left(\lim\limits_{x \to 1} \left(3x^9 - 3\sqrt[3]{x}\right)\right)^6 = (3 - 1)^6 = 64$

39. $\lim\limits_{x \to 0} \cos \pi(x + |x|) = \cos\left(\lim\limits_{x \to 0} \pi(x + |x|)\right) = \cos 0 = 1$

41. Self-explanatory

43. $\lim\limits_{x \to 0} \dfrac{2x}{\sin x} = 2 \left/ \lim\limits_{x \to 0} (\sin x / x) \right. = 2$

45. $\lim\limits_{x \to 0} \dfrac{\sin x}{\sin 3x} = \lim\limits_{x \to 0} \left(\dfrac{\sin x}{x} \left/ \dfrac{\sin 3x}{x}\right.\right) = 1/3$

47. $\lim\limits_{x \to 0} x \csc 3x = \lim\limits_{x \to 0} \dfrac{x}{\sin 3x} = 1 \left/ \lim\limits_{x \to 0} (\sin 3x / x) \right. = 1/3$

49. $f(x) = 9 - x^2 = 0 \Longleftrightarrow x = -3, 3$. The table below indicates the sign changes for $f(x)$.

53

Interval	$(-\infty, -3)$	$(-3, 3)$	$(3, +\infty)$
Sign f	$-$	$+$	$-$

$\implies$ $\quad f > 0$ on $(-3, 3)$
$\quad\quad f < 0$ on $(-\infty, -3)$, $(3, +\infty)$

51. $f(x) = x^3 + 2x^2 - x - 2 = x^2(x + 2) - (x + 2) = (x + 2)(x - 1)(x + 1) = 0 \iff x = -2, -1, 1$. The table below indicates the sign changes of $f(x)$.

Interval	$(-\infty, -2)$	$(-2, -1)$	$(-1, 1)$	$(1, +\infty)$
Sign f	$-$	$+$	$-$	$+$

$\implies$ $\quad f > 0$ on $(-2, -1)$, $(1, +\infty)$
$\quad\quad f < 0$ on $(-\infty, -2)$, $(-1, 1)$

53. $f(x) = \cos(x + \pi) = 0 \iff x + \pi = (k + 1/2)\pi \iff x = (k - 1/2)\pi$, k an integer. The table below indicates the sign changes of $f(x)$.

Interval	$\cdots$	$(-3\pi/2, -\pi/2)$	$(-\pi/2, \pi/2)$	$(\pi/2, 3\pi/2)$	$\cdots$
Sign f	$\cdots$	$+$	$-$	$+$	$\cdots$

$\implies$ $\quad f > 0$ on the intervals $((2k + 1/2)\pi, (2k + 3/2)\pi)$, k an integer
$\quad\quad f < 0$ on the intervals $((2k - 1/2)\pi, (2k + 1/2)\pi)$, k an integer

55. Let $f(x) = x(x + 6) + 8$. Then $x(x + 6) > -8 \iff f > 0$. Now, $f(x) = x^2 + 6x + 8 = (x + 4)(x + 2) = 0 \iff x = -4, -2$. The table below indicates the sign changes of $f(x)$.

Interval	$(-\infty, -4)$	$(-4, -2)$	$(-2, +\infty)$
Sign f	$+$	$-$	$+$

Hence $f > 0$ on $(-\infty, -4)$ or $(-2, +\infty)$ and therefore $x(x + 6) > -8 \iff x < -4$ or $x > -2$.

57. $f(x) = x^3 + x^2 - 2x = x(x^2 + x - 2) = x(x + 2)(x - 1) = 0 \iff x = -2, 0, 1$.

Interval	$(-\infty, -2)$	$(-2, 0)$	$(0, 1)$	$(1, +\infty)$
Sign f	$-$	$+$	$-$	$+$

Hence $x^3 + x^2 - 2x > 0 \iff -2 < x < 0$ or $x > 1$.

59. $f(x) = x^4 - 9x^2 = x^2(x^2 - 9) = 0 \iff x = -3, 3$.

Interval	$(-\infty, -3)$	$(-3, 0)$	$(0, 3)$	$(3, +\infty)$
Sign f	$+$	$-$	$-$	$+$

Hence $x^4 - 9x^2 > 0 \iff x < -3$ or $x > 3$.

61. Let $f(x) = \sin x \cos x$, where $-2\pi < x < 2\pi$. Then $f(x) = 0 \iff x = -3\pi/2, -\pi, -\pi/2, 0, \pi/2, \pi, 3\pi/2$.

Interval	$(-2\pi, -3\pi/2)$	$(-3\pi/2, -\pi)$	$(-\pi, -\pi/2)$	$(-\pi/2, 0)$	$(0, \pi/2)$	$(\pi/2, \pi)$	$(\pi, 3\pi/2)$	$(3\pi/2, 2\pi)$
Sign f	$+$	$-$	$+$	$-$	$+$	$-$	$+$	$-$

Hence $\sin x \cos x > 0 \iff -2\pi < x < -3\pi/2$ or $-\pi < x < -\pi/2$ or $0 < x < \pi/2$ or $\pi < x < 3\pi/2$.

63. $f(1) = 1$, $f(3/2) = 9/4 = 2.25 \implies f(c) = c^2 = 2$ for some $c \in [1, 3/2]$. Hence $c = \sqrt{2}$ is between 1 and 3/2. Similarly, $f(3/2) = 2.25$ and $f(2) = 4 \implies f(d) = d^2 = 3$ for some $d \in [3/2, 2] \implies d = \sqrt{3}$ is between 3/2 and 2.

65. Let $f(x) = x + \sin x - 4$. Then $f(\pi) = \pi - 4 < 0$ and $f(2\pi) = 2\pi - 4 > 0$. Hence f has a zero between π and 2π; *i.e.*, the equation $x + \sin x = 4$ has at least one solution in the interval $[\pi, 2\pi]$.

67. Let $x = a$ be any number. Then $\lim_{x \to a} f(x) = \lim_{x \to a} c = c = f(a)$. Hence the constant function $f(x) = c$ is continuous at all x.

69. (ii) Since $(cf)(a) = cf(a)$, cf is defined at $x = a$. Moreover,

$$
\begin{aligned}
\lim_{x \to a} (cf)(x) &= \lim_{x \to a} cf(x) && \text{definition of } cf \\
&= c \lim_{x \to a} f(x) && \text{Theorem 1(ii)} \\
&= cf(a) && \text{since } f \text{ is continuous at } x = a \\
&= (cf)(a) && \text{definition of } f.
\end{aligned}
$$

Hence cf is continuous at $x = a$.

(iii) Since $(fg)(a) = f(a)g(a)$, fg is defined at $x = a$. Moreover,

$$
\begin{aligned}
\lim_{x \to a} (fg)(x) &= \lim_{x \to a} f(x)g(x) && \text{definition of } fg \\
&= \left(\lim_{x \to a} f(x) \right) \left(\lim_{x \to a} g(x) \right) && \text{Theorem 1(iii)} \\
&= f(a)g(a) && \text{since } f, g \text{ are continuous at } x = a \\
&= (fg)(a) && \text{definition of } fg.
\end{aligned}
$$

Hence fg is continuous at $x = a$.

(iv) Since $(f/g)(a) = f(a)/g(a)$ and $g(a) \neq 0$, f/g is defined at $x = a$. Moreover,

$$
\begin{aligned}
\lim_{x \to a} (f/g)(x) &= \lim_{x \to a} f(x)/g(x) && \text{definition of } f/g \\
&= \frac{\lim_{x \to a} f(x)}{\lim_{x \to a} g(x)} && \text{Theorem 1(iv)} \\
&= f(a)/g(a) && \text{since } f, g \text{ are continuous at } x = a \text{ and } g(a) \neq 0 \\
&= (f/g)(a) && \text{definition of } f/g.
\end{aligned}
$$

Hence f/g is continuous at $x = a$.

71. Let $\varepsilon > 0$ be a given number and set $\delta = \varepsilon^{m/n}$. Then

$$
0 < x < \delta \implies x^{n/m} < \delta^{n/m} = \varepsilon \implies |x^{n/m} - 0| < \varepsilon.
$$

Hence $\lim_{x \to 0^+} x^{n/m} = 0$.

73. $c_{14} = 2.28680$, $f(c_{14}) = -0.00033$, $(b_{14} - a_{14})/2 = 0.00006$.

75. $c_{14} = 2.53516$, $f(c_{14}) = 0.00009$, $(b_{14} - a_{14})/2 = 0.00006$

Review Exercises – Chapter 2

1. $\lim\limits_{x \to 2}(x^2 - x + 2) = 4$

3. $\lim\limits_{x \to 3}(x^4 - x - 1) = 77$

5. $\lim\limits_{x \to 3}(4 - 4x)^{1/3} = (-8)^{1/3} = -2$

7. $\lim\limits_{x \to 3/2} \dfrac{4x^2 - 9}{2x - 3} = \lim\limits_{x \to 3/2} \dfrac{(2x - 3)(2x + 3)}{2x - 3} = \lim\limits_{x \to 3/2}(2x + 3) = 6$

9. $\lim\limits_{x \to -2} \dfrac{x^2 + x - 2}{x + 2} = \lim\limits_{x \to -2} \dfrac{(x + 2)(x - 1)}{x + 2} = \lim\limits_{x \to -2}(x - 1) = -3$

11. $\lim\limits_{x \to 1} \dfrac{x^2 + 2x - 3}{x^2 + x - 2} = \lim\limits_{x \to 1} \dfrac{(x + 3)(x - 1)}{(x + 2)(x - 1)} = \lim\limits_{x \to 1} \dfrac{x + 3}{x + 2} = \dfrac{4}{3}$

13. $\lim\limits_{x \to 1^-} \sqrt{\dfrac{1 - x^2}{1 + x}} = 0$

15. $\lim\limits_{x \to 7^+} \sqrt{\dfrac{x - 7}{x + 2}} = 0$

17. $\lim\limits_{x \to 0} \dfrac{\tan x}{\sin x} = \lim\limits_{x \to 0} \dfrac{1}{\cos x} = 1$

19. $\lim\limits_{x \to 0} \dfrac{(2 + x)^2 - 4}{x} = \lim\limits_{x \to 0} \dfrac{x^2 + 4x}{x} = \lim\limits_{x \to 0}(x + 4) = 4$

21. $\lim\limits_{x \to 0} \dfrac{|x|}{2x}$ does not exist

23. $\lim\limits_{x \to 0} 3x \csc 4x = \lim\limits_{x \to 0} \dfrac{3x}{\sin 4x} = \lim\limits_{x \to 0} \dfrac{3}{(\sin 4x)/x} = 3/4$

25. $\lim\limits_{x \to 1} \dfrac{x^3 - 1}{x - 1} = \lim\limits_{x \to 1} \dfrac{(x - 1)(x^2 + x + 1)}{x - 1} = \lim\limits_{x \to 1}(x^2 + x + 1) = 3$

27. $\lim\limits_{x \to 4^-} \sqrt{8 - 2x} = 0$

29. $\lim\limits_{x \to 1} \dfrac{|x - 1|}{x - 1}$ does not exist

31. $\lim\limits_{x \to 3}\left(2[\![x/2]\!] + 3\right) = 5$

33. $\lim\limits_{x \to 1^+} \dfrac{|x - 1|}{x - 1} = 1$

35. $\lim\limits_{x \to 0^+} \dfrac{2 + \sqrt{x}}{2 - \sqrt{x}} = 1$

37. $\lim\limits_{x \to 1^+} \sqrt{1 - [\![x]\!]} = 0$

39. $\lim\limits_{x \to 0} \dfrac{\sqrt[3]{x+1}-1}{x} = \lim\limits_{x \to 0} \dfrac{\sqrt[3]{x+1}-1}{x} \cdot \dfrac{\sqrt[3]{(x+1)^2}+\sqrt[3]{x+1}+1}{\sqrt[3]{(x+1)^2}+\sqrt[3]{x+1}+1} = \lim\limits_{x \to 0} \dfrac{1}{\sqrt[3]{(x+1)^2}+\sqrt[3]{x+1}+1} = \dfrac{1}{3}$

41. The slope of the tangent line at $x_0 = 1$ is

$$m = \lim_{h \to 0} \frac{f(1+h) - f(1)}{h} = \lim_{h \to 0} \frac{[3(1+h)^2 + 4] - 7}{h} = \lim_{h \to 0}(6 + 3h) = 6.$$

Therefore the equation of the tangent line at $(1, 7)$ is $y - 7 = 6(x-1)$, or $y = 6x + 1$.

43. The slope of the tangent line at $x_0 = 2$ is

$$m = \lim_{h \to 0} \frac{f(2+h) - f(2)}{h} = \lim_{h \to 0} \frac{\frac{1}{2+h} - \frac{1}{2}}{h} = \lim_{h \to 0} \frac{2 - (2+h)}{2h(2+h)} = \lim_{h \to 0} \frac{-1}{2(2+h)} = -\frac{1}{4}.$$

Therefore the equation of the tangent line at $(2, 1/2)$ is $y - 1/2 = (-1/4)(x - 2)$.

45. The slope of the tangent line at $x_0 = 1$ is

$$m = \lim_{h \to 0} \frac{f(1+h) - f(1)}{h} = \lim_{h \to 0} \frac{(2+h)^4 - 16}{h} = \lim_{h \to 0}(32 + 24h + 8h^2 + h^3) = 32.$$

Therefore the equation of the tangent line at $(1, 16)$ is $y - 16 = 32(x - 1)$.

47. $(-\infty, -2), (-2, 3), (3, +\infty)$ since f is continuous at all x for which $x^2 - x - 6 = (x-3)(x+2) \neq 0$.

49. $[0, \pi/2), (\pi/2, 3\pi/2), (3\pi/2, 2\pi]$ since y is continuous at all x in the interval $[0, 2\pi]$ for which $\cos x \neq 0$, namely $x \neq \pi/2$ and $x \neq 3\pi/2$.

51. $(-\infty, -1], [4, +\infty)$ since f is continuous at all x for which $x^2 - 3x - 4 = (x-4)(x+1) \geq 0$.

53. $(-\infty, \pi/4], (\pi/4, +\infty)$ since $\lim\limits_{x \to \pi/4^-} y = \sqrt{2}/2 = y(\pi/4)$ but $\lim\limits_{x \to \pi/4^+} y = 1 - \sqrt{2}/2 \neq y(\pi/4)$.

55. $(-\infty, -2), [-2, +\infty)$ since $\lim\limits_{x \to -2^-} f(x) = 1 - (-2) = 3 \neq f(-2)$ but $\lim\limits_{x \to -2^+} f(x) = -(-2) = 2 = f(-2)$, while $\lim\limits_{x \to 1^-} f(x) = -1 = \lim\limits_{x \to 1^+} f(x) = f(1)$.

57. $(-\infty, +\infty)$ since $\lim\limits_{x \to -5} f(x) = \lim\limits_{x \to -5} \dfrac{(x-2)(x+5)}{x+5} = \lim\limits_{x \to -5}(x - 2) = -7 = f(-5)$.

59. (a) $\lim\limits_{x \to -1^-} f(x) = \lim\limits_{x \to -1^-} 1 = 1$

   (b) $\lim\limits_{x \to -1^+} f(x) = \lim\limits_{x \to -1^+} \sqrt{x+2} = \sqrt{1} = 1$

   (c) $\lim\limits_{x \to -1} f(x) = 1$

   (d) $\lim\limits_{x \to 2^-} f(x) = \lim\limits_{x \to 2^-} \sqrt{x+2} = \sqrt{4} = 2$

   (e) $\lim\limits_{x \to 2^+} f(x) = \lim\limits_{x \to 2^+} \cos \pi x = \cos 2\pi = 1$

   (f) $\lim\limits_{x \to 2} f(x)$ does not exist since the one-sided limits are not equal

   (g) Yes, since $\lim\limits_{x \to -1} f(x) = 1 = f(-1)$.

   (h) No, since $\lim\limits_{x \to 2} f(x)$ does not exist.

61. $\displaystyle\lim_{x\to 1^-} f(x) = \lim_{x\to 1^-} (4 - x + x^2) = 4 = \lim_{x\to 1^+} f(x) = \lim_{x\to 1^+} (9 - ax^2) = 9 - a \implies a = 5$

63. $\displaystyle\lim_{x\to 0}(1 - |x|) = 1 \leq \lim_{x\to 0} f(x) \leq \lim_{x\to 0}(\sec x) = 1 \implies \lim_{x\to 0} f(x) = 1$

65. Let $\varepsilon > 0$ be given. Then

$$|(4x - 3) - 1| = 4|x - 1| < \varepsilon \iff |x - 1| < \varepsilon/4.$$

Thus, if we set $\delta = \varepsilon/4$, it follows that

$$0 < |x - 1| < \delta \implies |(4x - 3) - 1| < \varepsilon$$

and hence $\displaystyle\lim_{x\to 1}(4x - 3) = 1$.

67. Let $\varepsilon > 0$ be given and let x be sufficiently close to 4, say within one unit. Then $1/|x| < 1/3$ and hence

$$\left|\frac{1}{x} - \frac{1}{4}\right| = \frac{|x - 4|}{4|x|} < \frac{|x - 4|}{12} < \varepsilon \iff |x - 4| < 12\varepsilon.$$

Thus, if we let δ be the smaller of 1 and 12ε, it follows that

$$0 < |x - 4| < \delta \implies \left|\frac{1}{x} - \frac{1}{4}\right| < \varepsilon$$

and hence $\displaystyle\lim_{x\to 4}(1/x) = 1/4$.

69. Let $\varepsilon > 0$ be given and let $\delta = \varepsilon$. Then

$$0 < |x + 3| < \delta \implies \big||x + 3|\big| = |x + 3| < \varepsilon$$

and therefore $\displaystyle\lim_{x\to -3} |x + 3| = 0$.

71. Let $\varepsilon > 0$ be given. Then, if x is sufficiently close to -2, say within one unit, $|x - 1| < 2$ and hence

$$\left|(x^2 + x + 1) - 3\right| = |x + 2|\,|x - 1| < 2|x + 2| < \varepsilon \iff |x + 2| < \varepsilon/2.$$

Thus, if we let δ be the smaller of 1 and $\varepsilon/2$, it follows that

$$0 < |x + 2| < \delta \implies \left|(x^2 + x + 1) - 3\right| < \varepsilon$$

and therefore $\displaystyle\lim_{x\to -2}(x^2 + x + 1) = 3$.

73. $\displaystyle\lim_{x\to a}(x^3 - 4x^2 + 3x - 2) = a^3 - 4a^2 + 3a - 2 = -10 \implies a^3 - 4a^2 + 3a + 8 = (a + 1)(a^2 - 5a + 8) = 0$
$\implies a = -1$ since the equation $a^2 - 5a + 8$ has no real solutions.

75. False. Let $f(x) = -2$, $g(x) = 2$, and $h(x) = x/|x|$ for $x \neq 0$, $h(0) = 0$. Then f and g are continuous at $x = 0$ and $f(x) \leq h(x) \leq g(x)$ for all x, but h is not continuous at $x = 0$.

77. Suppose that $\displaystyle\lim_{x\to a} f(x) = M < L$ and let $\varepsilon > 0$ be a number such that $\varepsilon < L - M$. Since $\displaystyle\lim_{x\to a} f(x) = M$ and $\displaystyle\lim_{x\to a} g(x) = L$, there are numbers δ_1, δ_2 such that

$$0 < |x - a| < \delta_1 \implies |f(x) - M| < \varepsilon/2$$
$$0 < |x - a| < \delta_2 \implies |g(x) - L| < \varepsilon/2.$$

Let δ be the smaller of δ_1 and δ_2. Then, if x is any number for which $0 < |x - a| < \delta$, we have that

$$f(x) < M + \varepsilon/2 < M + (1/2)(L - M) = L - (1/2)(L - M) < L - \varepsilon/2 < g(x),$$

which contradicts the fact that $g(x) \leq f(x)$. Hence $\displaystyle\lim_{x\to a} f(x) \geq L$.

Chapter 3

The Derivative

3.1   The Derivative as a Function

1. a $\leftrightarrow$ iv, b $\leftrightarrow$ v, c $\leftrightarrow$ iii, d $\leftrightarrow$ i, e $\leftrightarrow$ ii.

3. $f(x) = 6x + 5 \implies$

$$f'(x) = \lim_{h \to 0} \frac{f(x+h) - f(x)}{h} = \lim_{h \to 0} \frac{[6(x+h)+5] - [6x+5]}{h} = \lim_{h \to 0} 6 = 6$$

5. $f(x) = 2x^3 + 3 \implies$

$$f'(x) = \lim_{h \to 0} \frac{f(x+h) - f(x)}{h} = \lim_{h \to 0} \frac{[2(x+h)^3 + 3] - [2x^3 + 3]}{h} = \lim_{h \to 0} \left(6x^2 + 6xh + 2h^2\right) = 6x^2$$

7. $f(x) = ax^3 + bx^2 + cx + d \implies$

$$f'(x) = \lim_{h \to 0} \frac{f(x+h) - f(x)}{h} = \lim_{h \to 0} \frac{\left[a(x+h)^3 + b(x+h)^2 + c(x+h) + d\right] - \left[ax^3 + bx^2 + cx + d\right]}{h}$$

$$= \lim_{h \to 0} \left(3ax^2 + 3axh + ah^2 + 2bx + bh + c\right) = 3ax^2 + 2bx + c$$

9. $f(x) = \dfrac{1}{2x + 3} \implies$

$$f'(x) = \lim_{h \to 0} \frac{f(x+h) - f(x)}{h} = \lim_{h \to 0} \frac{\dfrac{1}{2(x+h)+3} - \dfrac{1}{2x+3}}{h}$$

$$= \lim_{h \to 0} \frac{-2}{(2x+3)(2x+2h+3)} = \frac{-2}{(2x+3)^2}$$

11. $f(x) = \sqrt{x + 1} \implies$

$$f'(x) = \lim_{h \to 0} \frac{f(x+h) - f(x)}{h} = \lim_{h \to 0} \frac{\sqrt{x+h+1} - \sqrt{x+1}}{h}$$

$$= \lim_{h \to 0} \frac{\sqrt{x+h+1} - \sqrt{x+1}}{h} \cdot \frac{\sqrt{x+h+1} + \sqrt{x+1}}{\sqrt{x+h+1} + \sqrt{x+1}} = \lim_{h \to 0} \frac{1}{\sqrt{x+h+1} + \sqrt{x+1}} = \frac{1}{2\sqrt{x+1}}$$

13. $f(x) = \dfrac{1}{\sqrt{x+1}} \implies$

$$f'(x) = \lim_{h \to 0} \frac{f(x+h) - f(x)}{h} = \lim_{h \to 0} \frac{\dfrac{1}{\sqrt{x+h+1}} - \dfrac{1}{\sqrt{x+1}}}{h}$$

$$= \lim_{h \to 0} \frac{\sqrt{x+1} - \sqrt{x+h+1}}{h\sqrt{x+1}\sqrt{x+h+1}} \cdot \frac{\sqrt{x+1} + \sqrt{x+h+1}}{\sqrt{x+1} + \sqrt{x+h+1}}$$

$$= \lim_{h \to 0} \frac{-1}{\sqrt{x+1}\sqrt{x+h+1}\left(\sqrt{x+1} + \sqrt{x+h+1}\right)} = \frac{-1}{2(x+1)\sqrt{x+1}}$$

15. $f(x) = 1/x^2 \implies$

$$f'(x) = \lim_{h \to 0} \frac{f(x+h) - f(x)}{h} = \lim_{h \to 0} \frac{\dfrac{1}{(x+h)^2} - \dfrac{1}{x^2}}{h} = \lim_{h \to 0} \frac{-2x - h}{x^2(x+h)^2} = -\frac{2}{x^3}$$

17. $f(x) = (x+3)^3 \implies$

$$f'(x) = \lim_{h \to 0} \frac{f(x+h) - f(x)}{h} = \lim_{h \to 0} \frac{(x+h+3)^3 - (x+3)^3}{h}$$

$$= \lim_{h \to 0} \frac{(x+3)^3 + 3(x+3)^2 h + 3(x+3)h^2 + h^3 - (x+3)^3}{h}$$

$$= \lim_{h \to 0} \left[3(x+3)^2 + 3(x+3)h + h^2\right] = 3(x+3)^2$$

19. $f(x) = \dfrac{1}{\sqrt{x+5}} \implies$

$$f'(x) = \lim_{h \to 0} \frac{f(x+h) - f(x)}{h} = \lim_{h \to 0} \frac{\dfrac{1}{\sqrt{x+h+5}} - \dfrac{1}{\sqrt{x+5}}}{h}$$

$$= \lim_{h \to 0} \frac{\sqrt{x+5} - \sqrt{x+h+5}}{h\sqrt{x+5}\sqrt{x+h+5}} \cdot \frac{\sqrt{x+5} + \sqrt{x+h+5}}{\sqrt{x+5} + \sqrt{x+h+5}}$$

$$\lim_{h \to 0} \frac{-1}{\sqrt{x+5}\sqrt{x+h+5}\left(\sqrt{x+5} + \sqrt{x+h+5}\right)} = \frac{-1}{2(x+5)\sqrt{x+5}}$$

21. $f(x) = \dfrac{3}{(x-1)^2} \implies$

$$f'(x) = \lim_{h \to 0} \frac{f(x+h) - f(x)}{h} = \lim_{h \to 0} \frac{\dfrac{3}{(x+h-1)^2} - \dfrac{3}{(x-1)^2}}{h}$$

$$= \lim_{h \to 0} \frac{3(x-1)^2 - 3(x+h-1)^2}{h(x-1)^2(x+h-1)^2} = \lim_{h \to 0} \frac{-3(2x+h-2)}{(x-1)^2(x+h-1)^2} = \frac{-6}{(x-1)^3}$$

23. By Exercise 9, $y' = -2/(2x+3)^2$. Hence the slope of the tangent line at $(0, 1/3)$ is $y'(0) = -2/9$ and the equation is $y - 1/3 = (-2/9)x$.

25. By Exercise 4, $f'(x) = 2x$. Hence the slope of the tangent line at $P = (3, 2)$ is $f'(3) = 6$. Therefore the slope of the normal is $-1/6$ and the equation is $y - 2 = (-1/6)(x - 3)$.

27. Here $f'(x) = 1' - (x^2)' = -2x$. Hence the slopes of the tangent lines at $x = x_1$ and $x = x_2$ are $-2x_1$ and $-2x_2$. Now, if the triangle is equilateral, then each of the angles is $60°$ and hence

$$\text{slope of tangent at } x_1 = f'(x_1) = \tan 60° \implies -2x_1 = \sqrt{3} \implies x_1 = -\sqrt{3}/2.$$

Therefore $x_2 = \sqrt{3}/2$ since x_1 and x_2 are symmetric about the y-axis.

29. $y = 2 - ax^2 \implies y' = 2' - a(x^2)' = -2ax$. Hence the slope of the tangent line is 6 at $x = -1$ when $-2a(-1) = 6 \implies a = 3$.

31. $y = ax^2 + b \implies y' = a(x^2)' + (b)' = 2ax$. Now, if $y = 4x$ is the tangent line at $(1, 4)$, then the slope is 4 and hence $y'(1) = 2a = 4 \implies a = 2$. Since $y(1) = a + b = 4$, $b = 4 - a = 2$.

33. Let $m_h = \big(f(x + h) - f(x)\big)/h$ stand for the difference quotient. The values of m_h for $f(x) = \cos x$ and $h = \pm 0.5$, ± 0.1, ± 0.05, and ± 0.01 are listed in the table below, together with the estimate of the slope.

h	0	$\pi/4$	$\pi/2$	$3\pi/4$	π	$5\pi/4$	$3\pi/2$	$7\pi/4$	2π
0.5000	-0.2448	-0.8511	-0.9588	-0.5049	0.2448	0.8511	0.9589	0.5049	-0.2448
0.1000	-0.0500	-0.7413	-0.9983	-0.6706	0.0500	0.7413	0.9983	0.6706	-0.0500
0.0500	-0.0250	-0.7245	-0.9996	-0.6891	0.0250	0.7245	0.9996	0.6891	-0.0250
0.0100	-0.0050	-0.7106	-1.0000	-0.7036	0.0050	0.7106	1.0000	0.7036	-0.0050
-0.0100	0.0050	-0.7036	-1.0000	-0.7106	-0.0050	0.7036	1.0000	0.7106	0.0050
-0.0500	0.0250	-0.6891	-0.9996	-0.7245	-0.0250	0.6891	0.9996	0.7245	0.0250
-0.1000	0.0500	-0.6706	-0.9983	-0.7412	-0.0500	0.6706	0.9983	0.7413	0.0500
-0.5000	0.2448	-0.5049	-0.9588	-0.8511	-0.2448	0.5049	0.9589	0.8511	0.2448
Estimate	0	-0.707	-1	-0.707	0	0.707	1	0.707	0

The last row of the table gives the estimate for the slope of the tangent line. It appears that the slope is equal to $-\sin x$ for each x.

35. $f(x) = x^2 - 6x + 2 \implies f'(x) = (x^2)' + (-6x + 2)' = 2x - 6$. Hence $f' < 0 \iff x < 3$.

37. (a) No, $f'(1)$ does not exist since

$$\lim_{h \to 0^+} \frac{f(1 + h) - f(1)}{h} = \lim_{h \to 0^+} \frac{[(1 + h)^2 - (1 + h)] - 0}{h} = \lim_{h \to 0^+} (1 + h) = 1$$

but

$$\lim_{h \to 0^-} \frac{f(1 + h) - f(1)}{h} = \lim_{h \to 0^-} \frac{[2(1 + h) - 2] - 0}{h} = \lim_{h \to 0^-} 2 = 2.$$

(b) Yes, $\lim_{x \to 1} f(x) = 0 = f(1)$ since

$$\lim_{x \to 1^+} f(x) = \lim_{x \to 1^+} (x^2 - x) = 0 \quad \text{and} \quad \lim_{x \to 1^-} f(x) = \lim_{x \to 1^-} (2x - 2) = 0.$$

39. $f'(0) = \lim_{h \to 0} \dfrac{\sin(h + 0) - \sin 0}{h} = \lim_{h \to 0} \dfrac{\sin h}{h} = 1.$

41. (a) False; $f(a)$ must also exist and equal the limit. For example, if $f(x) = (x^2 - 1)/(x - 1)$, then $\lim_{x \to 1} f(x) = 2$, but f is not continuous at $x = 1$ since $f(1)$ does not exist.

(b) True; continuity means $\lim_{x \to a} f(x)$ exists and is equal to $f(a)$.

(c) False; if $f(x) = |x|$, then $\lim_{x \to 0} f(x) = 0$ but f is not differentiable at $x = 0$ (Example 5 in the text).

(d) True; if f is differentiable at $x = a$, then, by Theorem 2, f is continuous at $x = a$ and hence $\lim_{x \to a} f(x) = f(a)$.

(e) False; $f(x) = |x|$ is continuous but not differentiable at $x = 0$.

(f) True; this is Theorem 2.

43. $m_h(x) = [2(x + h) + 1 - (2x + 1)]/h = 2h/h = 2$. Range: $[-4.7, 4.7] \times [-3.1, 3.1]$. This graph, below on the left, is the true picture of the derivative for this function.

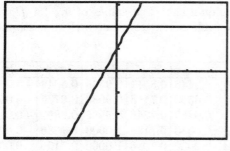

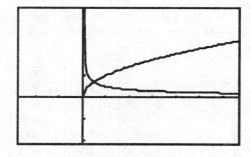

Exercise 43

Exercise 45

45. $m_h(x) = (\sqrt{x + h} - \sqrt{x})/h$. Range: $[-2.8, 6.7] \times [-2.2, 4.1]$ with $h = 0.01$. The graph is shown above on the right. Note that the graph of $m_h(x)$ is an approximation to the true graph of $f'(x)$, but as h gets smaller and smaller, the approximation becomes better.

47. $f(x) = x^2 - 2x + 1$, $a = 1$. The table to the right summarizes the values of m_h for $h = 2$, -1.5, -1, 0.1, -0.1, and 0.001. The graphs are shown for $h = 2$, -1 with range $[-3.8, 5.7] \times [-3.2, 3.1]$. $f(x)$ is differentiable at $a = 1$.

h	m_h
2	2
-1.5	-1.5
-1	-1
-0.1	0.1
0.001	0.001

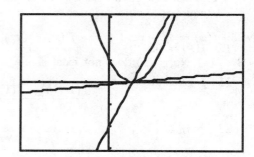

49. $f(x) = (\sin x)/x$, $a = 0$. The table to the right summarizes the values of m_h for $h = -2$, -1.5, 1, -0.1, 0.01, and 0.001. The graphs are shown for $h = 1$, 0.001 with range $[-2.35, 2.35] \times [-0.55, 2.55]$. $f(x)$ is differentiable at $a = 0$.

h	m_h
-2	0.2727
-1.5	0.2233
1	-0.1585
-0.1	0.0167
0.01	-0.0017
0.001	-0.0002

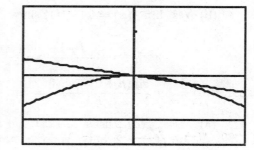

51. $f'(x) = 1$

53. $f'(x) = 2$

55. $f'(x) = 2x^2 + 2$

57. $f'(x) = 3x^2 - 10x$

59. $f'(x) = -1/(x+1)^2$

61. $f'(x) = -2\sin 2x$

3.2 Rules for Calculating Derivatives

1. $f(x) = 8x^3 - x^2 \implies f'(x) = 24x^2 - 2x$

3. $f(x) = ax^3 + bx \implies f'(x) = 3ax^2 + b$

5. $f(x) = (x^2 + 5)/2 = (1/2)x^2 + 5/2 \implies f'(x) = x$

7. $f(x) = (x-1)(x+2) = x^2 + x - 2 \implies f'(x) = 2x + 1$

9. $f(x) = (3x^2 - 8x)(x^2 + 2) \implies f'(x) = (3x^2 - 8x)(2x) + (6x - 8)(x^2 + 2) = 12x^3 - 24x^2 + 12x - 16$

11. $f(x) = \left(x^3 - x\right)^2 = x^6 - 2x^4 + x^2 \implies f'(x) = 6x^5 - 8x^3 + 2x$

13. $f(x) = 5x^2 + 2x + \dfrac{3}{x} - \dfrac{4}{x^2} = 5x^2 + 2x + 3x^{-1} - 4x^{-2} \implies f'(x) = 10x + 2 - 3x^{-2} + 8x^{-3}$

15. $f(x) = \dfrac{6}{3-x} \implies f'(x) = \dfrac{(3-x)(0) - 6(-1)}{(3-x)^2} = \dfrac{6}{(3-x)^2}$

17. $f(x) = \dfrac{x^4 + 4x + 4}{1 - x^3} \implies f'(x) = \dfrac{(1-x^3)(4x^3 + 4) - (x^4 + 4x + 4)(-3x^2)}{(1-x^3)^2} = \dfrac{-x^6 + 12x^3 + 12x^2 + 4}{(1-x^3)^2}$

19. $f(x) = 5x^{-3} - 2x^{-5} \implies f'(x) = -15x^{-4} + 10x^{-6}$

21. $f(x) = (x-2)\left(x + \dfrac{1}{x}\right) = x^2 - 2x + 1 - 2x^{-1} \implies f'(x) = 2x - 2 + 2x^{-2}$

23. $f(x) = \left(1 + \dfrac{3}{x}\right)^2 = 1 + 6x^{-1} + 9x^{-2} \implies f'(x) = -6x^{-2} - 18x^{-3}$

25. $f(x) = \left(\dfrac{x+1}{x-1}\right)^2 = \dfrac{x^2 + 2x + 1}{x^2 - 2x + 1} \implies$

$$f'(x) = \dfrac{(x^2 - 2x + 1)(2x + 2) - (x^2 + 2x + 1)(2x - 2)}{(x^2 - 2x + 1)^2} = \dfrac{-4x^2 + 4}{(x-1)^4} = \dfrac{-4(x+1)}{(x-1)^3}$$

27. $f(x) = \left(x^5 + x^{-2}\right)\left(x^3 - x^{-7}\right) = x^8 + x - x^{-2} - x^{-9} \implies f'(x) = 8x^7 + 1 + 2x^{-3} + 9x^{-10}$

29. $f(x) = \dfrac{x^{-3} - x^4}{x^5} = x^{-8} - x^{-1} \implies f'(x) = -8x^{-9} + x^{-2}$

31. By the Product Rule: $f'(x) = (x^2 + 2)(2x) + (x^2 - 2)(2x) = 4x^3$
Expanding first: $f(x) = (x^2 + 2)(x^2 - 2) = x^4 - 4 \implies f'(x) = 4x^3$

33. By the Product Rule: $f'(x) = (x^3 + 7)(12x^3 + 1) + (3x^4 + x + 9)(3x^2) = 21x^6 + 88x^3 + 27x^2 + 7$
Expanding first: $f(x) = (x^3 + 7)(3x^4 + x + 9) = 3x^7 + 22x^4 + 9x^3 + 7x + 63 \implies f'(x) = 21x^6 + 88x^3 + 27x^2 + 7$

35. By the Product Rule: $f(x) = \left(\dfrac{1}{x+1}\right)\left(\dfrac{2}{x+2}\right) \implies$

$$f'(x) = \frac{1}{x+1}\frac{(x+2)(0) - 2(1)}{(x+2)^2} + \frac{2}{x+2}\frac{(x+1)(0) - (1)(1)}{(x+1)^2} = \frac{-4x - 6}{(x+1)^2(x+2)^2}$$

Expanding first: $f(x) = \dfrac{2}{x^2 + 3x + 2} \implies f'(x) = \dfrac{(x^2 + 3x + 2)(0) - 2(2x + 3)}{(x^2 + 3x + 2)^2} = \dfrac{-4x - 6}{(x^2 + 3x + 2)^2}$

37. By the Product Rule: $f'(u) = (u^2 + u + 1)(2u - 1) + (u^2 - u - 1)(2u + 1) = 4u^3 - 2u - 2$
Expanding first: $f(u) = (u^2 + u + 1)(u^2 - u - 1) = u^4 - u^2 - 2u - 1 \implies f'(u) = 4u^3 - 2u - 2$

39. Using the Product Rule, we obtain

$$(fgh)'(x) = \big[(fg)h\big]'(x) = \big[(fg)'h + (fg)h'\big](x)$$

$$= \big[(f'g + fg')h + fgh'\big](x) = f'(x)g(x)h(x) + f(x)g'(x)h(x) + f(x)g(x)h'(x).$$

41. $f(s) = (2s - 1)(s - 3)(s^2 + 4) \implies$

$$f'(s) = (2)(s - 3)(s^2 + 4) + (2s - 1)(1)(s^2 + 4) + (2s - 1)(s - 3)(2s) = 8s^3 - 21s^2 + 22s - 28$$

43. $f(x) = \left(\dfrac{1}{x}\right)\left(\dfrac{1}{x+1}\right)\left(\dfrac{1}{x+2}\right) \implies$

$$f'(x) = \left(\frac{-1}{x^2}\right)\left(\frac{1}{x+1}\right)\left(\frac{1}{x+2}\right) + \left(\frac{1}{x}\right)\left(\frac{-1}{(x+1)^2}\right)\left(\frac{1}{x+2}\right) + \left(\frac{1}{x}\right)\left(\frac{1}{x+1}\right)\left(\frac{-1}{(x+2)^2}\right)$$

$$= \frac{-3x^2 - 6x - 2}{x^2(x+1)^2(x+2)^2}$$

45. $f(x) = (2x^3 - 6x + 9)^3 = (2x^3 - 6x + 9)(2x^3 - 6x + 9)(2x^3 - 6x + 9) \implies$

$$f'(x) = (6x^2 - 6)(2x^3 - 6x + 9)^2 + (6x^2 - 6)(2x^3 - 6x + 9)^2 + (6x^2 - 6)(2x^3 - 6x + 9)^2$$

$$= 3(6x^2 - 6)(2x^3 - 6x + 9)^2$$

47. Using the derivative for $f(x)$ found in Exercise 11, $f'(2) = 2(11)(6) = 132$.

49. $f(x) = \left(\dfrac{x - 1}{x}\right)^3 = (1 - x^{-1})^3 \implies f'(x) = 3(1 - x^{-1})^2(x^{-2})$

51. $f(x) = \dfrac{1}{(x - 6)^3} \implies f'(x) = \dfrac{(x - 6)^3(0) - (1)\left[3(x - 6)^2(1)\right]}{(x - 6)^6} = \dfrac{-3}{(x - 6)^4}$

53. $f(x) = (x + 2)(x - 1)^2 \implies f'(x) = (1)(x - 1)^2 + (x + 2)[2(x - 1)(1)] = 3x^2 - 3 = 3(x - 1)(x + 1)$

55. $f(x) = \dfrac{x-1}{x+1} \implies f'(x) = \dfrac{(x+1)(1)-(x-1)(1)}{(x+1)^2} = \dfrac{2}{(x+1)^2} \implies f'(1) = 1/2$. Hence the equation of the tangent line at $(1,0)$ is $y = (1/2)(x-1)$.

57. $f(x) = \left(1 - \dfrac{1}{x}\right)^2 = (1-x^{-1})^2 \implies f'(x) = 2(1-x^{-1})(x^{-2}) \implies f'(1) = 0$. Hence the equation of the tangent line at $(1,0)$ is $y = 0$.

59. $f(x) = 2x^3 + 1 \implies f'(x) = 6x^2 \implies f'(1) = 6$. Hence the slope of the normal line to the graph at $(1,3)$ is $-1/6$ and the equation is $y - 3 = (-1/6)(x-1)$.

61. $f(x) = 2/x^2 = 2x^{-2} \implies f'(x) = -4x^{-3} \implies f'(2) = -1/2$. Hence the slope of the normal line to the graph at $(2, 1/2)$ is 2 and the equation is $y - 1/2 = 2(x-2)$.

63. $f(x) = \dfrac{3x}{2x-4} \implies$

$$f'(x) = \frac{(2x-4)(3)-(3x)(2)}{(2x-4)^2} = \frac{-12}{(2x-4)^2} = -3$$

$$\implies \quad (2x-4)^2 = 4 \quad \implies \quad 2x - 4 = \pm 2 \quad \implies \quad x = 3, 1.$$

Hence there are two points where the slope of the tangent line is -3: $(3, 9/2)$ and $(1, -3/2)$.

65. $y = b/x^2 = bx^{-2} \implies y' = -2bx^{-3} \implies y'(-2) = b/4$. Since the slope of the line $4y - bx - 21 = 0$ is also $b/4$, this line is parallel to the tangent line at $x = -2$ for all values of b. If $x = -2$, $y = b/(-2)^2 = b/4$ and hence this line touches the graph when $4(b/4) - b(-2) - 21 = 0 \implies b = 7$.

67. If $f(x) = x^{1/n}$, then $f^n(x) = x$ and hence $nf^{n-1}(x)f'(x) = 1$. Therefore

$$f'(x) = \frac{1}{nf^{n-1}(x)} = \frac{1}{nx^{(n-1)/n}} = (1/n)x^{(1/n)-1}.$$

69. The slope of the secant line through $(1,1)$ and $(2,8)$ is 7. Hence, $f'(x) = 3x^2 = 7 \implies x = \pm\sqrt{7/3}$. The graph of $f(x)$, the secant line and tangent line are shown below.

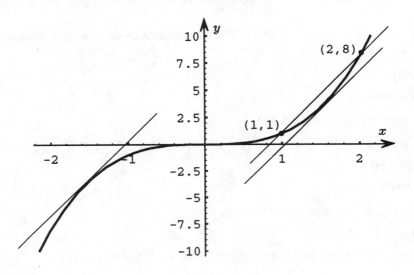

71. (a) By definition, the difference quotient of $1/g$ at x_0 is

$$\frac{\dfrac{1}{g(x_0+h)} - \dfrac{1}{g(x_0)}}{h} = \frac{g(x_0) - g(x_0+h)}{hg(x_0)g(x_0+h)}.$$

(b) By definition,

$$\left(\frac{1}{g}\right)'(x) = \lim_{h \to 0} \frac{\dfrac{1}{g(x+h)} - \dfrac{1}{g(x)}}{h} = \lim_{h \to 0} \frac{g(x) - g(x+h)}{hg(x)g(x+h)}$$

$$= \lim_{h \to 0} \frac{-[g(x+h) - g(x)]}{h} \frac{1}{g(x)} \frac{1}{g(x+h)} = -g'(x)\frac{1}{g(x)} \frac{1}{g(x)} = -\frac{g'(x)}{[g(x)]^2}.$$

(c) Using the Product Rule, we obtain

$$\left(\frac{f}{g}\right)'(x) = \left[f\left(\frac{1}{g}\right)\right]'(x) = f'(x)\left(\frac{1}{g(x)}\right) + f(x)\left(\frac{1}{g}\right)'(x)$$

$$= \frac{f'(x)}{g(x)} + f(x)\frac{-g'(x)}{[g(x)]^2} = \frac{f'(x)}{g(x)} - \frac{f(x)g'(x)}{[g(x)]^2} = \frac{f'(x)g(x) - f(x)g'(x)}{[g(x)]^2}.$$

3.3 The Derivative as a Rate of Change

1. Average rate of change from $t = 2$ to $t = 3$ is

$$\frac{s(3) - s(2)}{3 - 2} = \frac{16 - 9}{1} = 7.$$

Since $s'(t) = 2(1 + t)$, the instantaneous rate of change at $t_0 = 5/2$ is $2(1 + 5/2) = 7$.

3. Average rate of change from $t = 0$ to $t = 1$ is

$$\frac{f(1) - f(0)}{1 - 0} = \frac{0 - 0}{1} = 0.$$

Since $f'(t) = \left[1/(2\sqrt{t})\right](1 - t^3) + \sqrt{t}(-3t^2)$, the instantaneous rate of change at $t_0 = 1$ is $f'(1) = 0 + (-3) = -3$.

5. (a) $v(t) = s'(t) = 3$

(b) $v(t)$ is never zero

(c) $[0, +\infty)$ since $v(t) > 0$ for all $t \geq 0$.

7. (a) $v(t) = s'(t) = \left[(1+t)(0) - (1)(1)\right]/(1+t)^2 = -1/(1+t)^2$

(b) $v(t)$ is never zero

(c) $v(t)$ is never positive

9. (a) $v(t) = s'(t) = 3t^2 - 18t + 24 = 3(t-4)(t-2)$

(b) $v(t) = 0 \iff t = 2, 4$

(c) $[0, 2), (4, +\infty)$ since, for positive t, $v(t) > 0 \iff 0 < t < 2$ or $t > 4$.

11. (a) $v(t) = s'(t) = 3t^2 - 12t + 9 = 3(t-3)(t-1)$

    (b) $v(t) = 0 \iff t = 1, 3$

    (c) $[0,1), (3,+\infty)$ since, for positive t, $v(t) > 0 \iff 0 < t < 1$ or $t > 3$.

13. $s(t) = \dfrac{t^2+2}{t+1} \implies v(t) = s'(t) = \dfrac{(t+1)(2t) - (t^2+2)(1)}{(t+1)^2} = \dfrac{t^2+2t-2}{(t+1)^2}$. Hence $v(3) = (9+6-2)/4^2$
    $= 13/16$.

15. Let $V(t)$ stand for the volume of the snowball at time t. Then $V(t) = (4/3)\pi r^3 = (4/3)\pi(4-t^2)^3$.

    (a) The average rate of change of the volume from $t = 0$ to $t = 1$ is

    $$\frac{V(1) - V(0)}{1-0} = (4/3)\pi(27) - (4/3)\pi(64) = (4/3)\pi(-37) \approx -154.99 \text{ in}^3/\text{min}.$$

    The average rate of change of the volume from $t = 1$ to $t = 2$ is

    $$\frac{V(2) - V(1)}{2-1} = (4/3)\pi(0) - (4/3)\pi(27) = -36\pi \approx -113.1 \text{ in}^3/\text{min}.$$

    (b) $V'(t) = (4/3)\pi\left[3(4-t^2)^2(-2t)\right] = -8\pi t(4-t^2)^2$ (see Exercise 46(b), Section 3.2). Hence the instantaneous rate of change of the volume at $t = 1$ is $V'(1) = -8\pi(1)(9) = -72\pi \approx -226.2$ in^3/min.

17. (a) $v(t) = s'(t) = 6 - 2t \implies v_0 = v(0) = 6$.

    (b) $v(t) = 0 \iff 6 - 2t = 0 \iff t = 3$. Since $v(t) > 0$ for $t < 3$ and $v(t) < 0$ for $t > 3$, velocity changes sign at $t = 3$ and hence the particle changes direction at $t = 3$.

    (c) The particle is at the origin when $s(t) = 0 \implies t(6-t) = 0 \implies t = 0, 6$. Thus, the second time it is at the origin occurs when $t = 6$ and its velocity is $v(6) = 6 - 2(6) = -6$.

19. Let $s(t)$ stand for the position of the object after t seconds. Then $s(t) = -4.9t^2$ (see Example 2 in the text).

    (a) $v(t) = s'(t) = -9.8t = -49 \implies t = 5$. Hence the object fell for 5 seconds.

    (b) $s(5) = -4.9(5)^2 = -122.5$. Hence the object was dropped from 122.5 meters.

    (c) Average velocity from $t = 0$ to $t = 5$ is

    $$\frac{s(5) - s(0)}{5-0} = \frac{-122.5 - 0}{5} = -24.5 \text{ m/s}.$$

21. (a) $s(t) = 40 + 400t - 4.9t^2$

    (b) $v(t) = s'(t) = 400 - 9.8t = 0 \implies t = 40.82 \implies$ Maximum height $= s(40.82) = 8203.3$ meters.

    (c) For t positive,

    $$s(t) = 40 + 400t - 4.9t^2 = 0 \implies t = \frac{-400 - \sqrt{(400)^2 - 4(40)(-4.9)}}{2(-4.9)} = 81.73.$$

    Thus, the time to impact is 81.73 seconds. Impact speed $= |v(81.73)| = 400.95$ m/s.

23. (a) The graph of $s(t)$ is shown to the right.

    (b) $v(t) = s'(t) = 800 - 32t$.

(c) The object is rising $\iff v(t) > 0 \iff 800 - 32t > 0$
    $\iff 0 < t < 25$.

(d) The object is falling $\iff v(t) < 0 \iff 800 - 32t < 0$
    $\iff 25 < t < 50$.

(e) The object is at rest $\iff v(t) = 0 \iff 800 - 32t = 0$
    $\iff t = 25$.

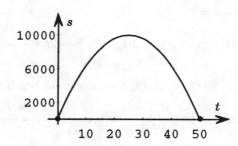

(f) Maximum height $= s(25) = 800(25) - 16(25)^2 = 10{,}000$ feet.

(g) $s(t) = 9600 \implies 800t - 16t^2 = 9600 \implies t^2 - 50t + 600 = (t - 20)(t - 30) = 0 \implies t = 20, 30$.
    Thus, it takes 20 seconds to reach a height of 9600 feet.

(h) For positive t, $s(t) = 0 \implies 800t - 16t^2 = t(800 - 16t) = 0 \implies t = 50$. Hence impact occurs at 50 seconds.

3.4 Derivatives of the Trigonometric Functions

1. $f(x) = 4\cos x \implies f'(x) = -4\sin x$

3. $f(x) = x^3 \tan x \implies f'(x) = (3x^2)\tan x + x^3(\sec^2 x)$

5. $f(x) = (x^3 - 2)\cot x \implies f'(x) = (3x^2)\cot x + (x^3 - 2)(-\csc^2 x)$

7. $f(x) = \sin x \sec x = \tan x \implies f'(x) = \sec^2 x$

9. $f(x) = x\cos x - x\sin x = x(\cos x - \sin x) \implies f'(x) = (1)(\cos x - \sin x) + x(-\sin x - \cos x)$

11. $f(x) = \sec x \tan x \implies f'(x) = (\sec x \tan x)\tan x + (\sec x)(\sec^2 x) = \sec x(\tan^2 x + \sec^2 x)$

13. $f(x) = \dfrac{x}{2 + \sin x} \implies f'(x) = \dfrac{(2 + \sin x)(1) - x(\cos x)}{(2 + \sin x)^2}$

15. $f(x) = \dfrac{\sin x - \cos x}{1 + \tan x} \implies f'(x) = \dfrac{(1 + \tan x)(\cos x + \sin x) - (\sin x - \cos x)(\sec^2 x)}{(1 + \tan x)^2}$

17. $f(x) = \dfrac{x^2 + 4\cot x}{x + \tan x} \implies f'(x) = \dfrac{(x + \tan x)(2x - 4\csc^2 x) - (x^2 + 4\cot x)(1 + \sec^2 x)}{(x + \tan x)^2}$

19. $f(x) = \dfrac{3\csc x}{4x^2 - 5\tan x} \implies$

$$f'(x) = \frac{(4x^2 - 5\tan x)(-3\csc x \cot x) - (3\csc x)(8x - 5\sec^2 x)}{(4x^2 - 5\tan x)^2}$$

21. $y = x\sin x \implies y' = \sin x + x\cos x \implies y'(\pi) = \sin \pi + \pi \sin \pi = -\pi$. Hence the slope of the tangent line at $(\pi, 0)$ is $-\pi$ and the equation is $y = -\pi(x - \pi)$.

23. The tangent line is horizontal $\iff y' = \sec x \tan x = 0 \iff \tan x = 0 \iff x = 0, \pi, 2\pi, 3\pi, 4\pi$ in the interval $[0, 4\pi]$.

25. The piston changes direction when its velocity changes sign. Now, $s(t) = 5 - 4\sin t \implies v(t) = s'(t) = -4\cos t = 0 \implies t = \pi/2, 3\pi/2, 5\pi/2, \ldots$.

Sign v	$-$	$+$	$-$
Interval	$[0, \pi/2]$	$[\pi/2, 3\pi/2]$	$[3\pi/2, 5\pi/2]$
Direction of piston	up	down	up

Hence the piston changes direction when $t = \pi/2, 3\pi/2, 5\pi/2, \ldots$.

27. (a) The tangent lines are parallel when the derivatives are equal:

$$\frac{d}{dx}(\sin x) = \frac{d}{dx}(\cos x) \implies \cos x = -\sin x \implies \tan x = -1 \implies x = 3\pi/4, 7\pi/4.$$

Note that if $\cos x = -\sin x$, then $\cos x \neq 0$ since, if $\cos x = 0$, $\sin x = \pm 1$, and hence we may divide by $\cos x$.

(b) The tangent lines are perpendicular when the product of the derivatives is equal to -1:

$$\left[\frac{d}{dx}(\sin x)\right]\left[\frac{d}{dx}(\cos x)\right] = -1 \implies (\cos x)(-\sin x) = -1 \implies (1/2)\sin 2x = 1 \implies \sin 2x = 2.$$

There are no such numbers x since $|\sin x| \leq 1$ for all x. Hence there are no points where the tangent lines are perpendicular.

29. $\dfrac{d}{dx}\cot x = \dfrac{d}{dx}\left(\dfrac{\cos x}{\sin x}\right) = \dfrac{(\sin x)(-\sin x) - (\cos x)(\cos x)}{\sin^2 x} = \dfrac{-1}{\sin^2 x} = -\csc^2 x$

$\dfrac{d}{dx}\sec x = \dfrac{d}{dx}\left(\dfrac{1}{\cos x}\right) = \dfrac{(\cos x)(0) - (1)(-\sin x)}{\cos^2 x} = \dfrac{1}{\cos x}\dfrac{\sin x}{\cos x} = \sec x \tan x$

$\dfrac{d}{dx}\csc x = \dfrac{d}{dx}\left(\dfrac{1}{\sin x}\right) = \dfrac{(\sin x)(0) - (1)(\cos x)}{\sin^2 x} = -\dfrac{1}{\sin x}\dfrac{\cos x}{\sin x} = -\csc x \cot x$

31. (a) $v(t) = s'(t) = -10\sin t$

(b) The mass changes direction when $v(t)$ changes sign, as indicated in the table below.

Sign v	$-$	$+$
Interval	$[0, \pi]$	$[\pi, 2\pi]$
Direction of motion	left	right

Hence the mass changes its direction of motion at $t = \pi, 2\pi, \ldots$.

3.5   The Chain Rule

1. $y = f(u) = u^3 + 1 \implies dy/du = 3u^2$; $u = g(x) = 1 - x^2 \implies du/dx = -2x$. Hence $y = f(g(x)) = (1 - x^2)^3 + 1$ and

$$\frac{dy}{dx} = \frac{dy}{du}\frac{du}{dx} = (3u^2)(-2x) = 3(1 - x^2)^2(-2x) = -6x(1 - x^2)^2.$$

3. $y = f(u) = u(1 - u^2) = u - u^3 \implies dy/du = 1 - 3u^2$; $u = g(x) = 1/x \implies du/dx = -1/x^2$. Hence $y = f(g(x)) = 1/x - 1/x^3$ and

$$\frac{dy}{dx} = \frac{dy}{du}\frac{dy}{du} = (1 - 3u^2)\left(-\frac{1}{x^2}\right) = \left(1 - \frac{3}{x^2}\right)\left(-\frac{1}{x^2}\right).$$

5. $y = f(u) = \tan u \implies dy/du = \sec^2 u$; $u = g(x) = 1 + x^4 \implies du/dx = 4x^3$. Hence $y = f(g(x)) = \tan(1 + x^4)$ and

$$\frac{dy}{dx} = \frac{dy}{du}\frac{du}{dx} = (\sec^2 u)(4x^3) = 4x^3 \sec^2(1 + x^4).$$

7. $y = f(u) = 1/(1 + u) \implies dy/du = -1/(1 + u)^2$; $u = g(x) = \sin^2 x \implies du/dx = 2\sin x \cos x$ (see Exercise 8, Section 3.4). Hence $y = f(g(x)) = 1/(1 + \sin^2 x)$ and

$$\frac{dy}{dx} = \frac{dy}{du}\frac{du}{dx} = \frac{-1}{(1 + u)^2}(2\sin x \cos x) = \frac{-2\sin x \cos x}{(1 + \sin^2 x)^2}.$$

9. $y = f(u) = \cos u \implies dy/du = -\sin u$; $u = g(x) = \sin x \implies du/dx = \cos x$. Hence $y = f(g(x)) = \cos(\sin x)$ and

$$\frac{dy}{dx} = \frac{dy}{du}\frac{du}{dx} = (-\sin u)(\cos x) = -\sin(\sin x)\cos x.$$

11. $y = f(u) = \sin u$. Hence

$$\frac{dy}{dx} = \frac{dy}{du}\frac{du}{dx} = (\cos u)(2x + 1) = [\cos(x^2 + x)](2x + 1).$$

13. $y = f(u) = (1 + u)/(1 - u)$. Hence

$$\frac{dy}{dx} = \frac{dy}{du}\frac{du}{dx} = \frac{(1 - u)(1) - (1 + u)(-1)}{(1 - u)^2}(\cos x) = \frac{2\cos x}{(1 - \sin x)^2}.$$

15. $y = f(u) = u^4$. Hence

$$\frac{dy}{dx} = \frac{dy}{du}\frac{du}{dx} = (4u^3)\frac{(3x - 7)(3) - (3x + 7)(3)}{(3x - 7)^2} = 4\left(\frac{3x + 7}{3x - 7}\right)^3\left[\frac{-42}{(3x - 7)^2}\right] = \frac{-168(3x + 7)^3}{(3x - 7)^5}.$$

17. $y = f(u) = [(u + 7)/(u - 7)]^4$. To find dy/du, let $v = (u + 7)/(u - 7)$. Then $y = v^4$ and hence

$$\frac{dy}{du} = \frac{dy}{dv}\frac{dv}{du} = (4v^3)\frac{(u - 7)(1) - (u + 7)(1)}{(u - 7)^2} = 4\left(\frac{u + 7}{u - 7}\right)^3\left[\frac{-14}{(u - 7)^2}\right].$$

Therefore

$$\frac{dy}{dx} = \frac{dy}{du}\frac{du}{dx} = 4\left(\frac{u + 7}{u - 7}\right)^3\left[\frac{-14}{(u - 7)^2}\right](3) = \frac{-168(3x + 7)^3}{(3x - 7)^5}.$$

19. $y = (x^2 + 4)^3 \implies dy/dx = 3(x^2 + 4)^2(2x) = 6x(x^2 + 4)^2$

21. $y = (\cos x - x)^6 \implies dy/dx = 6(\cos x - x)^5(-\sin x - 1)$

23. $y = \sin(x^3 + 3x) \implies dy/dx = [\cos(x^3 + 3x)](3x^2 + 3)$

25. $y = \tan(x^2 + x) \implies dy/dx = [\sec^2(x^2 + x)](2x + 1)$

27. $y = x(x^4 - 5)^3 \implies dy/dx = x[3(x^4 - 5)^2(4x^3)] + (x^4 - 5)^3(1) = (x^4 - 5)^2(13x^4 - 5)$

29. $y = \dfrac{1}{(x^2 - 9)^3} = (x^2 - 9)^{-3} \implies dy/dx = -3(x^2 - 9)^{-4}(2x) = \dfrac{-6x}{(x^2 - 9)^4}$

31. $y = (3\tan x - 2)^4 \implies dy/dx = 4(3\tan x - 2)^3(3\sec^2 x)$

33. $y = \left(\dfrac{x-3}{x+3}\right)^4 \implies$

$$\frac{dy}{dx} = 4\left(\frac{x-3}{x+3}\right)^3 \frac{(x+3)(1)-(x-3)(1)}{(x+3)^2} = 4\left(\frac{x-3}{x+3}\right)^3 \frac{6}{(x+3)^2} = \frac{24(x-3)^3}{(x+3)^5}$$

35. $y = x\cos(1-x^2) \implies dy/dx = x\left[(-\sin(1-x^2))(-2x)\right] + (1)\cos(1-x^2) = 2x^2\sin(1-x^2)+\cos(1-x^2)$

37. $y = \dfrac{\tan^2 x + 1}{1-x} \implies$

$$\frac{dy}{dx} = \frac{(1-x)\left[2(\tan x)(\sec^2 x)\right] - (\tan^2 x + 1)(-1)}{(1-x)^2} = \frac{2(1-x)\tan x\sec^2 x + \tan^2 x + 1}{(1-x)^2}$$

39. $y = \dfrac{\tan^2(\pi/4) + 1}{\sec(0)} \implies dy/dx = 0$ since y is constant.

41. $y = \dfrac{x}{(x^2+x+1)^6} \implies$

$$\frac{dy}{dx} = \frac{(x^2+x+1)^6(1) - x\left[6(x^2+x+1)^5(2x+1)\right]}{(x^2+x+1)^{12}} = \frac{-11x^2-5x+1}{(x^2+x+1)^7}$$

43. $y = \left(\dfrac{ax+b}{cx+d}\right)^4 \implies dy/dx = 4\left(\dfrac{ax+b}{cx+d}\right)^3 \dfrac{(cx+d)(a)-(ax+b)(c)}{(cx+d)^2} = \dfrac{4(ax+b)^3(ad-bc)}{(cx+d)^5}$

45. $y = \dfrac{1}{1+\cos^3 x} = (1+\cos^3 x)^{-1} \implies dy/dx = -(1+\cos^3 x)^{-2}3(\cos^2 x)(-\sin x) = \dfrac{3\cos^2 x\sin x}{(1+\cos^3 x)^2}$

47. $y = \dfrac{1}{(1+x^4)^3} = (1+x^4)^{-3} \implies dy/dx = -3(1+x^4)^{-4}(4x^3) = \dfrac{-12x^3}{(1+x^4)^4}$

49. $y = \tan(6x) - 6\tan x \implies dy/dx = \left[\sec^2(6x)\right](6) - 6\sec^2 x = 6\sec^2(6x) - 6\sec^2 x$

51. $y = \sec^3 4x \implies dy/dx = 3(\sec^2 4x)(\sec 4x\tan 4x)(4) = 12\sec^3 4x\tan 4x$

53. $y = \tan^2(\pi - x^2) \implies dy/dx = 2[\tan(\pi - x^2)][\sec^2(\pi - x^2)](-2x) = -4x\tan(\pi - x^2)\sec^2(\pi - x^2)$

55. $y = (\sin\pi x - \cos\pi x)^4 \implies dy/dx = 4(\sin\pi x - \cos\pi x)^3[\pi\cos\pi x + \pi\sin\pi x]$

57. $y = \left[1 + (x^2-3)^4\right]^6 \implies dy/dx = 6\left[1 + (x^2-3)^4\right]^5\left[4(x^2-3)^3(2x)\right] = 48x(x^2-3)^3\left[1 + (x^2-3)^4\right]^5$

59. $y = \cos^2\left(\dfrac{1+x}{1-x}\right) \implies$

$$\frac{dy}{dx} = 2\left[\cos\left(\frac{1+x}{1-x}\right)\right]\left[-\sin\left(\frac{1+x}{1-x}\right)\right]\left[\frac{(1-x)(1)-(1+x)(-1)}{(1-x)^2}\right]$$

$$= \frac{-4}{(1-x)^2}\cos\left(\frac{1+x}{1-x}\right)\sin\left(\frac{1+x}{1-x}\right)$$

61. $y = 1 + \left[1 + (1 + x^2)^2\right]^2 \implies$

$$\frac{dy}{dx} = 2\left[1 + (1 + x^2)^2\right]\left[2(1 + x^2)\right](2x) = 8x(1 + x^2)\left[1 + (1 + x^2)^2\right]$$

63. $y = \tan\sqrt{1 + x^2} \implies$

$$\frac{dy}{dx} = \left[\sec^2\sqrt{1 + x^2}\right]\left[\frac{1}{2\sqrt{1 + x^2}}\right](2x) = \frac{x}{\sqrt{1 + x^2}}\sec^2\sqrt{1 + x^2}$$

65. $y = \left(\dfrac{x}{x + 1}\right)^4 \implies$

$$y' = 4\left(\frac{x}{x + 1}\right)^3\frac{(x + 1)(1) - x(1)}{(x + 1)^2} = 4\left(\frac{x}{x + 1}\right)^3\frac{1}{(x + 1)^2} \implies y'(0) = 0.$$

Hence the slope of the tangent line at $(0, 0)$ is 0 and its equation is $y = 0$.

67. $y = x\cos(\pi + x^2) \implies y' = x\left[-\sin(\pi + x^2)\right](2x) + \cos(\pi + x^2) \implies y'(0) = \cos\pi = -1$. Hence the slope of the tangent line at $(0, 0)$ is -1 and its equation is $y = -x$.

69. $y = \tan\left(\dfrac{\pi/4 - x}{1 + x}\right) \implies$

$$y' = \left[\sec^2\left(\frac{\pi/4 - x}{1 + x}\right)\right]\frac{(1 + x)(-1) - (\pi/4 - x)(1)}{(1 + x)^2} = \sec^2\left(\frac{\pi/4 - x}{1 + x}\right)\frac{-1 - \pi/4}{(1 + x)^2}$$

$$\implies y'(0) = \left[\sec^2(\pi/4)\right](-1 - \pi/4) = -2(1 + \pi/4).$$

Hence the slope of the tangent line at $(0, 1)$ is $-2(1 + \pi/4)$ and its equation is $y - 1 = -2(1 + \pi/4)x$.

71. (a) $(f \circ g)'(1) = f'\big(g(1)\big)g'(1) = f'(2)g'(1) = (5)(2) = 10$

 (b) $(g \circ f)'(1) = g'\big(f(1)\big)f'(1) = g'(1)f'(1) = (2)(-2) = -4$

73. $h(t) = \sin g(t) \implies h'(t) = [\cos g(t)]g'(t) \implies$

$$h'(2) = [\cos g(2)]g'(2) = [\cos(3\pi/4)]\sqrt{2} = (-\sqrt{2}/2)\sqrt{2} = -1$$

75. $f(x) = \sin x^2 \implies f'(x) = (\cos x^2)(2x) \implies$

$$f'(\sqrt{\pi}/2) = \left[\cos(\sqrt{\pi}/2)^2\right](2\sqrt{\pi}/2) = [\cos(\pi/4)]\sqrt{\pi} = (\sqrt{2}/2)\sqrt{\pi} = \sqrt{\pi/2}.$$

Hence the slope of the tangent line at $(\sqrt{\pi}/2, \sqrt{2}/2)$ is $\sqrt{\pi/2}$ and hence the equation is $y - \sqrt{2}/2 = \sqrt{\pi/2}\,(x - \sqrt{\pi}/2)$.

77. $v(t) = s'(t) = 5[\cos(4t)](4) = 20\cos(4t)$ cm/s.

79. Let $V(t)$ and $r(t)$ stand for the volume and radius of the balloon after t seconds, respectively. Then $V(t) = (4/3)\pi r(t)^3 \implies dV/dt = 4\pi r(t)^2(dr/dt)$. From the given information, $dr/dt = 0.5$. Hence, when $r = 5$ inches,

$$\left.\frac{dV}{dt}\right|_{r=5} = 4\pi(5)^2(0.5) = 50\pi.$$

Thus, when the radius of the balloon is 5 inches the volume is increasing at a rate of 50π in^3/s.

81. (a) $|x|^2 = |x^2| = x^2$ since $x^2 \geq 0 \implies |x| = \sqrt{x^2}$ since $|x| \geq 0$.

(b) $f(x) = \sqrt{x^2} = (x^2)^{1/2} \implies$

$$f'(x) = \frac{1}{2}(x^2)^{-1/2}(2x) = \frac{x}{\sqrt{x^2}} = \frac{x}{|x|}.$$

(c) Because $\sqrt{x^2} = g(h(x))$, where $g(x) = \sqrt{x}$ and $h(x) = x^2$, and $g'(h(0)) = g'(0)$ does not exist.

83. (c) We have that

$$\begin{aligned}
(g^{n+1})'(x) &= (g \cdot g^n)'(x) \\
&= (g' \cdot g^n)(x) + g \cdot (g^n)'(x) && \text{by the Product Rule} \\
&= (g' \cdot g^n)(x) + g \cdot (ng^{n-1} \cdot g')(x) && \text{by the induction hypothesis} \\
&= g'(x)g^n(x) + ng^n(x)g'(x) \\
&= (n+1)g^n(x)g'(x).
\end{aligned}$$

85. (a) The graph with range $[-1,1] \times [-2,2]$ is shown below on the left. Note that since $\lim\limits_{x \to 0} \frac{f(x)}{x} = 0$,

$$f'(0) = \lim_{x \to 0} \frac{f(x) - f(0)}{x - 0} = \lim_{x \to 0} \frac{f(x)}{x} = 0.$$

Therefore $f(x)$ is differentiable at $x = 0$.

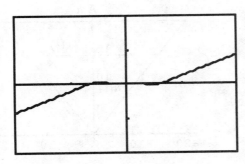

(a)

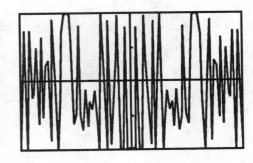

(b)

(b) $f'(x) = \begin{cases} 2x\sin(1/x) - \cos(1/x), & \text{if } x \neq 0 \\ 0, & \text{if } x = 0 \end{cases}$. The graph with the range $[-.01, .01] \times [-1,1]$ is shown above on the right. It appears that $f'(x)$ is not continuous at $x = 0$.

(c) $f'(x) = \frac{2}{n\pi}\sin(n\pi) - \cos(n\pi) = (-1)^{n+1}$. Hence $f'(x)$ is alternatively 1 and -1 for an infinite number of points arbitrarily close to 0.

3.6 Related Rates

1. $f(t) = 2[g(t)]^3 + 5 \implies f'(t) = 6[g(t)]^2 g'(t) \implies$

$$f'(1) = 6[g(1)]^2 g'(1) = 6(3)^2(-2) = -108.$$

73

3. $f(t) = \dfrac{1}{1 + g(t)} = [1 + g(t)]^{-1} \implies f'(t) = -[1 + g(t)]^{-2}g'(t) \implies$

$$f'(2) = -[1 + g(2)]^{-2}g'(2) = -[1 + 3]^{-2}(-2) = 1/8.$$

5. $\sin\big(f(t)\big) = [g(t)]^2 \implies \big[\cos\big(f(t)\big)\big]f'(t) = 2g(t)g'(t) \implies \big[\cos\big(f(0)\big)\big]f'(0) = 2g(0)g'(0) \implies$

$$[\cos(\pi/6)]f'(0) = 2(1)(-2) \implies (\sqrt{3}/2)f'(0) = -4 \implies f'(0) = -8\sqrt{3}/3.$$

7. Let $D(x)$ stand for the length of the diagonal of a cube of edge length x. Then $D(x)^2 = x^2 + x^2 + x^2 = 3x^2 \implies D(x) = x\sqrt{3} \implies dD/dt = \sqrt{3}\,(dx/dt)$. From the given information, $dx/dt = 2$. Hence $dD/dt = 2\sqrt{3}$. Thus, the diagonal is increasing at the rate of $2\sqrt{3}$ cm/s.

9. Let $V(t)$ and $r(t)$ stand for the volume and radius of the snowball after t seconds, respectively. Then $V(t) = (4/3)\pi r(t)^3 \implies dV/dt = 4\pi r(t)^2(dr/dt)$. From the given information, $dr/dt = -1$. Hence, when $r = 6$,

$$\left.\frac{dV}{dt}\right|_{r=6} = 4\pi(6)^2(-1) = -144\pi.$$

Thus, when the radius is 6 cm, the volume is decreasing at a rate of 144π cm^3/s.

11. Let $V(t)$, $h(t)$, and $r(t)$ stand for the volume, depth, and radius of water in the tank after t minutes, as shown in figures below.

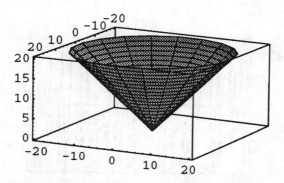
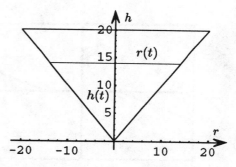

Then $V(t) = (1/3)\pi[r(t)]^2h(t)$. To eliminate $r(t)$ from this equation, observe that, by similar triangles,

$$\frac{r(t)}{h(t)} = \frac{20}{20} \implies r(t) = h(t).$$

Hence

$$V(t) = \frac{1}{3}\pi[h(t)]^2h(t) = \frac{\pi}{3}[h(t)]^3 \implies \frac{dV}{dt} = \pi[h(t)]^2\frac{dh}{dt}.$$

From the given information, $dV/dt = 40$. Hence, when $h = 8$,

$$\left.\frac{dV}{dt}\right|_{h=8} = \pi(8)^2\frac{dh}{dt} = 40 \implies \frac{dh}{dt} = \frac{5}{8\pi} \approx 0.199.$$

Thus, when the water is 8 meters deep, it is rising at the rate of approximately 0.199 m/min.

13. Let $A(t)$ stand for the area of a rectangle of width $y(t)$ and length $2y(t)$ at time t. Then

$$A(t) = [2y(t)][y(t)] = 2[y(t)]^2 \implies \frac{dA}{dt} = 4y(t)\frac{dy}{dt} = 8 \implies \frac{dy}{dt} = \frac{2}{y(t)}.$$

Hence, when $y = 5$, $dy/dt = 2/5$ and therefore the rate of increase of the length $2y$ is

$$\frac{d}{dt}[2y(t)] = 2\left.\frac{dy}{dt}\right|_{y=5} = 2\left(\frac{2}{5}\right) = \frac{4}{5} \text{ cm/s.}$$

15. Let $x(t)$ be the horizontal distance from the shore to the boat and $s(t)$ the length of rope at time t. Then

$$[x(t)]^2 + 6^2 = [s(t)]^2 \implies 2x(t)\frac{dx}{dt} = 2s(t)\frac{ds}{dt} \implies \frac{dx}{dt} = \frac{s(t)}{x(t)}\frac{ds}{dt}.$$

From the given information, $ds/dt = -5$. Hence, when $x = 8$, $s = \sqrt{8^2 + 6^2} = 10$ and therefore

$$\left.\frac{dx}{dt}\right|_{x=8} = \frac{10}{8}(-5) = -6.25.$$

Thus, the boat is approaching the shore at 6.25 m/min.

17. Let $x(t)$ be the distance from the lamp post to the woman and $s(t)$ the length of her shadow at time t, as illustrated in the figure below. Then, by similar triangle,

$$\frac{10}{1.6} = \frac{x(t) + s(t)}{s(t)}$$

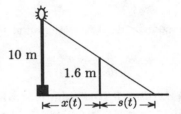

$$\implies 10s(t) = 1.6x(t) + 1.6s(t)$$

$$\implies s(t) = \frac{4}{21}x(t) \implies \frac{ds}{dt} = \frac{4}{21}\frac{dx}{dt}.$$

Observe that the formula for ds/dt does not involve x, only dx/dt, and hence the rate of change of the woman's shadow is independent of how far she is from the lamp post. From the given information, $dx/dt = 1.2$. Hence

$$\frac{ds}{dt} = \frac{4}{21}(1.2) \approx 0.23.$$

Thus, her shadow is increasing at a rate of approximately 0.23 m/s regardless of how far she is from the lamp post.

19. $N(t) = I(t) + S(t) \implies dN/dt = dI/dt + dS/dt$. From the given information, $dI/dt = 24$ and $dS/dt = -20$. Hence $dN/dt = 4$. Thus, the population is increasing at a rate of 4 persons per day.

21. Let $x(t)$ be the distance east from Columbus of the first truck, $y(t)$ the distance north of the second truck, and $D(t)$ the distance between the trucks t hours after noon. Note that $y(t) = 0$ for $0 \le t \le 1$ and $60(t - 1)$ for $t > 1$. Now,

$$D(t)^2 = x(t)^2 + y(t)^2 \implies 2D(t)\frac{dD}{dt} = 2x(t)\frac{dx}{dt} + 2y(t)\frac{dy}{dt}.$$

75

From the given information, $dx/dt = 40$ and $dy/dt = 60$. Hence

$$2D(t)\frac{dD}{dt} = 2x(t)(40) + 2y(t)(60) \implies \frac{dD}{dt} = \frac{40x(t) + 60y(t)}{D(t)}.$$

At 2:00 pm, $t = 2$, $x = 40(2) = 80$, $y = 60(1) = 60$, and $D = \sqrt{80^2 + 60^2}$. Hence

$$\left.\frac{dD}{dt}\right|_{t=2} = \frac{40(80) + 60(60)}{100} = 68.$$

Thus, at 2:00 pm the distance between the trucks is increasing at the rate of 68 km/h.

23. In this case, $x = h/\sqrt{3}$ and hence

$$V(t) = \frac{1}{2}(2x)(h)(400) = \frac{400h^2}{\sqrt{3}} \implies \frac{dV}{dt} = \frac{800h}{\sqrt{3}}\frac{dh}{dt} \implies \frac{dh}{dt} = \frac{\sqrt{3}}{800h}\frac{dV}{dt}.$$

Hence, when $h = 10$,

$$\left.\frac{dh}{dt}\right|_{h=10} = \frac{\sqrt{3}}{800(10)}(9000) = \frac{9}{8}\sqrt{3} \approx 1.95.$$

Thus, when the water is 10 cm deep it is rising at the rate of approximately 1.95 cm/s.

25. Let $\big(x(t), y(t)\big)$ stand for the position of the point at time t. Then

$$x^2 + y^2 = 1 \implies 2x\frac{dx}{dt} + 2y\frac{dy}{dt} = 0.$$

(a) If the coordinates are changing at the same rate, then $dx/dt = dy/dt$ and hence

$$2x\frac{dx}{dt} + 2y\frac{dx}{dt} = 2(x+y)\frac{dx}{dt} = 0 \implies x + y = 0 \quad\text{or}\quad \frac{dx}{dt} = \frac{dy}{dt} = 0.$$

Since $dx/dt \neq 0$ (the point is moving), it follows that the x- and y-coordinates are changing at the same rate at the points where $y = -x$, namely at $\left(-\sqrt{2}/2, \sqrt{2}/2\right)$ and $\left(\sqrt{2}/2, -\sqrt{2}/2\right)$.

(b) If the coordinates are changing at opposite rates, $dx/dt = -dy/dt$ and hence

$$2x\frac{dx}{dt} - 2y\frac{dx}{dt} = 2(x-y)\frac{dx}{dt} = 0.$$

Thus, if the coordinates are changing at opposite rates, $x - y = 0$, or $x = y$, and hence is at $\left(\sqrt{2}/2, \sqrt{2}/2\right)$ and $\left(-\sqrt{2}/2, -\sqrt{2}/2\right)$.

27. Let $s(\theta)$ stand for the length of the shadow when the sun is θ radians above the horizon. Then

$$s(\theta) = 30\cot\theta \implies \frac{ds}{dt} = -30\csc^2\theta\,\frac{d\theta}{dt}.$$

From the given information, $d\theta/dt = -15(\pi/180)$. Hence, when $\theta = 30° = \pi/6$,

$$\left.\frac{ds}{dt}\right|_{\theta=\pi/6} = -30\csc^2(\pi/6)\left(\frac{-15\pi}{180}\right) = 10\pi.$$

Thus, the shadow length is increasing at the rate of 10π m/h.

29. Let $V(t)$, $r(t)$, and $h(t)$ be the volume, radius, and height of the coffee in the bottom container. Since coffee fills only the bottom portion of the cone,

$$
\begin{aligned}
V(t) &= \text{Total volume of cone} - \text{Volume of unfilled portion of cone} \\
&= \frac{1}{3}\pi(4)^2(10) - \frac{1}{3}\pi r^2(10 - h) \\
&= \frac{1}{3}\pi\left[160 - r^2(10 - h)\right].
\end{aligned}
$$

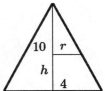

Now, $r/4 = (10 - h)/10 \implies r = (2/5)(10 - h)$. Hence

$$
V(t) = \frac{1}{3}\pi\left[160 - \frac{4}{25}(10 - h)^3\right]
$$

and therefore

$$
\frac{dV}{dt} = -\frac{\pi}{3}\frac{4}{25}\left[3(10 - h)^2\left(-\frac{dh}{dt}\right)\right] = 4 \implies \frac{dh}{dt} = \frac{25}{\pi(10 - h)^2}.
$$

Hence, when $h = 4$,

$$
\left.\frac{dh}{dt}\right|_{h=4} = \frac{25}{\pi(36)} \approx 0.22.
$$

Thus, when the depth of coffee in the bottom container is 4 cm, it is rising at the rate of approximately 0.22 cm/sec.

31. Let $x(t)$ be the distance along the shore from the piling to the light beam and $\theta(t)$ the angle between the light beam and the piling on shore. Then

$$
x = 200\tan\theta \implies \frac{dx}{dt} = 200\left(\sec^2\theta\right)\frac{d\theta}{dt}.
$$

From the given information, $d\theta/dt = 2$ rev/min $= 2(2\pi)$ rad/min $= 4\pi$ rad/min. Hence, when $x = 200$, $\tan\theta = 1 \implies \sec^2\theta = 2$ and therefore

$$
\frac{dx}{dt} = 200(2)(4\pi) = 1600\pi \approx 5026.5.
$$

Thus, when the light beam is 200 m from the piling, it is moving at the rate of 1600π m/min, or equivalently, $80\pi/3$ m/s.

33. $(\sin\alpha)/(\sin\beta) = C_a \implies \sin\beta = (1/C_a)\sin\alpha \implies (\cos\beta)(d\beta/dt) = (1/C_a)(\cos\alpha)(d\alpha/dt)$. From the given information, $C_a = 1.33$, $d\alpha/dt = 0.2$, and hence

$$
\cos\beta\frac{d\beta}{dt} = \frac{1}{1.33}(\cos\alpha)(0.2) = 0.15\cos\alpha \implies \frac{d\beta}{dt} = 0.15\frac{\cos\alpha}{\cos\beta}\text{ rad/s}.
$$

35. (a) Here $L = 25$, $r = 0$, $dR/dt = 0.2$. Therefore

$$
\nu = \frac{\alpha}{25}R^2 \implies \frac{d\nu}{dt} = \frac{\alpha}{25}\left(2R\frac{dR}{dt}\right) = \frac{\alpha}{25}(2R)(0.2) = 0.016\alpha R.
$$

When $R = 10$,

$$
\left.\frac{d\nu}{dt}\right|_{R=10} = 0.016\alpha(10) = 0.16\alpha.
$$

Thus, when $R = 10$, the acceleration at the center is 0.16α cm/min^2.

(b) In this case $R = 10$, $r = 0$, $dL/dt = 0.5$. Therefore

$$\nu = \frac{\alpha}{L}(10^2) \implies \frac{d\nu}{dL} = -\frac{100\alpha}{L^2}\frac{dL}{dt} = -\frac{100\alpha}{L^2}(0.5) = -\frac{50\alpha}{L^2}.$$

When $L = 25$,

$$\left.\frac{d\nu}{dL}\right|_{L=25} = -\frac{50\alpha}{25^2} = -0.08\alpha.$$

Thus, when $L = 25$, the acceleration at the center is -0.08α cm/s^2.

3.7 Linear Approximation

1. (a) $f(x) = \sqrt{x} \implies f'(x) = 1/(2\sqrt{x})$. Hence, if $x_0 = 4$,

$$f(x_0) + f'(x_0)\Delta x = \sqrt{4} + \frac{1}{2\sqrt{4}}\Delta x = 2 + \frac{1}{4}\Delta x.$$

(b) Here $\Delta x = 0.1$. Hence

$$\sqrt{4.1} \approx 2 + \frac{1}{4}(0.1) = 2.025.$$

3. (a) $f(x) = 1/x \implies f'(x) = -1/x^2$. Hence, if $x_0 = 2$,

$$f(x_0) + f'(x_0)\Delta x = \frac{1}{2} - \frac{1}{2^2}\Delta x = \frac{1}{2} - \frac{1}{4}\Delta x.$$

(b) Since $13/30 = 1/(30/13)$, $\Delta x = 30/13 - 2 = 0.308$ and hence

$$\frac{13}{30} \approx \frac{1}{2} - \frac{1}{4}(0.308) = 0.423.$$

5. (a) $f(x) = x^3 \implies f'(x) = 3x^2$. Hence, if $x_0 = 1/2$,

$$f(x_0) + f'(x_0)\Delta x = \left(\frac{1}{2}\right)^3 + 3\left(\frac{1}{2}\right)^2\Delta x = \frac{1}{8} + \frac{3}{4}\Delta x.$$

(b) Here $\Delta x = 0.48 - 0.5 = -0.02$. Hence

$$(0.48)^3 \approx \frac{1}{8} + \frac{3}{4}(-0.02) = 0.11.$$

7. (a) $f(x) = 2\sin x \implies f'(x) = 2\cos x$. Hence, if $x_0 = 0$,

$$f(x_0) + f'(x_0)\Delta x = 2\sin 0 + 2\cos 0(\Delta x) = 2\Delta x.$$

(b) Here $2° = 2(\pi/180) = 0.035$ radian and $\Delta x = 0.035$. Hence $2\sin 2° \approx 2(0.035) = 0.07$.

9. (a) $f(x) = \cos x \implies f'(x) = -\sin x$. Hence, if $x_0 = \pi/6$,

$$f(x_0) + f'(x_0)\Delta x = \cos(\pi/6) - [\sin(\pi/6)]\Delta x = \frac{\sqrt{3}}{2} - \frac{1}{2}\Delta x.$$

(b) Here $31° = 31(\pi/180) = 0.541$ radian and $\Delta x = 0.541 - \pi/6 = 0.017$. Hence

$$\cos 31° \approx \frac{\sqrt{3}}{2} - \frac{1}{2}(0.017) \approx 0.858.$$

11. Let $f(x) = \sqrt{x}$. Then $f'(x) = 1/(2\sqrt{x})$ and hence

$$f(x) \approx f(x_0) + f'(x_0)\Delta x = \sqrt{x_0} + \frac{1}{2\sqrt{x_0}}\,\Delta x.$$

Setting $x = 48$ and $x_0 = 49$, $\Delta x = -1$ and hence

$$\sqrt{48} \approx \sqrt{49} + \frac{1}{2\sqrt{49}}(-1) \approx 6.929.$$

13. Let $f(x) = 1/x$. Then $f'(x) = -1/x^2$ and hence

$$f(x) \approx f(x_0) + f'(x_0)\Delta x = \frac{1}{x_0} - \frac{1}{x_0^2}\,\Delta x.$$

Setting $x = 4.1$ and $x_0 = 4$, $\Delta x = 0.1$ and hence

$$\frac{1}{4.1} \approx \frac{1}{4} - \frac{1}{4^2}(0.1) = 0.244.$$

15. Let $f(x) = \cos x$. Then $f'(x) = -\sin x$ and hence

$$f(x) \approx f(x_0) + f'(x_0)\Delta x = \cos x_0 - (\sin x_0)\Delta x.$$

Setting $x = 59°$ and $x_0 = 60°$, $\Delta x = -1° = -\pi/180 = -0.017$ radian and hence

$$\cos 59° \approx \cos 60° - (\sin 60°)(-0.017) \approx 0.515.$$

17. Let $f(x) = x^3$. Then $f'(x) = 3x^2$ and hence

$$f(x) \approx f(x_0) + f'(x_0)\Delta x = x_0^3 + 3x_0^2\Delta x.$$

Setting $x = 2.97$ and $x_0 = 3$, $\Delta x = -0.03$ and hence

$$(2.97)^3 \approx 3^3 + 3(3)^2(-0.03) = 26.19.$$

19. The table below summarizes the calculator value, linear approximation, relative error, and percentage error for the approximations in Exercises 11, 13, 15, and 17.

		calculator value y	linear approx y_1	relative error $|(y - y_1)/y|$	% error
Ex. 11	$\sqrt{48}$	6.9282	6.929	0.000115	0.0115%
Ex. 13	$1/4.1$	0.2439	0.244	0.0004	0.04%
Ex. 15	$\cos 59°$	0.51504	0.515	0.000074	0.0074%
Ex. 17	$(2.97)^3$	26.198073	26.19	0.00031	0.031%

21. Relative error $= |(y - y_1)/y| = |(\sqrt{2}/2 - 0.6701)/\sqrt{2}/2| = 0.0523$, or 5.23%.

23. $f(36) \approx f(39) + f'(39)(36 - 39) = 7 + 0.65(-3) = 5.05$

25. Let $V(r)$ stand for the volume of the bearing of radius r and length $\ell = 5$. Then

$$V(r) = \pi r^2(5) = 5\pi r^2 \quad \Longrightarrow \quad \Delta V \approx dV = 10\pi r dr.$$

If $r = 2$ and $\Delta r = dr = 0.02$,

$$\Delta V \approx 10\pi(2)(0.02) = 0.4\pi \approx 1.26.$$

Thus, the volume varies by approximately 1.26 cm^3.

27. (a) $a(\theta) = -32 \sin \theta \Longrightarrow a'(\theta) = -32 \cos \theta \Longrightarrow a'(0) = -32 \Longrightarrow a(\theta) = a(0) + a'(0)(\theta - 0) = -32\theta$.

(b) $a(\pi/6) \approx -32(\pi/6) \approx -16.76$.

(c) Let y = actual value = $a(\pi/6) = -16$ and y_1 = linear approximation = $a(0) + a'(0)(\pi/6 - 0) = -16.76$. Then the relative error in the linear approximation is equal to

$$\left|\frac{y - y_1}{y}\right| = \frac{0.76}{16} = 0.0475.$$

Remark: The relative error in approximating the acceleration $a(\pi/6)$, calculated above, should not be confused with the relative error in approximating the *change* in acceleration, as discussed in Example 4 of the text. The relative error in approximating the change in acceleration is equal to

$$\left|\frac{\Delta a - da}{\Delta a}\right| = \left|\frac{-16 - (-16.76)}{-16}\right| = 0.0475.$$

Although the two relative errors are the same in this case, they are conceptually different ideas and, in general, are not equal. See also Exercise 95 in the review exercises for Chapter 3.

29. $f(x) = \sqrt{x} \Longrightarrow f'(x) = 1/(2\sqrt{x})$.

(a) If $x_0 = 49$, $\Delta x = 75 - 49 = 8$ and hence

$$\sqrt{57} = f(49 + 8) \approx f(49) + f'(49)(8) = \sqrt{49} + \frac{1}{2\sqrt{49}}(8) = 7.57.$$

(b) If $x_0 = 64$, $\Delta x = 75 - 64 = -7$ and hence

$$\sqrt{57} = f(64 + (-7)) \approx f(64) + f'(64)(-7) = \sqrt{64} + \frac{1}{2\sqrt{64}}(-4) = 7.56.$$

(c) The value in (a) is larger.

(d) The graph of $y = \sqrt{x}$ is shown below with the tangent lines at $x_0 = 49$ and $x_0 = 64$.

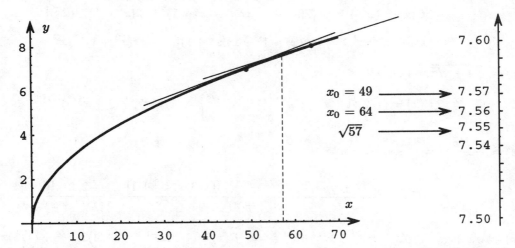

(e) $\sqrt{57}$ lies to the left (below) both of the approximations, as shown above on the number line on the right.

31. (a) $f'(x) = -100x/(1+10x^2)^2 \iff f'(1/2) = -200/49$. Therefore $t_1(x) = -200(x-1/2)/49 + 10/7$.

    (b) The graph with range $[0,2] \times [0,5]$ is shown below.

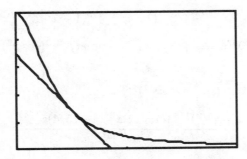

    (c) $t_1(x)$ is the best linear approximation to $f(x)$ at the point $(1/2, 10/7)$.

    (d) $h = 0.03$.

    (e) $f'(1) = -100/121 \iff t_2(x) = -100(x-1)/121 + 5/11$. The value of h for $t_2(x)$ is 0.09.

    (f) The function $f(x)$ is flatter near $x = 1$ than it is near $x = 1/2$.

3.8 Implicit Differentiation and Rational Power Functions

1. $f(x) = (x+2)^{4/3} \implies f'(x) = (4/3)(x+2)^{1/3}$

3. $f(x) = \sqrt{x} + \dfrac{1}{\sqrt{x}} = x^{1/2} + x^{-1/2} \implies f'(x) = (1/2)x^{-1/2} - (1/2)x^{-3/2} = \dfrac{x-1}{2x\sqrt{x}}$

5. $f(x) = x^{2/3} + x^{-2/3} \implies f'(x) = (2/3)x^{-1/3} - (2/3)x^{-5/3}$

7. $f(x) = (x^2 - 1)^{-2}(x^2 + 1)^2 \implies$

$$f'(x) = (x^2 - 1)^{-2}\left[2(x^2 + 1)(2x)\right] + (x^2 + 1)^2\left[-2(x^2 - 1)^{-3}(2x)\right]$$

$$= (x^2 - 1)^{-3}(x^2 + 1)\left[4x(x^2 - 1) - 4x(x^2 + 1)\right] = -8x(x^2 - 1)^{-3}(x^2 + 1)$$

9. $f(x) = \sqrt{1 + \sqrt[3]{x}} = (1 + x^{1/3})^{1/2} \implies$

$$f'(x) = (1/2)(1 + x^{1/3})^{-1/2}\left[(1/3)x^{-2/3}\right] = (1/6)x^{-2/3}(1 + x^{1/3})^{-1/2}$$

11. $f(x) = \dfrac{\sqrt[3]{x}}{\sqrt[3]{x} + x} = \dfrac{x^{1/3}}{x^{1/3} + x} \implies$

$$f'(x) = \frac{(x^{1/3} + x)(1/3)x^{-2/3} - x^{1/3}\left[(1/3)x^{-2/3} + 1\right]}{(x^{1/3} + x)^2} = \frac{(-2/3)x^{1/3}}{(x^{1/3} + x)^2}$$

13. $f(x) = \sqrt{x} + \sqrt[4]{x} + \sqrt[8]{x} = x^{1/2} + x^{1/4} + x^{1/8} \implies f'(x) = (1/2)x^{-1/2} + (1/4)x^{-3/4} + (1/8)x^{-7/8}$

15. $f(x) = \dfrac{(x^2 + 1)^{2/3}}{9 - x^2)^{4/3}} = (1 + x^2)^{2/3}(1 - x^2)^{-4/3} \implies$

$$f'(x) = (1 + x^2)^{2/3}\left[(-4/3)(1 - x^2)^{-7/3}(-2x)\right] + (1 - x^2)^{-4/3}\left[(2/3)(1 + x^2)^{-1/3}(2x)\right]$$

$$= (4/3)x(1 + x^2)^{-1/3}(1 - x^2)^{-7/3}\left[2(1 + x^2) + (1 - x^2)\right]$$

$$= (4/3)x(1 + x^2)^{-1/3}(1 - x^2)^{-7/3}(x^2 + 3)$$

17. $f(x) = \sqrt[3]{\sin x^2} = (\sin x^2)^{1/3} \implies f'(x) = (1/3)(\sin x^2)^{-2/3}(\cos x^2)(2x)$

19. $f(x) = \dfrac{x^{1/4}(x^{4/3} - x)}{(\sin^3 x - \sqrt{x})^{1/3}} = \dfrac{x^{19/12} - x^{5/4}}{(\sin^3 x - x^{1/2})^{1/3}} \implies$

$$f'(x) = \frac{(\sin^3 x - x^{1/2})^{1/3}\left[(19/12)x^{7/12} - (5/4)x^{1/4}\right]}{(\sin^3 x - x^{1/2})^{2/3}}$$

$$- \frac{(x^{19/12} - x^{5/4})(1/3)(\sin^3 x - x^{1/2})^{-2/3}\left[3\sin^2 x \cos x - (1/2)x^{-1/2}\right]}{(\sin^3 x - x^{1/2})^{2/3}}$$

$$= \frac{(\sin^3 x - \sqrt{x})\left(\frac{19}{12}x^{7/12} - \frac{5}{4}x^{1/4}\right) - \frac{1}{3}(x^{19/12} - x^{5/4})\left(3\sin^3 x \cos x - \frac{1}{2}x^{-1/2}\right)}{(\sin^3 x - \sqrt{x})^{4/3}}$$

21. $f(x) = \sin^{5/2}(x^{2/3}) \implies$

$$f'(x) = (5/2)\sin^{3/2}(x^{2/3})\left[\cos(x^{2/3})\right](2/3)x^{-1/3} = (5/3)x^{-1/3}\sin^{3/2}(x^{2/3})\cos x^{2/3}$$

23. $f(x) = \sin\sqrt[3]{x} + \sqrt[3]{\sin x} = \sin(x^{1/3}) + (\sin x)^{1/3} \implies$

$$f'(x) = \left[\cos(x^{1/3})\right](1/3)x^{-2/3} + (1/3)(\sin x)^{-2/3}\cos x$$

25. $f(x) = \sec \sqrt[4]{3x+1} = \sec(3x+1)^{1/4} \Longrightarrow$

$$f'(x) = \left[\sec(3x+1)^{1/4}\tan(3x+1)^{1/4}\right](1/4)(3x+1)^{-3/4}(3)$$

$$= (3/4)(3x+1)^{-3/4}\sec(3x+1)^{1/4}\tan(3x+1)^{1/4}$$

27. $x^2 + y^2 = 25 \Longrightarrow 2x + 2y(dy/dx) = 0 \Longrightarrow dy/dx = -x/y$

29. $x^{1/2} + y^{1/2} = 4 \Longrightarrow (1/2)x^{-1/2} + (1/2)y^{-1/2}(dy/dx) = 0 \Longrightarrow dy/dx = -\sqrt{y/x}$

31. $x = \tan y \Longrightarrow 1 = (\sec^2 y)(dy/dx) \Longrightarrow dy/dx = \cos^2 y$

33. $x \sin y = y \cos x \Longrightarrow$

$$x(\cos y)\frac{dy}{dx} + \sin y = y(-\sin x) + (\cos x)\frac{dy}{dx}$$

$$\Longrightarrow \frac{dy}{dx}(x\cos y - \cos x) = -\sin y - y\sin x$$

$$\Longrightarrow \frac{dy}{dx} = -\frac{y\sin x + \sin y}{x\cos y - \cos x}$$

35. $(xy)^{1/2} = xy - x \Longrightarrow$

$$(1/2)(xy)^{-1/2}\left[x\frac{dy}{dx} + y\right] = x\frac{dy}{dx} + y - 1$$

$$\Longrightarrow \frac{dy}{dx}\left[(1/2)x(xy)^{-1/2} - x\right] = y - 1 - (1/2)y(xy)^{-1/2}$$

$$\Longrightarrow \frac{dy}{dx} = \frac{y - 1 - (1/2)y(xy)^{-1/2}}{(1/2)x(xy)^{-1/2} - x} = \frac{2(y-1)\sqrt{xy} - y}{x(1 - 2\sqrt{xy})}$$

37. $x^3 + x^2y + xy^2 + y^3 = 15 \Longrightarrow$

$$3x^2 + x^2\frac{dy}{dx} + 2xy + x\left(2y\frac{dy}{dx}\right) + y^2 + 3y^2\frac{dy}{dx} = 0$$

$$\Longrightarrow (x^2 + 2xy + 3y^2)\frac{dy}{dx} = -3x^2 - 2xy - y^2$$

$$\Longrightarrow \frac{dy}{dx} = -\frac{3x^2 + 2xy + y^2}{x^2 + 2xy + 3y^2}$$

39. $\sqrt{x+y} = xy - x \Longrightarrow$

$$(1/2)(x+y)^{-1/2}\left(1 + \frac{dy}{dx}\right) = x\frac{dy}{dx} + y - 1$$

$$\Longrightarrow \frac{dy}{dx}\left[(1/2)(x+y)^{-1/2} - x\right] = y - 1 - (1/2)(x+y)^{-1/2}$$

$$\Longrightarrow \frac{dy}{dx} = \frac{y - 1 - (1/2)(x+y)^{-1/2}}{(1/2)(x+y)^{-1/2} - x} = \frac{2(y-1)\sqrt{x+y} - 1}{1 - 2x\sqrt{x+y}}$$

83

41. $\cot y = 3x^2 + \cot(x + y) \implies$

$$-\left(\csc^2 y\right)\frac{dy}{dx} = 6x - \left[\csc^2(x+y)\right]\left(1 + \frac{dy}{dx}\right)$$

$$\implies \frac{dy}{dx}\left[-\csc^2 y + \csc^2(x+y)\right] = 6x - \csc^2(x+y)$$

$$\implies \frac{dy}{dx} = \frac{6x - \csc^2(x+y)}{\csc^2(x+y) - \csc^2 y}$$

43. $\sin(xy) = 1/2 \implies [\cos(xy)][x(dy/dx)+y] = 0 \implies dy/dx = -y/x$. Note that $\cos(xy) \neq 0$; for if $\cos(xy) = 0$, then $\sin(xy) = \pm 1$, not 1/2.

45. $y^4 = x^5 \implies 4y^3(dy/dx) = 5x^4 \implies dy/dx = 5x^4/4y^3 = 5x^5/4xy^3 = 5y^4/4xy^3 = 5y/4x$

47. $y^2 + 3 = x\sec y \implies$

$$2y\frac{dy}{dx} = x(\sec y \tan y)\frac{dy}{dx} + \sec y$$

$$\implies \frac{dy}{dx}(2y - x\sec y \tan y) = \sec y$$

$$\implies \frac{dy}{dx} = \frac{\sec y}{2y - x\sec y \tan y}$$

49. $x = \sin 2xy \implies 1 = (\cos 2xy)[2x(dy/dx) + 2y] \implies dy/dx = (\sec 2xy - 2y)/2x$

51. $xy = 9 \implies$

$$x\frac{dy}{dx} + y = 0 \quad \implies \quad \frac{dy}{dx} = -\frac{y}{x} \quad \implies \quad \left.\frac{dy}{dx}\right|_{(3,3)} = -1.$$

Hence the slope of the tangent line is -1 and the equation is $y - 3 = -(x - 3)$, or $y = -x + 6$.

53. $x^3 + y^3 = 16 \implies$

$$3x^2 + 3y^2\frac{dy}{dx} = 0 \quad \implies \quad \frac{dy}{dx} = -\frac{x^2}{y^2} \quad \implies \quad \left.\frac{dy}{dx}\right|_{(2,2)} = -1.$$

Hence the slope of the tangent line is -1 and the equation is $y - 2 = -(x - 2)$, or $y = -x + 4$.

55. $\dfrac{x+y}{x-y} = 4 \implies x + y = 4(x - y) \implies 3x = 5y \implies dy/dx = 3/5$. Thus the slope of the tangent line is $3/5$ at all points. Hence the equation of the tangent line at $(5, 3)$ is $y - 3 = (3/5)(x - 5)$, or $y = 3x/5$.

57. $y^4 + 3x - x^2\sin y = 3 \implies$

$$4y^3\frac{dy}{dx} + 3 - x^2\cos y\frac{dy}{dx} - 2x\sin y = 0 \quad \implies \quad \frac{dy}{dx} = \frac{2x\sin y - 3}{4y^3 - x^2\cos y} \quad \implies \quad \left.\frac{dy}{dx}\right|_{(1,0)} = 3.$$

Hence the slope of the tangent line is 3 and the equation is $y = 3(x - 1)$.

59. By definition, the equation of the tangent line at the point (x_0, y_0) has the form

$$y - y_0 = \left.\frac{dy}{dx}\right|_{(x_0,y_0)} (x - x_0).$$

It follows that if the tangent line is horizontal, then $y - y_0 = 0$ and hence $\left.\dfrac{dy}{dx}\right|_{(x_0,y_0)} = 0.$

61. $x^2 + y^2 = 2 \implies 2x + 2y(dy/dx) = 0 \implies dy/dx = -x/y$. Hence

$$\frac{dy}{dx} = -\frac{x}{y} = 0 \implies x = 0 \implies y = \pm\sqrt{2}$$

$$\implies \text{ horizontal tangents: } y = \sqrt{2} \text{ at } (0, \sqrt{2}) \text{ and } y = -\sqrt{2} \text{ at } (0, -\sqrt{2});$$

$$\frac{dx}{dy} = -\frac{y}{x} = 0 \implies y = 0 \implies x = \pm\sqrt{2}$$

$$\implies \text{ vertical tangents: } x = \sqrt{2} \text{ at } (\sqrt{2}, 0) \text{ and } x = -\sqrt{2} \text{ at } (-\sqrt{2}, 0).$$

63. $x^2 - y^2 = 1 \implies 2x - 2y(dy/dx) = 0 \implies dy/dx = x/y$. Now, if $dy/dx = 0$, then $x = 0$, in which case $-y^2 = 1$, which has no real solutions. Thus, there are no points where the tangent line is horizontal. If $dx/dy = y/x = 0$, then $y = 0$ and hence $x = \pm 1$. Thus, the graph has vertical tangents $x = \pm 1$ at $(\pm 1, 0)$.

65. $x^2 + y^2 + 2x + 4y = -4 \implies 2x + 2y(dy/dx) + 2 + 4(dy/dx) = 0 \implies dy/dx = -(2x + 2)/(2y + 4) = -(x+1)/(y+2)$. Hence

$$\frac{dy}{dx} = -\frac{x+1}{y+2} = 0 \implies x = -1 \implies y^2 + 4y + 3 = (y+3)(y+1) = 0 \implies y = -1, -3$$

$$\implies \text{ horizontal tangents: } y = -1 \text{ at } (-1, -1) \text{ and } y = -3 \text{ at } (-1, -3)$$

and

$$\frac{dx}{dy} = -\frac{y+2}{x+1} = 0 \implies y = -2 \implies x^2 + 2x = x(x+2) = 0 \implies x = 0, -2$$

$$\implies \text{ vertical tangents: } x = 0 \text{ at } (0, -2) \text{ and } x = -2 \text{ at } (-2, -2).$$

67. $a = 1$. To obtain this value, we first find the intersection points of the two circles by solving their equations simultaneously. Subtracting the equations, we find that:

$$(x-a)^2 - (x+a)^2 = 0 \implies (x^2 - 2ax + a^2) - (x^2 + 2ax + a^2) = -4ax = 0 \implies x = 0.$$

Note that $a \neq 0$ since, if $a = 0$, there is only one circle. Now, if $x = 0$, then $a^2 + y^2 = 2 \implies y = \pm\sqrt{2 - a^2}$. Thus, the two circles intersect at the points $(0, \sqrt{2 - a^2})$ and $(0, -\sqrt{2 - a^2})$. Now,

$$(x-a)^2 + y^2 = 2 \implies 2(x-a) + 2y\frac{dy}{dx} = 0 \implies \frac{dy}{dx} = -\frac{x-a}{y} \implies \frac{dy}{dx}\bigg|_{(0, \sqrt{2-a^2})} = \frac{a}{\sqrt{2-a^2}}$$

and

$$(x+a)^2 + y^2 = 2 \implies 2(x+a) + 2y\frac{dy}{dx} = 0 \implies \frac{dy}{dx} = -\frac{x+a}{y} \implies \frac{dy}{dx}\bigg|_{(0, \sqrt{2-a^2})} = \frac{-a}{\sqrt{2-a^2}}.$$

Thus, the slopes of the tangent lines at $(0, \sqrt{2 - a^2})$ are $a/\sqrt{2 - a^2}$ and $-a/\sqrt{2 - a^2}$. If the tangent lines are perpendicular, then

$$\frac{a}{\sqrt{2-a^2}} \cdot \frac{-a}{\sqrt{2-a^2}} = -1 \implies a^2 = 2 - a^2 \implies a = \pm 1.$$

At the point $(0, -\sqrt{2 - a^2})$, the slopes of the tangent lines are the same numbers, except in reverse order:

$$\frac{a}{-\sqrt{2-a^2}} \quad \text{and} \quad \frac{a}{\sqrt{2-a^2}}.$$

Thus, if the tangent lines at the intersection points of the two circles are perpendicular, then $a = \pm 1$. But the two circles for $a = 1$ are the same as the two circles for $a = -1$. Hence $a = 1$.

69. $x^2 + y^2 = a^2 \implies 2x + 2y(dy/dx) = 0 \implies dy/dx = -x/y$. Thus the slope of the tangent line at a typical point (x_0, y_0) on the curve is

$$\left.\frac{dy}{dx}\right|_{(x_0, y_0)} = -\frac{x_0}{y_0}.$$

Therefore the slope of the normal line is y_0/x_0 and the equation is $y - y_0 = (y_0/x_0)(x - x_0)$. Letting $x = 0$, we find that $y - y_0 = -y_0 \implies y = 0$. Hence the normal line passes through the origin $(0, 0)$.

71. $5x^2 + 6xy + 5y^2 = 8 \implies$

$$10x + 6x\frac{dy}{dx} + 6y + 10y\frac{dy}{dx} = 0 \quad \implies \quad \frac{dy}{dx} = \frac{-10x - 6y}{6x + 10y} = \frac{-5x - 3y}{3x + 5y}.$$

Now, the slope of the line $x - y = 1$ is 1. Hence

$$\frac{dy}{dx} = \frac{-5x - 3y}{3x + 5y} = 1 \quad \implies \quad -5x - 3y = 3x + 5y \quad \implies \quad y = -x.$$

Thus, if (x, y) is a point on the curve at which the tangent line is parallel to $x - y = 1$, then $y = -x$. In this case,

$$5x^2 + 6xy + 5y^2 = 5x^2 + 6x(-x) + 5(-x)^2 = 4x^2 = 8 \quad \implies x = \pm\sqrt{2} \quad \implies \quad y = \mp\sqrt{2}.$$

Hence there are two points on the curve at which the tangent line is parallel to $x - y = 1$, namely $(\sqrt{2}, -\sqrt{2})$ and $(-\sqrt{2}, \sqrt{2})$.

73. $xy^2 + x^2y = 4x^2 \implies x(y^2 + xy - 4x) = 0$. Thus, the graph of $xy^2 + x^2y = 4x^2$ consists of all points on the graph of $y^2 + xy = 4x$ together with those points at which $x = 0$, namely the y-axis. Now,

$$y^2 + xy - 4x = y^2 + x(y - 4) = 0 \quad \implies \quad x = \frac{y^2}{4 - y}.$$

Hence, as $y \to 0$, $x \to 0$. Moreover,

$$\frac{dx}{dy} = \frac{(4 - y)(2y) - y^2(-1)}{(4 - y)^2} = \frac{8y - y^2}{(4 - y)^2}.$$

Hence, as $y \to 0$, $dx/dy \to 0$. In other words, the tangent lines are approaching the vertical for x near zero.

3.9 Higher Order Derivatives

1. $f(x) = 2x^3 - 6x \implies f'(x) = 6x^2 - 6 \implies f''(x) = 12x$

3. $f(x) = x^5 - 3x^{-2} \implies f'(x) = 5x^4 + 6x^{-3} \implies f''(x) = 20x^3 - 18x^{-4}$

5. $f(x) = 1/(1 + x) \implies (1 + x)^{-1} \implies f'(x) = -(1 + x)^{-2} \implies f''(x) = 2(1 + x)^{-3}$

7. $f(x) = \sin^2 x \implies f'(x) = 2\sin x \cos x \implies f''(x) = (2\sin x)(-\sin x) + (2\cos x)\cos x = -2\sin^2 x + 2\cos^2 x$

86

9. $f(x) = 1/(1 + \cos x) = (1 + \cos x)^{-1} \Longrightarrow$

$$\begin{aligned} f'(x) &= -(1 + \cos x)^{-2}(-\sin x) = (\sin x)(1 + \cos x)^{-2} \\ \Longrightarrow \quad f''(x) &= (\sin x)\left[(-2)(1 + \cos x)^{-3}(-\sin x)\right] + (\cos x)(1 + \cos x)^{-2} \\ &= (1 + \cos x)^{-2}\left[2\sin^2 x\,(1 + \cos x)^{-1} + \cos x\right] \\ &= (1 + \cos x)^{-2}\left[2(1 - \cos^2 x)(1 + \cos x)^{-1} + \cos x\right] \\ &= (1 + \cos x)^{-2}(2 - \cos x) \end{aligned}$$

11. $f(x) = \tan^2 x \Longrightarrow f'(x) = 2(\tan x)\sec^2 x \Longrightarrow$

$$f''(x) = (2\tan x)[2\sec x(\sec x \tan x)] + (2\sec^2 x)(\sec^2 x) = 2\sec^2 x(2\tan^2 x + \sec^2 x)$$

13. $f(x) = x - x\sec x \Longrightarrow f'(x) = 1 - (x\sec x \tan x + \sec x) = 1 - (\sec x)(x\tan x + 1) \Longrightarrow$

$$\begin{aligned} f''(x) &= -(\sec x)\left[x\sec^2 x + \tan x\right] - (\sec x \tan x)(x\tan x + 1) \\ &= -x\sec^3 x - 2\sec x \tan x - x\sec x \tan^2 x \end{aligned}$$

15. $f(x) = (x^3 + 1)^5 \Longrightarrow f'(x) = 5(x^3 + 1)^4(3x^2) = 15x^2(x^3 + 1)^4 \Longrightarrow$

$$f''(x) = (15x^2)\left[4(x^3 + 1)^3(3x^2)\right] + (30x)(x^3 + 1)^4 = 30x(x^3 + 1)^3(7x^3 + 1)$$

17. $y = x^{-2} - 2x^{-4} \Longrightarrow dy/dx = -2x^{-3} + 8x^{-5} \Longrightarrow d^2y/dx^2 = 6x^{-4} - 40x^{-6}$

19. $dy/dx = \sec x \tan x \Longrightarrow d^2y/dx^2 = (\sec x)(\sec^2 x) + (\sec x \tan x)\tan x = (\sec x)(\sec^2 x + \tan^2 x)$

21. $y = x^3 \cos x \Longrightarrow dy/dx = x^3(-\sin x) + (3x^2)\cos x = x^2(-x\sin x + 3\cos x) \Longrightarrow$

$$\frac{d^2y}{dx^2} = x^2[(-x)\cos x + (-1)\sin x - 3\sin x] + (2x)(-x\sin x + 3\cos x) = -x^3 \cos x - 6x^2 \sin x + 6x\cos x$$

23. $y = 1/(1 - x^2) = (1 - x^2)^{-1} \Longrightarrow dy/dx = -(1 - x^2)^{-2}(-2x) = 2x(1 - x^2)^{-2} \Longrightarrow$

$$\frac{d^2y}{dx^2} = (2x)\left[-2(1 - x^2)^{-3}(-2x)\right] + (2)(1 - x^2)^{-2} = 2(1 - x^2)^{-3}(3x^2 + 1)$$

25. $f(x) = \dfrac{x}{x + 2} \Longrightarrow f'(x) = \dfrac{(x + 2)(1) - x(1)}{(x + 2)^2} = 2(x + 2)^{-2} \Longrightarrow f''(x) = -4(x + 2)^{-3}$

27. $y = (x^2 - 1)^4 \Longrightarrow dy/dx = 4(x^2 - 1)^3(2x) = 8x(x^2 - 1)^3 \Longrightarrow$

$$\frac{d^2y}{dx^2} = (8x)\left[3(x^2 - 1)^2(2x)\right] + 8(x^2 - 1)^3 = 8(x^2 - 1)^2(7x^2 - 1)$$

$$\frac{d^3y}{dx^3} = 8(x^2 - 1)^2(14x) + \left[16(x^2 - 1)(2x)\right](7x^2 - 1) = 48(x^2 - 1)(7x^3 - 3x)$$

$$\frac{d^4y}{dx^4} = 48(x^2 - 1)(21x^2 - 3) + (96x)(7x^3 - 3x) = 48(35x^4 - 30x^2 + 3)$$

29. $y^2 = 4x \Longrightarrow 2y(dy/dx) = 4 \Longrightarrow dy/dx = 2y^{-1} \Longrightarrow d^2y/dx^2 = -2y^{-2}(dy/dx) = -2y^{-2}(2y^{-1}) = -4y^{-3}$

31. $y^2 - xy = 4 \implies$

$$2y\frac{dy}{dx} - x\frac{dy}{dx} - y = 0 \implies \frac{dy}{dx} = \frac{y}{2y - x}$$

$$\implies \frac{d^2y}{dx^2} = \frac{(2y - x)(dy/dx) - y[2(dy/dx) - 1]}{(2y - x)^2} = \frac{-x[y/(2y - x)] + y}{(2y - x)^2} = \frac{2y(y - x)}{(2y - x)^3}$$

33. Because the derivative of a polynomial is a polynomial. Note that if $p(x)$ has degree n, then $p^{(k)}(x) = 0$ for all $k > n$.

35. (a) $v(t) = s'(t) = 3t^2 - 12t - 30 = 3(t^2 - 4t - 10)$

 (b) $v(t) = 0 \iff t = (4 \pm \sqrt{56})/2 = 2 \pm \sqrt{14}$. The table below summarizes the sign changes for $v(t)$.

Sign $v(t)$	+	−	+
Interval	$(-\infty, 2 - \sqrt{14})$	$(2 - \sqrt{14}, 2 + \sqrt{14})$	$(2 + \sqrt{14}, +\infty)$

 Velocity is negative when $2 - \sqrt{14} < t < 2 + \sqrt{14}$.

 (c) $a(t) = v'(t) = 6t - 12 = 6(t - 2)$

 (d) Acceleration is positive when $t > 2$.

37. $s(t) = 10 + 5\cos t \implies v(t) = s'(t) = -5\sin t \implies a(t) = v'(t) = -5\cos t$. The graph below shows that when $v(t) > 0$, $s(t)$ is increasing, while when $a(t) > 0$, $v(t)$ is increasing.

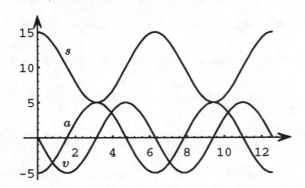

Exercise 37

39. For any integer $k \geq 0$, $f^{(4k)}(x) = -\sin x$. Therefore $f^{(161)}(x) = \left(f^{(160)}(x)\right)' = (-\sin x)' = -\cos x$.

41. $f(x) = x\cos x \implies f'(x) = -x\sin x + \cos x \implies f''(x) = -x\cos x - 2\sin x \implies f'''(x) = x\sin x - 3\sin s$
$\implies f^{(4)}(x) = x\cos x + 4\sin x \implies f^{(8)}(x) = (x\cos x)^{(4)} + 4(\sin x)^{(4)} = x\cos x + 4\sin x + 4\sin x = x\cos x + 8\sin x$. In general, $f^{(4k)}(x) = x\cos x + 4k\sin x$. Therefore

$$f^{(50)}(x) = (x\cos + 48\sin x)'' = -x\cos x - 2\sin x + 48(-\sin x) = -x\cos x - 50\sin x.$$

88

43. First recall that $d(|x|)/dx = |x|/x$. Hence, by the Chain Rule,

$$f'(x) = \frac{|x^2-1|}{x^2-1}(2x) = \begin{cases} 2x, & x < -1 \\ -2x, & -1 < x < 1 \\ 2x, & x > 1 \end{cases}$$

and

$$f''(x) = \frac{|x^2-1|}{x^2-1}(2) + (2x)\frac{(x^2-1)\left[|x^2-1|/(x^2-1)\right](2x) - |x^2-1|(2x)}{(x^2-1)^2}$$

$$= 2\frac{|x^2-1|}{x^2-1} = \begin{cases} 2, & x < -1 \\ -2, & -1 < x < 1 \\ 2, & x > 1. \end{cases}$$

The domain of $f(x)$ is all real numbers, while the domain of $f'(x)$ and $f''(x)$ is all $x \neq \pm 1$.

45. Let $f(x) = ax^2 + bx + c$. Then

$$f(2) = 4a + 2b + c = 2$$
$$f'(x) = 2ax + b \implies f'(2) = 4a + b = 4$$
$$f''(x) = 2a \implies f''(2) = 2a = 6.$$

Therefore $a = 3$, $b = -8$, $c = 6$. Hence $f(x) = 3x^2 - 8x + 6$.

47. The table below summarizes the first four derivatives of $\tan x$, $\cot x$, $\sec x$ and $\csc x$. Clearly, there is no apparent pattern to the derivatives.

$f(x)$	$f'(x)$	$f''(x)$	$f'''(x)$	$f^{(4)}(x)$
$\tan x$	$\sec^2 x$	$2\sec^2 x \tan x$	$2\sec^2 x(\sec^2 x + 2\tan^2 x)$	$8\sec^2 x \tan x(2\sec^2 x + \tan^2 x)$
$\cot x$	$-\csc^2 x$	$2\csc^2 x \cot x$	$-2\csc^2 x(\csc^2 x + 2\cot^2 x)$	$8\csc^2 x \cot x(\cot^2 x + 2\csc^2 x)$
$\sec x$	$\sec x \tan x$	$\sec x(\sec^2 x + \tan^2 x)$	$\sec x \tan x(5\sec^2 x + \tan^2 x)$	$\sec x(5\sec^4 x + 18\sec^2 x \tan^2 x + \tan^4 x)$
$\csc x$	$-\csc x \cot x$	$\csc x(\csc^2 x + \cot^2 x)$	$-\csc x \cot x(\cot^2 x + 5\csc^2 x)$	$\csc x(\cot^4 x + 18\cot^2 x \csc^2 x + 5\csc^4 x)$

3.10 Newton's Method

1. $f(x) = x^2 - (7/2)x + 3/4 \implies f(0) = 3/4$ and $f(2) = -9/4$. Hence $f(x)$ has a zero between $a = 0$ and $b = 2$. Using $x_0 = 0.5$ as initial guess, set

$$x_{n+1} = x_n - \frac{f(x_n)}{f'(x_n)} = x_n - \frac{x_n^2 - (7/2)x_n + 3/4}{2x_n - 7/2}$$

for $n \geq 0$. The table below summarizes the successive iterations and shows that the root, to within six decimal places, is 0.229309.

n	x_n	$f(x_n)$	$f'(x_n)$
0	0.8	-1.41	-1.9
1	0.0578947	0.55072	-3.38421
2	0.220627	0.0264818	-3.05875
3	0.229285	0.0000749562	-3.04143
4	0.229309	6.07370×10^{-10}	-3.04138
5	0.229309		

n	x_n	$f(x_n)$	$f'(x_n)$
0	0.5	0.25	$-2.$
1	0.625	-0.015625	-2.25
2	0.6180556	-0.0000482253	-2.23611
3	0.618034	-4.65118×10^{-10}	-2.23607
4	0.618034		

Exercise 1

Exercise 3

3. $f(x) = 1 - x - x^2 \implies f(0) = 1$ and $f(1) = -1$. Hence $f(x)$ has a zero between $a = 0$ and $b = 1$. Using $x_0 = 0.5$ as initial guess, set

$$x_{n+1} = x_n - \frac{f(x_n)}{f'(x_n)} = x_n - \frac{1 - x_n - x_n^2}{-1 - 2x_n}$$

for $n \geq 0$. The table below summarizes the successive iterations and shows that the root, to within six decimal places, is 0.618034.

5. $f(x) = \sqrt{x + 3} - x \implies f(1) = 1$ and $f(3) = \sqrt{6} - 3 < 0$. Hence $f(x)$ has a zero between $a = 1$ and $b = 3$. Using $x_0 = 1.5$ as initial guess, set

$$x_{n+1} = x_n - \frac{f(x_n)}{f'(x_n)} = x_n - \frac{\sqrt{x_n + 3} - x_n}{(1/2\sqrt{x_n + 3}) - 1}$$

for $n \geq 0$. The table below summarizes the successive iterations and shows that the root, to within six decimal places, is 2.302776.

n	x_n	$f(x_n)$	$f'(x_n)$
0	1.5	0.62132	−0.764298
1	2.31293	−0.00795038	−0.783078
2	2.302777	-1.05315×10^{-6}	−0.782871
3	2.302776	-1.85249×10^{-14}	−0.782871
4	2.302776		

n	x_n	$f(x_n)$	$f'(x_n)$
0	−1.5	0.0625	−13.5
1	−1.49537	0.000288757	−13.3754
2	−1.495349	6.25312×10^{-9}	−13.3748
3	−1.495349		

Exercise 5 · · · Exercise 7

7. $f(x) = x^4 - 5 \implies f(-2) = 11$ and $f(-1) = -4$. Hence $f(x)$ has a zero between $a = -2$ and $b = -1$. Using $x_0 = -1.5$ as initial guess, set

$$x_{n+1} = x_n - \frac{f(x_n)}{f'(x_n)} = x_n - \frac{x_n^4 - 5}{4x_n^3}$$

for $n \geq 0$. The table below summarizes the successive iterations and shows that the root, to within six decimal places, is −1.495349.

9. The two graphs intersect when $\sqrt{x + 3} = x$ or, equivalently, when $\sqrt{x + 3} - x = 0$. By Exercise 5, one of the solutions of this equation is 2.302776. Hence one of the intersection points is $(2.302776, 2.302776)$.

11. Let $f(x) = x^2 - 37$. Then $f(x)$ has a root between 6 and 7, namely $\sqrt{37}$. Using $x_0 = 6.5$ as initial guess, set

$$x_{n+1} = x_n - \frac{f(x_n)}{f'(x_n)} = x_n - \frac{x_n^2 - 37}{2x_n}$$

for $n \geq 0$. Then we find that

$$x_1 = 6.096154, \quad x_2 = 6.082777, \quad x_3 = 6.082763, \quad x_4 = 6.082763.$$

Hence, to six decimals, $\sqrt{37} \approx 6.082763$.

13. Let $f(x) = x^5 - 18$. Then $f(x)$ has a root between 1 and 2, namely $\sqrt[5]{18}$. Using $x_0 = 1.5$ as initial guess, set

$$x_{n+1} = x_n - \frac{f(x_n)}{f'(x_n)} = x_n - \frac{x_n^5 - 18}{5x_n^4}$$

for $n \geq 0$. Then we find that

$$x_1 = 1.911111, \quad x_2 = 1.798761, \quad x_3 = 1.78289, \quad x_4 = 1.782603, \quad x_5 = 1.782602, \quad x_6 = 1.782602.$$

Hence, to six decimals, $\sqrt[5]{18} \approx 1.782602$.

15. The graph of $f(x) = 4x^3 - 3x^2 - 11x + 7$, shown below, indicates that the function has three zeros; one between -1 and -2, another between 0 and 1, and another between 1 and 2. Choose $x_0 = 1.5$, 0.5, and -1.5 as initial guesses and, for $n \geq 0$, set

$$x_{n+1} = x_n - \frac{f(x_n)}{f'(x_n)} = x_n - \frac{4x_n^3 - 3x_n^2 - 11x_n + 7}{12x_n^2 - 6x_n - 11}.$$

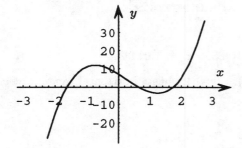

0	1.5	0.5	−1.5
1	1.892857	0.6136364	−1.63
2	1.76893	0.6180257	−1.618139
3	1.750408	0.618034	−1.618034
4	1.75	0.618034	−1.618034
5	1.75	0.618034	−1.618034

The table on the right summarizes the successive iterations using the three initial guesses and shows that, to six decimal places, the roots are 1.75, 0.618034, and −1.618034.

17. The graph of $f(x) = x^4 + 4x^3 + 3x^2 - 2x - 1$, shown below, indicates that the function has four zeros; one between -3 and -2, another between -2 and -1, another between -1 and 0, and one between 0 and 1. Choose $x_0 = -2.5$, -1.5, -0.5, and 0.5 as initial guesses and, for $n \geq 0$, set

$$x_{n+1} = x_n - \frac{f(x_n)}{f'(x_n)} = x_n - \frac{x^4 + 4x^3 + 3x^2 - 2x - 1}{4x^3 + 12x^2 + 6x - 2}.$$

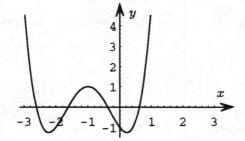

0	−2.5	−1.5	−0.5	0.5
1	−2.65278	−1.625	−0.375	0.652778
2	−2.61999	−1.61805	−0.381954	0.619985
3	−2.61804	−1.61803	−0.381966	0.618041
4	−2.61803	−1.61803	−0.381966	0.618034
5	−2.61803	−1.61803	−0.381966	0.618034

The table on the right summarizes the successive iterations using the three initial guess and shows that, to five decimal places, the roots are −2.61803, −1.61803, −0.38197, and 0.61803.

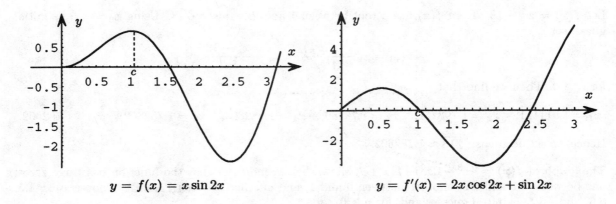

$$y = f(x) = x \sin 2x \qquad\qquad y = f'(x) = 2x \cos 2x + \sin 2x$$

19. $f(x) = x \sin 2x \implies f'(x) = 2x \cos 2x + \sin 2x$. The graph of $y = f'(x)$ is shown below on the right. The smallest positive number c such that $f'(c) = 0$ is approximately 1. Using $x_0 = 1$ and setting

$$x_{n+1} = x_n - \frac{f'(x_n)}{f''(x_n)} = x_n - \frac{2x_n + \sin 2x_n}{-4x \sin 2x + 4 \cos 2x},$$

for $n \geq 0$, we find that

$$x_0 = 1, \quad x_1 = 1.01452, \quad x_2 = 1.01438, \quad x_3 = 1.01438.$$

Hence $c \approx 1.01438$. Geometrically, the graph of $y = f(x) = x \sin 2x$ has a horizontal tangent at c.

21. Let $f(x) = x^4$. Using $x_1 = 0.01$ and setting

$$x_{n+1} = x_n - \frac{f(x_n)}{f'(x_n)} = x_n - \frac{x_n^4}{4x_n^3} = 0.75x_n,$$

we find that

x_1	$= 0.01$	1 decimal place accuracy
x_2	$= 0.0075$	1 decimal place accuracy
x_3	$= 0.005625$	1 decimal place accuracy
x_4	$= 0.00421875$	2 decimal place accuracy
x_5	$= 0.00316406$	2 decimal place accuracy
x_6	$= 0.00237305$	2 decimal place accuracy
x_7	$= 0.00177979$	2 decimal place accuracy
x_8	$= 0.00133484$	2 decimal place accuracy
x_9	$= 0.00100113$	2 decimal place accuracy
x_{10}	$= 0.000750847$	2 decimal place accuracy.

If convergence had been quadratic, one would expect x_{10} to be accurate to 2^9, or 512 decimal places.

23. Here $x_1 = 3$ and

$$x_{n+1} = x_n - \frac{(x-2)^{1/3}}{(1/3)(x-2)^{-2/3}} = x_n - 3(x_n - 2) = 6 - 2x_n.$$

Hence

$$x_1 = 3, \quad x_2 = 0, \quad x_3 = 6, \quad x_4 = -6, \quad x_5 = 18, \quad x_6 = -30, \quad x_7 = 66.$$

The iterates continue to diverge as n increases. The graph of $y = f(x)$ on the following page shows that the tangent lines are becoming flatter as x increases, thus forcing the x-intercept to increase without bound rather than converge.

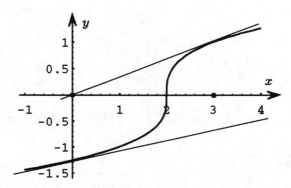

Exercise 23

25. (a) $x_1 = 0$, $x_2 = 1.258229874$, $x_3 = 2.768911636$, $x_4 = 4.235147453$.

(b) The roots are near the asymptotes of $t = \tan 2x$. These asymptotes are $x = \pi/4 + k\pi/2$, $k = 0, 1, 2, \ldots$.

Review Exercises – Chapter 3

1. $f(x) = \dfrac{x}{3x - 7} \implies f'(x) = \dfrac{(3x - 7)(1) - x(3)}{(3x - 7)^2} = \dfrac{-7}{(3x - 7)^2}$

3. $y = \dfrac{6x^3 - x^2 + x}{3x^5 + x^3} \implies$

$$y' = \dfrac{(3x^5 + x^3)(18x^2 - 2x + 1) - (6x^3 - x^2 + x)(15x^4 + 3x^2)}{(3x^5 + x^3)^2} = \dfrac{-36x^4 + 9x^3 - 12x^2 + x - 2}{x^3(3x^2 + 1)^2}$$

5. $f(t) = \sqrt{t} \sin t \implies f'(t) = \sqrt{t} \cos t + \dfrac{\sin t}{2\sqrt{t}}$

7. $y = \dfrac{1}{x + x^2 + x^3} = (x + x^2 + x^3)^{-1} \implies y' = -(x + x^2 + x^3)^{-2}(1 + 2x + 3x^2)$

9. $g(t) = (t^{-2} - t^{-3})^{-1} \implies g'(t) = -(t^{-2} - t^{-3})^{-2}(-2t^{-3} + 3t^{-4})$

11. $y = x^2 \sin x^2 \implies y' = x^2[(\cos x^2)(2x)] + 2x \sin x^2 = 2x(x^2 \cos x^2 + \sin x^2)$

13. $y = 1/(x^4 + 4x^2)^3 = (x^4 + 4x^2)^{-3} \implies y' = -3(x^4 + 4x^2)^{-4}(4x^3 + 8x) = \dfrac{-12(x^2 + 2)}{x^7(x^2 + 4)^4}$

15. $f(x) = \dfrac{2x - 7}{\sqrt{x^2 + 1}} = \dfrac{(x^2 + 1)^{1/2}(2) - (2x - 7)\left[(1/2)(x^2 + 1)^{-1/2}(2x)\right]}{x^2 + 1} = \dfrac{7x + 2}{(x^2 + 1)^{3/2}}$

17. $y = \left[(x^2 + 1)^3 - 7\right]^5 \implies y' = 5\left[(x^2 + 1)^3 - 7\right]^4 \left[3(x^2 + 1)^2(2x)\right] = 30x(x^2 + 1)^2 \left[(x^2 + 1)^3 - 7\right]^4$

93

19. $f(x) = \sqrt{\dfrac{3x-9}{x^2+3}} \implies$

$$f'(x) = \frac{1}{2}\left(\frac{3x-9}{x^2+3}\right)^{-1/2}\frac{(x^2+3)(3)-(3x-9)(2x)}{(x^2+3)^2} = \frac{1}{2}\left(\frac{3x-9}{x^2+3}\right)^{-1/2}\frac{-3x^2+18x+9}{(x^2+3)^2}$$

21. $f(s) = \tan^2 s \cot^5 s = \cot^3 s \implies f'(s) = 3\cot^2 s\,(-\csc^2 s) = -3\cot^2 s\,\csc^2 s$

23. $y = \dfrac{\sqrt{1+\sin x}}{\cos x} = \dfrac{(1+\sin x)^{1/2}}{\cos x} \implies$

$$y' = \frac{(\cos x)\left[(1/2)(1+\sin x)^{-1/2}(\cos x)\right]-(1+\sin x)^{1/2}(-\sin x)}{\cos^2 x} = \frac{(1/2)\cos^2 x + \sin x + \sin^2 x}{(\cos^2 x)\sqrt{1+\sin x}}$$

25. $f(t) = t^{9/2} - 6t^{5/2} \implies f'(t) = (9/2)t^{7/2} - 15t^{3/2}$

27. $h(t) = \cos^2(t^2-1) \implies h'(t) = 2\left[\cos(t^2-1)\right]\left[-\sin(t^2-1)\right](2t) = -4t\cos(t^2-1)\sin(t^2-1)$

29. $y = \sqrt{x^3}\,(x^{-3}-2x^{-1}+2)^2 = x^{3/2}(x^{-3}-2x^{-1}+2)^2 \implies$

$$y' = x^{3/2}\left[2(x^{-3}-2x^{-1}+2)(-3x^{-4}+2x^{-2})\right] + (3/2)x^{1/2}(x^{-3}-2x^{-1}+2)^2$$

$$= x^{1/2}(x^{-3}-2x^{-1}+2)\left[(-9/2)x^{-3}+x^{-1}+3\right]$$

$$= (1/2)x^{-11/2}(12x^6-8x^5-4x^4-12x^3+20x^2-9)$$

31. $f(x) = x\sin x \implies$

$$f'(x) = x\cos x + \sin x$$

$$f''(x) = x(-\sin x) + \cos x + \cos x = -x\sin x + 2\cos x$$

33. $f(x) = \sec x \tan x \implies$

$$f'(x) = (\sec x)(\sec^2 x) + (\tan x)(\sec x \tan x) = (\sec x)(\sec^2 x + \tan^2 x)$$

$$f''(x) = (\sec x)\left[2(\sec x)(\sec x \tan x) + 2(\tan x)(\sec^2 x)\right] + (\sec x \tan x)(\sec^2 x + \tan^2 x)$$

$$= \sec x \tan x(5\sec^2 x + \tan^2 x)$$

35. $s(t) = t^2 - 1 + t^{1/2} \implies v(t) = 2t + (1/2)t^{-1/2} \implies a(t) = 2 - (1/4)t^{-3/2}.$

37. $s(t) = \sqrt{2+t^2} = (2+t^2)^{1/2} \implies$

$$v(t) = (1/2)(2+t^2)^{-1/2}(2t) = t(2+t^2)^{-1/2}$$

$$a(t) = t\left[(-1/2)(2+t^2)^{-3/2}(2t)\right] + (2+t^2)^{-1/2} = 2(2+t^2)^{-3/2}$$

39. $y = 6x - 2 \implies$

$$\frac{dy}{dx} = \lim_{h\to 0}\frac{f(x+h)-f(x)}{h} = \lim_{h\to 0}\frac{[6(x+h)-2]-[6x-2]}{h} = \lim_{h\to 0} 6 = 6$$

41. $y = \sqrt{x+1} \implies$

$$\frac{dy}{dx} = \lim_{h \to 0} \frac{f(x+h) - f(x)}{h} = \lim_{h \to 0} \frac{\sqrt{x+h+1} - \sqrt{x+1}}{h}$$

$$= \lim_{h \to 0} \frac{\sqrt{x+h+1} - \sqrt{x+1}}{h} \cdot \frac{\sqrt{x+h+1} + \sqrt{x+1}}{\sqrt{x+h+1} + \sqrt{x+1}}$$

$$= \lim_{h \to 0} \frac{1}{\sqrt{x+h+1} + \sqrt{x+1}} = \frac{1}{2\sqrt{x+1}}$$

43. $y = x^2 + x + 1 \implies$

$$\frac{dy}{dx} = \lim_{h \to 0} \frac{f(x+h) - f(x)}{h} = \lim_{h \to 0} \frac{[(x+h)^2 + (x+h) + 1] - [x^2 + x + 1]}{h}$$

$$= \lim_{h \to 0} \frac{2xh + h^2 + h}{h} = \lim_{h \to 0}(2x + h + 1) = 2x + 1$$

45. $6x^2 - xy - 4y^2 = 0 \implies$

$$12x - x\frac{dy}{dx} - y - 8y\frac{dy}{dx} = 0$$

$$\implies \quad \frac{dy}{dx}(-x - 8y) = y - 12x$$

$$\implies \quad \frac{dy}{dx} = \frac{12x - y}{x + 8y}$$

47. $x = y + y^2 + y^3 \implies$

$$1 = \frac{dy}{dx} + 2y\frac{dy}{dx} + 3y^2\frac{dy}{dx} = \frac{dy}{dx}(1 + 2y + 3y^2) \quad \implies \quad \frac{dy}{dx} = \frac{1}{1 + 2y + 3y^2}$$

49. $xy = \cot xy \implies$

$$x\frac{dy}{dx} + y = (-\csc^2 xy)\left(x\frac{dy}{dx} + y\right) = -x(\csc^2 xy)\frac{dy}{dx} - y\csc^2 xy$$

$$\implies \quad \frac{dy}{dx}(x + x\csc^2 xy) = -y - y\csc^2 xy$$

$$\implies \quad \frac{dy}{dx} = \frac{-y - \csc^2 xy}{x + x\csc^2 xy} = \frac{-y(1 + \csc^2 xy)}{x(1 + \csc^2 xy)} = -\frac{y}{x}$$

51. $x\sin y = y \implies x\cos(dy/dx) + \sin y = dy/dx \implies dy/dx = -(\sin y)/(x\cos y - 1)$

53. $x^2 + y^2 = 4 \implies$

$$2x + 2y\frac{dy}{dx} = 0 \quad \implies \quad \frac{dy}{dx} = -\frac{x}{y}$$

$$\implies \quad \frac{d^2y}{dx^2} = -\frac{y - x(dy/dx)}{y^2} = -\frac{y - x(-x/y)}{y^2} = -\frac{y^2 + x^2}{y^3} = -\frac{4}{y^3}$$

55. $\dfrac{1}{x} + \dfrac{1}{y} = 4 \implies$

$$-\frac{1}{x^2} - \frac{1}{y^2}\frac{dy}{dx} = 0 \implies \frac{dy}{dx} = -\frac{y^2}{x^2} \implies \frac{dy}{dx}\bigg|_{(1/2,1/2)} = -\frac{(1/2)^2}{(1/2)^2} = -1.$$

Therefore the slope of the tangent line at $(1/2, 1/2)$ is -1 and the equation is $y - 1/2 = -(x - 1/2)$, or $y = -x + 1$.

57. $f(x) = \dfrac{\cos x}{\sin x} = \cot x \implies f'(x) = -\csc^2 x \implies f'(\pi/4) = -\csc^2(\pi/4) = -2$. Therefore the slope of the tangent line at $(\pi/4, 1)$ is -2 and the equation is $y - 1 = -2(x - \pi/4)$, or $y = -2x + 1 + \pi/2$.

59. $f(x) = \sec x \tan x \implies$

$$f'(x) = (\sec x)\sec^2 x + (\tan x)(\sec x \tan x) \implies f'(\pi/4) = \sec^3(\pi/4) + \sec(\pi/4)\tan^2(\pi/4) = 3\sqrt{2}.$$

Therefore the slope of the tangent line is $3\sqrt{2}$ and the equation is $y - \sqrt{2} = 3\sqrt{2}\,(x - \pi/4)$.

61. $y = x \sin x \implies dy = (dy/dx)dx = (x\cos x + \sin x)dx$

63. $y = \sqrt{1 + x^2} \implies dy = (dy/dx)dx = (1/2)(1 + x^2)^{-1/2}(2x)dx = x(1 + x^2)^{-1/2}dx$

65. $y = \cos\sqrt{x} \implies dy = (dy/dx)dx = (-\sin\sqrt{x})(1/2)x^{-1/2}dx$

67. Let $f(x) = x^{2/3}$. Then $f'(x) = (2/3)x^{-1/3}$ and hence

$$f(x) \approx f(x_0) + f'(x_0)\Delta x = x_0^{2/3} + (2/3)x_0^{-1/3}\Delta x.$$

Setiing $x_0 = 8$ and $x = 8.2$, $\Delta x = 0.2$ and hence

$$(8.2)^{2/3} \approx 8^{2/3} + (2/3)8^{-1/3}(0.2) = 4.0667.$$

69. Let $f(x) = x^3$. Then $f'(x) = 3x^2$. and hence

$$f(x) \approx f(x_0) + f'(x_0)\Delta x = x_0^3 + 3x_0^2\Delta x.$$

Setting $x_0 = 1$ and $x = 0.96$, $\Delta x = -0.04$ and hence

$$(0.96)^3 \approx 1^3 + 3(1)^2(-0.04) = 0.88.$$

71. No;

$$\lim_{h \to 0^+} \frac{f(2 + h) - f(2)}{h} = \lim_{h \to 0^+} \frac{|(2+h)^2 - 6(2+h) + 8|}{h} = \lim_{h \to 0^+} \frac{-h(h-2)}{h} = 2$$

but

$$\lim_{h \to 0^-} \frac{f(2 + h) - f(2)}{h} = \lim_{h \to 0^-} \frac{|(2+h)^2 - 6(2+h) + 8|}{h} = \lim_{h \to 0^-} \frac{h(h-2)}{h} = -2.$$

73. No;

$$\lim_{h \to 0^+} \frac{f(2 + h) - f(2)}{h} = \lim_{h \to 0^+} \frac{2(2+h) - 4}{h} = 2$$

but

$$\lim_{h \to 0^-} \frac{f(2 + h) - f(2)}{h} = \lim_{h \to 0^-} \frac{4 - (2+h)^2}{h} = \lim_{h \to 0^-} (-4 - h) = -4.$$

75. a ↔ v, b ↔ viii, c ↔ i

77. Let $A(t)$ be the area of a hexagon whose side is $s(t)$ meters at time t. In the figure below, each central angle is 60° and hence each triangle is equilateral. Hence

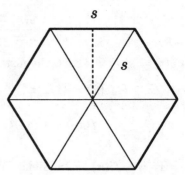

$$A = 6 \text{ Area(triangle)}$$

$$= 6\left[\left(\frac{1}{2}s\right)\left(\frac{\sqrt{3}}{2}s\right)\right]$$

$$= \frac{3s^2\sqrt{3}}{2}$$

$$\implies \frac{dA}{dt} = 3s\sqrt{3}\frac{ds}{dt}.$$

From the given information, $ds/dt = 2$. Hence

$$\left.\frac{dA}{dt}\right|_{s=10} = 3(10)\sqrt{3}\,(2) = 60\sqrt{3}.$$

Thus, when the side is 10 cm, the area is increasing at the rate of $60\sqrt{3}$ cm^2/s.

79. Let $x(t)$ stand for the horizontal displacement of the base of the ladder from the wall and $y(t)$ the vertical displacement of the top, as illustrated in the figure below. Then

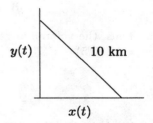

$$A = \frac{1}{2}xy = \frac{1}{2}x\sqrt{100 - x^2}$$

$$\implies \frac{dA}{dt} = \left[\left(\frac{x}{2}\right)\left(\frac{1}{2}\right)(100 - x^2)^{-1/2}(-2x) + \frac{1}{2}\sqrt{100 - x^2}\right]\frac{dx}{dt}$$

$$= \frac{50 - x^2}{\sqrt{100 - x^2}}\frac{dx}{dt}.$$

From the given information, $dx/dt = 1$. Hence, when $x = 6$,

$$\left.\frac{dA}{dt}\right|_{x=6} = \frac{50 - 6^2}{\sqrt{100 - 6^2}}(1) = 1.75.$$

Thus, when the base of the ladder is 6 feet from the wall, the area of the triangle is increasing at a rate of 1.75 m^2/s.

81. Let $V(t)$ and $r(t)$ be the volume and radius of the snowball at time t. From the given information, $dV/dt = -k(4\pi r^2)$ for some constant $k > 0$, where $4\pi r^2$ is the surface area of the snowball. Since

$$V = \frac{4}{3}\pi r^3 \implies \frac{dV}{dt} = 4\pi r^2\frac{dr}{dt} = -k(4\pi r^2) \implies \frac{dr}{dt} = -k,$$

it follows that the radius of the snowball decreases at a constant rate.

83. $y = \sin x^2 \implies y' = (\cos x^2)(2x) \implies y'\left(\sqrt{\pi/4}\right) = [\cos(\pi/4)]\left(2\sqrt{\pi/4}\right) = \sqrt{\pi/2}$. Therefore the slope of the normal line at $\left(\sqrt{\pi/4}, \sqrt{2}/2\right)$ is $-\sqrt{2/\pi}$ and the equation is $y - \sqrt{2}/2 = -\sqrt{2/\pi}\left(x - \sqrt{\pi/4}\right)$.

85. $y = x^3 - 2x \implies y' = 3x^2 - 2 \implies y'(0) = -2$. Therefore the slope of the normal line at $(0,0)$ is $1/2$ and the equation is $y = (1/2)x$.

87. The average velocity during the time interval $0 \le t \le \pi$ is

$$\frac{s(\pi) - s(0)}{\pi - 0} = \frac{[-4\cos\pi] - [-4\cos 0]}{\pi - 0} = \frac{8}{\pi}.$$

Since $v(t) = s'(t) = 4\sin t$, the instantaneous velocity at $t = \pi/2$ is $v(\pi/2) = 4\sin(\pi/2) = 4$.

89. $f(x) = \sqrt{x} \implies f'(x) = (1/2)x^{-1/2}$. Hence, if $x_0 = 8$,

$$f(x_0) + f'(x_0)\Delta x = \sqrt{x_0} + \frac{1}{2\sqrt{x_0}}(x - x_0) = \sqrt{8} + \frac{1}{2\sqrt{8}}(x - 8).$$

91. $f(x) = x^2 \implies f'(x) = 2x$. Hence, if $x_0 = 1$,

$$f(x_0) + f'(x_0)\Delta x = x_0^2 + 2x_0(x - x_0) = 1 + 2(x - 1) = 2x - 1.$$

93. $h(x) = f(g(x)) \implies h'(x) = f'(g(x))g'(x) \implies h'(3) = f'(g(3))g'(3) = f'(4)g'(3) = (5)(2) = 10$.

95. Let $V(r)$ be the volume of a sphere of radius r. Then

$$V = (4/3)\pi r^3 \implies dV = 4\pi r^2 dr.$$

If $r_0 = 20$ and $r = 20.5$, $\Delta r = 0.5$ and hence

$$dV = \text{approximate change in volume} = 4\pi(20)^2(0.5) = 2513.27.$$

Thus, the volume increases by approximately 2513.27 cm^3. Since the actual change in volume is $\Delta V = V(20.5) - V(20) = (4/3)\pi(20.5^3 - 20^3) = 2576.63$, the relative error is

$$\left| \frac{\Delta V - dV}{\Delta V} \right| = \frac{2576.63 - 2513.27}{2576.63} = 0.0246$$

and the percentage error is 2.46 %

Remark: Note that the problem asks for the relative error in approximating the *change* in volume when r increases from 20 cm to 20.5 cm, not the relative error in approximating $V(20.5)$. The relative error in approximating $V(20.5)$ is

$$\left| \frac{y - y_1}{y} \right| = 0.0018,$$

where y = actual volume = $V(20.5) = 36086.9512$ and y_1 = linear approximation to $V(20.5) = 36023.5957$.

97. $y = 4x^3 - 4x + 4 \implies y' = 12x^2 - 4$. Since the slope of the line $y - 8x + 6 = 0$ is 8, the tangent line is parallel to this line $\iff 12x^2 - 4 = 8 \iff x^2 = 1 \iff x = \pm 1$. Hence the tangent line is parallel to $y - 8x + 6 = 0$ at two points: $(1, 4)$ and $(-1, 4)$.

99. $f(x) = (8/5)x^{5/4}$

101. (a) $v(t) = s'(t) = t^3 - 9t^2 + 24t - 20$

    (b) $a(t) = v'(t) = 3t^2 - 18t + 24$

(c) To find sign changes for $v(t)$, we first solve $v(t) = 0$. Observe that $t = 2$ is a root of $v = 0$. Hence

$$v(t) = (t-2)(t^2 - 7t + 10) = (t-2)(t-5)(t-2) = 0 \implies t = 2, 5.$$

Sign v	$-$	$-$	$+$
Interval	$(-\infty, 2)$	$(2, 5)$	$(5, +\infty)$

Hence the particle is moving to the right when $t > 5$.

(d) From the table in part (c), the particle is moving to the left when $t < 2$ and when $2 < t < 5$.

(e) By part (c), $v(t) = 0 \iff t = 2, 5$.

(f) Only when $t = 5$.

(g) $a(t) = 3t^2 - 18t + 24 = 3(t-4)(t-2) = 0 \iff t = 2, 4$.

103. $s(t) = t^3 + 3t + 3 \implies v(t) = 3t^2 + 3 \implies a(t) = 6t$. Hence

$$|v(t)| = |a(t)| \implies 3t^2 + 3 = \pm 6t \implies 3(t^2 \mp 2t + 1) = 3(t \mp 1)^2 = 0 \implies t = \pm 1.$$

Thus, velocity and acceleration have the same magnitude when $t = \pm 1$.

105. Let $f(x)$ be an odd function. Then

$$f'(-x) = \lim_{h \to 0} \frac{f(-x + h) - f(-x)}{h}.$$

Replacing h by $-h$, and noting that $h \to 0 \iff -h \to 0$,

$$
\begin{aligned}
f'(-x) &= \lim_{h \to 0} \frac{f(-x - h) - f(-x)}{-h} \\
&= \lim_{h \to 0} \frac{-f(x + h) + f(x)}{-h} \qquad \text{since } f \text{ is odd} \\
&= \lim_{h \to 0} \frac{f(x + h) - f(x)}{h} \\
&= f'(x).
\end{aligned}
$$

Hence $f'(x)$ is an even function.

Chapter 4

Applications of the Derivative

4.1   Extreme Values

1. a) Critical numbers: x_1, x_2.
    b) Numbers at which f has a relative minimum on $[a, b]$: a, b.
    c) Numbers at which f has an absolute maximum on $[a, b]$: x_2.
    d) Numbers at which f has an absolute extremum on $[a, b]$: a, x_2.

3. a) x_1, x_2, x_3
    b) a, b
    c) x_3
    d) x_3, b

5. $f(x) \;=\; x^2 - 4x + 3 \qquad [0, 4] \qquad c = 2$
    $f'(x) \;=\; 2x - 4$

    a) $f'(c) = f'(2) = 2(2) - 4 = 0$

    b)

Exercise 5. b)

c) $f(c) = f(2) = -1$ is the absolute minimum of f on $[0, 4]$.

7. $f(x) \;=\; x^2 - 7 \qquad [-3, 4] \qquad c = 0$
    $f'(x) \;=\; 2x$

    a) $f'(0) = (2)(0) = 0$

100

b)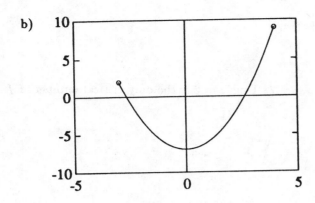

Exercise 7. b)

c) $f(0) = -7$ is the absolute minimum of f on $[-3, 4]$.

9. $f(x) = \cos(\pi x)$ $[0, 2]$ $c = 1$
$f'(x) = -\pi \sin(\pi x)$

    a) $f'(1) = -\pi \sin(\pi) = 0$

    b)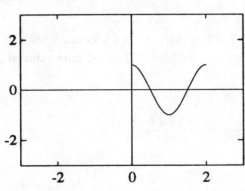

Exercise 9. b)

c) $f(1) = -1$ is the absolute minimum of f on $[0, 2]$.

In Exercises 11–28, find all critical numbers and the maximum and minimum values for f on the given interval.

11. $f(x) = x^2(x - 4), \quad x \in [-1, 3]$
     $f'(x) = 2x(x - 4) + x^2 = 3x^2 - 8x$

Let $f'(x) = 0$, we have $3x^2 - 8x = 0$
$$\Rightarrow \quad x(3x - 8) = 0$$

Thus, the critical numbers are 0 and $\dfrac{8}{3}$.

$f(-1) = -5$
$f(0) = 0$
$f\left(\dfrac{8}{3}\right) = -\dfrac{256}{27}$
$f(3) = -9$

Hence, $f(0) = 0$ is the maximum value of f on $[-1, 3]$

and $f\left(\dfrac{8}{3}\right) = -\dfrac{256}{27}$ is the minimum value of f on $[-1, 3]$.

101

13. $f(x) = \dfrac{1}{x(x-4)}, \quad x \in [1,3]$

$f'(x) = \dfrac{-(x-4+x)}{x^2(x-4)^2} = \dfrac{-2x+4}{x^2(x-4)^2}$

Let $f'(x) = 0$, we have: $\dfrac{-2x+4}{x^2(x-4)^2} = 0 \Rightarrow x = 2$. Thus $x = 2$ is the only critical number of f.

$f(1) = -\frac{1}{3}$

$f(2) = -\frac{1}{4}$

$f(3) = -\frac{1}{3}$

Therefore $f(1) = f(3) = -\frac{1}{3}$ is the minimum value of f on $[1,3]$

and $f(2) = -\frac{1}{4}$ is the maximum value of f on $[1,3]$.

15. $f(x) = x^2(x-2), \quad x \in [-1,2]$
$f'(x) = 2x(x-2) + x^2 = 3x^2 - 4x$
$3x^2 - 4x = 0 \Rightarrow x = 0, \frac{4}{3}$.
Therefore the critical numbers are 0 and $\frac{4}{3}$.

$f(-1) = -3$ (minimum value of f on $[-1,2]$) $\quad f(0) = 0$ (maximum value of f on $[-1,2]$)

$f\left(\frac{4}{3}\right) = -\frac{32}{27}$ $\qquad\qquad\qquad\qquad\qquad f(2) = 0$ (maximum value of f on $[-1,2]$)

17. $f(x) = 1 - \tan^2 x, \quad x \in \left[-\frac{\pi}{4}, \frac{\pi}{4}\right]$

$f'(x) = -2\tan x \sec^2 x = -2\dfrac{\sin x}{\cos x}\dfrac{1}{\cos^2 x} = \dfrac{-2\sin x}{\cos^3 x}$

$f'(x) = 0 \Rightarrow -2\sin x = 0$
$\qquad\qquad \Rightarrow \sin x = 0$
$\qquad\qquad \Rightarrow x = 0$

$f'(x)$ is undefined implies $\cos^3 x = 0$
$\qquad\qquad\qquad\qquad\qquad \Rightarrow \cos x = 0.$

Since $\cos x \neq 0$ in $\left[-\frac{\pi}{4}, \frac{\pi}{4}\right]$, the only critical number is 0.

$f\left(-\frac{\pi}{4}\right) = 0 \longleftarrow$ minimum value of f on $\left[-\frac{\pi}{4}, \frac{\pi}{4}\right]$

$f(0) = 1 \longleftarrow$ maximum value of f on $\left[-\frac{\pi}{4}, \frac{\pi}{4}\right]$

$f\left(\frac{\pi}{4}\right) = 0 \longleftarrow$ minimum value of f on $\left[-\frac{\pi}{4}, \frac{\pi}{4}\right]$

19. $f(x) = \sin x + \cos x, \quad x \in [0, \pi]$
$f'(x) = \cos x - \sin x$
$\cos x - \sin x = 0$ when $\cos x = \sin x$
$\qquad\qquad\qquad\qquad \Rightarrow x = \frac{\pi}{4}.$

Therefore the only critical number is $\frac{\pi}{4}$.

$f(0) = 1$

$f\left(\frac{\pi}{4}\right) = \sqrt{2} \longleftarrow$ maximum value of f on $[0, \pi]$

$f(\pi) = -1 \longleftarrow$ minimum value of f on $[0, \pi]$

21. $f(x) = x + \frac{1}{x}, \quad x \in \left[\frac{1}{2}, 2\right]$

$f'(x) = 1 - \frac{1}{x^2} = \frac{x^2 - 1}{x^2}$

$\frac{x^2 - 1}{x^2} = 0$ when $x^2 - 1 = 0 \Rightarrow (x+1)(x-1) = 0 \Rightarrow x = \pm 1.$

Therefore the only critical number in $\left[\frac{1}{2}, 2\right]$ is 1.

$f\left(\frac{1}{2}\right) = \frac{5}{2} \longleftarrow$ maximum value of f on $\left[\frac{1}{2}, 2\right]$

$f(1) = 2 \longleftarrow$ minimum value of f on $\left[\frac{1}{2}, 2\right]$

$f(2) = \frac{5}{2} \longleftarrow$ maximum value of f on $\left[\frac{1}{2}, 2\right]$

23. $f(x) = \sec x, \quad x \in \left[-\frac{\pi}{4}, \frac{\pi}{4}\right]$

$f'(x) = \sec x \tan x = \frac{\sin x}{\cos^2 x}$

$f'(x) = 0$ when $\sin x = 0 \Rightarrow x = 0$ in $\left[-\frac{\pi}{4}, \frac{\pi}{4}\right].$

$f'(x)$ is undefined when $\cos^2 x = 0 \Rightarrow \cos x = 0.$

Since $\cos x \neq 0$ in $\left[-\frac{\pi}{4}, \frac{\pi}{4}\right]$, the only critical number is 0.

$f\left(-\frac{\pi}{4}\right) = \sqrt{2} \longleftarrow$ maximum value of f on $\left[-\frac{\pi}{4}, \frac{\pi}{4}\right]$

$f(0) = 1 \longleftarrow$ minimum value of f on $\left[-\frac{\pi}{4}, \frac{\pi}{4}\right]$

$f\left(\frac{\pi}{4}\right) = \sqrt{2} \longleftarrow$ maximum value of f on $\left[-\frac{\pi}{4}, \frac{\pi}{4}\right]$

25. $f(x) = \frac{x^3}{2+x}, \quad x \in [-1, 1]$

$f'(x) = \frac{3x^2(2+x) - x^3}{(2+x)^2} = \frac{2x^3 + 6x^2}{(2+x)^2} = \frac{2x^2(x+3)}{(2+x)^2}$

$f'(x) = 0$ when $2x^2(x+3) = 0 \Rightarrow x = 0$ or $x = -3.$

Therefore the only critical number in $[-1, 1]$ is 0.

$f(-1) = -1 \longleftarrow$ minimum value of f on $[-1, 1]$

$f(0) = 0$

$f(1) = \frac{1}{3} \longleftarrow$ maximum value of f on $[-1, 1]$

27. $f(x) = 8x^{1/3} - 2x^{4/3}, \quad x \in [-1, 8]$

$f'(x) = \frac{8}{3x^{2/3}} - \frac{8x^{1/3}}{3} = \frac{8 - 8x}{3x^{2/3}}$

$f'(x) = 0$ when $8 - 8x = 0 \Rightarrow x = 1.$

$f'(x)$ is undefined when $3x^{2/3} = 0 \Rightarrow x = 0.$ Therefore the critical numbers are $0, 1.$

$f(-1) = -10$ $\qquad\qquad\qquad f(0) = 0$

$f(1) = 6$ (maximum value of f on $[-1, 8]$) $\quad f(8) = -16$ (minimum value of f on $[-1, 8]$)

In Exercises 29–32, find the maximum and minimum values for f on the given interval.

29. $f(x) = \left(\sqrt{x} - x\right)^2, \quad x \in [0, 4]$

$f'(x) = 2\left(\sqrt{x} - x\right)\left(\frac{1}{2\sqrt{x}} - 1\right) = 2\sqrt{x}\left(1 - \sqrt{x}\right)\left(\frac{1 - 2\sqrt{x}}{2\sqrt{x}}\right) = \left(1 - \sqrt{x}\right)\left(1 - 2\sqrt{x}\right)$

$f'(x) = 0$ when $1 - \sqrt{x} = 0 \Rightarrow x = 1$ or when $1 - 2\sqrt{x} = 0 \Rightarrow x = \frac{1}{4}$.
Therefore the critical numbers are 1 and $\frac{1}{4}$.

$f(0) = 0$ (minimum value of f on $[0,4]$) $f(1) = 0$ (minimum value of f on $[0,4]$)
$f\left(\frac{1}{4}\right) = \frac{1}{16}$ $f(4) = 4$ (maximum value of f on $[0,4]$)

31. $f(x) = \frac{\sqrt{x}}{1 + x}, \quad x \in [0, 4]$

$f'(x) = \frac{\frac{1}{2\sqrt{x}}(1 + x) - \sqrt{x}}{(1 + x)^2} = \frac{1 - x}{2\sqrt{x}(1 + x)^2}$

$f'(x) = 0$ when $x = 1$ and $f'(x)$ is undefined when $x = 0$.

$f(0) = 0 \quad \longleftarrow \quad$ minimum value of f on $[0,4]$

$f(1) = \frac{1}{2} \quad \longleftarrow \quad$ maximum value of f on $[0,4]$

$f(4) = \frac{2}{5}$

33. a) $g'(x) = c\,f'(x)$ and $c \neq 0$ imply that $g'(x) = 0$ exactly when $f'(x) = 0$; so: the critical numbers of f and g are the same.

   b) $h'(x) = f'(x + c)$ implies that x is a critical number for h if and only if $x + c$ is a critical number for f. So: the critical numbers for h are those for f shifted c units to the left.

   c) $k'(x) = c\,f'(cx)$ and $c \neq 0$ imply that x is a critical number for k if and only if cx is a critical number for f. So: the critical numbers for k are those for f divided by c.

35. We are told that $f(3) = -7$ is the minimum value of $f(x) = x^2 + bx + c$ on $[0, 5]$. Since $x = 3$ is an interior point, $f'(3) = 0$. Now $f'(x) = 2x + b$ implies

$$6 + b = f'(3) = 0$$

and $f(x) = x^2 + bx + c$ implies

$$9 + 3b + c = f(3) = -7$$

So $b = -6$ and $c = -7 - 9 - 3(-6) = 2$.

37. a) $f(x) = \sin x - \cos x \Rightarrow f'(x) = \cos x + \sin x$. Now $f'(x) = 0 \Rightarrow \sin x = -\cos x \Rightarrow \tan x = -1 \Rightarrow x = \frac{3\pi}{4}$ or $x = \frac{7\pi}{4}$ in $[0, 2\pi]$. Evaluating f at these critical numbers: $f\left(\frac{3\pi}{4}\right) = \sqrt{2}$ and $f\left(\frac{7\pi}{4}\right) = -\sqrt{2}$. Evaluating f at the endpoints: $f(0) = f(2\pi) = -1$.
   So: maximum value of f is $\sqrt{2}$, minimum value of f is $-\sqrt{2}$.

   b) Notice (from trigonometry) that

$$f(x) = 2\sin\left(\frac{\pi}{2} - x\right) = 2\cos x$$
$$\text{so } f'(x) = -2\sin x.$$

   Now $f'(x) = 0 \Rightarrow x \in \{0, \pi, 2\pi\}$ in $[0, 2\pi]$. Evaluating: $f(0) = f(2\pi) = 2$ and $f(\pi) = -2$.
   So: maximum value of f is 2, minimum value of f is -2.

c) The double angle formula yields ·

$$f(x) = 2 \sin x \cos x = \sin 2x$$

which has period π; so we need only look at $[0, \pi]$. Here,

$$f'(x) = 2 \cos 2x$$
$$\text{so } f'(x) = 0 \Rightarrow x \in \left\{ \frac{\pi}{4}, \frac{3\pi}{4} \right\} \text{ in } [0, \pi].$$

Evaluating: $f(0) = f(\pi) = 0$, $f\left(\frac{\pi}{4}\right) = 1$ and $f\left(\frac{3\pi}{4}\right) = -1$.
So: maximum value of f is 1, minimum value of f is -1.

39. $f(x) = |x - 3|$, $x \in [-2, 6]$
$x - 3 = 0 \Leftrightarrow x = 3$. Let $g(x) = x - 3$; $g'(x) = 1 \neq 0$. Thus $g(x)$ has no critical numbers.
$$\begin{array}{lll} f(3) & = & 0 \quad \longleftarrow \text{ minimum value} \\ f(-2) & = & 5 \quad \longleftarrow \text{ maximum value} \\ f(6) & = & 3 \end{array}$$

41. $f(x) = |x^2 - x - 6|$, $x \in [-3, 5]$
$x^2 - x - 6 = (x - 3)(x + 2) = 0 \Leftrightarrow x = 3, -2$. Let $g(x) = x^2 - x - 6$; $g'(x) = 2x - 1 = 0$ at $x = \frac{1}{2}$.
$$\begin{array}{lll} f(3) & = & 0 \quad \longleftarrow \text{ minimum value} \qquad f(-3) = 6 \\ f(-2) & = & 0 \quad \longleftarrow \text{ minimum value} \qquad f(5) = 14 \quad \longleftarrow \text{ maximum value} \\ f\left(\frac{1}{2}\right) & = & \frac{25}{4} \end{array}$$

43. $f(x) = \dfrac{|x|}{4 + x}$; $x \in [-2, 4]$
Note that $4 + x > 0$ on $[-2, 4]$. Therefore $f(x)$ can be rewritten as $\left| \dfrac{x}{4+x} \right|$. $\dfrac{x}{4 + x} = 0 \Leftrightarrow x = 0$. Let
$g(x) = \dfrac{x}{4 + x}$, $g'(x) = \dfrac{4}{(4 + x)^2} > 0$ for $x \in (-2, 4)$. Thus $g(x)$ has no critical numbers on $(-2, 4)$.
$$\begin{array}{lll} f(0) & = & 0 \quad \longleftarrow \text{ minimum value} \\ f(-2) & = & 1 \quad \longleftarrow \text{ maximum value} \\ f(4) & = & \frac{1}{2} \end{array}$$

45. Let $p_2(x) = ax^2 + bx + c$ and let $a \neq 0$. In fact, suppose first that $a > 0$. From $p_2'(x) = 2ax + b$ we see that p_2 has exactly one critical number: namely, $x = -\frac{b}{2a}$. If $x < -\frac{b}{2a}$ than $p_2(x) < 0$ whilst if $x > -\frac{b}{2a}$ then $p_2(x) > 0$; so p_2 has a relative minimum at $x = -\frac{b}{2a}$. This is actually an absolute minimum. The easiest way to see this is by completing the square:

$$\begin{aligned} p_2(x) \quad + \quad & a \left(x^2 + \frac{b}{a} x + \frac{c}{a} \right) \\ = \quad & a \left\{ \left(x + \frac{b}{2a} \right)^2 + \frac{c}{a} - \frac{b^2}{4a^2} \right\} \\ = \quad & a \left(x + \frac{b}{2a} \right)^2 + \frac{4ac - b^2}{4a} \end{aligned}$$

which assumes a minimum value of $\dfrac{4ac - b^2}{4a}$ when $x = -\dfrac{b}{2a}$.

Similarly, if $a < 0$ then p_2 has an absolute maximum value of $\dfrac{4ac - b^2}{4a}$ at its unique critical number $x = -\dfrac{b}{2a}$.

105

47. a) $p'^+(0) = \lim_{h \to 0^+} \frac{p(0+h) - p(0)}{h} = \lim_{h \to 0^+} \frac{h^{3/2} - 0}{h}$
$= \lim_{h \to 0^+} h^{1/2} = 0.$

The function $p(x) = x^{3/2}$ can be said to have a critical number at 0 since its right-hand derivative is zero at 0.

b) $f'^+(a) = \lim_{h \to 0^+} \frac{f(a+h) - f(a)}{h}$
$= \lim_{h \to 0^+} \frac{g(a+h) - g(a)}{h}$
$= g'(a)$

since g is assumed differentiable at a and since $f(x) = g(x)$ if $x \in [a, b]$.

c) In the notation of part b) we let $f(x) = x$ and $g(x) = p(x) = |x|$. Since $f(x) = g(x)$ for $x \geq 0$ and since $f'(0)$ is surely 1, we deduce that $p'^+(0) = g'^+(0) = f'(0) = 1$.

d) The left-hand derivative $f'^-(b)$ of f at b is equal to

$$\lim_{h \to 0^-} \frac{f(b+h) - f(b)}{h}.$$

e) $p(x) = x^4 - 2x^2$
$\Rightarrow p'(x) = 4x^3 - 4x$
$= 4x(x^2 - 1)$
$\Rightarrow p'(x) = 0$ when $x = 0, +1, -1.$

Thus, p has critical numbers at $0, +1, -1$. Strictly speaking, we deal with $x = -1$ using part b) and with $x = +1$ using the corresponding argument based on part c), since these are endpoints of $[-1, +1]$.

49. absolute minimum: $(-3, -12 - \sqrt[3]{7})$
relative minimum: $(0.46987, -2.63767)$
relative minimum: $(1.22175, -3.15492)$
relative maximum: $(-1.25314, 2.77862)$
relative maximum: $(0.532, -2.10943)$
absolute maximum: $(3, 12 + \sqrt[3]{5})$

51. absolute minimum: $(-3, -18)$
relative minimum: $(1, -2)$
relative maximum: $(-1, 2)$
absolute maximum: $(3, 6)$

4.2 Applied Maximum-Minimum Problems

1. Let the two nonnegative numbers be x and y; thus $x + y = 20$ and so $y = 20 - x$.

a) The product of the numbers is

$$P = xy = x(20 - x)$$

therefore
$$\frac{dP}{dx} = 20 - 2x$$

which is zero if $x = 10$, $P(10) = 100$, $P(0) = 0$, $P(20) = 0$. Thus the product is least when $x = 0$, $y = 20$.

b) and c) From

$$S = x^2 + y^2 = x^2 + (20 - x)^2$$

$$= 2x^2 - 40x + 400$$

we deduce that

$$\frac{dS}{dx} = 4x - 40$$

which is zero when $x = 10$. Now, $x = 10 \Rightarrow S = 200$, $x = 0$ (one endpoint) $\Rightarrow S = 400$ and $x = 20$ (other endpoint) $\Rightarrow S = 400$. So: the sum of the squares is

    a) least when the two numbers are both 10.

    b) greatest when one is 0 and the other 20.

3. Let the three sides of the fence have lengths (in meters) x, y and y, so that $x + 2y = 20$. The function to be optimized is the area enclosed, which is

$$A = xy = (20 - 2y)y$$

therefore
$$\frac{dA}{dy} = 20 - 4y$$

which is zero if $y = 5$. Also, A is zero at the endpoints of the relevant interval $[0, 10]$. Thus: the area enclosed is greatest when the rectangle is $10\,\text{m}$ (measured paralled to the house) by $5\,\text{m}$ (measured at right angles to the house).

5. With the notation of the figure in the textbook, the total length of fence used is

$$120 = 6l + 3w$$
therefore
$$w = 40 - 2l.$$

Now, the total area of the pen is the function to optimize, namely

$$A = 3lw = 6l(20 - l)$$

therefore
$$\frac{dA}{dl} = 120 - 12l$$

which is zero if $l = 10$. Also, $A = 0$ at the endpoints $l = 0$ and $l = 20$ of $[0, 20]$. Thus: A is greatest when $l = 10$ and $w = 20$.

7. The box has height x and has base of length $24 - 2x$ and width $16 - 2x$; its volume is thus given by

$$\begin{aligned} V &= x(16 - 2x)(24 - 2x) \\ &= 4(x^3 - 20x^2 + 96x) \end{aligned}$$

where $0 \leq x \leq 8$ (see figure in textbook). Now

$$V'(x) = 4(3x^2 - 40x + 96)$$

which is zero when $x = \frac{40 \pm 8\sqrt{7}}{6} = \frac{20 \pm 4\sqrt{7}}{3}$. The critical number in $[0, 8]$ is $x = \frac{20 - 4\sqrt{7}}{3}$, at which $V > 0$; at the endpoints, $V = 0$. Consequently, the volume of the box is a maximum when the box has height $\frac{20 - 4\sqrt{7}}{3}$, width $\frac{8}{3}(1 + \sqrt{7})$ and length $\frac{8}{3}(4 + \sqrt{7})$.

9. From the figure in the textbook, note that $l + 2h = 10$ and $2h + 2w = 16$, whilst the box has volume $V = wlh$. Choose one dimension in terms of which to express the other two: say h. Then $l = 10 - 2h$ and $w = 8 - h$, so $V = (8 - h)(10 - 2h)h = 2h(5 - h)(8 - h) = 2(h^3 - 13h^2 + 40h)$; the appropriate range of values for h is $0 \leq h \leq 5$. Now $\frac{dV}{dh} = 2(3h^2 - 26h + 40) = 2(h - 2)(3h - 20)$ so V has critical numbers $h = 2$ and $h = \frac{20}{3}$, of which only $h = 2$ lies in $[0, 5]$. At the endpoints, $V = 0$; so the volume of the box is a maximum of $72\,\text{cm}^3$ when $h = 2$, $l = 6$, $w = 6$.

11. Bisect the triangle vertically as in the accompanying figure. Let the whole rectangle have width $2x$ and height y (in centimeters). Since the large triangle and the lower right triangle are similar, we have

$$\frac{y}{3-x} = \frac{4}{3}$$

therefore

$$y = 4 - \frac{4}{3}x.$$

The area A of the whole rectangle is the function to optimize, namely

$$A = 2xy = 2x\left(4 - \frac{4}{3}x\right)$$

so

$$\frac{dA}{dx} = 8 - \frac{16}{3}x$$

which is zero when $x = \frac{3}{2}$. Also, $A = 0$ at the endpoints of the relevant interval $[0,3]$ for x. Thus the rectangle's area is greatest when its width is 3cm and its height 2 cm.

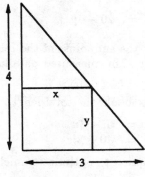

Exercise 11

13. Let the rectangle have width $2x$ and height y so that $y = 4 - x^2$ as in the figure. Its area A is then the function to optimize, given by

$$A = 2xy = 2x(4 - x^2)$$

therefore

$$\frac{dA}{dx} = 8 - 6x^2$$

which is 0 if $x = \frac{2}{\sqrt{3}}$. Also, $A = 0$ at the endpoints of the relevant interval $0 \le x \le 2$. Thus the area of the rectangle is greatest when its width is $\frac{4}{\sqrt{3}}$ units and its height $\frac{8}{3}$ units.

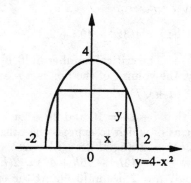

Exercise 13

108

15. The fixed perimeter of the sector is given by $P = 2r + r\theta$ and its area (the function to optimize) by

$$A = \frac{1}{2}r^2\theta = \frac{1}{2}r(P - 2r).$$

Now

$$\frac{dA}{dr} = \frac{1}{2}P - 2r$$

which is zero when $r = \frac{1}{4}P$. At the endpoints of the relevant interval $\left[0, \frac{P}{2}\right]$ for r, the area is zero. Thus the sectorial area is maximized when $r = \frac{1}{4}P$ and $\theta = \frac{P}{r} - 2 = 2$.

17. Let T be the number of trees per acre and Y the yield per tree; we are told that $\frac{dY}{dT} = -15$ and that $Y = 495$ when $T = 25$, so

$$\begin{aligned} Y - 495 &= (-15)(T - 25) \\ Y &= 870 - 15T. \end{aligned}$$

therefore

If A (the function to optimize) is the yield per acre then

$$\begin{aligned} A &= TY = T(870 - 15T) \\ \frac{dA}{dT} &= 870 - 30T \end{aligned}$$

therefore

which is zero when $T = 29$. The relevant interval is $0 \leq T \leq 58$, at the endpoints of which A is zero. Thus, yield per acre will be maximized by planting four more trees per acre.

19. Let the northeast vertex of the rectangle have coordinates (x, y) when the circle is centered at the origin, so that $x^2 + y^2 = 16$. The area of the rectangle is the function to maximize, namely

$$A = 4xy = 4x\sqrt{16 - x^2}$$

therefore

$$\frac{DA}{dx} = \frac{8(8 - x^2)}{\sqrt{16 - x^2}}.$$

If $x = 2\sqrt{2}$ then $\frac{dA}{dx} = 0$; at the endpoints of the relevant interval $0 \leq x \leq 4$, $A = 0$. Thus the area of the rectangle is greatest when the rectangle is a square of side $4\sqrt{2}$ and area 32 square units.

21. Let the inscribed cylinder have base radius r and height h (in centimeters). The accompanying figure shows a vertical section of half the cone through its axis. The large triangle is similar to that on the lower right, so

$$\frac{h}{3 - r} = \frac{5}{3} \quad \text{therefore } h = 5 - \frac{5}{3}r.$$

The cylinder has volume (the function to maximize) given by

$$\begin{aligned} V &= \frac{1}{3}\pi r^2 h = \frac{1}{3}\pi r^2 \left(5 - \frac{5}{3}r\right) \\ &= \frac{5}{3}\pi \left(r^2 - \frac{1}{3}r^3\right) \end{aligned}$$

therefore

$$\frac{dV}{dr} = \frac{5}{3}\pi(2r - r^2)$$

which is zero when $r = 0$ and $r = 2$. Also, $V = 0$ at the endpoints of the relevant interval $0 \leq r \leq 3$. Thus the volume is maximized when the cylinder has base radius 2 cm and height $\frac{5}{3}$ cm.

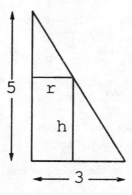

Exercise 21

23. With the given data, the growth rate is

$$\frac{dy}{dt} = \alpha + \beta\gamma\pi\cos(\gamma\pi t)$$
$$= 3 + \frac{2}{3}\pi\cos\left(\frac{1}{3}\pi t\right)$$

therefore

$$\frac{d}{dt}\left(\frac{dy}{dt}\right) = -\frac{2\pi^2}{9}\sin\left(\frac{1}{3}\pi t\right)$$

which (for $0 \leq t \leq 9$) is zero when $t = 0,\ 3,\ 6,\ 9$.

t	0	3	6	9
$\frac{dy}{dt}$	$3 + \frac{2}{3}\pi$	$3 - \frac{2}{3}\pi$	$3 + \frac{2}{3}\pi$	$3 - \frac{2}{3}\pi$

From the above table, minimum growth occurs at 3 months and 9 months whilst maximum growth occurs at birth and 6 months.

25. Let the northeast vertex of the rectangle be at (x, y), so that $x^2 + 4y^2 = 4$ and the area of the rectangle is the function to optimize, given by

$$A = 4xy = 8y\sqrt{1 - y^2}.$$

Now

$$\frac{dA}{dy} = 8\frac{(1 - 2y^2)}{\sqrt{1 - y^2}}$$

which is zero if $y = \frac{1}{\sqrt{2}}$. Further, $A = 0$ at the endpoints of the appropriate interval $0 \leq y \leq 1$. Thus: the rectangle of maximum area has height $2y = \sqrt{2}$ units and width $2x = 2\sqrt{2}$ units.

27. Let the circle have radius r and the square have side s (in centimeters). The respective perimeters are then $2\pi r$ (circle) and $4s$ (square) and we are given that $2\pi r + 4s = 50$. Now, the area enclosed by the circle and square together is to be optimized, given by

$$A = \pi r^2 + s^2$$
$$= \pi r^2 + \left(\frac{25}{2} - \frac{\pi r}{2}\right)^2$$

110

$$= \left(\pi + \frac{\pi^2}{4}\right) r^2 - \frac{25\pi}{2} r + \frac{625}{4}$$

so

$$\frac{dA}{dr} = \left(2\pi + \frac{\pi^2}{2}\right) r - \frac{25\pi}{2}$$

which is zero when $r = \frac{25}{4+\pi}$ and then $A = \frac{625}{4+\pi}$. The physical limits to the problem are $r = 0$ (no circle: $A = \frac{625}{4}$) and $r = \frac{25}{\pi}$ (no square: $A = \frac{625}{\pi}$). Inspecting these three values for A, the figures will enclose

  a) the greatest area $\left(\frac{625}{\pi} \text{ cm}^2\right)$ when $r = \frac{25}{\pi}$ (do not cut the wire, form a circle) and

  b) the least $\left(\frac{625}{4+\pi} \text{ cm}^2\right)$ when $r = \frac{25}{4+\pi}$ (cut the wire at $\frac{50\pi}{4+\pi}$ cm).

29. With the notation of the figure in the textbook, the route shown covers a distance $300 - x$ on land and a distance $\sqrt{100^2 + x^2}$ by sea, and so takes a time

$$T = \frac{300 - x}{5} + \frac{\sqrt{100^2 + x^2}}{3},$$

which is the function to minimize. Now

$$\frac{dT}{dx} = -\frac{1}{5} + \frac{x}{3\sqrt{100^2 + x^2}}$$

which is zero when

$$3\sqrt{100^2 + x^2} = 5x$$

therefore

$$9(100^2 + x^2) = 25x^2$$

therefore

$$9 \times 100^2 = 16x^2$$

therefore

$$x = 75.$$

If $x = 75$ then $T = \frac{260}{3}$, if $x = 0$ then $T = \frac{280}{3}$ and if $x = 300$ then $T = \frac{100\sqrt{10}}{3}$. Thus: in order to minimize time, the swimmer should aim to reach the shore $300 - 75 = 225$ meters from the person in distress.

31. Now monthly profit P (to be maximized) is given by

$$\begin{aligned} P &= \text{revenue} - \text{cost} \\ &= px - (500 + 10x) \\ &= 3000p - 100p^2 - 20500 \end{aligned}$$

so

$$\frac{dP}{dp} = 3000 - 200p$$

which is zero if $p = 15$. In addition, $P = 0$ at the endpoints of the appropriate interval $[15 - 2\sqrt{5},\ 15 + 2\sqrt{5}]$. So: monthly profits are maximized when $p = 15$.

33. Let the number of rooms rented per day be y and the daily rate be $\$x$ per room. We are told that
$$y - 200 = (-4)(x - 40)$$
therefore
$$y = 360 - 4x.$$

Now daily revenue R (to be maximized) is given by
$$R = xy = x(360 - 4x)$$
therefore
$$\frac{dR}{dx} = 360 - 8x$$

which is zero when $x = 45$. Checking endpoints of the appropriate interval $0 \le x \le 90$ as usual, we see that this rental rate of $\$45$ maximizes revenue.

35. If S is the survival rate at distance x then $S = kdp$ for some constant of proportionality k. With the given data, this function to maximize is
$$S = \frac{1}{10} k \frac{x}{1 + \left(\frac{x}{5}\right)^2}$$
therefore
$$\frac{dS}{dx} = \frac{1}{10} k \frac{\left(1 - \frac{x^2}{25}\right)}{\left(1 + \frac{x^2}{25}\right)}$$

which is zero when $x = 5$. Now, $x = 5 \Rightarrow S = \frac{k}{4}$, $x = 0$ (one end of the given interval) $\Rightarrow S = 0$ and $x = 10$ (the other end) $\Rightarrow S = \frac{k}{5}$; so survival rate is a maximum when $x = 5$: at a distance of $5\,\text{m}$ from the trunk.

37. a) $v = \dfrac{(\rho - 100)^2}{100} \Rightarrow \dfrac{dv}{d\rho} = \dfrac{\rho - 100}{50}$

which is zero when $\rho = 100$. Since $\rho = 0 \Rightarrow v = 100$ and $\rho = 100 \Rightarrow v = 0$, it follows that velocity will be a maximum when $\rho = 0$ (!).

  b) $q = \dfrac{\rho(\rho - 100)^2}{100} \Rightarrow \dfrac{dq}{d\rho} = \dfrac{3\rho^2 - 400\rho + 10000}{100}$

which is zero when $\rho = 100$ and when $\rho = \frac{100}{3}$. Now, $\rho = \frac{100}{3} \Rightarrow q = \frac{40000}{27}$, $\rho = 100 \Rightarrow q = 0$ and $\rho = 0 \Rightarrow q = 0$; of these values for q, the greatest is $\frac{40000}{27}$. So: maximum flow rate occurs when $\rho = \frac{100}{3}$, so that there are 100 automobiles in every 3 km.

39. a) From the diagram $\tan\theta = \frac{8 - h}{1.5}$, so $h = 8 - 1.5\tan\theta$. Also $\cos\theta = \frac{1.5}{l}$ so $l = 1.5\sec\theta$. Therefore,
$$\begin{aligned} C(\theta) &= 275(1.5\sec\theta) + 185(8 - 1.5\tan\theta) \\ &= 412.5\sec\theta - 277.5\tan\theta + 1480. \end{aligned}$$

  b) The domain is $0 \le \theta \le \theta_0$ where $\tan\theta_0 = \frac{8}{1.5}$. So $\theta = 1.385448$ radians or $79.380345°$.

  c) Graph using the range $[0, 1.4] \times [1600, 2400]$. The minimum is at $\theta = 0.7368$ with $C(\theta) = 1785.21$.

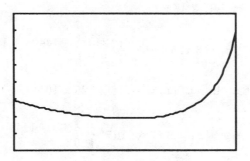

Exercise 39. c)

d) $C'(\theta)$ $=$ $412.5 \sec\theta\tan\theta - 277.5\sec^2\theta$
$=$ $\sec\theta(412.5\tan\theta - 277.5\sec\theta)$.

Zooming in we find $C'(\theta) = 0$ for $\theta = 0.737889$. For this value of θ the cost is $C(0.737889) = 1785.20$. A more precise value of θ can be found by solving

$$412.5\tan\theta - 277.5\sec\theta = 0$$

or

$$\sin\theta = \frac{277.5}{412.5} = 0.6727272727.$$

Therefore $\theta_* = 0.7378886885$ radians or $42.27791°$. This gives a distance of $l = 1.5\sec\theta_* = 2.02733$ miles under water and $h = 8 - 1.5\tan\theta_* = 6.63616$ miles on land.

41. a) Below is the graph of the function **using the range:** $[0,6] \times [-0.00025, 0]$.

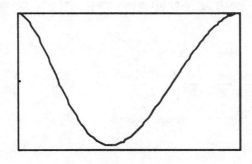

Exercise 41. a)

From the graph the maximum deflection is 0.000241 at about $x = 2.53$.

b) Note the $y(3) = -0.0002278125$ and $\lim\limits_{x\to 3^+} y(x) = -0.0002278125$ so $y(x)$ is continuous at $x = 3$. Also note the following computations:

$$y'(x) = 3.125 \times 10^{-7} \begin{cases} 20x^4 - 240x^3 + 972x^2 - 1242x & \text{if } x \le 3 \\ -108x^2 + 918x - 1620 & \text{if } x > 3. \end{cases}$$

$y'(3) = 0.000050625 = \lim\limits_{x\to 3^+} y'(x).$

$$y''(x) = 3.125 \times 10^{-7} \begin{cases} 80x^3 - 720x^2 + 1944x - 1242 & \text{if } x \le 3 \\ -216x + 918 & \text{if } x > 3. \end{cases}$$

$$y''(3) = 0.000084375 = \lim_{x \to 3^+} y''(x).$$

$$y^{(3)}(x) = 3.125 \times 10^{-7} \begin{cases} 240x^2 - 1440x + 1944 & \text{if } x \le 3 \\ -216 & \text{if } x > 3. \end{cases}$$

$$y^{(3)}(3) = -0.0000675 = \lim_{x \to 3^+} y'''(x) = -216(3.125 \times 10^{-7}).$$

$$y^{(4)}(x) = 3.125 \times 10^{-7} \begin{cases} 480x - 1440 & \text{if } x \le 3 \\ 0 & \text{if } x > 3. \end{cases}$$

$$y^{(4)}(3) = 0 = \lim_{x \to 3^+} y^{(4)}(x).$$

$$y^{(5)}(x) = 3.125 \times 10^{-7} \begin{cases} 480 & \text{if } x \le 3 \\ 0 & \text{if } x > 3. \end{cases}$$

$y^{(5)}(3) = 480 \ne \lim_{x \to 3^+} y^{(5)}(x) = 0.$ These computations show that $y(x)$, $y'(x)$, $y^{(3)}(x)$, and $y^{(4)}(x)$ are all continuous, but $y^{(5)}(x)$ is not continuous.

4.3   The Mean Value Theorem

1. $f(x) = x^2 - 2x + 1$ on $[0, 2]$. Clearly, f is continuous on $[0, 2]$ and differentiable on $(0, 2)$. Also, $f(0) = f(2) = 1$. $f'(x) = 2x - 2$; $f'(c) = 2c - 2 = 0$ when $c = 1$.

3. $f(x) = \sin 3x$ on $[0, 2\pi]$. Clearly, f is continuous on $[0, 2\pi]$ and differentiable on $(0, 2\pi)$.
$$\begin{aligned} f(0) &= f(2\pi) = 0 \\ f'(x) &= 3\cos 3x; \\ f'(c) &= 3\cos 3c = 0 \text{ when } \cos 3c = 0 \\ &\Rightarrow 3c = \frac{\pi}{2}, \frac{3\pi}{2}, \frac{5\pi}{2}, \frac{7\pi}{2}, \frac{9\pi}{2}, \frac{11\pi}{2} \Rightarrow c = \frac{\pi}{6}, \frac{\pi}{2}, \frac{5\pi}{6}, \frac{7\pi}{6}, \frac{3\pi}{2}, \frac{11\pi}{6}. \end{aligned}$$

5. $f(x) = 3x + 1$, $x \in [0, 3]$. Clearly, f is continuous on $[0, 3]$ and differentiable on $(0, 3)$.
$$f'(x) = 3; \ f'(c) = 3.$$
$$\frac{f(3) - f(0)}{3 - 0} = \frac{9}{3} = 3 = f'(c)$$
Thus all numbers in $(0, 3)$ will satisfy the equation $f'(c) = \dfrac{f(b) - f(a)}{b - a}$.

7. $f(x) = \cos x$, $x \in \left[\dfrac{\pi}{4}, \dfrac{7\pi}{4}\right]$. Clearly, f is continuous on $\left[\dfrac{\pi}{4}, \dfrac{7\pi}{4}\right]$ and differentiable on $\left(\dfrac{\pi}{4}, \dfrac{7\pi}{4}\right)$.
$$f'(x) = -\sin x, \ f'(c) = -\sin c$$
$$-\sin c = \frac{f\left(\frac{7\pi}{4}\right) - f\left(\frac{\pi}{4}\right)}{\frac{7\pi}{4} - \frac{\pi}{4}} = \frac{\frac{\sqrt{2}}{2} - \frac{\sqrt{2}}{2}}{\frac{3\pi}{2}} = 0$$
Thus $c = \pi$.

9. $f(x) = x^2 + 2x$, $x \in [0, 4]$. Clearly, f is continuous on $[0, 4]$ and differentiable on $(0, 4)$.

$$f'(x) = 2x + 2; \ f'(c) = 2c + 2$$

$$2c + 2 = \frac{f(4) - f(0)}{4 - 0} = \frac{24 - 0}{4} = 6$$

Thus $c = 2$.

11. $f(x) = x^2 + 2x - 3$, $x \in [-3, 0]$. f is continuous on $[-3, 0]$ and differentiable on $(-3, 0)$.

$$f'(x) = 2x + 2, \ f'(c) = 2c + 2$$

$$2c + 2 = \frac{f(0) - f(-3)}{0 - (-3)} = \frac{-3 - 0}{3} = -1$$

Thus $c = -\frac{3}{2}$.

13. $f(x) = x^3 + x^2 + x$, $x \in [1, 3]$. f is continuous on $[1, 3]$ and differentiable on $(1, 3)$.

$$f'(c) = 3c^2 + 2c + 1 = \frac{f(3) - f(1)}{3 - 1} = \frac{39 - 3}{2} = 18$$

Thus $c = \dfrac{-1 + 2\sqrt{13}}{3}$.

15.

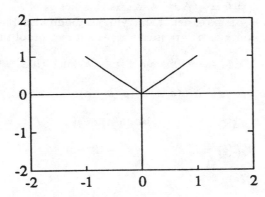

Graph of $f(x) = |x|$ on $[-1, 1]$.

The Mean Value Theorem does not apply for f on $[-1, 1]$ since f is not differentiable at $x = 0$.

17. Let $f(x) = x^3 + 3x^2 + 6x + 1$. Then f is continuous on $[-1, 0]$. $f(-1) = -3$ and $f(0) = 1$. By the Intermediate Value Theorem, there exists a $c \in (-1, 0)$ with $f(c) = 0$. Thus $x^3 + 3x^2 + 6x + 1 = 0$ has a solution between -1 and 0.

$$f'(x) = 3x^2 + 6x + 6 = 3(x^2 + 2x + 1) + 3$$
$$= 3(x + 1)^2 + 3 > 0 \text{ for all } x.$$

By Rolle's Theorem, c is the only solution to $x^3 + 3x^2 + 6x + 1 = 0$ since if there were a d such that $f(d) = 0 = f(c)$, then $f'(z)$ must equal to zero for some z between c and d, but we have shown that $f'(x) > 0$ for all x.

19. According to the MVT, there is a c in (a, b) such that $f(b) - f(a) = f'(c)(b - a)$ so

$$|f(b) - f(a)| = |f'(c)| \, |b - a| \leq M(b - a)$$

since it is given that $|f'(c)| \leq M$.

21. Take $f(x) = \sin x$ so that $f'(x) = \cos x$ and certainly $|f'(x)| \leq 1$ for all x. Since $|f'(t)| \leq 1$ for all t in the interval (x, y) (assuming, for the sake of argument, that $x < y$) it follows at once from Exercise 19 (with $M = 1$) that $|\sin y - \sin x| \leq |y - x|$.

23. Let $f(t)$ be the distance (in km) traveled after t hours so that $f'(t)$ is the speed (in km/h) after t hours. The total distance traveled after 6 hours is $f(6) - f(0)$ and we are given that $f'(t)$ lies betweeen 75 and 90 for $0 < t < 6$. Applying Exercise 22 with $a = 0$, $b = 6$, $m = 75, M = 90$, we deduce that

$$450 = 75 \times 6 \ \leq \ f(6) - f(0) \ \leq \ 90 \times 6 \ = \ 540.$$

Thus: the automobile travels between 450 km and 540 km.

25. Let $f(x) = x^{1/3}$ on the interval $[a, b] = [1, 1 + \Delta x]$ say. From $f'(x) = \frac{1}{3} x^{-2/3}$ it follows that if $x \geq 1$ then $0 < f'(x) \leq \frac{1}{3}$. Putting $M = \frac{1}{3}$ in Exercise 20, we have

$$\sqrt[3]{1 + \Delta x} \ = \ f(1 + \Delta x) \ \leq \ f(1) + M(1 + \Delta x - 1)$$
$$= \ 1 + \frac{1}{3}\Delta x.$$

Notice that here the inequality is weak ($\leq$) rather than strong ($<$). To see that $\sqrt[3]{1 + \Delta x} < 1 + \frac{1}{3}\Delta x$, we must see that $\sqrt[3]{1 + \Delta x} \neq 1 + \frac{1}{3}\Delta x$. However, if $\sqrt[3]{1 + \Delta x} = 1 + \frac{1}{3}\Delta x$ then (cube both sides) $1 + \Delta x = 1 + \Delta x + \frac{1}{3}(\Delta x)^2 + \frac{1}{27}(\Delta x)^3$ so that $0 = \frac{1}{3}(\Delta x)^2 + \frac{1}{27}(\Delta x)^3$ which is absurd since $\Delta x > 0$. [Remark: This last step, getting from $\leq$ to $<$, seems easier than trying to modify Exercise 20's conclusion. Notice also that with a little work, the same closing argument gives a direct proof that $\sqrt[3]{1 + \Delta x} < 1 + \frac{1}{3}\Delta x$.]

27. Let $h = g - f$. According to the MVT, there exists a c in (a, b) such that

$$h(b) - h(a) = h'(c)(b - a).$$

Now, $h'(c) = g'(c) - f'(c) > 0$ since $g'(c) > f'(c)$; also, $h(a) = g(a) - f(a) = 0$ since $g(a) = f(a)$. So

$$h(b) = h(a) + h'(c)(b - a) > 0.$$

But $h(b) = g(b) - f(b)$; so $g(b) > f(b)$.

29. $f(x) = \tan x \Rightarrow f'(x) = \sec^2 x$, $f(0) = 0$, $f\left(\frac{\pi}{4}\right) = 1$. So: we're looking for a $c \in \left(0, \frac{\pi}{4}\right)$ such that

$$1 \ = \ f\left(\frac{\pi}{4}\right) - f(0) = f'(c)\left(\frac{\pi}{4} - 0\right) = \frac{\pi}{4} f'(c)$$

or $\qquad 1 \ = \ \frac{\pi}{4} \sec^2 c$ or $\cos^2 c = \frac{\pi}{4}$.

Let $g(x) = \cos^2 x - \frac{\pi}{4}$ so $g'(x) = -2\cos x \sin x = -\sin 2x$.

n	x_n	$g(x_n)$	$g'(x_n)$	x_{n+1}
1	0.50000	-0.01525	-0.84147	0.48188
2	0.48188	-0.00018	-0.82134	0.48166
3	0.48166	0.00000		

(The initial guess was made noting that $\cos^2 c \approx \frac{3}{4} \Rightarrow \cos c \approx \frac{\sqrt{3}}{2} \Rightarrow c \approx \frac{\pi}{6} \approx \frac{1}{2}$). So: an approximation to c is $c \approx 0.48166$.

116

31. $f(x) = ax^2 + bx + c \Rightarrow f'(x) = 2ax + b$. Now if we wish $d \in (x_1, x_2)$ to satisfy

$$f(x_2) - f(x_1) \;=\; f'(d)(x_2 - x_1)$$

then

$$ax_2^2 + bx_2 - ax_1^2 - bx_1 \;=\; (2ad + b)(x_2 - x_1)$$

or

$$a(x_1 + x_2)(x_2 - x_1) + b(x_2 - x_1) \;=\; (2ad + b)(x_2 - x_1)$$

or (cancelling $x_2 - x_1 \neq 0$ and subtracting b)

$$a(x_1 + x_2) \;=\; 2ad$$

whece $d = \frac{x_1 + x_2}{2}$ so long as $a \neq 0$. [If $a = 0$ then any point d between x_1 and x_2 works in the MVT].

For Exercises 32–33, the two lines given are parallel.

33. Slope: $m = 1 - \dfrac{2}{3\pi} = 0.7877934092$
    Point of intersection: $c = -1.070795768$
    $l_1:$ $\quad y = 0.7877934092 \left(x + \dfrac{\pi}{2} \right) - \dfrac{\pi}{2}$
    $l_2:$ $\quad y = 0.7877934092 \left(x + 1.070795768 \right) - 2.068290875.$

35. Slope: $m = -0.1958003507 = \dfrac{1 - \sqrt[3]{4}}{3}$. There is no point of intersection. The Mean-Value Theorem requires $f'(x)$ to exist on the open interval (a, b) in order to guarantee that $f'(c) = m$ for some number c between a and b. If $f'(x)$ fails to exist then there may not be such a number.

4.4 Increasing and Decreasing Functions

1. $f(x) = x^2 - 4x + 6$
   $f'(x) = 2x - 4 = 0$ at $x = 2$.
   Therefore the only critical number is 2.

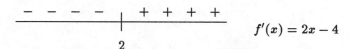

$$f'(x) = 2x - 4$$

Thus f is decreasing on $(-\infty, 2]$ and increasing on $[2, \infty)$. f has a relative minimum and absolute minimum at $(2, 2)$.

3. $f(x) = 2x^3 - 3x^2 - 12x$
   $f'(x) = 6x^2 - 6x - 12 = 6(x - 2)(x + 1) = 0$ at $x = 2, -1$.
   Therefore the critical numbers are 2 and -1.

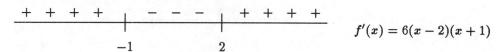

$$f'(x) = 6(x - 2)(x + 1)$$

Thus f is increasing on $(-\infty, -1]$ and $[2, \infty)$ and decreasing on $[-1, 2]$. f has a relative minimum at $(2, -20)$ and a relative maximum at $(-1, 7)$.

5. $f(x) = 4 + x^{2/3}$

$f'(x) = \dfrac{2}{3x^{1/3}} \neq 0$ for all x. $f'(x)$ is undefined when $3x^{1/3} = 0 \Rightarrow x = 0$.

Therefore the only critical number is 0.

$-\quad-\quad-\quad-\ |\ +\ +\ +\ +$

0

$f'(x) = \dfrac{2}{3x^{1/3}}$

Thus f is decreasing on $(-\infty, 0]$ and increasing on $[0, \infty)$. f has a relative minimum and absolute minimum at $(0, 4)$.

7. $f(x) = x^3 + 4$
$f'(x) = 3x^2 = 0$ at $x = 0$.
Therefore the only critical number is 0.

$+\ +\ +\ +\ |\ +\ +\ +\ +$

0

$f'(x) = 3x^2$

Thus f is increasing on $(-\infty, \infty)$. f has no relative extrema.

9. $f(x) = |4 - x^2|$. Note that $f(x) = \begin{cases} 4 - x^2 & \text{if } -2 \leq x \leq 2 \\ x^2 - 4 & \text{if } -\infty < x \leq -2 \text{ and } 2 \leq x < \infty. \end{cases}$

$\displaystyle \lim_{x \to -2^-} \frac{f(x) - f(-2)}{x + 2} = \lim_{x \to -2^-} \frac{x^2 - 4 - 0}{x + 2} = \lim_{x \to -2^-} (x - 2) = -4$

$\displaystyle \lim_{x \to -2^+} \frac{f(x) - f(-2)}{x + 2} = \lim_{x \to -2^+} \frac{4 - x^2 - 0}{x + 2} = \lim_{x \to -2^+} (2 - x) = 4$

Since $\displaystyle \lim_{x \to -2^-} \frac{f(x) - f(-2)}{x + 2} \neq \lim_{x \to -2^+} \frac{f(x) - f(-2)}{x + 2}$, $f'(-2)$ does not exist.

$\displaystyle \lim_{x \to 2^-} \frac{f(x) - f(2)}{x - 2} = \lim_{x \to 2^-} \frac{4 - x^2 - 0}{x - 2} = \lim_{x \to 2^-} -(2 + x) = -4$

$\displaystyle \lim_{x \to 2^+} \frac{f(x) - f(2)}{x - 2} = \lim_{x \to 2^+} \frac{x^2 - 9 - 0}{x - 2} = \lim_{x \to 2^+} (x + 2) = 4$

Since $\displaystyle \lim_{x \to 2^+} \frac{f(x) - f(2)}{x - 2} \neq \lim_{x \to 2^-} \frac{f(x) - f(2)}{x - 2}$, $f'(2)$ does not exist.

Therefore we have: $f'(x) = \begin{cases} -2x & \text{if } -2 < x < 2 \\ 2x & \text{if } -\infty < x < -2 \text{ and } 2 < x < \infty \\ \text{does not exist} & \text{if } x = \pm 2. \end{cases}$

The critical numbers are $0, -2, 2$.

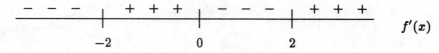

$-\quad-\quad-\ |\ +\ +\ +\ |\ -\quad-\quad-\ |\ +\ +\ +$

$-2 \qquad\quad 0 \qquad\quad 2$

$f'(x)$

Thus f is increasing on $[-2, 0]$ and $[2, \infty)$ and decreasing on $(-\infty, -2]$ and $[0, 2]$. f has a relative maximum at $(0, 4)$ and a relative (absolute) minima at $(\pm 2, 0)$.

11. $f(x) = \sin\left(x + \frac{\pi}{4}\right)$, $0 \leq x \leq 2\pi$

$f'(x) = \cos\left(x + \frac{\pi}{4}\right) = 0$ when $x + \frac{\pi}{4} = \frac{\pi}{2}, \frac{3\pi}{2} \Rightarrow x = \frac{\pi}{4}, \frac{5\pi}{4}$.

Therefore the critical numbers are $\frac{\pi}{4}$ and $\frac{5\pi}{4}$.

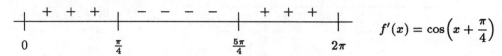

f is increasing on $\left[0, \frac{\pi}{4}\right]$ and $\left[\frac{5\pi}{4}, 2\pi\right]$ and decreasing on $\left[\frac{\pi}{4}, \frac{5\pi}{4}\right]$. f has a relative maximum at $\left(2\pi, \frac{\sqrt{2}}{2}\right)$, an absolute maximum at $\left(\frac{\pi}{4}, 1\right)$, a relative minimum at $\left(0, \frac{\sqrt{2}}{2}\right)$, and an absolute minimum at $\left(\frac{5\pi}{4}, -1\right)$.

13. $f(x) = x^4 - 1$

$f'(x) = 4x^3 = 0$ at $x = 0$.

Therefore the only critical number is 0.

$$\underset{0}{\underline{\quad - \quad - \quad - \quad | \quad + \quad + \quad +\quad}} \qquad f'(x) = 4x^3$$

Thus f is decreasing on $(-\infty, 0]$ and increasing on $[0, \infty)$. f has a relative minimum and absolute minimum at $(0, -1)$.

15. $f(x) = \dfrac{2}{x + 1}$. Note that the domain of f is $(-\infty, -1) \cup (-1, \infty)$.

$f'(x) = \dfrac{-2}{(x+1)^2} < 0$ for all x in the domain.

Thus f is decreasing on $(-\infty, -1) \cup (-1, \infty)$ and f has no relative extrema.

17. $f(x) = \begin{cases} 4 - x^2 & \text{if } -\infty < x \leq 1 \\ x + 2 & \text{if } 1 < x < \infty. \end{cases}$

$\displaystyle \lim_{x \to 1^-} \frac{f(x) - f(1)}{x - 1} = \lim_{x \to 1^-} \frac{4 - x^2 - 3}{x - 1} = \lim_{x \to 1^-} -(1 + x) = -2$

$\displaystyle \lim_{x \to 1^+} \frac{f(x) - f(1)}{x - 1} = \lim_{x \to 1^+} \frac{x + 2 - 3}{x - 1} = 1$

Since $\displaystyle \lim_{x \to 1^-} \frac{f(x) - f(1)}{x - 1} \neq \lim_{x \to 1^+} \frac{f(x) - f(1)}{x - 1}$, $f'(1)$ does not exist.

We have $f'(x) = \begin{cases} -2x & \text{if } -\infty < x < 1 \\ 1 & \text{if } 1 < x < \infty \\ \text{does not exist} & \text{if } x = 1. \end{cases}$

Therefore the critical numbers are 0 and 1.

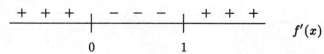

Thus f is increasing on $(-\infty, 0]$ and $[1, \infty)$ and decreasing on $[0, 1]$. f has a relative maximum at $(0, 4)$ and a relative minimum at $(1, 3)$.

19. $f(x) = 4x^3 + 9x^2 - 12x + 7$
$f'(x) = 12x^2 + 18x - 12 = 6(2x - 1)(x + 2) = 0$ at $x = \frac{1}{2}, -2$.
Therefore the critical numbers are $\frac{1}{2}$ and -2.

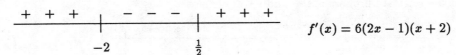

$f'(x) = 6(2x - 1)(x + 2)$

f is increasing on $(-\infty, -2]$ and $\left[\frac{1}{2}, \infty\right)$ and decreasing on $\left[-2, \frac{1}{2}\right]$. f has a relative maximum at $(-2, 35)$ and a relative minimum at $\left(\frac{1}{2}, \frac{15}{4}\right)$.

21. $f(x) = x + \sin x$
$f'(x) = 1 + \cos x = 0$ when $\cos x = -1 \Rightarrow x = \pi + 2n\pi$ for $n \in \mathcal{Z}$.
Therefore f has infinitely many critical numbers.

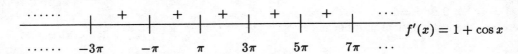

$f'(x) = 1 + \cos x$

f is increasing on $(-\infty, \infty)$ and f has no relative extrema.

23. $f(x) = \dfrac{x}{1 + x}$. Note that the domain of f is $(-\infty, -1) \cup (-1, \infty)$.

$f'(x) = \dfrac{(1 + x) - x}{(1 + x)^2} = \dfrac{1}{(1 + x)^2} > 0$ for all x in the domain.
Thus f is increasing for all x in the domain and f has no relative extrema.

25. $f(x) = \dfrac{1}{1 + x^2}$. Note that the domain of f is $(-\infty, \infty)$.

$f'(x) = \dfrac{-2x}{(1 + x^2)^2} = 0$ when $x = 0$.
Therefore the only critical number is 0.

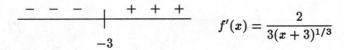

$f'(x) = \dfrac{-2x}{(1 + x^2)^2}$

Thus f is increasing on $(-\infty, 0]$ and decreasing on $[0, \infty)$. f has a relative (absolute) maximum at $(0, 1)$.

27. $f(x) = (x + 3)^{2/3}$

$f'(x) = \dfrac{2}{3(x + 3)^{1/3}} \neq 0$ for all x.
$f'(x)$ is undefined when $3(x + 3)^{1/3} = 0 \Rightarrow x = -3$.
Therefore the only critical number is -3.

$f'(x) = \dfrac{2}{3(x + 3)^{1/3}}$

Thus f is decreasing on $(-\infty, -3]$ and increasing on $[-3, \infty)$. f has a relative (absolute) minimum at $(-3, 0)$.

29. $f(x) = \frac{1}{3} x^3 - 3x^2 - 7x + 5$
$f'(x) = x^2 - 6x - 7 = (x - 7)(x + 1) = 0$ at $x = -1, 7$.
Therefore the critical numbers are -1 and 7.

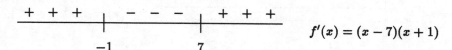

$$f'(x) = (x - 7)(x + 1)$$

Thus f is increasing on $(-\infty, -1]$ and on $[7, \infty)$ and decreasing on $[-1, 7]$. f has a relative maximum at $\left(-1, \frac{26}{3}\right)$ and a relative minimum at $\left(7, \frac{-230}{3}\right)$.

31. $f(x) = x^3 + 6x^2 + 9x + 1$
$f'(x) = 3x^2 + 12x + 9 = 3(x + 3)(x + 1) = 0$ at $x = -1, -3$.
Therefore the critical numbers are -1 and -3.

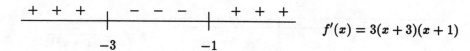

$$f'(x) = 3(x + 3)(x + 1)$$

Thus f is increasing on $(-\infty, -3]$ and on $[-1, \infty)$ and decreasing on $[-3, -1]$. f has a relative maximum at $(-3, 1)$ and a relative minimum at $(-1, -3)$.

33. $f(x) = x^3 + 3x^2 + 6x - 1$
$f'(x) = 3x^2 + 6x + 6 = 3(x^2 + 2x + 2) = 3\left[(x + 1)^2 + 1\right] > 0$ for all x.
Thus f is increasing on $(-\infty, \infty)$ and f has no relative extrema.

35. $f(x) = x^4 + 4x^3 + 2x^2 + 1$

$f'(x) = 4x^3 + 12x^2 + 4x = 4x(x^2 + 3x + 1) = 0$ when $x = 0, \dfrac{-3 \pm \sqrt{5}}{2}$. $\left(\dfrac{-3 \pm \sqrt{5}}{2} \approx -.38, -2.62\right)$

Therefore the critical numbers are $0, \dfrac{-3 \pm \sqrt{5}}{2}$.

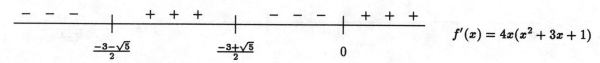

$$f'(x) = 4x(x^2 + 3x + 1)$$

Thus f is decreasing on $\left(-\infty, \dfrac{-3 - \sqrt{5}}{2}\right]$ and on $\left[\dfrac{-3 + \sqrt{5}}{2}, 0\right]$ and

increasing on $\left[\dfrac{-3 - \sqrt{5}}{2}, \dfrac{-3 + \sqrt{5}}{2}\right]$ and on $[0, \infty)$.

f has a relative maximum at $\left(\dfrac{-3 + \sqrt{5}}{2}, 1.09\right)$ and a relative minimum at $(0, 1)$ and a relative

(absolute) minimum at $\left(\dfrac{-3 - \sqrt{5}}{2}, -10.09\right)$.

37. $f(x) = 6x^{5/2} - 70x^{3/2} + 15$. Note that the domain of f is $[0, \infty)$.
$f'(x) = 15x^{3/2} - 105x^{1/2} = 15x^{1/2}(x-7) = 0$ when $x = 0, 7$.

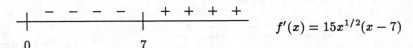

Thus f is decreasing on $[0, 7]$ and increasing on $[7, \infty)$. f has a relative (absolute) minimum at $(7, -503.6)$.

39. $f(x) = \dfrac{x^2}{1 + x^2}$. Note that the domain of f is $(-\infty, \infty)$.

$$f'(x) = \frac{2x(1 + x^2) - x^2(2x)}{(1 + x^2)^2} = \frac{2x}{(1 + x^2)^2} = 0 \text{ at } x = 0.$$

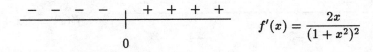

Thus f is decreasing on $(-\infty, 0]$ and increasing on $[0, \infty)$. f has a relative (absolute) minimum at $(0, 0)$.

41. $f(x) = \dfrac{x^{1/3}}{x^{2/3} - 4}$. Note that $x^{2/3} - 4 = 0$ when $x = \pm 8$.
Therefore the domain of f is $(-\infty, -8) \cup (-8, 8) \cup (8, \infty)$.

$$f'(x) = \frac{\frac{1}{3x^{2/3}}(x^{2/3} - 4) - x^{1/3}(\frac{2}{3x^{1/3}})}{(x^{2/3} - 4)^2} = \frac{x^{2/3} - 4 - 2x^{2/3}}{3x^{2/3}(x^{2/3} - 4)^2} = \frac{-(x^{2/3} + 4)}{3x^{2/3}(x^{2/3} - 4)^2} \neq 0$$

for all x in the domain.

$f'(x)$ is undefined at $x = 0$. Therefore, the only critical number is 0.

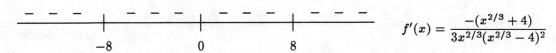

Thus f is decreasing on $(-\infty, -8) \cup (-8, 8) \cup (8, \infty)$. f has no relative extrema.

43. $f(x) = \sec^2 x$, $0 \le x \le 2\pi$. Note that the domain of f is $\left[0, \frac{\pi}{2}\right) \cup \left(\frac{\pi}{2}, \frac{3\pi}{2}\right) \cup \left(\frac{3\pi}{2}, 2\pi\right]$.
$f'(x) = 2\sec x \sec x \tan x = \dfrac{2\sin x}{\cos^3 x} = 0$ when $\sin x = 0 \Rightarrow x = \pi$ for $x \in (0, 2\pi)$.

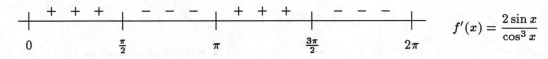

Thus f is increasing on $\left[0, \frac{\pi}{2}\right)$ and $\left[\pi, \frac{3\pi}{2}\right)$ and decreasing on $\left(\frac{\pi}{2}, \pi\right]$ and $\left(\frac{3\pi}{2}, 2\pi\right]$. f has a relative (absolute) minima at $(0, 1)$, $(\pi, 1)$ and $(2\pi, 1)$.

45. $f(x) = |\sin x|$

Since $(\sin x)$ is a periodic function with period 2π we only need to analyze $f(x)$ on $[0, 2\pi]$.

$$f(x) = \begin{cases} \sin x & \text{if } 0 \le x \le \pi \\ -\sin x & \text{if } \pi \le x \le 2\pi. \end{cases} \quad \text{Then } f'(x) = \begin{cases} \cos x & \text{if } 0 < x < \pi \\ -\cos x & \text{if } \pi < x < 2\pi \\ \text{does not exist} & \text{if } x = \pi, 0, 2\pi. \end{cases}$$

$f'(x) = 0$ when $x = \frac{\pi}{2}, \frac{3\pi}{2}$.

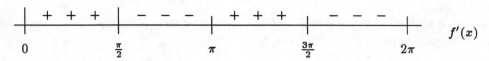

Thus f is increasing on $\left[0, \frac{\pi}{2}\right]$ and $\left[\pi, \frac{3\pi}{2}\right]$ and decreasing on $\left[\frac{\pi}{2}, \pi\right]$ and $\left[\frac{3\pi}{2}, 2\pi\right]$. f has a relative (absolute) minimum at $(\pi, 0)$ and relative (absolute) maxima at $\left(\frac{\pi}{2}, 1\right)$ and $\left(\frac{3\pi}{2}, 1\right)$.

In general, f is increasing on $\left[n\pi, (2n+1)\frac{\pi}{2}\right]$ and decreasing on $\left[(2n-1)\frac{\pi}{2}, n\pi\right]$ for $n \in \mathcal{Z}$. f has relative (absolute) minima at $(n\pi, 0)$ and relative (absolute) maxima at $\left((2n+1)\frac{\pi}{2}, 1\right)$ for $n \in \mathcal{Z}$.

47. $f(x) = x^{5/3} - 5x^{2/3} + 3$

$f'(x) = \frac{5}{3} x^{2/3} - \frac{10}{3x^{1/3}} = \frac{5(x-2)}{3x^{1/3}} = 0$ at $x = 2$.

$f'(x)$ is undefined when $3x^{1/3} = 0 \Rightarrow x = 0$.
Therefore the critical numbers are 0 and 2.

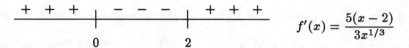

Thus f is increasing on $(-\infty, 0]$ and $[2, \infty)$ and decreasing on $[0, 2]$. f has a relative maximum at $(0, 3)$ and a relative minimum at $(2, -1.76)$.

49. $f(x) = \frac{1}{4} x^4 - 2x^3 + \frac{3}{2} x^2 + 10x - 8$

$f'(x) = x^3 - 6x^2 + 3x + 10 = (x-2)(x-5)(x+1) = 0$ when $x = -1, 2, 5$.

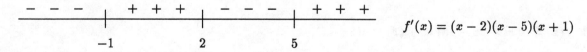

Thus f is decreasing on $(-\infty, -1]$ and $[2, 5]$ and increasing on $[-1, 2]$ and $[5, \infty)$. f has a relative maximum at $(2, 6)$ and a relative (absolute) minima at $\left(-1, \frac{-57}{4}\right)$ and $\left(5, \frac{-57}{4}\right)$.

51. $f(x) = x + |\sin x|$, $0 \le x \le 2\pi$

$$f(x) = \begin{cases} x + \sin x & \text{if } 0 \le x \le \pi \\ x - \sin x & \text{if } \pi \le x \le 2\pi. \end{cases}$$

$$\text{Then } f'(x) = \begin{cases} 1 + \cos x & \text{if } 0 < x < \pi \\ 1 - \cos x & \text{if } \pi < x < 2\pi \\ \text{does not exist} & \text{if } x = \pi, 0, 2\pi. \end{cases}$$

123

$f'(x) \neq 0$ for $x \in (0, 2\pi)$.

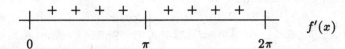

Thus f is increasing on $[0, 2\pi]$. f has a relative minimum at $(0, 0)$ and a relative maximum at $(2\pi, 2\pi)$.

53. $f(x) = (x^2 - 4x)^3$
$f'(x) = 3(x^2 - 4x)^2(2x - 4) = 6(x^2)(x - 4)^2(x - 2) = 0$ when $x = 0, 4$ and 2.
Therefore the critical numbers are $0, 4$ and 2.

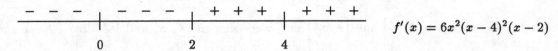

Thus f is decreasing on $(-\infty, 2]$ and increasing on $[2, \infty)$. f has a relative (absolute) minimum at $(2, -64)$.

55. $f(x) = 2x^3 - 3ax^2 + 6$
$f'(x) = 6x^2 - 6ax = 6x(x - a) < 0$ if and only if x lies between 0 and a. Thus $a = 3$.

57. $f(x) = \frac{1}{3}x^3 - x^2 + qx + 10 \Rightarrow f'(x) = x^2 - 2x + q$. We are given that f should be increasing on $(-\infty, -3]$ and $[5, \infty)$ and decreasing on $[-3, 5]$. In particular, $f'(-3)$ and $f'(5)$ should be zero. Since $f'(-3) = 15 + q = f'(5)$ it is clear that we should take $q = -15$, so that

$$f'(x) = x^2 - 2x - 15 = (x - 5)(x + 3)$$

which is indeed positive on $(-\infty, -3)$ and $(5, \infty)$ and negative on $(-3, 5)$.

59. $f(x) = x^2 + bx + c \Rightarrow f'(x) = 2x + b$. Notice that $f'\left(-\frac{b}{2}\right) = 0$ and that $f'(x) < 0$ if $x < -\frac{b}{2}$ whilst $f'(x) > 0$ if $x > -\frac{b}{2}$; so f has a relative minimum at $x = -\frac{b}{2}$. Of course, f cannot have a relative maximum at any point, since f has only one critical number: a (relative) minimum.

61. $f(x) = x^2 - ax + b \Rightarrow f'(x) = 2x - a$. Given that f has a relative minimum at $x = 2$ we deduce that

$$0 = f'(2) = 4 - a$$

so $a = 4$.

63. $f(x) = a(x^2 - bx + 16) \Rightarrow f'(x) = a(2x - b)$.

   a) Since f has a relative extremum at $x = 5$, we have $0 = f'(5) = a(10 - b)$ so that $b = 10$. Of course, we exclude the possibility that $a = 0$!

   b) In order that $(5, f(5))$ be a relative minimum, we ask that $f'(x)$ be negative if $x < 5$ and positive if $x > 5$. Since $f'(x) = a(2x - 10)$ it is clear that a should be positive.

65. a) if the length of the rectangle is l, then its perimeter is $2x + 2l$; given that this equals 24, it follows that $l = 12 - x$. The area A of the rectangle is now $xl = x(12 - x) = A(x)$.

   b) $y = x(12 - x) = 12x - x^2$ implies that $\frac{dy}{dx} = 12 - 2x$; this is positive if $x < 6$ and negative if $x > 6$. So: the largest subinterval of $[0, 12]$ on which y is increasing is $[0, 6]$.

   c) At the endpoints $x = 0$ and $x = 12$, y is zero. From $\frac{dy}{dx} = 12 - 2x$ we see that y has a local extremum at $x = 6$, where $y = 6(12 - 6) = 36$. So: the maximum value of y on $[0, 12]$ is 36.

67. a) $s(t) = -t^2 + 6t - 8 \Rightarrow s'(t) = -2t + 6$; so $s'(t) > 0$ for $t < 3$ and $s'(t) < 0$ for $t > 3$, from which we deduce the largest interval on which s is increasing to be $[0, 3]$. During this time, the motion is to the right of the horizontal number line.

    b) At time $t = 3$, the direction of motion changes: the particle comes to a halt $\big(s'(3) = 0\big)$ and then moves to the left $\big(s'(t) < 0$ for $t > 3\big)$.

69. Let the box have height y and square base of side x; its (fixed) volume V is then given by $V = x^2 y$ and its surface area is $A = 2x^2 + 4xy$ (counting four sides, top and bottom). Solve for y in terms of x and V to get $y = \frac{V}{x^2}$ and substitute into A to get $A = 2x^2 + \frac{4V}{x}$. Now

$$\frac{dA}{dx} = 4x - \frac{4V}{x^2} = \frac{4(x^3 - V)}{x^2}$$

which is zero when $x^3 = V$ so that $x = \sqrt[3]{V}$ and $y = \frac{V}{x^2} = \sqrt[3]{V}$ also. If $x < \sqrt[3]{V}$ then $\frac{dA}{dx} < 0$ whilst if $x > \sqrt[3]{V}$ then $\frac{dA}{dx} > 0$; so A has a minimum value when x and y equal $\sqrt[3]{V}$. This shows that the box has least surface area when it is actually a cube.

71. Let the bunker have height y and square base of side x; its volume is then given by $96 = x^2 y$. The base has area x^2 and so costs $10x^2$; the top has area x^2 and so costs $6x^2$; and the four sides have total area $4xy$ and so cost $24xy$. The total cost is therefore

$$C = 16x^2 + 24xy.$$

Solving for y in terms of x from $96 = x^2 y$ and substituting into C gives

$$C = 16x^2 + \frac{24 \times 96}{x}.$$

Now

$$\begin{aligned}\frac{dC}{dx} &= 32x - \frac{24 \times 96}{x^2} \\ &= \frac{32(x^3 - 72)}{x^2}\end{aligned}$$

which is zero when $x = 2\sqrt[3]{9}$, negative when $x < 2\sqrt[3]{9}$ and positive when $x > 2\sqrt[3]{9}$. Thus the cost of the bunker is least when $x = 2\sqrt[3]{9}$ and $y = \frac{96}{x^2} = \frac{8}{3}\sqrt[3]{9}$.

73. The idea here is to minimize the distance D from (x_0, y_0) to an arbitrary point (x, y) on the given line; as in **Example 10** and **Remark 3**, it is convenient to minimize $S = D^2$ first. If $b = 0$ then the given line is $x = -\frac{c}{a}$, and the solution is obvious: the least distance from (x_0, y_0) to the line $x = -\frac{c}{a}$ is the horizontal one. So suppose $b \neq 0$ and substitute $y = -\frac{c + ax}{b}$ into

$$S = (x - x_0)^2 + (y - y_0)^2$$

to get

$$S = (x - x_0)^2 + \left(\frac{c + ax}{b} + y_0\right)^2$$

whence

125

$$\frac{dS}{dx} = 2(x - x_0) + 2\frac{a}{b}\left(\frac{c + ax}{b} + y_0\right)$$

which is zero when (and only when)

$$(a^2 + b^2)x = b(bx_0 - ay_0) - ac$$

and so (from $y = -\dfrac{c + ax}{b}$ or from symmetry)

$$(a^2 + b_2)y = a(ay_0 - bx_0) - bc.$$

Checking the sign of S' shows that these values of x and y produce a minimum value for S. To calculate this minimum value, note that

$$\begin{aligned}(a^2 + b^2)(x - x_0) &= (a^2 + b^2)x - (a^2 + b^2)x_0 \\ &= b(bx_0 - ay_0) - ac - a^2 x_0 - b^2 x_0 \\ &= -a(ax_0 + by_0 + c)\end{aligned}$$

and similarly

$$(a^2 + b^2)(y - y_0) = -b(ax_0 + by_0 + c).$$

Thus

$$\begin{aligned}(a^2 + b^2)^2 S &= \left((a^2 + b^2)(x - x_0)\right)^2 + \left((a^2 + b^2)^2(y - y_0)\right)^2 \\ &= (a^2 + b^2)(ax_0 + by_0 + c)\end{aligned}$$

whence cancellation of a factor $a^2 + b^2$ and taking the square root yields

$$D = \frac{|ax_0 + by_0 + c|}{\sqrt{a^2 + b^2}}.$$

75. The blue lines in the figure indicated the length L available to the rod when it makes an angle θ with the South wall of the corridor; from the figure, this available length is given by

$$L = 4\csc\theta + 4\sec\theta.$$

Now

$$\frac{dL}{d\theta} = -4\csc\theta\cot\theta + 4\sec\theta\tan\theta = 4\left(-\frac{\cos\theta}{\sin^2\theta} + \frac{\sin\theta}{\cos^3\theta}\right)$$

which is zero when $\cos^3\theta = \sin^3\theta$ or $\cos\theta = \sin\theta$ or $\theta = \frac{\pi}{4}$ (note that $0 \leq \theta \leq \frac{\pi}{2}$ from the figure). If $0 < \theta < \frac{\pi}{4}$ then $\frac{dL}{d\theta} < 0$ whilst if $\frac{\pi}{4} < \theta < \frac{\pi}{2}$ then $\frac{dL}{d\theta} > 0$: thus, the available length L is least when $\theta = \frac{\pi}{4}$; of course, this agrees with intuition. Since the rod is required to negotiate the corner completely, its length must be at most

$$L\left(\frac{\pi}{4}\right) = 4\csc\frac{\pi}{4} + 4\sec\frac{\pi}{4} = 8\sqrt{2},$$

which is therefore the length (in feet) of the longest rod which can be carried horizontally around the corner.

126

77. The value of the conical cup is

$$36\pi = \frac{1}{3}(\pi r^2)h$$

so that

$$108 = r^2 h.$$

Substituting $h = \frac{108}{r^2}$ into the formula for the surface area yields

$$S = \pi r\sqrt{r^2 + h^2} = \pi r\sqrt{r^2 + \frac{(108)^2}{r^4}}.$$

Bearing in mind **Remark 3**, we find it convenient to minimize S^2, where

$$S^2 = \pi^2 r^2 \left(r^2 + \frac{(108)^2}{r^4}\right)$$

$$= \pi^2 \left(r^4 + \frac{(108)^2}{r^2}\right).$$

From

$$\frac{d(S^2)}{dr} = \pi^2 \left(4r^3 - \frac{2(108)^2}{r^3}\right)$$

we see that $\frac{d(S^2)}{dr} = 0$ when

$$4r^6 = 2(108)^2 \quad \text{therefore} \quad r^6 = 2^3 3^6 \quad \text{therefore} \quad r = 3\sqrt{2}$$

and that $r < 3\sqrt{2} \Rightarrow \frac{d(S^2)}{dr} < 0$ whereas $r > 3\sqrt{2} \Rightarrow \frac{d(S^2)}{dr} > 0$. So the area of the conical cup is a minimum when the cup has radius $r = 3\sqrt{2}$ and height $h = 6$ (in centimetres).

79. $f(x) = x + \frac{1}{x} \Rightarrow f'(x) = 1 - \frac{1}{x^2}$

a) $f'(x) > 0 \Leftrightarrow 1 > \frac{1}{x^2} \Leftrightarrow x^2 > 1 \Leftrightarrow x < -1$ or $x > 1$ so: f is increasing on $(-\infty, -1]$ and $[1, \infty)$.

b) On $[1, \infty)$ the function f is increasing, so surely $f(x) \geq f(1) = 2$ for $x \geq 1$; on $(0, 1]$ the function f is decreasing, so also $f(x) \geq f(1) = 2$ if $0 < x \leq 1$. Consequently, $x + \frac{1}{x} \geq 2$ for $x > 0$.

c) From $(x - 1)^2 \geq 0$ we deduce that $x^2 - 2x + 1 \geq 0$ so $x^2 + 1 \geq 2x$ and therefore (dividing by positive x) $1 + \frac{1}{x} \geq 2$.

If $x < 0$ then $x + \frac{1}{x} = f(x) \leq f(-1) = -2$. This can be seen either by modifying the above argument or noting that f is an odd function.

81. $r = \dfrac{1/w\,C}{\sqrt{R^2 + \left(1/w\,C\right)^2}} = \dfrac{1}{\sqrt{R^2 w^2 C^2 + 1}}$ implies that $\frac{dr}{dw} = -\frac{1}{2}(2R^2 C^2 w)(1 + R^2 w^2 C^2)^{-3/2}$

$= -R^2 C^2 w(1 + R^2 C^2 w^2)^{-3/2}$, which is negative if $w > 0$; so r is a decreasing function of w on $(0, \infty)$.

83. Let x_1 and x_2 lie in $[a, c]$ with $x_1 < x_2$. If both x_1 and x_2 lie in $[a, b]$ then $f(x_1) < f(x_2)$ since f is increasing on $[a, b]$; likewise, $f(x_1) < f(x_2)$ if both x_1 and x_2 lie in $[b, c]$. So we may suppose $x_1 \in [a, b]$ and $x_2 \in [b, c]$. If in fact $x_1 < b$ and $b \leq x_2$ then $f(x_1) < f(b)$ and $f(b) \leq f(x_2)$ so that $f(x_1) < f(x_2)$; here, we again use the fact that f increases on $[a, b]$ and $[b, c]$. If instead $x_1 \leq b$ and $b < x_2$ then, similarly, $f(x_1) \leq f(b)$ and $f(b) < f(x_2)$ whence $f(x_1) < f(x_2)$ again. Having covered all contingencies, the proof is complete.

85. (ii) Either: proceed exactly as in the text's proof of (i) but reverse all inequalities and interchange the words 'increasing' and 'decreasing;' or apply part (i) to the function $g = -f$.

(iii) Since $f'(x) > 0$ for all $x \in (a, c)$ and f is continuous on $[a, c]$ we conclude that f is increasing on $[a, c]$ by Therorem 6; similarly, f is increasing on $[c, b]$. Now apply the result of Exercise 83: since f is increasing on $[a, c]$ and $[c, b]$, it is increasing on $[a, b]$.

(iv) Let $g = -f$; since $f' < 0$ on $(a, c) \cup (c, b)$ it follows that $g' = -f' > 0$ on $(a, c) \cup (c, b)$; according to part (iii) it follows that g is increasing on $[a, b]$, therefore $f = -g$ is decreasing on $[a, b]$.

87. It will be enough to use the technique of Exercise 86 for $0 < x \le \frac{\pi}{2}$ since $|\sin x| \le 1 < \frac{\pi}{2}$ for all x. Let $f(x) = x$ and $g(x) = \sin x$, and note that $f(0) = 0 = g(0)$. If $0 < x < \frac{\pi}{2}$ then

$$g'(x) = \cos x < 1 = f'(x)$$

so that $(f - g)'(x) > 0$. According to **Theorem 6** (applied to $I = \left[0, \frac{\pi}{2}\right)$) it follows that if $0 < x < \frac{\pi}{2}$ then $(f - g)(x) > (f - g)(0) = 0$, whence $\sin x = g(x) < f(x) = x$ for such x.

89. Exercise 87 tells us that if $x > 0$ then $\sin x < x$. To handle the inequality $x - \frac{x^3}{6} < \sin x$ for $x > 0$, let $f(x) = \sin x$ and $g(x) = x - \frac{x^3}{6}$ so that $f'(x) = \cos x$ and $g'(x) = 1 - \frac{x^2}{2}$. Clearly $f(0) = g(0) = 0$, whilst Exercise 88 implies that if $x > 0$ then

$$(f - g)'(x) \;=\; f'(x) - g'(x) \;=\; \cos x - \left(1 - \frac{x^2}{2}\right) \;>\; 0$$

so that (by **Theorem 6**) if $x > 0$ then

$$(f - g)(x) \;>\; (f - g)(0) \;=\; 0$$

whence

$$x \;>\; 0 \;\Rightarrow\; \sin x \;>\; x - \frac{x^3}{6}.$$

[Incidentally, if we divide through the inequalities $x - \frac{x^3}{6} < \sin x < x$ by $x > 0$ we obtain $1 - \frac{x^2}{6} < \frac{\sin x}{x} < 1$; these last inequalitites also hold for $x < 0$ since sine is odd. By an application of the "Pinching Theorem" it follows that $\lim\limits_{x \to 0} \frac{\sin x}{x} = 1$ once again.

91. a) $s(x) = x^2 \Rightarrow s'(x) = 2x$, which is positive if $x > 0$; by **Theorem 6**, it follows that s is increasing on $[0, \infty)$.

b) To say that $f(c)$ is a (relative) minimum for f is to say that $f(c) \le f(x)$ for all x (near c). Since s is increasing on $[0, \infty)$ and f is nonnegative, the inequality $f(c) \le f(x)$ is equivalent to $f(c)^2 \le f(x)^2$ and this in turn means that $f(c)^2$ is a minimum for f^2.

c) In the notation of **Example 10** and **Remark 3**,

$$S(x) \;=\; x^2 - 7x + 16$$
$$\Rightarrow \quad S'(x) \;=\; 2x - 7$$

which is zero when $x = \frac{7}{2}$. As in **Example 10**, this is the only critical number for S in $[0, \infty)$ and in fact yields a minimum for S, hence a minimum for D by part b).

d) As suggested in Remark 3, the difference between this approach and that of **Example 10** is that this approach circumvents the unsightly fractional powers.

128

93. $f(x) = 4x^5 + 5x^4 + 2 \Rightarrow f'(x) = 20x^3(x+1)$. Now, $f' < 0$ on $(-1, 0)$ and $f' > 0$ on $(-\infty, -1)$ and $(0, \infty)$; thus f is decreasing on $[-1, 0]$ and increasing on the intervals $(-\infty, -1]$ and $[0, \infty)$. We have $f(-2) = -46$, $f(-1) = 3$, and $f(0) = 2$. Therefore, f has exactly one root, and it lies between -2 and -1. In fact, the root is approximately -1.38564.

95.
$$\begin{aligned}
f(x) &= 40x^7 - 140x^6 + 168x^5 - 70x^4 + 1 \\
\Rightarrow f'(x) &= 280x^6 - 840x^5 + 840x^4 - 280x^3 \\
&= 280x^3(x^3 - 3x^2 + 3x - 1) \\
&= 280x^3(x-1)^3.
\end{aligned}$$

Here, $f' > 0$ on $(-\infty, 0)$ and $(1, \infty)$ whilst $f' < 0$ on $(0, 1)$. Thus, f is increasing on $(-\infty, 0]$ and $[1, \infty)$ and decreasing on $[0, 1]$. Since $f(-1) = -417$, $f(0) = 1$, $f(1) = -1$, and $f(2) = 417$, we know that f has exactly three roots— one in $(-1, 0)$, one in $(0, 1)$, and one in $(1, 2)$. Their approximate values are -0.29465, 0.5, and 1.29465.

4.5   Signficance of the Second Derivative: Concavity

1.   a) The intervals on which the function is increasing: $[1, 2]$, $[4, 6]$ and $[8, 9]$.

   b) The intervals on which the graph is concave up: $[3, 6]$ and $[6, 9]$.

   c) The relative minima: $x = 1$, $x = 4$ and $x = 8$.

   d) The relative maxima: $x = 2$, $x = 6$ and $x = 9$.

   e) The inflection point: $x = 3$.

3. $f(x) = x^2 - 4x - 5$
   $f'(x) = 2x - 4$
   $f''(x) = 2 > 0$ for all x.
   Thus f is concave up on $(-\infty, \infty)$ and f has no inflection points.

5. $f(x) = |1 - x^2| = \begin{cases} 1 - x^2 & \text{if } -1 \le x \le 1 \\ x^2 - 1 & \text{if } x \le -1 \text{ and } x \ge 1. \end{cases}$

   $f'(x) = \begin{cases} -2x & \text{if } -1 < x < 1 \\ 2x & \text{if } x < -1 \text{ and } x > 1 \\ \text{does not exist} & \text{if } x = \pm 1. \end{cases}$ $f''(x) = \begin{cases} -2 & \text{if } -1 < x < 1 \\ 2 & \text{if } x < -1 \text{ and } x > 1 \\ \text{does not exist} & \text{if } x = \pm 1. \end{cases}$

   Thus f is concave up on $(-\infty, -1]$ and $[1, \infty)$ and concave down on $[-1, 1]$.
   f has inflection points at $(\pm 1, 0)$.

7. $f(x) = \cos x$, $0 \le x \le 2\pi$
   $f'(x) = -\sin x$
   $f''(x) = -\cos x = 0$ at $x = \frac{\pi}{2}$, $\frac{3\pi}{2}$.

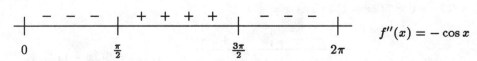

Thus f is concave down on $\left[0, \frac{\pi}{2}\right]$ and $\left[\frac{3\pi}{2}, 2\pi\right]$ and concave up on $\left[\frac{\pi}{2}, \frac{3\pi}{2}\right]$. f has inflection points at $\left(\frac{\pi}{2}, 0\right)$ and $\left(\frac{3\pi}{2}, 0\right)$.

9. $f(x) = (x+3)^3$

   $f'(x) = 3(x+3)^2$

   $f''(x) = 6(x+3) = 0$ at $x = -3$.

$$\begin{array}{c} - \ - \ - \quad | \quad + \ + \ + \\ \hline \\ -3 \end{array} \qquad f''(x) = 6(x+3)$$

f is concave down on $(-\infty, -3]$ and concave up on $[-3, \infty)$. f has an inflection point at $(-3, 0)$.

11. $f(x) = \sqrt{x+2}$. (**Note:** domain of f: $[-2, \infty)$)

    $f'(x) = \frac{1}{2}(x+2)^{-1/2}$

    $f''(x) = \dfrac{-1}{4(x+2)^{3/2}} < 0$ for all x in $(-2, \infty)$.

    Thus f is concave down on $[-2, \infty)$ and f has no inflection points.

13. $f(x) = \dfrac{x}{x+1}$. (**Note:** domain of f: $(-\infty, -1) \cup (-1, \infty)$)

    $f'(x) = \dfrac{(x+1) - x}{(x+1)^2} = \dfrac{1}{(x+1)^2}$

    $f''(x) = \dfrac{-2}{(x+1)^3} \neq 0$ for all x in the domain.

$$\begin{array}{c} + \ + \ + \quad | \quad - \ - \ - \\ \hline \\ -1 \end{array} \qquad f''(x) = \dfrac{-2}{(x+1)^3}$$

f is concave up on $(-\infty, -1)$ and concave down on $(-1, \infty)$ and f has no inflection points.

15. $f(x) = (2x+1)^3$

    $f'(x) = 6(2x+1)^2$

    $f''(x) = 24(2x+1) = 0$ at $x = -\frac{1}{2}$.

$$\begin{array}{c} - \ - \ - \quad | \quad + \ + \ + \\ \hline \\ -\frac{1}{2} \end{array} \qquad f''(x) = 24(2x+1)$$

f is concave down on $\left(-\infty, -\frac{1}{2}\right]$ and concave up on $\left[-\frac{1}{2}, \infty\right)$. f has an inflection point at $\left(-\frac{1}{2}, 0\right)$.

17. $f(x) = 2x^3 - 3x^2 + 18x - 12$

    $f'(x) = 6x^2 - 6x + 18$

    $f''(x) = 12x - 6 = 6(2x-1) = 0$ at $x = \frac{1}{2}$

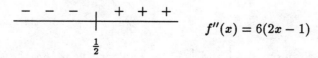

$$\begin{array}{c} - \ - \ - \quad | \quad + \ + \ + \\ \hline \\ \frac{1}{2} \end{array} \qquad f''(x) = 6(2x-1)$$

f is concave down on $\left(-\infty, \frac{1}{2}\right]$ and concave up on $\left[\frac{1}{2}, \infty\right)$. f has an inflection point at $\left(\frac{1}{2}, -\frac{7}{2}\right)$.

19. $f(x) = x^4 + 2x^3 - 36x^2 + 24x - 6$
$f'(x) = 4x^3 + 6x^2 - 72x + 24$
$f''(x) = 12x^2 + 12x - 72 = 12(x^2 + x - 6) = 12(x + 3)(x - 2) = 0$ at $x = -3, 2$.

$$+ + + \quad | \quad - - - \quad | \quad + + +$$
$$ -3 2 f''(x) = 12(x + 3)(x - 2)$$

f is concave up on $(-\infty, -3]$ and $[2, \infty)$ and concave down on $[-3, 2]$. f has inflection points at $(-3, -375)$ and $(2, -70)$.

21. $f(x) = x^4 + 6x^3 + 6x^2 + 12x + 1$
$f'(x) = 4x^3 + 18x^2 + 12x + 12$

$f''(x) = 12x^2 + 36x + 12 = 12(x^2 + 3x + 1) = 0$ where $x = \dfrac{-3 \pm \sqrt{5}}{2} \approx -.382, \ -2.62$.

$$+ + + + \quad | \quad - - - - \quad | \quad + + + +$$
$$ \frac{-3-\sqrt{5}}{2} \frac{-3+\sqrt{5}}{2} f''(x) = 12(x^2 + 3x + 1)$$

Thus f is concave up on $\left(-\infty, \frac{-3-\sqrt{5}}{2}\right]$ and $\left[\frac{-3+\sqrt{5}}{2}, \infty\right)$ and concave down on $\left[\frac{-3-\sqrt{5}}{2}, \frac{-3+\sqrt{5}}{2}\right]$.

f has inflection points at $\left(\frac{-3-\sqrt{5}}{2}, -50.04\right)$ and $\left(\frac{-3+\sqrt{5}}{2}, -3.02\right)$.

23. $f(x) = 3x^5 - 25x^4 + 60x^3 - 60x^2 + 2x + 1$
$f'(x) = 15x^4 - 100x^3 + 180x^2 - 120x + 2$
$f''(x) = 60x^3 - 300x^2 + 360x - 120 = 60(x^3 - 5x^2 + 6x - 2) = 60(x - 1)(x^2 - 4x + 2) = 0$ when
$x = 1, 2 \pm \sqrt{2}$. $[2 \pm \sqrt{2} \approx 3.41, \ .586]$.

$$- - - \quad | \quad + + + \quad | \quad - - - \quad | \quad + + +$$
$$2 - \sqrt{2} 1 2 + \sqrt{2} f''(x) = 60(x - 1)(x^2 - 4x + 2)$$

Thus f is concave down on $\left(-\infty, 2-\sqrt{2}\right]$ and $\left[1, 2+\sqrt{2}\right]$ and concave up on $\left[2-\sqrt{2}, 1\right]$ and $\left[2+\sqrt{2}, \infty\right)$.
f has inflection points at $x = 2 - \sqrt{2}, \ 1$ and $2 + \sqrt{2}$.

25. $f(x) = x^{2/3}(119 - x^5)$

$f'(x) = \frac{2}{3} x^{-1/3}(119 - x^5) + x^{2/3}(-5x^4) = \dfrac{17(14 - x^5)}{3x^{1/3}}$

$f''(x) = \dfrac{17}{3}\left[\dfrac{-5x^4(x^{1/3}) - (14 - x^5)\frac{1}{3}(x^{-2/3})}{x^{2/3}}\right] = \dfrac{-238}{9}\left(\dfrac{x^5 + 1}{x^{4/3}}\right) = 0$ at $x = -1$.

$f''(x)$ is undefined when $x = 0$.

$$+ + + \quad | \quad - - - \quad | \quad - - -$$
$$ -1 0 f''(x) = \dfrac{-238}{9}\left(\dfrac{x^5 + 1}{x^{4/3}}\right)$$

Thus f is concave up on $(-\infty, -1]$ and concave down on $[-1, 0]$ and $[0, \infty)$. f has an inflection point at $(-1, 120)$.

27. $f(x) = x^{5/3} - 5x^{2/3} + 3$

$f'(x) = \dfrac{5}{3} x^{2/3} - \dfrac{10}{3} x^{-1/3}$

$f''(x) = \dfrac{10}{9} x^{-1/3} + \dfrac{10}{9} x^{-4/3} = \dfrac{10}{9}\left(\dfrac{x+1}{x^{4/3}}\right) = 0$ at $x = -1$. $f''(x)$ is undefined when $x = 0$.

$$\begin{array}{ccccc} - & - & - & + & + & + & + & + & + \\ & & & | & & & | & & \\ & & & -1 & & & 0 & & \end{array} \qquad f''(x) = \dfrac{10}{9}\left(\dfrac{x+1}{x^{4/3}}\right)$$

f is concave down on $(-\infty, -1]$ and concave up on $[-1, 0]$ and $[0, \infty)$. f has an inflection point at $(-1, -3)$.

29. $f(x) = \sin^2 x, \ x = \dfrac{\pi}{2}$

$f'(x) = 2\sin x \cos x = \sin 2x; \ f'\left(\dfrac{\pi}{2}\right) = 0$

$f''(x) = 2\cos 2x; \ f''\left(\dfrac{\pi}{2}\right) = 2\cos \pi = -2 < 0$. Thus f has a relative maximum at $x = \dfrac{\pi}{2}$.

31. $f(x) = x^4 - 4x^3 - 48x^2 + 24x + 20, \ x = 4$
$f'(x) = 4x^3 - 12x^2 - 96x + 24; \ f'(4) = -296 \neq 0$
Thus $x = 4$ is not a critical number. Therefore f does not have a relative extremum at $x = 4$.

33. $f(x) = x^4 - 4x^3 + 6x^2 - 4x, \ x = 1$
$f'(x) = 4x^3 - 12x^2 + 12x - 4; \ f'(1) = 0$
$f''(x) = 12x^2 - 24x + 12; \ f''(1) = 0$. Thus the Second Derivative Test fails. We'll use the First Derivative Test.

$$\begin{array}{cccccccc} - & - & - & - & + & + & + & + \\ & & & | & & & & \\ & & & 1 & & & & \end{array} \qquad f'(x) = 4(x - 1)^3$$

Thus f has a relative (absolute) minimum at $x = 1$.

35. $f'(x) = \dfrac{x - 1}{x + 1}, \ x = 1$

$f'(1) = 0$

$$\begin{array}{ccccccc} & - & - & - & + & + & + \\ | & & & | & & & \\ -1 & & & 1 & & & \end{array} \qquad f'(x) = \dfrac{x - 1}{x + 1}$$

Thus f has a relative minimum at $x = 1$.

37. $f'(x) = (x - 1)(x + 2), \ x = -2$.
$f'(-2) = 0$

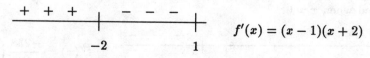

$$\begin{array}{ccccccc} + & + & + & - & - & - & \\ & & | & & & | & \\ & & -2 & & & 1 & \end{array} \qquad f'(x) = (x - 1)(x + 2)$$

Thus f has a relative maximum at $x = -2$.

39. $f(x) = x^3 - 3x^2 + 6$
$f'(x) = 3x^2 - 6x = 3x(x - 2) = 0$ at $x = 0, 2$.

132

$f''(x) = 6x - 6 = 6(x - 1) = 0$ at $x = 1$.
$f''(0) = -6 < 0$. Thus f has a relative maximum at $(0, 6)$.
$f''(2) = 6 > 0$. Thus f has a relative minimum at $(2, 2)$.

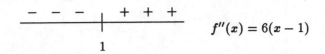

$$f''(x) = 6(x - 1)$$

f is concave down on $(-\infty, 1]$ and concave up on $[1, \infty)$. f has an inflection point at $(1, 4)$.

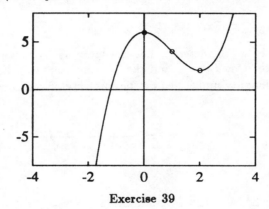

Exercise 39

41. $y = |9 - x^2| = \begin{cases} 9 - x^2 & \text{if } -3 \le x \le 3 \\ x^2 - 9 & \text{if } x \le -3 \text{ and } x \ge 3 \end{cases}$

$y' = \begin{cases} -2x & \text{if } -3 < x < 3 \\ 2x & \text{if } x < -3 \text{ and } x > 3 \\ \text{does not exist} & \text{if } x = \pm 3. \end{cases}$

$y' = 0$ at $x = 0$.

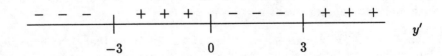

Thus y has a relative maximum at $(0, 9)$ and relative (absolute) minima at $(\pm 3, 0)$.

$y'' = \begin{cases} -2 & \text{if } -3 < x < 3 \\ 2 & \text{if } x < -3 \text{ and } x > 3 \\ \text{does not exist} & \text{if } x = \pm 3. \end{cases}$

Thus y is concave up on $(-\infty, -3]$ and $[3, \infty)$ and concave down on $[-3, 3]$. y has inflection points at $(\pm 3, 0)$.

133

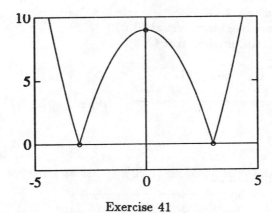

Exercise 41

43. Let the distance from P_1 to Q be x (in meters). Let the lamp at P_2 be twice as strong as that at P_1. The illumination at Q due to P_1 is $I_1 = \frac{k}{x^2}$ and the illumination at Q due to P_2 is $\frac{2k}{(20-x)^2}$ where $k > 0$ is the constant of proportionality multiplied by the intensity of the lamp at P_1. Thus: the total illumination at Q is

$$I = I_1 + I_2 = \frac{k}{x^2} + \frac{2k}{(20-x)^2}$$

therefore

$$\frac{dI}{dx} = -k\left(\frac{2}{x^3} + \frac{4}{(x-20)^3}\right)$$

and

$$\frac{d^2I}{dx^2} = k\left(\frac{6}{x^4} + \frac{12}{(x-20)^4}\right).$$

Now, $\frac{dI}{dx} = 0$ when $\left(\frac{20-x}{x}\right)^3 = 2$ therefore $\frac{20}{x} - 1 = \sqrt[3]{2}$ so $x = \frac{20}{1+\sqrt[3]{2}}$. Since $\frac{d^2I}{dx^2}$ is clearly positive (for $0 < x < 20$) it follows that I is a minimum when $x = \frac{20}{1+\sqrt[3]{2}}$.

45. f is increasing on $[1,3]$ and $[5,6]$. f is concave down on $[2,4]$ and $[7,8]$. The function f has an endpoint relative maximum at $x = 0$ and an endpoint relative minimum at $x = 8$. It also has relative minima at $x = 1$ and $x = 5$ and relative maxima at $x = 3$ and $x = 6$.

47. $f(x) = ax^2 + bx + c \Rightarrow f'(x) = 2ax + b \Rightarrow f''(x) = 2a$. In order that f be concave up for all x, it is necessary and sufficient that a be positive: for if $a > 0$ then $f''(x) > 0$ for all x and we can apply **Theorem 8**; whilst if $a \le 0$ then $f''(x) \le 0$ for all x so that f' is not increasing.

49. Let $f(x) = x^4$ for all real x. Then $x \ne 0 \Rightarrow f(x) = x^4 > 0 = f(0)$, so that $f(0)$ is a relative (in fact, absolute) minimum, but $f''(x) = 12x^2$ so that $f''(0) = 0$.

51. If $n = 2$ then f can have no inflection points, since f'' is a constant and so cannot change sign. If $n = 3$ then f has one inflection point: f'' is a linear function, and so crosses the x-axis exactly once. If $n = k$ then f'' is a polynomial of degree $k-2$ and so has at most $k-2$ real roots; thus, f has at most $k-2$ inflection points (but may have fewer: for example, $f(x) = x^4$ has none).

53. An example: $f(x) = -x^{1/3}$ (for which $f'(0)$ and $f''(0)$ are undefined). A sketch:

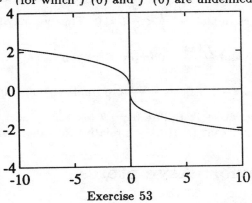

Exercise 53

55.

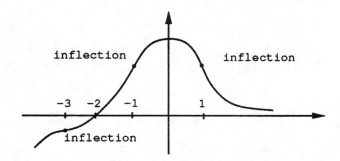

Exercise 55

57. Let $a < b$ be points of I. Since the graph of f lies above the tangent at $\big(a, f(a)\big)$ we have

$$f(b) > f(a) + f'(a)(b - a)$$

whilst since the graph of f lies above the tangent at $\big(b, f(b)\big)$ we have

$$f(a) > f(b) + f'(b)(a - b).$$

The first of these inequalities implies

$$f'(a) < \frac{f(b) - f(a)}{b - a}$$

whilst the second implies

$$f'(b) > \frac{f(b) - f(a)}{b - a}$$

in view of $a < b$. It follows at once that $f'(a) < f'(b)$. Thus: f' is increasing on I.

59. a) As suggested, let $f''(c) = L < 0$.

   b) Since $f'(c) = 0$ we have

$$\frac{f'(x)}{x - c} = \frac{f'(x) - f'(c)}{x - c}$$

   which (by definition) approaches $f''(c) = L$ as $x \to c$.

c) In the ϵ-δ definition of limit, let $\epsilon = \frac{1}{2}|L| = -\frac{1}{2}L$; there exists a $\delta > 0$ such that $|x - c| < \delta$ implies $\left|\dfrac{f'(x)}{x - c} - L\right| < \frac{1}{2}|L|$, whence in fact $\dfrac{f'(x)}{x - c} < \frac{1}{2}L < 0$. Thus we may take $(a, b) = (c - \delta,\ c + \delta)$.

d) If $x \in (a, c)$ then $x - c < 0$ and $\dfrac{f'(x)}{x - c} < 0$ imply $f'(x) > 0$.

e) Similarly, $x \in (c, b)$ implies $x - c > 0$, so $\dfrac{f'(x)}{x - c} < 0$ implies $f'(x) < 0$.

f) According to the First Derivative Test, **Theorem 7** (especially part (i)) in Section 4.4, it follows at once from d) and e) above that $\big(c, f(c)\big)$ is a relative maximum as desired.

4.6 Large-Scale Behavior: Asymptotes

1. $\displaystyle\lim_{x\to\infty} \frac{3x^2 + 2}{10x^2 - 3x} = \lim_{x\to\infty} \frac{3x^2 + 2}{10x^2 - 3x}\,\frac{\left(1/x^2\right)}{\left(1/x^2\right)} = \lim_{x\to\infty} \frac{3 + \frac{2}{x^2}}{10 - \frac{3}{x}} = \frac{3 + 0}{10 - 0} = \frac{3}{10}$

3. $\displaystyle\lim_{x\to\infty} \frac{x(4 - x^3)}{3x^4 + 2x^2} = \lim_{x\to\infty} \frac{x(4 - x^3)}{3x^4 + 2x^2}\,\frac{\left(1/x^4\right)}{\left(1/x^4\right)} = \lim_{x\to\infty} \frac{\frac{4}{x^3} - 1}{3 + \frac{2}{x^2}} = \frac{0 - 1}{3 + 0} = -\frac{1}{3}$

5. $\displaystyle\lim_{x\to-\infty} x^{-1/3} = \lim_{x\to-\infty} \frac{1}{x^{1/3}} = 0$

7. $\displaystyle\lim_{x\to\infty} \frac{3x^2 + 7x}{1 - x^4} = \lim_{x\to\infty} \frac{3x^2 + 7x}{1 - x^4}\,\frac{\left(1/x^4\right)}{\left(1/x^4\right)} = \lim_{x\to\infty} \frac{\frac{3}{x^2} + \frac{7}{x^3}}{\frac{1}{x^4} - 1} = \frac{0 + 0}{0 - 1} = 0$

9. $\displaystyle\lim_{x\to-\infty} \frac{\sin x}{x}$. Note that $-\dfrac{1}{x} \le \dfrac{\sin x}{x} \le \dfrac{1}{x}$.
   Since $\displaystyle\lim_{x\to-\infty} -\frac{1}{x} = \lim_{x\to-\infty} \frac{1}{x} = 0$, $\displaystyle\lim_{x\to-\infty} \frac{\sin x}{x} = 0$.

11. $\displaystyle\lim_{x\to-\infty} \frac{x^4 - 3\sin x}{3x + 5x^5} = \lim_{x\to-\infty} \left(\frac{x^4 - 3\sin x}{3x + 5x^5}\right)\frac{\left(1/x^5\right)}{\left(1/x^5\right)} = \lim_{x\to-\infty} \frac{\frac{1}{x} - \frac{3\sin x}{x^5}}{\frac{3}{x^4} + 5}$

$$= \lim_{x\to-\infty} \frac{\frac{1}{x} - 3\left(\frac{\sin x}{x}\right)\left(\frac{1}{x^4}\right)}{\frac{3}{x^4} + 5} = \frac{0 - 3(0)(0)}{0 + 5} = 0$$

Note that $\displaystyle\lim_{x\to-\infty} \frac{\sin x}{x} = 0$ (See problem 9.)

13. $\displaystyle\lim_{x\to\infty} \frac{|2 - x^2|}{x^2 + 1} = \lim_{x\to\infty} \frac{|2 - x^2|}{x^2 + 1}\,\frac{\left(1/x^2\right)}{\left(1/x^2\right)} = \lim_{x\to\infty} \frac{\left|\frac{2}{x^2} - 1\right|}{1 + \frac{1}{x^2}} = \frac{|0 - 1|}{1 + 0} = \frac{1}{1} = 1$

15. $\displaystyle\lim_{x\to\infty} \frac{\sqrt{x - 1}}{x^2} = \lim_{x\to\infty} \frac{\sqrt{x - 1}}{x^2}\,\frac{\left(1/x^2\right)}{\left(1/x^2\right)} = \lim_{x\to\infty} \frac{\sqrt{\frac{x-1}{x^4}}}{1} = \lim_{x\to\infty} \frac{\sqrt{\frac{1}{x^3} - \frac{1}{x^4}}}{1} = \frac{0}{1} = 0$

17. $\lim\limits_{x\to\infty} \dfrac{x^{2/3}+x}{1+x^{3/4}} = \lim\limits_{x\to\infty} \dfrac{x^{2/3}+x}{1+x^{3/4}} \dfrac{\left(1/x^{3/4}\right)}{\left(1/x^{3/4}\right)} = \lim\limits_{x\to\infty} \dfrac{\frac{1}{x^{1/12}}+x^{1/4}}{\frac{1}{x^{3/4}}+1}$

numerator: $0 + \lim\limits_{x\to\infty} x^{1/4} = \infty$

denominator: $0 + 1 = 1$

Therefore $\lim\limits_{x\to\infty} \dfrac{\frac{1}{x^{1/12}}+x^{1/4}}{\frac{1}{x^{3/4}}+1} = +\infty.$

19. $\lim\limits_{x\to\infty} x\sin x$

As x runs through the odd multiples of $\frac{\pi}{2}$, $(\sin x)$ takes the values $+1$ and -1 alternately. So at these values, $(x\sin x)$ is alternately $+x$ and $-x$. Since these values of x can be arbitrarily large, $(x\sin x)$ assumes arbitrarily large positive and negative values as $x \to \infty$. Thus $\lim\limits_{x\to\infty} (x\sin x)$ does not exist.

(See Figure 1.5 on page 193.)

21. $\lim\limits_{x\to 2^+} \dfrac{1}{x-2} = \infty$

23. $\lim\limits_{x\to\frac{\pi}{2}^-} \tan x = \lim\limits_{x\to\frac{\pi}{2}^-} \dfrac{\sin x}{\cos x} = \infty$ (since $\sin x \to 1$ as $x \to \frac{\pi}{2}^-$ and $\cos x \to 0^+$ as $x \to \frac{\pi}{2}^-$)

25. $\lim\limits_{x\to\frac{\pi}{2}} \tan x$: does not exist (See problems 23 and 24.)

27. $\lim\limits_{x\to 0^-} \dfrac{|x|}{x} = \lim\limits_{x\to 0^-} \dfrac{-x}{x} = \lim\limits_{x\to 0^-} (-1) = -1$

29. $\lim\limits_{x\to 5^-} \dfrac{x^2+1}{x+5} = \dfrac{25+1}{5+5} = \dfrac{26}{10} = \dfrac{13}{5}$

31. $\lim\limits_{x\to 5^-} \dfrac{x^2-25}{x-5} = \lim\limits_{x\to 5^-} \dfrac{(x+5)(x-5)}{x-5} = \lim\limits_{x\to 5^-} (x+5) = 10$

33. $\lim\limits_{x\to 5^-} \dfrac{x-7}{x-5} = \infty$ (since as $x \to 5^-$, $x-7 \to -2$ and $x-5 \to 0^-$).

35. $\lim\limits_{x\to 1^+} \dfrac{1}{(x-1)^{1/3}} = \infty$ (since as $x \to 1^+$, $(x-1)^{1/3} \to 0^+$).

37. $\lim\limits_{x\to 1^+} \dfrac{1}{(x-1)^{2/3}} = \infty$

39. $y = \dfrac{1}{1+x}$

$\lim\limits_{x\to\infty} \dfrac{1}{1+x} = \lim\limits_{x\to\infty} \dfrac{1}{1+x} \dfrac{\left(1/x\right)}{\left(1/x\right)} = 0$

$\lim\limits_{x\to-\infty} \dfrac{1}{1+x} = 0$

Therefore $y = 0$ is the horizontal asymptote.

41. $y = \dfrac{2x^2}{x^2 + 1}$

$$\lim_{x \to \infty} \frac{2x^2}{x^2 + 1} = 2; \quad \lim_{x \to -\infty} \frac{2x^2}{x^2 + 1} = 2$$

Therefore $y = 2$ is the horizontal asymptote.

43. $y = \dfrac{2x^2}{(x^2 + 1)^2}$

$$\lim_{x \to \infty} \frac{2x^2}{(x^2 + 1)^2} = \lim_{x \to \infty} \frac{2x^2}{(x^2 + 1)^2} \frac{\left(1/x^4\right)}{\left(1/x^4\right)} = \lim_{x \to \infty} \frac{\frac{2}{x^2}}{\left(1 + \frac{1}{x^2}\right)^2} = 0$$

Similarly, $\displaystyle\lim_{x \to -\infty} \frac{2x^2}{(x^2 + 1)^2} = 0$.

Therefore $y = 0$ is the horizontal asymptote.

45. $y = 3 + \dfrac{\sin x}{x}$

$$\lim_{x \to \infty} \left(3 + \frac{\sin x}{x}\right) = 3 + \lim_{x \to \infty} \frac{\sin x}{x} = 3 + 0 = 3$$

Similarly, $\displaystyle\lim_{x \to -\infty} \left(3 + \frac{\sin x}{x}\right) = 3$.

Therefore $y = 3$ is the horizontal asymptote.

47. $y = \dfrac{4 - 3x}{\sqrt{1 + 2x^2}}$

Note that for $x > 0$, $\sqrt{x^2} = x$ and for $x < 0$, $\sqrt{x^2} = -x$.

$$\lim_{x \to \infty} \frac{4 - 3x}{\sqrt{1 + 2x^2}} = \lim_{x \to \infty} \frac{4 - 3x}{\sqrt{1 + 2x^2}} \frac{\left(1/x\right)}{\left(1/x\right)} = \lim_{x \to \infty} \frac{\frac{4}{x} - 3}{\sqrt{\frac{1}{x^2} + 2}} = -\frac{3}{\sqrt{2}} = \frac{-3\sqrt{2}}{2}$$

$$\lim_{x \to -\infty} \frac{4 - 3x}{\sqrt{1 + 2x^2}} = \lim_{x \to -\infty} \frac{4 - 3x}{\sqrt{1 + 2x^2}} \frac{\left(-1/x\right)}{\left(-1/x\right)} = \lim_{x \to -\infty} \frac{-\frac{4}{x} + 3}{\sqrt{\frac{1}{x^2} + 2}} = \frac{3}{\sqrt{2}} = \frac{3\sqrt{2}}{2}$$

Thus $y = \pm \dfrac{3\sqrt{2}}{2}$ are the horizontal asymptotes.

49. $3 = \displaystyle\lim_{x \to \pm\infty} \frac{ax + 7}{4 - x} = \lim_{x \to \pm\infty} \frac{ax + 7}{4 - x} \frac{\left(1/x\right)}{\left(1/x\right)} = \lim_{x \to \pm\infty} \frac{a + \frac{7}{x}}{\frac{4}{x} - 1} = -a$

Thus $a = -3$.

51. $y = \dfrac{1}{(x + 2)^2}$

$$\lim_{x \to -2^+} \frac{1}{(x + 2)^2} = \infty \text{ and } \lim_{x \to -2^-} \frac{1}{(x + 2)^2} = \infty$$

Thus $x = -2$ is the vertical asymptote.

53. $y = \dfrac{1}{(x - 1)^{2/3}}; \quad \displaystyle\lim_{x \to 1^+} \frac{1}{(x - 1)^{2/3}} = \lim_{x \to 1^-} \frac{1}{(x - 1)^{2/3}} = \infty.$

Thus $x = 1$ is the vertical asymptote.

55. $y = \dfrac{x^2}{1-x}$

$\lim\limits_{x\to 1^+} \dfrac{x^2}{1-x} = -\infty$ and $\lim\limits_{x\to 1^-} \dfrac{x^2}{1-x} = \infty$.

Thus $x = 1$ is the vertical asymptote.

57. $\lim\limits_{x\to\pm\infty} \dfrac{\pi + ax^r}{1 - 3x^{2/3}} = -2$ if and only if $\dfrac{ax^r}{-3x^{2/3}} = -2$ which implies $r = \frac{2}{3}$ and $\frac{a}{-3} = -2$. Hence $a = 6$.

59. $y = (x^2 + ax + b)^{-1} = \dfrac{1}{x^2 + ax + b}$ has vertical asymptotes of $x = 3$ and $x = 5$ if and only if $x^2 + ax + b = (x-3)(x-5) = x^2 - 8x + 15$.
Thus $a = -8$ and $b = 15$.

61. a)

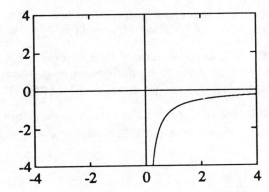

Exercise 61. a)

b) $f(x) = -\frac{1}{x}$ for $x > 0$

c) No.
   Counterexample: $f(x) = -1 - \frac{1}{x}$
   $\lim\limits_{x\to\infty} f(x) = \lim\limits_{x\to\infty}\left(-1 - \dfrac{1}{x}\right) = -1 \neq 0$

63.

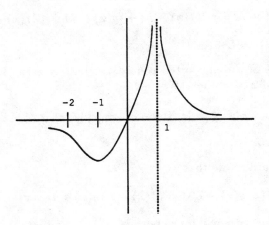

Exercise 63

65. A convenient definition of "$\lim_{x \to -\infty} f(x) = L$" is "given $\epsilon > 0$ there is an integer N so that if $x < -N$ then $|f(x) - L| < \epsilon$."

   a) $f(x) = \dfrac{1}{x}$, $L = 0$. Given $\epsilon > 0$ pick any integer $N > \frac{1}{\epsilon}$. If $x < -N$ then $|x| = -x > N$ so that $|f(x) - L| = \left| \frac{1}{x} - 0 \right| = \frac{1}{|x|} < \frac{1}{N} < \epsilon$.

   b) $f(x) = \dfrac{x}{2x+1}$, $L = \frac{1}{2}$. Here, $|f(x) - L| = \left| \frac{x}{2x+1} - \frac{1}{2} \right| = \left| \frac{-1}{4x+2} \right| = \frac{1}{|4x+2|}$. For $x < -\frac{1}{2}$ we have $4x + 2 < 0$ so $|4x + 2| = -4x - 2$. Now, given $\epsilon > 0$ pick any integer N so that $4N - 2 > \frac{1}{\epsilon}$ (that is, $N > \frac{1}{2} + \frac{1}{4\epsilon}$). If $x < -N$ then $4x + 2 < -4N + 2$ so $-4x - 2 > 4N - 2 > \frac{1}{\epsilon}$ and therefore $|f(x) - L| = \frac{1}{|4x+2|} = -\frac{1}{4x+2} < \epsilon$.

   c) $f(x) = \dfrac{6x-2}{x+1}$, $L = 6$. Here, $|f(x) - L| = \left| \frac{6x-2}{x+1} - 6 \right| = \left| \frac{-8}{x+1} \right| = \frac{8}{|x+1|}$. For $x < -1$ we have $x + 1 < 0$ so $|x + 1| = -(x + 1)$. Now, given $\epsilon > 0$ select any integer N such that $N > 1 + \frac{8}{\epsilon}$. If $x < -N$ then $x + 1 < -N + 1$ so $-(x+1) > N - 1 > \frac{8}{\epsilon}$ so that $|f(x) - L| = \frac{-8}{x+1} < \epsilon$.

67. "The limit of $f(x)$ as $x \to a^-$ is ∞" means that if N is any real number then there is a $\delta > 0$ so that $f(x) > N$ whenever $a - \delta < x < a$. Here, N may be assumed to be positive without harm. Now let $f(x) = \tan x$ and $a = \frac{\pi}{2}$. As in the previous question, $\sin x > \frac{\sqrt{2}}{2}$ whenever $x \in \left(-\frac{\pi}{2}, \frac{\pi}{2} \right)$. If $N > 0$ is given then pick $\delta > 0$ so that $\left| x - \frac{\pi}{2} \right| < \delta$ implies $|\cos x| < \frac{\sqrt{2}}{2N}$; this is again possible since cosine is continuous and $\cos \frac{\pi}{2} = 0$. If we once more stipulate that δ is at most $\frac{\pi}{4}$, then $\frac{\pi}{2} - \delta < x < \frac{\pi}{2}$ implies that $\sin x > \frac{\sqrt{2}}{2}$ and $0 < \cos x < \frac{\sqrt{2}}{2N}$, whence

$$\tan x = \frac{\sin x}{\cos x} > N$$

as required.

69. Let $\lim_{x \to \infty} f(x) = L$ and $\lim_{x \to \infty} g(x) = M$.

   i) Given $\epsilon > 0$ choose N large enough so that both $|f(x) - L| < \frac{1}{2}\epsilon$ and $|g(x) - M| < \frac{1}{2}\epsilon$ if $x > N$; this ensures that if $x > N$ then

$$|(f(x) + g(x)) - (L + M)| = |(f(x) - L) + (g(x) - M)| \leq |f(x) - L| + |g(x) - M| < \epsilon$$

   so that $\lim_{x \to \infty} (f(x) + g(x)) = L + M$.

   ii) If $c = 0$ then both sides of the purported equality are surely zero. Let $c \neq 0$ and, given $\epsilon > 0$, choose N so large that if $x > N$ then $|f(x) - L| < \frac{\epsilon}{|c|}$. Then

$$x > N \Rightarrow |c f(x) - cL| = |c| \, |f(x) - L| < \epsilon$$

   so that $\lim_{x \to \infty} (c f(x)) = cL$.

   iii) This is less straightforward. Note that

$$f(x) g(x) - LM = (f(x) - L) g(x) + L(g(x) - M).$$

By hypothesis, large values of x will make both $|f(x) - L|$ and $|g(x) - M|$ small: so the right side will be small if we can control the term $g(x)$ multiplying $(f(x) - L)$; we can first pick N_1 so that

140

$x > N_1 \Rightarrow |g(x) - M| < 1$ (say: any positive number would do in place of 1); this is possible since $\lim_{x \to \infty} g(x) = M$. Note now that if $x > N_1$ then

$$|g(x)| = |g(x) - M + M| \leq |g(x) - M| + |M| < 1 + |M|.$$

Now, given $\epsilon > 0$ we pick N_2 so that $x > N_2 \Rightarrow |f(x) - L| < \frac{\epsilon}{2(1+|M|)}$ and pick N_3 so that $x > N_3 \Rightarrow |g(x) - M| < \frac{\epsilon}{2(1+|L|)}$. Finally, let N be any number greater than N_1, N_2 and N_3. If $x > N$ then

$$
\begin{aligned}
|f(x)\,g(x) - LM| &= |(f(x) - L)\,g(x) + L(g(x) - M)| \\
&\leq |f(x) - L|\,|g(x)| + |L|\,|g(x) - M| \\
&\leq \frac{\epsilon}{2(1+|M|)}\,(1 + |M|) + |L|\frac{\epsilon}{2(1+|L|)} \\
&< \epsilon \qquad \text{as required.}
\end{aligned}
$$

iv) Suppose we can show $\lim_{x \to \infty} \frac{1}{g(x)} = \frac{1}{M}$. Then applying (iii) with $\frac{1}{g}$ in place of g will give

$$
\begin{aligned}
\lim_{x \to \infty} \left(\frac{f(x)}{g(x)}\right) &= \lim_{x \to \infty} \left(f(x) \cdot \frac{1}{g(x)}\right) \\
&= L \cdot \frac{1}{M} = \frac{L}{M}.
\end{aligned}
$$

So we need only show $\lim_{x \to \infty} \frac{1}{g(x)} = \frac{1}{M}$, for which the calculations are less involved. Note that

$$
* \qquad \left|\frac{1}{g(x)} - \frac{1}{M}\right| = \left|\frac{M - g(x)}{M\,g(x)}\right| = \frac{|g(x) - M|}{|M|\,|g(x)|}.
$$

Since $M \neq 0$ we can find an N_1 so that $x > N_1 \Rightarrow |g(x) - M| < \frac{1}{2}|M| \Rightarrow |g(x)| > \frac{1}{2}|M|$; this ensures that the denominator of (∗) does not cause difficulties by being zero. Now, given $\epsilon > 0$, choose N_2 so that $x > N_2 \Rightarrow |g(x) - M| < \frac{1}{2}|M|^2\epsilon$. Letting N be any number greater than N_1 and N_2, if $x > N$ then $|M|\,|g(x)| > \frac{1}{2}|M|^2$ and $|g(x) - M| < \frac{1}{2}|M|^2\epsilon$, so

$$
\left|\frac{1}{g(x)} - \frac{1}{M}\right| < \frac{\frac{1}{2}|M|^2\epsilon}{\frac{1}{2}|M|^2} = \epsilon
$$

so that $\lim_{x \to \infty} \frac{1}{g(x)} = \frac{1}{M}$ as required.

Remark: These proofs are of course similar to the proofs of **Theorem 1**; see Section 2.4 and Appendix II of the text. Also, part (ii) is a special case of part (iii) given by letting $g(x) = c$ for all x.

For Exercises 70–73, HA denotes horizontal asymptote and VA denotes vertical asymptote.

71. VA: $x = \pm 3, x = 0$
    HA: $y = 0$
    Relative Maximum: $(1.26229, -0.59827)$
    Relative Minimum: $(-1.26229, 0.59827)$

73. VA: $x = -0.222102$

    HA: $y = \frac{1}{2}$

    This function has an infinite number of relative extrema.

    In the interval $[-2, 2]$ we have:

    Relative Maxima: $(-1.16199, 1.31915)$

    $\qquad\qquad\qquad$ $(0.55592, 8.22225)$

    $\qquad\qquad\qquad$ $(1.71356, 0.51648)$

    Relative Minima: $(-0.61225, 0.69910)$

    $\qquad\qquad\qquad$ $(1.43590, 0.48692)$

75. Oblique asymptote: $y = x$.

77. For $x > 0$ and large $f(x)$ looks like the horizontal line $y = 7.3891$.

4.7   Curve Sketching

1. $f(x) = x^2 - 2x - 8$

   1) Domain: $(-\infty, \infty)$

   2) $x^2 - 2x - 8 = (x - 4)(x + 2) = 0$ when $x = 4$, $x = -2$.

   3) f has no vertical or horizontal asymptotes.

   4) $f'(x) = 2x - 2 = 0$ when $x = 1$.

   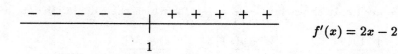

   $f'(x) = 2x - 2$

   Thus f is decreasing on $(-\infty, 1]$ and increasing on $[1, \infty)$. The point $(1, -9)$ is the minimum.

   5) $f''(x) = 2 > 0$. Thus f is concave up on $(-\infty, \infty)$.

   6) $f(0) = -8$

   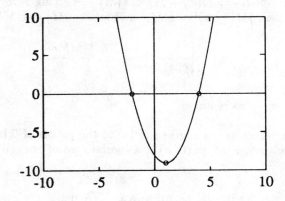

   Exercise 1

3. $f(x) = 2x^3 - 3x^2$

   1) Domain: $(-\infty, \infty)$

142

2) $2x^3 - 3x^2 = x^2(2x - 3) = 0$ when $x = 0$ and $x = \frac{3}{2}$.

3) f has no vertical or horizontal asymptotes.

4) $f'(x) = 6x^2 - 6x = 6x(x - 1) = 0$ when $x = 0, 1$.

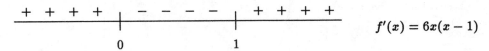

f is increasing on $(-\infty, 0]$ and $[1, \infty)$ and decreasing on $[0, 1]$. The point $(0, 0)$ is the relative maximum and $(1, -1)$ is the relative minimum.

5) $f''(x) = 12x - 6 = 0$ when $x = \frac{1}{2}$.

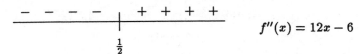

f is concave down on $\left(-\infty, \frac{1}{2}\right]$ and concave up on $\left[\frac{1}{2}, \infty\right)$. The point $\left(\frac{1}{2}, -\frac{1}{2}\right)$ is the inflection point.

6)

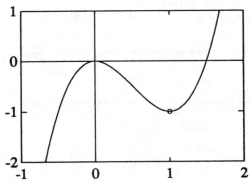

Exercise 3

5. $f(x) = x^4 - 8x^3 + 24x^2 - 32x + 19$

1) Domain: $(-\infty, \infty)$

2) f has no vertical or horizontal asymptotes.

3) $f'(x) = 4x^3 - 24x^2 + 48x - 32 = 4(x^3 - 6x^2 + 12x - 8) = 4(x - 2)^3 = 0$ at $x = 2$.

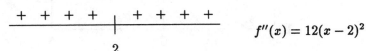

f is decreasing on $(-\infty, 2]$ and increasing on $[2, \infty)$. f has a minimum at $(2, 3)$.

4) $f''(x) = 12(x - 2)^2 = 0$ at $x = 2$.

$$f''(x) = 12(x - 2)^2$$

f is concave up on $(-\infty, \infty)$.

5) $f(0) = 19$ and $f(4) = 19$.

6)

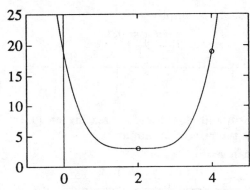

Exercise 5

7. $f(x) = x^3 + x^2 - 8x + 8$

1) Domain: $(-\infty, \infty)$

2) f has no vertical or horizontal asymptotes.

3) $f'(x) = 3x^2 + 2x - 8 = (3x - 4)(x + 2) = 0$ at $x = \frac{4}{3}, -2$.

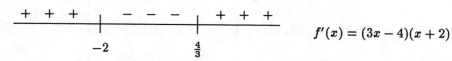

$$f'(x) = (3x - 4)(x + 2)$$

f is increasing on $(-\infty, -2]$ and $\left[\frac{4}{3}, \infty\right)$ and decreasing on $\left[-2, \frac{4}{3}\right]$. f has a relative maximum at $(-2, 20)$ and a relative minimum at $\left(\frac{4}{3}, \frac{40}{27}\right)$.

4) $f''(x) = 6x + 2 = 2(3x + 1) = 0$ at $x = -\frac{1}{3}$.

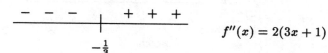

$$f''(x) = 2(3x + 1)$$

f is concave down on $\left(-\infty, -\frac{1}{3}\right]$ and concave up on $\left[-\frac{1}{3}, \infty\right)$. The inflection point is $\left(-\frac{1}{3}, \frac{290}{27}\right)$.

5) $f(0) = 8$ and $f(-4) = -8$

6)

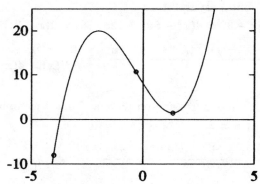

Exercise 7

144

9. $f(x) = x^3 - 7x + 6$

1) Domain: $(-\infty, \infty)$

2) $x^3 - 7x + 6 = (x-1)(x+3)(x-2) = 0$ at $x = -3, 1, 2$.

3) f has no vertical or horizontal asymptotes.

4) $f'(x) = 3x^2 - 7 = 0$ at $x = \pm\frac{\sqrt{21}}{3}$.

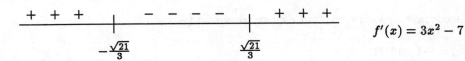

$$f'(x) = 3x^2 - 7$$

f is increasing on $\left(-\infty, -\frac{\sqrt{21}}{3}\right]$ and $\left[\frac{\sqrt{21}}{3}, \infty\right)$ and decreasing on $\left[-\frac{\sqrt{21}}{3}, \frac{\sqrt{21}}{3}\right]$. f has a relative maximum at $\left(-\frac{\sqrt{21}}{3}, 13.13\right)$ and a relative minimum at $\left(\frac{\sqrt{21}}{3}, -1.13\right)$.

5) $f''(x) = 6x = 0$ at $x = 0$.

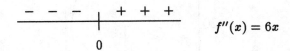

$$f''(x) = 6x$$

f is concave up on $[0, \infty)$ and concave down on $(-\infty, 0]$. $(0, 6)$ is an inflection point.

6)

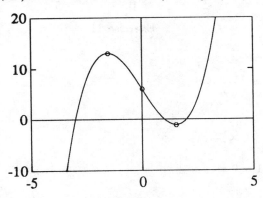

Exercise 9

11. $f(x) = \dfrac{x+4}{x-4}$

1) $x - 4 = 0$ at $x = 4$, therefore Domain: $(-\infty, 4) \cup (4, \infty)$.

2) $f(x) = 0$ when $x + 4 = 0 \Rightarrow x = -4$.

3) $\lim\limits_{x \to \pm\infty} f(x) = 1$, therefore $y = 1$ is the horizontal asymptote.

$\lim\limits_{x \to 4^-} f(x) = -\infty$ and $\lim\limits_{x \to 4^+} f(x) = \infty$; therefore $x = 4$ is the vertical asymptote.

4) $f'(x) = \dfrac{(x-4) - (x+4)}{(x-4)^2} = \dfrac{-8}{(x-4)^2} < 0$ for all x in the domain.

Therefore f is decreasing for all x in the domain.

145

5) $f''(x) = \dfrac{16}{(x-4)^3}$

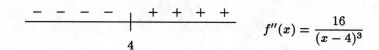

$f''(x) = \dfrac{16}{(x-4)^3}$

f is concave up on $(4,\infty)$ and concave down on $(-\infty,4)$.

6) $f(0) = -1$ and $f(5) = 9$.

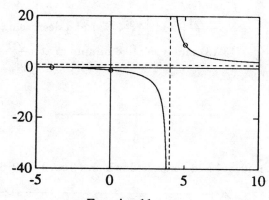

Exercise 11

13. $f(x) = 9x - x^{-1} = 9x - \dfrac{1}{x} = \dfrac{9x^2-1}{x}$

1) Domain: $(-\infty,0) \cup (0,\infty)$

2) $f(x) = 0$ when $9x^2 - 1 = 0 \Rightarrow x = \pm\frac{1}{3}$.

3) $f(x)$ has no horizontal asymptote.
   $\lim\limits_{x\to 0^-} f(x) = +\infty$ and $\lim\limits_{x\to 0^+} f(x) = -\infty$; thus, $x = 0$ is the vertical asymptote. Also, note that $y = 9x$ is an oblique asymptote of $f(x)$.

4) $f'(x) = 9 + x^{-2} = 9 + \dfrac{1}{x^2} = \dfrac{9x^2+1}{x^2} > 0$ for all x in the domain. Thus f is increasing for all x in the domain.

$$f''(x) = \dfrac{-2}{x^3}$$

f is concave up on $(-\infty,0)$ and concave down on $(0,\infty)$.

146

6)

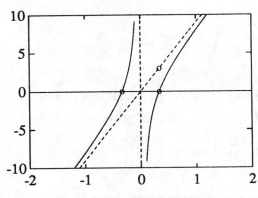

Exercise 13

15. $f(x) = |4 - x^2|$

To graph $f(x)$, first we will graph $g(x) = 4 - x^2$. Then reflect the portion of the graph of $g(x)$ extending below the x-axis through the x-axis.

- $g(x) = 4 - x^2$

  1) Domain: $(-\infty, \infty)$

  2) $g(x) = 0$ at $x = \pm 2$.

  3) $g(x)$ has no vertical or horizontal asymptotes.

  4) $g'(x) = -2x = 0$ at $x = 0$.

  g is increasing on $(-\infty, 0]$ and decreasing on $[0, \infty)$. g has a maximum at $(0, 4)$.

  5) $g''(x) = -2 < 0$. Thus g is concave down for all x.

6)

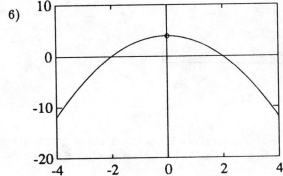

For the graph of $f(x)$, see Figure 5.6 on **page 240** of the textbook.

The graph of $g(x) = 4 - x^2$
Exercise 15

17. $f(x) = x^2 + \frac{2}{x}$

   1) Domain: $(-\infty, 0) \cup (0, \infty)$

   2) $f(x) = x^2 + \frac{2}{x} = \frac{x^3 + 2}{x} = 0$ when $x^3 + 2 = 0 \Rightarrow x = \sqrt[3]{-2}$.

   3) $\lim\limits_{x \to \pm\infty} f(x) = \infty$, therefore f has no horizontal asymptotes.

      $\lim\limits_{x \to 0^-} f(x) = -\infty$ and $\lim\limits_{x \to 0^+} f(x) = \infty$, therefore $x = 0$ is the vertical asymptote.

   4) $f'(x) = 2x - \frac{2}{x^2} = \frac{2(x^3 - 1)}{x^2} = 0$ at $x = 1$.

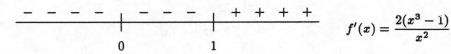

$$f'(x) = \frac{2(x^3 - 1)}{x^2}$$

   f is increasing on $[1, \infty)$ and decreasing on $(-\infty, 0)$ and $(0, 1]$. f has a relative minimum at $(1, 3)$.

   5) $f''(x) = 2 + \frac{4}{x^3} = \frac{2(x^3 + 2)}{x^3} = 0$ at $x = \sqrt[3]{-2}$.

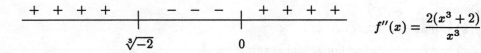

$$f''(x) = \frac{2(x^3 + 2)}{x^3}$$

   f is concave up on $\left(-\infty, \sqrt[3]{-2}\right]$ and $(0, \infty)$ and concave down on $\left[\sqrt[3]{-2}, 0\right)$. $\left(\sqrt[3]{-2}, 0\right)$ is an inflection point.

   6) Note that $y = x^2$ is an oblique asymptote for f.

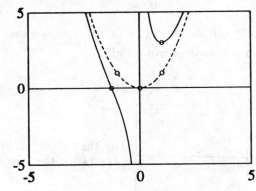

Exercise 17

19. $f(x) = |x^2 - 3x + 2|$

To graph $f(x)$, first we will graph $g(x) = x^2 - 3x + 2$. Then reflect the portion of the graph extending below the x-axis through the x-axis. (See problem 38 in Section 4.1.) Consider $g(x) = x^2 - 3x + 2$.

   1) Domain: $(-\infty, \infty)$

   2) $g(x) = x^2 - 3x + 2 = (x - 1)(x - 2) = 0$ at $x = 1, 2$.

   3) g has no vertical or horizontal asymptote.

148

4) $g'(x) = 2x - 3 = 0$ at $x = \frac{3}{2}$.

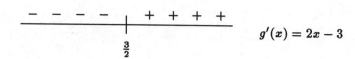

$g'(x) = 2x - 3$

g is decreasing on $\left(-\infty, \frac{3}{2}\right]$ and increasing on $\left[\frac{3}{2}, \infty\right)$. g has a minimum at $\left(\frac{3}{2}, -\frac{1}{4}\right)$.

5) $g''(x) = 2 > 0$. Thus g is concave up on $(-\infty, \infty)$.

6) $g(0) = 2$

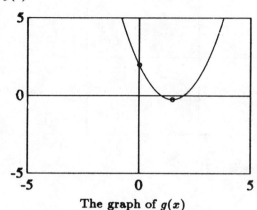

The graph of $g(x)$

The graph of $f(x)$

21. $f(x) = \dfrac{1}{x(x-4)} = \dfrac{1}{x^2 - 4x}$

1) $x(x-4) = 0$ at $x = 0, 4$; therefore Domain: $(-\infty, 0) \cup (0, 4) \cup (4, \infty)$

2) $f(x) \neq 0$ for all x in the domain.

3) $\lim\limits_{x \to \pm\infty} f(x) = 0$; therefore $y = 0$ is the horizontal asymptote.

   $\lim\limits_{x \to 0^-} f(x) = \infty$ and $\lim\limits_{x \to 0^+} f(x) = -\infty$, therefore $x = 0$ is a vertical asymptote.

   $\lim\limits_{x \to 4^-} f(x) = -\infty$ and $\lim\limits_{x \to 4^+} f(x) = \infty$, therefore $x = 4$ is a vertical asymptote.

4) $f'(x) = \dfrac{-(2x-4)}{x^2(x-4)^2} = \dfrac{2(2-x)}{x^2(x-4)^2} = 0$ at $x = 2$.

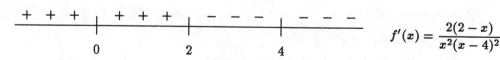

$f'(x) = \dfrac{2(2-x)}{x^2(x-4)^2}$

f is increasing on $(-\infty, 0)$ and $(0, 2]$ and decreasing on $[2, 4)$ and $(4, \infty)$. f has a relative maximum at $\left(2, -\frac{1}{4}\right)$.

5) $f''(x) = \dfrac{-2(x^2)(x-4)^2 - (2)(2-x)[2x(x-4)^2 + 2(x-4)x^2]}{x^4(x-4)^4}$

$= \dfrac{-2x(x-4) + 2(x-2)[2(x-4) + 2x]}{x^3(x-4)^3} = \dfrac{2(3x^2 - 12x + 16)}{x^3(x-4)^3}$

$$= \frac{2[3(x-2)^2+4]}{x^3(x-4)^3} \neq 0 \text{ for all } x \text{ in the domain.}$$

$$\underset{0 \qquad\qquad 4}{\underset{+\ +\ +\ |\ -\ -\ -\ |\ +\ +\ +}{\rule{8cm}{0.4pt}}} \qquad f''(x) = \frac{2[3(x-2)^2+4]}{x^3(x-4)^3}$$

f is concave up on $(-\infty, 0)$ and $(4, \infty)$ and concave down on $(0, 4)$. f has no inflection points.

6) $f(-1) = \frac{1}{5}$ and $f(5) = \frac{1}{5}$.

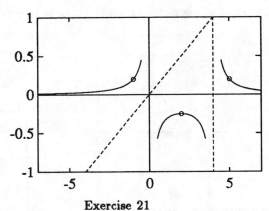

Exercise 21

23. $f(x) = \dfrac{x}{(2x+1)^2}$

1) $2x+1 = 0$ at $x = -\frac{1}{2}$; therefore Domain: $\left(-\infty, -\frac{1}{2}\right) \cup \left(-\frac{1}{2}, \infty\right)$.

2) $f(x) = 0$ when $x = 0$.

3) $\displaystyle\lim_{x \to \pm\infty} f(x) = 0$; therefore $y = 0$ is the horizontal asymptote.

$\displaystyle\lim_{x \to -\frac{1}{2}^-} f(x) = \lim_{x \to -\frac{1}{2}^+} f(x) = -\infty$; therefore $x = -\frac{1}{2}$ is the vertical asymptote.

4) $f'(x) = \dfrac{(2x+1)^2 - x(2)(2x+1)(2)}{(2x+1)^4} = \dfrac{1-2x}{(2x+1)^3} = 0$ at $x = \frac{1}{2}$.

$$\underset{-\frac{1}{2} \qquad\qquad \frac{1}{2}}{\underset{-\ -\ -\ |\ +\ +\ +\ |\ -\ -\ -}{\rule{8cm}{0.4pt}}} \qquad f'(x) = \frac{1-2x}{(2x+1)^3}$$

f is decreasing on $\left(-\infty, -\frac{1}{2}\right)$ and $\left[\frac{1}{2}, \infty\right)$ and increasing on $\left(-\frac{1}{2}, \frac{1}{2}\right]$.

f has a maximum at $\left(\frac{1}{2}, \frac{1}{8}\right)$.

5) $f''(x) = \dfrac{-2(2x+1)^3 - (1-2x)\, 3\, (2x+1)^2 (2)}{(2x+1)^6}$

$$= \frac{-2(2x+1) + 6(2x-1)}{(2x+1)^4} = \frac{8(x-1)}{(2x+1)^4} = 0 \text{ at } x = 1$$

$$\underset{-\frac{1}{2} \qquad\qquad 1}{\underset{-\ -\ -\ |\ -\ -\ -\ |\ +\ +\ +}{\rule{8cm}{0.4pt}}} \qquad f''(x) = \frac{8(x-1)}{(2x+1)^4}$$

150

f is concave down on $\left(-\infty, -\frac{1}{2}\right)$ and $\left(-\frac{1}{2}, 1\right]$ and concave up on $[1, \infty)$. f has an inflection point at $\left(1, \frac{1}{9}\right)$.

6.) $f(-1) = -1$

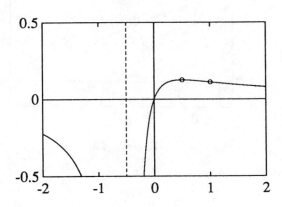

Exercise 23

25. $f(x) = (x-3)^{2/3} + 1$

1) Domain: $(-\infty, \infty)$

2) $f(x) > 0$ for all x since $(x-3)^{2/3} > 0$ for all x.

3) f has no vertical or horizontal asymptote.

4) $f'(x) = \frac{2}{3}(x-3)^{-1/3} \neq 0$ for all x.
   $f'(x)$ is undefined when $x - 3 = 0 \Rightarrow x = 3$.
   Therefore $x = 3$ is a critical number.

$$\frac{- \quad - \quad - \ | \ + \ + \ +}{3} \qquad f'(x) = \frac{2}{3(x-3)^{1/3}}$$

f is decreasing on $(-\infty, 3]$ and increasing on $(3, \infty)$. f has a minimum at $(3, 1)$.

5) $f''(x) = -\frac{2}{9}(x-3)^{-4/3} = \frac{-2}{9(x-3)^{4/3}} \neq 0$ for all x.
   $f''(x)$ is undefined at $x = 3$.

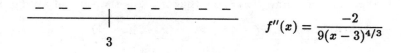

$f''(x) = \frac{-2}{9(x-3)^{4/3}}$

f is concave down on $(-\infty, 3]$ and $[3, \infty)$.

6) $f(0) \approx 3.08$

151

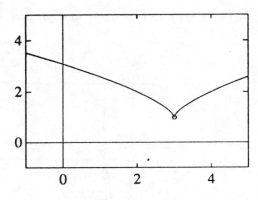

Exercise 25

27. $f(x) = \frac{1}{3}x^3 - x^2 - 3x + 4$

   1) Domain: $(-\infty, \infty)$

   2) f has no vertical or horizontal asymptotes.

   3) $f'(x) = x^2 - 2x - 3 = (x - 3)(x + 1) = 0$ at $x = -1, 3$.

$$\underset{\underset{-1}{}}{+\ +\ +} \quad \underset{\underset{3}{}}{-\ -\ -} \quad +\ +\ + \qquad f'(x) = (x-3)(x+1)$$

   f is increasing on $(-\infty, -1]$ and $[3, \infty)$ and decreasing on $[-1, 3]$. f has a relative maximum at $(-1, 6^{1/3})$ and a relative minimum at $(3, -5)$.

   4) $f''(x) = 2x - 2 = 2(x - 1) = 0$ at $x = 1$.

$$\underset{\underset{1}{}}{-\ -\ -} \quad +\ +\ + \qquad f''(x) = 2(x-1)$$

   f is concave down on $(-\infty, 1]$ and concave up on $[1, \infty)$. f has an inflection point at $\left(1, \frac{1}{3}\right)$.

   5) $f(0) = 4$

   6) (See Figure 4.9 on p.226 of the textbook.)

29. $f(x) = \dfrac{x^2 - 4x + 5}{x - 2} = \dfrac{(x^2 - 4x + 4) + 1}{x - 2} = \dfrac{(x - 2)^2 + 1}{x - 2}$

   1) $x - 2 = 0$ at $x = 2$; therefore Domain: $(-\infty, 2) \cup (2, \infty)$.

   2) $(x - 2)^2 + 1 > 0$ for all x in the domain. Thus $f(x) \neq 0$ for all x in the domain.

   3) f has no horizontal asymptotes.
   $\lim\limits_{x \to 2^-} f(x) = -\infty$ and $\lim\limits_{x \to 2^+} f(x) = \infty$; thus $x = 2$ is the vertical asymptote.

   4) $f'(x) = \dfrac{(2x - 4)(x - 2) - (x^2 - 4x + 5)}{(x - 2)^2} = \dfrac{(x - 3)(x - 1)}{(x - 2)^2} = 0$

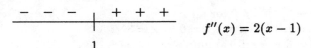

$$\underset{\underset{1}{}}{+\ +\ +} \quad \underset{\underset{2}{}}{-\ -\ -} \quad \underset{\underset{3}{}}{-\ -\ -} \quad +\ +\ + \qquad f'(x) = \dfrac{(x-3)(x-1)}{(x-2)^2}$$

152

f is increasing on $(-\infty, 1]$ and $[3, \infty)$ and decreasing on $[1, 2)$ and $(2, 3]$. f has a relative maximum at $(1, -2)$ and a relative minimum at $(3, 2)$.

5) $f''(x) = \dfrac{(2x-4)(x-2)^2 - (x^2 - 4x + 3)\, 2\,(x-2)}{(x-2)^4} = \dfrac{2}{(x-2)^3} \neq 0$ for all x in the domain.

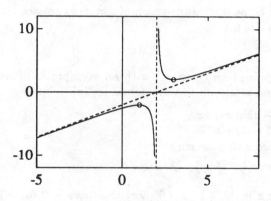

$f''(x) = \dfrac{2}{(x-2)^3}$

f is concave down on $(-\infty, 2)$ and concave up on $(2, \infty)$.

6) $f(x) = \dfrac{(x-2)^2 + 1}{x - 2} = (x - 2) + \dfrac{1}{x - 2}$

Note that $y = x - 2$ is an oblique asymptote for $f(x)$.

7) $f(0) = -\dfrac{5}{2}$

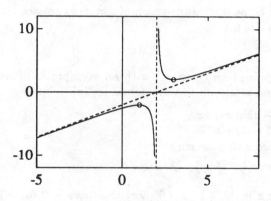

Exercise 29

31. $f(x) = x + \cos x$.

1)–5) This function is defined on $(-\infty, \infty)$ and has neither horizontal nor vertical asymptotes.

$$\begin{aligned}
f'(x) &= 1 - \sin x \\
f'(x) &= 0 \Rightarrow \sin x = 1 \Rightarrow x = (4n+1)\tfrac{\pi}{2}\text{(for any integer } n) \\
&= \ldots\ldots, -\frac{3\pi}{2}, \frac{\pi}{2}, \frac{5\pi}{2}, \ldots\ldots \\
f'(x) &> 0 \text{ for all other values of } x.
\end{aligned}$$

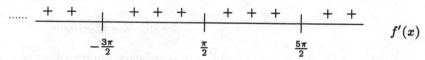

So f is increasing on $(-\infty, \infty)$ with critical numbers $x = (4n+1)\tfrac{\pi}{2}$ and no relative extrema.

$f''(x) = -\cos x$, so

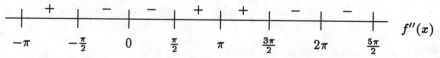

So f is concave up on $\left((4n+1)\tfrac{\pi}{2},\ (4n+3)\tfrac{\pi}{2}\right)$ and concave down on $\left((4n-1)\tfrac{\pi}{2},\ (4n+1)\tfrac{\pi}{2}\right)$ with inflection points where x is any odd multiple of $\tfrac{\pi}{2}$.

6)

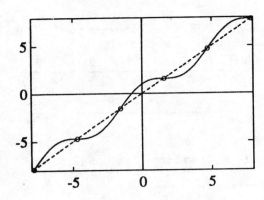

Exercise 31

(Dashed line $y = x$ helps sketching: graph of $y = f(x)$ crosses this line when x is an odd multiple of $\frac{\pi}{2}$ since then $\cos x = 0$.)

33. $f(x) = \sin x + \cos^2 x$

1)–5) This function is defined on $(-\infty, \infty)$ with no asymptotes of either kind. Since it is periodic of period 2π we confine attention to the interval $[0, 2\pi]$.

$$
\begin{aligned}
f'(x) &= \cos x - 2 \sin x \cos x \\
&= \cos x (1 - 2 \sin x) \\
f'(x) &= 0 \Rightarrow \cos x = 0 \text{ or } \sin x = \tfrac{1}{2} \\
&\Rightarrow x = \tfrac{\pi}{2}, \tfrac{3\pi}{2}, \tfrac{\pi}{6}, \tfrac{5\pi}{6}
\end{aligned}
$$

Since $\cos x > 0$ for x in $\left[0, \frac{\pi}{2}\right) \cup \left(\frac{3\pi}{2}, 2\pi\right]$ and $\cos x < 0$ for x in $\left(\frac{\pi}{2}, \frac{3\pi}{2}\right)$ while $\sin x > \frac{1}{2}$ for x in $\left(\frac{\pi}{6}, \frac{5\pi}{6}\right)$ and $\sin x < \frac{1}{2}$ for x in $\left[0, \frac{\pi}{6}\right) \cup \left(\frac{5\pi}{6}, 2\pi\right)$ we see:

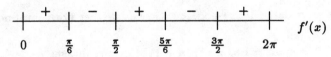

So f is increasing on the intervals $\left[0, \frac{\pi}{6}\right]$, $\left[\frac{\pi}{2}, \frac{5\pi}{6}\right]$, $\left[\frac{3\pi}{2}, 2\pi\right]$ and decreasing on the intervals $\left[\frac{\pi}{6}, \frac{\pi}{2}\right]$, $\left[\frac{5\pi}{6}, \frac{3\pi}{2}\right]$, and has relative extrema at

$$\left(\frac{\pi}{6}, \frac{5}{4}\right) \quad \longleftarrow \text{ absolute maximum}$$

$$\left(\frac{\pi}{2}, 1\right) \quad \longleftarrow \text{ relative minimum}$$

$$\left(\frac{5\pi}{6}, \frac{5}{4}\right) \quad \longleftarrow \text{ absolute maximum}$$

$$\left(\frac{3\pi}{2}, -1\right) \quad \longleftarrow \text{ absolute minimum}$$

$$
\begin{aligned}
f''(x) &= -\sin x - 2 \cos^2 x + 2 \sin^2 x \\
&= 4 \sin^2 x - \sin x - 2
\end{aligned}
$$

By the quadratic formula, $f''(x) = 0 \Rightarrow \sin x = \frac{1 \pm \sqrt{33}}{8}$.

Approximate solutions (in degrees): $x_1 \approx 57\frac{1}{2}^\circ$, $x_2 \approx 122\frac{1}{2}^\circ$, $x_3 \approx 216\frac{1}{3}^\circ$, $x_4 \approx 323\frac{2}{3}^\circ$.

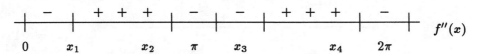

[To determine the sign of f'' on (for example) the interval (x_2, x_3) simply substitute a specific value of x such as: then $f''(\pi) = 4\sin^3\pi - \sin\pi - 2 = -2 < 0$.]

So f is concave up on the intervals $[x_1, x_2]$, $[x_3, x_4]$ and concave down on the intervals $[0, x_1]$, $[x_2, x_3]$, $[x_4, 2\pi]$ and has inflection points at x_1, x_2, x_3, x_4 found above (approximately).

To see where the graph crosses the x-axis, use the quadratic formula again:
$$f(x) = \sin x + \cos^2 x = \sin x + 1 - \sin^2 x.$$
Therefore $f(x) = 0 \Rightarrow \sin x = \frac{1 \pm \sqrt{5}}{2}$.

Of these, only $\sin x = \frac{1 - \sqrt{5}}{2}$ is possible, yielding (approximately, in degrees)

$$x \approx 218^\circ \text{ and } x \approx 322^\circ.$$

6)

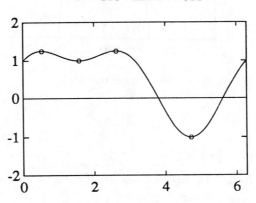

Exercise 33

35. $f(x) = 4x^2(1 - x^2)$. Note that f is an even function.
Therefore, f is symmetric with respect to the y-axis.

1) Domain: $(-\infty, \infty)$

2) $f(x) = 0$ when $x = 0, \pm 1$.

3) f has no vertical or horizontal asymptotes.

4) $f'(x) = 8x(1 - x^2) + 4x^2(-2x) = 8x(1 - 2x^2) = 0$ at $x = 0, \pm\frac{\sqrt{2}}{2}$

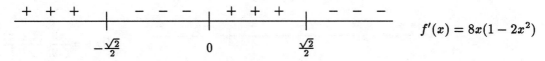

f is increasing on $\left(-\infty, -\frac{\sqrt{2}}{2}\right]$ and $\left[0, \frac{\sqrt{2}}{2}\right]$ and decreasing on $\left[-\frac{\sqrt{2}}{2}, 0\right]$ and $\left[\frac{\sqrt{2}}{2}, \infty\right)$.
f has a relative minimum at $(0, 0)$ and maxima at $\left(\pm\frac{\sqrt{2}}{2}, 1\right)$.

155

5) $f''(x) = 8(1 - 2x^2) + 8x(-4x) = 8(1 - 6x^2) = 0$ at $x = \pm\frac{\sqrt{6}}{6}$.

$$f''(x) = 8(1 - 6x^2)$$

f is concave up on $\left[-\frac{\sqrt{6}}{6}, \frac{\sqrt{6}}{6}\right]$ and concave down on $\left(-\infty, -\frac{\sqrt{6}}{6}\right]$ and $\left[\frac{\sqrt{6}}{6}, \infty\right)$. The inflection points are $\left(\pm\frac{\sqrt{6}}{6}, \frac{5}{9}\right)$.

6)

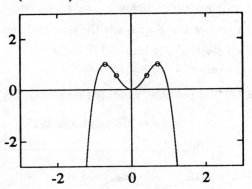

Exercise 35

37. $f(x) = x^{2/3} - \frac{1}{5} x^{5/3}$

1) Domain: $(-\infty, \infty)$

2) $x^{2/3} - \frac{1}{5}x^{5/3} = x^{2/3}\left(1 - \frac{1}{5}x\right) = 0$ at $x = 0, 5$.

3) f has no vertical or horizontal asymptotes.

4) $f'(x) = \frac{2}{3}x^{-1/3} - \frac{1}{3}x^{2/3} = \frac{2-x}{3x^{1/3}} = 0$ at $x = 2$ and f' is undefined at $x = 0$. The critical numbers are $0, 2$.

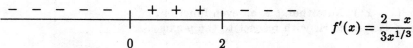

$$f'(x) = \frac{2-x}{3x^{1/3}}$$

f is increasing on $[0, 2]$ and decreasing on $(-\infty, 0]$ and $[2, \infty)$. f has a relative minimum at $(0, 0)$ and a relative maximum at $\left(2, \frac{3\sqrt[3]{4}}{5}\right)$.

5) $f''(x) = -\frac{2}{9}x^{-4/3} - \frac{2}{9}x^{-1/3} = -\frac{2}{9}\left(\frac{1+x}{x^{4/3}}\right) = 0$ at $x = -1$. $f''(x)$ is undefined at $x = 0$.

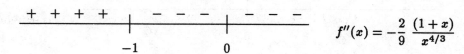

$$f''(x) = -\frac{2}{9}\frac{(1+x)}{x^{4/3}}$$

f is concave up on $(-\infty, -1]$ and concave down on $[-1, 0)$ and $[0, \infty)$. $\left(-1, \frac{6}{5}\right)$ is an inflection point.

156

6) See Figure 5.8 on page 241 of the textbook.

39. $f(x) = \dfrac{x^2}{9 - x^2}$

1) $9 - x^2 = 0$ when $x = \pm 3$; therefore Domain: $(-\infty, -3) \cup (-3, 3) \cup (3, \infty)$.

2) $f(x) = 0$ when $x = 0$.

3) $\lim\limits_{x \to \pm\infty} f(x) = -1$; therefore $y = -1$ is the horizontal asymptote.
   $\lim\limits_{x \to -3^-} f(x) = -\infty$ and $\lim\limits_{x \to -3^+} f(x) = +\infty$; therefore $x = -3$ is a vertical asymptote.
   $\lim\limits_{x \to 3^-} f(x) = +\infty$ and $\lim\limits_{x \to 3^+} f(x) = -\infty$; therefore $x = 3$ is a vertical asymptote.

4) $f'(x) = \dfrac{2x(9 - x^2) - x^2(-2x)}{(9 - x^2)^2} = \dfrac{18x}{(9 - x^2)^2} = 0$ at $x = 0$.

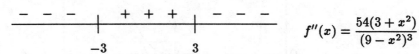

$f'(x) = \dfrac{18x}{(9 - x^2)^2}$

f is decreasing on $(-\infty, -3)$ and $(-3, 0]$ and increasing on $[0, 3)$ and $(3, \infty)$. f has a relative minimum at $(0, 0)$.

5) $f''(x) = \dfrac{18(9 - x^2)^2 - 18x(2)(9 - x^2)(-2x)}{(9 - x^2)^4} = \dfrac{18(9 - x^2 + 4x^2)}{(9 - x^2)^3} = \dfrac{54(3 + x^2)}{(9 - x^2)^3} \neq 0$ for all x in the domain.

$f''(x) = \dfrac{54(3 + x^2)}{(9 - x^2)^3}$

f is concave down on $(-\infty, -3)$ and $(3, \infty)$ and concave up on $(-3, 3)$.

6)

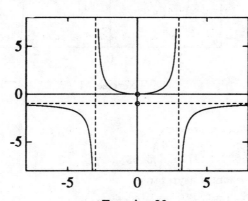

Exercise 39

41. $f(x) = 5x^3 - x^5$. Note that f is an odd function. Therefore the graph of f is symmetric with respect to the origin.

1) Domain: $(-\infty, \infty)$

2) $5x^3 - x^5 = x^3(5 - x^2) = 0$ at $x = 0$, $\pm\sqrt{5}$.

3) f has no vertical or horizontal asymptotes.

4) $f'(x) = 15x^2 - 5x^4 = 5x^2(3 - x^2) = 0$ at $x = 0, \pm\sqrt{3}$.

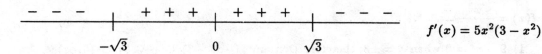

$f'(x) = 5x^2(3 - x^2)$

f is increasing on $\left[-\sqrt{3}, \sqrt{3}\right]$ and decreasing on $\left(-\infty, -\sqrt{3}\right]$ and $\left[\sqrt{3}, \infty\right)$. f has a relative minimum at $\left(-\sqrt{3}, -6\sqrt{3}\right)$ and a relative maximum at $\left(\sqrt{3}, 6\sqrt{3}\right)$.

5) $f''(x) = 30x - 20x^3 = 10x(3 - 2x^2) = 0$ at $x = 0, \pm\frac{\sqrt{6}}{2}$.

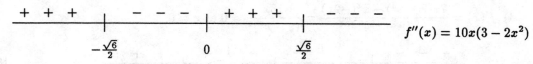

$f''(x) = 10x(3 - 2x^2)$

f is concave up on $\left(-\infty, -\frac{\sqrt{6}}{2}\right]$ and $\left[0, \frac{\sqrt{6}}{2}\right]$ and concave down on $\left[-\frac{\sqrt{6}}{2}, 0\right]$ and $\left[\frac{\sqrt{6}}{2}, \infty\right)$. The inflection points are $\left(-\frac{\sqrt{6}}{2}, \frac{-21\sqrt{6}}{8}\right)$, $(0, 0)$ and $\left(\frac{\sqrt{6}}{2}, \frac{21\sqrt{6}}{8}\right)$.

6)

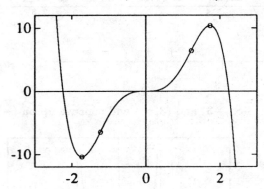

Exercise 41

43. $f(x) = 16 - 20x^3 + 3x^5$

1) Domain: $(-\infty, \infty)$

2) $f(x) = 0$ when $x \approx -2.6377, .9774, 2.5163$

3) f has no vertical or horizontal asymptotes.

4) $f'(x) = -60x^2 + 15x^4 = 15x^2(x^2 - 4) = 0$ at $x = 0, \pm 2$.

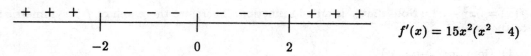

$f'(x) = 15x^2(x^2 - 4)$

f is increasing on $(-\infty, -2]$ and $[2, \infty)$ and decreasing on $[-2, 2]$. f has a relative maximum at $(-2, 80)$ and a relative minimum at $(2, -48)$.

5) $f''(x) = -120x + 60x^3 = 60x(x^2 - 2) = 0$ at $x = 0, \pm\sqrt{2}$.

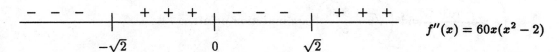

$$f''(x) = 60x(x^2 - 2)$$

f is concave down on $\left(-\infty, -\sqrt{2}\right]$ and $\left[0, \sqrt{2}\right]$ and concave up on $\left[-\sqrt{2}, 0\right]$ and $\left[\sqrt{2}, \infty\right)$. The inflection points are $\left(-\sqrt{2},\ 16 + 28\sqrt{2}\right)$, $(0, 16)$ and $\left(\sqrt{2},\ 16 - 28\sqrt{2}\right)$.

6)

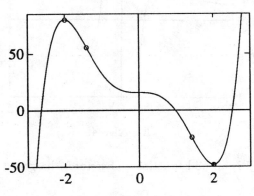

Exercise 43

45. $f(x) = \dfrac{1 - x^2}{x^3}$

1) Domain: $(-\infty, 0) \cup (0, \infty)$

2) $f(x) = 0$ when $1 - x^2 = (1 - x)(1 + x) = 0 \Rightarrow x = \pm 1$.

3) $\displaystyle\lim_{x \to \pm\infty} f(x) = 0 \Rightarrow y = 0$ is the horizontal asymptote.

$\displaystyle\lim_{x \to 0^+} f(x) = \infty$ and $\displaystyle\lim_{x \to 0^-} f(x) = -\infty$. Thus $x = 0$ is the vertical asymptote.

4) $f'(x) = \dfrac{-2x(x^3) - (1 - x^2)(3x^2)}{x^6} = \dfrac{x^2 - 3}{x^4} = 0$ when $x = \pm\sqrt{3}$.

$$f'(x) = \frac{x^2 - 3}{x^4}$$

f is increasing on $\left(-\infty, -\sqrt{3}\right]$ and $\left[\sqrt{3}, \infty\right)$ and decreasing on $\left[-\sqrt{3}, 0\right)$ and $\left(0, \sqrt{3}\right]$. f has a relative minimum at $\left(\sqrt{3}, \dfrac{-2\sqrt{3}}{9}\right)$ and a relative maximum at $\left(-\sqrt{3}, \dfrac{2\sqrt{3}}{9}\right)$.

5) $f''(x) = \dfrac{2x(x^4) - 4x^3(x^2 - 3)}{x^8} = \dfrac{2(6 - x^2)}{x^5} = 0$ at $x = \pm\sqrt{6}$.

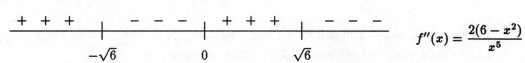

$$f''(x) = \frac{2(6 - x^2)}{x^5}$$

159

f is concave up on $\left(-\infty, -\sqrt{6}\right]$ and $\left(0, \sqrt{6}\right]$ and concave down on $\left[-\sqrt{6}, 0\right)$ and $\left[\sqrt{6}, \infty\right)$. $\left(-\sqrt{6}, \frac{5\sqrt{6}}{36}\right)$ and $\left(\sqrt{6}, \frac{-5\sqrt{6}}{36}\right)$ are inflection points.

6)

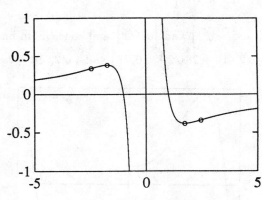

Exercise 45

47. $f(x) = \dfrac{x}{x^2 + 3}$. Note that f is an odd function. Therefore, the graph of f is symmetric with respect to the origin.

1) $(x^2 + 3)$ is always positive. Therefore Domain: $(-\infty, \infty)$.

2) $f(x) = 0$ when $x = 0$.

3) $\lim\limits_{x \to \pm\infty} f(x) = 0 \Rightarrow y = 0$ is the horizontal asymptote. $f(x)$ has no vertical asymptotes.

4) $f'(x) = \dfrac{x^2 + 3 - x(2x)}{(x^2 + 3)^2} = \dfrac{3 - x^2}{(x^2 + 3)^2} = 0$ when $x = \pm\sqrt{3}$.

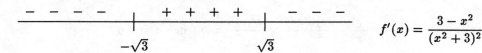

$$f'(x) = \frac{3 - x^2}{(x^2 + 3)^2}$$

f is decreasing on $\left(-\infty, -\sqrt{3}\right]$ and $\left[\sqrt{3}, \infty\right)$ and increasing on $\left[-\sqrt{3}, \sqrt{3}\right]$.
The point $\left(-\sqrt{3}, \frac{-\sqrt{3}}{6}\right)$ is a relative minimum and $\left(\sqrt{3}, \frac{\sqrt{3}}{6}\right)$ is a relative maximum.

5) $f''(x) = \dfrac{-2x(x^2 + 3)^2 + (x^2 - 3)(2)(x^2 + 3)(2x)}{(x^2 + 3)^4} = \dfrac{2x(x + 3)(x - 3)}{(x^2 + 3)^3} = 0$ when $x = 0, -3, 3$.

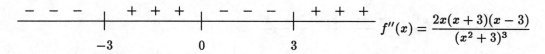

$$f''(x) = \frac{2x(x + 3)(x - 3)}{(x^2 + 3)^3}$$

f is concave up on $[-3, 0]$ and $[3, \infty)$ and concave down on $(-\infty, -3]$ and $[0, 3]$.
The points $(0, 0)$, $\left(-3, -\frac{1}{4}\right)$ and $\left(3, \frac{1}{4}\right)$ are inflection points.

6)

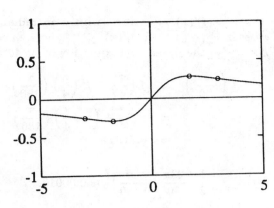

Exercise 47

4.8   Linear Approximation Revisited

In Exercises 1 through 4, we take $f(x) = x^2$ so that $f'(x) = 2x$ and $f''(x) = 2$; consequently, we take $M = 2$ in each case. [Note: in each case, the estimated error is the actual error; this is typical for $f(x) = x^2$, as can be seen from the binomial expansion of $(x_0 + \Delta x)^2$.]

1. Here, with $x_0 = 2$ and $\Delta x = 0.02$, the linear approximation to $(2.02)^2$ is

$$f(2.02) \approx f(2) + f'(2)(0.02) = 4.08$$

with error at most

$$\frac{1}{2} M (\Delta x)^2 = 0.0004.$$

3. With $x_0 = 2$ and $\Delta x = -0.03$, the linear approximation to $(1.97)^2$ is

$$f(1.97) \approx f(2) + f'(2)(-0.03) = 3.88$$

with error at most

$$\frac{1}{2}(2)(-0.03)^2 = 0.0009.$$

5. $f(x) = x^4 \Rightarrow f'(x) = 4x^3 \Rightarrow f''(x) = 12x^2$. The maximum value of f'' on $[2, 2.03]$ is assumed at the right endpoint, so we take $M = 12(2.03)^2 = 49.4508$. With $x_0 = 2$ and $\Delta x = 0.03$, the linear approximation to $(2.03)^4$ is

$$
\begin{aligned}
f(2.03) &\approx f(2) + f'(2)(0.03) \\
&= 16 + (32)(0.03) = 16.96
\end{aligned}
$$

with error at most

$$\frac{1}{2}(49.4508)(0.03)^2 = 0.02225286.$$

Remark: It might be more in keeping with the level of calculations in the earlier problems to handle the error a little differently. The maximum value of f'' on $[2, 2.03]$ is surely less than $f''(3) = (12)(9) = 108$, so the error in problem 5 is surely less than $\frac{1}{2}(108)(0.03)^2 = 0.0486$; this error estimate is easier to calculate than the (closer) estimate provided by taking $M = f''(2.03)$.]

7. $f(x) = \sin x \Rightarrow f'(x) = \cos x \Rightarrow f''(x) = -\sin x$ (with x in radians). Using the approximation $|\sin x| \leq |x|$ we see that if x lies in $\left[\frac{\pi}{4}, \frac{23\pi}{90}\right]$ then $|f''(x)| \leq \frac{23\pi}{90}$; here, of course, $\frac{23\pi}{90}$ radians corresponds to 46°. With $x_0 = \frac{\pi}{4}$ and $\Delta x = \frac{\pi}{180}$, the linear approximation to $\sin 46° = \sin\frac{23\pi}{90}$ is

$$
\begin{aligned}
f\left(\frac{23\pi}{90}\right) &\approx f\left(\frac{\pi}{4}\right) + f'\left(\frac{\pi}{4}\right)\left(\frac{\pi}{180}\right) \\
&= \sin\frac{\pi}{4} + \left(\cos\frac{\pi}{4}\right)\left(\frac{\pi}{180}\right) \\
&\approx 0.71945
\end{aligned}
$$

with error at most

$$\frac{1}{2}\left(\frac{23\pi}{90}\right)\left(\frac{\pi}{180}\right)^2 \approx 0.00012.$$

9. $f(x) = \sin x \Rightarrow f'(x) = \cos x \Rightarrow f''(x) = -\sin x$. Here, $|f''(x)| = |\sin x| \leq |x|$ so that $|f''(x)| \leq \frac{5\pi}{180} = \frac{\pi}{36} = M$ for x in $\left[0, \frac{5\pi}{180}\right]$. Consequently, if we let $x_0 = 0$ and $\Delta x = \frac{5\pi}{180}$ then the corresponding linear approximation to $\sin 5° = \sin\frac{5\pi}{180}$ is

$$
\begin{aligned}
f\left(\frac{\pi}{36}\right) &\approx f(0) + f'(0)\frac{\pi}{36} \\
&= \frac{\pi}{36} \approx 0.08727
\end{aligned}
$$

with error at most

$$\frac{1}{2}M(\Delta x)^2 = \frac{1}{2}\left(\frac{\pi}{36}\right)\left(\frac{\pi}{36}\right)^2 \approx 0.00033.$$

11. Let $f(x) = x^{1/2}$, so $f'(x) = \frac{1}{2}x^{-1/2}$ and $f''(x) = -\frac{1}{4}x^{-3/2}$. With $x_0 = 36$ and $\Delta x = 1$, the linear approximation to $\sqrt{37} = f(37)$ is

$$
\begin{aligned}
f(37) &\approx f(36) + f'(36)(1) \\
&= 6\frac{1}{12} = 6.08\dot{3}.
\end{aligned}
$$

To estimate the error, note this $|f''(x)| = -|\frac{1}{4}x^{-3/2}|$ is decreasing for $x \in [36, 37]$; thus, we may take $M = |f''(36)| = \frac{1}{864}$ and find that the error is at most

$$\frac{1}{2}M(\Delta x)^2 \approx 0.000579.$$

13. Let $f(x) = \sqrt{x}$ so that $f'(x) = \frac{1}{2\sqrt{x}}$ and $f''(x) = -\frac{1}{4\sqrt{x^3}}$. With $x_0 = 49$ we get

$$f(50) \approx f(49) + f'(49)(1) = 7\frac{1}{14} \approx 7.07143$$

and with $x_0 = 50.41\left(= (7.1)^2\right)$ we get

$$
\begin{aligned}
f(50) &\approx 7.1 + f'(50.41)(-0.41) \\
&\approx 7.07113.
\end{aligned}
$$

Since $|f''(x)| = \frac{1}{4\sqrt{x^3}}$ is decreasing on $[49, 50]$ we estimate the error (according to **Theorem 11**) in the second approximation as being at most

$$
\begin{aligned}
\frac{1}{2}|f''(49)|(-0.41)^2 &= \frac{1}{2}\left(\frac{1}{4 \times 343}\right)(-0.41)^2 \\
&= 0.00006;
\end{aligned}
$$

162

this makes it clear that 7.1 is a better approximation to $\sqrt{50}$ than is 7. [Alternatively, note that

$$\left(\sqrt{50} - 7\right)\left(\sqrt{50} + 7\right) = 50 - 49 = 1$$

so

$$\sqrt{50} - 7 = \frac{1}{\sqrt{50} + 7}$$

whereas

$$\left(7.1 - \sqrt{50}\right)\left(7.1 + \sqrt{50}\right) = 50.41 - 50 = 0.41$$

so

$$7.1 - \sqrt{50} = \frac{0.41}{\sqrt{50} + 7.1}.$$

Since $0.41 < 1$ and $\sqrt{50} + 7.1 > \sqrt{50} + 7$ it follows that $7.1 - \sqrt{50} < \sqrt{50} - 7$.]

15. $f(x) = \sin x \Rightarrow f''(x) = -\sin x$. Since $|\sin x| \leq 1$ we shall take $M = 1$ in **Theorem 11**; see the naïve estimation in **Example 3** of the text. According to **Theorem 11**, the error made in the linear approximation

$$f(x) \approx f(0) + f'(0)x = x$$

is at most

$$\frac{1}{2} M x^2 = \frac{1}{2} x^2.$$

If we wish to make this error at most 0.005 (see **Remark 2** of the text) then we take $x^2 \leq 0.01$ so that $|x| \leq 0.1$. An interval as required is therefore $[-0.1, 0.1]$. [Needless to say, this is not the largest such interval.]

17. Graph with the range: $[-3, 2] \times [-3, 6]$.

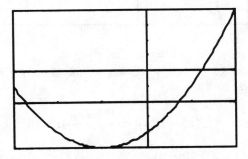

Exercise 17

Since $f''(x) = 2$, the linear approximation has the same accuracy at all points.

19. Graph with the range: $[0, \pi] \times [-1, 2.5]$.

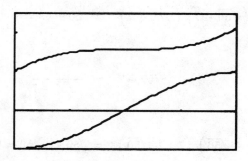

Exercise 19

Since $f''(x) = -\cos x$, the linear approximation is most accurate at $x = \frac{\pi}{2}$ where $f''(x) = 0$ and least accurate at $x = 0$ and $x = \pi$ where $|f''(x)|$ is the largest.

21. $f'(x) = \dfrac{3x^2}{8} - 1$

   a) $f(-2) = 1$ and $f'(-2) = \frac{1}{2}$ so $T_1(x) = 1 + \frac{x+2}{2} = \frac{x}{2} + 2$. Take $h_1 = 0.11$.

   b) $f(0) = 0$ and $f'(0) = -1$ so $T_2(x) = 0 - x = -x$. Take $h_2 = 0.42$.

   c) $f''(x) = \frac{3x}{4}$ so $|f''(-2)| = \frac{3}{2}$ and $|f''(0)| = 0$. The error terms $|f(x) - T_1(x)|$ and $|f(x) - T_2(x)|$ depend on the size of $f''(x)$ near $x = -2$ and $x = 0$. This explains why the value for h_1 in part a is larger than the value in part b.

   d) Graph with the range: $[-5, 5] \times [-3, 3]$.

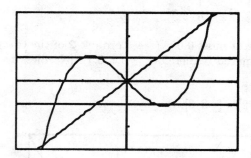

Exercise 21. d)

Graph $f(x)$, $f''(x)$, and $y = \pm 1$. From the graph we see that $|f''(x)| \leq 1$ for $-\frac{4}{3} \leq x \leq \frac{4}{3}$. For any point, a, in this interval we have

$$|f(x) - T(x)| = \frac{|f''(z_x)|}{2} |x - a|^2$$

where $T(x) = f(a) + f'(a)(x - a)$. Since the maximum of $|f''(x)| = 1$ on the interval $-\frac{4}{3} \leq x \leq \frac{4}{3}$ then

$$|f(x) - T(x)| \leq \frac{1}{2} |x - a|^2.$$

164

23. Graph with the range: $[-\pi, \pi] \times [-4, 4]$, $h = 0.01$.

$$S_h(x) = \frac{\sin(2(x+h)) - 2\sin 2x + \sin(2(x-h))}{h^2}$$

$$= \frac{8 \sin x \cos x (\cos^2 h - 1)}{h^2}$$

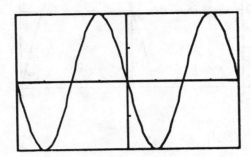

Exercise 23

25. Graph with the range: $[-2, 2] \times [-2, 1]$, $h = 0.01$.

$$S_h(x) = \frac{1}{h^2} \left[\frac{1}{1 + (x+h)^2} - \frac{2}{1 + x^2} + \frac{1}{1 + (x-h)^2} \right].$$

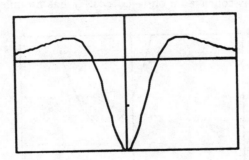

Exercise 25

Review Exercises—Chapter 4

1. $f(x) = 4x - x^2$

   1) $f'(x) = 4 - 2x = 0$ at $x = 2$.

$$+ \;+\;+ \quad | \quad -\;-\;- $$
$$\underline{\hspace{6cm}} \qquad f'(x) = 4 - 2x$$
$$2$$

f is increasing on $(-\infty, 2]$ and decreasing on $[2, \infty)$. f has an absolute maximum at $(2, 4)$.

165

2) $f''(x) = -2 < 0$. Thus f is concave down on $(-\infty, \infty)$. f has no inflection points.

3) $f(x) = 0$ at $x = 0, 4$.

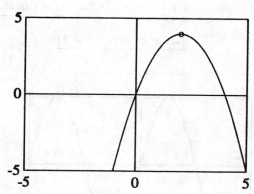

Exercise 1

3. $f(x) = x^2 - 2x + 3$

1) $f'(x) = 2x - 2 = 0$ at $x = 1$.

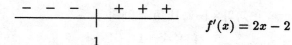

$f'(x) = 2x - 2$

f is decreasing on $(-\infty, 1]$ and increasing on $[1, \infty)$. f has an absolute minimum at $(1, 2)$.

2) $f''(x) = 2 > 0$. Thus f is concave up on $(-\infty, \infty)$. f has no inflection points.

3) $f(0) = 3$

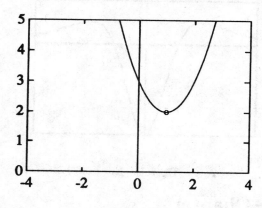

Exercise 3

5. $f(x) = 2\sin(\pi x)$. Since $\sin \pi x$ has period 2, we will need to analyze $f(x)$ on $[0, 2]$ only.

1) $f(0) = f(2) = 0$

2) $f'(x) = 2\pi \cos(\pi x) = 0$ when $\cos(\pi x) = 0 \Rightarrow \pi x = \frac{\pi}{2}, \frac{3\pi}{2} \Rightarrow x = \frac{1}{2}, \frac{3}{2}$.

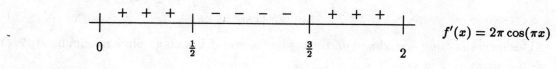

$f'(x) = 2\pi \cos(\pi x)$

f is increasing on $\left[0, \frac{1}{2}\right]$ and $\left[\frac{3}{2}, 2\right]$ and decreasing on $\left[\frac{1}{2}, \frac{3}{2}\right]$. f has an absolute maximum at $\left(\frac{1}{2}, 2\right)$ and an absolute minimum at $\left(\frac{3}{2}, -2\right)$.

3) $f''(x) = -2\pi^2 \sin \pi x = 0$ when $\sin \pi x = 0 \Rightarrow \pi x = \pi \Rightarrow x = 1$.

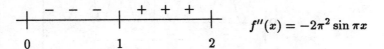

f is concave down on $[0, 1]$ and concave up on $[1, 2]$. f has an inflection point at $(1, 0)$.

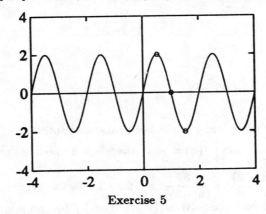

Exercise 5

7. $f(t) = 2\cos^2(2t) = 2\left[\frac{1}{2}(1 + \cos 4t)\right] = 1 + \cos 4t$. Since $\cos 4t$ has period $\frac{\pi}{2}$, we'll need to analyze $f(t)$ on $\left[0, \frac{\pi}{2}\right]$ only.

1) $f(t) = 0$ when $\cos 4t = -1 \Rightarrow 4t = \pi \Rightarrow t = \frac{\pi}{4}$.

2) $f'(t) = -4\sin 4t = 0$ when $\sin 4t = 0 \Rightarrow 4t = \pi \Rightarrow t = \frac{\pi}{4}$.

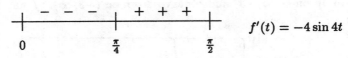

f is decreasing on $\left[0, \frac{\pi}{4}\right]$ and increasing on $\left[\frac{\pi}{4}, \frac{\pi}{2}\right]$. f has an absolute minimum at $\left(\frac{\pi}{4}, 0\right)$. f has absolute maxima at $(0, 2)$ and $\left(\frac{\pi}{2}, 2\right)$.

3) $f''(t) = -16\cos 4t = 0$ when $\cos 4t = 0 \Rightarrow 4t = \frac{\pi}{2}, \frac{3\pi}{2} \Rightarrow t = \frac{\pi}{8}, \frac{3\pi}{8}$.

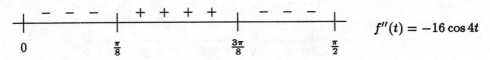

f is concave down on $\left[0, \frac{\pi}{8}\right]$ and $\left[\frac{3\pi}{8}, \frac{\pi}{2}\right]$ and concave up on $\left[\frac{\pi}{8}, \frac{3\pi}{8}\right]$. f has points of inflection at $\left(\frac{\pi}{8}, 1\right)$ and $\left(\frac{3\pi}{8}, 1\right)$.

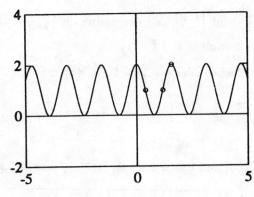

Exercise 7

9. $f(x) = \dfrac{x-3}{x+3}$

1) Domain: $(-\infty, -3) \cup (-3, \infty)$

2) $f(x) = 0$ when $x = 3$.

3) $\displaystyle\lim_{x \to \pm\infty} f(x) = 1$, therefore $y = 1$ is the horizontal asymptote.

   $\displaystyle\lim_{x \to -3^-} f(x) = \infty$ and $\displaystyle\lim_{x \to -3^+} f(x) = -\infty$, therefore $x = -3$ is the vertical asymptote.

4) $f'(x) = \dfrac{(x+3) - (x-3)}{(x+3)^2} = \dfrac{6}{(x+3)^2} > 0$ for all x in the domain.

   Thus f is increasing on $(-\infty, -3) \cup (-3, \infty)$ and f has no relative extrema.

5) $f''(x) = \dfrac{-12}{(x+3)^3} \neq 0$ for all x in the domain.

$$\underset{\underset{-3}{\big|}}{\overline{+\ \ +\ \ +\ \ -\ \ -\ \ -}} \qquad f''(x) = \dfrac{-12}{(x+3)^3}$$

f is concave up on $(-\infty, -3)$ and concave down on $(-3, \infty)$. f has no inflection points.

6) $f(0) = -1$

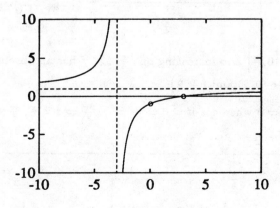

Exercise 9

11. $f(x) = x\sqrt{16 - x^2}$

168

1) Domain: $[-4, 4]$

2) $f(x) = 0$ when $x = 0, \pm 4$.

3) $f'(x) = (16 - x^2)^{1/2} + x\left(\frac{1}{2}\right)(16 - x^2)^{-1/2}(-2x) \stackrel{\ast}{=} \dfrac{2(8 - x^2)}{(16 - x^2)^{1/2}} = 0$ when $x = \pm 2\sqrt{2}$.

$f'(x) = \dfrac{2(8 - x^2)}{(16 - x^2)^{1/2}}$

f is decreasing on $\left[-4, -2\sqrt{2}\right]$ and $\left[2\sqrt{2}, 4\right]$ and increasing on $\left[-2\sqrt{2}, 2\sqrt{2}\right]$. f has an absolute minimum at $\left(-2\sqrt{2}, -8\right)$ and an absolute maximum at $\left(2\sqrt{2}, 8\right)$.

4) $f''(x) = \dfrac{-4x(16 - x^2)^{1/2} - 2(8 - x^2)\left(\frac{1}{2}\right)(16 - x^2)^{-1/2}(-2x)}{(16 - x^2)} = \dfrac{2x(x^2 - 24)}{(16 - x^2)^{3/2}} = 0$ when $x = 0$.

Note that $x^2 - 24 \neq 0$ on $[-4, 4]$.

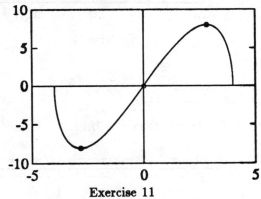

$f''(x) = \dfrac{2x(x^2 - 24)}{(16 - x^2)^{3/2}}$

f is concave up on $[-4, 0]$ and concave down on $[0, 4]$. f has an inflection point at $(0, 0)$.

Exercise 11

13. $f(t) = \dfrac{t^2}{t^2 + 9}$

1) Domain: $(-\infty, \infty)$

2) $f(t) = 0$ when $t = 0$.

3) $\lim\limits_{t \to \pm\infty} f(t) = 1$; therefore $y = 1$ is the horizontal asymptote.

4) $f'(t) = \dfrac{2t(t^2 + 9) - t^2(2t)}{(t^2 + 9)^2} = \dfrac{18t}{(t^2 + 9)^2} = 0$ at $t = 0$.

$f'(t) = \dfrac{18t}{(t^2 + 9)^2}$

f is decreasing on $(-\infty, 0]$ and increasing on $[0, \infty)$.
f has an absolute minimum at $(0, 0)$.

5) $f''(t) = \dfrac{18(t^2 + 9)^2 - (18t)(2)(t^2 + 9)(2t)}{(t^2 + 9)^4} = \dfrac{54(3 - t^2)}{(t^2 + 9)^3} = 0$ when $t = \pm\sqrt{3}$.

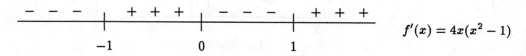

$$f''(t) = \frac{54(3 - t^2)}{(t^2 + 9)^3}$$

f is concave down on $(-\infty, -\sqrt{3}]$ and $[\sqrt{3}, \infty)$ and concave up on $[-\sqrt{3}, \sqrt{3}]$. f has points of inflection at $\left(\pm\sqrt{3}, \frac{1}{4}\right)$.

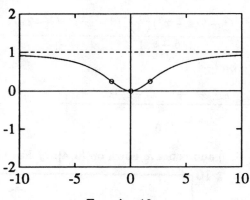

Exercise 13

15. $f(x) = x^4 - 2x^2$

1) $f(x) = x^2(x^2 - 2) = 0$ when $x = 0, \pm\sqrt{2}$.

2) $f'(x) = 4x^3 - 4x = 4x(x^2 - 1) = 0$ when $x = 0, \pm 1$.

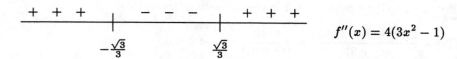

$$f'(x) = 4x(x^2 - 1)$$

$f(x)$ is decreasing on $(-\infty, -1]$ and $[0, 1]$ and increasing on $[-1, 0]$ and $[1, \infty)$. f has a relative maximum at $(0, 0)$ and absolute minima at $(\pm 1, -1)$.

3) $f''(x) = 12x^2 - 4 = 4(3x^2 - 1) = 0$ at $x = \pm\frac{\sqrt{3}}{3}$.

$$f''(x) = 4(3x^2 - 1)$$

f is concave up on $\left(-\infty, -\frac{\sqrt{3}}{3}\right]$ and on $\left[\frac{\sqrt{3}}{3}, \infty\right)$ and concave down on $\left[-\frac{\sqrt{3}}{3}, \frac{\sqrt{3}}{3}\right]$. f has inflection points at $\left(\pm\frac{\sqrt{3}}{3}, -\frac{5}{9}\right)$.

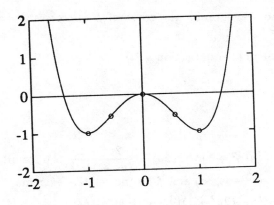

Exercise 15

17. $f(x) = x - \cos x$

This function is defined on the whole real line and has neither horizontal nor vertical asymptotes.

1) $f'(x) = 1 + \sin x$

$f'(x) = 0 \Rightarrow \sin x = -1 \Rightarrow x = (4n+3)\frac{\pi}{2}$ (any integer n) = $\cdots\cdots, -\frac{\pi}{2}, \frac{3\pi}{2}, \frac{7\pi}{2}, \cdots\cdots$.

$f'(x) > 0$ for all other values of x.

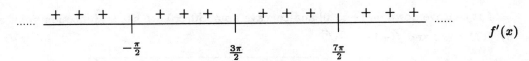

So f is increasing on $(-\infty, \infty)$ with critical numbers $x = (4n+3)\frac{\pi}{2}$ and no relative extrema.

2) $f''(x) = \cos x$

This being a familiar function, we see:

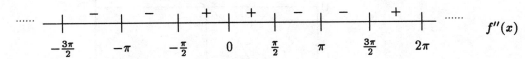

So f is concave up on the intervals $\left((4n-1)\frac{\pi}{2}, (4n+1)\frac{\pi}{2}\right)$ and concave down on the intervals $\left((4n+1)\frac{\pi}{2}, (4n+3)\frac{\pi}{2}\right)$ and has inflection points where x is any odd multiple of $\frac{\pi}{2}$.

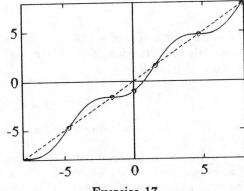

Exercise 17

171

(Dashed line helps to draw graph: the graph of $f(x)$ crosses $y = x$ when x is an odd multiple of $\frac{\pi}{2}$ since then $\cos x = 0$).

19. $f(x) = x^2 + \frac{2}{x}$. (See problem (17) in Section 4.7.)

21. $f(u) = u\sqrt{u^2 + 1}$

   1) Domain: $(-\infty, \infty)$

   2) $f(u) = 0$ when $u = 0$.

   3) $f'(u) = \sqrt{u^2 + 1} + u(\frac{1}{2})(u^2 + 1)^{-1/2}(2u) = \dfrac{2u^2 + 1}{(u^2 + 1)^{1/2}} > 0$ for all x.

   Thus f is increasing on $(-\infty, \infty)$ and f has no relative extrema.

   4) $f''(u) = \dfrac{4u(u^2 + 1)^{1/2} - (2u^2 + 1)(\frac{1}{2})(u^2 + 1)^{-1/2}(2u)}{u^2 + 1} = \dfrac{u[4(u^2 + 1) - (2u^2 + 1)]}{(u^2 + 1)^{3/2}}$

   $= \dfrac{u(2u^2 + 3)}{(u^2 + 1)^{3/2}} = 0$ at $u = 0$.

$$- \quad - \quad - \quad | \quad + \quad + \quad +$$
$$0$$

   $f''(u) = \dfrac{u(2u^2 + 3)}{(u^2 + 1)^{3/2}}$

   f is concave down on $(-\infty, 0]$ and concave up on $[0, \infty)$. f has an inflection point at $(0, 0)$.

   5) $f(1) = \sqrt{2}$ and $f(-1) = -\sqrt{2}$.

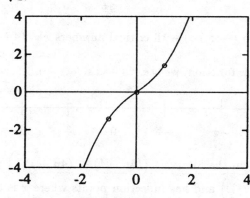

Exercise 21

23. $f(x) = \cos x + \sin^2 x$

   This can be treated along the same lines as Exercise Set 4.7 number 33 $\cdots\cdots$ or reduced to this very problem by two simple trigonometric substitutions. From $\cos\left(x - \frac{\pi}{2}\right) = \cos x \cos \frac{\pi}{2} + \sin x \sin \frac{\pi}{2} = \sin x$ and similarly $\sin\left(x - \frac{\pi}{2}\right) = -\cos x$, it follows that

$$f\left(x - \frac{\pi}{2}\right) = \sin x + (-\cos x)^2$$
$$= \sin x + \cos^2 x$$

   which is the very function of number 33 in Exercise Set 4.7. This means that in order to obtain the graph of $f(x) = \cos x + \sin^2 x$, all we need do is shift the graph in 4.7 number 33 by $\frac{\pi}{2}$ units to the left. Notice that the resulting graph is symmetric about the vertical line $x = \pi$. If we had drawn it

on all of $(-\infty, \infty)$ we would have **seen that the graph is also** symmetric about the y-axis (after all, $f(x) = \cos x + \sin^2 x$ is an even function).

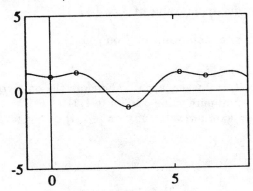

Exercise 23

By shifting the solution to 4.7 number **33 a total of $\frac{\pi}{2}$** radians (or 90°) to the left, we find:

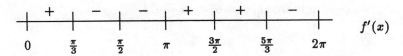

So: f increases on $\left[0, \frac{\pi}{3}\right]$, $\left[\pi, \frac{5\pi}{3}\right]$ and decreases on $\left[\frac{\pi}{3}, \pi\right]$, $\left[\frac{5\pi}{3}, 2\pi\right]$ and has relative extrema at

$$(0, 1) \quad \longleftarrow \quad \text{relative minimum}$$
$$\left(\frac{\pi}{3}, \frac{5}{4}\right) \quad \longleftarrow \quad \text{absolute maximum}$$
$$(\pi, -1) \quad \longleftarrow \quad \text{absolute minimum}$$
$$\left(\frac{5\pi}{3}, \frac{5}{4}\right) \quad \longleftarrow \quad \text{absolute maximum}$$
$$(2\pi, 1) \quad \longleftarrow \quad \text{relative minimum}$$

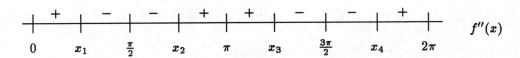

Here (approximately, in degrees):

$$x_1 \approx 32\frac{1}{2}^\circ, \quad x_2 \approx 126\frac{1}{3}^\circ, \quad x_3 \approx 233\frac{2}{3}^\circ, \quad x_4 \approx 302\frac{1}{2}^\circ$$

So f is concave up on $[0, x_1]$, $[x_2, x_3]$, $[x_4, 2\pi]$ and concave down on $[x_1, x_2]$, $[x_3, x_4]$, with inflection points at x_1, x_2, x_3, x_4 found above.

25. $f(x) = \dfrac{1}{x^2 - 4}$, $x \in [-1, 1]$

$f'(x) = \dfrac{-2x}{(x^2 - 4)^2} = 0$ at $x = 0$. Therefore the only critical number of f on $[-1, 1]$ is 0.

173

$$f(-1) \;=\; -\tfrac{1}{3} \;\longleftarrow\; \text{minimum value of } f \text{ on } [-1,1]$$

$$f(0) \;=\; -\tfrac{1}{4} \;\longleftarrow\; \text{maximum value of } f \text{ on } [-1,1]$$

$$f(1) \;=\; -\tfrac{1}{3} \;\longleftarrow\; \text{minimum value of } f \text{ on } [-1,1]$$

27. $y = x^3 - 3x^2 + 1,\; x \in [-1,1]$
$y' = 3x^2 - 6x = 3x(x-2) = 0$ when $x = 0$. Therefore the only critical number of f on $[-1,1]$ is 0.
$$y(-1) \;=\; -3 \;\longleftarrow\; \text{minimum value of } y \text{ on } [-1,1]$$
$$y(0) \;=\; 1 \;\longleftarrow\; \text{maximum value of } y \text{ on } [-1,1]$$
$$y(1) \;=\; -1$$

29. $f(x) = x + x^{2/3},\; x \in [-1,1]$

$$f'(x) = 1 + \frac{2}{3x^{1/3}} = \frac{3x^{1/3}+2}{3x^{1/3}} = 0 \text{ at } x = -\tfrac{8}{27}.$$

$f'(x)$ is undefined when $x = 0$. Therefore the critical numbers of f are 0 and $-\tfrac{8}{27}$.

$$f(-1) \;=\; 0 \;\longleftarrow\; \text{minimum value of } f \text{ on } [-1,1]$$

$$f(-\tfrac{8}{27}) \;=\; \tfrac{4}{27}$$

$$f(0) \;=\; 0 \;\longleftarrow\; \text{minimum value of } f \text{ on } [-1,1]$$

$$f(1) \;=\; 2 \;\longleftarrow\; \text{maximum value of } f \text{ on } [-1,1]$$

31. $y = x - \sqrt{1-x^2},\; x \in [-1,1]$
$$y' = 1 - \tfrac{1}{2}(1-x^2)^{1/2}(-2x) = \frac{(1-x^2)^{1/2}+x}{(1-x^2)^{1/2}} = 0 \text{ when } x = -\tfrac{\sqrt{2}}{2}.$$

$$y(-1) \;=\; -1$$

$$y\!\left(-\tfrac{\sqrt{2}}{2}\right) \;=\; -\sqrt{2} \;\longleftarrow\; \text{minimum value of } y \text{ on } [-1,1]$$

$$y(1) \;=\; 1 \;\longleftarrow\; \text{maximum value of } y \text{ on } [-1,1]$$

33. $f(x) = x - 4x^{-2},\; x \in [-3,-1]$

$$f'(x) = 1 + 8x^{-3} = \frac{x^3+8}{x^3} = 0 \text{ when } x = -2.$$
$$f(-3) \;=\; -\tfrac{31}{9}$$

$$f(-2) \;=\; -3 \;\longleftarrow\; \text{maximum value of } f \text{ on } [-3,-1]$$

$$f(-1) \;=\; -5 \;\longleftarrow\; \text{minimum value of } f \text{ on } [-3,-1]$$

35. $y = x^4 - 2x^2,\; x \in [-2,2]$
$y' = 4x^3 - 4x = 4x(x^2-1) = 0$ when $x = 0, \pm 1$.
$$y(-2) \;=\; 8 \;\longleftarrow\; \text{maximum value of } y \text{ on } [-2,2]$$
$$y(-1) \;=\; -1 \;\longleftarrow\; \text{minimum value of } y \text{ on } [-2,2]$$
$$y(0) \;=\; 0$$
$$y(1) \;=\; -1 \;\longleftarrow\; \text{minimum value of } y \text{ on } [-2,2]$$
$$y(2) \;=\; 8 \;\longleftarrow\; \text{maximum value of } f \text{ on } [-2,2]$$

37. $f(x) = \dfrac{2 + \cos x}{2 - \cos x}$, $x \in [-\pi, \pi]$

$f'(x) = \dfrac{-\sin x(2 - \cos x) - (2 + \cos x)(\sin x)}{(2 - \cos x)^2} = \dfrac{-4\sin x}{(2 - \cos x)^2} = 0$ when $-4\sin x = 0 \Rightarrow x = 0$.

Therefore the only critical number of f on $[-\pi, \pi]$ is 0.

$\begin{aligned} f(-\pi) &= \tfrac{1}{3} &\longleftarrow \text{ minimum value of } f \text{ on } [-\pi, \pi] \\ f(0) &= 3 &\longleftarrow \text{ maximum value of } f \text{ on } [-\pi, \pi] \\ f(\pi) &= \tfrac{1}{3} &\longleftarrow \text{ maximum value of } f \text{ on } [-\pi, \pi] \end{aligned}$

39. $y = \dfrac{x + 2}{\sqrt{x^2 + 1}}$, $x \in [0, 2]$

$y' = \dfrac{(x^2 + 1)^{1/2} - (x + 2)(\frac{1}{2})(x^2 + 1)^{-1/2}(2x)}{x^2 + 1} = \dfrac{1 - 2x}{(x^2 + 1)^{3/2}} = 0$ when $x = \tfrac{1}{2}$.

$\begin{aligned} y(0) &= 2 \\ y\left(\tfrac{1}{2}\right) &= \sqrt{5} &\longleftarrow \text{ maximum value of } y \text{ on } [0, 2] \\ y(2) &= \tfrac{4\sqrt{5}}{5} &\longleftarrow \text{ minimum value of } y \text{ on } [0, 2] \end{aligned}$

41. $f(x) = |x^3 + x^2 - x|$, $x \in \left[-\tfrac{3}{2}, 1\right]$

$f(x) = \begin{cases} x^3 + x^2 - x & \text{if } -\dfrac{3}{2} \le x \le 0 \text{ and } \dfrac{-1 + \sqrt{5}}{2} \le x \le 1 \\ -(x^3 + x^2 - x) & \text{if } 0 \le x \le \dfrac{-1 + \sqrt{5}}{2} \end{cases}$

$f'(x) = \begin{cases} 3x^2 + 2x - 1 & \text{if } -\dfrac{3}{2} < x < 0 \text{ and } \dfrac{-1 + \sqrt{5}}{2} < x < 1 \\ -(3x^2 + 2x - 1) & \text{if } 0 < x < \dfrac{-1 + \sqrt{5}}{2} \\ \text{does not exist} & \text{if } x = -\dfrac{3}{2}, \ 0, \ \dfrac{-1 + \sqrt{5}}{2}, \ 1 \end{cases}$

$f'(x) = 0$ when $x = -1, \tfrac{1}{3}$.

$\begin{aligned} f\left(-\tfrac{3}{2}\right) &= \tfrac{3}{8} \\ f(-1) &= 1 &\longleftarrow \text{ maximum value of } f \text{ on } \left[-\tfrac{3}{2}, 1\right] \\ f(0) &= 0 &\longleftarrow \text{ minimum value of } f \text{ on } \left[-\tfrac{3}{2}, 1\right] \\ f\left(\tfrac{1}{3}\right) &= \tfrac{5}{27} \\ f\left(\dfrac{-1 + \sqrt{5}}{2}\right) &= 0 &\longleftarrow \text{ minimum value of } f \text{ on } \left[-\tfrac{3}{2}, 1\right] \\ f(1) &= 1 &\longleftarrow \text{ maximum value of } f \text{ on } \left[-\tfrac{3}{2}, 1\right] \end{aligned}$

43. $\displaystyle\lim_{x \to -\infty} \frac{3x^2 + x + 1}{x^2 + 1} = \lim_{x \to -\infty} \frac{3 + \frac{1}{x} + \frac{1}{x^2}}{1 + \frac{1}{x^2}} = 3$

45. $\displaystyle\lim_{x \to -1^+} \frac{x^2 + x + 4}{x^2 - 2x - 3} = \lim_{x \to -1^+} \frac{x^2 + x + 4}{(x - 3)(x + 1)} = -\infty$

47. $\lim_{x \to 7} \dfrac{x-8}{x^2-8x+7}$ = does not exist since $\lim_{x \to 7+} \dfrac{x-8}{x^2-8x+7} = \lim_{x \to 7+} \dfrac{x-8}{(x-7)(x-1)} = -\infty$

and $\lim_{x \to 7-} \dfrac{x-8}{x^2-8x+7} = \infty$.

49. $\lim_{x \to 7} \dfrac{x-8}{x^2-14x+49} = \lim_{x \to 7} \dfrac{x-8}{(x-7)^2} = -\infty$

51. $\lim_{x \to \infty} \cos\left(\dfrac{1}{x}\right) = \cos 0 = 1$

53. $\lim_{\theta \to \frac{\pi}{2}-} \dfrac{\sec \theta}{\tan \theta} = \lim_{\theta \to \frac{\pi}{2}-} \dfrac{1}{\sin \theta} = 1$

55.

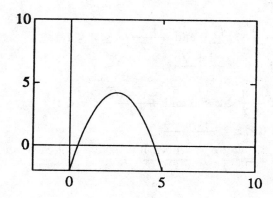

Exercise 55

57.

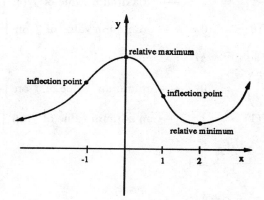

Exercise 57

176

59.

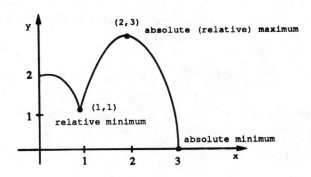

Exercise 59

61.

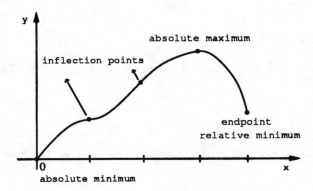

Exercise 61

63. Let the foot of the ladder be a distance x (in meters) from the foot of the fence, so (Pythagoras) the distance along the ladder from its foot to the top of the fence is $\sqrt{x^2 + 9}$. An application of similar triangles shows that the length L of the ladder is given by

$$\frac{L}{\sqrt{x^2 + 9}} = \frac{x + 1}{x}$$

therefore
$$L = \left(1 + \frac{1}{x}\right)\sqrt{x^2 + 9}$$

therefore
$$\frac{dL}{dx} = \frac{x^3 - 9}{x^2\sqrt{x^2 + 9}}$$

which is zero if $x = \sqrt[3]{9}$, negative if $x < \sqrt[3]{9}$ and positive if $x > \sqrt[3]{9}$. The shortest ladder for the job therefore has length $(1 + 9^{-1/3})(9^{2/3} + 9)^{1/2}$ meters.

65. Let s = number of subscriptions sold per year and let r = yearly subscription rate. The marketing department predicts $\frac{ds}{dr} = -50$, so the (linear!) equation relating r and s is

$$s - 2000 = (-50)(r - 40)$$

therefore

$$s = 4000 - 50r.$$

Now, total revenue R is

$$R \;=\; rs \;=\; r(4000 - 50r)$$

therefore

$$\frac{dR}{dr} \;=\; 4000 - 100r$$

which is zero if $r = 40$, < 0 if $r > 40$ and > 0 if $r < 40$. So: the subscription rate to maximize revenue is \$40 per year.

67. $$f(t) \;=\; t^3 - 6t^2 + 14t + 5$$
$$\Rightarrow \quad f'(t) \;=\; 3t^2 - 12t + 14$$
$$=\; 3\left(t^2 - 4t + \frac{14}{3}\right)$$
$$=\; 3\left((t-2)^2 + \frac{2}{3}\right) \;\geq\; 2 \text{ for all } t.$$

Since $f'(t) > 0$ for all real t, it follows that f is increasing on $(-\infty, \infty)$ by Theorem 6.

69. $$f(x) \;=\; \frac{x}{1 + ax^2}$$
$$\Rightarrow \quad f'(x) \;=\; \frac{1 - ax^2}{(1 + ax^2)^2}.$$

Now, since f is to assume its minima at the interior points $x = \pm 3$, we must have $f'(\pm 3) = 0$; thus $9a = 1$ and so $a = \frac{1}{9}$.

71. $f(x) = x^n + ax + b \Rightarrow f'(x) = n\,x^{n-1} + a$. If n is even then f' is increasing (look at the sign of f''). In particular, $f'(x) = 0$ for just one value of x. Now, if $p < q < r$ are three zeros of f then (by Rolle's Theorem) there exist $s \in (p, q)$ and $t \in (q, r)$ such that $f'(s) = 0$ and $f'(t) = 0$: this contradicts the finding of the previous sentence, and shows that f has at most two zeros. If n is odd then f can have three zeros [for example, look at $f(x) = x^3 - x = (x+1)\,x\,(x-1)$].

73. The surface area of the pool is $S = \frac{1}{2}r^2\theta$ (constant) and its perimeter P is given by $P = 2r + r\theta$. Solve $S = \frac{1}{2}r^2\theta$ for $\theta = \frac{2S}{r^2}$ and substitute into P to get

$$P \;=\; 2r + \frac{2S}{r}$$

therefore

$$\frac{dP}{dr} \;=\; 2 - \frac{2S}{r^2}$$

which is 0 if $r = \sqrt{S}$, < 0 if $r < \sqrt{S}$ and > 0 if $r > \sqrt{S}$. Thus the perimeter P is minimized when $r = \sqrt{S}$ and $\theta = 2$ (radians).

75. Let $S = D^2$ where D is the distance from $(0, 0)$ to the point (x, y) on the hyperbola $x^2 - x + \frac{5}{4} - y^2 = 0$, so that

$$S \;=\; x^2 + y^2 \;=\; 2x^2 - x + \frac{5}{4}.$$

Then $\frac{dS}{dx} = 4x - 1$, which is zero if $x = \frac{1}{4}$, negative if $x < \frac{1}{4}$ and positive if $x > \frac{1}{4}$; so S is a minimum when $x = \frac{1}{4}$ and so (Remark 3 of Section 4.4) D is a minimum when $x = \frac{1}{4}$. From $y^2 = x^2 - x + \frac{5}{4}$ we deduce that $y = \pm\frac{\sqrt{17}}{4}$. Thus: the sought for points are $\left(\frac{1}{4}, \pm\frac{\sqrt{17}}{4}\right)$.

77. Let the log have diameter D (given as fixed) and let the beam have width w and depth d, so (Pythagoras) $D^2 = w^2 + d^2$. The strength S of the beam is given by $S = kwd^2$ where k is some constant of proportionality; by substitution, it follows that

$$S = kw(D^2 - w^2)$$

therefore

$$\frac{dS}{dw} = kD^2 - 3kw^2$$

which is zero when $w = \frac{D}{\sqrt{3}}$. An examination of the sign of $\frac{dS}{dw}$ on either side of $w = \frac{D}{\sqrt{3}}$ shows that $w = \frac{D}{\sqrt{3}}$ maximizes S. Now, $w = \frac{D}{\sqrt{3}} \Rightarrow d = \sqrt{D^2 - w^2} = D\sqrt{2/3}$ so that the optimal ratio of d to w is $\frac{d}{w} = \sqrt{2}$.

79. Let the cylindrical portion of the tank have length l and radius r. If the cost of the material for this portion is k per unit area then the total construction cost C is the sum of $2\pi rlk$ (for cylinder) and $(4\pi r^2)(3k)$ (for caps) so

$$C = 2\pi rlk + 12\pi r^2 k.$$

The total volume of the tank is fixed at

$$V = \pi r^2 h + \frac{4}{3}\pi r^3$$

so

$$l = \frac{V - \frac{4}{3}\pi r^3}{\pi r^2}.$$

Thus

$$C = 2k\left[\frac{V}{r} + \frac{14}{3}\pi r^2\right]$$

so

$$\frac{dC}{dr} = 2k\left[\frac{-V}{r^2} + \frac{28}{3}\pi r\right]$$

which is zero when $r = \sqrt[3]{\frac{3V}{28\pi}}$. Checking the sign of $\frac{dC}{dr}$ for r less than or greater than this value shows that cost C is minimized when

$$r = \sqrt[3]{\frac{3V}{28\pi}} \quad \text{and} \quad l = 8r.$$

81. Given $x_1, \cdots\cdots, x_n$ let us write

$$S = (\overline{x} - x_1)^2 + \cdots\cdots + (\overline{x} - x_n)^2.$$

Then

$$\frac{dS}{d\overline{x}} = 2(\overline{x} - x_1) + \cdots\cdots + 2(\overline{x} - x_n)$$

which is zero when

$$\overline{x} = \frac{x_1 + \cdots\cdots + x_n}{n};$$

checking the sign of $\frac{dS}{d\overline{x}}$ as usual shows that S is minimized.

83. Let the box have width w, length $2w$, height h (in centimeters). Its volume is then

$$1024 = w \times 2w \times h$$

so

$$h = \frac{512}{w^2}.$$

Now, the total cost of construction (in cents) is

$$
\begin{aligned}
C &= \underset{\text{(top)}}{12 \times 2w^2} + \underset{\text{(bottom)}}{6 \times 2w^2} + \underset{\text{(sides)}}{6 \times 6wh} \\
&= 36(w^2 + wh) \\
&= 36\left(w^2 + \frac{512}{w}\right).
\end{aligned}
$$

Thus

$$\frac{dC}{dw} = 36\left(2w - \frac{512}{w^2}\right)$$

which is zero when $w^3 = 256$ so $w = 4\sqrt[3]{4}$. If $w > 4\sqrt[3]{4}$ then $\frac{dC}{dw} > 0$ whilst if $0 < w < 4\sqrt[3]{4}$ then $\frac{dC}{dw} < 0$. Thus, cost C is least when $w = 4\sqrt[3]{4}$ and $h = 2\sqrt[3]{256}$.

85.
$$
\begin{aligned}
f(x) &= x^3 + qx^2 + 5x - 6 \\
\Rightarrow \quad f'(x) &= 3x^2 + 2qx + 5.
\end{aligned}
$$

With $c = 2$, $a = 0$ and $b = 2$, the Mean Value Theorem applied to f yields

$$
\begin{aligned}
f(2) - f(0) &= 2f'(2) \\
\text{or} \qquad (12 + 4q) - (-6) &= 2(17 + 4q) \\
\text{or} \qquad q &= -4.
\end{aligned}
$$

87. Let x be the number of times per year that parts are ordered, and we assume that all orders are equal in size. Then the cost function (in dollars per year) is

$$
\begin{aligned}
C(x) &= (.05)\,3000 + 20x + \left(\frac{.48}{x}\right)\left(\frac{1500}{x}\right)x \\
&= 150 + 20x + \frac{720}{x}.
\end{aligned}
$$

Thus

$$C'(x) = 20 - \frac{720}{x^2}.$$

To find the critical numbers, we solve $C'(x) = 0$. That is,

$$
\begin{aligned}
\frac{720}{x^2} &= 20 \\
36 &= x^2,
\end{aligned}
$$

and $x = 6$ is the critical number in the interval $[1, \infty)$. Note that $x = 6$ corresponds to a minimum because

$$C''(x) = \frac{1440}{x^3},$$

which implies that the graph of C is concave up on $[1, \infty)$.

89. This problem is similar to Exercise 87. Let x be the number of times per year that shirts are ordered. Then the cost function is

$$
\begin{aligned}
C(x) &= (.20)\,2000 + 50x + \left(\frac{3}{x}\right)\left(\frac{1000}{x}\right)x \\
&= 400 + 50x + \frac{3000}{x}.
\end{aligned}
$$

Thus

$$
C'(x) = 50 - \frac{3000}{x^2},
$$

and the critical numbers are obtained by solving $C'(x) = 0$. We obtain

$$
x^2 = 60,
$$

which yields the critical number $\sqrt{60}$. If we want an integral number for x, we use the fact that $\sqrt{60}$ lies between 7 and 8, and we calculate

$$
\begin{aligned}
C(7) &= 1178.59 \\
C(8) &= 1175.
\end{aligned}
$$

Note that $C''(x) > 0$ on $[1, \infty)$, so the graph of C is concave upward on $[1, \infty)$. So shirts should be ordered 8 times per year.

Chapter 5

Antidifferentiation

5.1 Antiderivatives

In Exercises 1–18, find the indefinite integral.

1. $\displaystyle \int \left(2x^2 + 1\right) dx = \frac{2x^3}{3} + x + C$

3. $\displaystyle \int \left(x^{2/3} + x^{5/2}\right) dx = \frac{3}{5}\,x^{5/3} + \frac{2}{7}\,x^{7/2} + C$

5. $\displaystyle \int \left(2\cos t - \sin t\right) dt = 2\sin t + \cos t + C$

7. $\displaystyle \int \left(\sin x + \sec^2 x\right) dx = \int \sin x\, dx + \int \sec^2 x\, dx = -\cos x + \tan x + C$

9. $\displaystyle \int \frac{t^2 + 1}{\sqrt{t}}\, dt = \int \left(t^2 + 1\right) \cdot t^{-1/2}\, dt = \int t^{3/2} + t^{-1/2}\, dt = \frac{2}{5}\,t^{5/2} + 2t^{1/2} + C$

11. $\displaystyle \int (t^2 + 1)(t + 2)\, dt = \int \left(t^3 + t + 2t^2 + 2\right) dt = \frac{t^4}{4} + \frac{2t^3}{3} + \frac{t^2}{2} + 2t + C$

13. $\displaystyle \int \frac{1}{\sqrt[3]{x}}\, dx = \int x^{-1/3}\, dx = \frac{3}{2}\,x^{2/3} + C$

15. $\displaystyle \int \left(\tan^2 x + 1\right) dx = \int \left((\sec^2 x - 1) + 1\right) dx = \int \sec^2 x\, dx = \tan x + C$

17. $\displaystyle \int \left(3x^2 - \sin x + 2\sec^2 x\right) dx = 3\int x^2\, dx - \int \sin x\, dx + 2\int \sec^2 x\, dx$

$\displaystyle \qquad\qquad = 3\,\frac{x^3}{3} - -\cos x + 2\tan x + C = x^3 + \cos x + 2\tan x + C$

In Exercises 19–26, find the position function s corresponding to the given velocity function and initial condition.

19. $v(t) = \cos t,\ s(0) = 2$

$$s(t) = \int \cos t \, dt,$$
$$s(t) = \sin t + C$$

$$s(0) = \sin(0) + C = 2$$
$$0 + C = 2$$
$$C = 2$$
$$s(t) = \sin t + 2$$

21. $v(t) = 2t^2,\ s(0) = 4$

$$s(t) = \int 2t^2 \, dt,$$
$$s(t) = \frac{2t^3}{3} + C$$

$$s(0) = 0 + C = 4$$
$$C = 4$$
$$s(t) = \frac{2t^3}{3} + 4$$

23. $v(t) = \sqrt{t}(t + 4),\ s(0) = 0$

$$s(t) = \int \sqrt{t}(t + 4) \, dt$$
$$= \int t^{1/2}(t + 4) \, dt$$
$$= \int \left(t^{3/2} + 4t^{1/2}\right) dt$$
$$s(t) = \frac{2}{5} t^{5/2} + \frac{8}{3} t^{3/2} + C$$

$$s(0) = 0 + C = 0$$
$$C = 0$$
$$s(t) = \frac{2}{5} t^{5/2} + \frac{8}{3} t^{3/2}$$

25. $v(t) = \left(\sqrt{t} + 4\right)\left(\sqrt{t} - 4\right),\ s(2) = 3$

$$s(t) = \int \left(\sqrt{t} + 4\right)\left(\sqrt{t} - 4\right) dt$$
$$= \int (t - 16) \, dt$$
$$s(t) = \frac{t^2}{2} - 16t + C$$

$$s(2) = \frac{4}{2} - 16 \cdot 2 + C = 3$$
$$2 - 32 + C = 3$$
$$-30 + C = 3$$
$$C = 33$$
$$s(t) = \frac{t^2}{2} - 16t + 33$$

In Exercises 27–37, find the velocity function v and position function s corresponding to the given acceleration function and initial conditions.

27. $a(t) = 2,\ v(0) = 3,\ s(0) = 0$

$$v(t) = \int a(t) \, dt = \int 2 \, dt = 2t + C_1$$
$$v(t) = 2t + C_1$$
$$v(0) = 0 + C_1 = 3,\ C_1 = 3$$
$$v(t) = 2t + 3$$

$$s(t) = \int v(t) \, dt$$
$$= \int (2t + 3) \, dt = t^2 + 3t + C_2$$
$$s(t) = t^2 + 3t + C_2$$
$$s(0) = 0 + C_2 = 0$$
$$C_2 = 0$$
$$s(t) = t^2 + 3t$$

29. $a(t) = 3t,\ v(0) = 0,\ s(0) = 20$

$$v(t) = \int 3t\, dt$$

$$v(t) = \frac{3t^2}{2} + C_1$$

$$v(0) = 0 + C_1 = 0$$

$$C_1 = 0$$

$$v(t) = \frac{3t^2}{2}$$

$$s(t) = \int \frac{3t^2}{2}\, dt$$

$$s(t) = \frac{3t^3}{6} + C_2$$

$$s(t) = \frac{t^3}{2} + C_2$$

$$s(0) = 0 + C_2 = 20$$

$$C_2 = 20$$

$$s(t) = \frac{t^3}{2} + 20$$

31. $a(t) = 4t + 4$, $v(0) = 8$, $s(0) = 12$

$$v(t) = \int (4t + 4)\, dt$$

$$v(t) = 2t^2 + 4t + C_1$$

$$v(0) = 0 + 0 + C_1 = 8$$

$$C_1 = 8$$

$$v(t) = 2t^2 + 4t + 8$$

$$s(t) = \int (2t^2 + 4t + 8)\, dt$$

$$s(t) = \frac{2t^3}{3} + \frac{4t^2}{2} + 8t + C_2$$

$$s(t) = \frac{2t^3}{3} + 2t^2 + 8t + C_2$$

$$s(0) = 0 + 0 + 0 + C_2 = 12$$

$$C_2 = 12$$

$$s(t) = \frac{2t^3}{3} + 2t^2 + 8t + 12$$

33. $a(t) = \sin t$, $v(0) = 0$, $s(0) = 2$

$$v(t) = \int \sin t\, dt$$

$$v(t) = -\cos t + C_1$$

$$v(0) = -\cos(0) + C_1 = 0$$

$$-1 + C_1 = 0$$

$$C_1 = 1$$

$$v(t) = -\cos t + 1$$

$$s(t) = \int (-\cos t + 1)\, dt$$

$$s(t) = -\sin t + t + C_2$$

$$s(0) = -\sin(0) + 0 + C_2 = 2$$

$$0 + 0 + C_2 = 2$$

$$C_2 = 2$$

$$s(t) = -\sin t + t + 2$$

35. $a(t) = -2\sin t$, $v(\pi) = -3$, $s(\pi) = -2\pi$

$$v(t) = \int -2\sin t\, dt$$

$$v(t) = 2\cos t + C_1$$

$$v(\pi) = 2\cos(\pi) + C_1 = -3$$

$$2(-1) + C_1 = -3$$

$$-2 + C_1 = -3$$

$$C_1 = -1$$

$$v(t) = 2\cos t - 1$$

$$s(t) = \int (2\cos t - 1)\, dt$$

$$s(t) = 2\sin t - t + C_2$$

$$s(\pi) = 2\sin \pi - \pi + C_2 = -2\pi$$

$$0 - \pi + C_2 = -2\pi$$

$$C_2 = -\pi$$

37. $a(t) = 0$, $v(1) = 1$, $s(2) = 3$

$$v(t) = \int 0\, dt$$

$$v(t) = C_1$$

$$v(1) = C_1 = 1$$

$$C_1 = 1$$

$$v(t) = 1$$

$$s(t) = \int 1\, dt$$

$$s(t) = t + C_2$$

$$s(2) = 2 + C_2 = 3$$

$$C_2 = 1$$

$$s(t) = t + 1$$

39. $v(0) = 15 \, m/s$

$$a = -9.8 \, m/s^2$$
$$v(t) = \int -9.8 \, dt$$
$$v(t) = -9.8t + C_1$$
$$v(0) = 0 + C_1 = 15$$

$$C_1 = 15$$
$$v(t) = -9.8t + 15$$
velocity $= 0$ when $-9.8t + 15 = 0$
$$15 = 9.8t$$
$$\frac{15}{9.8} = t$$

(a) Stops rising at $t = \frac{15}{9.8} \approx 1.53$ secs.

(b) Strikes the ground at $t = 2 \cdot \frac{15}{9.8} \approx 3.06$ secs.

41. A particle moves along a line with constant acceleration $a = 3 m/s^2$.

(a) How fast is it moving after $6 \, s$ if its initial velocity is $v(0) = 10 m/s$?

$$v(t) = \int 3 \, dt$$
$$= 3t + C_1$$
$$v(0) = 0 + C_1 = 10$$
$$C_1 = 10$$

$$v(t) = 3t + 10 \, m/s$$
$$v(6) = 3.6 + 10$$
$$= 28 \, m/s$$

(b) What is its initial velocity $v(0)$ if its speed after $3 \, s$ is $15 m/s$?

$$v(3) = 3 \cdot 3 + C_1 = 15$$
$$9 + C_1 = 15$$
$$C_1 = 6$$
$$v(t) = 3t + 6 \, m/s$$
Find $v(0)$,
$$v(0) = 3 \cdot 0 + 6 = 6 \, m/s$$

43. What constant acceleration will enable the driver of an automobile to increase its speed from $20 \, m/s$ to $25 \, m/s$ in $10 \, s$? Find $a(t) = K$.

$$\left. \begin{array}{l} \text{if } v(0) = 20 \, m/s \\ v(10) = 25 \, m/s \end{array} \right\} \quad \begin{array}{l} v(t) = \int K \, dt \\ v(t) = Kt + C_1 \end{array}$$

$$v(0) = 0 + C_1 = 20$$
$$C_1 = 20$$
$$v(t) = Kt + 20$$

$$v(10) = 10K + 20 = 25$$
$$10K = 5$$
$$K = \frac{1}{2}$$

$$\text{Therefore } a(t) = \frac{1}{2} \, m/s^2.$$

45. Figures a–d illustrate the graphs $y = f(x)$ of four particular functions, and Figures i–iv represent the slope portraits for their antiderivatives F. Match each function f with the slope portrait of its

185

antiderivative F. Then, on each of the four slope portraits, illustrate the antidifferentiation process by sketching the graphs $y = F(x)$ of three different antiderivatives.

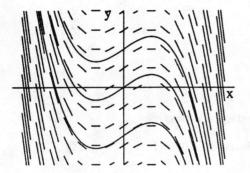

Exercise 45(a)
Matches ii

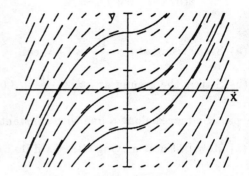

Exercise 45(b)
Matches i

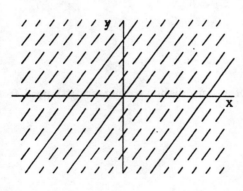

Exercise 45(c)
Matches iv

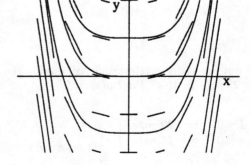

Exercise 45(d)
Matches iii

In each of Exercises 46–49, sketch the slope portrait for the antiderivatives F of the given function f. Then, on each portrait, sketch the graphs $y = F(x)$ of three antiderivatives of f. Finally, using the formula for $f(x)$ calculate the indefinite integral of f, and reconcile the graphs $y = F(x)$ with the integral.

47. $f(x) = 1 - x$

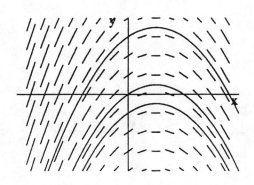

m	
0	$1 - x = 0, \; x = 1$
1	$1 - x = 1, \; x = 0$
2	$1 - x = 2, \; x = -1$
-1	$1 - x = -1, \; x = 2$

$$\int (1 - x)\, dx = x - \frac{x^2}{2} + C$$

Exercise 47

49. $f(x) = \sin x$

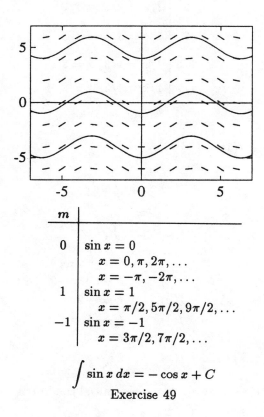

m	
0	$\sin x = 0$
	$x = 0, \pi, 2\pi, \dots$
	$x = -\pi, -2\pi, \dots$
1	$\sin x = 1$
	$x = \pi/2, 5\pi/2, 9\pi/2, \dots$
-1	$\sin x = -1$
	$x = 3\pi/2, 7\pi/2, \dots$

$$\int \sin x\, dx = -\cos x + C$$

Exercise 49

51. Let $f(x) = 1/x$. From the sketch, the $\lim\limits_{x \to 0} \dfrac{1}{x}$ appears to be $-\infty$.

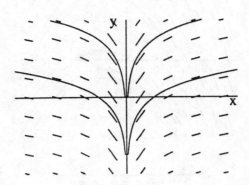

m	
0	$1/x = 0$ undefined
1	$1/x = 1$, $x = 1$
2	$1/x = 2$, $x = 1/2$
3	$1/x = 3$, $x = 1/3$
1/3	$1/x = 1/3$, $x = 3$
1/2	$1/x = 1/2$, $x = 2$

Exercise 51

53. $f''(x) = 0 \qquad f(0) = -2 \qquad f(1) = 1$

$$
\begin{aligned}
f'(x) &= \int 0\,dx \\
&= C_1 \\
f(x) &= \int C_1\,dx \\
&= C_1 x + C_2 \\
\text{Now } f(0) &= 0 + C_2 = -2, \; C_2 = -2 \\
f(x) &= C_1 x - 2 \\
\text{and } f(1) &= C_1 - 2 = 1 \\
C_1 &= 3. \\
\text{Therefore } f(x) &= 3x - 2.
\end{aligned}
$$

Only one such function that satisfies the two given initial conditions.

188

55. $f''(x) = 2 \sin x$

$$
\begin{aligned}
f'(x) &= 2 \int \sin x \, dx \\
&= -2 \cos x + C_1 \\
f(x) &= (-2 \cos x + C_1) \, dx \\
&= -2 \sin x + C_1 x + C_2
\end{aligned}
$$

57. F is continuous on $[-2, 12]$; $F(0) = 0$; and F is the antiderivative of f.

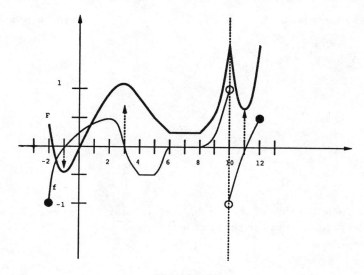

Exercise 57

59. Range: $[-2\pi, 2\pi] \times [-1.5, 1.5]$.

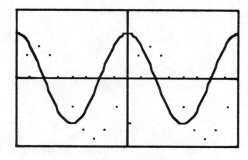

Graph for $n = 20$

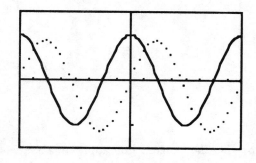

Graph for $n = 50$

61. Range: $[-2, 2] \times [0, 3]$.

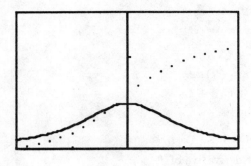

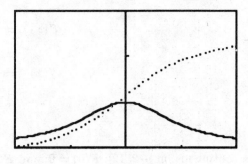

Graph for $n = 20$ Graph for $n = 50$

63.   a) The graph to the left is that of the points $(t, r(t))$ using the range $[0, 30] \times [0, 35]$. The graph to the right is the prediction of the shape of $n(t)$.

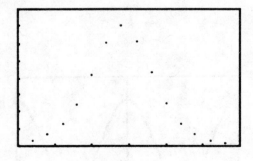

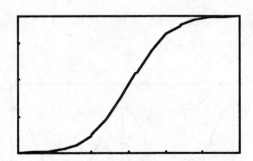

   b) Use the tangent line approximation with $h = 2$ and $n(0) = 1$. This takes the form of $n(t_i) = n(t_{i-1}) + hr(t_{i-1})$ where $t_0 = 0$, $t_1 = 2$, $t_2 = 4, \ldots$, etc. The values of $n(t_i)$ are given in the following table.

t	$n(t)$
0	1.0000
2	2.4798
4	5.5572
6	11.8760
8	24.5174
10	48.5335
12	89.9374
14	150.4876
16	220.4014
18	281.9954
20	324.6750
22	349.6395
24	362.8414
26	369.4562
28	372.6816
30	374.2334

After 30 days all 375 persons have the disease.

5.2 Finding Antiderivatives by Substitution

In Exercises 1–6, use the designated substitution to find the specified antiderivatives.

1. $\displaystyle \int x^5 \sin\left(x^6 + 2\right) dx = \frac{1}{6} \int \sin u \, du \qquad \begin{aligned} u &= x^6 + 2 \\ du &= 6x^5 \, dx \end{aligned}$

    $\displaystyle \qquad\qquad\qquad\qquad = \frac{1}{6}(-\cos u) + C$

    $\displaystyle \qquad\qquad\qquad\qquad = \frac{-\cos(x^6 + 2)}{6} + C$

3. $\displaystyle \int \frac{\cos \sqrt{x}}{\sqrt{x}} \, dx = 2 \int \cos u \, du \qquad u = \sqrt{x} = x^{1/2}$

    $\displaystyle \qquad\qquad\qquad = 2 \sin u + C$

    $\displaystyle \qquad\qquad\qquad = 2 \sin \sqrt{x} + C \qquad du = \frac{1}{2} x^{-1/2} \, dx$

    $\displaystyle \qquad\qquad\qquad\qquad\qquad\qquad\qquad\qquad du = \frac{1}{2\sqrt{x}} \, dx$

5. $\displaystyle \int x\sqrt{x + 2} \, dx = \int (u - 2)u^{1/2} \, du \qquad \begin{aligned} u &= x + 2 \\ du &= dx \\ x &= u - 2 \end{aligned}$

    $\displaystyle \qquad\qquad\qquad = \int \left(u^{3/2} - 2u^{1/2} \right) du$

    $\displaystyle \qquad\qquad\qquad = \frac{2}{5} u^{5/2} - 2 \cdot \frac{2}{3} u^{3/2} + C$

    $\displaystyle \qquad\qquad\qquad = \frac{2}{5}(x + 2)^{5/2} - \frac{4}{3}(x + 2)^{3/2} + C$

In Exercises 7–14, use the method of substitution to find the antiderivative. Check your answer by differentiating.

191

7. $\displaystyle \int x\sqrt{x^2+1}\, dx \;=\; \frac{1}{2}\int u^{1/2}\, du$ $u = x^2+1$
     $du = 2x\, dx$

$\displaystyle =\; \frac{1}{2}\cdot\frac{2}{3}\, u^{3/2} + C$

$\displaystyle =\; \frac{1}{3}(x^2+1)^{3/2} + C$

9. $\displaystyle \int \sec 2\theta \tan 2\theta \, d\theta \;=\; \frac{1}{2}\int \sec u \tan u\, du$ $u = 2\theta$
     $du = 2\, d\theta$

$\displaystyle =\; \frac{1}{2}\sec u + C$

$\displaystyle =\; \frac{1}{2}\sec 2\theta + C$

11. $\displaystyle \int x\csc^2(x^2)\, dx \;\Rightarrow\; \frac{1}{2}\int \csc^2 u\, du$ $u = x^2$
     $du = 2x\, dx$

$\displaystyle =\; -\frac{1}{2}\cot u + C$

$\displaystyle =\; -\frac{1}{2}\cot(x^2) + C$

13. $\displaystyle \int \left(1-t^2\right)\sqrt{3t^3-9t+9}\, dt \;=\; -\frac{1}{9}\int u^{1/2}\, du$ $u = 3t^3-9t+9$
     $du = 9t^2 - 9\, dt$
     $= 9(t^2-1)\, dt$
     $= -9(1-t^2)\, dt$

$\displaystyle =\; -\frac{1}{9}\cdot\frac{2}{3}\, u^{3/2} + C$

$\displaystyle =\; -\frac{2}{27}\left(3t^3-9t+9\right)^{3/2} + C$

In Exercises 15–24, find the antiderivative.

15. $\displaystyle \int \sec^2(x-\pi)\, dx \;=\; \int \sec^2 u\, du$ $u = x-\pi$
     $du = dx$

$\displaystyle =\; \tan u + C$

$\displaystyle =\; \tan(x-\pi) + C \text{ or } \tan x + C$

17. $\displaystyle \int \frac{(2\sqrt{x}+3)^2}{\sqrt{x}}\, dx \;=\; \int u^2\, du$ $u = 2\sqrt{x}+3$

$\displaystyle =\; \frac{u^3}{3} + C$ $du = 2\cdot\frac{1}{2}\, x^{-1/2}\, dx$

$\displaystyle =\; \frac{(2\sqrt{x}+3)^3}{3} + C$ $du = \frac{1}{\sqrt{x}}\, dx$

19. $\displaystyle \int (x^2+1)^2\, dx \;=\; \int (x^4+2x^2+1)\, dx$

$\displaystyle =\; \frac{x^5}{5} + \frac{2x^3}{3} + x + C$

21. $\displaystyle \int x\sqrt{x-1}\sqrt{x+1}\, dx \;=\; \int x\sqrt{x^2-1}\, dx$ $u = x^2-1$
     $du = 2x\, dx$

$\displaystyle =\; \frac{1}{2}\int u^{1/2}\, du$

$\displaystyle =\; \frac{1}{2}\cdot\frac{2}{3}u^{3/2} + C$

$\displaystyle =\; \frac{1}{3}(x^2-1)^{3/2} + C$

23. $\displaystyle\int x\sec(\pi - x^2)\tan(\pi - x^2)\,dx \;=\; -\frac{1}{2}\int \sec u \tan u\,du$ $\qquad u = \pi - x^2$

$\qquad\qquad\qquad\qquad\qquad\qquad\qquad\qquad du = -2x\,dx$

$\displaystyle =\; -\frac{1}{2}\sec u + C$

$\displaystyle =\; -\frac{1}{2}\sec(\pi - x^2) + C$

25. $\displaystyle\int (x^5 - 2x^3)(x^6 - 3x^4)^{5/2}\,dx \;=\; \frac{1}{6}\int u^{5/2}\,du$ $\qquad u = x^6 - 3x^4$

$\qquad\qquad\qquad\qquad\qquad\qquad\qquad du = 6x^5 - 12x^3\,dx$

$\displaystyle =\; \frac{1}{6}\cdot\frac{2}{7}u^{7/2} + C$ $\qquad\qquad = 6(x^5 - 2x^3)\,dx$

$\displaystyle =\; \frac{1}{21}(x^6 - 3x^4)^{7/2} + C$

27. $\displaystyle\int x^2(1 - x^4)^2\,dx \;=\; \int x^2(1 - 2x^4 + x^8)\,dx$

$\displaystyle =\; \int x^2 - 2x^6 + x^{10}\,dx$

$\displaystyle =\; \frac{x^3}{3} - \frac{2x^7}{7} + \frac{x^{11}}{11} + C$

29. $\displaystyle\int \frac{t^2 + 2}{\sqrt{t^3 + 6t}}\,dt \;=\; \frac{1}{3}\int \frac{1}{\sqrt{u}}\,du$ $\qquad u = t^3 + 6t$

$\qquad\qquad\qquad\qquad\qquad\qquad\qquad du = 3t^2 + 6\,dt$

$\displaystyle =\; \frac{1}{3}\int u^{-1/2}\,du$ $\qquad\qquad\qquad = 3(t^2 + 2)\,dt$

$\displaystyle =\; \frac{1}{3}\cdot 2u^{1/2} + C$

$\displaystyle =\; \frac{2}{3}\sqrt{t^3 + 6t} + C$

31. $\displaystyle\int \frac{1}{(1 - 4x)^{2/3}}\,dx \;=\; -\frac{1}{4}\int u^{-2/3}\,du$ $\qquad u = 1 - 4x$

$\qquad\qquad\qquad\qquad\qquad\qquad\qquad du = -4\,dx$

$\displaystyle =\; -\frac{1}{4}\cdot 3u^{1/3} + C$

$\displaystyle =\; -\frac{3}{4}(1 - 4x)^{1/3} + C$

33. $\displaystyle\int x^3\sqrt{x^2 + 1}\,dx \;=\; \frac{1}{2}\int (u - 1)u^{1/2}\,du$ $\qquad u = x^2 + 1$

$\qquad\qquad\qquad\qquad\qquad\qquad\qquad du = 2x\,dx$

$\displaystyle =\; \frac{1}{2}\int u^{3/2} - u^{1/2}\,du$ $\qquad\qquad x^2 = u - 1$

$\displaystyle =\; \frac{1}{2}\cdot\left(\frac{2}{5}u^{5/2} - \frac{2}{3}u^{3/2}\right) + C$

$\displaystyle =\; \frac{1}{5}(x^2 + 1)^{5/2} - \frac{1}{3}(x^2 + 1)^{3/2} + C$

In Exercises 35–37, state a substitution appropriate to each part and find the antiderivative.

35. $\quad u = 3x + 4$
$\quad du = 3\,dx$

(a) $\displaystyle\int (3x + 4)^{10}\,dx = \frac{1}{3}\int u^{10}\,du = \frac{1}{3}\cdot\frac{u^{11}}{11} + C = \frac{1}{33}(3x + 4)^{11} + C$

(b) $\displaystyle\int \sqrt{3x + 4}\,dx = \frac{1}{3}\int u^{1/2}\,du = \frac{1}{3}\cdot\frac{2}{3}u^{3/2} + C = \frac{2}{9}(3x + 4)^{3/2} + C$

(c) $\int \sin(3x + 4)\, dx = \frac{1}{3} \int \sin u \, du = \frac{1}{3}(-\cos u) + C = -\frac{1}{3} \cos(3x + 4) + C$

(d) $\displaystyle \int \sin^2(3x+4) \cos(3x+4)\, dx$

$\begin{aligned}
&= \int \sin^2 u \cos u \, du & w &= \sin u \\
& & dw &= \cos u \, du \\
&= \frac{1}{3} \int w^2 \, dw \\
&= \frac{1}{3} \frac{w^3}{3} + C \\
&= \frac{\sin^3 u}{9} + C \\
&= \frac{\sin^3(3x+4)}{9} + C
\end{aligned}$

37. $\quad u = x^2 + 6$
$\quad du = 2x \, dx$

(a) $\int (x^2+6)^{21} x \, dx = \frac{1}{2} \int u^{21} \, du = \frac{1}{2} \frac{u^{22}}{22} + C = \frac{(x^2+6)^{22}}{44} + C$

(b) $\int \frac{x}{\sqrt{x^2+6}} \, dx = \frac{1}{2} \int u^{-1/2} \, du = \frac{1}{2} \cdot 2 u^{1/2} + C = \sqrt{x^2+6} + C$

(c) $\int x \sec^2(x^2+6) \, dx = \frac{1}{2} \int \sec^2 u \, du = \frac{1}{2} \tan u + C = \frac{1}{2} \tan(x^2+6) + C$

39.   I. $\quad \displaystyle \int \sin x \cos x \, dx$

$\begin{aligned}
&= \int u \, du & u &= \sin x \\
& & du &= \cos x \, dx \\
&= \frac{u^2}{2} + C_1 \\
&= \frac{\sin^2 x}{2} + C_1
\end{aligned}$

II. $\sin 2x = 2 \sin x \cos x$

$\begin{aligned}
\int \sin x \cos x \, dx &= \frac{1}{2} \int \sin 2x \, dx & u &= 2x \\
& & du &= 2 \, dx \\
&= \frac{1}{4} \int \sin u \, du \\
&= \frac{1}{4}(-\cos u) + C_2 \\
&= -\frac{\cos(2x)}{4} + C_2
\end{aligned}$

III. Use $\cos 2x = \cos^2 x - \sin^2 x$ and $\cos^2 x = 1 - \sin^2 x$ in answer II.

$\begin{aligned}
-\frac{\cos(2x)}{4} + C_2 &= \frac{-(\cos^2 x - \sin^2 x)}{4} + C_2 \\
&= \frac{-\cos^2 x + \sin^2 x}{4} + C_2 \\
&= \frac{-(1 - \sin^2 x) + \sin^2 x}{4} + C_2 \\
&= \frac{-1 + 2\sin^2 x}{4} + C_2 \\
&= \frac{\sin^2 x}{2} - \frac{1}{4} + C_2
\end{aligned}$

Yes, answers are the same if $C_1 = -\dfrac{1}{4} + C_2$.

41. Four of the six antiderivatives can be calculated using the methods discussed in Sections 5.1 and 5.2.

 I. Which are they? (a, c, d, and e). Three of these four can be determined using a substitution. Determine which three and indicate the substitution.

    (a) $\displaystyle\int \sec^2(2\theta)\,d\theta$ $\begin{aligned} u &= 2\theta \\ du &= 2\,d\theta \end{aligned}$ substitution

    (d) $\displaystyle\int u(u^2+1)^{10}\,du$ $\begin{aligned} w &= u^2+1 \\ dw &= 2u\,du \end{aligned}$ substitution

    (e) $\displaystyle\int \frac{t+1}{\sqrt{t^2+2t}}\,dt$ $\begin{aligned} u &= t^2+2t \\ du &= (2t+2)\,dt \\ &= 2(t+1)\,dt \end{aligned}$ substitution

 II. How do you calculate the fourth?

    $[(c)]\displaystyle\int (u^2+1)^{10}\,du$

    Since substitution won't work, we can multiply it out and then integrate.

 III. One of the three amenable to the method of substitution can also be determined without using substitution. Which one? How can this antiderivative be determined if substitution is not used?

    $[(d)]\displaystyle\int u(u^2+1)^{10}\,du$

    It can be multiplied out.

5.3   Differential Equations

In Exercises 1–14, find the general solution of the differential equation.

1. $\begin{aligned} \frac{dy}{dx} &= x-1 \\ y &= \int (x-1)\,dx \\ y &= \frac{x^2}{2} - x + C \end{aligned}$

3. $\begin{aligned} \frac{dy}{dx} &= x^2 - \frac{1}{x^2} \\ y &= \int \left(x^2 - x^{-2}\right)dx \\ &= \frac{x^3}{3} - \frac{x^{-1}}{-1} + C \\ y &= \frac{x^3}{3} + \frac{1}{x} + C \end{aligned}$

5. $\begin{aligned} \frac{dy}{dx} &= \sin x \cos x \sqrt{1+\sin^2 x} \\ y &= \int \sin x \cos x \sqrt{1+\sin^2 x}\,dx \\ y &= \frac{1}{2}\int u^{1/2}\,du \\ y &= \frac{1}{2}\cdot\frac{2}{3}\,u^{3/2} + C \\ y &= \frac{1}{3}(1+\sin^2 x)^{3/2} + C \end{aligned}$ $\begin{aligned} u &= 1+\sin^2 x \\ du &= 2\sin x \cos x\,dx \end{aligned}$

195

7. $dy = (x - \sqrt{x})\,dx$

   $y = \displaystyle\int x - x^{1/2}\,dx$

   $y = \dfrac{x^2}{2} - \dfrac{2}{3}\,x^{3/2} + C$

9. $dy = \dfrac{x}{y}\,dx$

   $\displaystyle\int y\,dy = \int x\,dx$

   $\dfrac{y^2}{2} = \dfrac{x^2}{2} + C$

   $\dfrac{y^2}{2} - \dfrac{x^2}{2} = C$

   $y^2 - x^2 = C$ or $y^2 = x^2 + C$

11. $dy = -4xy^2\,dx$

    $\dfrac{1}{y^2}\,dy = -4x\,dx$

    $\displaystyle\int y^{-2}\,dy = -4\int x\,dx$

    $\dfrac{y^{-1}}{-1} = -\dfrac{4x^2}{2} + C$

    $-\dfrac{1}{y} = -2x^2 + C$

    $y = \dfrac{1}{2x^2 + C}$

13. $\dfrac{dy}{dx} = -\dfrac{x}{y}$

    $\displaystyle\int y\,dy = -\int x\,dx$

    $\dfrac{y^2}{2} = -\dfrac{x^2}{2} + C$

    $\dfrac{y^2}{2} + \dfrac{x^2}{2} = C$

    $x^2 + y^2 = C$ or $y^2 = C - x^2$

In Exercises 15–26, find the solution of the initial value problem.

15. $\dfrac{dy}{dx} = \dfrac{x}{y},\quad y(1) = 4$

    $y\,dy = x\,dx$

    $\dfrac{y^2}{2} = \dfrac{x^2}{2} + C$

    $y^2 = x^2 + C$

    $y = \pm\sqrt{x^2 + C}$

    Now $y(1) = \pm\sqrt{1^2 + C} = 4$
    implies that we must
    use the positive radical.

    $16 = 1^2 + C$

    $15 = C$

    $y = \sqrt{x^2 + 15}$

17. $\dfrac{dy}{dx} = -\dfrac{x}{y},\quad y\left(\sqrt{3}\right) = 1$

    $\displaystyle\int y\,dy = -\int x\,dx$

    $\dfrac{y^2}{2} = -\dfrac{x^2}{2} + C$

    $y^2 = -x^2 + C$

    $(1)^2 = -\left(\sqrt{3}\right)^2 + C$

    Now $y\left(\sqrt{3}\right) = \pm\sqrt{-3 + C} = 1$
    implies we must
    use the positive radical.

    $C = 4$

    $y = \sqrt{-x^2 + 4}$

19. $dy = \dfrac{x^2}{\left(1 + x^3\right)^2}\,dx,\quad y(0) = \dfrac{1}{2}$

$$y = \int \frac{x^2}{(1+x^3)^2}\,dx \qquad u = 1+x^3$$
$$du = 3x^2\,dx$$

$$y = \frac{1}{3}\int \frac{1}{u^2}\,du$$

$$y = \frac{1}{3}\int u^{-2}\,du$$

$$y = \frac{1}{3}\frac{u^{-1}}{-1} + C$$

$$y = -\frac{1}{3u} + C$$

$$y = \frac{-1}{3(1+x^3)} + C$$

$$\frac{1}{2} = \frac{-1}{3(1+0)} + C$$

$$\frac{1}{2} = \frac{-1}{3} + C$$

$$\frac{5}{6} = C$$

$$y = \frac{-1}{3(1+x^3)} + \frac{5}{6}$$

21. $dy = y^2(1+x^2)\,dx, \quad y(0) = 1$

$$\int y^{-2}\,dy = \int (1+x^2)\,dx$$

$$\frac{y^{-1}}{-1} = x + \frac{x^3}{3} + C$$

$$-\frac{1}{y} = \frac{3x + x^3 + C}{3}$$

$$y = -\frac{3}{x^3 + 3x + C}$$

$$1 = -\frac{3}{0+0+C}$$

$$C = -3$$

$$y = -\frac{3}{x^3 + 3x - 3}$$

23. $\dfrac{d^2y}{dx^2} = x^2, \quad y(0) = 2, \quad y'(0) = 4$

$$y' = \frac{dy}{dx} = \frac{x^3}{3} + C_1 \Rightarrow \begin{aligned} 4 &= 0 + C_1 \\ 4 &= C_1 \end{aligned}$$

$$y' = \frac{x^3}{3} + 4$$

$$y = \frac{x^4}{12} + 4x + C_2$$

$$\begin{aligned} 2 &= 0 + 0 + C_2 \\ 2 &= C_2 \end{aligned}$$

$$y = \frac{x^4}{12} + 4x + 2$$

25. $\dfrac{d^2y}{dx^2} = -16\sin 4x, \quad y(\pi) = \pi + 1, \quad y'(\pi) = 1$

$$y' = -16\int \sin 4x\,dx \qquad u = 4x$$
$$du = 4dx$$

$$y' = -4\int \sin u\,du$$

$$y' = -4(-\cos u) + C_1$$
$$y' = 4\cos u + C_1$$
$$y' = 4\cos(4x) + C_1$$
$$1 = 4\cos 4\pi + C_1$$
$$1 = 4 + C_1$$
$$-3 = C_1$$
$$y' = 4\cos(4x) - 3$$

$$y' = 4\cos 4x - 3$$
$$y = 4\int \cos 4x\,dx - \int 3\,dx$$
$$y = 4\cdot\frac{1}{4}\int \cos u\,du - 3x + C_2$$
$$y = \sin u - 3x + C_2$$
$$y = \sin 4x - 3x + C_2$$
$$\pi + 1 = \sin 4\pi - 3\pi + C_2$$
$$\pi + 1 = 0 - 3\pi + C_2$$
$$4\pi + 1 = C_2$$

Therefore $y = \sin 4x - 3x + 4\pi + 1.$

27. $\dfrac{dy}{dx} = x - \sqrt{x}, \quad y(1) = 2$

197

$$y = \int \left(x - x^{1/2} \right) dx$$

$$y = \frac{x^2}{2} - \frac{2}{3} x^{3/2} + C$$

$$2 = \frac{1}{2} - \frac{2}{3} + C$$

$$2 = -\frac{1}{6} + C$$

$$\frac{13}{6} = C$$

$$y = \frac{x^2}{2} - \frac{2}{3} x^{3/2} + \frac{13}{6}$$

29. $\dfrac{dy}{dx} = 2$ for $x \leq 1$, $\dfrac{dy}{dx} = x + 1$ for $x > 1$

$$y = \int 2 \, dx$$

$$y = 2x + C_1 \text{ for } x \leq 1$$

$$y(0) = 1$$

$$y(0) = 1 = C_1$$

$$y = \int (x + 1) \, dx$$

$$y = \frac{x^2}{2} + x + C_2 \text{ for } x > 1$$

Therefore $y = 2x + 1$ for $x \leq 1$.

If the curve is continuous then $\displaystyle\lim_{x \to 1+} \frac{x^2}{2} + x + C_2 = f(1) = 2(1) + 1 = 3$

$$\frac{1}{2} + 1 + C_2 = 3$$

$$C_2 = \frac{3}{2}$$

$$y = \frac{x^2}{2} + x + \frac{3}{2} \text{ for } x > 1$$

$$\text{and} \quad y(2) = \frac{2^2}{2} + 2 + \frac{3}{2}$$

$$y(2) = \frac{11}{2}.$$

31. $\dfrac{dP}{dt} = \dfrac{900}{P^2}$, $P(0) = 10$

$$\int P^2 \, dP = 900 \int dt$$

$$\frac{P^3}{3} = 900t + C_1$$

$$\frac{10^3}{3} = 0 + C_1, \; C_1 = \frac{10^3}{3}$$

$$\frac{P^3}{3} = 900t + \frac{10^3}{3}$$

$$P^3 = 2700t + 10^3$$

$$P(t) = (2700t + 1000)^{1/3}$$

33. $\dfrac{dP}{dt} = 4\sqrt{P}$, $P(0) = 10$

$$\int P^{-1/2} \, dP = 4 \int dt$$

$$2\sqrt{P} = 4t + C$$

$$\sqrt{P} = 2t + C$$

$$P = (2t + C)^2$$

$$P(0) = 10 = C^2, \; C = \sqrt{10}$$

$$P(t) = \left(2t + \sqrt{10} \right)^2$$

35. Use the initial condition $y(-2) = -2$. Range: $[-2, 2] \times [-2, 0]$. For $n = 10$ and $n = 30$ the value of y_i eventually overflows due to round-off error. If you try $n = 60$ or $n = 120$ then a graph is produced.

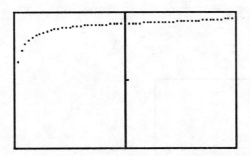

Graph for $n = 60$ Graph for $n = 120$

37. Range: $[0, 2\pi] \times [0, 1]$. Graphs for all values of n.

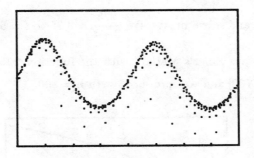

39.   a) Now

$$\frac{dv}{dt} = -kv^2$$

so

$$\int \frac{dv}{v^2}\, dt = -k \int dt.$$

Therefore

$$\frac{-1}{v} = kt + c$$

or

$$v = \frac{-1}{kt + c}.$$

Now $v(0) = 75$ so $75 = -\frac{1}{c}$, i.e., $c = -\frac{1}{75}$. Therefore

$$v(t) = \frac{-1}{kt - \frac{1}{75}} = \frac{75}{1 - 75kt}.$$

199

b) Since $v(2) = 45$ we have that $45 = \dfrac{75}{1 - 75k(2)}$ and so $1 - 150k = \dfrac{75}{45}$ implying that $k = -\dfrac{1}{225}$. So

$$v(t) = \frac{75}{1 + \frac{75}{225}\,t} = \frac{225}{3 + t}.$$

Graph with the range: $[0, 100] \times [0, 75]$.

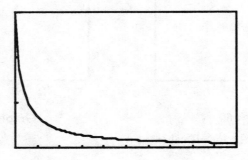

Now 5% of 75 is 3.75 and so we must solve $\dfrac{225}{3 + t} = 3.75$ so $t = 57$.

c) Now $\dfrac{ds}{dt} = \dfrac{225}{3 + t}$. So use Euler's Method with the function $\mathbf{225/(3+X)}$. The graph below has the range $[0, 57] \times [0, 700]$ and was produced with $n = 300$.

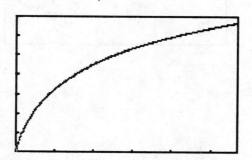

The last value computed by the programs is $s(57)$ and this value is 667.35 feet. More accurate values of this can be found by increasing the value of n. For example, if $n = 1000$, $s(57) \approx 672.02$ and if $n = 2000$, $s(57) \approx 673.03$.

Review Exercises—Chapter 5

In Exercises 1–18, find the indefinite integral for the given function.

1. $f(x) = 6x^2 - 2x + 1$

$$\int \left(6x^2 - 2x + 1\right) dx = 2x^3 - x^2 + x + C$$

3. $y = \dfrac{\sin \sqrt{t}}{\sqrt{t}}$

$$
\begin{aligned}
\int \frac{\sin \sqrt{t}}{\sqrt{t}} \, dt &= 2 \int \sin u \, du & u &= \sqrt{t} \\
&= 2(-\cos u) + C & du &= \frac{1}{2} t^{-1/2} \, dt \\
&= -2 \cos \sqrt{t} + C & &= \frac{1}{2\sqrt{t}} \, dt
\end{aligned}
$$

5. $f(x) = 3\sqrt{x} + \dfrac{3}{\sqrt{x}}$

$$
\int \left(3\sqrt{x} + \frac{3}{\sqrt{x}} \right) dx = \int \left(3x^{1/2} + 3x^{-1/2} \right) dx = 3 \cdot \frac{2}{3} x^{3/2} + 3 \cdot 2x^{1/2} + C = 2x^{3/2} + 6x^{1/2} + C
$$

7. $y = x \sec^2 x^2$

$$
\begin{aligned}
\int x \sec^2 x^2 \, dx &= \frac{1}{2} \int \sec^2 u \, du & u &= x^2 \\
& & du &= 2x \, dx \\
&= \frac{1}{2} \tan u + C \\
&= \frac{1}{2} \tan x^2 + C
\end{aligned}
$$

9. $f(x) = \dfrac{x^3 - 7x^2 + 6x}{x}$

$$
\int \frac{x^3 - 7x^2 + 6x}{x} \, dx = \int x^2 - 7x + 6 \, dx = \frac{x^3}{3} - \frac{7x^2}{2} + 6x + C
$$

11. $f(t) = t^5 \sqrt{t^2 + 1}$

$$
\begin{aligned}
\int t^5 \sqrt{t^2 + 1} \, dt &= \int t^4 \sqrt{t^2 + 1} \cdot t \, dt & u &= t^2 + 1 \\
&= \frac{1}{2} \int (u - 1)^2 u^{1/2} \, du & du &= 2t \, dt \\
&= \frac{1}{2} \int \left(u^2 - 2u + 1 \right) u^{1/2} \, du & t^2 &= u - 1 \\
&= \frac{1}{2} \int \left(u^{5/2} - 2u^{3/2} + u^{1/2} \right) du \\
&= \frac{1}{2} \left(\frac{2}{7} u^{7/2} - 2 \cdot \frac{2}{5} u^{5/2} + \frac{2}{3} u^{3/2} \right) + C \\
&= \frac{1}{7} u^{7/2} - \frac{2}{5} u^{5/2} + \frac{1}{3} u^{3/2} + C \\
&= \frac{1}{7} (t^2 + 1)^{7/2} - \frac{2}{5} (t^2 + 1)^{5/2} + \frac{1}{3} (t^2 + 1)^{3/2} + C
\end{aligned}
$$

13. $f(x) = x \cos(1 + x^2)$

$$
\begin{aligned}
\int x \cos(1 + x^2) \, dx &= \frac{1}{2} \int \cos u \, du & u &= 1 + x^2 \\
& & du &= 2x \, dx \\
&= \frac{1}{2} \sin u + C \\
&= \frac{1}{2} \sin(1 + x^2) + C
\end{aligned}
$$

201

15. $y = (2x - 1)(2x + 1)$

$$\int (2x - 1)(2x + 1)\, dx = \int (4x^2 - 1)\, dx = \frac{4}{3} x^3 - x + C$$

17. $f(x) = |x|$

$$\int |x|\, dx \Rightarrow \begin{cases} \displaystyle\int x\, dx & \text{if } x \geq 0 \\[2mm] \displaystyle\int -x\, dx & \text{if } x < 0 \end{cases} \qquad f(x) = \begin{cases} \dfrac{x^2}{2} + C & x \geq 0 \\[2mm] -\dfrac{x^2}{2} + C & x < 0 \end{cases}$$

In Exercises 19–24, find the particular function satisfying the stated conditions.

19. $f'(x) = 1 + \cos x, \quad f(0) = 3$

$$\begin{aligned} f(x) &= \int (1 + \cos x)\, dx \\ f(x) &= x + \sin x + C \\ f(0) &= 0 + 0 + \ C \ = \ 3 \\ C &= 3 \\ f(x) &= x + \sin x + 3 \end{aligned}$$

21. $f''(x) = 3, \quad f'(1) = 6, \quad f(0) = 4$

$$\begin{aligned} f'(x) &= \int 3\, dx \\ f'(x) &= 3x + C_1 \\ f'(1) &= 3 + \ C_1 \ = \ 6 \\ C_1 &= 3 \\ f'(x) &= 3x + 3 \\ f(x) &= \int (3x + 3)\, dx \end{aligned}$$

$$\begin{aligned} f(x) &= \frac{3x^2}{2} + 3x + C_2 \\ f(0) &= 0 + 0 + \ C_2 \ = \ 4 \\ C_2 &= 4 \\ f(x) &= \frac{3x^2}{2} + 3x + 4 \end{aligned}$$

23. $f'(x) = x \sec x^2 \tan x^2, \quad f\left(\sqrt{\pi/4}\,\right) = 1$

$$\begin{aligned} f(x) &= \int x \sec x^2 \tan x^2\, dx \qquad u = x^2 \\ &\qquad\qquad\qquad\qquad\qquad du = 2x\, dx \\ f(x) &= \frac{1}{2} \int \sec u \tan u\, du \\ &= \frac{1}{2} \sec u + C \\ &= \frac{1}{2} \sec x^2 + C \end{aligned}$$

$$\begin{aligned} f\left(\sqrt{\frac{\pi}{4}}\right) &= \frac{1}{2} \sec \left(\sqrt{\frac{\pi}{4}}\right)^2 + C = 1 \\ &= \frac{1}{2} \sec \frac{\pi}{4} + C = 1 \\ &= \frac{1}{2} \cdot \sqrt{2} + C = 1 \\ C &= 1 - \frac{\sqrt{2}}{2} = \frac{2 - \sqrt{2}}{2} \\ f(x) &= \frac{1}{2} \sec(x^2) + \frac{2 - \sqrt{2}}{2} \end{aligned}$$

25. $\dfrac{dy}{dx} = 2x, \quad y(2) = 9$

$$\begin{aligned} y &= 2 \int x\, dx \\ y &= x^2 + C \\ y(2) &= 2^2 + C = 9 \end{aligned}$$

$$\begin{aligned} 4 + C &= 9 \\ C &= 5 \\ y &= x^2 + 5 \end{aligned}$$

27. Find s if $a = v'(t) = s''(t) = -9.8 m/s^2$, $v(0) = 0$, $s(0) = 0$.

$$v(t) = -\int 9.8 \, dt$$
$$= -9.8t + C_1$$
$$v(0) = 0 + C_1 = 0$$
$$C_1 = 0$$
$$v(t) = -9.8t$$
$$-9.8t = -49$$
$$t = \frac{49}{9.8} = 5$$

$$s(t) = \int -9.8t \, dt$$
$$s(t) = -\frac{9.8t^2}{2} + C_2$$
$$s(0) = 0 + C_2 = 0$$
$$C_2 = 0$$
$$s(t) = -4.9t^2$$
$$s(5) = -4.9(5)^2$$
$$\approx -122.5m$$

Therefore the initial height is $\approx 122.5m$.

In Exercises 28–31, find the solution of the initial value problem.

29. $\dfrac{dy}{dx} = \dfrac{x+1}{y}$, $y(1) = 2$

$$\int y \, dy = \int (x+1) \, dx$$
$$\frac{y^2}{2} = \frac{x^2}{2} + x + C$$
$$y^2 = x^2 + 2x + C$$
$$y(1) = 2$$

$$2^2 = 1^2 + 2(1) + C$$
$$4 = 1 + 2 + C$$
$$1 = C$$
$$y^2 = x^2 + 2x + 1$$
$$y^2 = (x+1)^2$$
$$y = x+1$$

31. $dy = \dfrac{\sqrt{x+1}}{\sqrt{y}} \, dx$, $y(0) = 1$

$$\int y^{1/2} \, dy = \int (x+1)^{1/2} \, dx \qquad u = x+1$$
$$\qquad\qquad\qquad\qquad\qquad du = dx$$
$$\frac{2}{3} y^{3/2} = \int u^{1/2} \, du$$
$$\frac{2}{3} y^{3/2} = \frac{2}{3} u^{3/2} + C$$
$$y^{3/2} = u^{3/2} + C$$
$$y^{3/2} = (x+1)^{3/2} + C$$

$$y(0) = 1$$
$$(0+1)^{3/2} + C = 1$$
$$C = 0$$
$$y^{3/2} = (x+1)^{3/2}$$
$$y = x+1$$

33. $s(t) = t^3 - 6t^2 + 9t - 4$

   (a) $s(0) = 0 - 0 + 0 - 4$
       $s(0) = -4$ or 4 units to the left of the origin.

   (b) $v(t) = 3t^2 - 12t + 9$
       $v(0) = 0 - 0 + 9 = 9$ units moving in a positive direction, i.e. right.

   (c) Changes direction when velocity = 0.

$$v(t) = 0, \ 3t^2 - 12t + 9 = 0$$
$$3(t^2 - 4t + 3) = 0$$
$$3(t-3)(t-1) = 0$$
$$t = 1, 3$$

     It changes direction at time $t = 1$ and at $t = 3$.

   (d) It changes direction twice.

Chapter 6

The Definite Integral

6.1   Area and Summation

1. $\displaystyle\sum_{j=1}^{10} j = 1 + 2 + \cdots + 9 + 10$

   [Use special formula] [Eq (4)]

   $\displaystyle = \frac{10(10+1)}{2} = 55.$

3. $\displaystyle\sum_{j=1}^{12} j^3 = 1^3 + 2^3 + \cdots + 11^3 + 12^3$

   [Use special formula] [Eq(6)]

   $\displaystyle = (12)^2\,\frac{(12+1)^2}{4}$

   $\displaystyle = 6084.$

5. $\displaystyle\sum_{j=1}^{5} \sin\left(\frac{j\pi}{2}\right) = \sin\left(\frac{1\pi}{2}\right) + \sin\left(\frac{2\pi}{2}\right) + \sin\left(\frac{3\pi}{2}\right) + \sin\left(\frac{4\pi}{2}\right) + \sin\left(\frac{5\pi}{2}\right) = 1 + 0 - 1 + 0 + 1 = 1.$

7. $\displaystyle\sum_{j=2}^{200}(-1)^j = (-1)^2 + (-1)^3 + \cdots + (-1)^{198} + (-1)^{199} + (-1)^{200} = (+1-1) + \cdots + (+1-1) + (1) =$
   $0 + \cdots + 0 + 1 = 1.$ Method: group the terms in pairs, each of which sums to zero.

9. $2 + 4 + 6 + 8 + 10$
    $= 2(1) + 2(2) + 2(3) + 2(4) + 2(5)$
    $= 2(1 + 2 + 3 + 4 + 5)$
    $\displaystyle = 2\sum_{j=1}^{5} j = \sum_{j=1}^{5} 2j$

11. $4 + 9 + 16 + 25 + 36$

$\qquad = \ 2^2 + 3^2 + 4^2 + 5^2 + 6^2$

$\qquad = \ \displaystyle\sum_{j=2}^{6} j^2$

13. Since $\Delta x = \frac{b-a}{n} = \frac{4-0}{4} = 1$ then $x_0 = 0$, $x_1 = 1$, $x_2 = 2$, $x_3 = 3$, $x_4 = 4$. Since $f(x) = x$ increases, then for the lower sum, the c_j are the left endpoints. So

$$\underline{S}_4 \ = \ \sum_{j=1}^{4} f(c_j)\Delta x$$
$$= \ 0(1) + 1(1) + 2(1) + 3(1)$$
$$= \ 6.$$

15. Since $\Delta x = \frac{b-a}{n} = \frac{4-0}{4} = 1$, then $x_0 = 0$, $x_1 = 1$, $x_2 = 2$, $x_3 = 3$, $x_4 = 4$. Since $f(x) = 2x^2 + 3$ increases on $[0,4]$, then the c_j are the left endpoints for a lower sum. So

$$\underline{S}_4 \ = \ \sum_{j=1}^{4} f(c_j)\Delta x$$
$$= \ [2(0^2) + 3](1) + [2(1)^2 + 3](1) + [2(2)^2 + 3](1) + [2(3)^2 + 3](1)$$
$$= \ 40.$$

17. Since $\Delta x = \frac{b-a}{n} = \frac{\pi - 0}{4} = \frac{\pi}{4}$ then $x_0 = 0$, $x_1 = \frac{\pi}{4}$, $x_2 = \frac{2\pi}{4}$, $x_3 = \frac{3\pi}{4}$, $x_4 = \frac{4\pi}{4}$. Since $f(x) = \sin(x)$ increases on $\left[0, \frac{\pi}{2}\right]$ and decreases on $\left[\frac{\pi}{2}, \pi\right]$, then for a lower sum the endpoints are $c_1 = 0$, $c_2 = \frac{\pi}{4}$, $c_3 = \frac{3\pi}{4}$, $c_4 = \frac{4\pi}{4}$. So

$$\underline{S}_4 \ = \ \sum_{j=1}^{4} f(c_j)\Delta x$$
$$= \ [\sin(0)]\frac{\pi}{4} + \left[\sin\left(\frac{\pi}{4}\right)\right]\frac{\pi}{4} + \left[\sin\left(\frac{3\pi}{4}\right)\right]\frac{\pi}{4} + \left[\sin\left(\frac{4\pi}{4}\right)\right]\frac{\pi}{4}$$
$$= \ 0 + \left(\sqrt{\frac{1}{2}}\right)\frac{\pi}{4} + \left(\sqrt{\frac{1}{2}}\right)\frac{\pi}{4} + 0 \ = \ \frac{\pi\sqrt{2}}{4}.$$

19. Since $\Delta x = \frac{b-a}{n} = \frac{3-1}{200} = \frac{1}{100}$, then $x_0 = 1$, $x_1 = 1.01, \cdots$, $x_{199} = 2.99$, $x_{200} = 3$. Since $f(x) = 4x + 1$ increases on $[1,3]$, then the c_j are the left endpoints for a lower sum: $\quad c_j \ = \ 1 + \left(\dfrac{j}{100}\right) - \left(\dfrac{1}{100}\right)$

$$= \ \frac{99}{100} + \frac{j}{100}.$$

So

$$\underline{S}_{200} \ = \ \sum_{j=1}^{200} f(c_j)\Delta x \ = \ \sum_{j=1}^{200}\left[f\left(\frac{99}{100} + \frac{j}{100}\right)\right]\frac{1}{100}$$

See 29 a) b).

$$= \frac{1}{100}\left\{\sum_{j=1}^{200} 4\left(\frac{99}{100}+\frac{j}{100}\right)+1\right\} = \frac{1}{100}\left\{\sum_{j=1}^{200}\left(\frac{99}{25}+1\right)+\frac{j}{25}\right\}$$

$$= \frac{1}{100}\left\{\frac{124}{25}\left(\sum_{j=1}^{200} 1\right)+\frac{1}{25}\left(\sum_{j=1}^{200} j\right)\right\}$$

[Use special formula for $\sum n$ and $\sum j$.] [Eq(3)] [Eq(4)]

$$= \frac{1}{100}\left\{\left(\frac{124}{25}\right)(200)+\frac{1}{25}\left(\frac{(200)(200+1)}{2}\right)\right\} = 17.96.$$

21. Since $\Delta x = \frac{b-a}{n} = \frac{3-0}{3} = 1$, then $x_0 = 0$, $x_1 = 1$, $x_2 = 2$, $x_3 = 3$. Since $f(x) = 2x$ increases on $[0,3]$, then the c_j for an upper sum are the right endpoints. So

$$\overline{S}_3 = \sum_{j=1}^{3} f(c_j)\Delta x = 2(1)(1)+2(2)(1)+2(3)(1) = 12.$$

23. Since $\Delta x = \frac{b-a}{n} = \frac{3+1}{4} = 1$, then $x_0 = -1$, $x_1 = 0$, $x_2 = 1$, $x_3 = 2$, $x_4 = 3$. Since $f(x) = 3x^2 + 10$ decreases on $[-1,0]$ and increases on $[0,3]$, then

$$\overline{S}_4 = f(-1)(1)+f(1)(1)+f(2)(1)+f(3)(1) = 13+13+22+37 = 85.$$

25. Since $\Delta x = \frac{b-a}{n} = (\frac{\pi}{2}+\frac{\pi}{2})/4 = \frac{\pi}{4}$, then $x_0 = -\frac{2\pi}{4}$, $x_1 = -\frac{\pi}{4}$, $x_2 = 0$, $x_3 = \frac{\pi}{4}$, $x_4 = \frac{2\pi}{4}$. Since $f(x) = \cos(x)$ increases on $\left[-\frac{\pi}{2},\ 0\right]$ and decreases on $\left[0,\ \frac{\pi}{2}\right]$, then

$$\overline{S}_4 = \sum_{j=1}^{4} f(c_j)\Delta x = \left[\cos\left(-\frac{\pi}{4}\right)\right]\left(\frac{\pi}{4}\right)+\left[\cos(0)\right]\left(\frac{\pi}{4}\right)+\left[\cos(0)\right]\left(\frac{\pi}{4}\right)+\left[\cos\left(\frac{\pi}{4}\right)\right]$$

$$= \left(\sqrt{\frac{1}{2}}\right)\frac{\pi}{4}+\frac{\pi}{4}+\frac{\pi}{4}+\left(\sqrt{\frac{1}{2}}\right)\frac{\pi}{4} = \frac{\pi}{2}+\frac{\pi\sqrt{2}}{4} = \pi\left(\frac{2+\sqrt{2}}{4}\right).$$

27. Since $\Delta x = \frac{b-a}{n} = \frac{4-1}{300} = \frac{1}{100}$, then $x_0 = 1$, $x_1 = 1.01, \cdots, x_{299} = 3.99$, $x_{300} = 4$. Since $f(x) = 4x - 1$ increases, then the c_j for an upper sum are the right endpoints: $c_j = 1+\frac{j}{100}$. So

$$\overline{S}_{300} = \sum_{j=1}^{300} f(c_j)\Delta x = \sum_{j=1}^{300} f\left(1+\frac{j}{100}\right)\cdot\Delta x = \sum_{j=1}^{300}\left[4\left(1+\frac{j}{100}\right)-1\right]\left(\frac{1}{100}\right)$$

$$= \sum_{j=1}^{300}\left[3+\frac{j}{25}\right]\left(\frac{1}{100}\right) = \frac{1}{100}\left\{3\left(\sum_{j=1}^{300} 1\right)+\frac{1}{25}\left(\sum_{j=1}^{300} j\right)\right\}$$

[Use special formula: Eq (3) for $\sum n$ and Eq (4) for $\sum j$.]

$$= \frac{1}{100}\left\{300(3)+\frac{1}{25}\frac{300(300+1)}{2}\right\} = 27.06.$$

29. a) $\displaystyle\sum_{j=1}^{n}(x_j + y_j)$. The summation notation means $(x_1 + y_1) + (x_2 + y_2) + \cdots + (x_n + y_n)$. Since addition is associative and commutative, this can be commuted and associated (regrouped) as $(x_1 + x_2 + \cdots + x_n) + (y_1 + y_2 + \cdots + y_n)$. The summation notation for this is $\displaystyle\sum_{j=1}^{n} x_j + \sum_{j=1}^{n} y_j$.

b) The summation notation $\displaystyle\sum_{j=1}^{n} cx_j$ means $cx_1 + cx_2 + \cdots + cx_n$. By the distributive law, this is $c(x_1 + x_2 + \cdots + x_n)$. The summation notation for this is $c\displaystyle\sum_{j=1}^{n} x_j$.

c) This is the same as a), where we replace y_j with $-y_j$ as follows:

$$\sum_{j=1}^{n} x_j - y_j \;=\; \sum_{j=1}^{n} x_j + (-1)y_j.$$

By 29 a), b) this is equal to

$$\sum x_j + \sum(-1)(y_j) \;=\; \left[\sum x_j\right] + (-1)\left[\sum y_j\right]$$
$$= \sum x_j - \sum y_j.$$

d) $\displaystyle\sum_{j=1}^{n} c$ means $c + c + \cdots + c$ where c is summed n times. By the distributive law, this equals nc.

31.
$$
\begin{aligned}
\sum_{j=1}^{7} 3j^2 &= 3\sum_{j=1}^{7} j^2 \quad \text{[29 b)]}\\
&= 3\,\frac{7(7+1)(2\cdot 7 + 1)}{6} \quad \text{[Eq(4)]}\\
&= 420.
\end{aligned}
$$

33.
$$
\begin{aligned}
\sum_{j=1}^{6} 2j^2 + 3j + 5 &= \sum_{j=1}^{6} 2j^2 + \sum_{j=1}^{6} 3j + \sum_{j=1}^{6} 5 \quad \text{[29 a)]}\\
&= 2\sum_{j=1}^{6} j^2 + 3\sum_{j=1}^{6} j + \sum_{j=1}^{6} 5 \quad \text{[29 b)]}\\
&= 2\,\frac{(6)(6+1)(2\cdot 6 + 1)}{6} + 3\,\frac{(6)(6+1)}{2} + 6\cdot 5 \quad \text{By [Eq(3),(4),(5)]}\\
&= 275.
\end{aligned}
$$

35.
$$
\begin{aligned}
\sum_{j=1}^{n} (6j^2 - 4j) &= 6\cdot\sum_{j=1}^{n} j^2 - 4\cdot\sum_{j=1}^{n} j \quad \text{[By 29 b), c)]}\\
&= 6\left[\frac{n(n+1)(2n+1)}{6}\right] - 4\left[\frac{n(n+1)}{2}\right] \quad \text{[By Eq (4), (5)]}\\
&= (2n^3 + 3n^2 + n) - (2n^2 + 2n)\\
&= 2n^3 + n^2 - n.
\end{aligned}
$$

37. If $120 = \sum\limits_{j=4}^{n} 2j$ then

$$120 = 2\left[\sum_{j=4}^{n}(j)\right] = 2\left[\sum_{j=1}^{n} j - \sum_{j=1}^{3} j\right] = 2\left[\frac{n(n+1)}{2} - (1+2+3)\right] = n(n+1) - 2(6) = n^2 + n - 12.$$

Solve $120 = n^2 + n - 12$ for n:

$$\begin{aligned} n^2 + n - 132 &= 0 \\ (n-11)(n+12) &= 0 \\ n &= 11, -12. \end{aligned}$$

Solution: $n = 11$.

39. If $\sum\limits_{j=n}^{15}(1-2j) = -209$ then

$$\begin{aligned} -209 &= \sum_{j=1}^{15}(1-2j) - \sum_{j=1}^{n-1}(1-2j) = \left[\sum_{j=1}^{15} 1 - 2\sum_{j=1}^{15} j\right] - \left[\sum_{j=1}^{n-1} 1 - 2\sum_{j=1}^{n-1} j\right] \\ &= \left[15 - 2\left(\frac{15(15+1)}{2}\right)\right] - \left[(n-1) - 2\frac{(n-1)(n)}{2}\right] \\ &= [15 - 240] - [-n^2 + 2n - 1] = n^2 - 2n - 224. \end{aligned}$$

Solve $-209 = n^2 - 2n - 224$ for n:

$$\begin{aligned} 0 &= n^2 - 2n - 15 \\ 0 &= (n+3)(n-5) \end{aligned}$$
Our solution is $n = 5$.

41. $$\begin{aligned} 13 + 15 + \cdots + 241 + 243 &= \sum_{j=1}^{116}(11 + 2j) = \sum 11 + 2\sum j \\ &= [116(11)] + 2\left[\frac{(116)(116+1)}{2}\right] \quad \text{[By Eq(4)]} \\ &= 1276 + (116)(117) = 14,848. \end{aligned}$$

43. $$\begin{aligned} \left(\frac{1}{n^2}\right)\sum_{j=1}^{n}(2j+1) &= \frac{1}{n^2}\left[2\left(\sum_{j=1}^{n} j\right) + \left(\sum_{j=1}^{n} 1\right)\right] = \frac{1}{n^2}\left[2\left(\frac{(n)(n+1)}{2}\right) + n \cdot 1\right] \\ &= \frac{1}{n^2}[n^2 + 2n] = 1 + \frac{2}{n} \to 1 \text{ as } n \to \infty. \end{aligned}$$

45. $$\begin{aligned} \left(\frac{1}{n^3}\right)\sum_{j=1}^{n} 6j^2 &= \left(\frac{1}{n^3}\right) 6\sum j^2 = \left(\frac{6}{n^3}\right)\frac{(n(n+1)(2n+1))}{6} = \frac{n(n+1)(2n+1)}{n \cdot n \cdot n} \\ &= (1)\left(1 + \frac{1}{n}\right)\left(2 + \frac{1}{n}\right) \to 1 \cdot 1 \cdot 2 = 2 \text{ as } n \to \infty. \end{aligned}$$

47. a) Since $\Delta x = \frac{b-a}{n} = \frac{2-0}{4} = \frac{1}{2}$, then $x_0 = 0$, $x_1 = \frac{1}{2}$, $x_2 = \frac{2}{2}$, $x_3 = \frac{3}{2}$, $x_4 = \frac{4}{2}$. Since f increases, then for a lower sum, the c_j are the left endpoints. So

$$\begin{aligned} \underline{S}_4 &= \sum_{j=1}^{4} f(c_j)\Delta x = \left(\frac{1}{2}\right)\sum f(c_j) = \left(\frac{1}{2}\right)\left[f(0) + f\left(\frac{1}{2}\right) + f(1) + f\left(\frac{3}{2}\right)\right] \\ &= \left(\frac{1}{2}\right)\left(2 + \frac{7}{2} + \frac{10}{2} + \frac{13}{2}\right) = 8.5. \end{aligned}$$

b) $\Delta x = \frac{2}{100} = \frac{1}{50}$ and $c_j = 0 + \frac{j-1}{50}$. So

$$\underline{S}_{100} = \sum_{j=1}^{100} f(c_j)\Delta x = \left(\frac{1}{50}\right)\sum f\left(\frac{j-1}{50}\right) = \left(\frac{1}{50}\right)\sum\left[3\left(\frac{j-1}{50}\right)+2\right]$$

$$= \frac{1}{50}\left\{\frac{3}{50}\left(\sum j\right) + \left(2 - \frac{3}{50}\right)\left(\sum 1\right)\right\}$$

$$= \frac{1}{50}\left\{\frac{3}{50}\left(\frac{100(100+1)}{2}\right) + \frac{97}{50}(100)\right\} = 9.94$$

c) $\Delta x = \frac{2-0}{n} = \frac{2}{n}$ and $c_j = 0 + (\Delta x)(j-1) = \frac{2(j-1)}{n}$. So

$$\underline{S}_n = \sum f(c_j)\Delta x = \sum_{j=1}^{n}\left[3\left(\frac{2(j-1)}{n}\right)+2\right]\left(\frac{2}{n}\right)$$

$$= \sum\left[\frac{6}{n}(j-1)\left(\frac{2}{n}\right)+2\left(\frac{2}{n}\right)\right] = \sum\left[\frac{12}{n^2}(j-1)+\frac{4}{n}\right]$$

$$= \frac{12}{n^2}\sum j - \frac{12}{n^2}\sum 1 + \frac{4}{n}\sum 1 = \frac{12}{n^2}\cdot\frac{n(n+1)}{2} - \frac{12}{n^2}\cdot\frac{n}{1} + \frac{4}{n}\frac{n}{1}$$

$$= \frac{12}{1}\left(\frac{n}{n}\right)\frac{1}{2}\left(\frac{n+1}{n}\right) - \frac{12}{n} + 4 = \frac{6}{1}\left(1+\frac{1}{n}\right) - \frac{12}{n} + 4 = 10 - \frac{6}{n}.$$

d) $\underline{S} = 10 - \frac{6}{n} \to 10$ as $n \to \infty$.

e) Use the right endpoints: $c_j = 0 + \frac{1}{50}j$.

$$\overline{S}_{100} = \sum f(c_j)\Delta x = \sum_{j=1}^{100}\left[3\left(\frac{j}{50}\right)+2\right]\left(\frac{1}{50}\right)$$

$$= \sum\left(\frac{1}{50}\cdot\frac{3}{50}\right)j + \sum\frac{2}{50}$$

$$= \frac{3}{50(50)}\left(\sum j\right) + \frac{2}{50}\left(\sum 1\right)$$

$$= \frac{1}{50}\left(\frac{3}{50}\right)\frac{(100)(100+1)}{2} + \frac{2}{50}(100) = 10.06.$$

f) $\Delta x = \frac{2-0}{n}$,
   $c_j = 0 + \left(\frac{2}{n}\right)j$.

So $\quad \overline{S}_n = \sum f(c_j)\Delta x = \sum_{j=1}^{100}\left[3\left(\frac{2}{n}j\right)+2\right]\left(\frac{2}{n}\right)$

$$= 3\left(\frac{2}{n}\right)\left(\frac{2}{n}\right)\sum j + 2\left(\frac{2}{n}\right)\sum 1 = \left(\frac{12}{n^2}\right)\frac{n(n+1)}{2} + \left(\frac{4}{n}\right)\cdot n$$

$$= \frac{12}{2}\left(\frac{n}{n}\right)\left(\frac{n+1}{n}\right) + 4 = (6)\left(1+\frac{1}{n}\right) + 4 = 10 + \frac{6}{n}.$$

g) $\overline{S}_n = 10 + \frac{6}{n} \to 10$ as $n \to \infty$.

h) The area is $10 = \lim \overline{S}_n = \lim \underline{S}_n$ by **Definition 1**.

49. a) Since $\Delta x = \frac{b-0}{n} = \frac{b}{n}$ and $f(x) = x^2$ increases on $[0, b]$, then c_j is the left endpoint for a lower sum: $c_j = 0 + (\Delta x)j - \Delta x = \left(\frac{b}{n}\right)j - \left(\frac{b}{n}\right)$. So

$$\underline{S}_n = \sum_{j=1}^{n} f(c_j)\Delta x = \sum \left[\frac{b}{n}(j-1)\right]^2 \left(\frac{b}{n}\right) = \sum \left(\frac{b}{n}\right)^2 (j^2 - 2j + 1)\left(\frac{b}{n}\right)$$

$$= \left(\frac{b}{n}\right)^3 \left\{\sum j^2 - 2\sum j + \sum 1\right\} = \frac{b^3}{n^3}\left\{\frac{(n)(n+1)(2n+1)}{6} - \frac{2}{1}\frac{n(n+1)}{2} + n\right\}$$

$$= b^3 \left\{\left(\frac{n}{n}\right)\left(\frac{n+1}{n}\right)\left(\frac{2n+1}{n}\right)\frac{1}{6} - \left(\frac{n}{n}\right)\left(\frac{n+1}{n}\right)\left(\frac{1}{n}\right) + \frac{n}{n}\frac{1}{n^2}\right\}$$

$$= b^3\left\{\frac{1}{6}\left(1 + \frac{1}{n}\right)\left(2 + \frac{1}{n}\right) - \left(1 + \frac{1}{n}\right)\left(\frac{1}{n}\right) + \frac{1}{n^2}\right\} = b^3\left\{\frac{1}{3} - \frac{1}{2}\frac{1}{n} + \frac{1}{6}\frac{1}{n^2}\right\}.$$

b) $\underline{S}_n \to b^3\left\{\frac{1}{3} - 0 + 0\right\} = \frac{b^3}{3}$ as $n \to \infty$.

c) $\Delta x = \frac{b-a}{n}$, and $c_j = a + (\Delta x)j - \Delta x$. So

$$\underline{S}_n = \sum_{j=1}^{n} f(c_j)\Delta x = \sum \left[a + \left(\frac{b-a}{n}\right)(j-1)\right]^2 \left(\frac{b-a}{n}\right)$$

$$= \sum \left[a^2 + 2a\left(\frac{b-a}{n}\right)(j-1) + \left(\frac{b-a}{n}\right)^2 (j-1)^2\right]\left(\frac{b-a}{n}\right)$$

$$= \left\{\sum \left[a^2 - 2a\left(\frac{b-a}{n}\right) + \left(\frac{b-a}{n}\right)^2\right] + \left[2a\left(\frac{b-a}{n}\right) - 2\left(\frac{b-a}{n}\right)^2\right]j \right.$$

$$\left. + \left[\left(\frac{b-a}{n}\right)^2\right]j^2\right\}\left(\frac{b-a}{n}\right)$$

$$= \left\{\left[a^2 - \frac{2a(b-a)}{n} + \left(\frac{b-a}{n}\right)^2\right]n + \left[2a\left(\frac{b-a}{n}\right) - 2\left(\frac{b-a}{n}\right)^2\right]\frac{n(n+1)}{2}\right.$$

$$\left. + \left[\left(\frac{b-a}{n}\right)^2\right]\frac{n(n+1)(2n+1)}{6}\right\}\left(\frac{b-a}{n}\right)$$

$$= \left[(b-a)a^2\frac{n}{n}\right] + \left[\frac{-2a(b-a)2n}{n^2}\right] + \left[(b-a)^3\left(\frac{n}{n^3}\right)\right] + \left[\frac{2a(b-a)^2}{2}\frac{n(n+1)}{n\cdot n}\right]$$

$$+ \left[\frac{-2(b-a)^3}{2}\frac{(n)}{n}\frac{(n+1)}{n}\frac{1}{n}\right] + \left[\frac{(b-a)^3}{6}\frac{(n)}{n}\frac{(n+1)}{n}\frac{(2n+1)}{n}\right]$$

This expression is convenient for taking the limit, in 49 d).

d) Take the limit of each term of $\underline{S}_n$ from 49 c):

$$\underline{S}_n \to (b-a)a^2 + 0 + 0 + a(b-a)^2 + 0 + \frac{(b-a)^3}{6}\left(\frac{2}{1}\right)$$

$$= a^2b - a^3 + a^3 - 2a^2b + ab^2 + \frac{b^3 - 3b^2a + 3ba^2 - a^3}{3} = \frac{-a^3}{3} + \frac{b^3}{3}.$$

e) The area A0b from 0 to b is equal to the sum of the area A0a from 0 to a and the area Aab from a to b: A0b = A0a + Aab. So Aab = A0b − A0a. And A0a = $\frac{a^3}{3}$ from 49 b). So Aab = $\frac{b^3}{3} - \frac{a^3}{3}$.

f) The area from 1 to 5 is the area from 0 to 5 minus the area from 0 to 1: $\frac{5^3}{3} - \frac{1^3}{3} = \frac{125-1}{3} = \frac{124}{3}$.

51. Use the method pointed out in 49 e): find the area from $x = 0$ to $x = 3$, then from $x = 0$ to $x = 2$, then subtract the second area.

(1) The function increases, so use left endpoints for a lower sum: $c_j = 0 + \left(\frac{3}{n}\right)(j-1)$.

$$
\begin{aligned}
\underline{S}_n &= \sum_{j=1}^{n} f(c_j)\Delta x = \sum \left[3\left(\frac{3}{n}(j-1)\right)^2 + 7 \right]\left(\frac{3}{n}\right) \\
&= \sum \left[3\left\{\left(\frac{3}{n}\right)^2 (j^2 - 2j + 1)\right\} + 7 \right]\left(\frac{3}{n}\right) \\
&= 3\left(\frac{3}{n}\right)^2 \left(\frac{3}{n}\right)\left(\sum j^2\right) - 3\left(\frac{3}{n}\right)^2 \left(\frac{3}{n}\right)(2)\left(\sum j\right) + \left[3\left(\frac{3}{n}\right)^2 + 7\right]\left(\frac{3}{n}\right)\left(\sum 1\right) \\
&= 3\left(\frac{3^3}{n^3}\right)\left(\frac{(n)(n+1)(2n+1)}{6}\right) - (2)(3)\left(\frac{3}{n}\right)^3 \left(\frac{n(n+1)}{2}\right) + \left[3\left(\frac{3}{n}\right)^3 + 7\left(\frac{3}{n}\right)\right]\left(\frac{n}{1}\right) \\
&= \left(\frac{3}{6}\right)\frac{3^3}{1}\left(\frac{n}{n}\right)\left(\frac{n+1}{n}\right)\left(\frac{2n+1}{n}\right) - (3)\frac{3^3}{n}\left(\frac{n}{n}\frac{n+1}{n}\right) + \left(3\left(\frac{3^3}{n^2}\right) + \frac{3(7)}{1}\right)
\end{aligned}
$$

which approaches $\frac{3^3}{2}\left(\frac{2}{1}\right) - (0) + (0 + 21) = 48$ as $n \to \infty$.

(2) Replacing 3 by 2 above, we get.

$$
\underline{S}_n = \left(\frac{3}{6}\right) 2^3 \left(\frac{n}{n} \frac{n+1}{n} \frac{2n+1}{n}\right) - 3\frac{2^3}{n} \frac{n}{n} \frac{n+1}{n} + \left(3\frac{2^3}{n^2} + 2(7)\right)
$$

which approaches $22 = 8 - 0 + 14$ as $n \to \infty$. The difference between the areas is $48 - 22 = 26$.

53. Now $\Delta x = \frac{1-0}{n} = \frac{1}{n}$, and since f increases on $[0,1]$, then the c_j are the left endpoints for a lower sum: $c_j = 0 + (\Delta x)(j-1) = \left(\frac{j}{n}\right) - \left(\frac{1}{n}\right)$. So

$$
\begin{aligned}
\underline{S}_n &= \sum_{j=1}^{n} f(c_j)\Delta x = \sum \left[3\left[\frac{j-1}{n}\right]^2 + 6 \right]\left(\frac{1}{n}\right) = \sum \left[3(j^2 - 2j + 1)\left(\frac{1}{n}\right)^2 + 6 \right]\left(\frac{1}{n}\right) \\
&= 3\left(\frac{1}{n}\right)^3 \left(\sum j^2\right) - 3(2)\left(\frac{1}{n}\right)^3 \left(\sum j\right) + \left[3\left(\frac{1}{n}\right)^3 + 6\frac{1}{n}\right]\left(\sum 1\right) \\
&= 3\frac{1}{n^3}\frac{n(n+1)(2n+1)}{6} - 3(2)\frac{1}{n^3}\cdot\frac{n(n+1)}{2} + \left[3\frac{1}{n^3} + 6\frac{1}{n}\right]\frac{n}{1} \\
&= \frac{1}{2}\left(\frac{n}{n}\right)\left(1+\frac{1}{n}\right)\left(2+\frac{1}{n}\right) - \left(\frac{3}{1}\right)\frac{n}{n}\left(1+\frac{1}{n}\right)\left(\frac{1}{n}\right) + \left[\frac{3}{n^2} + 6\right]
\end{aligned}
$$

which approaches $\frac{1}{2}(1)(1)(2) - 3(1)(1)(0) + (0) + 6 = 7$ as $n \to \infty$.

55. Since $f(x) = \text{ABS}(1 - x^2)$ decreases from 1 to 0 on $[0,1]$ and increases from 0 to 1 on $[1,2]$, we will divide the sum into two parts, each with n terms, each covering one of those intervals.

In the first sum, the c_j are the right endpoints: $c_j = 0 + (\Delta x)j = \frac{j}{n}$, and $1 - x^2 \geq 0$.

$$
\underline{S}_n \text{ on}[0,1] = \sum_{j=1}^{n} f(c_j)\Delta x = \sum \text{ABS}\left(1 - \left(\frac{j}{n}\right)^2\right)\left(\frac{1}{n}\right) = \frac{-1}{n^3}\sum j^2 + \frac{1}{n}\sum 1
$$

211

$$= \left(\frac{-1}{n^3}\right)\frac{n(n+1)(2n+1)}{6} + \left(\frac{1}{n}\right)n = -\frac{1}{6}\frac{n}{n}\frac{n+1}{n}\frac{2n+1}{n} + 1 \to -\frac{1}{6}\frac{2}{1} + 1$$

$$= \frac{2}{3} \text{ as } n \to \infty.$$

In the second sum, on $[1,2]$, the c_j are the left endpoints, $c_j = 1 + (\Delta x)(j-1)$, and $1 - x^2 \le 0$. So

$$
\begin{aligned}
\underline{S}_n \text{ on } [1,2] &= \sum_{j=1}^{n} f(c_j)\Delta x = \sum \left\{ \text{ABS}\left(1 - \left[1 + \frac{j-1}{n}\right]^2\right)\right\}\left(\frac{1}{n}\right) \\
&= \sum\left\{\left[1 + 2\left(\frac{j-1}{n}\right) + \frac{j^2 - 2j + 1}{n^2}\right] - 1\right\}\left(\frac{1}{n}\right) \\
&= \left(\frac{1}{n^3}\right)\left(\sum j^2\right) + \left(\frac{2}{n^2} - \frac{2}{n^3}\right)\left(\sum j\right) + \left(-\frac{2}{n^2} + \frac{1}{n^3}\right)\sum 1 \\
&= \left[\frac{n(n+1)(2n+1)}{n \cdot n \cdot n \cdot 6}\right] + \left[\left(\frac{2}{n^2} - \frac{2}{n^3}\right)\frac{n(n+1)}{2}\right] + \left[\left(-\frac{2}{n^2} + \frac{1}{n^3}\right)\frac{n}{1}\right] \\
&\to \left[\frac{2}{1}\frac{1}{6}\right] + [1 - 0] + [0 + 0] \\
&= \frac{4}{3} \text{ as } n \to \infty.
\end{aligned}
$$

So the entire sum approaches $\frac{2}{3} + \frac{4}{3} = 2$.

57. a) If $\frac{x^2}{9} + \frac{y^2}{4} = 1$, then

$$\frac{y^2}{4} = 1 - \frac{x^2}{9}, \quad \frac{y}{2} = \pm\sqrt{1 - \frac{x^2}{9}}, \quad \text{and} \quad y = \pm 2\sqrt{1 - \frac{x^2}{9}}.$$

b) The graph is symmetric about the x-axis and y-axis, so we can do only the part in Quadrant I, then multiply by 4 to get the total area.

c) The function decreases in Quadrant I, so use the right endpoints for a lower sum.
$\underline{S}_3 : c_1 = 1, \ c_2 = 2, \ c_3 = 3$. Also $\Delta x = \frac{3-0}{3} = 1$. So

$$\underline{S}_3 = \sum_{j=1}^{3} f(c_j)\Delta x = 2\sqrt{1 - \frac{1^2}{9}} + 2\sqrt{1 - \frac{2^2}{9}} + 2\sqrt{1 - \frac{3^2}{9}} \approx 3.37633.$$

So $\underline{S}_3 \cdot 4 \approx 13.50532$.

d) If $n = 6$, $\Delta x = \frac{3-0}{6} = \frac{1}{2}$ then $c_1 = \frac{1}{2}, c_2 = \frac{2}{2}, c_3 = \frac{3}{2}, \cdots, c_6 = \frac{6}{2}$, and

$$\underline{S}_6 = \sum_{j=1}^{6} f(c_j)\Delta x \approx 4.09297.$$

So $\underline{S}_6 \cdot 4 \approx 16.37190$.

59. a) We have divided the entire circle, measuring $2\pi(rad)$, into $2n$ equal angles, so each measures $\frac{2\pi}{2n} = \frac{\pi}{n}(rad)$.

b) As shown in the figure, the base of each triangle is $r\cos\left(\frac{\pi}{n}\right)$ and the height is $r\sin\left(\frac{\pi}{n}\right)$, from basic trigonometry. So the area is $\left(\frac{1}{2}\right)$ (base) (height) $= \left(\frac{r^2}{2}\right)\sin\left(\frac{\pi}{n}\right)\cos\left(\frac{\pi}{n}\right) = \left(\frac{r^2}{4}\right)\left[2\sin\left(\frac{\pi}{n}\right)\cos\left(\frac{\pi}{n}\right)\right] = \left(\frac{r^2}{4}\right)\sin\left(\frac{2\pi}{n}\right)$ by the double-angle identity for sin.

c) To get A_n, sum the areas of the triangles:

$$
\begin{aligned}
A_n &= \sum_{t=1}^{2n} \frac{r^2}{4}\sin\left(\frac{2\pi}{n}\right) \\
&= \frac{r^2}{4}\sin\left(\frac{2\pi}{n}\right)\left(\sum 1\right) \\
&= \frac{r^2}{4}\left[\sin\left(\frac{2\pi}{n}\right)\right](2n) \\
&= \frac{nr^2}{2}\sin\left(\frac{2\pi}{n}\right).
\end{aligned}
$$

d)

Table 1.2

r	n	A_n	$A = \pi r^2$
1	5		
1	20	3.09017	
1	50		
4	5	38.04226	50.26548
4	20		
4	50		
4	200	50.25721	
10	5		314.15927
10	20		
10	50		
10	200	314.10759	
10	500	314.15100	
10	1000		

61. The triangles in Ex. 59 are inscribed in the circle, so the sum of their areas, $\left(\frac{nr^2}{2}\right)\sin\left(\frac{2\pi}{n}\right)$, is a lower sum for the area. The triangles in Ex. 60 are circumscribed, so $nr^2\tan\left(\frac{\pi}{n}\right)$ is an upper sum for the area. Take the limits of the two sums:

$$\frac{nr^2}{2}\sin\left(\frac{2\pi}{n}\right) = \pi r^2\left(\frac{n}{2\pi}\right)\sin\left(\frac{2\pi}{n}\right) = \pi r^2\frac{\sin(u)}{u}\left[u = \frac{2\pi}{n}\right]$$

which approaches $\pi r^2 \cdot 1$ as $n \to \infty$ since $\frac{2\pi}{n} = u \to 0$ as $n \to \infty$.

Also $nr^2\tan\left(\frac{\pi}{n}\right) = \pi r^2\left(\frac{n}{\pi}\right)\tan\left(\frac{\pi}{n}\right) = \pi r^2\frac{\tan(w)}{w}\left[w = \frac{\pi}{n}\right]$ which approaches $\pi r^2 \cdot 1$ as $n \to \infty$, since $w = \frac{\pi}{n} \to 0$ as $n \to \infty$.

63. Assume Eq(3) and Ex. 62. Then

$$
\begin{aligned}
\sum_{j=1}^{n} j &= \sum\left(\frac{1}{2}(2j-1)+\frac{1}{2}\right) = \frac{1}{2}\sum(2j-1)+\frac{1}{2}\sum 1 \\
&= \frac{1}{2}n^2 + \frac{1}{2}n = \frac{1}{2}(n^2+n) = \frac{1}{2}n(n+1).
\end{aligned}
$$

65. a) Using addition identitites: $\sin(\alpha+\beta) - \sin(\alpha-\beta) = [\sin\alpha\cos\beta + \cos\alpha\sin\beta] - [\sin\alpha\cos\beta - \cos\alpha\sin\beta] = 2\cos\alpha\sin\beta$.

b) To show

$$\sum_{j=0}^{n-1}\cos(j\gamma) = \frac{1}{2} + \frac{\sin\left(n-\frac{1}{2}\right)\gamma}{2\sin\left(\frac{1}{2}\gamma\right)}$$

for all n, show it for (1) $n = 1$, then show (2) that if it is true for n, then it is also true for $n+1$.

(1) $\displaystyle\sum_{j=0}^{1-1}\cos(j\gamma) = \cos(0\cdot\gamma) = 1$ and

$$\frac{1}{2}\frac{\sin\left(1-\frac{1}{2}\right)\gamma}{2\sin\left(\frac{1}{2}\gamma\right)} = \frac{1}{2} + \frac{\sin\left(\frac{1}{2}r\right)}{2\sin\left(\frac{1}{2}r\right)} = \frac{1}{2} + \frac{1}{2} = 1.$$

213

(2) Assume the equation is true for n. Then

$$\sum_{j=0}^{(n+1)-1} \cos j\gamma = \left[\sum_{j=0}^{n-1} \cos(j\gamma)\right] + \cos(n\gamma) = \left[\frac{1}{2} + \frac{1}{2}\frac{\sin\left(n - \frac{1}{2}\right)\gamma}{\sin\left(\frac{1}{2}\right)\gamma}\right] + \cos(n\gamma).$$

We want to show that this is equal to $\left[\dfrac{1}{2} + \dfrac{1}{2}\dfrac{\sin\left(n + 1 - \frac{1}{2}\right)\gamma}{\sin\left(\frac{1}{2}\gamma\right)}\right]$. This is equivalent to showing that $\cos(n\gamma)$ is equal to the difference of the last two expressions in brackets. So take the difference and simplify it:

$$\left[\frac{1}{2} + \frac{1}{2}\frac{\sin\left(n + \frac{1}{2}\right)\gamma}{\sin\left(\frac{1}{2}\gamma\right)}\right] - \left[\frac{1}{2} + \frac{1}{2}\frac{\sin\left(n - \frac{1}{2}\right)\gamma}{\sin\left(\frac{1}{2}\gamma\right)}\right] =$$

$$\left[\frac{1}{2} - \frac{1}{2}\right] + \left[\frac{1}{2\sin\left(\frac{1}{2}\gamma\right)}\right]\left[\sin\left(n + \frac{1}{2}\right)\gamma - \sin\left(n - \frac{1}{2}\right)\gamma\right]$$

$$= 0 + \frac{1}{2\sin\left(\frac{1}{2}\gamma\right)}\left[\sin\left(n\gamma + \frac{\gamma}{2}\right) - \sin\left(n\gamma - \frac{\gamma}{2}\right)\right]$$

[Use part a) now, where $\alpha = n\gamma$ and $\beta = \dfrac{\gamma}{2}$.]

$$= \frac{1}{2\sin\left(\frac{1}{2}\gamma\right)}\left[2\cos(n\gamma)\sin\left(\frac{\gamma}{2}\right)\right] = \cos(n\gamma)$$

after cancelling.

67. **Graph using the range $[-1, 2] \times [0, 2]$ and $n = 16$.**

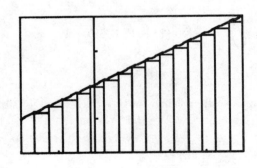 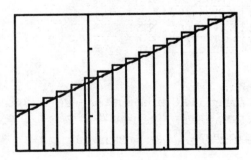

Table:

n	4	8	16	64
$\underline{S}_n$	3.1875	3.46875	3.609375	3.71484375
$\overline{S}_n$	4.3125	4.03125	3.890625	3.78515625

69. **Graph using the range $[0, 1] \times [0, 1]$ and $n = 16$.**

214

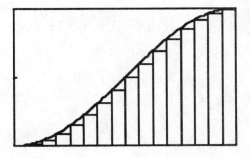 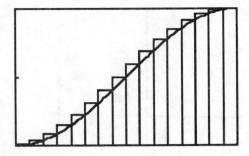

Table:

n	4	8	16	64
$\underline{S}_n$	0.375	0.4375	0.46875	0.4921875
$\overline{S}_n$	0.625	0.5625	0.53125	0.5078125

71. Table:

n	$\overline{S}_n$	$\underline{S}_n$	$\overline{S}_n - \underline{S}_n$
1	4	2	2
2	3.6	2.6	1
4	3.3811765	2.8811765	0.5
8	3.2639885	3.0139885	0.25
16	3.20344161	3.07844161	0.125
32	3.17267989	3.11017989	0.0625
64	3.15717696	3.12592696	0.3125

From the table it appears that $\overline{S}_n - \underline{S}_n = \frac{1}{2}^{m-1}$ where m is chosen so that $2^m = n$. So for $n = 1$, $m = 0$; $n = 2$, $m = 1$; $n = 4$, $m = 2$; $n = 8$, $m = 3$; etc. Therefore if $\overline{S}_n - \underline{S}_n < 10^{-5}$ then we must choose m so that $10^{-5} < \frac{1}{2}^{m-1}$. So solving the equation $2^{1-m} = 10^{-5}$ gives $m = 17.6096$. Thus take $m = 18$. Then $\frac{1}{2}^{17} = 7.6295 \times 10^{-6} < 10^{-5}$. Thus $n = 2^{18} = 262144$ yields the prescribed error.

6.2 Riemann Sums: The Definite Integral

1. $f(x) = x + 2$, $P_3 = \{1, 2, 3, 5\}$

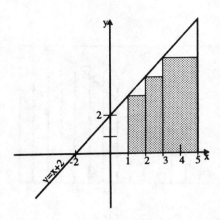

$$R_3 = \sum_{j=1}^{3} f(t_j)\Delta x_j$$

$$= f(1)(2-1) + f(2)(3-2)$$

$$+ f(3)(5-3)$$

$$= 3(1) + 4(1) + 5(2) = 17.$$

Exercise 1

3. $f(x) = 3 - x$, $P_4 = \left\{0, 1, \frac{3}{2}, 2, 4\right\}$

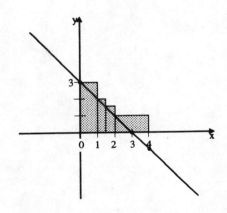

$$R_4 = \sum_{j=1}^{4} f(t_j)\Delta x_j$$

$$= f(0)(1-0) + f(1)\left(\frac{3}{2}-1\right)$$

$$+ f\left(\frac{3}{2}\right)\left(2-\frac{3}{2}\right) + f(2)(4-2)$$

$$= 3(1) + 2\left(\frac{1}{2}\right)$$

$$+ \frac{3}{2}\left(\frac{1}{2}\right) + 1(2)$$

$$= 6\frac{3}{4}.$$

Exercise 3

5. $f(x) = x^2 - x$, $P_3 = \{0, 1, 2, 4\}$.

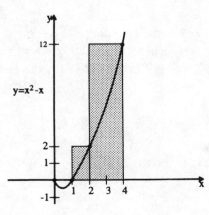

$$R_3 = \sum_{j=1}^{3} f(t_j)\Delta x_j$$

$$= f(1)(1-0) + f(2)(2-1)$$

$$+ f(4)(4-2)$$

$$= 0(1) + 2(1) + 12(2)$$

$$= 26.$$

Exercise 5

216

7. $f(x) = \cos x$, $P_4 = \left\{ 0, \frac{\pi}{2}, \pi, 2\pi, \frac{5\pi}{2} \right\}$.

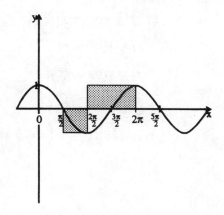

Exercise 7

$$
\begin{aligned}
R_4 &= \sum_{j=1}^{4} f(t_j)\Delta x_j \\
&= f\left(\tfrac{\pi}{2}\right)\left(\tfrac{\pi}{2} - 0\right) + f(\pi)\left(\pi - \tfrac{\pi}{2}\right) \\
&\quad + f(2\pi)(2\pi - \pi) \\
&\quad + f\left(\tfrac{5\pi}{2}\right)\left(\tfrac{5\pi}{2} - 2\pi\right) \\
&= 0\left(\tfrac{\pi}{2}\right) + (-1)\left(\tfrac{\pi}{2}\right) \\
&\quad + 1(\pi) + 0\left(\tfrac{\pi}{2}\right) \\
&= \tfrac{\pi}{2}.
\end{aligned}
$$

9. $f(x) = \sqrt{x}$, $P_3 = \{0, 2, 6, 12\}$.

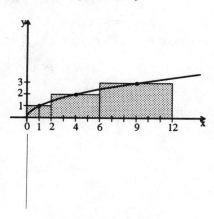

Exercise 9

$$
\begin{aligned}
R_3 &= \sum_{j=1}^{3} f(t_j)\Delta x_j \\
&= f(1)(2 - 0) + f(4)(6 - 2) \\
&\quad + f(9)(12 - 6) \\
&= 1(2) + 2(4) + 3(6) \\
&= 28.
\end{aligned}
$$

11. $f(x) = \sin x$, $P_4 = \left\{ 0, \frac{\pi}{2}, \pi, \frac{3\pi}{2}, 2\pi \right\}$.

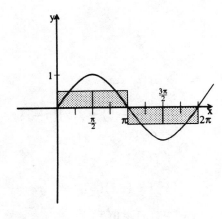

$$R_4 = \sum_{j=1}^{4} f(t_j)\Delta x_j$$

$$= f\left(\tfrac{\pi}{4}\right)\left(\tfrac{\pi}{2} - 0\right) + f\left(\tfrac{3\pi}{4}\right)\left(\pi - \tfrac{\pi}{2}\right)$$

$$+ f\left(\tfrac{5\pi}{4}\right)\left(\tfrac{3\pi}{2} - \pi\right)$$

$$+ f\left(\tfrac{7\pi}{4}\right)\left(2\pi - \tfrac{3\pi}{2}\right)$$

$$+ \sqrt{\tfrac{1}{2}}\left(\tfrac{\pi}{2}\right) + \sqrt{\tfrac{1}{2}}\left(\tfrac{\pi}{2}\right)$$

$$+ \left(-\sqrt{\tfrac{1}{2}}\right)\left(\tfrac{\pi}{2}\right) + \left(-\sqrt{\tfrac{1}{2}}\right)\left(\tfrac{1}{2}\right)$$

$$= 0.$$

Exercise 11

13. $f(x) = 4 - x^2$, $[a, b] = [-2, 2]$
$P_4 = \{-2, -1, 0, 1, 2\}$

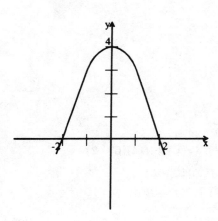

$$\text{largest } R_n = f(-1)(-1 - (-2))$$

$$+ f(0)(0 - (-1))$$

$$+ f(0)(1 - 0)$$

$$+ f(1)(2 - 1)$$

$$= 3(1) + 4(1) + 4(1) + 3(1)$$

$$= 14.$$

$$\text{smallest } R_n = f(-2)(1) + f(-1)(1)$$

$$+ f(1)(1) + f(2)(1)$$

$$= 0 + 3 + 3 + 0$$

$$= 6.$$

Exercise 13

15. $f(x) = \sin x$, $[a, b] = [0, 2\pi]$

$P_4 = \left\{0, \tfrac{\pi}{2}, \pi, \tfrac{3\pi}{2}, 2\pi\right\}$, $\Delta x_j = \tfrac{\pi}{2}$.

$$\text{largest } R_n = f\left(\tfrac{\pi}{2}\right) \cdot \tfrac{\pi}{2} + f\left(\tfrac{\pi}{2}\right) \cdot \tfrac{\pi}{2} + f(\pi)\left(\tfrac{\pi}{2}\right) + f(2\pi)\left(\tfrac{\pi}{2}\right) = (1 + 1 + 0 + 0)\left(\tfrac{\pi}{2}\right) = \pi.$$

$$\text{smallest } R_n = \left[f(0) + f(\pi) + f\left(\tfrac{3\pi}{2}\right) + f\left(\tfrac{3\pi}{2}\right)\right]\left(\tfrac{\pi}{2}\right) = (0 + 0 - 1 - 1)\left(\tfrac{\pi}{2}\right) = -\pi.$$

17. $\displaystyle\int_0^5 6\,dx = 6(5 - 0) = 30$ by Example 5.

19. $\displaystyle\int_{-2}^{4}(3x-2)\,dx \;=\; 3\int_{-2}^{4}x\,dx - 2\int_{-2}^{4}1\,dx$

$$= \; 3\left[\frac{1}{2}\left(4^2-(-2)^2\right)\right] - 2[4-(-2)] \;=\; 3[6]-2[6] \;=\; 6.$$

21. $\displaystyle\int_{4}^{0}(9-2x)\,dx \;=\; 9\int_{4}^{0}1\,dx - 2\int_{4}^{0}x\,dx \;=\; 9(0-4)-2\left[\frac{1}{2}\left(0^2-4^2\right)\right]$

$$= \; -36-2[-8] \;=\; -20.$$

23. $\displaystyle\int_{3}^{1}(6-x^2)\,dx \;=\; \int_{3}^{1}6\,dx - \int_{3}^{1}x^2\,dx \;=\; 6(1-3)-\left[\frac{1}{3}\left(1^3-3^3\right)\right]$

$$= \; -12-\left(-\frac{26}{3}\right) \;=\; \frac{-36+26}{3} \;=\; -\frac{10}{3}.$$

25. $\displaystyle\int_{0}^{1}(x^2-x+2)\,dx \;=\; \int_{0}^{1}x^2\,dx - \int_{0}^{1}x\,dx + 2\int_{0}^{1}1\,dx$

$$= \; \frac{1}{3}\left(1^3-0^3\right)-\frac{1}{2}\left(1^2-0^2\right)+2(1-0) \;=\; \frac{1}{3}-\frac{1}{2}+2 \;=\; \frac{11}{6}.$$

27. $\displaystyle\int_{0}^{4}x(2+x)\,dx \;=\; \int_{0}^{4}(x^2+2x)\,dx \;=\; \int_{0}^{4}x^2\,dx + 2\int_{0}^{4}x\,dx$

$$= \; \frac{1}{3}\left(4^3-0^3\right)+2\frac{1}{2}\left(4^2-0^2\right) \;=\; \frac{1}{3}64+16 \;=\; \frac{64+48}{3} \;=\; \frac{112}{3}.$$

29. $\displaystyle\int_{-4}^{-1}2x(1-2x)\,dx \;=\; \int(-4x^2+2x)\,dx \;=\; -4\int x^2\,dx + 2\int x\,dx$

$$= \; (-4)\frac{1}{3}\left[(-1)^3-(-4)^3\right]+2\left[\frac{1}{2}\left((-1)^2-(-4)^2\right)\right]$$

$$= \; \left(-\frac{4}{3}\right)(63)+(-15) \;=\; -84-15 \;=\; -99.$$

31. a) $\displaystyle\int_{2}^{0}f(x)\,dx = -\int_{0}^{2}f(x)\,dx = -(3).$

b) $\displaystyle\int_{0}^{5}f(x)\,dx = \int_{0}^{2}f(x)\,dx + \int_{2}^{5}f(x)\,dx = 3-2 = 1.$

c) $\displaystyle\int_{2}^{8}f(x)\,dx = \int_{2}^{5} + \int_{5}^{8} = -2+5 = 3.$

d) $\displaystyle\int_{0}^{8}f(x)\,dx = \int_{0}^{2} + \int_{2}^{5} + \int_{5}^{8} = 3-2+5 = 6.$

e) $\displaystyle\int_{5}^{0}f(x)\,dx = -\int_{0}^{5} = -1$ from (b).

f) $\displaystyle\int_{8}^{2}f(x)\,dx = -\int_{2}^{8} = -3$ from (c).

219

33. a) $\int_1^3 [f(x) + g(x)]\, dx = \int_1^3 f(x)\, dx + \int_1^3 g(x)\, dx = 5 - 2 = 3.$

  b) $\int_1^3 [6g(x) - 2f(x)]\, dx = 6\int_1^3 g(x)\, dx - 2\int_1^3 f(x)\, dx = 6(-2) - 2(5) = -22.$

  c) $\int_3^1 2f(x)\, dx - \int_1^3 g(x)\, dx = 2\left(-\int_1^3 f(x)\, dx\right) - (-2) = 2(-5) + 2 = -8.$

  d) $\int_1^3 [2f(x) + 3]\, dx = 2\int_1^3 f(x)\, dx + 3\int_1^3 1\, dx = 2(5) + 3(3 - 1) = 16.$

  e) $\int_3^1 [g(x) - 4f(x) + 5]\, dx = -\int_1^3 g(x)\, dx - 4\left(-\int_1^3 f(x)\, dx\right) + 5\left(-\int_1^3 1\, dx\right)$
  $= -(-2) - 4(-5) + 5(-2) = 12.$

  f) $\int_1^3 [1 - 4g(x) + 3f(x)]\, dx = \int_1^3 1\, dx - 4\int_1^3 g(x)\, dx + 3\int_1^3 f(x)\, dx$
  $= 2 - 4(-2) + 3(5) = 25.$

35.

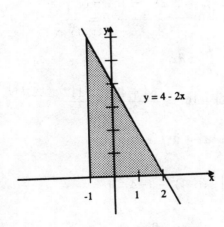

y = 4 - 2x

Exercise 35

$\int_{-1}^{2} (4 - 2x)\, dx$ = area of triangle

$= \tfrac{1}{2}bh$

$= \tfrac{1}{2}(3)(6)$

$= 9.$

37.

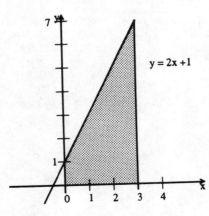

y = 2x + 1

Exercise 37

$\int_0^3 (2x + 1)\, dx$ = area of trapezoid

$= \tfrac{1}{2}(b_1 + b_2)(h)$

$= \tfrac{1}{2}(1 + 7)(3)$

$= 12.$

39.

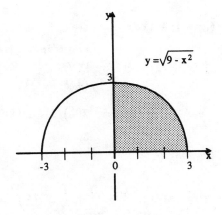

Exercise 39

$$\int_{+0}^{3} \sqrt{9 - x^2} \, dx \quad = \quad \text{area quarter circle}$$

$$= \quad \tfrac{1}{4} \pi r^2$$

$$= \quad \tfrac{9\pi}{4}.$$

41.

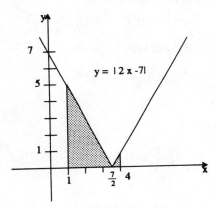

Exercise 41

$$\int_{1}^{4} |2x - 7| \, dx \quad = \quad \text{sum of areas}$$

$$\text{of triangles}$$

$$= \quad \tfrac{1}{2} b_1 h_1 + \tfrac{1}{2} b_2 h_2$$

$$= \quad \tfrac{1}{2} \left(\tfrac{5}{2}\right)(5) + \tfrac{1}{2} \left(\tfrac{1}{2}\right)(1)$$

$$= \quad \tfrac{25+1}{4}$$

$$= \quad \tfrac{13}{2}.$$

43.

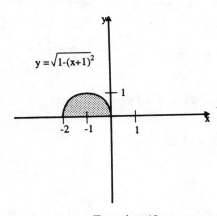

Exercise 43

$$\int_{-2}^{0} \sqrt{1 - (x + 1)^2} \, dx \quad = \quad \text{area semicircle}$$

$$= \quad \tfrac{1}{2} \pi r^2$$

$$= \quad \tfrac{1}{2} \pi 1^2$$

$$= \quad \tfrac{\pi}{2}.$$

221

45.

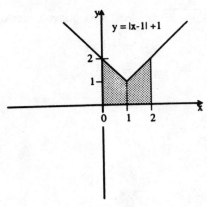

Exercise 45

$\int_0^2 (|x - 1| + 1)\, dx$ = area of

two trapezoids

= twice

$\left(\frac{1}{2}(b_1 + b_2)h\right)$

= $2(\frac{1}{2})(2 + 1)(1)$

= 3.

47.

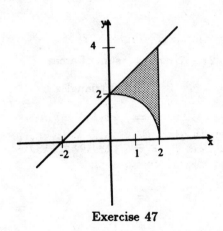

Exercise 47

$\int_0^2 \left(x + 2 - \sqrt{4 - x^2}\right) dx$

= area trapezoid −

   area quarter circle

= $\frac{1}{2}(b_1 + b_2)h - \frac{1}{4}\pi r^2$

= $\frac{1}{2}(2 + 4)(2) - \frac{1}{4}\pi 2^2$

= $6 - \pi$.

49.

$$\int_0^2 f(x)\, dx = \int_0^1 (1 - x)\, dx + \int_1^2 (x^2 - 1)\, dx$$

$$= \int_0^1 1\, dx - \int_0^1 x\, dx + \int_1^2 x^2\, dx - \int_1^2 1\, dx = (1 - 0) - \frac{1}{2}(1^2 - 0^2)$$

$$+ \frac{1}{3}(2^3 - 1^3) - (2 - 1) = 1 - \frac{1}{2} + \frac{7}{3} - 1 = \frac{14 - 3}{6} = \frac{11}{6}.$$

51. $\int_0^2 f(x)\, dx = \int_0^1 (x^2 - 1)\, dx + \int_1^2 (x - 1)\, dx = \int_0^1 x^2\, dx - \int_0^1 1\, dx + \int_1^2 x\, dx - \int_1^2 1\, dx$

$$= \frac{1}{3}(1^3 - 0^3) - (1 - 0) + \frac{1}{2}(2^2 - 1^2) - (2 - 1) = \frac{1}{3} - 1 + \frac{3}{2} - 1$$

$$= \frac{2 - 6 + 9 - 6}{6} = -\frac{1}{6}.$$

53.   a) Suppose f is continuous on $[a, b]$. Then it has a max M and a min m on $[a, b]$; so $m \leq f(x) \leq M$

for $x \in [a, b]$. Hence for any Riemann sum R_n,

$$\sum_{j=1}^{n} m \Delta x_j \;\leq\; \sum_{j=1}^{n} f(t_j) \Delta x_j \;\leq\; \sum_{j=1}^{n} M \Delta x_j.$$

Factor m and M out of each sum:

$$m \left(\sum \Delta x_j \right) \;\leq\; R_n \;\leq\; M \left(\sum \Delta x_j \right).$$

So

$$m(b - a) \leq R_n \;\leq\; M(b - a),$$

since $\displaystyle\sum_{j=1}^{n} \Delta x_j = b - a$. So each R_n is sandwiched between the constants $m(b-a)$ and $M(b-a)$,

and therefore its limit, $\int_a^b f(x)\, dx$, must also lie in the interval $[m(b-a),\; M(b-a)]$.

b) If $1 \leq x \leq 4$, then $1 \leq \sqrt{x} \leq 2$, so $\frac{1}{1} \geq \frac{1}{\sqrt{x}} \geq \frac{1}{2}$, and the min and max of $f(x) = \frac{1}{\sqrt{x}}$ on $[1, 4]$ are $\frac{1}{2}$ and 1. So by # 53 a),

$$\frac{1}{2}(4 - 1) \;\leq\; \int_1^4 \frac{1}{\sqrt{x}}\, dx \;\leq\; 1(4 - 1), \quad \text{or} \quad \frac{3}{2} \leq \int_1^4 \frac{1}{\sqrt{x}}\, dx \leq 3.$$

55. **By Theorem 2,** we can use the partitions with equal-length, since the limit is independent of which we use, so long as $||P_n|| \to 0$. Then $\Delta x_j = \frac{b-a}{n} = \Delta x$, and $x_j = a + j\Delta x$ is the jth point of the partition $P_n = \{x_0, x_1, \cdots, x_n\}$. Choose the right endpoint of each subinterval:

$$t_j = x_j = a + j\Delta x \text{ for } j = 1, 2, \cdots, n.$$

Then

$$\begin{aligned}
R_n &= \sum_{j=1}^{n} f(t_j) \Delta x_j = \sum_{j=1}^{n} (a + j\Delta x)\Delta x = \sum a\Delta x + \sum j(\Delta x)^2 \\
&= a\Delta x \sum 1 + (\Delta x)^2 \sum j = a\Delta x \cdot n + (\Delta x)^2 \frac{n(n+1)}{2} \\
&= \frac{a(b-a)}{n}\frac{n}{1} + \frac{(b-a)^2}{n^2}\frac{n(n+1)}{2} = a(b-a) + \frac{1}{2}(b-a)^2 \left(\frac{n}{n}\right)\left(\frac{n+1}{n}\right).
\end{aligned}$$

As $n \to \infty$, this approaches

$$\begin{aligned}
a(b-a) + \frac{1}{2}(b-a)^2(1)(1) &= ab - a^2 + \frac{1}{2}(b^2 - 2ab + a^2) \\
&= ab - a^2 + \frac{1}{2}b^2 - ab + \frac{1}{2}a^2 = \frac{1}{2}b^2 - \frac{1}{2}a^2 = \frac{1}{2}(b^2 - a^2).
\end{aligned}$$

57. a) $\displaystyle\sum_{j=1}^{n} \frac{j}{n^2} = \sum_{j=1}^{n} \left(\frac{j}{n}\right)\left(\frac{1}{n}\right)$. If we partition $[0, 1]$ into n equal parts, then

$$\Delta x = \frac{1}{n}, \text{ and } x_j = j\Delta x, \; j = 0, 1, 2, \cdots, n.$$

If we then let t_j be the right endpoint of the jth subintervals,

$$t_j \;=\; x_j \;=\; \frac{j}{n},\; j = 1, 2, \cdots, n,$$

then

$$\sum_{j=1}^{n} \left(\frac{j}{n}\right)\left(\frac{1}{n}\right) \;=\; \sum_{j=1}^{n} t_j \Delta x \;=\; \sum_{j=1}^{n} f(t_j) \Delta x_j$$

which is a Riemann sum R_n for the function $f(x) = x$ over the interval $[0, 1]$.

b) Since $R_n \to \int_0^1 f(x)\, dx$ as $n \to \infty$, then

$$\lim_{n\to\infty} \sum_{j=1}^{n} \frac{j}{n^2} = \lim_{n\to\infty} (R_n) = \int_0^1 f(x)\, dx = \frac{1}{2}\left(1^2 - 0^2\right) = \frac{1}{2}.$$

59. $$\lim_{n\to\infty} \sum_{j=1}^{n} \frac{\sqrt{n^2 - j^2}}{n^2} \;=\; \lim_{n\to\infty} \sum \frac{\sqrt{n^2 - j^2}}{n} \cdot \left(\frac{1}{n}\right) \;=\; \lim_{n\to\infty} \sum \sqrt{1 - \left(\frac{j}{n}\right)^2}\left(\frac{1}{n}\right)$$

$$=\; \lim_{n\to\infty} R_n \;=\; \int_0^1 \left(\sqrt{1 - x^2}\right) dx$$

$$=\; \text{area of a quarter circle} = \frac{1}{4}\pi 1^2 \;=\; \frac{1}{4}\pi.$$

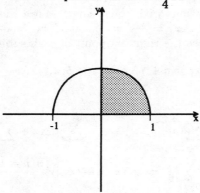

Exercise 59

61. For $\int_0^b \cos x\, dx$, we use the partition $P_n = \{x_j\}$ where $x_j = j\left(\frac{b}{n}\right)$, $j = 0, 1, \cdots, n$, and the left endpoints $t_j = x_{j-1} = (j-1)\left(\frac{b}{n}\right)$ for the Riemann sum:

$$R_n \;=\; \sum_{j=1}^{n} f(t_j) \Delta x_j \;=\; \sum_{j=1}^{n} \cos\left(\frac{(j-1)b}{n}\right)\left(\frac{b}{n}\right)$$

$$=\; \left(\frac{b}{n}\right) \sum_{j=1}^{n} \cos\left[(j-1)\left(\frac{b}{n}\right)\right] \;=\; \left(\frac{b}{n}\right) \sum_{k=0}^{n-1} \cos\left(k\left(\frac{b}{n}\right)\right)$$

where we substitute $k = j - 1$ in the summation for convenience. So

$$R_n = \left(\frac{b}{2n}\right) + \left[\frac{\left(\frac{b}{2n}\right)}{\sin\left(\frac{b}{2n}\right)}\right] \sin\left[\left(n - \frac{1}{2}\right)\left(\frac{b}{n}\right)\right]$$

as $n \to \infty$, $\left(\frac{b}{2n}\right) \to 0$, and $\left(n - \frac{1}{2}\right)\left(\frac{b}{n}\right) = b - \frac{b}{2n} \to b$ so $\sin\left[\left(n - \frac{1}{2}\right)\frac{b}{n}\right] \to \sin(b)$ since $\sin x$ is continuous. Also $\frac{\sin x}{x} \to 1$ as $x \to 0$, so substituting $x = \frac{b}{2n}$, we get

$$R_n \to 0 + [1][\sin(b)] = \sin(b).$$

So $\int_0^b \cos(x)\,dx = \sin(b)$.

63. The text shows that if C is between A and B, then $\int_A^B = \int_A^C + \int_C^A$. Now we use that fact to show that $\int_a^b = \int_a^c + \int_c^b$, no matter what the order among a, b, c.

The six possible orders are

(i) $a < c < b$

(ii) $c < a < b$ (already discussed in the text)

(iii) $a < b < c$

(iv) $b < a < c$

(v) $b < c < a$ and

(vi) $c < b < a$.

We use the same method for each order, starting with the fact that the equation above is true for any $A < C < B$, and **Definition 4**.

(iii) We know from the proof, $\int_a^c = \int_a^b + \int_b^c$, since b is between a and c. [Hence $a = A$, $b = C$, $c = B$].

Hence $\int_a^b = + \int_a^c - \int_b^c = \int_a^c + \int_c^b$ by **Definition 4**.

(iv) a is between b and c, so $\int_b^c = \int_b^a + \int_a^c$. Hence $-\int_b^a = \int_a^c - \int_b^c$, so $\int_a^b = \int_a^c + \int_c^b$ by **Definition 4**.

(v) c is between b and a, so $\int_b^a = \int_b^c + \int_c^a$. Hence $-\int_b^a = -\int_c^a - \int_b^c$, so $\int_a^b = \int_a^b + \int_c^b$.

(vi) b is between c and a, so $\int_c^a = \int_c^b + \int_b^a$. Hence $-\int_b^a = -\int_c^a + \int_c^b$, so $\int_a^b = \int_a^c + \int_c^b$.

65. a) [As with several problems in this set, the statement of the problem leaves nothing else to be said.]

b) $f(x) = g(x)$ for $0 \le x < 1$. So if we never choose 1 for one of the t_j in a Riemann sum for g, then the sums for f and g over $[0, 1]$ will be equal if we choose the same P and t_j.

c) **Theorem 2** tells us that the Riemann sums for f approach a limit, the same limit, regardless of our choice of t_j, so long as $||P|| \to 0$. Hence we can avoid $t_n = 1$ in our sums for f over the interval $[0, 1]$, and still be sure that they approach $\int_0^1 f(x)\,dx$ as $n \to \infty$; $||P|| \to 0$. By (b), if we use the same P and t_j for our sums for g, they will be the same as the sums for f. Hence those sums for g approach the same limit: $\int_0^1 x\,dx = \frac{1}{2}(1^2 - 0^2) = \frac{1}{2}$.

But to show g is integrable, we also must show that we get the same limit in case t_n can be equal to 1 in the sums. We argue as follows. If $t_n = 1$ in some R_n, then we compute R_n' with the same P and t_j, except $t_n' \ne 1$. Then the difference between R_n and R_n' is no larger than $(g(1) - g(t_n'))\Delta x_n \le 2||P||$. As $||P|| \to 0$, then, this difference approaches zero. So the sums R_n with $t_n = 1$ are larger than the R_n' sum, but they all approach the same limit, which we now can call $\int_0^1 g(x)\,dx$.

67. The same procedure as used in Ex. 66 can be applied to any finite number of discontinuities—but not necessarily an infinite number. If g is continuous on $[a, b]$, except at the points c_j, $j = 1, 2, \cdots, n$. $a = c_0 < c_1 < c_2 < \cdots < c_n < c_{n+1} = b$ then on each subinterval $[c_j, c_{j+1}]$, $j = 0, 1, \cdots, n$ we define

$$f_j(x) = \begin{cases} \lim_{x \to c_j^+} g(x) & \text{if } x = c_j \\ g(x) & \text{if } c_j < x < c_{j+1} \\ \lim_{x \to c_j^-} g(x) & \text{if } x = c_{j+1}. \end{cases}$$

Then each f_j is continuous, and a Riemann sum $(R\,g)_m$ for g over $[a, b]$ is equal to the sum of n different sums

$$(R\,g)_m = (R\,f_1)_{m_1} + \cdots (R\,f_n)_{m_n}$$

where $m_1 + m_2 + \cdots + m_n = m$ and $(R\,f_j)m_j$ is a sum for f_j over $[c_j, c_{j+1}]$, and where we do **not** allow any t_k to be equal to any c_j, and where each c_j is in the partition P.

Such sums approach $\int_a^{c_1} f_0 + \int_{c_1}^{c_2} f_1 + \cdots + \int_{c_n}^b f_n$ as $||P|| \to 0$. If any $t_k = c_j$, then as before (Ex. 66) we form a sum $(R\,g)'_m$ such that in $(R\,g)'_m$ no t_k is equal to any c_j, but

$$|(R\,g)'_m - (R\,g)_m| \le B\,||P||,$$

where now B equals the **finite** sum of $(\text{Max}_1 - \text{min}_1) + \cdots + (\text{Max}_n - \text{min}_n)$ where Max_j, min_j are the max and min of f_j on $[c_j, c_{j+1}]$. So, as before, all Riemann sums for g over $[a, b]$ approach the same limit, so g is integrable over $[a, b]$, and

$$\int_a^b g = \int_a^{c_1} f_0 + \cdots + \int_{c_n}^b f_n.$$

[NOTE: If there were an infinite number of points c_j, we could not conclude that necessarily the sum $(\text{Max}_1 - \text{min}_1 + \cdots) = B$ was finite.]

69. $h(x) = \begin{cases} \frac{1}{x} & \text{if } x \ne 0 \\ 0 & \text{if } x = 0. \end{cases}$ If we can show that any particular sequence of sums R_n does not approach a (finite) limit, then we have shown that it is not true all sums R_n approach a definite limit, and so we have shown that h is not integrable.

Let $P = \{0, \frac{1}{n}, \frac{2}{n}, \cdots, 1\}$ and choose $t_1 = \frac{1}{n^2}$, $t_2 = \frac{1}{n}, t_3 = \frac{2}{n}, \cdots, t_n = \frac{n-1}{n}$. Then

$$\begin{aligned} R_n &= \sum h(t_j)\Delta x_j = \left[\frac{1}{(\frac{1}{n^2})} + \frac{1}{\frac{1}{n}} + \cdots + \frac{1}{\frac{n-1}{n}} \right] \left(\frac{1}{n} \right) \\ &= \left[n^2 + n + \frac{1}{2}n + \frac{1}{3}n + \cdots + \left(\frac{1}{n-1} \right)n \right] \left(\frac{1}{n} \right) \\ &= n + \left(1 + \frac{1}{2} + \frac{1}{3} + \cdots + \frac{1}{n-1} \right) > n. \end{aligned}$$

So as $n \to \infty$, $R_n \to \infty$, and h is not integrable over $[0, 1]$.

71. $\int_1^4 \frac{1}{x}\,dx \approx \sum_{j=1}^4 f(m_j)\Delta x_j = \frac{1}{\frac{3}{2}}(1) + \frac{1}{\frac{5}{2}}(1) + \frac{1}{\frac{7}{2}}(1) = \frac{2}{3} + \frac{2}{5} + \frac{2}{7} = \frac{70 + 42 + 30}{3 \cdot 5 \cdot 7} = \frac{142}{105}.$

73. $\int_0^4 \sqrt{x^3 + 1}\,dx \approx \sum_{j=1}^4 f(m_j)\Delta x_j = f\left(\frac{1}{2} \right)(1) + f\left(\frac{3}{2} \right)(1) + f\left(\frac{5}{2} \right)(1) + f\left(\frac{7}{2} \right)(1) \approx 13.85350749.$

75. $\int_0^4 \dfrac{1}{1+x^2}\,dx \approx \sum_{j=1}^{4} f(m_j)\Delta x_j = f\left(\dfrac{1}{2}\right)\cdot(1) + f\left(\dfrac{3}{2}\right)\cdot(1) + f\left(\dfrac{5}{2}\right)\cdot(1) + f\left(\dfrac{7}{2}\right)\cdot(1) \approx 1.321095040.$

77. $\int_0^1 \sqrt{x^4+1}\,dx \approx \sum_{j=1}^{6} f(m_j)\Delta x_j \approx 1.087792562$

   where $\{m_j\} = \left\{\dfrac{1}{12}, \dfrac{3}{12}, \dfrac{5}{12}, \dfrac{7}{12}, \dfrac{9}{12}, \dfrac{11}{12}\right\}$ and $\Delta x_j = \dfrac{1}{6}.$

79. $\int_0^3 \sin\sqrt{x}\,dx \approx \sum_{j=1}^{6} f(m_j)\Delta x_j \approx 2.551760163$ where $\{m_j\} = \left\{\dfrac{1}{4}, \dfrac{3}{4}, \dfrac{5}{4}, \dfrac{7}{4}, \dfrac{9}{4}, \dfrac{11}{4}\right\}$ and $\Delta x_j = \dfrac{1}{2}.$

81.

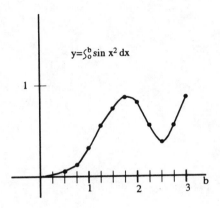

Exercise 81

b	$\int_0^b \sin(x^2)\,dx$ midpoint rule, $n = 5$
0.25	0.005154896
0.5	0.041077014
0.75	0.136284354
1.0	0.308439081
1.25	0.545833991
1.5	0.785324320
1.75	0.912511721
2.0	0.824278992
2.25	0.557850284
2.5	0.367701578
2.75	0.529437490
3.0	0.889646736

83. Graph using the range $[-4, 5] \times [-1.5, 3]$ and $n = 16$.

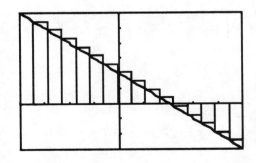
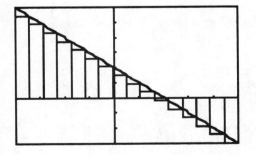

Table:

n	4	8	16	32	64
L_n	11.8125	9.28125	8.015625	7.3828125	7.06640625
R_n	1.6875	4.21875	5.484375	6.1171875	6.43359375

85. Graph using the range $[0,4] \times [0,5]$ and $n = 16$.

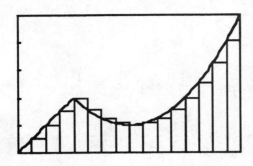 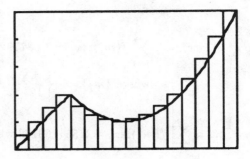

Table:

n	4	8	16	32	64
L_n	5	5.875	6.40625	6.6953125	6.8457031
R_n	10	8.375	7.65625	7.3203125	7.1582031

In Exercises 87–90, we approximate to 3 significant digits:

87. $L_n = 3.1425795$, $R_n = 3.1405795$, $n = 2000$.

89. $L_n = 1.8500817$, $R_n = 1.8472228$, $n = 1000$.

6.3   The Fundamental Theorem of Calculus

1. $\displaystyle\int_{-2}^{2} (x^3 - 1)\, dx = \frac{x^4}{4} - x \bigg]_{-2}^{2} = \left(\frac{16}{4} - 2\right) - \left(\frac{16}{4} - (-2)\right) = -4.$

3. $\displaystyle\int_{1}^{2} \left(\frac{2}{x^3} + 5x\right) dx = -\frac{1}{x^2} + \frac{5}{2}x^2 \bigg]_{1}^{2} = \left(-\frac{1}{4} + 10\right) - \left(-\frac{1}{1} + \frac{5}{2}\right) = \frac{33}{4}.$

5. $\displaystyle\int_{1}^{3} (t^2 + 2)^2\, dt = \int_{1}^{3} (t^4 + 4t^2 + 4)\, dt = \frac{1}{5}t^5 + \frac{4}{3}t^3 + 4t \bigg]_{1}^{3} = \frac{243}{5} + 36 + 12 - \left(\frac{1}{5} + \frac{4}{3} + 4\right) = 91\frac{1}{15}.$

7. $\displaystyle\int_{1}^{2} \frac{1-x}{x^3}\, dx = -\frac{1}{2x^2} + \frac{1}{x} \bigg]_{1}^{2} = -\frac{1}{8} + \frac{1}{2} - \left(-\frac{1}{2} + \frac{1}{1}\right) = -\frac{1}{8}.$

9. $\displaystyle\int_{2}^{0} x\left(\sqrt{x} - 1\right) dx = \int_{2}^{0} (x^{3/2} - x)\, dx = \frac{2}{5}x^{5/2} - \frac{x^2}{2} \bigg]_{2}^{0} = -\frac{2}{5}\left(4\sqrt{2}\right) + \frac{4}{2} = 2 - \frac{8}{5}\sqrt{2} = -0.26274.$

11. $\displaystyle\int_{0}^{4} |x - 3|\, dx = -\int_{0}^{3} (x-3)\, dx + \int_{3}^{4} (x-3)\, dx = -\left[\frac{x^2}{2} - 3x\right]_{0}^{3} + \left[\frac{x^2}{2} - 3x\right]_{3}^{4}$

$\displaystyle = -\left(\frac{9}{2} - 9\right) + \left[\left(\frac{16}{2} - \frac{9}{2}\right) - 3(4-3)\right] = \frac{9}{2} + \frac{7}{2} - 3 = 5.$

13. $\displaystyle\int_0^4 |x - \sqrt{x}|\,dx = -\int_0^1 (x - \sqrt{x})\,dx + \int_1^4 (x - \sqrt{x})\,dx$

$$= -\left[\frac{x^2}{2} - \frac{2}{3}x^{3/2}\right]_0^1 + \left[\frac{x^2}{2} - \frac{2}{3}x^{3/2}\right]_1^4$$

$$= -\left(\frac{1}{2} - \frac{2}{3}\right) + \left(\left(\frac{16}{2} - \frac{1}{2}\right) - \frac{2}{3}(8 - 1)\right) = 3.$$

15. $\displaystyle\int_4^9 x^{1/2}(1 - x^{3/2})\,dx = \int_4^9 (x^{1/2} - x^2)\,dx = \frac{2}{3}x^{3/2} - \frac{x^3}{3}\bigg]_4^9 = \frac{2}{3}(27 - 8) - \frac{1}{3}(729 - 64) = -209.$

17. $\displaystyle\int_{-1}^4 |x^2 - x - 2|\,dx = \quad x^2 - x - 2 = (x - 2)(x + 1) \Rightarrow x^2 - x - 2 = 0$ at $x = -1$, $x = 2$

$$-\int_{-1}^2 (x^2 - x - 2)\,dx + \int_2^4 (x^2 - x - 2)\,dx = -\left[\frac{x^3}{3} - \frac{x^2}{2} - 2x\right]_{-1}^2 + \left[\frac{x^3}{3} - \frac{x^2}{2} - 2x\right]_2^4$$

$$= -\left(\frac{8}{3} - \frac{4}{2} - 4\right) + \left(-\frac{1}{3} - \frac{1}{2} + 2\right)$$

$$+ \left(\frac{64}{3} - \frac{16}{2} - 8\right) - \left(\frac{8}{3} - \frac{4}{2} - 4\right) = \frac{79}{6}.$$

19. $\displaystyle\int_0^\pi \sin u\,du = -\cos u\bigg]_0^\pi = -(-1 - 1) = 2.$

21. $\displaystyle\int_{-\pi/3}^{\pi/3} \sec^2\theta\,d\theta = \tan\theta\bigg]_{-\pi/3}^{\pi/3} = \sqrt{3} - (-\sqrt{3}) = 2\sqrt{3}.$

23. $\displaystyle\int_{-2}^3 f(x)\,dx$, where $f(x) = \begin{cases} x^2 - 1 & \text{for } x \le 1 \\ x - 1 & \text{for } x > 1 \end{cases}$

$$= \int_{-2}^1 (x^2 - 1)\,dx + \int_1^3 (x - 1)\,dx$$

$$= \frac{x^3}{3} - x\bigg]_{-2}^1 + \frac{x^2}{2} - x\bigg]_1^3 = \frac{1}{3}(1 + 8) - (1 + 2) + \frac{1}{2}(9 - 1) - (3 - 1)$$

$$= 3 - 3 + 4 - 2 = 2.$$

25. $\displaystyle\int_0^\pi \sin 1\,dx = (\sin 1)x\bigg]_0^\pi = \pi\sin 1 \approx 2.6436.$

27. $f(x) = x^2 + x + 1$, $[0, 1]$

$$\bar{f} = \frac{1}{1 - 0}\int_0^1 (x^2 + x + 1)\,dx = \frac{x^3}{3} + \frac{x^2}{2} + x\bigg]_0^1 = \frac{1}{3} + \frac{1}{2} + 1 = \frac{11}{6}.$$

29. $f(x) = \sec^2 x$, $\left[0, \dfrac{\pi}{4}\right]$

229

$$\bar{f} = \frac{1}{\frac{\pi}{4} - 0} \int_0^{\pi/4} \sec^2 x \, dx = \frac{4}{\pi} \tan x \Big]_0^{\pi/4} = \frac{4}{\pi}(1 - 0) = \frac{4}{\pi}.$$

31. $F(x) = \int_1^x t \, dt = \dfrac{t^2}{2}\Big]_1^x = \dfrac{x^2 - 1}{2}.$ $\qquad$ $F'(x) = \dfrac{2x}{2} = x.$

33. $F(x) = \int_0^x \cos t \, dt = \sin t \Big]_0^x = \sin x - 0 = \sin x.$

$F'(x) = \cos x.$

35. $F(x) = \int_2^x t^2 \sin t^2 \, dt = \int_2^x f(t) \, dt.$ $\qquad$ $F'(x) = f(x) = x^2 \sin x^2.$

37. $F(x) = \int_x^1 t^3 \cos^2 t \, dt = -\int_1^x t^3 \cos^2 t \, dt = -\int_1^x f(t) \, dt.$

$F'(x) = -f(x) = -x^3 \cos^2 x.$

39. $F(x) = \int_0^{3x} \sqrt{1 + \sin t} \, dt = \int_0^{g(x)} f(t) \, dt.$ $\qquad$ $F'(x) = f[g(x)] \, g'(x) = 3\sqrt{1 + \sin 3x}.$

41. $F(x) = \int_{x^2}^x \cos^3(t + 1) \, dt = \int_{g(x)}^x f(t) \, dt.$

$F'(x) = f(x) - f[g(x)]g'(x) = \cos^3(x + 1) - 2x\left(\cos^3(x^2 + 1)\right).$

43. $F(x) = x \int_3^{x^2} (t^2 + 1)^{-3} \, dt = x \int_3^{g(x)} f(t) \, dt.$

$F'(x) = \int_3^{g(x)} f(t) \, dt + x \, f[g(x)]g'(x) = \int_3^{x^2} (t^2 + 1)^{-3} \, dt + 2x^2(x^4 + 1)^{-3}.$

45. a) $\dfrac{d}{dx}\left[\int_a^x f(t) \, dt\right] = f(x).$ $\qquad\qquad$ b) $\dfrac{d}{dx}\left[\int_x^b f(t) \, dt\right] = -f(x).$

c) $\dfrac{d}{dx}\left[\int_a^b f(t) \, dt\right] = 0.$ $\qquad\qquad$ d) $\dfrac{d}{dx}\left[\int f(x) \, dx\right] = f(x).$

e) $\int_a^b f'(x) \, dx = f(b) - f(a).$ $\qquad\qquad$ f) $\int f'(x) \, dx = f(x) + C.$

47. a) $\int_1^4 |x - 2| \, dx = (4 - 1)f(c) = 3|c - 2|$

$\int_1^2 (2 - x) \, dx + \int_2^4 (x - 2) \, dx = 3|c - 2|$

$2x - \dfrac{x^2}{2}\Big]_1^2 + \dfrac{x^2}{2} - 2x\Big]_2^4 = 3|c - 2|$

$$\frac{1}{2} + 2 = 3|c - 2|$$

So we want c such that $|c - 2| = \frac{5}{6}$. Thus $c = \frac{7}{6}$ or $\frac{17}{6}$.

b) $f(x) = 4 - x^2$, $[-2, 3]$.

$$\int_{-2}^{3} (4 - x^2) \, dx = (3 + 2) f(c) = 5(4 - c^2).$$

$$4x - \frac{x^3}{3} \Big]_{-2}^{3} = 4(3 + 2) - \frac{1}{3}(27 + 8) = 20 - \frac{35}{3} = \frac{25}{3}.$$

$$\frac{25}{3} = 5(4 - c^2); \quad 4 - c^2 = \frac{5}{3}; \quad c^2 = 4 - \frac{5}{3} = \frac{7}{3}.$$

$$c = \pm\sqrt{\frac{7}{3}} \approx 1.5275.$$

49. $f(x) = 2x + 3$, $[1, b]$; $\bar{f} = 11$.

$$\bar{f} = \frac{1}{b - 1} \int_{1}^{b} (2x + 3) \, dx = \frac{1}{b - 1} \left[x^2 + 3x \right]_{1}^{b} = \frac{1}{b - 1}(b^2 + 3b - 4) = 11.$$

$b^2 + 3b - 4 = 11b - 11$; $\quad b^2 - 8b + 7 = 0$.
$(b - 1)(b - 7) = 0$; choose $b = 7$.

51. $A(x) = \displaystyle\int_{0}^{x} f(t) \, dt$, $0 \le x \le 8$

    a) $A'(x) = f(x)$, which is greater than zero on $[0, 4]$ and $[7, 8]$.

    b) A is constant on the interval $[6, 7]$.

    c) $A'(x) = 0 \Rightarrow f(x) = 0$, which occurs at $x = 4$, and $6 \le x \le 7$.
       A relative maximum of A occurs at $x = 4$ and at the endpoint $x = 8$.

    d) $A(0) = \displaystyle\int_{0}^{0} f(t) \, dt = 0$.

53. $\bar{f} = \dfrac{1}{b - a} \displaystyle\int_{a}^{b} f(x) \, dx$; $\quad \bar{g} = \dfrac{1}{b - a} \displaystyle\int_{a}^{b} g(x) \, dx$.

$\bar{f} - \bar{g} = \dfrac{1}{b - a} \displaystyle\int_{a}^{b} [f(x) - g(x)] \, dx$. If $f(x) - g(x) \ge 0$ on $[a, b]$, then $\displaystyle\int_{a}^{b} [f(x) - g(x)] \, dx \ge 0$ and $\bar{f} - \bar{g} \ge 0$.

55. $\bar{f} = \dfrac{1}{b - a} \displaystyle\int_{a}^{b} f(x) \, dx \le \dfrac{1}{b - a} \displaystyle\int_{a}^{b} M \, dx$.

$\bar{f}(b - a) = M(b - a) \Rightarrow \bar{f} = M$.

If $\bar{f} = M$, $\bar{f}(b - a) = \displaystyle\int_{a}^{b} f(x) \, dx \le \displaystyle\int_{a}^{b} M \, dx \Rightarrow f(x) = M$ for all $x \in [a, b]$.

57. $\displaystyle\lim_{n \to \infty} \left(\frac{x}{n}\right) \sum_{j=1}^{n} \left(\sin \frac{jx}{n}\right) = \lim_{n \to \infty} \sum_{j=1}^{n} \sin y_j \, \Delta y$, where $y_j = \dfrac{x}{n} j$ and $\Delta y = \dfrac{x}{n}$.

$$\int_{0}^{x} \sin y \, dy = -\cos y \Big]_{0}^{x} = -\cos x + 1 = 1 - \cos x.$$

59.   a) Graph using the range $[0, 10] \times [-50, 50]$.

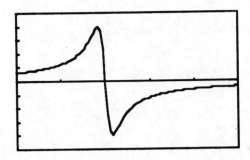

Since $\frac{dp}{dt} = 0$ when $t = 4$ and $\frac{dp}{dt} > 0$ for $t < 4$, $\frac{dp}{dt} < 0$ for $t > 4$, then at $t = 4$, $p(t)$ has a maximum.

   b) Using the Midpoint Rule programs for drawing an antiderivative (Appendix V) with **K=32, N=20**, and the range $[0, 8] \times [0, 135]$ the following graph is drawn.

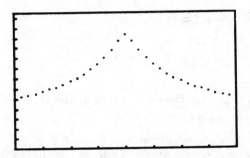

   c) From the graph in part b) it appears that the maximum number of rabbits is approximately 114.

   d) Using the Midpoint Rule programs for drawing an antiderivative (Appendix V) with **K=50** and **N=6**, and the range $[0, 50] \times [-20, 135]$ the following graph is drawn.

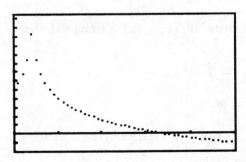

The graph crosses the axis at approximately $t = 34$. So the rabbits will die out after 34 years.

6.4 Substitution in Definite Integrals

1. $\int_0^{\pi/4} \cos 2x\, dx, \ u = 2x$

$du = 2x\, dx$

if $x = 0, \ u = 0$; if $x = \frac{\pi}{4}, \ u = \frac{\pi}{2}$

$$\frac{1}{2}\int_0^{\pi/2} \cos u\, du = \frac{1}{2}\sin u\Big]_0^{\pi/2} = \frac{1}{2}(1-0) = \frac{1}{2}.$$

3. $\int_0^{\pi/6} \sec(2x)\tan(2x)\, dx, \ u = 2x$

$du = 2\, dx$

if $x = 0, \ u = 0$; if $x = \frac{\pi}{6}, \ u = \frac{\pi}{3}$

$$\frac{1}{2}\int_0^{\pi/3} \sec u\tan u\, du = \frac{1}{2}\sec u\Big]_0^{\pi/3} = \frac{1}{2}(2-1) = \frac{1}{2}.$$

5. $\int_1^2 (2-x^2)(x^3-6x+7)^{50}\, dx, \ u = x^3 - 6x + 7$

$du = (3x^2 - 6)\, dx = -3(2-x^2)\, dx$

if $x = 1, \ u = 2$; if $x = 2, \ u = 3$

$$-\frac{1}{3}\int_2^3 u^{50}\, du = -\frac{1}{3(51)}u^{51}\Big]_2^3 = -\frac{1}{153}(3^{51} - 2^{51}) \approx -1.40764 \times 10^{22}.$$

7. $\int_0^{\pi/4} (1 - \cos 2x)\, dx = x - \frac{1}{2}\sin 2x\Big]_0^{\pi/4} = \frac{\pi}{4} - \frac{1}{2} \approx 0.2854.$

9. $\int_0^1 2x(x^2+1)^2\, dx$

Let $u = x^2 + 1, \ du = 2x\, dx$

if $x = 0, \ u = 1$; if $x = 1, \ u = 2$

$$\int_1^2 u^2\, du = \frac{1}{3}u^3\Big]_1^2 = \frac{1}{3}(8-1) = \frac{7}{3}.$$

11. $\int_1^2 x\sqrt{4-x^2}\, dx$

Let $u = 4 - x^2, \ du = -2x\, dx$

if $x = 1, \ u = 3$; if $x = 2, \ u = 0$

$$-\frac{1}{2}\int_3^0 \sqrt{u}\, du = \frac{1}{2}\left(\frac{2}{3}\right)u^{3/2}\Big]_0^3 = \frac{1}{3}(3^{3/2}) = \sqrt{3}.$$

13. $\int_0^{\sqrt{\pi/2}} t\sin(\pi - t^2)\, dt$

Let $u = \pi - t^2, \ du = -2t\, dt$

233

if $t = 0$, $u = \pi$; if $t = \frac{\sqrt{\pi}}{2}$, $u = \frac{3\pi}{4}$

$$-\frac{1}{2} \int_{\pi}^{3\pi/4} \sin u \, du = \frac{1}{2} \cos u \bigg]_{\pi}^{3\pi/4} = \frac{1}{2} \left(-\frac{\sqrt{2}}{2} + 1 \right) = \frac{2 - \sqrt{2}}{4}.$$

15. $\displaystyle\int_{-\pi/4}^{\pi/4} \frac{\sin x}{\cos^2 x} \, dx = \int_{-\pi/4}^{\pi/4} (\cos x)^{-2} \sin x \, dx = \frac{1}{\cos x} \bigg]_{-\pi/4}^{\pi/4} = \sec x \bigg]_{-\pi/4}^{\pi/4}$

$\qquad\qquad = \sqrt{2} - \sqrt{2} = 0.$

17. $\displaystyle\int_{0}^{4} \frac{dt}{\sqrt{2t + 1}}$

Let $u = 2t + 1$, $du = 2 \, dt$

if $t = 0$, $u = 1$; if $t = 4$, $u = 9$.

$$\frac{1}{2} \int_{1}^{9} \frac{du}{\sqrt{u}} = \sqrt{u} \bigg]_{1}^{9} = 3 - 1 = 2.$$

19. $\displaystyle\int_{-1}^{1} \frac{x}{(1 + x^2)^3} \, dx$

Let $u = 1 + x^2$, $du = 2x \, dx$

if $x = -1$, $u = 2$; if $x = 1$, $u = 2$

$\displaystyle\frac{1}{2} \int_{2}^{2} \frac{du}{u^3} = 0.$

21. $\displaystyle\int_{\pi^2/4}^{\pi^2} \frac{\sin \sqrt{x}}{\sqrt{x}} \, dx$

Let $u = \sqrt{x}$, $du = \frac{1}{2\sqrt{x}} \, dx$

if $x = \frac{\pi^2}{4}$, $u = \frac{\pi}{2}$; if $x = \pi^2$, $u = \pi$

$$2 \int_{\pi/2}^{\pi} \sin u \, du = -2 \cos u \bigg]_{\pi/2}^{\pi} = -2(-1 - 0) = 2.$$

23. $\displaystyle\int_{1}^{2} \frac{x}{(2x^2 - 1)^3} \, dx$

Let $u = 2x^2 - 1$, $du = 4x \, dx$

if $x = 1$, $u = 1$; if $x = 2$, $u = 7$

$$\frac{1}{4} \int_{1}^{7} \frac{du}{u^3} = -\frac{1}{8} \frac{1}{u^2} \bigg]_{1}^{7} = \frac{1}{8} \left(1 - \frac{1}{49} \right) = \frac{6}{49}.$$

25. $\displaystyle\int_{1}^{4} s|2 - s|^{12} \, ds = \int_{1}^{4} s(2 - s)^{12} \, ds$

Let $u = 2 - s$, $du = -ds$, $s = 2 - u$. If $s = 1$, $u = 1$; if $s = 4$, $u = -2$.

$$-\int_{1}^{-2} (2 - u)u^{12} \, du = \int_{-2}^{1} 2 \, u^{12} \, du - \int_{-2}^{1} u^{13} \, du$$

$$= \frac{2}{13} u^{13} \Big]_{-2}^{1} - \frac{1}{14} u^{14} \Big]_{-2}^{1} = \frac{2}{13}(1 + 2^{13}) - \frac{1}{14}(1 - 2^{14})$$

$$= \frac{2}{13} - \frac{1}{14} + 2^{14} \left(\frac{1}{13} + \frac{1}{14} \right) = \frac{15}{182} + \frac{27 \cdot 2^{14}}{182} = \frac{442,383}{182} \approx 2430.68.$$

27. $\int_{0}^{1} \frac{1}{(4-x)^2} \, dx = \frac{1}{4-x} \Big]_{0}^{1} = \frac{1}{3} - \frac{1}{4} = \frac{1}{12}.$

29. $\int_{0}^{\pi} \cos^2 x \, dx = \int_{0}^{\pi} \left(\frac{1}{2} + \frac{1}{2} \cos 2x \right) dx = \frac{x}{2} + \frac{1}{4} \sin 2x \Big]_{0}^{\pi} = \frac{\pi}{2} + \frac{1}{4}(0) = \frac{\pi}{2}.$

31. $\int_{0}^{\pi/4} \sec^2 u \sqrt{\tan u} \, du = \frac{2}{3} (\tan u)^{3/2} \Big]_{0}^{\pi/4} = \frac{2}{3}.$

(Let $z = \tan u$, etc.)

33. $\int_{0}^{\sqrt{\pi/2}} t \cos^3(t^2) \sin(t^2) \, dt$

Let $u = t^2$, $du = 2t \, dt$

if $t = 0$, $u = 0$; if $t = \sqrt{\frac{\pi}{2}}$, $u = \frac{\pi}{2}$

$$\frac{1}{2} \int_{0}^{\pi/2} \cos^3 u \sin u \, du = -\frac{1}{8} \cos^4(u) \Big]_{0}^{\pi/2} = \frac{1}{8}.$$

(Let $z = \cos u$. etc.)

35. $\int_{0}^{\pi/4} \tan^2 x \sec^2 x \, dx$

Let $u = \sec x$, $du = \sec x \tan x \, dx$; $\tan x = \sqrt{\sec^2 x - 1} = \sqrt{u^2 - 1}$

if $x = 0$, $u = 1$; if $x = \frac{\pi}{4}$, $x = \sqrt{2}$

$$\int_{1}^{\sqrt{2}} \sqrt{u^2 - 1} \, u \, du = \frac{1}{2} \cdot \frac{2}{3} (u^2 - 1)^{3/2} \Big]_{1}^{\sqrt{2}} = \frac{1}{3}(1 - 0) = \frac{1}{3}.$$

(Let $z = u^2 - 1$, $dz = 2u \, du$, etc.)

37. $f(x) = x\sqrt{x^2 + 1}$, $[0, 2]$

$$\bar{f} = \frac{1}{2-0} \int_{0}^{2} x\sqrt{x^2 + 1} \, dx = \frac{1}{2} \left(\frac{1}{3} \right) (x^2 + 1)^{3/2} \Big]_{0}^{2} = \frac{1}{6}(5^{3/2} - 1) \approx 1.6967.$$

(Let $u = x^2 + 1$, $du = 2x \, dx$, etc;)

39. $f(x) = x \cos x^2$, $\left[0, \frac{\sqrt{\pi}}{2} \right]$

$$\bar{f} = \frac{1}{\frac{\sqrt{\pi}}{2} - 0} \int_{0}^{\sqrt{\pi}/2} x \cos x^2 \, dx$$

Let $u = x^2$, $du = 2x\,dx$

if $x = 0$, $u = 0$; if $x = \frac{\sqrt{\pi}}{2}$, $u = \frac{\pi}{4}$

$$\bar{f} = \frac{2}{\sqrt{\pi}} \cdot \frac{1}{2} \int_0^{\pi/4} \cos u\,du = \frac{1}{\sqrt{\pi}} \sin u \Big]_0^{\pi/4} = \frac{1}{\sqrt{\pi}} \frac{\sqrt{2}}{2} = \frac{1}{\sqrt{2\pi}} \approx 0.3989.$$

41. $F(x) = \displaystyle\int_1^x \frac{t}{\sqrt{1 + t^2}}\,dt = \sqrt{1 + t^2}\,\Big]_1^x = \sqrt{1 + x^2} - \sqrt{2}.$

$\ \ F'(x) = \dfrac{2x}{2\sqrt{1 + x^2}} = \dfrac{x}{\sqrt{1 + x^2}}.$

43. $F(x) = \displaystyle\int_x^0 \cos^2 t \sin t\,dt = -\frac{1}{3} \cos^3 t\,\Big]_x^0 = -\frac{1}{3} + \frac{1}{3} \cos^3 x.\ .$

$\ \ F'(x) = \dfrac{1}{3}(3\cos^2 x)(-\sin x) = -\cos^2 x \sin x.$

6.5 Finding Areas By Integration

1. $f(x) = 2x + 5$, $a = 0$, $b = 2$.

$\quad A = \displaystyle\int_0^2 (2x + 5)\,dx = x^2 + 5x\,\Big]_0^2 = 4 + 10 = 14.$

3. $f(x) = \dfrac{1}{\sqrt{x - 2}}$, $a = 3$, $b = 5$.

$\quad A = \displaystyle\int_3^5 \frac{1}{\sqrt{x - 2}}\,dx = 2\sqrt{x - 2}\,\Big]_3^5 = 2(\sqrt{3} - 1) \approx 1.4641.$

5. $f(x) = \dfrac{x + 1}{(x^2 + 2x)^2}$, $a = 1$, $b = 2$.

$\quad A = \displaystyle\int_1^2 \frac{x + 1}{(x^2 + 2x)^2}\,dx = -\frac{1}{2}\left(\frac{1}{x^2 + 2x}\right)\Big]_1^2 = -\frac{1}{2}\left(\frac{1}{8} - \frac{1}{3}\right) = \frac{5}{48}.$

7. $f(x) = \sqrt{5 + x}$, $a = -4$, $b = 4$.

$\quad A = \displaystyle\int_{-4}^4 \sqrt{5 + x}\,dx = \frac{2}{3}(5 + x)^{3/2}\,\Big]_{-4}^4 = \frac{2}{3}(27 - 1) = \frac{52}{3}.$

9. $f(x) = x + 1$, $g(x) = -2x + 1$, $0 \le x \le 2$.

$$A = \int_0^2 [x + 1 - (-2x + 1)]\, dx$$

$$= \int_0^2 3x\, dx$$

$$= \left. \frac{3}{2} x^2 \right]_0^2$$

$$= 6.$$

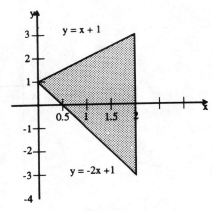

Exercise 9

11. $f(x) = \sqrt{x},\ g(x) = -x^2,\ 0 \le x \le 4.$

$$A = \int_0^4 [\sqrt{x} - (-x^2)]\, dx$$

$$= \left. \frac{2}{3} x^{3/2} + \frac{x^3}{3} \right]_0^4$$

$$= \frac{2}{3}(8) + \frac{64}{3}$$

$$= \frac{80}{3}.$$

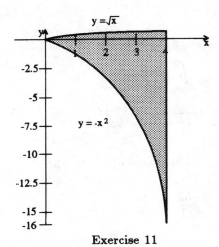

Exercise 11

13. $f(x) = \sin x,\ g(x) = \cos x,\ 0 \le x \le 2\pi.$

$\cos x \ge \sin x$ on $\left[0, \frac{\pi}{4}\right] \cup \left[\frac{5\pi}{4}, 2\pi\right]$

$\sin x \ge \cos x$ on $\left[\frac{\pi}{4}, \frac{5\pi}{4}\right]$

$$A = \int_0^{\pi/4} (\cos x - \sin x)\, dx + \int_{\pi/4}^{5\pi/4} (\sin x - \cos x)\, dx + \int_{5\pi/4}^{2\pi} (\cos x - \sin x)\, dx$$

$$= \sin x + \cos x \Big]_0^{\pi/4} + (-\cos x - \sin x) \Big]_{\pi/4}^{5\pi/4} + \sin x + \cos x \Big]_{5\pi/4}^{2\pi}$$

$$= \frac{\sqrt{2}}{2} + \left(\frac{\sqrt{2}}{2} - 1\right) + \left(\frac{\sqrt{2}}{2} + \frac{\sqrt{2}}{2}\right) + \left(\frac{\sqrt{2}}{2} + \frac{\sqrt{2}}{2}\right) + \left(0 + \frac{\sqrt{2}}{2} + 1 + \frac{\sqrt{2}}{2}\right)$$

$$= 4\sqrt{2} \approx 5.6569.$$

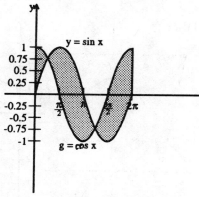

Exercise 13

15. $f(x) = x - 3$, $g(x) = x^2$, $-2 \leq x \leq 2$.

$$A = \int_{-2}^{2} [x^2 - (x - 3)]\, dx$$

$$= \frac{x^3}{3} - \frac{x^2}{2} + 3x \Big]_{-2}^{2}$$

$$= \frac{1}{3}(8 + 8) - \frac{1}{2}(4 - 4) + 3(2 + 2)$$

$$= \frac{52}{3}.$$

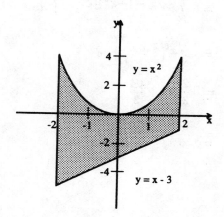

Exercise 15

17. $f(x) = |4 - x^2|$, $g(x) = 5$, $-3 \leq x \leq 3$.

$$A = \int_{-3}^{3} (5 - |4 - x^2|)\, dx = \int_{-3}^{-2} (5 - x^2 + 4)\, dx + \int_{-2}^{2} (5 - 4 + x^2)\, dx + \int_{2}^{3} (5 - x^2 + 4)\, dx$$

$$= 9x - \frac{x^3}{3}\Big]_{-3}^{-2} + x + \frac{x^3}{3}\Big]_{-2}^{2} + 9x - \frac{x^3}{3}\Big]_{2}^{3}$$

$$= 9(-2 + 3) - \frac{1}{3}(-8 + 27) + (2 + 2) + \frac{1}{3}(8 + 8) + 9(3 - 2) - \frac{1}{3}(27 - 8)$$

$$= \frac{44}{3}.$$

238

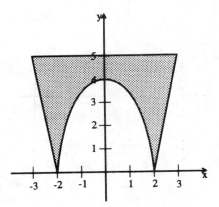

Exercise 17

19. $f(x) = \dfrac{x^2 - 1}{x^2}$, $g(x) = \dfrac{1 - x^2}{x^2}$, $1 \le x \le 2$.

$$
\begin{aligned}
A &= \int_1^2 \left(\frac{x^2 - 1}{x^2} - \frac{1 - x^2}{x^2} \right) dx \\[2mm]
&= \int_1^2 \left(2 - \frac{2}{x^2} \right) dx \\[2mm]
&= \left. 2x + \frac{2}{x} \right]_1^2 \\[2mm]
&= 2(2 - 1) + 2 \left(\frac{1}{2} - 1 \right) \\[2mm]
&= 1.
\end{aligned}
$$

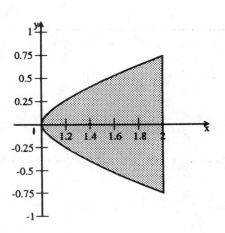

Exercise 19

21. $y = 4 - x^2$, $y = x - 2$.

$4 - x^2 = x - 2 \Rightarrow x^2 + x - 6 = 0$;

$(x + 3)(x - 2) = 0$; $x = 2, -3$

$$
\begin{aligned}
A &= \int_{-3}^2 [4 - x^2 - (x - 2)]\, dx \\[2mm]
&= \int_{-3}^2 (-x^2 - x + 6)\, dx \\[2mm]
&= \left. -\frac{x^3}{3} - \frac{x^2}{2} + 6x \right]_{-3}^2 \\[2mm]
&= -\frac{1}{3}(8 + 27) - \frac{1}{2}(4 - 9) + 6(2 + 3) \\[2mm]
&= \frac{125}{6} \\[2mm]
&= 20.833.
\end{aligned}
$$

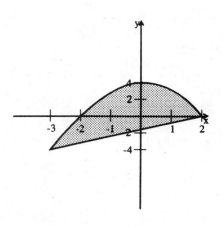

Exercise 21

23. $y = x^2$, $y = x^3$.

$$x^2 = x^3 \Rightarrow x^2(x-1) = 0 \Rightarrow x = 0, 1$$

$$\begin{aligned} A &= \int_0^1 (x^2 - x^3)\, dx \\ &= \left. \frac{x^3}{3} - \frac{x^4}{4} \right]_0^1 \\ &= \frac{1}{3} - \frac{1}{4} \\ &= \frac{1}{12}. \end{aligned}$$

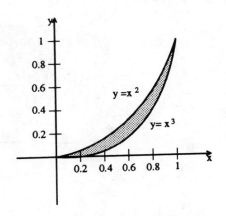

Exercise 23

25. $x + y^2 = 4$, $y = x + 2$.

$$4 - y^2 = y - 2 \Rightarrow y^2 + y - 6 = 0 \Rightarrow (y+3)(y-2) = 0$$
$$y = 2, -3$$

$$\begin{aligned} A &= \int_{-3}^2 [4 - y^2 - (y-2)]\, dy \\ &= \int_{-3}^2 (-y^2 - y + 6)\, dy \\ &= \left. -\frac{y^3}{3} - \frac{y^2}{2} + 6y \right]_{-3}^2 \\ &= -\frac{1}{3}(8 + 27) - \frac{1}{2}(4 - 9) + 6(2 + 3) \\ &= \frac{125}{6} \\ &\approx 20.833. \end{aligned}$$

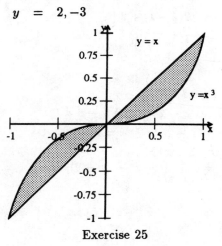

Exercise 25

27. $2x = y^2$, $x + 2 = y^2$.

$$2x = x + 2 \Rightarrow x = 2; \; y = \pm 2$$

240

$$A = \int_{-2}^{2} \left[\frac{1}{2} y^2 - (y^2 - 2) \right] dy$$

$$= 2 \int_{0}^{2} \left(-\frac{y^2}{2} + 2 \right) dy$$

$$= 2 \left[-\frac{y^3}{6} + 2y \right]_{0}^{2}$$

$$= 2 \left(-\frac{8}{6} + 4 \right)$$

$$= \frac{16}{3}.$$

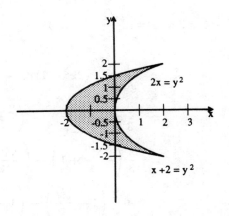

Exercise 27

29. $y = x^{2/3}$, $y = x^2$.

$x^{2/3} = x^2 \Rightarrow x = 0, \pm 1$

$$A = \int_{-1}^{1} (x^{2/3} - x^2) \, dx$$

$$= 2 \int_{0}^{1} (x^{2/3} - x^2) \, dx$$

$$= 2 \left[\frac{3}{5} x^{5/3} - \frac{x^3}{3} \right]_{0}^{1}$$

$$= 2 \left(\frac{3}{5} - \frac{1}{3} \right)$$

$$= \frac{8}{15}$$

$$\approx 0.5333.$$

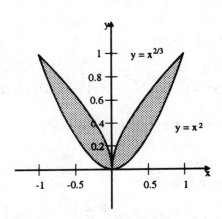

Exercise 29

31. $f(x) = 3 - x$, $[0, 4]$.

a) $\int_{0}^{4} (3 - x) \, dx = 3x - \frac{x^2}{2} \Big]_{0}^{4} = 12 - 8 = 4.$

b) $3 - x = 0 \Rightarrow x = 3$

$$A = \int_{0}^{3} (3 - x) \, dx - \int_{3}^{4} (3 - x) \, dx$$

$$= 3x - \frac{x^2}{2} \Big]_{0}^{3} - \left(3x - \frac{x^2}{2} \right) \Big]_{3}^{4}$$

$$= 9 - \frac{9}{2} - (12 - 8) + \left(9 - \frac{9}{2} \right) = 5.$$

33. $f(x) = 2x^2 - 1$, $[-1, 1]$.

a) $\int_{-1}^{1} (2x^2 - 1) \, dx = \frac{2}{3} x^3 - x \Big]_{-1}^{1} = \frac{2}{3} (1 + 1) - (1 + 1) = -\frac{2}{3}.$

b) $2x^2 - 1 = 0 \Rightarrow x = \pm\dfrac{\sqrt{2}}{2}$

$$A = \int_{-1}^{-\sqrt{2}/2}(2x^2-1)\,dx - \int_{-\sqrt{2}/2}^{\sqrt{2}/2}(2x^2-1)\,dx + \int_{\sqrt{2}/2}^{1}(2x^2-1)\,dx$$

$$= \frac{2}{3}x^3 - x\Big]_{-1}^{-\sqrt{2}/2} - \left(\frac{2}{3}x^3 - x\right)\Big]_{-\sqrt{2}/2}^{\sqrt{2}/2} + \frac{2}{3}x^3 - x\Big]_{\sqrt{2}/2}^{1}$$

$$= \frac{2}{3}\left(-\frac{\sqrt{2}}{4}+1\right) - \left(-\frac{\sqrt{2}}{2}+1\right) - \left(\frac{2}{3}\cdot\frac{\sqrt{2}}{4} - \frac{\sqrt{2}}{2} + \frac{2}{3}\cdot\frac{\sqrt{2}}{4} - \frac{\sqrt{2}}{2}\right)$$

$$\quad + \frac{2}{3}\left(1 - \frac{\sqrt{2}}{4}\right) - \left(1 - \frac{\sqrt{2}}{2}\right)$$

$$= \frac{2}{3}(2\sqrt{2}-1) \approx 1.2190.$$

35. $f(x) = x^2 + 6$, $[0,3]$.

  a) $\displaystyle\int_0^3 (x^2+6)\,dx = \frac{x^3}{3} + 6x\Big]_0^3 = 9 + 18 = 27.$

  b) Since $x^2 + 6 > 0$ on $[0,3]$, $A = 27$.

37. $y = x^{2/3}$, $y = 1$
  $x^{2/3} = 1 \Rightarrow x = \pm 1$

  $$A = \int_{-1}^{1}(1 - x^{2/3})\,dx = x - \frac{3}{5}x^{5/3}\Big]_{-1}^{1} = 1 + 1 - \frac{3}{5}(1+1) = \frac{4}{5}.$$

39. $y = x^3$, $y = -x^3$, $y = 1$, $y = -1$
  $y = x^3 \Rightarrow x = y^{1/3}$
  By symmetry the area is
  $4 \times$ area in quadrant I.

  $$A = 4\int_0^1 (y^{1/3} - 0)\,dy$$

  $$= \frac{4\cdot 3}{4}y^{4/3}\Big]_0^1$$

  $$= 3.$$

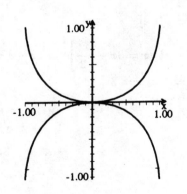

Exercise 39

41. $y = 4 - x^2$, an inverted parabola with x-intercepts ± 2. For the areas above and below $y = a$ to be equal,

$$\int_0^a \sqrt{4-y}\,dy = \int_a^4 \sqrt{4-y}\,dy$$

$$-\frac{2}{3}(4-y)^{3/2}\Big]_0^a = -\frac{2}{3}(4-y)^{3/2}\Big]_a^4$$

$$(4-a)^{3/2} - 8 = 0 - (4-a)^{3/2}$$

$$(4-a)^{3/2} = 4$$

$$4-a = 4^{2/3}$$

$$a = 4 - 4^{2/3} \approx 1.4802.$$

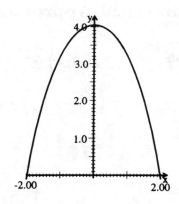

Exercise 41

43. a) $y = x$, $x = 3$, $y = -1$

$$A = \int_{-1}^3 [x - (-1)]\,dx = \frac{x^2}{2} + x\Big]_{-1}^3 = \frac{1}{2}(9-1) + 3 + 1 = 8.$$

The figure is a triangle with base of length 4 and altitude of length 4. Its area $= \frac{1}{2}(4\cdot4) = 8$.

b) $A = \int_1^3 [(2x+2) - (-2x-1)]\,dx = 2x^2 + 3x\Big]_1^3 = 2(9-1) + 3(3-1) = 22.$

The figure is a parallelogram with base at $x = 3$ of length $8 - (-7) = 15$ and base at $x = 1$ of length $4 - (-3) = 7$. The altitude is $3 - 1 = 2$. Area $= \left(\dfrac{b_1 + b_2}{2}\right) \times$ altitude $= \left(\dfrac{15 + 7}{2}\right) \cdot 2 = 22$.

45. $f(x) = \sin\sqrt{x}$, $[0, 4\pi^2]$

$$\Delta x = \frac{4\pi^2}{8} = \frac{\pi^2}{2} \approx 4.9348$$

i	x_i	$\sin\sqrt{x_i}$
1	2.4674	1.0000
2	7.4022	0.4086
3	12.3370	-0.3624
4	17.2718	-0.8491
5	22.2066	-1.0000
6	27.1414	-0.8788
7	32.0762	-0.5807
8	37.0110	-0.1982

$$\text{Area} \approx \Delta x \sum_{i=1}^8 |\sin\sqrt{x_i}|$$

$$\approx 4.9348(5.2778)$$
$$\approx 26.045.$$

If $n = 50$, we obtain the approximation: area ≈ 25.159.

6.6 Numerical Approximation of the Definite Integral

1. $\displaystyle\int_1^5 \frac{1}{x}\,dx \;\approx\; \left\{\sum_{j=1}^{8} f(m_j)\right\}\Delta x \;=\; M_8$

$\qquad = \left\{f\left(\frac{5}{4}\right) + f\left(\frac{7}{4}\right) + f\left(\frac{9}{4}\right) + \cdots + f\left(\frac{19}{4}\right)\right\}\left(\frac{5-1}{8}\right)$

$\qquad = \left(\frac{4}{5} + \frac{4}{7} + \frac{4}{9} + \cdots + \frac{4}{19}\right)\left(\frac{1}{2}\right) \approx 1.599844394.$

3. $\displaystyle\int_0^8 \sqrt{1+x^3}\,dx \;\approx\; M_4 = \left\{\sum_{j=1}^{4} f(m_j)\right\}\Delta x$

$\qquad = \left\{\sqrt{1+1^3} + \sqrt{1+3^3} + \sqrt{1+5^3} + \sqrt{1+7^3}\right\}\left(\frac{8-0}{4}\right) \approx 72.95585067.$

5. $\displaystyle\int_0^4 \sqrt{x^3+2}\,dx \;\approx\; T_4 = \left(\frac{4-0}{4}\right)\left(\frac{1}{2}\right)[f(0) + 2f(1) + 2f(2) + 2f(3) + f(4)]$

$\qquad = (1)\left(\frac{1}{2}\right)\left[\sqrt{2} + 2\sqrt{1+2} + 2\sqrt{8+2} + 2\sqrt{27+2} + \sqrt{64+2}\right] \approx 15.04861925.$

7. $\displaystyle\int_1^5 \frac{1}{x}\,dx \;\approx\; T_8$

$\qquad = \left(\frac{5-1}{8}\right)\left(\frac{1}{2}\right)\left[\left(\frac{2}{2}\right) + 2\left(\frac{2}{3}\right) + 2\left(\frac{2}{4}\right) + 2\left(\frac{2}{5}\right) + 2\left(\frac{2}{6}\right) + 2\left(\frac{2}{7}\right) + \left(\frac{2}{8}\right)\right]$

$\qquad \approx 1.628968254.$

9. $\displaystyle\int_0^8 \sqrt{1+x^3}\,dx \;\approx\; T_4$

$\qquad = \left(\frac{8-0}{4}\right)\left(\frac{1}{2}\right)\left[\sqrt{1} + 2\sqrt{1+2^3} + 2\sqrt{1+4^3} + 2\sqrt{1+6^3} + \sqrt{1+8^3}\right]$

$\qquad \approx 75.23585852.$

11. $\displaystyle\int_1^5 \frac{1}{x}\,dx \;\approx\; S_8$

$\qquad = \left(\frac{5-1}{8}\right)\left(\frac{1}{3}\right)\left[\left(\frac{2}{2}\right) + 4\left(\frac{2}{3}\right) + 2\left(\frac{2}{4}\right) + 4\left(\frac{2}{5}\right) + 2\left(\frac{2}{6}\right) + 4\left(\frac{2}{7}\right)\right.$

$\qquad\qquad \left. + 2\left(\frac{2}{8}\right) + 4\left(\frac{2}{9}\right) + \left(\frac{2}{10}\right)\right]$

$\qquad \approx 1.610846561.$

13. $\displaystyle\int_1^5 \sqrt{x^3+1}\,dx \;\approx\; S_4$

$$= \left(\frac{5-1}{4}\right)\left(\frac{1}{3}\right)\left[\sqrt{1^3+1}+4\sqrt{2^3+1}+2\sqrt{3^3+1}+4\sqrt{4^3+1}+\sqrt{5^3+1}\right]$$

$$\approx 22.49040732.$$

15. $\displaystyle\int_0^2 \sin(x^2)\,dx \;\approx\; T_{10}$

$$= \left(\frac{2-0}{10}\right)\left(\frac{1}{2}\right)\left[\sin(0)+2\sin\left(\frac{1}{5}\right)^2+\cdots+2\sin\left(\frac{9}{5}\right)^2+\sin\left(\frac{10}{5}\right)\right]$$

$$\approx 0.795924733.$$

17. $\displaystyle\int_0^1 \frac{1}{1+x^2}\,dx \;\approx\; T_{10} = \left(\frac{1-0}{10}\right)\left(\frac{1}{2}\right)\left[\frac{1}{1+0^2}+\frac{2}{1+(.1)^2}+\cdots+\frac{2}{1+(.9)^2}+\frac{1}{1+(1.0)^2}\right]$

$$\approx 0.784981497.$$

19. $\displaystyle\int_0^2 \frac{1}{\sqrt{1+x^3}}\,dx \;\approx\; S_{10} = \left(\frac{2-0}{10}\right)\left(\frac{1}{3}\right)\left[\frac{1}{\sqrt{1}}+\frac{4}{\sqrt{1+\left(\frac{1}{5}\right)^3}}+\frac{2}{\sqrt{1+\left(\frac{2}{5}\right)^3}}+\cdots\right.$

$$\left.+\frac{4}{\sqrt{1+\left(\frac{9}{5}\right)^3}}+\frac{1}{\sqrt{1+2^3}}\right] \approx 1.402206274.$$

21. $\displaystyle\int_1^2 \frac{1}{x}\,dx \;\approx\; S_{10} = \left(\frac{2-1}{10}\right)\left(\frac{1}{3}\right)\left[\frac{1}{1}+\frac{4}{1.1}+\frac{2}{1.2}+\cdots+\frac{4}{1.9}+\frac{1}{2}\right]$

$$\approx 0.693150231.$$

23. $\displaystyle\int_0^1 \frac{4}{1+x^2}\,dx \;\approx\; M_{10} = 3.142425985 \approx T_{10} = 3.139925989$

$$M_{100} \approx 3.141600987$$

$$T_{100} \approx 3.141575987.$$

25. Since $f(x)=\sqrt{x^3+1}$ is positive and $f''(x)=\dfrac{3}{4}\dfrac{x}{(x^3+1)^{3/2}}[x^3+1]=\dfrac{3}{4}\left(\dfrac{x}{\sqrt{x^3+1}}\right)$ which is positive for $0<x<8$, then $y=f(x)$ is concave up between 0 and 8. So by **Theorem 10**,

$$M_4 \;\le\; \int_0^8 f(x)\,dx \;\le\; T_4,$$

so

$$72.9558 \;\le\; \int_0^8 \;\le\; 75.2359$$

27. Since no measurement is shown in Fig 6.12 to the left of the line measuring 12 ft, we will use 0 ft as $f(x_0)$ and 13 ft as $f(x_5)$. Then the area is approximately

$$T_5 = (5 \text{ ft})\left(\frac{1}{2}\right)[0 + 2(12) + 2(15) + 2(15) + 2(20) + 2(18) + 13]\,(\text{ft})$$

$$= \frac{5}{2}[173]\,(\text{ft}^2) = 432.5\,\text{ft}^2$$

[Note: if we use 0 for $f(x_0)$ and 13 for $f(x_6)$, then we get $T_6 = \left(\frac{5}{2}\right)[0 + 2(12) + \cdots + 2(18) + 13] = 432.5$.]

29. We will use trapezoids to estimate the number of autos in each subinterval. Note that the 5 subintervals are not equal in length.

$$A_1 = [(9-5)(\text{hrs})]\left(\frac{1}{2}\right)\left[(10 + 100)\left(\frac{\text{auto}}{\text{hr}}\right)\right] = 220 \text{ autos}$$

$$A_2 = [4 \text{ hrs}]\left(\frac{1}{2}\right)\left[(100 + 50)\left(\frac{\text{auto}}{\text{hr}}\right)\right] = 300 \text{ autos}$$

$$A_3 = [2 \text{ hrs}]\left(\frac{1}{2}\right)\left[(50 + 60)\left(\frac{\text{auto}}{\text{hr}}\right)\right] = 110 \text{ autos}$$

$$A_4 = [2 \text{ hrs}]\left(\frac{1}{2}\right)\left[(60 + 110)\left(\frac{\text{auto}}{\text{hr}}\right)\right] = 170 \text{ autos}$$

$$A_5 = [6 \text{ hrs}]\left(\frac{1}{2}\right)\left[(110 + 20)\left(\frac{\text{auto}}{\text{hr}}\right)\right] = 390 \text{ autos}$$

$$A_6 = [(11-5)(\text{hrs})]\left(\frac{1}{2}\right)\left[(20 + 10)\left(\frac{\text{auto}}{\text{hr}}\right)\right] = 90 \text{ autos}$$

$$\sum A_j = 1280 \text{ autos}.$$

31. Graph the antiderivative with the range $[1, 5] \times [0, 2]$. Take $F(1) = 0$, $k = 50$, and $n = 6$. So using the programs set **FA=0**, **K=50**, and **N=6**. Typical graph will look like the following.

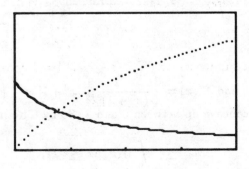

Exercise 31

33. Graph the antiderivative with the range $[-4, 4] \times [0, 3]$. Take $F(-4) = 0$, $k = 50$, and $n = 6$. So using the programs set **FA=0**, **K=50**, and **N=6**. Typical graph will look like the following.

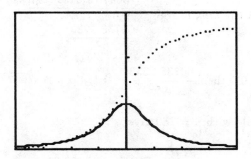

Exercise 33

35. a) Graph with the range $[1, 365] \times [-10, 100]$.

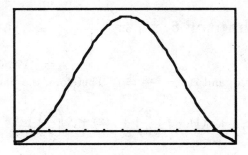

Exercise 35

Hottest months: June, July, and August.

Coldest months: January and December.

Now the minimum is -6.884 and the maximum is 92.983 so

$$-6.884 < \frac{1}{364} \int_1^{365} T(t)\, dt < 92.983.$$

Therefore the approximate average value is $\dfrac{92.983 - 6.884}{2} = 43.0495$.

b) Now $T'(t) = \dfrac{102\pi}{365} \cos\left(\dfrac{2\pi}{365}(t - 95.2)\right)$, and $T''(t) = \dfrac{-204\pi^2}{133225} \sin\left(\dfrac{2\pi}{365}(t - 95.2)\right)$.

So the maximum of $|T''|$ occurs where $\sin\left(\dfrac{2\pi}{365}(t - 95.2)\right) = 1$ or when

$$\begin{aligned}
\frac{2\pi}{365}(t - 95.2) &= \frac{\pi}{2} \\
t - 95.2 &= \frac{365}{4} \\
t &= 186.45.
\end{aligned}$$

Therefore the maximum of $|T''| = \dfrac{204\pi^2}{133225} = 0.01511277$. The error for the Trapezoidal Rule is then

$$|\text{error}| \leq \frac{(b-a)^3 M}{12n^2}$$

247

$$= \frac{(364)^2(0.01511277)}{12n^2}$$

$$= \frac{60738.9233}{n^2}.$$

So if we require $|\text{error}| < 0.1$ then $\dfrac{60738.9233}{n^2} < 0.1$. So $n > \sqrt{607389.233} = 779.352$. Thus take $n = 780$.

c) Using the Trapezoidal Rule with $n = 780$ gives

$$\int_1^{365} T(t)\, dt \approx 15702.91.$$

So the average value is approximately $\dfrac{15702.91}{364} = 43.14.$

Review Exercises—Chapter 6

1. $f(x) = 3x + 1$, $x \in [0, 3]$, $n = 6$.

   Note that f is increasing on $[0,3]$ and $\Delta x = \frac{3-0}{6} = \frac{1}{2}$. Therefore, the partition is $\left\{0, \frac{1}{2}, 1, \frac{3}{2}, 2, \frac{5}{2}, 3\right\}$.

   a) $\underline{S}_6 \;=\; \dfrac{1}{2}\left(f(0) + f\left(\dfrac{1}{2}\right) + f(1) + f\left(\dfrac{3}{2}\right) + f(2) + f\left(\dfrac{5}{2}\right)\right)$

   $\qquad = \dfrac{1}{2}\left(1 + \dfrac{5}{2} + 4 + \dfrac{11}{2} + 7 + \dfrac{17}{2}\right)$

   $\qquad = \dfrac{57}{4}.$

   b) $\overline{S}_6 \;=\; \dfrac{1}{2}\left(f\left(\dfrac{1}{2}\right) + f(1) + f\left(\dfrac{3}{2}\right) + f(2) + f\left(\dfrac{5}{2}\right) + f(3)\right)$

   $\qquad = \dfrac{1}{2}\left(\dfrac{5}{2} + 4 + \dfrac{11}{2} + 7 + \dfrac{17}{2} + 10\right)$

   $\qquad = \dfrac{75}{4}.$

3. $f(x) = \dfrac{1}{1 + x^2}$, $x \in [-1, 1]$, $n = 6$.

   Note that f is increasing on $[-1, 0]$ and decreasing on $[0, 1]$. We have $\Delta x = \frac{1-(-1)}{6} = \frac{1}{3}$. The partition is

   $$x_i \;=\; -1 + i(\Delta x)$$
   $$=\; -1 + \frac{i}{3}.$$

   That is, the partition is $\left\{-1, -\frac{2}{3}, -\frac{1}{3}, 0, \frac{1}{3}, \frac{2}{3}, 1\right\}$.

a) $\underline{S}_6 = \Delta x \left(f(x_0) + f(x_1) + f(x_2) + f(x_4) + f(x_5) + f(x_6) \right)$

$$= \frac{1}{3} \left(\frac{1}{2} + \frac{9}{13} + \frac{9}{10} + \frac{9}{10} + \frac{9}{13} + \frac{1}{2} \right)$$

$$= \frac{272}{195}.$$

b) $\overline{S}_6 = \Delta x \left(f(x_1) + f(x_2) + f(x_3) + f(x_3) + f(x_4) + f(x_5) \right)$

$$= \frac{1}{3} \left(\frac{9}{13} + \frac{9}{10} + 1 + 1 + \frac{9}{10} + \frac{9}{13} \right)$$

$$= \frac{337}{195}.$$

5. $f(x) = \cos \pi x$, $x \in \left[-\frac{1}{2}, \frac{1}{2} \right]$, $n = 4$.

Note that f is increasing on $\left[-\frac{1}{2}, 0 \right]$ and decreasing on $\left[0, \frac{1}{2} \right]$. We have $\Delta x = \frac{\frac{1}{2} - \left(-\frac{1}{2} \right)}{4} = \frac{1}{4}$. The partition is

$$x_i = -\frac{1}{2} + i(\Delta x)$$
$$= -\frac{1}{2} + \frac{i}{4}.$$

In other words, the partition is $\left\{ -\frac{1}{2}, -\frac{1}{4}, 0, \frac{1}{4}, \frac{1}{2} \right\}$.

a) $\underline{S}_4 = \Delta x (f(x_0) + f(x_1) + f(x_3) + f(x_4))$

$$= \frac{1}{4} \left(0 + \frac{\sqrt{2}}{2} + \frac{\sqrt{2}}{2} + 0 \right)$$

$$= \frac{\sqrt{2}}{4}.$$

b) $\overline{S}_4 = \Delta x (f(x_1) + f(x_2) + f(x_2) + f(x_3))$

$$= \frac{1}{4} \left(\frac{\sqrt{2}}{2} + 1 + 1 + \frac{\sqrt{2}}{2} \right)$$

$$= \frac{1}{2} + \frac{\sqrt{2}}{4}.$$

7. $f(x) = 3 - x$, $x \in [0, 2]$, $n = 100$

Note that f is decreasing on $[0, 2]$. Also, we have $\Delta x = \frac{2-0}{100} = \frac{1}{50}$. The partition is

$$x_i = 0 + i(\Delta x)$$
$$= \frac{i}{50}.$$

In other words, the partition is $\left\{ 0, \frac{1}{50}, \frac{1}{25}, \ldots, \frac{49}{25}, \frac{99}{50}, 2 \right\}$.

a) $\underline{S}_{100} = \Delta x \left(\sum_{i=1}^{100} f(x_i) \right)$

$= \dfrac{1}{50} \left(\sum_{i=1}^{100} (3 - x_i) \right)$

$= \dfrac{1}{50} \left(\sum_{i=1}^{100} \left(3 - \dfrac{i}{50} \right) \right)$

$= \dfrac{1}{50} \left(300 - \dfrac{1}{50} \sum_{i=1}^{100} i \right)$

$= 6 - \dfrac{1}{50^2} \left(\dfrac{100(101)}{2} \right)$

$= 6 - \dfrac{101}{50} = 3\dfrac{49}{50} = 3.98.$

b) $\overline{S}_{100} = \Delta x \left(\sum_{i=0}^{99} f(x_i) \right)$

$= \dfrac{1}{50} \left(\sum_{i=0}^{99} (3 - x_i) \right)$

$= \dfrac{1}{50} \left(\sum_{i=0}^{99} \left(3 - \dfrac{i}{50} \right) \right)$

$= \dfrac{1}{50} \left(300 - \dfrac{1}{50} \sum_{i=0}^{99} i \right)$

$= 6 - \dfrac{1}{50^2} \left(\dfrac{99(100)}{2} \right)$

$= 6 - \dfrac{99}{50} = 4\dfrac{1}{50} = 4.02.$

9. $f(x) = \cos x$, $\left[-\dfrac{\pi}{2}, \pi \right]$, $n = 6$

Note that f is positive on $\left[-\dfrac{\pi}{2}, \dfrac{\pi}{2} \right]$ and negative on $\left[\dfrac{\pi}{2}, \pi \right]$. Also, f is increasing on $\left[-\dfrac{\pi}{2}, 0 \right]$ and decreasing on $[0, \pi]$. We have $\Delta x = \dfrac{\pi - \left(-\frac{\pi}{2} \right)}{6} = \dfrac{\pi}{4}$. So the partition is

$$x_i = -\dfrac{\pi}{2} + i \dfrac{\pi}{4},$$

which is $\left\{ -\dfrac{\pi}{2}, -\dfrac{\pi}{4}, 0, \dfrac{\pi}{4}, \dfrac{\pi}{2}, \dfrac{3\pi}{4}, \pi \right\}$.

i	x_i	$f(x_i)$
0	$-\pi/2$	0
1	$-\pi/4$	$\sqrt{2}/2$
2	0	1
3	$\pi/4$	$\sqrt{2}/2$
4	$\pi/2$	0
5	$3\pi/4$	$-\sqrt{2}/2$
6	π	-1

The smallest Riemann sum for $n = 6$ is

$$\Delta x(f(x_0) + f(x_1) + f(x_3) + f(x_4) + f(x_5) + f(x_6)) = \dfrac{\pi}{4} \left(0 + \dfrac{\sqrt{2}}{2} + \dfrac{\sqrt{2}}{2} + 0 - \dfrac{\sqrt{2}}{2} - 1 \right)$$

$$= \dfrac{\pi}{8} \left(\sqrt{2} - 2 \right).$$

The largest Riemann sum for $n = 6$ is

$$\Delta x(f(x_1) + f(x_2) + f(x_2) + f(x_3) + f(x_4) + f(x_5)) = \dfrac{\pi}{4} \left(\dfrac{\sqrt{2}}{2} + 1 + 1 + \dfrac{\sqrt{2}}{2} + 0 - \dfrac{\sqrt{2}}{2} \right)$$

$$= \dfrac{\pi}{8} \left(\sqrt{2} + 4 \right).$$

11. $f(x) = x^2 + 1$, $x \in [-1, 2]$, $n = 6$

Note that f is positive on $[-1, 2]$. Moreover, it is decreasing on $[-1, 0]$ and increasing on $[0, 2]$. We have $\Delta x = \frac{2-(-1)}{6} = \frac{1}{2}$. So the partition is

$$\begin{aligned} x_i &= -1 + i(\Delta x) \\ &= -1 + \frac{i}{2}. \end{aligned}$$

In other words, the partition is $\left\{ -1, -\frac{1}{2}, 0, \frac{1}{2}, 1, \frac{3}{2}, 2 \right\}$.

The smallest Riemann sum is

$$\begin{aligned} \underline{S}_6 &= \frac{1}{2} \left(f(x_1) + f(x_2) + f(x_2) + f(x_3) + f(x_4) + f(x_5) \right) \\ &= \frac{1}{2} \left(\frac{5}{4} + 1 + 1 + \frac{5}{4} + 2 + \frac{13}{4} \right) \\ &= \frac{39}{8}. \end{aligned}$$

The largest Riemann sum is

$$\begin{aligned} \overline{S}_6 &= \frac{1}{2} \left(f(x_0) + f(x_1) + f(x_3) + f(x_4) + f(x_5) + f(x_6) \right) \\ &= \frac{1}{2} \left(2 + \frac{5}{4} + \frac{5}{4} + 2 + \frac{13}{4} + 5 \right) \\ &= \frac{59}{8}. \end{aligned}$$

13. $y = 2x - 2$, $[2, 4]$

Let $\Delta x = \frac{4-2}{n} = \frac{2}{n}$, $x_i = 2 + \left(\frac{2}{n} \right) i$

f is increasing on $[2, 4]$.

$$\begin{aligned} \underline{S}_n &= \Delta x \sum_{i=0}^{n-1} f(x_i) = \frac{2}{n} \sum_{i=0}^{n-1} 4 + \left(\frac{4}{n} \right) i - 2 \\ &= \frac{2}{n} \sum_{i=0}^{n-1} 2 + \frac{4i}{n} = \frac{2}{n} (2n) + \frac{8}{n^2} \sum_{i=0}^{n-1} i \\ &= 4 + \frac{8}{n^2} \frac{(n-1)(n)}{2}. \\ \lim_{n \to \infty} \underline{S}_n &= 4 + 4 = 8. \end{aligned}$$

15. $\displaystyle\int_0^9 \sqrt{x}\, dx = \frac{2}{3} x^{3/2} \Big]_0^9 = \frac{2}{3} \cdot 27 = 18$

17. $\displaystyle\int_0^1 (x^{2/3} - x^{1/2})\, dx = \frac{3}{5} x^{5/3} - \frac{2}{3} x^{3/2} \Big]_0^1 = \frac{3}{5} - \frac{2}{3} = -\frac{1}{15}$

19. $\displaystyle\int_0^1 x^3(x+1)\, dx = \frac{x^5}{5} + \frac{x^4}{4} \Big]_0^1 = \frac{1}{5} + \frac{1}{4} = \frac{9}{20}$

21. $\int_0^1 (3x + 2)\, dx = \dfrac{3}{2}\, x^2 + 2x \bigg]_0^1 = \dfrac{3}{2} + 2 = \dfrac{7}{2}$

23. $\int_{-3}^5 (x^2 + 2)\, dx = \dfrac{x^3}{3} + 2x \bigg]_{-3}^5 = \dfrac{1}{3}(125 + 27) + 2(5 + 3) = \dfrac{152}{3} + 16 \approx 66.667$

25. $\int_1^4 \sqrt{x}\, dx = \dfrac{2}{3}\, x^{3/2} \bigg]_1^4 = \dfrac{2}{3}(8 - 1) = \dfrac{14}{3}$

27. $\int_1^4 \left(\sqrt{x} - \dfrac{1}{\sqrt{x}} \right) dx = \dfrac{2}{3}\, x^{3/2} - 2\sqrt{x} \bigg]_1^4 = \dfrac{2}{3}(8 - 1) - 2(2 - 1) = \dfrac{14}{3} - 2 = \dfrac{8}{3}$

29. $\int_0^2 (x + 7)(2x + 2)\, dx = \int_0^2 (2x^2 + 16x + 14)\, dx = \dfrac{2}{3}\, x^3 + 8x^2 + 14x \bigg]_0^2 = \dfrac{16}{3} + 32 + 28 = \dfrac{196}{3} \approx 65.333$

31. $\int_0^{\pi/4} \sin x\, dx = -\cos x \bigg]_0^{\pi/4} = -\dfrac{\sqrt{2}}{2} + 1 = \dfrac{2 - \sqrt{2}}{2} \approx 0.29289$

33. $\int_0^{\pi/4} \sin(2x)\, dx = -\dfrac{1}{2} \cos 2x \bigg]_0^{\pi/4} = -\dfrac{1}{2}(0 - 1) = \dfrac{1}{2}$

35. $\int_1^2 (x + 4)^{10}\, dx = \dfrac{1}{11}(x + 4)^{11} \bigg]_1^2 = \dfrac{1}{11}(6^{11} - 5^{11}) \approx 28.543 \times 10^6$

37. $\int_2^4 \dfrac{t^2 - 2t}{5}\, dt = \dfrac{t^3}{15} - \dfrac{t^2}{5} \bigg]_2^4 = \dfrac{1}{15}(64 - 8) - \dfrac{1}{5}(16 - 4) = \dfrac{56}{15} - \dfrac{12}{5} = \dfrac{4}{3}$

39. $\int_0^3 \dfrac{dt}{(t + 1)^2} = -\dfrac{1}{t + 1} \bigg]_0^3 = -\left(\dfrac{1}{4} - 1 \right) = \dfrac{3}{4}$

41. $\int_0^{\pi/4} \dfrac{1}{\cos^2 x}\, dx = \int_0^{\pi/4} \sec^2 x\, dx = \tan x \bigg]_0^{\pi/4} = 1$

43. $\int_0^1 (x - \sqrt{x})^2\, dx = \int_0^1 (x^2 - 2x\sqrt{x} + x)\, dx = \dfrac{x^3}{3} - \dfrac{4}{5}\, x^{5/2} + \dfrac{x^2}{2} \bigg]_0^1 = \dfrac{1}{3} - \dfrac{4}{5} + \dfrac{1}{2} = \dfrac{1}{30}$

45. $\int_1^8 t \left(\sqrt[3]{t} - 2t \right) dt = \int_1^8 (t^{4/3} - 2t^2)\, dt = \dfrac{3}{7}\, t^{7/3} - \dfrac{2}{3}\, t^3 \bigg]_1^8 = \dfrac{3}{7}(128 - 1) - \dfrac{2}{3}(512 - 1) = -\dfrac{6011}{21} \approx -286.238$

47. $\int_1^2 \dfrac{1}{(1 - 2x)^3}\, dx$

    Let $u = 1 - 2x$, $du = -2\, dx$
    if $x = 1$, $u = -1$; if $x = 2$, $u = -3$.

$$-\dfrac{1}{2} \int_{-1}^{-3} \dfrac{du}{u^3} = \dfrac{1}{4} \cdot \dfrac{1}{u^2} \bigg]_{-1}^{-3} = \dfrac{1}{4} \left(\dfrac{1}{9} \cdot \dfrac{1}{1} \right) = -\dfrac{2}{9} \approx -0.2223$$

49. $\int_0^1 \dfrac{x^3 + 8}{x + 2}\, dx = \int_0^1 (x^2 - 2x + 4)\, dx = \dfrac{x^3}{3} - x^2 + 4x \bigg]_0^1 = \dfrac{1}{3} - 1 + 4 = \dfrac{10}{3}$

51. $\displaystyle\int_0^{\pi/2} \sec\left(\frac{u}{2}\right)\tan\left(\frac{u}{2}\right)du = 2\sec\left(\frac{u}{2}\right)\Bigg]_0^{\pi/2} = 2(\sqrt{2}-1) \approx 0.8284$

53. $\displaystyle\int_{-1}^{3}|x-x^2|\,dx = -\int_{-1}^{0}(x-x^2)\,dx + \int_0^1 (x-x^2)\,dx - \int_1^3 (x-x^2)\,dx = -\left(\frac{x^2}{2}-\frac{x^3}{3}\right)\Bigg]_{-1}^{0} + \frac{x^2}{2}-\frac{x^3}{3}\Bigg]_0^1 -$

$\left(\frac{x^2}{2}-\frac{x^3}{3}\right)\Bigg]_1^3 = -\left(0-\frac{1}{2}-0-\frac{1}{3}\right) + \left(\frac{1}{2}-\frac{1}{3}\right) - \left(\frac{9}{2}-\frac{1}{2}-9+\frac{1}{3}\right) = \frac{5}{6}+\frac{1}{6}+\frac{14}{3} = \frac{17}{3} \approx 5.6667$

55. $\displaystyle\int_0^1 x\sqrt{x+1}\,dx$

Let $u = x+1$, $du = dx$, $x = u-1$
if $x = 0$, $u = 1$; if $x = 1$, $u = 2$.

$\displaystyle\int_1^2 (u-1)\sqrt{u}\,du = \int_1^2 \left(u^{3/2}-\sqrt{u}\right)du = \frac{2}{5}u^{5/2}-\frac{2}{3}u^{3/2}\Bigg]_1^2 = \frac{2}{5}\left(4\sqrt{2}-1\right)-\frac{2}{3}\left(2\sqrt{2}-1\right)$

$\displaystyle = \frac{4}{15}\sqrt{2}+\frac{4}{15} = \frac{4}{15}\left(\sqrt{2}+1\right) \approx 0.6438$

57. $\displaystyle 2\sum_{i=1}^{m}i - 2\sum_{i=1}^{10}i = 1782.$

$$\begin{aligned}
\frac{2(m)(m+1)}{2} - 2\frac{(10)(11)}{2} &= 1782 \\
m(m+1) &= 1782 + 110 = 1892 \\
m^2 + m - 1892 &= 0 \\
m &= \frac{-1\pm\sqrt{1+7568}}{2} = \frac{-1\pm 87}{2} = \frac{86}{2} = 43 \\
n &= 2m = 86
\end{aligned}$$

59. $f(x) = \sqrt{x-1}$, $1 \le x \le 5$
$A = \displaystyle\int_1^5 \sqrt{x-1}\,dx = \frac{2}{3}(x-1)^{3/2}\Bigg]_1^5 = \frac{2}{3}(8) = \frac{16}{3}$

61. $f(x) = (x^2+2)^2$, $0 \le x \le 1$
$A = \displaystyle\int_0^1 (x^2+2)^2\,dx = \int_0^1 (x^4+4x^2+4)\,dx = \frac{x^5}{5}+\frac{4x^3}{3}+4x\Bigg]_0^1 = \frac{1}{5}+\frac{4}{3}+4 = \frac{83}{15} \approx 5.533$

63. $f(x) = 4x-x^2$, $4 \le x \le 5$
$A = -\displaystyle\int_4^5 (4x-x^2)\,dx = -\left(2x^2-\frac{x^3}{3}\right)\Bigg]_4^5 = -\left(50-\frac{125}{3}\right) + \left(32-\frac{64}{3}\right) = -18+\frac{61}{3} = \frac{7}{3}$

65. $f(x) = 3+2x-x^2$, $1 \le x \le 4$
$f \ge 0$ on $[1,3]$; $f \le 0$ on $[3,4]$.

$$\begin{aligned}
A &= \int_1^3 (3+2x-x^2)\,dx - \int_3^4 (3+2x-x^2)\,dx = 3x+x^2-\frac{x^3}{3}\Bigg]_1^3 - \left(3x+x^2-\frac{x^3}{3}\right)\Bigg]_3^4 \\
&= 9+9-9-\left(3+1-\frac{1}{3}\right)-\left(12+16-\frac{64}{3}\right)+(9+9-9) = \frac{23}{3} \approx 7.667
\end{aligned}$$

67. $f(x) = x^2 - x - 2,\ 2 \le x \le 4$

$$A = \int_2^4 (x^2 - x - 2)\, dx = \frac{x^3}{3} - \frac{x^2}{2} - 2x \bigg]_2^4 = \frac{64}{3} - \frac{8}{3} - \frac{16}{2} + \frac{4}{2} - 8 + 4 = \frac{26}{3} \approx 8.667$$

69. $f(x) = \sqrt{1-x},\ -1 \le x \le 1$

$$A = \int_{-1}^1 \sqrt{1-x}\, dx = -\frac{2}{3}(1-x)^{3/2}\bigg]_{-1}^1 = -\frac{2}{3}(0 - 2^{3/2}) = \frac{4\sqrt{2}}{3} \approx 1.8856$$

71. $f(x) = (ax^2 + 3)^2,\ 0 \le x \le 3.$

$$A = \int_0^3 (ax^2 + 3)^2\, dx = \int_0^3 (a^2 x^4 + 6ax^2 + 9)\, dx$$

$$= \frac{1}{5}a^2 x^5 + 2ax^3 + 9x\bigg]_0^3 = \frac{1}{5}a^2(243) + 2a(27) + 27 = \frac{243}{5}a^2 + 54a + 27$$

73. $f(x) = x^2 + 2x - 3,\ -3 \le x \le 1$

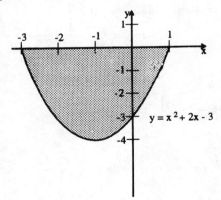

$y = x^2 + 2x - 3$

Exercise 73

$$A = -\int_{-3}^1 (x^2 + 2x - 3)\, dx = -\left[\frac{x^3}{3} + x^2 - 3x\right]_{-3}^1 = -\left(\frac{1}{3} + 1 - 3\right) + (-9 + 9 + 9) = \frac{32}{3}$$

75. $f(x) = 2x - 4,\ 0 \le x \le 4$

254

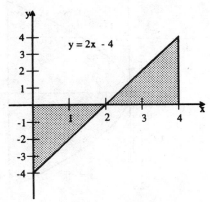

Exercise 75

$$A = -\int_0^2 (2x-4)\,dx + \int_2^4 (2x-4)\,dx = -\left[x^2 - 4x\right]_0^2 + \left[x^2 - 4x\right]_2^4 = -(4-8)+(16-16)-(4-8) = 8$$

77. $y = x^3$, $y = 0$, $x = -2$, $x = 2$

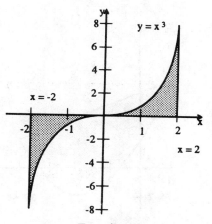

Exercise 77

$$A = -\int_{-2}^0 x^3\,dx + \int_0^2 x^3\,dx = 2\int_0^2 x^3\,dx = \frac{1}{2}x^4\Big]_0^2 = 8$$

79. $y = \sqrt{x}$, $y = -\sqrt{x}$, $x = 0$, $x = 4$

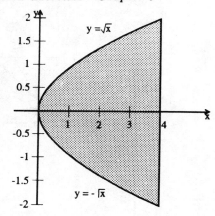

Exercise 79

$$A = \int_0^4 \left[\sqrt{x} - (-\sqrt{x}) \right] dx = \frac{4}{3} x^{3/2} \Big]_0^4 = \frac{4}{3}(8) = \frac{32}{3}$$

81. $y = x^2 - 4x + 2, \; x + y = 6$

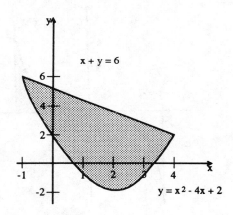

Exercise 81

Find intersections: $x^2 - 4x + 2 = 6 - x \Rightarrow x^2 - 3x - 4 = 0; \; (x+1)(x-4) = 0; \; x = -1, 4.$

$$A = \int_{-1}^4 [6 - x - (x^2 - 4x + 2)] \, dx = 4x + \frac{3}{2} x^2 - \frac{x^3}{3} \Big]_{-1}^4 = 4(4+1) + \frac{3}{2}(16-1) - \frac{1}{3}(64+1) = \frac{125}{6} \approx 20.833$$

83. $y = \dfrac{x - 2}{\sqrt{x^2 - 4x + 8}}, \; x = 0, \; y = 0; \; y = f(x) < 0 \text{ on } [0, 2].$

256

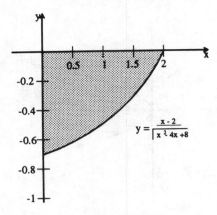

Exercise 83

$$A = \int_0^2 -\frac{(x-2)}{\sqrt{x^2-4x+8}}\,dx$$

Let $u = x^2 - 4x + 8$, $du = (2x - 4)\,dx$
if $x = 0$, $u = 8$; if $x = 2$, $u = 4$.

$$A = \int_8^4 \frac{-du}{2\sqrt{u}} = -\sqrt{u}\Big]_8^4 = \sqrt{8} - 2 = 2\left(\sqrt{2} - 1\right) \approx 0.8284$$

85. $x + y^2 = 0$, $x + y + 2 = 0$

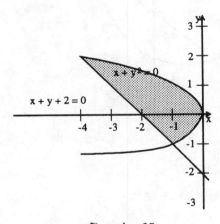

Exercise 85

Find intersections: $-y^2 = -y - 2 \Rightarrow y^2 - y - 2 = 0$; $(y-2)(y+1) = 0$; $y = -1, 2$.

$$A = \int_{-1}^2 [-y^2 - (-y-2)]\,dy = -\frac{y^3}{3} + \frac{y^2}{2} + 2y\Big]_{-1}^2 = -\frac{1}{3}(8+1) + \frac{1}{2}(4-1) + 2(2+1) = \frac{9}{2}$$

87. $x = y^{2/3}$, $x = y^2$

257

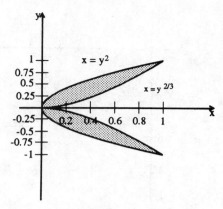

Exercise 87

Find intersections: $y^{2/3} = y^2 \Rightarrow y^{2/3}(1 - y^{4/3}) = 0 \Rightarrow y = 0, -1, 1$

$$A = \int_{-1}^{1} (y^{2/3} - y^2)\, dy = 2\left[\frac{3}{5}y^{5/3} - \frac{y^3}{3}\right]_{0}^{1} = 2\left(\frac{3}{5} - \frac{1}{3}\right) = \frac{8}{15}.$$

89. $y = x^2 + x - 2$, and the **line through** $(-1, -2)$ and $(1, 0)$.

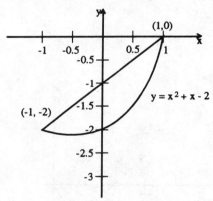

Exercise 89

Slope of line: $\quad m = \dfrac{0 - (-2)}{1 - (-1)} = 1$

$\dfrac{y - 0}{x - 1} = 1 \Rightarrow y = x - 1.$

Find intersections: $x^2 + x - 2 = x - 1 \Rightarrow x^2 - 1 = 0 \Rightarrow x = -1, 1.$

$$A = \int_{-1}^{1} [x - 1 - (x^2 + x - 2)]\, dx = x - \frac{x^3}{3}\Big]_{-1}^{1} = (1 + 1) - \frac{1}{3}(1 + 1) = \frac{4}{3}$$

91. $3x + 5y = 23,\; 5x - 2y = 28,\; 2x - 7y = -26$

258

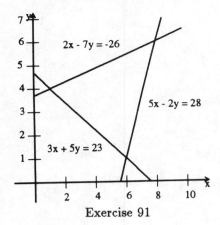

Exercise 91

Find intersections:

$$
\begin{array}{rcl}
2x - 7y &=& -26 \\
3x + 5y &=& 23
\end{array}
\quad \rightarrow \quad
\begin{array}{rcl}
6x - 21y &=& -78 \\
6x + 10y &=& 46 \\
\hline
-31y &=& -124 \Rightarrow \quad y = 4, \ x = 1
\end{array}
$$

$$
\begin{array}{rcl}
2x - 7y &=& -26 \\
5x - 2y &=& 28
\end{array}
\quad \rightarrow \quad
\begin{array}{rcl}
10x - 35y &=& -130 \\
10x - 4y &=& 56 \\
\hline
-31y &=& -186 \Rightarrow \quad y = 6, \ x = 8
\end{array}
$$

$$
\begin{array}{rcl}
3x + 5y &=& 23 \\
5x - 2y &=& 28
\end{array}
\quad \rightarrow \quad
\begin{array}{rcl}
15x + 25y &=& 115 \\
15x - 6y &=& 84 \\
\hline
31y &=& 31 \Rightarrow \quad y = 1, \ x = 6
\end{array}
$$

$$
\begin{aligned}
A &= \int_1^6 \left(\frac{2x+26}{7} - \frac{23-3x}{5} \right) dx + \int_6^8 \left(\frac{2x+26}{7} - \frac{5x-28}{2} \right) dx \\
&= x^2 \left(\frac{1}{7} + \frac{3}{10} \right) + x \left(\frac{26}{7} - \frac{23}{5} \right) \Big]_1^6 + x^2 \left(\frac{1}{7} - \frac{5}{4} \right) + x \left(\frac{26}{7} + 14 \right) \Big]_6^8 \\
&= \frac{31}{70}(36 - 1) - \frac{31}{35}(6 - 1) - \frac{31}{28}(64 - 36) + \frac{124}{7}(8 - 6) \\
&= \frac{31}{2} - \frac{31}{7} - 31 + \frac{248}{7} = \frac{217}{14} = \frac{31}{2}
\end{aligned}
$$

93. $y = \sin x \cos x$ on $\left[0, \frac{\pi}{2} \right]$.

$$
\overline{f} = \frac{1}{\frac{\pi}{2} - 0} \int_0^{\pi/2} \sin x \cos x \, dx = \frac{2}{\pi} \cdot \frac{1}{2} \sin^2 x \Big]_0^{\pi/2} = \frac{1}{\pi}(1 - 0) = \frac{1}{\pi}
$$

95. $y = \dfrac{x}{\sqrt{1 + x^2}}$ on $[0, 2]$.

$$
\overline{f} = \frac{1}{2 - 0} \int_0^2 \frac{x}{\sqrt{1 + x^2}} \, dx = \frac{1}{2} \sqrt{1 + x^2} \Big]_0^2 = \frac{1}{2} \left(\sqrt{5} - 1 \right) \approx 0.6180
$$

259

97. $y = \sin x$ on $[0, 10\pi]$.

$$\overline{f} = \frac{1}{10\pi - 0} \int_0^{10\pi} \sin x \, dx = -\frac{1}{10\pi} \cos x \Big]_0^{10\pi} = -\frac{1}{10\pi}(1 - 1) = 0$$

99. $f(x) = \sin^2 x$ on $[0, n\pi]$.

$$\overline{f} = \frac{1}{n\pi} \int_0^{n\pi} \sin^2 x \, dx = \frac{1}{2n\pi} \int_0^{n\pi} (1 - \cos 2x) \, dx$$

$$= \frac{1}{2n\pi} \left[x - \frac{1}{2} \sin 2x \right]_0^{n\pi} = \frac{1}{2n\pi} \left(n\pi - \frac{1}{2}(0) \right) = \frac{1}{2}$$

101. $F(x) = \int_0^x t^2 \sqrt{1+t} \, dt$ for $x > -1$.

    a) $F(0) = \int_0^0 t^2 \sqrt{1+t} \, dt = 0$

    b) $F'(x) = x^2 \sqrt{1+x}$

    c) $F'(3) = 9\sqrt{4} = 18$

    d) $F(2x) = \int_0^{2x} t^2 \sqrt{1+t} \, dt$

        $F'(2x) = (2x)^2 \sqrt{1+2x} = 4x^2 \sqrt{1+2x}$

103. $\dfrac{d}{dx}\left[\int_a^x f(t) \, dt \right] = f(x)$.

$$\int_a^x \frac{d}{dt}[f(t)] \, dt = \int_a^x f'(t) \, dt$$

$$= f(x) - f(a).$$

So the equality does not hold if $f(a) \neq 0$.

105. If $a \neq b$, then we know that there exists a number c between a and b such that

$$f(c) = \frac{\int_a^b f(x) \, dx}{b - a} = 0$$

by the Mean Value Theorem for Integrals. So the statement is true if $a \neq b$. If $a = b$, then $\int_a^b f(x) \, dx = 0$ regardless of the value of $f(a)$.

107. $\dfrac{d}{dx}\left[\int_{2x}^0 \sec t \, dt \right] = -\dfrac{d}{dx}\left[\int_0^{2x} \sec t \, dt \right] = -\sec(2x)\dfrac{d}{dx}(2x) = -2\sec(2x)$

109. $\int_a^b [f(x)]^n f'(x) \, dx$; assume that $f'(x)$ is continuous on $[a, b]$. Let $f(x) = u$, $f'(x) \, dx = du$.

$$\int u^n \, du = \frac{1}{n+1} u^{n+1}; \quad \frac{1}{n+1}\left[f(x)^{n+1} \right]_a^b = \frac{1}{n+1}\left[f(b)^{n+1} - f(a)^{n+1} \right] \quad (n \neq -1)$$

260

111. $\displaystyle\int_{-1}^{1} \sqrt{1 - x^2}\, dx = \text{area of semicircle of radius } 1 = \frac{\pi}{2}.$

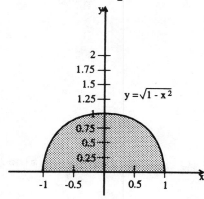

Exercise 111

113. $\displaystyle\int_{-3}^{3} |3 - x|\, dx$

  =   area of triangle of base 6

   and altitude 6.

 Area = $\frac{1}{2}(6 \cdot 6)$

    =   18.

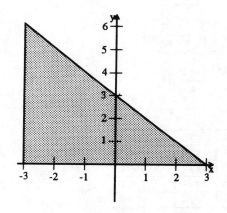

Exercise 113

115. $\displaystyle\int_{-1}^{1} (3 + \sqrt{1 - x^2})\, dx$

  =   area of semicircle
   of radius 1
   plus area of rectangle

   (2×3)

  = $\frac{1}{2}\pi + 6$

  $\approx$ 7.5708.

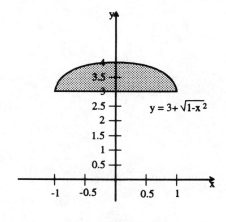

Exercise 115

117. $\displaystyle\int_0^4 \sqrt{1+x^2}\, dx$, $n = 4$

$\Delta x = \dfrac{4-0}{4} = 1$, $x_i = i$

i	x_i	$f(x_i)$
0	0	1.0000
1	1	1.4142
2	2	2.2361
3	3	3.1623
4	4	4.1231

$$\int_0^4 f(x)\,dx \;\approx\; \tfrac{1}{2}[1.0000 + 2(1.4142)$$
$$+\; 2(2.2361) + 2(3.1623)$$
$$+\;4.1231]$$
$$=\; 9.3742.$$

119. $\overline{f} = \dfrac{1}{10-0}\displaystyle\int_0^{10} f(x)\,dx$

i	x_i	$f(x_i)$
0	0	72
1	2	82
2	4	90
3	6	94
4	8	91
5	10	85

$$\int_0^{10} f(x)\,dx \;=\; \tfrac{\Delta x}{2}[72 + 2(82) + 2(90)$$
$$+\;2(94) + 2(91) + 85]$$
$$=\; \tfrac{2}{2}(871)$$
$$=\; 871$$

$$\overline{f} \;=\; \text{average temperature}$$
$$=\; \tfrac{1}{10}(871)$$
$$=\; 87.1.$$

121. $\displaystyle\int_0^{2\pi} \dfrac{\sin x}{x}\, dx$, $n = 4$

$\Delta x = \dfrac{2\pi - 0}{4} = \dfrac{\pi}{2}$, $x_i = \dfrac{\pi}{4} + \dfrac{\pi}{2}i - \Delta x$

i	x_i	$f(x_i)$
1	$\pi/4$	0.9003
2	$3\pi/4$	0.3001
3	$5\pi/4$	-0.1801
4	$7\pi/4$	-0.1286

$$\int_0^{2\pi} f(x)\,dx \;\approx\; \tfrac{\pi}{2}\,(0.9003 + 0.3001$$
$$-\;0.1801 - 0.1286)$$
$$=\; 0.4458\pi$$
$$\approx\; 1.4007$$

Chapter 7

Applications of the Definite Integral

7.1 Calculating Volumes by Slicing

1. $f(x) = 2x + 1,\ 1 \le x \le 4$

   a) $V = \displaystyle\int_1^4 \pi (2x+1)^2\, dx$

   b) $V = \dfrac{\pi}{6}(2x+1)^3\Big]_1^4 = \dfrac{\pi}{6}(729 - 27) = 117\pi$

3. $f(x) = \sin x,\ 0 \le x \le \pi$

   a) $V = \displaystyle\int_0^\pi \pi \sin^2 x\, dx$

   b) $V = \dfrac{\pi}{2}\displaystyle\int_0^\pi (1 - \cos 2x)\, dx = \dfrac{\pi}{2}\left(x - \dfrac{\sin 2x}{2}\right)\Big]_0^\pi = \dfrac{\pi}{2}(\pi) = \dfrac{\pi^2}{2}$

5. $f(x) = x(x^3 - 1)^2,\ 0 \le x \le 1$

   a) $V = \displaystyle\int_0^1 \pi x^2 (x^3 - 1)^4\, dx$

   b) $V = \dfrac{\pi}{15}(x^3 - 1)^5\Big]_0^1 = \dfrac{\pi}{15}(0 + 1) = \dfrac{\pi}{15}$

7. $f(x) = \sqrt{4 - x^2},\ 1 \le x \le 2$

   a) $V = \displaystyle\int_1^2 \pi (4 - x^2)\, dx$

   b) $V = \pi\left(4x - \dfrac{x^3}{3}\right)\Big]_1^2 = \pi\left[8 - \dfrac{8}{3} - \left(4 - \dfrac{1}{3}\right)\right] = \dfrac{5\pi}{3}$

9. $f(x) = \tan x,\ 0 \le x \le \dfrac{\pi}{4}$

   $$V = \int_0^{\pi/4} \pi \tan^2 x\, dx = \pi \int_0^{\pi/4} (\sec^2 x - 1)\, dx = \pi(\tan x - x)\Big]_0^{\pi/4} = \pi\left(1 - \dfrac{\pi}{4}\right) = \dfrac{\pi}{4}(4 - \pi)$$

263

11. $f(x) = \sec x$, $0 \le x \le \dfrac{\pi}{4}$

$$V = \int_0^{\pi/4} \pi \sec^2 x\, dx = \pi \tan x \bigg]_0^{\pi/4} = \pi$$

13. $f(x) = |3x - 6|$, $0 \le x \le 5$

$$V = \int_0^5 \pi(3x - 6)^2\, dx = \frac{\pi}{9}(3x - 6)^3 \bigg]_0^5 = \frac{\pi}{9}(729 + 216) = 105\pi$$

15. $f(x) = -x^2 + 6$, $g(x) = 2$

a) Find intersections: $-x^2 + 6 = 2 \Rightarrow x^2 = 4 \Rightarrow x = \pm 2$.

$$V = \int_{-2}^2 \pi[(-x^2 + 6)^2 - 4]\, dx = 2\pi \int_0^2 [(-x^2 + 6)^2 - 4]\, dx$$

b) $V = 2\pi \left[\dfrac{x^5}{5} - \dfrac{12}{3}x^3 + 36x - 4x \right]_0^2 = 2\pi \left(\dfrac{32}{5} - 32 + 72 - 8 \right) = \dfrac{384}{5}\pi \approx 241.274$

17. $f(x) = \dfrac{x^2}{4}$, $g(x) = x$

a) Find intersections: $\dfrac{x^2}{4} = x \Rightarrow x\left(\dfrac{x}{4} - 1\right) = 0 \Rightarrow x = 0, 4$

$$V = \int_0^4 \pi \left[x^2 - \dfrac{x^4}{16} \right] dx$$

b) $V = \pi \left[\dfrac{x^3}{3} - \dfrac{x^5}{80} \right]_0^4 = \pi \left(\dfrac{64}{3} - \dfrac{1024}{80} \right) = \pi \left(\dfrac{64}{3} - \dfrac{64}{5} \right) = \dfrac{128\pi}{15}$

19. $f(x) = \sin x$, $g(x) = \cos x$, $0 \le x \le \dfrac{\pi}{2}$

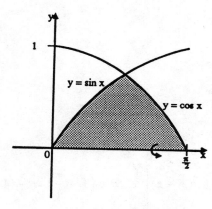

Exercise 19

$$
\begin{aligned}
V &= \int_0^{\pi/4} \pi(\cos^2 x - \sin^2 x)\, dx \\
&\quad + \int_{\pi/4}^{\pi/2} \pi(\sin^2 x - \cos^2 x)\, dx \\
&= 2\pi \int_0^{\pi/4} (\cos^2 x - \sin^2 x)\, dx \\
&= 2\pi \int_0^{\pi/4} \cos 2x\, dx \\
&= \pi(\sin 2x) \bigg]_0^{\pi/4} \\
&= \pi
\end{aligned}
$$

264

21. $y = x^2$, $y = 0$, $0 \leq x \leq 2$

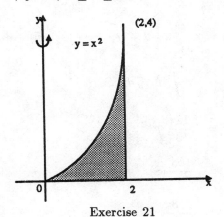

Exercise 21

a) $x = 2 \Rightarrow y = 2^2 = 4$

$V = \int_0^4 \pi[4 - y]\,dy$

b) $V = \pi\left[4y - \dfrac{y^2}{2}\right]_0^4$

$= (16 - 8)\pi$

$= 8\pi$

23. $y = x^3$, $x = 2$, $y = 0$

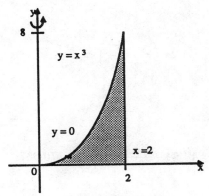

Exercise 23

a) $x = 2 \Rightarrow y = 2^3 = 8$

$V = \int_0^8 \pi[4 - y^{2/3}]\,dy$

b) $V = \pi\left[4y - \dfrac{3}{5}y^{5/3}\right]_0^8$

$= \pi\left(32 - \dfrac{96}{5}\right)$

$= \dfrac{64\pi}{5}$

25. $y = x^4$, $y = 0$, $x = 2$

$x = 2 \Rightarrow y = 2^4 = 16$

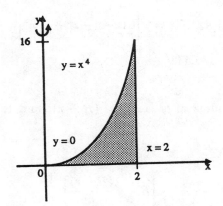

Exercise 25

$V = \int_0^{16} \pi(4 - \sqrt{y})\,dy$

$= \pi\left[4y - \dfrac{2}{3}y^{3/2}\right]_0^{16}$

$= \pi\left(64 - \dfrac{128}{3}\right)$

$= \dfrac{64}{3}\pi$

27. $y = (x - 2)^2$, $y = 4$

265

$$x - 2 = \pm\sqrt{y} \;\Rightarrow\; \begin{aligned}&x = 2 + \sqrt{y},\\ &x = 2 - \sqrt{y}\end{aligned}$$

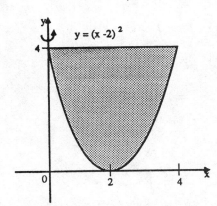

$y = (x-2)^2$

Exercise 27

$$V = \int_0^4 \pi \left[(2 + \sqrt{y})^2 - (2 - \sqrt{y})^2 \right] dy$$

$$= \pi \int_0^4 8\sqrt{y}\, dy$$

$$= \frac{16\pi}{3} y^{3/2} \Big]_0^4$$

$$= \frac{16\pi}{3}(8)$$

$$= \frac{128\pi}{3}$$

29. Consider rotating a semicircle of radius r about the y-axis.

$$V = 2\int_0^r \pi(r^2 - y^2)\, dy = 2\pi \left[r^2 y - \frac{y^3}{3} \right]_0^r = 2\pi \left(r^3 - \frac{r^3}{3} \right) = \frac{4\pi r^3}{3}$$

31. $A(x) = \frac{1}{2}(2y)^2 = 2y^2 = 2(16 - x^2)$

$$V = 2\int_0^4 2(16 - x^2)\, dx = 4\left(16x - \frac{x^3}{3} \right)\Big]_0^4 = 4\left(64 - \frac{64}{3} \right) = \frac{512}{3} \approx 170.667$$

33. $f(x) = x^2$, $g(x) = 8 - x^2$.
   Intersections: $x^2 = 8 - x^2 \Rightarrow 2x^2 = 8 \Rightarrow x = \pm 2$.
   At any value of $x \in [-2, 2]$, the base of the square has length $8 - x^2 - x^2 = 8 - 2x^2$. $A(x) = (8 - 2x^2)^2$.

$$V = \int_{-2}^2 (8 - 2x^2)^2\, dx = 2\int_0^2 (64 - 32x^2 + 4x^4)\, dx = 2\left[64x - \frac{32}{3}x^3 + \frac{4}{5}x^5 \right]_0^2$$

$$= 2\left(128 - \frac{32}{3}(8) + \frac{4}{5}(32) \right) = \frac{2048}{15} \approx 136.533$$

35. Consider circles perpendicular to the y-axis. The radius at y is $R - \frac{y}{h}(R - r)$, and the area is $\pi \left[R - y\frac{(R - r)}{h} \right]^2$.

$$V = \int_0^h \pi \left[R - y\frac{(R - r)}{h} \right]^2 dy = -\frac{\pi}{3}\frac{h}{(R - r)} \left[R - y\frac{(R - r)}{h} \right]^3 \Big]_0^h$$

$$= -\frac{\pi}{3}\frac{h}{(R - r)} [r^3 - R^3] = \frac{\pi h}{3}(R^2 + rR + r^2)$$

266

37. From Exercise 32, $\quad V = \pi \int_0^h (20y - y^2)\, dy$

$$\frac{dV}{dt} = \pi(20h - h^2)\frac{dh}{dt};\ 10 = \pi(20 \cdot 4 - 16)\frac{dh}{dt}$$

$$\frac{dh}{dt} = \frac{10}{\pi(64)} \approx 0.04974\ \text{m/min}$$

39. $y = 1 - x^2$ and x-axis about the line $y = -3$.

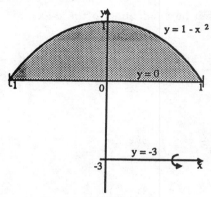

Exercise 39

$$V = \int_{-1}^1 \pi[(1 - x^2 - (-3))^2 - 9]\, dx = \pi \int_{-1}^1 (16 - 8x^2 + x^4 - 9)\, dx$$

$$= \pi \left[7x - \frac{8}{3}x^3 + \frac{x^5}{5}\right]_{-1}^1 = 2\pi\left(7 - \frac{8}{3} + \frac{1}{5}\right) = \frac{136}{15}\pi \approx 28.4838$$

41. $y = \sqrt{x}$, $y = 0$, and $x = 9$ about the line $y = -2$.

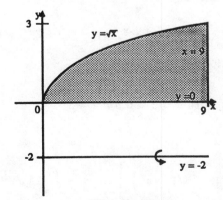

Exercise 41

$$V = \int_0^9 \pi[(\sqrt{x} + 2)^2 - 4]\, dx = \pi \int_0^9 (x + 4\sqrt{x})\, dx$$

$$= \pi \left[\frac{x^2}{2} + \frac{8}{3}x^{3/2}\right]_0^9 = \pi\left(\frac{81}{2} + \frac{8}{3} \cdot 27\right) = \frac{225}{2}\pi \approx 353.429$$

267

43. $\dfrac{x^2}{a^2} + \dfrac{y^2}{b^2} = 1$ about the y-axis.

$$V = \int_{-b}^{b} \pi x^2 \, dy = 2\pi \int_{0}^{b} a^2 \left(1 - \frac{y^2}{b^2}\right) dy = 2\pi a^2 \left[y - \frac{y^3}{b^2}\right]_{0}^{b} = \frac{4\pi a^2 b}{3}$$

45. Cross-sections taken perpendicular to the x-axis are rectangles of height $x \tan \theta$ and width $2y$, where y is given by $x^2 + y^2 = r^2$.

$$V = 2 \int_{0}^{r} (x \tan \theta)\sqrt{r^2 - x^2} \, dx = -\frac{2}{3} \tan \theta (r^2 - x^2)^{3/2} \Big]_{0}^{r} = \frac{2}{3} r^3 \tan \theta$$

Alternatively, cross-sections taken perpendicular to the y-axis are right triangles of base x and altitude $x \tan \theta$.

$$V = 2 \int_{0}^{r} \frac{1}{2} x \, (x \tan \theta) \, dy = \tan \theta \int_{0}^{r} (r^2 - y^2) \, dy = \tan \theta \left[r^2 y - \frac{y^3}{3}\right]_{0}^{r} = \frac{2}{3} r^3 \tan \theta$$

47. $f(x) = \dfrac{1}{\sqrt{1 + x^2}}$, $y = 0$, $0 \leq x \leq 4$

$$V = \int_{0}^{4} \pi \left(\frac{1}{1 + x^2}\right) dx = \pi \int_{0}^{4} \frac{dx}{1 + x^2} = \pi \int_{0}^{4} g(x) \, dx$$

$\Delta x = \frac{4}{12} = \frac{1}{3}$; $x_i = 0 + \frac{1}{3} i$

i	x_i	$g(x_i)$
0	0.0000	1.0000
1	0.3333	0.9000
2	0.6667	0.6923
3	1.0000	0.5000
4	1.3333	0.3600
5	1.6667	0.2647
6	2.0000	0.2000

i	x_i	$g(x_i)$
7	2.3333	0.1552
8	2.6667	0.1233
9	3.0000	0.1000
10	3.3333	0.0826
11	3.6667	0.0692
12	4.0000	0.0588

$$V = \pi \left(\frac{1}{9}\right) [f(x_0) + 4f(x_1) + 2f(x_2) + 4f(x_3) + \cdots + 2f(x_{10}) + 4f(x_{11}) + f(x_{12})]$$

$$= \frac{\pi}{9} (11.9316) \approx 4.1649$$

49. $\dfrac{x^2}{a^2} + \dfrac{y^2}{b^2} = 1$ about the line $y = -2$, with $a = 3$ and $b = 1$.

$$y^2 = \frac{9 - x^2}{9} \Rightarrow y = \pm \frac{1}{3} \sqrt{9 - x^2}$$

$$V = \int_{-3}^{3} \pi \left[\left(2 + \frac{1}{3}\sqrt{9 - x^2}\right)^2 - \left(2 - \frac{1}{3}\sqrt{9 - x^2}\right)^2\right] dx = \frac{\pi}{9} \int_{-3}^{3} 24\sqrt{9 - x^2} \, dx = \frac{8\pi}{3} \int_{-3}^{3} \sqrt{9 - x^2} \, dx$$

The integral $\int_{-3}^{3} \sqrt{9 - x^2} \, dx$ equals the area of a semicircle of radius 3, or $\frac{\pi}{2} (9)$.

$$\text{Volume} = \frac{8\pi}{3} \cdot \frac{9\pi}{2} = 12\pi^2.$$

268

7.2 Calculating Volumes by the Method of Cylindrical Shells

1. $x + y = 1$, $x = 0$, $y = 0$

a)

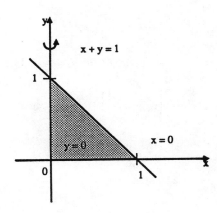

Exercise 1

b) $V = \displaystyle\int_0^1 2\pi x(1-x)\,dx$

$= 2\pi \displaystyle\int_0^1 (x - x^2)\,dx$

c) $V = 2\pi \left[\dfrac{x^2}{2} - \dfrac{x^3}{3} \right]_0^1$

$= 2\pi \left(\dfrac{1}{2} - \dfrac{1}{3} \right)$

$= \dfrac{\pi}{3}$

3. $y = x^3$, $y = 0$, $1 \le x \le 3$

a)

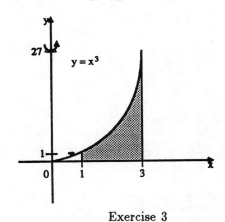

Exercise 3

b) $V = \displaystyle\int_1^3 2\pi x(x^3)\,dx$

$= 2\pi \displaystyle\int_1^3 x^4\,dx$

c) $V = \dfrac{2\pi}{5} x^5 \Big]_1^3$

$= \dfrac{2\pi}{5}(243 - 1)$

$= \dfrac{484\pi}{5}$

≈ 304.106

5. $y = \sqrt{x}$, $y = -x$, $x = 4$

269

a)

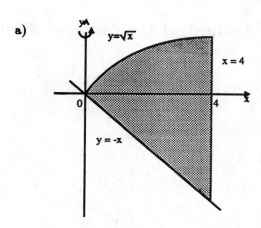

Exercise 5

b) $V = \int_0^4 2\pi x(\sqrt{x} + x)\, dx = 2\pi \int_0^4 (x^{3/2} + x^2)\, dx$

c) $V = 2\pi \left[\dfrac{2}{5} x^{5/2} + \dfrac{x^3}{3} \right]_0^4 = \dfrac{4\pi}{5}(32) + \dfrac{2\pi}{3}(64) = \dfrac{1024\pi}{15} \approx 214.467$

7. $y = -x^2 - 4x - 3, \ y = 0$

   a) $y = -(x^2 + 4x + 3) = -(x+2)^2 + 1$

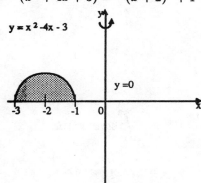

$y = 0 \Rightarrow (x+2)^2 = 1 \quad \Rightarrow \quad x + 2 = \pm 1$

$\Rightarrow \quad x = -3, -1$

Exercise 7

b) Since $x < 0$ on $[-3, -1]$, use $|x| = -x$ as the radius of the shell.

$$V = \int_{-3}^{-1} 2\pi(-x)(-x^2 - 4x - 3)\, dx = 2\pi \int_{-3}^{-1} (x^3 + 4x^2 + 3x)\, dx$$

c) $V = 2\pi \left[\dfrac{x^4}{4} + \dfrac{4}{3} x^3 + \dfrac{3}{2} x^2 \right]_{-3}^{-1} = 2\pi \left[\dfrac{1}{4}(1 - 81) + \dfrac{4}{3}(-1 + 27) + \dfrac{3}{2}(1 - 9) \right]$

$= 2\pi \left(\dfrac{8}{3} \right) = \dfrac{16\pi}{3} \approx 16.7552$

9. $y = \sin x^2, \ y = 1, \ 0 \le x \le \sqrt{\dfrac{\pi}{2}}$

270

a)

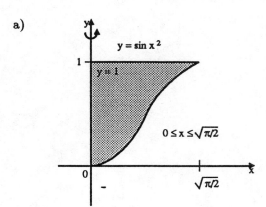

Exercise 9

b) $V = \int_0^{\sqrt{\pi/2}} 2\pi x(1 - \sin x^2)\, dx = \pi \int_0^{\sqrt{\pi/2}} 2x(1 - \sin x^2)\, dx$

c) $V = \pi(x^2 + \cos x^2)\Big]_0^{\sqrt{\pi/2}} = \pi\left(\dfrac{\pi}{2} - 1\right)$

11. $y = -x \cos x^3, \quad 0 \le x \le \sqrt[3]{\dfrac{\pi}{2}}$

a)

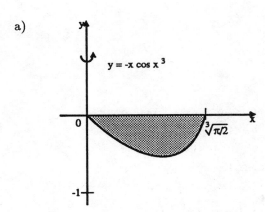

Exercise 11

b) $V = \int_0^{\sqrt[3]{\pi/2}} 2\pi x(0 + x \cos x^3)\, dx = 2\pi \int_0^{\sqrt[3]{\pi/2}} x^2 \cos x^3\, dx = \dfrac{2\pi}{3} \sin x^3 \Big]_0^{\sqrt[3]{\pi/2}} = \dfrac{2\pi}{3}$

13. $y = \sqrt{9 - x^2}, \; y = 0, \; 0 \le x \le 3$

a)

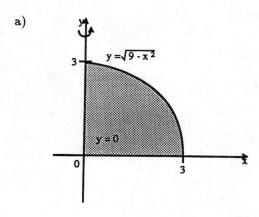

Exercise 13

b) $V = \int_0^3 2\pi x \sqrt{9 - x^2}\, dx = -\frac{2\pi}{3}(9 - x^2)^{3/2}\Big]_0^3 = -\frac{2\pi}{3}(0 - 27) = 18\pi$

15. $y = \dfrac{\sqrt{1 + x^{3/2}}}{\sqrt{x}},\ y = 0,\ 1 \le x \le 4$

a)

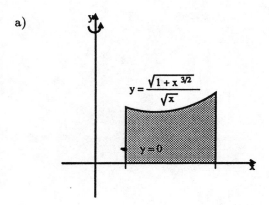

Exercise 15

b) $V = \int_1^4 2\pi x \left(\dfrac{\sqrt{1 + x^{3/2}}}{\sqrt{x}}\right) dx = 2\pi \int_1^4 \sqrt{1 + x^{3/2}}\, x^{1/2}\, dx$

$= \dfrac{8\pi}{9}(1 + x^{3/2})^{3/2}\Big]_1^4 = \dfrac{8\pi}{9}\left(27 - 2\sqrt{2}\right) \approx 67.500$

17. $x = y^2,\ x = 4$, about the line $x = -1$.

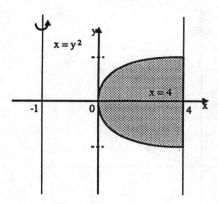

Exercise 17

$$V = \int_0^4 2\pi(x+1)\left(\sqrt{x}-(-\sqrt{x})\right)\,dx = 4\pi \int_0^4 \left(x^{3/2}+\sqrt{x}\right)\,dx = 4\pi \left[\frac{2}{5}\,x^{5/2}+\frac{2}{3}\,x^{3/2}\right]_0^4$$

$$= 4\pi\left(\frac{64}{5}+\frac{16}{3}\right) = \frac{1088\pi}{15} \approx 227.870$$

19. $y = x^2$, $y = \sqrt{x}$, about the line $y = -2$.

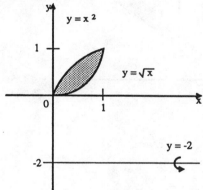

Exercise 19

$$x^2 = \sqrt{x} \Rightarrow x = 0,1; \; y = x^2 \Rightarrow x = \sqrt{y}; \; y = \sqrt{x} \Rightarrow x = y^2$$

$$V = \int_0^1 2\pi(y+2)(\sqrt{y}-y^2)\,dy = 2\pi \int_0^1 (-y^3 - 2y^2 + y^{3/2} + 2y^{1/2})\,dy$$

$$= 2\pi\left[-\frac{y^4}{4} - \frac{2}{3}\,y^3 + \frac{2}{5}\,y^{5/2} + \frac{4}{3}\,y^{3/2}\right]_0^1 = 2\pi\left(-\frac{1}{4} - \frac{2}{3} + \frac{2}{5} + \frac{4}{3}\right)$$

$$= \frac{49\pi}{30} \approx 5.1313$$

21. $y = x$, $y = x^3$, about the y-axis.

273

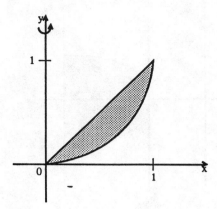

Exercise 21

$$V = 2\int_0^1 2\pi x(x - x^3)\,dx = 4\pi\int_0^1 (x^2 - x^4)\,dx = 4\pi\left[\frac{x^3}{3} - \frac{x^5}{5}\right]_0^1 = 4\pi\left(\frac{1}{3} - \frac{1}{5}\right) = \frac{8\pi}{15}$$

23. Triangle with vertices $(1, 3)$, $(1, 7)$, and $(4, 7)$ about the line $y = 1$.

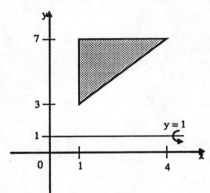

Exercise 23

Use cylinders parallel to the x-axis.

Equation of line: $\quad m = \dfrac{7 - 3}{4 - 1} = \dfrac{4}{3}$

$$\frac{y - 3}{x - 1} = \frac{4}{3}; \quad y = \frac{4}{3}x + \frac{5}{3} \Rightarrow x = \frac{3y - 5}{4}$$

$$V = \int_3^7 2\pi(y - 1)\left(\frac{3y - 5}{4} - 1\right)dy = \frac{3\pi}{2}\int_3^7 (y - 1)(y - 3)\,dy = \frac{3\pi}{2}\int_3^7 (y^2 - 4y + 3)\,dy$$

$$= \frac{3\pi}{2}\left[\frac{y^3}{3} - 2y^2 + 3y\right]_3^7 = \frac{3\pi}{2}\left[\frac{1}{3}(343 - 27) - 2(49 - 9) + 3(7 - 3)\right] = \frac{3\pi}{2}\left(\frac{112}{3}\right)$$

$$= 56\pi \approx 175.929$$

Alternatively, disks parallel to the y-axis may be used.

$$V = \int_1^4 \pi\left[6^2 - \left(\frac{4}{3}x + \frac{5}{3} - 1\right)^2\right]dx = \frac{16\pi}{9}\int_1^4 (20 - x - x^2)\,dx = \frac{16\pi}{9}\left[20x - \frac{x^2}{2} - \frac{x^3}{3}\right]_1^4$$

$$= \frac{16\pi}{9} \left[20(4-1) - \frac{1}{2}(16-1) - \frac{1}{3}(64-1) \right] = \frac{504\pi}{9} = 56\pi$$

25. $\frac{x^2}{a^2} + \frac{y^2}{b^2} = 1$ about the y-axis.

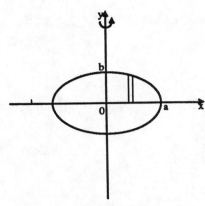

$$y = \frac{b}{a}\sqrt{a^2 - x^2}$$

Exercise 25

$$V = 2\int_0^a 2\pi x \left(\frac{b}{a}\sqrt{a^2 - x^2} \right) dx = \frac{4\pi b}{a} \int_0^a \left(\sqrt{a^2 - x^2} \right) x\, dx$$

$$= -\frac{4\pi b}{3a} (a^2 - x^2)^{3/2} \Big]_0^a = -\frac{4\pi b}{3a} (0 - a^3) = \frac{4\pi a^2 b}{3}$$

27. $y = \sin x$, x-axis, about the y-axis, for $0 \le x \le \pi$.

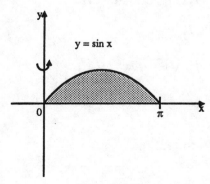

$y = \sin x$

Exercise 27

$$V = \int_0^\pi 2\pi x (\sin x)\, dx = 2\pi \int_0^\pi x \sin x\, dx$$

With $n = 8$, $\Delta x = \frac{\pi}{8}$ and $x_i = \frac{\pi}{8} i$. Let $f(x_i) = x_i \sin x_i$.

i	x_i	$f(x_i)$
0	0	0.0000
1	$\pi/8$	0.1503
2	$\pi/4$	0.5554
3	$3\pi/8$	1.0884
4	$\pi/2$	1.5708
5	$5\pi/8$	1.8140
6	$3\pi/4$	1.6661
7	$7\pi/8$	1.0520
8	π	0.0000

$$V = 2\pi\left(\frac{\pi}{16}\right)[f(x_0) + f(x_8)$$

$$+ 2\sum_{i=1}^{7} f(x_i)\Big]$$

$$= \frac{\pi^2}{8}(15.794)$$

$$\approx 19.485$$

29. $y = \sqrt{1 + x^4}$, $y = 0$, $x = 0$, $x = 2$, about the line $y = -1$.

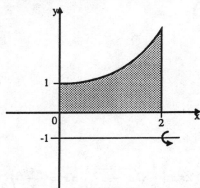

Exercise 29

$$V = \pi\int_0^2\left[\left(1 + \sqrt{1 + x^4}\right)^2 - 1^2\right]dx = \pi\int_0^2\left(2\sqrt{1 + x^4} + 1 + x^4\right)dx \approx 15.707\pi \approx 49.345$$

Simpson's Rule with $n = 16$ was used.

7.3   Arc Length and Surface Area

1. $y = 2x + 3$, from $(-3, -3)$ to $(2, 7)$.

    a) $L = \int_{-3}^{2}\sqrt{1 + (y')^2}\,dx = \int_{-3}^{2}\sqrt{1 + 4}\,dx = \sqrt{5}\int_{-3}^{2}dx$

    b) $L = \sqrt{5}\,x\Big]_{-3}^{2} = 5\sqrt{5}$

3. $y - 2 = (x + 2)^{3/2}$, from $(-2, 2)$ to $(2, 10)$.

    a) $L = \int_{-2}^{2}\sqrt{1 + (y')^2}\,dx = \int_{-2}^{2}\sqrt{1 + \left(\frac{3}{2}(x + 2)^{1/2}\right)^2}\,dx = \frac{1}{2}\int_{-2}^{2}\sqrt{4 + 9(x + 2)}\,dx$

       $= \frac{1}{2}\int_{-2}^{2}\sqrt{22 + 9x}\,dx$

276

b) $L = \frac{1}{2} \cdot \frac{2}{27} (22 + 9x)^{3/2} \Big]_{-2}^{2} = \frac{1}{27} (40^{3/2} - 4^{3/2}) = \frac{8}{27}(10^{3/2} - 1) \approx 9.0734$

Note that the graph in Exercise 3 is the same as that in Exercise 2, translated from $(0,0)$ to $(-2,2)$ at its origin.

5. $3y = 2(x+1)^{3/2}$, from $\left(0, \frac{2}{3}\right)$ to $\left(3, \frac{16}{3}\right)$

a) $L = \int_0^3 \sqrt{1 + (y')^2}\, dx = \int_0^3 \sqrt{1 + x + 1}\, dx = \int_0^3 \sqrt{2 + x}\, dx$

b) $L = \frac{2}{3}(2+x)^3 \Big]_0^3 = \frac{2}{3}\left[5^{3/2} - 2^{3/2}\right] \approx 5.568$

7. $(3y - 6)^2 = (x^2 + 2)^3$, for $0 \le x \le 3$

$(y - 2)^2 = \frac{1}{9}(x^2 + 2)^3$; $y - 2 = \pm\frac{1}{3}(x^2 + 2)^{3/2}$

$y' = \pm x(x^2 + 2)^{1/2}$; $(y')^2 = x^2(x^2 + 2)$

$L = \int_0^3 \sqrt{1 + (y')^2}\, dx = \int_0^3 \sqrt{1 + 2x^2 + x^4}\, dx = \int_0^3 (1 + x^2)\, dx = x + \frac{x^3}{3} \Big]_0^3 = 3 + 9 = 12$

Note that the graph has two branches, symmetrical about the line $y = 2$. Each arc has length 12.

9. $y = \frac{x^3}{6} + \frac{1}{2x}$, from $\left(1, \frac{2}{3}\right)$ to $\left(4, \frac{259}{24}\right)$.

$y' = \frac{x^2}{2} - \frac{1}{2x^2}$; $(y')^2 = \frac{1}{4}\left(x^4 - 2 + \frac{1}{x^4}\right)$

$L = \int_1^4 \sqrt{1 + (y')^2}\, dx = \frac{1}{2} \int_1^4 \sqrt{x^4 + 2 + \frac{1}{x^4}}\, dx = \frac{1}{2} \int_1^4 \left(x^2 + \frac{1}{x^2}\right) dx$

$= \frac{1}{2} \left[\frac{x^3}{3} - \frac{1}{x}\right]_1^4 = \frac{1}{2}\left[\frac{1}{3}(64 - 1) - \left(\frac{1}{4} - 1\right)\right] = \frac{87}{8} = 10.875$

11. $y = \frac{x^4 + 6x + 3}{6x}$, for $2 \le x \le 4$

$y' = \frac{x^2}{2} + 0 - \frac{1}{2x^2}$; $(y')^2 = \frac{1}{4}\left(x^4 - 2 + \frac{1}{x^4}\right)$

$L = \int_2^4 \sqrt{1 + (y')^2}\, dx = \frac{1}{2} \int_2^4 \sqrt{x^4 + 2 + \frac{1}{x^4}}\, dx = \frac{1}{2} \int_2^4 \left(x^2 + \frac{1}{x^2}\right) dx = \frac{1}{2}\left[\frac{x^3}{3} - \frac{1}{x}\right]_2^4$

$= \frac{1}{2}\left[\frac{1}{3}(64 - 8) - \left(\frac{1}{4} - \frac{1}{2}\right)\right] = \frac{227}{24} \approx 9.4583$

13. $y = 2x + 3$, x-axis, for $1 \le x \le 3$ about the x-axis.

$y' = 2$; $(y')^2 = 4$

$S = \int_1^3 2\pi y \sqrt{1 + (y')^2}\, dx = 2\pi \int_1^3 (2x + 3)\sqrt{5}\, dx = 2\pi\sqrt{5} \left[x^2 + 3x\right]_1^3$

$= 2\pi\sqrt{5}\left[(9 - 1) + 3(3 - 1)\right] = 2\pi\sqrt{5}\,(14) = 28\pi\sqrt{5} \approx 196.695$

15. $y = x^{1/3}$, y-axis, for $0 \le y \le 2$ about the y-axis.

$x = y^3$; $\dfrac{dx}{dy} = 3y^2$; $\left(\dfrac{dx}{dy}\right)^2 = 9y^4$

$$S = \int_0^2 2\pi x \sqrt{1 + (x')^2}\, dy = 2\pi \int_0^2 y^3 \sqrt{1 + 9y^4}\, dy = \frac{2\pi}{36} \cdot \frac{2}{3}(1 + 9y^4)^{3/2}\Big]_0^2$$

$$= \frac{\pi}{27}(145^{3/2} - 1) \approx 203.044$$

17. $y = \sqrt{x}$, x-axis, for $1 \le x \le 2$ about the x-axis.

$y' = \dfrac{1}{2\sqrt{x}}$; $(y')^2 = \dfrac{1}{4x}$

$$S = \int_1^2 2\pi y \sqrt{1 + (y')^2}\, dx = 2\pi \int_1^2 \sqrt{x}\sqrt{1 + \frac{1}{4x}}\, dx = \pi \int_1^2 \sqrt{4x + 1}\, dx = \frac{\pi}{6}(4x+1)^{3/2}\Big]_1^2$$

$$= \frac{\pi}{6}(9^{3/2} - 5^{3/2}) = \frac{\pi}{6}(27 - 5^{3/2}) \approx 8.2832$$

19. $12xy - 3y^4 = 4$, y-axis, for $1 \le y \le 3$ about the y-axis.

$x = \dfrac{1}{3y} + \dfrac{y^3}{4}$; $\dfrac{dx}{dy} = -\dfrac{1}{3y^2} + \dfrac{3y^2}{4}$

$$\left(\frac{dx}{dy}\right)^2 = \frac{1}{9y^4} - \frac{1}{2} + \frac{9y^4}{16}$$

$$S = \int_1^3 2\pi x \sqrt{1 + (x')^2}\, dy = 2\pi \int_1^3 \left(\frac{1}{3y} + \frac{y^3}{4}\right)\sqrt{\frac{1}{9y^4} + \frac{1}{2} + \frac{9y^4}{16}}\, dy$$

$$= 2\pi \int_1^3 \left(\frac{1}{3y} + \frac{y^3}{4}\right)\left(\frac{1}{3y^2} + \frac{3y^2}{4}\right) dy = 2\pi \int_1^3 \left(\frac{1}{9y^3} + \frac{y}{3} + \frac{3y^5}{16}\right) dy$$

$$= 2\pi\left[-\frac{1}{18y^2} + \frac{y^2}{6} + \frac{3y^6}{96}\right]_1^3 = 2\pi\left[-\frac{1}{18}\left(\frac{1}{9} - 1\right) + \frac{1}{6}(9 - 1) + \frac{1}{32}(729 - 1)\right]$$

$$= \frac{7819\pi}{162} \approx 151.630$$

21. $f(x) = \displaystyle\int_0^x \sqrt{2t^2 + t^4}\, dt$, $1 \le x \le 2$

$L = \displaystyle\int_1^2 \sqrt{1 + (y')^2}\, dx$; $y' = \sqrt{2x^2 + x^4}$

$$L = \int_1^2 \sqrt{1 + 2x^2 + x^4}\, dx = \int_1^2 (1 + x^2)\, dx = x + \frac{x^3}{3}\Big]_1^2 = (2 - 1) + \frac{1}{3}(8 - 1) = \frac{10}{3}$$

23. $y = \dfrac{rx}{\sqrt{l^2 - r^2}}$, $0 \le x \le \sqrt{l^2 - r^2}$, about the x-axis.

$y' = \dfrac{r}{\sqrt{l^2 - r^2}}$, $1 + (y')^2 = 1 + \dfrac{r^2}{l^2 - r^2} = \dfrac{l^2}{l^2 - r^2}$

$$S = \int_0^{\sqrt{l^2-r^2}} 2\pi y \sqrt{1 + (y')^2}\, dx = 2\pi \int_0^{\sqrt{l^2-r^2}} \frac{rx}{\sqrt{l^2 - r^2}} \cdot \frac{l}{\sqrt{l^2 - r^2}}\, dx$$

$$= \left. \frac{2\pi r l}{l^2 - r^2} \left(\frac{x^2}{2} \right) \right]_0^{\sqrt{l^2 - r^2}} = \frac{\pi r l}{l^2 - r^2} (l^2 - r^2) = \pi r l$$

25. Circle centered at $(1, 1)$ with radius 5. Arc joining $(4, 5)$ and $(5, 4)$ revolved about x-axis.

$(x - 1) + (y - 1)^2 = 25$; $y = 1 + \sqrt{25 - (x - 1)^2}$

$y' = \dfrac{-(x - 1)}{\sqrt{25 - (x - 1)^2}}$; $1 + (y')^2 = 1 + \dfrac{(x - 1)^2}{25 - (x - 1)^2} = \dfrac{25}{25 - (x - 1)^2}$

$$\begin{aligned} S &= \int_4^5 2\pi y \sqrt{1 + (y')^2} \, dx = 2\pi \int_4^5 \left[1 + \sqrt{25 - (x - 1)^2} \right] \left(\frac{5}{\sqrt{25 - (x - 1)^2}} \right) dx \\ &= 10\pi \int_4^5 \left(\frac{1}{\sqrt{25 - (x - 1)^2}} + 1 \right) dx \end{aligned}$$

Let $u = x - 1$, $du = dx$; $x = 4 \Rightarrow u = 3$, $x = 5 \Rightarrow u = 4$

$S = 10\pi \int_3^4 \left(\dfrac{1}{\sqrt{25 - u^2}} + 1 \right) du \approx 40.3316$ as in Exercise 24.

27. $f(x) = \sin x$ between $(0, 0)$ and $(\pi, 0)$ about the x-axis.

$f'(x) = \cos x$

$S = \int_0^\pi 2\pi y \sqrt{1 + (y')^2} \, dx = 2\pi \int_0^\pi \sqrt{1 + \cos^2 x} \sin x \, dx$

Let $n = 10$, $\Delta x = \dfrac{\pi}{10}$, $x_i = \dfrac{\pi}{20} + \left(\dfrac{\pi}{10} \right) (i - 1)$, $f(x_i) = \sqrt{1 + \cos^2 x_i} \sin x_i$

i	x_i	$f(x_i)$
1	$\pi/20$	0.2199
2	$3\pi/20$	0.6081
3	$5\pi/20$	0.8660
4	$7\pi/20$	0.9785
5	$9\pi/20$	0.9997

i	x_i	$f(x_i)$
6	$11\pi/20$	0.9997
7	$13\pi/20$	0.9785
8	$15\pi/20$	0.8660
9	$17\pi/20$	0.6081
10	$19\pi/20$	0.2199

$S \approx 2\pi h \sum_{i=1}^{10} f(x_i) \approx 14.497$

29. $\dfrac{x^2}{a^2} + \dfrac{y^2}{b^2} = 1$, for $-\dfrac{a}{2} \le x \le \dfrac{a}{2}$ about the x-axis.

$y = \dfrac{b}{a} \sqrt{a^2 - x^2}$; $y' = \dfrac{-bx}{a\sqrt{a^2 - x^2}}$

$1 + (y')^2 = 1 + \dfrac{b^2 x^2}{a^2(a^2 - x^2)}$

Let $a = 4$, $b = 3$.

$$S = \int_{-2}^2 2\pi \left(\frac{3}{4} \sqrt{16 - x^2} \right) \sqrt{1 + \frac{9x^2}{16(16 - x^2)}} \, dx = \frac{3\pi}{8} \int_{-2}^2 \sqrt{256 - 7x^2} \, dx$$

With $n = 10$, $\Delta x = 0.4$, $x_i = -2 + 0.4i$, $f(x_i) = \sqrt{256 - 7x_i^2}$

i	x_i	$f(x_i)$
0	-2.0	15.0997
1	-1.6	15.4298
2	-1.2	15.6818
3	-0.8	15.8594
4	-0.4	15.9650
5	0.0	16.0000

i	x_i	$f(x_i)$
6	0.4	15.9650
7	0.8	15.8594
8	1.2	15.6818
9	1.6	15.4298
10	2.0	15.0997

$$S \approx \frac{3\pi}{8}\left(\frac{h}{3}\right)[f(x_0) + 4f(x_1) + 2f(x_2) + \cdots + 4f(x_9) + f(x_{10})] \approx 74.000$$

7.4 Distance and Velocity

1. $v(t) = 9 - t^2$, $a = 0$, $b = 3$
   $v(t) \geq 0$ on $[0, 3]$.

$$D = \int_0^3 v(t)\,dt = \int_0^3 (9 - t^2)\,dt = 9t - \frac{t^3}{3}\Big]_0^3 = 27 - 9 = 18$$

3. $v(t) = \sin t$, $a = 0$, $b = \pi$

$$D = \int_0^\pi \sin t\,dt = -\cos t\Big]_0^\pi = -(-1 - 1) = 2$$

5. $v(t) = 2 + t^2$, $a = 0$, $b = 5$

$$D = \int_0^5 (2 + t^2)\,dt = 2t + \frac{t^3}{3}\Big]_0^5 = 10 + \frac{125}{3} = \frac{155}{3} \approx 51.667$$

7. $v(t) = \cos \pi t$, $a = \dfrac{1}{2}$, $b = \dfrac{3}{2}$

$\cos \pi t \;=\; 0$ at $t = \dfrac{1}{2}$, $t = \dfrac{3}{2}$

$\cos \pi t \;\geq\; 0$ on $\left[0, \dfrac{1}{2}\right]$

$\cos \pi t \;\leq\; 0$ on $\left[\dfrac{1}{2}, \dfrac{3}{2}\right]$

$$D = -\int_{1/2}^{3/2} \cos \pi t\,dt = -\frac{1}{\pi}\sin \pi t\Big]_{1/2}^{3/2} = -\frac{1}{\pi}(-1 - 1) = \frac{2}{\pi}$$

9. $v(t) = \sin^2 t \cos t$, $a = 0$, $b = \dfrac{\pi}{2}$

$$D = \int_0^{\pi/2} \sin^2 t \cos t\,dt = \frac{1}{3}\sin^3 t\Big]_0^{\pi/2} = \frac{1}{3}$$

280

11. $v(t) = (2t - 5)(t^2 - 5t + 6)^{1/3}$, $a = 3$, $b = 4$
$t^2 - 5t + 6 = (t - 2)(t - 3) \Rightarrow$ zeros at $t = 2, 3$.

$$D = \int_3^4 (t^2 - 5t + 6)^{1/3}(2t - 5)\, dt = \frac{3}{4}(t^2 - 5t + 6)^{4/3}\Big]_3^4 = \frac{3}{4}[2^{4/3} - 0] = \frac{3}{2^{2/3}} \approx 1.8899$$

13. $v(t) = 2 - t^2$, $t = 0$ to $t = 4$.
$2 - t^2 = 0 \Rightarrow t = \sqrt{2}$

$$D = \int_0^{\sqrt{2}} (2 - t^2)\, dt + \int_{\sqrt{2}}^4 (t^2 - 2)\, dt = 2t - \frac{t^3}{3}\Big]_0^{\sqrt{2}} + \frac{t^3}{3} - 2t\Big]_{\sqrt{2}}^4$$

$$= 2\sqrt{2} - \frac{2\sqrt{2}}{3} + \frac{64}{3} - 8 - \frac{2\sqrt{2}}{3} + 2\sqrt{2} = \frac{8\sqrt{2}}{3} + \frac{40}{3} \approx 17.105$$

15. $v(t) = \cos t \sin t$, $t = 0$ to $t = \frac{\pi}{2}$.

$\cos t \sin t \geq 0$ on $\left[0, \frac{\pi}{2}\right]$

$$D = \int_0^{\pi/2} \cos t \sin t\, dt = \frac{1}{2}\sin^2 t\Big]_0^{\pi/2} = \frac{1}{2}$$

17. $v(t) = (t - 1)^5$, $t = 0$ to $t = 2$.
$(t - 1)^5 = 0 \Rightarrow t = 1$

$$D = \int_0^1 -(t - 1)^5\, dt + \int_1^2 (t - 1)^5\, dt = -\frac{1}{6}(t - 1)^6\Big]_0^1 + \frac{1}{6}(t - 1)^6\Big]_1^2 = -\frac{1}{6}(-1) + \frac{1}{6}(1) = \frac{1}{3}$$

19. $v(t) = t(t^2 - 9)^5$, $t = 2$ to $t = 4$.
$t^2 - 9 = 0 \Rightarrow t = 3$

$$D = -\int_2^3 t(t^2 - 9)^5\, dt + \int_3^4 t(t^2 - 9)^5\, dt = -\frac{1}{12}(t^2 - 9)^6\Big]_2^3 + \frac{1}{12}(t^2 - 9)^6\Big]_3^4$$

$$= -\frac{1}{12}(0 - 15{,}625) + \frac{1}{12}(117{,}649) = \frac{133{,}274}{12} = \frac{66{,}637}{6} \approx 11{,}106.17$$

21. a) $v(t) = t - 4$, $t = 0$ to $t = 6$.
Distance $= 10$, calculated in Exercise 12.
Displacement $= \dfrac{t^2}{2} - 4t\Big]_0^6 = 18 - 24 = -6$.

b) $v(t) = 2 - t^2$, $t = 0$ to $t = 4$.
Distance $= \dfrac{8\sqrt{2} + 40}{3}$, calculated in Exercise 13.
Displacement $= 2t - \dfrac{t^3}{3}\Big]_0^4 = 8 - \dfrac{64}{3} = -\dfrac{40}{3}$.

c) $v(t) = \sin(2t)$, $t = 0$ to $t = \frac{\pi}{2}$.
Since $v(t) \geq 0$ on $\left[0, \frac{\pi}{2}\right]$, the distance and the displacement are the same. The value is 1, calculated in Exercise 14.

d) $v(t) = \cos t \sin t$, $t = 0$ to $t = \frac{\pi}{2}$.

Since $v(t) \geq 0$ on $\left[0, \frac{\pi}{2}\right]$, the distance and the displacement are the same. The value is $\frac{1}{2}$, calculated in Exercise 15.

23. $v(t) = -t^2 + 5t - 6$, $s(0) = 4$

a) $s(t) = \int (-t^2 + 5t - 6)\, dt = -\frac{t^3}{3} + \frac{5}{2}t^2 - 6t + C$

$s(0) = 4 \Rightarrow C = 4$

$s(t) = -\frac{t^3}{3} + \frac{5}{2}t^2 - 6t + 4$

b) $a(t) = v'(t) = -2t + 5$

c) $v(t) = 0 \Rightarrow (t-3)(t-2) = 0 \Rightarrow t = 2, 3$

d) $v(t) > 0$ for $t \in (2, 3)$

e) $v(t) < 0$ for $0 \leq t \leq 2$, $3 \leq t < \infty$

f) $D = -\int_0^2 (-t^2 + 5t - 6)\, dt + \int_2^3 (-t^2 + 5t - 6)\, dt$

$= \left. \frac{t^3}{3} - \frac{5}{2}t^2 + 6t \right]_0^2 - \left. \left(\frac{t^3}{3} - \frac{5}{2}t^2 + 6t \right) \right]_2^3$

$= \frac{8}{3} - 10 + 12 - \left(9 - \frac{45}{2} + 18\right) + \left(\frac{8}{3} - 10 + 12\right) = \frac{29}{6}$

g) Displacement $= \left. -\frac{t^3}{3} + \frac{5}{2}t^2 - 6t \right]_0^3 = -9 + \frac{45}{2} - 18 = -4.5$.

25. $a(t) = -9.8\,\mathrm{m/s^2}$, $v(0) = 12\,\mathrm{m/s}$

a) $v(t) = \int -9.8\, dt = -9.8t + C_1$

$v(0) = 12 \Rightarrow C_1 = 12;\ v(t) = -9.8t + 12$

b) $s(t) = \int (-9.8t + 12)\, dt = -4.9t^2 + 12t + C_2$

$s(0) = 10 \Rightarrow C_2 = 10;\ s(t) = -4.9t^2 + 12t + 10$

c) $v(t) = 0 \Rightarrow -9.8t + 12 = 0 \Rightarrow t = \frac{12}{9.8} \approx 1.2245$ seconds.

d) $s(t) = 0 \Rightarrow -4.9t^2 + 12t + 10 = 0$

$\Rightarrow t \approx 3.106$ seconds

e) Displacement $= -10$ meters.

Distance $= \int_0^{1.2245} (-9.8t + 12)\, dt - \int_{1.2245}^{3.106} (-9.8t + 12)\, dt$

$= \left. (-4.9t^2 + 12t) \right]_0^{1.2245} - \left. (-4.9t^2 + 12t) \right]_{1.2245}^{3.106}$

$= 24.694$

27. $a_1(t) = 1.0 \, \text{m/s}^2$, $a_2(t) = -1.0 \, \text{m/s}^2$, $v_{\text{max}} = 108 \, \text{km/h}$

$$v(t) = \int a(t)\, dt = 1.0t + v_0 = t, \text{ since } v_0 = 0$$

$$108 \, \text{km/h} = \frac{108{,}000}{3600} \, \text{m/s} = 30 \, \text{m/s}$$

$$v(T) = T = 30 \, \text{m/s} \Rightarrow T = 30 \, \text{s}$$

Therefore $v(t) = \begin{cases} t & 0 \le t \le 30 \\ 30 & 30 < t < 10(60) \end{cases}$

$$D = \int_0^{30} t\, dt + \int_{30}^{600} 30\, dt = \frac{t^2}{2}\Bigg]_0^{30} + 30t\Bigg]_{30}^{600} = 450 + 30(600 - 30) = 17{,}550 \, \text{m} = 17.55 \, \text{km}$$

29. If $v(t) \le 0$ for all $t \in [a, b]$, then

$$\int_a^b |v(t)|\, dt = \int_a^b -v(t)\, dt = \int_b^a v(t)\, dt$$

$= s(a) - s(b)$ if s is an antiderivative of v, that is, $s'(t) = v(t)$.

7.5 Hydrostatic Pressure

1.

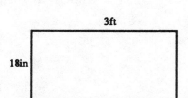

3ft

18in

Exercise 1

$$\begin{aligned} F &= \int_0^{1.5} (62.4)\, h\, (3)\, dh \\ &= 187.2 \, \frac{h^2}{2}\Bigg]_0^{1.5} \\ &= 187.2 \left(\frac{2.25}{2} \right) \\ &= 210.6 \, \text{lb} \end{aligned}$$

on each end panel

or $2(210.6) = 421.2 \, \text{lb}$

on both ends.

3.

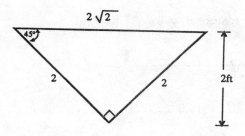

Exercise 3

$$F = \int_0^2 (62.4)h(2)(2-h)\,dh = 124.8\left[h^2 - \frac{h^3}{3}\right]_0^2 = 124.8\left(\frac{4}{3}\right) = 166.4\text{ lb}$$

5.

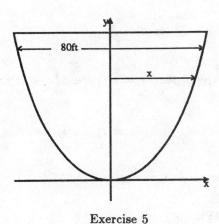

Exercise 5

$$y = \frac{x^2}{100} = \frac{40^2}{100} = 16\text{ ft depth.}$$

$x^2 = 100y \Rightarrow x = 10\sqrt{y}$, measured from the bottom. $x = 10\sqrt{16-y}$, measured from the top.

$$F = 2\int_0^{16} (62.4)\,h\left(10\sqrt{16-h}\right)\,dh = 1248\int_0^{16} h\sqrt{16-h}\,dh$$

Let $u = 16 - h$, $du = -dh$, $h = 16 - u$
if $h = 0$, $u = 16$; if $h = 16$, $u = 0$.

$$-1248\int_{16}^0 (16-u)\sqrt{u}\,du = -1248\left(\frac{32}{3}\right)u^{3/2}\Big]_{16}^0 + 1248\left(\frac{2}{5}\right)u^{5/2}\Big]_{16}^0$$
$$= 851{,}968 - 511{,}180.8 = 340{,}787.2\text{ lb}$$

284

7.

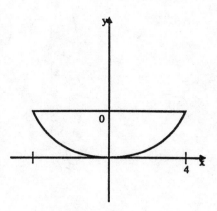

Exercise 7

Assume that the trough is full.

$$\frac{x^2}{16} + \frac{y^2}{4} = 1 \Rightarrow x = \pm 2\sqrt{4 - y^2} \text{ with coordinates centered at top of trough.}$$

$$F = 2\int_0^2 (62.4)\, h\left[2\sqrt{4 - h^2}\right]\, dh = 124.8\int_0^2 2h\sqrt{4 - h^2}\, dh = -124.8\left(\frac{2}{3}\right)(4 - h^2)^{3/2}\Big]_0^2$$

$$= -\frac{249.6}{3}(-8) = 665.6 \text{ lb}$$

9.

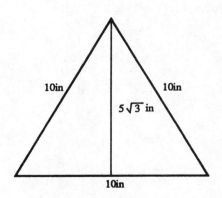

10in   10in

$5\sqrt{3}$ in

10in

Exercise 9

$$F = \int_{30}^{30 + \frac{5\sqrt{3}}{12}} (62.4)\, h\left[\frac{10}{12}\frac{h - 30}{\frac{5\sqrt{3}}{12}}\right]\, dh = \frac{124.8}{\sqrt{3}}\left[\frac{h^3}{3} - 15h^2\right]_{30}^{30 + \frac{5\sqrt{3}}{12}}$$

$$\approx \frac{124.8}{\sqrt{3}}\left[(9665.269 - 9000) - (14,157.332 - 13,500)\right] \approx 571.89 \text{ lb}$$

285

11.

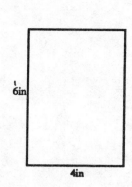

$$F = \int_0^{1/2} (72)\, h\left(\frac{1}{3}\right) dh$$

$$= 12h^2\Big]_0^{1/2}$$

$$= 12\left(\frac{1}{4}\right)$$

$$= 3 \text{ lb}$$

Exercise 11

13. $F = \int_0^{12} 58h\left(\frac{4}{3}\sqrt{36-(h-6)^2}\right) dh \approx 52,230$ lb with $n=20$ in Simpson's Rule.

(Exact answer is $16,704\pi$ lb $\approx 52,477$ lb)

15. $F = \int_2^{2+\frac{10}{12}} (62.4)\, h\left(\frac{15}{12}\right) dh = 39h^2\Big]_2^{2+\frac{10}{12}} = 39\left(\frac{289}{36}-4\right)$

$$= \frac{39(145)}{36} = \frac{1885}{12} \approx 157.083 \text{ lb}$$

Using the average depth gives $F = (62.4)\left(\frac{15.10}{144}\right)\left(2+\frac{5}{12}\right) = 65\left(\frac{29}{12}\right) = \frac{1885}{12}$, as calculated above.

17. $F = \int_7^9 (62.4)\, h\,(3)\, dh = 93.6h^2\Big]_7^9 = 93.6(81-49) = 2995.2$ lb

Using the average depth gives $F = (62.4)\,(6)\,(7+1) = 2995.2$ lb.

7.6   Work

1. $W = \int_0^{1/2} F(x)\, dx = \int_0^{1/2} 40x\, dx = 20x^2\Big]_0^{1/2} = \frac{20}{4} = 5$ ft-lb

3. $W = \int_{1/2}^1 40x\, dx = 20x^2\Big]_{1/2}^1 = 20-5 = 15$ ft-lb

5. $F = kx$; $50 = k\left(\frac{4}{12}\right)$; $k = 150$ lb/ft

$$W = \int_{1/3}^{2/3} 150x\, dx = 75x^2\Big]_{1/3}^{2/3} = 75\left(\frac{4}{9}-\frac{1}{9}\right) = 25 \text{ ft-lb}$$

7. True. $W_1 = \int_A^B F(x)\, dx$, $W_2 = \int_B^C F(x)\, dx$

$$W = W_1 + W_2 = \int_A^C F(x)\, dx$$

286

The property of definite integrals is

$$\int_A^B f(x)\,dx + \int_B^C f(x)\,dx = \int_A^C f(x)\,dx.$$

9. a) $F = kx$; $10 = k(2)$; $k = 5$ lb/ft

    b) $F = 5(3) = 15$ lb; additional weight $= 5$ lb.

    c) $W = \int_0^3 kx\,dx = \int_0^3 5x\,dx = \left. \frac{5}{2}x^2 \right]_0^3 = \frac{45}{2}$ ft-lb

11.

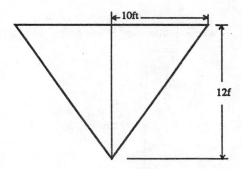

Exercise 11

r depends on h : $r = 10\left(1 - \dfrac{h}{12}\right)$

$$
\begin{aligned}
W &= \int_0^{12} (62.4)\pi \left[100\left(1 - \frac{h}{12}\right)^2\right] h\,dh \\
&= 6240\,\pi \int_0^{12} \left(h - \frac{h^2}{6} + \frac{h^3}{144}\right) dh \\
&= 6240\pi \left[\frac{h^2}{2} - \frac{h^3}{18} + \frac{h^4}{576}\right]_0^{12} \\
&= 6240\pi(72 - 96 + 36) \\
&= 74,880\pi \text{ ft-lb} \approx 235,242 \text{ ft-lb}
\end{aligned}
$$

13. $W = \int_0^{20} \pi r^2 \rho h\,dh = \left. 3120\pi h^2 \right]_0^{20} = 1,248,000\pi$ ft-lb

     $\approx 3,920,708$ ft-lb

15. $W = \int_0^{20} \pi r^2 \rho h\,dh = 62.4\pi \int_0^{20} 36\left(1 - \dfrac{h}{20}\right)^2 h\,dh = 2246.4\pi \int_0^{20}\left(h - \dfrac{h^2}{10} + \dfrac{h^3}{400}\right) dh$

     $= 2246.4\pi \left[\dfrac{h^2}{2} - \dfrac{h^3}{30} + \dfrac{h^4}{1600}\right]_0^{20} = 2246.4\pi\left(200 - \dfrac{8000}{30} + \dfrac{160,000}{1600}\right)$

     $= 74,880\pi$ ft-lb $\approx 235,242$ ft-lb

17.
$$\text{Volume of water} = \frac{\pi r^2 h}{3}.$$

$$r = \frac{4}{20}(6) = 1.2 \text{ ft}; \ V = \frac{\pi (1.2^2)(4)}{3} = 1.92\pi \text{ cu ft}$$

$$\text{Work to raise water 15 ft} = (62.4)(1.92\pi)(15)$$
$$= 1797.12\pi \text{ ft-lb.}$$
$$\text{Total work} = 2036.74\pi + 1797.12\pi \text{ ft-lb}$$
$$= 3833.86\pi \text{ ft-lb}$$
$$\approx 12,044 \text{ ft-lb.}$$

$$\text{Alternatively, } W = \int_{16}^{20} \pi r^2 \rho (h + 15)\, dh = 3833.86\pi \text{ ft-lb.}$$

19. $W = \int_0^{20} F(x)\, dx = \int_0^{20} 2(20 - x)\, dx = 40x - x^2 \Big]_0^{20} = 800 - 400 \text{ ft-lb} = 400 \text{ ft-lb}$

21.

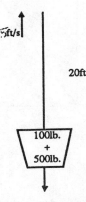

Exercise 21

$$W = \int_0^{20} F(x)\, dx$$

Leakage = 10 lb/s; $\frac{10}{5}$ = 2 lb/ft.

$$W = \int_0^{20} (100 + 500 - 2x)\, dx = 600x - x^2 \Big]_0^{20} = 12,000 - 400 = 11,600 \text{ ft-lb}$$

If the cable weighs 3 lb/ft,

$$W = \int_0^{20} (100 + 500 + 60 - (3 + 2)x)\, dx = 660x - \frac{5x^2}{2} \Big]_0^{20} = 13,200 - 1000 = 12,200 \text{ ft-lb.}$$

Alternatively, add

$$W = \int_0^{20} 3(20 - x)\, dx = 60x - \frac{3}{2}x^2 \Big]_0^{20} = 1200 - 600 = 600 \text{ ft-lb to original answer.}$$

23.   a) $F = K\frac{q_1 q_2}{r^2}$; $10 = \frac{K q^2}{4^2}$; $K = \frac{160}{q^2}\ \dfrac{\text{newton-meter}^2}{\text{coulomb}^2}$

288

b) $W = \int_0^2 F(x)\,dx = \int_0^2 \frac{160}{q^2}\frac{q^2}{(4-x)^2}\,dx = \frac{160}{(4-x)}\Big]_0^2 = 160\left(\frac{1}{2}-\frac{1}{4}\right) = 40$ joules

25.

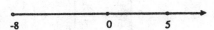

Exercise 25

$$W = \int_0^4 \left[\frac{160}{1}\frac{1}{(5-x)^2} - \frac{160}{1}\frac{1}{(8+x)^2}\right]dx = 160\left[\frac{1}{5-x}+\frac{1}{8+x}\right]_0^4$$

$$= 160\left[\left(1-\frac{1}{5}\right)+\left(\frac{1}{12}-\frac{1}{8}\right)\right] = 160\left(\frac{4}{5}-\frac{1}{24}\right) = \frac{364}{3} \approx 121.33 \text{ joules}$$

7.7 Moments and Centers of Mass

1. $\bar{x} = \dfrac{\sum_{j=1}^n x_j m_j}{\sum_{j=1}^n m_j} = \dfrac{1(-4)+31(2)+7(6)}{1+31+7} = \dfrac{100}{39} \approx 2.5641$

3. $80 \cdot 3 = 50x;\ x = \dfrac{24}{5} = 4.8$ ft

5. The moment of mass m_j about the y-axis is $x_j m_j$, independent of the value of y_j. Hence,

$$\sum_{j=1}^n (x_j - \bar{x})m_j = 0,$$

where $\bar{x}$ is the x-coordinate of the center of mass.

7. $\displaystyle\sum_{j=1}^n (x_j - \bar{x})m_j = 0 \Rightarrow \sum_{j=1}^n x_j m_j = \bar{x}\sum_{j=1}^n m_j \Rightarrow \bar{x} = \frac{\sum_{j=1}^n x_j m_j}{\sum_{j=1}^n m_j}$

Similarly, $\bar{y} = \dfrac{\sum_{j=1}^n y_j m_j}{\sum_{j=1}^n m_j}$.

9. $\bar{x} = 0 = \dfrac{2(1)+4(-2)+3x_3}{2+4+3} = \dfrac{-6+3x_3}{9}$ $\qquad\qquad \bar{y} = 1 = \dfrac{2(1)+4(3)+3y_3}{9} = \dfrac{14+3y_3}{9}$

$x_3 = \dfrac{6}{3} = 2$ $\qquad\qquad\qquad\qquad\qquad\qquad y_3 = \dfrac{9-14}{3} = -\dfrac{5}{3}$

11. $y = x^2$, $y = x^3$

Intersections: $x = 0, 1$

$$A = \int_0^1 (x^2 - x^3)\, dx = \frac{x^3}{3} - \frac{x^4}{4} \Big]_0^1 = \frac{1}{3} - \frac{1}{4} = \frac{1}{12}$$

$$\bar{x} = 6 \int_0^1 x\,(x^2 - x^3)\, dx = 6 \left[\frac{x^4}{4} - \frac{x^5}{5} \right]_0^1 = 12 \left(\frac{1}{4} - \frac{1}{5} \right) = \frac{3}{5}$$

$$\bar{y} = 6 \int_0^1 (x^4 - x^6)\, dx = 6 \left[\frac{x^5}{5} - \frac{x^7}{7} \right]_0^1 = 6 \left(\frac{1}{5} - \frac{1}{7} \right) = 6 \left(\frac{2}{35} \right) = \frac{12}{35}$$

$$(\bar{x}, \bar{y}) = \left(\frac{3}{5}, \frac{12}{35} \right)$$

13. $y = 4 - x$, $x = 0$, $y = 0$

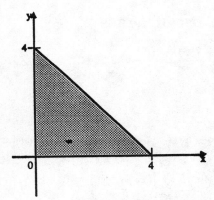

Exercise 13

$$A = \int_0^4 (4 - x)\, dx = 4x - \frac{x^2}{2} \Big]_0^4 = 16 - 8 = 8$$

$$\bar{x} = \frac{1}{8} \int_0^4 x(4 - x)\, dx = \frac{1}{8} \left[2x^2 - \frac{x^3}{3} \right]_0^4 = \frac{1}{8} \left(32 - \frac{64}{3} \right) = \frac{4}{3}$$

$$\bar{y} = \frac{4}{3} \text{ by symmetry.}$$

$$(\bar{x}, \bar{y}) = \left(\frac{4}{3}, \frac{4}{3} \right)$$

290

15. $y = \sqrt{4 - x^2}$, $y = 0$

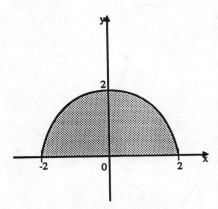

Exercise 15

A = area of semicircle of radius 2 = 2π.

$\overline{x}$ = 0 by symmetry.

$\overline{y}$ = $\dfrac{1}{4\pi} \displaystyle\int_{-2}^{2} (4 - x^2)\, dx = \dfrac{1}{4\pi} \left[4x - \dfrac{x^3}{3} \right]_{-2}^{2} = \dfrac{1}{2\pi} \left(8 - \dfrac{8}{3} \right) = \dfrac{8}{3\pi}$

$(\overline{x}, \overline{y})$ = $\left(0, \dfrac{8}{3\pi} \right)$

17. $y = 2x^2$, $y = x^2 + 1$

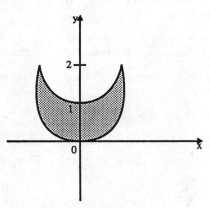

Exercise 17

Intersections: $2x^2 = x^2 + 1 \Rightarrow x^2 = 1 \Rightarrow x = \pm 1,\ y = 2$

A = $2 \displaystyle\int_{0}^{1} (x^2 + 1 - 2x^2)\, dx$

= $2 \displaystyle\int_{0}^{1} (1 - x^2)\, dx = 2x - \dfrac{2x^3}{3} \Big]_{0}^{1} = \dfrac{4}{3}.$

By symmetry, $\overline{x} = 0$.

$\overline{y}$ = $\dfrac{3}{8} \displaystyle\int_{-1}^{1} [(x^2 + 1)^2 - 4x^4]\, dx = \dfrac{3}{4} \displaystyle\int_{0}^{1} (-3x^4 + 2x^2 + 1)\, dx$

= $\dfrac{3}{4} \left[-\dfrac{3}{5} x^5 + \dfrac{2}{3} x^3 + x \right]_{0}^{1} = \dfrac{3}{4} \left(-\dfrac{3}{5} + \dfrac{2}{3} + 1 \right) = \dfrac{4}{5}$

291

$$(\overline{x}, \overline{y}) \;=\; \left(0, \frac{4}{5}\right)$$

19.

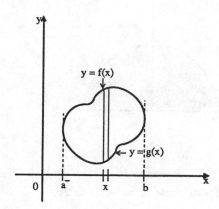

Exercise 19

a) Let R be a region bounded above by the graph of $y = f(x)$ and below by the graph of $y = g(x)$, for $a \leq x \leq b$. Assume that R is the face of a plate of uniform thickness c and of uniform density ρ.

b) Partition the interval $[a, b]$ into n subintervals of equal length $\Delta x = \frac{b-a}{n}$. Select one number t_j in each subinterval $[x_{j-1}, x_j]$.

c) Consider the section of R lying over the interval $[x_{j-1}, x_j]$ to be a slab of height $[f(t_j) - g(t_j)]$, width Δx, thickness c and density ρ. Its mass is approximated by $m_j \approx c\rho[f(t_j) - g(t_j)]\Delta x$.

d) The center of mass of the slab described in step c), with respect to the y-axis, must lie on the line $y = y_j = \frac{1}{2}[f(t_j) + g(t_j)]$.

e) An analogue of equation (6), for equilibrium, is

$$
\begin{aligned}
0 \;&=\; \sum_{j=1}^{n}(y_j - \overline{y})m_j \approx \sum_{j=1}^{n}(y_j - \overline{y})c\rho[f(t_j) - g(t_j)]\Delta x \\
&=\; \sum_{j=1}^{n}\left\{\frac{1}{2}[f(t_j) + g(t_j)] - \overline{y}\right\} c\rho[f(t_j) - g(t_j)]\Delta x \\
&=\; c\rho\sum_{j=1}^{n}\left[\frac{1}{2}[f(t_j)^2 - g(t_j)^2] - \overline{y}[f(t_j) - g(t_j)]\right]\Delta x
\end{aligned}
$$

Take the limit as $n \to \infty$, and divide through by $c\rho \neq 0$.

$$
\begin{aligned}
0 \;&=\; \int_a^b \frac{1}{2}[f(x)^2 - g(x)^2]\,dx - \overline{y}\int_a^b [f(x) - g(x)]\,dx \\
\overline{y} \;&=\; \frac{\frac{1}{2}\int_a^b [f(x)^2 - g(x)^2]\,dx}{\int_a^b = [f(x) - g(x)]\,dx} = \frac{1}{2A}\int_a^b [f(x)^2 - g(x)^2]\,dx
\end{aligned}
$$

21. $f(x) = x^3$, $g(x) = 0$, for $0 \leq x \leq 1$.

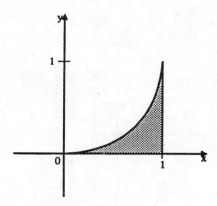

Exercise 21

$$A = \int_0^1 x^3\,dx = \frac{x^4}{4} = \frac{1}{4}$$

$$\overline{x} = 4\int_0^1 x(x^3)\,dx = \frac{4x^5}{5}\bigg]_0^1 = \frac{4}{5}$$

$$\overline{y} = 2\int_0^1 (x^6)\,dx = \frac{2x^7}{7}\bigg]_0^1 = \frac{2}{7}$$

$$(\overline{x}, \overline{y}) = \left(\frac{4}{5}, \frac{2}{7}\right)$$

23. $f(x) = x$, $g(x) = -x$, for $0 \le x \le 4$.

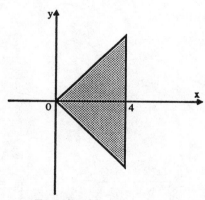

Exercise 23

$$A = \int_0^4 [x - (-x)]\,dx = x^2\bigg]_0^4 = 16$$

$$\overline{x} = \frac{1}{16}\int_0^4 x(2x)\,dx = \frac{1}{16}\cdot\frac{2}{3}x^3\bigg]_0^4 = \frac{1}{24}(64) = \frac{8}{3}$$

$$\overline{y} = 0, \text{ by symmetry.}$$

$$(\overline{x}, \overline{y}) = \left(\frac{8}{3}, 0\right)$$

25. $f(x) = x^2 + x + 1$, $g(x) = 0$, for $1 \le x \le 3$.

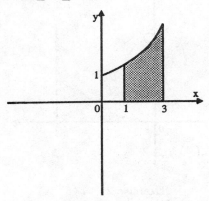

Exercise 25

$$A = \int_1^3 (x^2 + x + 1)\, dx = \frac{x^3}{3} + \frac{x^2}{2} + x \Big]_1^3 = 9 + \frac{9}{2} + 3 - \left(\frac{1}{3} + \frac{1}{2} + 1\right) = \frac{44}{3}$$

$$\bar{x} = \frac{3}{44} \int_1^3 x(x^2 + x + 1)\, dx = \frac{3}{44}\left[\frac{x^4}{4} + \frac{x^3}{3} + \frac{x^2}{2}\right]_1^3 = \frac{3}{44}\left[\frac{81}{4} + 9 + \frac{9}{2} - \left(\frac{1}{4} + \frac{1}{3} + \frac{1}{2}\right)\right] = \frac{49}{22}$$

$$\bar{y} = \frac{3}{88} \int_1^3 (x^2 + x + 1)^2\, dx = \frac{3}{88} \int_1^3 (x^4 + 2x^3 + 3x^2 + 2x + 1)\, dx$$

$$= \frac{3}{88}\left[\frac{x^5}{5} + \frac{x^4}{2} + x^3 + x^2 + x\right]_1^3 = \frac{3}{88}\left[\frac{243}{5} + \frac{81}{2} + 27 + 9 + 3 - \left(\frac{1}{5} + \frac{1}{2} + 3\right)\right] = \frac{933}{220}$$

$$(\bar{x}, \bar{y}) = \left(\frac{49}{22}, \frac{933}{220}\right)$$

27. $\rho(x) = 2 + \left(\dfrac{12 - 2}{10}\right) x = 2 + x$

$$\bar{x} = \frac{\int_0^{10} x\rho(x)\, dx}{\int_0^{10} \rho(x)\, dx} = \frac{\int_0^{10} x(2 + x)\, dx}{\int_0^{10} (2 + x)\, dx} = \frac{x^2 + \frac{x^3}{3} \Big]_0^{10}}{2x + \frac{x^2}{2} \Big]_0^{10}} = \frac{100 + \frac{1000}{3}}{20 + 50} = \frac{1300}{210} = \frac{130}{21}$$

29. $\rho(x) = 1 + x^2$

$$\bar{x} = \frac{\int_0^{10} x(1 + x^2)\, dx}{\int_0^{10} (1 + x^2)\, dx} = \frac{\frac{x^2}{2} + \frac{x^4}{4} \Big]_0^{10}}{x + \frac{x^3}{3} \Big]_0^{10}} = \frac{50 + 2500}{10 + \frac{1000}{3}} = \frac{7650}{1030} = \frac{765}{103}$$

31.

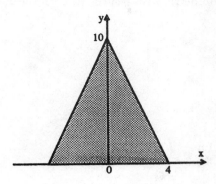

Exercise 31

$$\overline{y} = \frac{\int_0^h \rho y A(y)\,dy}{\int_0^h \rho A(y)\,dy} = \frac{\int_0^{10} y\pi r^2\,dy}{\int_0^{10} \pi r^2\,dy}$$

Use $r = \dfrac{4(10-y)}{10}$.

$$\overline{y} = \frac{\int_0^{10} y(10-y)^2\,dy}{\int_0^{10} (10-y)^2\,dy} = \frac{50y^2 - \frac{20}{3}y^3 + \frac{y^4}{4}\Big]_0^{10}}{100y - 10y^2 + \frac{y^3}{3}\Big]_0^{10}} = \frac{5000 - \frac{20,000}{3} + 2500}{1000 - 1000 + \frac{1000}{3}} = \frac{2500}{1000}$$

$$= \text{2.5 cm from the base.}$$

$$\overline{x} = 0, \text{ by symmetry.}$$

$$(\overline{x}, \overline{y}) = (0, 2.5)$$

33. $y = x^2$, $0 \le x \le 2$, about x-axis.

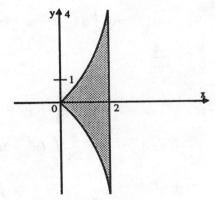

Exercise 33

$$\overline{x} = \frac{\int_0^2 x A(x)\,dx}{\int_0^2 A(x)\,dx} = \frac{\int_0^2 x\pi x^4\,dx}{\int_0^2 \pi x^4\,dx} = \frac{\frac{x^6}{6}\Big]_0^2}{\frac{x^5}{5}\Big]_0^2} = \frac{\frac{32}{3}}{\frac{32}{5}} = \frac{5}{3}$$

$$\overline{y} = 0, \text{ by symmetry.}$$

$$(\overline{x}, \overline{y}) = \left(\frac{5}{3}, 0\right)$$

35. A circle centered at the origin is symmetric with respect to both the x-axis and the y-axis. Therefore, $(\overline{x}, \overline{y})$ is the center of the circle.

37. $f(x) = 4 - x^2$, $f(x) = \begin{cases} -x - 2 & \text{for } -2 \leq x \leq 0 \\ x - 2 & \text{for } 0 \leq x \leq 2 \end{cases}$

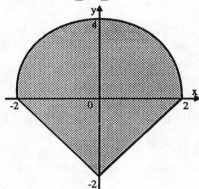

Exercise 37

$$A = 2 \int_0^2 [4 - x^2 - (x - 2)]\, dx = 2 \int_0^2 (6 - x - x^2)\, dx = 2 \left[6x - \frac{x^2}{2} - \frac{x^3}{3} \right]_0^2$$

$$= 2 \left(12 - 2 - \frac{8}{3} \right) = \frac{44}{3}$$

$$\overline{x} = 0, \text{ by symmetry.}$$

$$\overline{y} = \frac{3}{88} \int_{-2}^0 [(4 - x^2)^2 - (-x - 2)^2]\, dx + \frac{3}{88} \int_0^2 [(4 - x^2)^2 - (x - 2)^2]\, dx$$

$$= \frac{3}{88} \int_{-2}^0 (x^4 - 9x^2 - 4x + 12)\, dx + \frac{3}{88} \int_0^2 (x^4 - 9x^2 + 4x + 12)\, dx$$

$$= \frac{3}{44} \int_0^2 (x^4 - 9x^2 + 12)\, dx - \frac{3}{88} \int_{-2}^0 4x\, dx + \frac{3}{88} \int_0^2 4x\, dx$$

$$= \frac{3}{44} \left[\frac{x^5}{5} - 3x^3 + 12x \right]_0^2 - \frac{3}{44} x^2 \Big|_{-2}^0 + \frac{3}{44} x^2 \Big|_0^2$$

$$= \frac{3}{44} \left(\frac{32}{5} - 24 + 24 \right) - \frac{3}{44} (-4) + \frac{3}{44} (4) = \frac{3}{44} \left(\frac{72}{5} \right) = \frac{54}{55}$$

$$(\overline{x}, \overline{y}) = \left(0, \frac{54}{55} \right)$$

296

39.

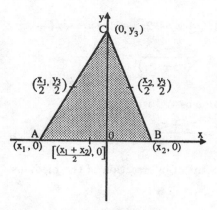

Exercise 39

Position the triangle with its base on the x-axis and vertex on the y-axis.

Area $= (x_2 - x_1)\frac{y_3}{2}$.

line AC: $\dfrac{y - 0}{x - x_1} = \dfrac{y_3 - 0}{0 - x_1}$; $y = \dfrac{-y_3(x - x_1)}{x_1}$

line BC: $\dfrac{y - 0}{x - x_2} = \dfrac{y_3 - 0}{0 - x_2}$; $y = \dfrac{-y_3(x - x_2)}{x_2}$

$$\begin{aligned}
\overline{x} &= \frac{2}{y_3(x_2 - x_1)}\left[\int_{x_1}^0 \frac{-xy_3(x - x_1)}{x_1}\,dx + \int_0^{x_2} \frac{-xy_3(x - x_2)}{x_2}\,dx\right] \\
&= \frac{2}{(x_2 - x_1)}\left\{\frac{-x^3}{3x_1} + \frac{x^2}{2}\bigg]_{x_1}^0 + \frac{-x^3}{3x_2} + \frac{x^2}{2}\bigg]_0^{x_2}\right\} = \frac{2}{(x_2 - x_1)}\left(\frac{x_1^2}{3} - \frac{x_1^2}{2} - \frac{x_2^2}{3} + \frac{x_2^2}{2}\right) \\
&= \frac{2(x_2^2 - x_1^2)}{6(x_2 - x_1)} = \frac{x_2 + x_1}{3} \\
\overline{y} &= \frac{1}{y_3(x_2 - x_1)}\left[\int_{x_1}^0 \frac{y_3^2(x - x_1)^2}{x_1^2}\,dx + \int_0^{x_2} \frac{y_3^2(x - x_2)^2}{x_2^2}\,dx\right] \\
&= \frac{y_3}{(x_2 - x_1)}\left\{\frac{(x - x_1)^3}{3x_1^2}\bigg]_{x_1}^0 + \frac{(x - x_2)^3}{3x_2^2}\bigg]_0^{x_2}\right\} = \frac{y_3}{(x_2 - x_1)}\left(-\frac{x_1}{3} - \frac{-x_2}{3}\right) = \frac{y_3}{3}
\end{aligned}$$

The equations of the median lines are:

From A: $\dfrac{y - 0}{x - x_1} = \dfrac{\frac{y_3}{2}}{\frac{x_2}{2} - x_1}$; $y = \dfrac{y_3(x - x_1)}{x_2 - 2x_1}$

From B: $\dfrac{y - 0}{x - x_2} = \dfrac{\frac{y_3}{2}}{\frac{x_1}{2} - x_2}$; $y = \dfrac{y_3(x - x_2)}{x_1 - 2x_2}$

From C: $\dfrac{y - y_3}{x - 0} = \dfrac{y_3}{0 - \frac{x_1 + x_2}{2}}$; $y = y_3 - \dfrac{2xy_3}{x_1 + x_2}$, $x_1 \neq -x_2$

Find intersections:

AB: $\dfrac{x - x_1}{x_2 - 2x_1} = \dfrac{x - x_2}{x_1 - 2x_2}$; $x(x_1 - 2x_2 - x_2 + 2x_1) = x_1^2 - x_2^2$

$$x = \frac{x_1^2 - x_2^2}{3(x_1 - x_2)} = \frac{x_1 + x_2}{3}$$

$$y = \frac{y_3\left(\frac{x_1 + x_2}{3} - x_1\right)}{x_2 - 2x_1} = \frac{y_3}{3}$$

297

AC: $\dfrac{x - x_1}{x_2 - 2x_1} = \dfrac{x_1 + x_2 - 2x}{x_1 + x_2}$; $x(x_1 + x_2 + 2x_2 - 4x_1) = -x_1^2 + x_2^2$

$$x = \frac{x_2^2 - x_1^2}{3(x_2 - x_1)} = \frac{x_1 + x_2}{3}$$

BC: $\dfrac{x - x_2}{x_1 - 2x_2} = \dfrac{x_1 + x_2 - 2x}{x_1 + x_2}$; same as AC with $x_1 \leftrightarrow x_2$

$$x = \frac{x_1 + x_2}{3}$$

Therefore the centroid equals the point of intersection of the medians.

7.8 The Theorem of Pappus

1. $x^2 + (y - 5)^2 = 9$, about the x-axis.

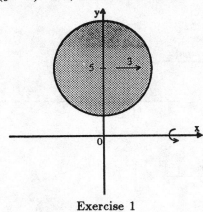

Exercise 1

$$\begin{aligned}
V &= 2\pi A \overline{y} \\
&= 2\pi(9\pi)\,5 \\
&= 90\pi^2
\end{aligned}$$

3. $y = x^2$, $y = \sqrt{x}$, about the y-axis.

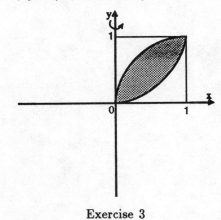

Exercise 3

$$\begin{aligned}
V &= 2\pi A \overline{x} \\
&= 2\pi \left(\frac{1}{3}\right)\left(\frac{9}{20}\right) \\
&= \frac{3\pi}{10}
\end{aligned}$$

5.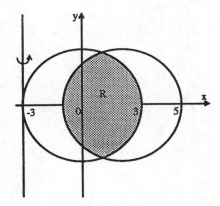

$$V = 2\pi A(\overline{x} + 3)$$
$$A = 2\pi A(1 + 3) = 8\pi A$$

Exercise 5

7. $\overline{y} = \dfrac{8}{2} = 4$

force $= (10)(8)(4)(62.4) = 19,968$ lb

9. $\overline{y} = 30 + \dfrac{4}{12}$ ft $= \dfrac{91}{3}$ ft

$A = (6)(8) + \pi(4^2) = 48 + 16\pi$ in^2

force $= A\rho\overline{y} = \dfrac{48 + 16\pi}{144} (62.4) \left(\dfrac{91}{3}\right) \approx 1291.65$ lb

11. $\quad V = \displaystyle\int_1^3 2\pi x \left[2\sqrt{1 - (x-2)^2}\right] dx$

Let $u = x - 2,\ du = dx,\ x = u + 2$
$\quad\ x = 1 \Rightarrow u = -1;\ x = 3 \Rightarrow u = 1$
$\quad\ V = 4\pi \displaystyle\int_{-1}^{1} (u+2)\sqrt{1 - u^2}\,du = 4\pi \int_{-1}^{1} u\sqrt{1 - u^2}\,du + 8\pi \int_{-1}^{1} \sqrt{1 - u^2}\,du$

$\qquad = -\dfrac{4\pi}{3}(1 - u^2)^{3/2}\Big]_{-1}^{1} + 8\pi\left(\dfrac{\pi}{2}\right)$ (area of semicircle)

$\qquad = -\dfrac{4\pi}{3}(0 - 0) + 4\pi^2 = 4\pi^2$

Review Exercises—Chapter 7

1. Rotate the region bounded by $y = \sqrt{r^2 - x^2}$ and the x-axis about the x-axis.

$$V = \int_{-r}^{r} \pi(r^2 - x^2)\,dx = 2\pi\left[r^2 x - \frac{x^3}{3}\right]_0^r = 2\pi\left(r^3 - \frac{r^3}{3}\right) = \frac{4\pi r^3}{3}$$

3. $y = \dfrac{1}{x},\ 1 \le x \le 4$, about the x-axis,

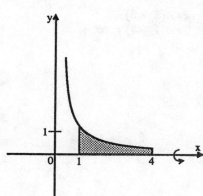

$$V = \int_1^4 \pi \left(\frac{1}{x}\right)^2 dx$$

$$= -\left.\frac{\pi}{x}\right]_1^4$$

$$= \pi\left(1 - \frac{1}{4}\right)$$

$$= \frac{3\pi}{4}$$

Exercise 3

5. $y = \sin x$, $0 \le x \le \dfrac{3\pi}{2}$, about the x-axis.

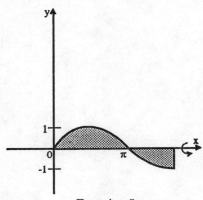

$$V = \int_0^{3\pi/2} \pi \sin^2 x \, dx$$

$$= \frac{\pi}{2} \int_0^{3\pi/2} (1 - \cos 2x) \, dx$$

$$= \frac{\pi}{2}\left[x - \frac{1}{2}\sin 2x\right]_0^{3\pi/2}$$

$$= \frac{\pi}{2}\left(\frac{3\pi}{2}\right)$$

$$= \frac{3\pi^2}{4}$$

Exercise 5

7. $f(x) = x^{2/3}$, $0 \le x \le 2$, about the y-axis.

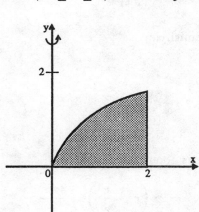

Exercise 7

$$V = \int_0^2 2\pi x(x^{2/3})\, dx = 2\pi \int_0^2 x^{5/3}\, dx = \left.\frac{3\pi}{4} x^{8/3}\right]_0^2 = \frac{3\pi}{4}\left(4 \cdot 2^{2/3}\right)$$

$$= (3\pi)2^{2/3} \approx 14.961$$

9. $y_1 = \sin x^2$, $y_2 = \cos x^2$, $0 \le x \le \dfrac{\sqrt{\pi}}{2}$, about the *y*-axis.

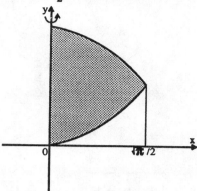

Exercise 9

$$V = \int_0^{\sqrt{\pi}/2} 2\pi x(\cos x^2 - \sin x^2)\, dx = \pi\left[\sin x^2 + \cos x^2\right]_0^{\sqrt{\pi}/2} = \pi\left(\frac{\sqrt{2}}{2} + \frac{\sqrt{2}}{2} - 1\right)$$

$$= \pi\left(\sqrt{2} - 1\right) \approx 1.3013$$

11. $y = x^{2/3} + 1$, $0 \le x \le 2$, about $y = -3$.

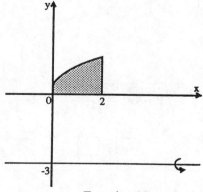

Exercise 11

$$V = \int_0^2 \pi[(x^{2/3} + 4)^2 - 3^2]\, dx = \pi\int_0^2 (x^{4/3} + 8x^{2/3} + 7)\, dx = \pi\left[\frac{3}{7}x^{7/3} + \frac{24}{5}x^{5/3} + 7x\right]_0^2$$

$$= \pi\left[\frac{12}{7}\sqrt[3]{2} + \frac{48}{5}\sqrt[3]{4} + 14\right] \approx 98.643$$

13. $y = 1 + x^{3/2}$ from $(0, 1)$ to $(4, 9)$.

301

$$y' = \frac{3}{2}x^{1/2};\ 1 + (y')^2 = 1 + \frac{9}{4}x$$

$$L = \int_0^4 \sqrt{1 + (y')^2}\,dx = \int_0^4 \sqrt{1 + \frac{9}{4}x}\,dx = \frac{8}{27}\left(1 + \frac{9}{4}x\right)^{3/2}\Bigg]_0^4$$

$$= \frac{8}{27}(10^{3/2} - 1) \approx 9.0734$$

15. $y^2 = (x+1)^3$ from $(0, 1)$ to $(1, \sqrt{8})$

$$y = (x+1)^{3/2};\ y' = \frac{3}{2}(x+1)^{1/2}$$

$$L = \int_0^1 \sqrt{1 + (y')^2}\,dx = \int_0^1 \sqrt{1 + \frac{9}{4}(x+1)}\,dx = \frac{4}{9}\cdot\frac{2}{3}\left[1 + \frac{9}{4}(x+1)\right]^{3/2}\Bigg]_0^1$$

$$= \frac{8}{27}\left[\left(\frac{11}{2}\right)^{3/2} - \left(\frac{13}{4}\right)^{3/2}\right] = \frac{8}{27}\left(\frac{11\sqrt{11}}{2\sqrt{2}} - \frac{13\sqrt{13}}{8}\right) = \frac{1}{27}\left(\frac{44\sqrt{11}}{\sqrt{2}} - 13\sqrt{13}\right)$$

$$= \frac{22\sqrt{22} - 13\sqrt{13}}{27} \approx 2.0858$$

17. $y = x^5 + \dfrac{1}{12x} + 1$ from $\left(1, \dfrac{25}{12}\right)$ to $\left(2, \dfrac{217}{24}\right)$

$$y' = 3x^2 - \frac{1}{12x^2}$$

$$L = \int_1^2 \sqrt{1 + (y')^2}\,dx = \int_1^2 \sqrt{1 + 9x^4 - \frac{1}{2} + \frac{1}{144x^4}}\,dx = \int_1^2 \sqrt{9x^4 + \frac{1}{2} + \frac{1}{144x^4}}\,dx$$

$$= \int_1^2 \left(3x^2 + \frac{1}{12x^2}\right)dx = x^3 - \frac{1}{12x}\Bigg]_1^2 = (8 - 1) - \frac{1}{12}\left(\frac{1}{2} - 1\right) = \frac{169}{24}$$

19. $A(y) = A\left(1 - \dfrac{y}{h}\right)^2$

$$V = \int_0^h A(y)\,dy = \int_0^h A\left(1 - \frac{y}{h}\right)^2 dy = -\frac{Ah}{3}\left(1 - \frac{y}{h}\right)^3\Bigg]_0^h$$

$$= -\frac{Ah}{3}(0 - 1) = \frac{Ah}{3}$$

21.

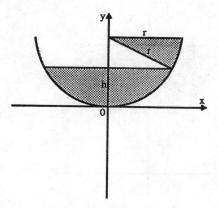

Exercise 21

At any value of h, $r^2 = (r - h)^2 + x^2$

$$V = \int_0^{r/2} \pi[r^2 - (r-h)^2]\,dh = \pi \int_0^{r/2} (2rh - h^2)\,dh = \pi\left[rh^2 - \frac{h^3}{3}\right]_0^{r/2}$$

$$= \pi\left(\frac{r^3}{4} - \frac{r^3}{24}\right) = \frac{5\pi r^3}{24}$$

$$\% \text{ of capacity} = \frac{\frac{5\pi r^3}{24}}{\frac{2\pi r^3}{3}}(100) = \frac{500}{16} = 31.25\%$$

23. $f(x) = \sin x$, $0 \le x \le \pi$, about the x-axis.

$$f'(x) = \cos x$$
$$S = \int_0^\pi 2\pi(\sin x)\sqrt{1 + \cos^2 x}\,dx$$

Using Simpson's Rule with $n = 16$ gives $S \approx 14.424$.

25. $F = kx$

$$10 = k(20);\ k = 0.5 \text{ newton/cm}$$
$$W = \int_0^{50} F(x)\,dx = \int_0^{50} 0.5x\,dx = \frac{x^2}{4}\Big]_0^{50}$$
$$= 625 \text{ newton-cm} = 6.25 \text{ newton-m or joules}$$
$$W = \int_{50}^{60} 0.5x\,dx = \frac{x^2}{4}\Big]_{50}^{60}$$
$$= \frac{1}{4}(3600 - 2500) = 275 \text{ newton-cm} = 2.75 \text{ newton-m or joules}$$

27.

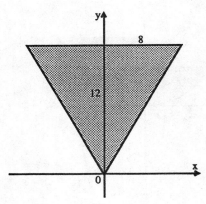

Exercise 27

From the figure, we see that

$$\frac{r}{h} = \frac{8}{12} = \frac{2}{3}.$$

So

$$
\begin{aligned}
W &= \int_0^{12} F(h)\,dh = \int_0^{12} \rho\pi r^2 (22 - h)\,dh \\
&= 62.4\pi \int_0^{12} \frac{4h^2}{9}(22 - h)\,dh \\
&= 62.4\pi(3328) \\
&= 207667.2\pi \text{ ft-lbs} \\
&= 652,405.75 \text{ ft-lbs}
\end{aligned}
$$

29. $W = \int_0^{20} F(x)\,dx = \int_0^{20} \frac{400}{20}x\,dx = 20\left.\frac{x^2}{2}\right]_0^{20} = 4000 \text{ ft-lb}$

31. $W = \int_0^{30} F(x)\,dx = \int_0^{30}(1000 + 6x)\,dx = 1000x + 3x^2\Big]_0^{30} = 30,000 + 2700 = 32,700 \text{ ft-lb}$

33.

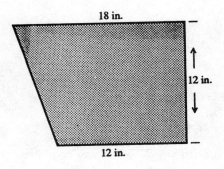

Exercise 33

$$F = \int_0^1 62.4(h)\,(w(h))\,dh$$

$$w(h) = 1.5 - \frac{h}{2}$$

$$F = 62.4 \int_0^1 h\left(1.5 - \frac{h}{2}\right) dh = 62.4 \left[\frac{3}{4}h^2 - \frac{1}{6}h^3\right]_0^1 = 62.4 \left(\frac{14}{24}\right) = 36.4\,\text{lb}$$

35.

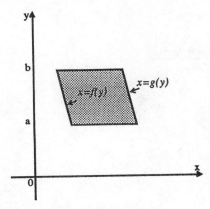

Exercise 35

Let the left-hand boundary of the face be given by $x = f(y)$ and the right-hand boundary by $x = g(y)$. Then, by integrating with respect to y,

$$\overline{y} = \frac{\int_a^b y[g(y) - f(y)]\, dy}{\int_a^b [g(y) - f(y)]\, dy} = \frac{\int_a^b hw(h)\, dh}{A}$$

where h is used in place of y and $w(h)$ represents the width of the face.

The force on the face of the plate is

$$\rho A\overline{y} = \rho \int_a^b hw(h)\, dh = \int_a^b \rho hw(h)\, dh,$$

which is the usual form for the hydrostatic force, except that h is usually measured as increasing downward. Let $h = -y$ and change the limits:

$$F = \int_b^a \rho hw(h)\, dh.$$

Usually $b = 0$, at the surface of the water, and

$$F = \int_0^a \rho hw(h)\, dh.$$

37. $\overline{x} = \dfrac{\int_0^{20} 16\pi x\rho(x)\, dx}{\int_0^{20} 16\pi\rho(x)\, dx}$

$$\rho(x) = 2 + \frac{6}{20}x$$

$$\overline{x} = \frac{\int_0^{20} x\left(2 + \frac{6}{20}x\right) dx}{\int_0^{20}\left(2 + \frac{6}{20}x\right) dx} = \frac{\int_0^{20}\left(2x + \frac{6}{20}x^2\right) dx}{\int_0^{20}\left(2 + \frac{6}{20}x\right) dx} = \frac{x^2 + \frac{2}{20}x^3\big]_0^{20}}{2x + \frac{3}{20}x^2\big]_0^{20}} = \frac{400 + 800}{40 + 60} = \frac{1200}{100} = 12\ \text{cm}$$

39. $\bar{x} = \dfrac{\int_0^3 \rho A(x)\,dx}{\int_0^3 \rho A(x)\,dx} = \dfrac{\int_0^3 \frac{20\left(10+\frac{90x}{3}\right)}{2\,(100^2)}\,x\,dx}{\int_0^3 \frac{20\left(10+\frac{90x}{3}\right)}{2\,(100^2)}\,dx} = \dfrac{\int_0^3 (10x+30x^2)\,dx}{\int_0^3 (10+30x)\,dx} = \dfrac{5x^2+10x^3 \Big]_0^3}{10x+15x^2 \Big]_0^3}$

$= \dfrac{45+270}{30+135} = \dfrac{315}{165} = \dfrac{21}{11}$ m

41. $\bar{x} = \dfrac{\sum_{j=1}^4 x_j w_j}{\sum_{j=1}^4 w_j} = \dfrac{m(-2-2+0+1)}{4m} = -\dfrac{3}{4}$

$\bar{y} = \dfrac{m(4-2+6-4)}{4m} = 1$

43. Answers will vary.

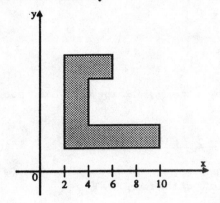 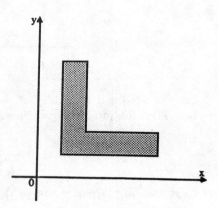

Exercise 43

45. $y = 16 - x^2$

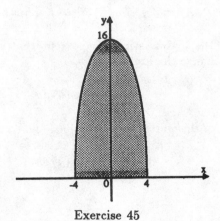

Exercise 45

$\bar{x} = 0$, by symmetry

$A = 2\int_0^4 (16-x^2)\,dx = 2\left[16x - \dfrac{x^3}{3}\right]_0^4 = 2\left(64 - \dfrac{64}{3}\right) = \dfrac{256}{3}$

$\bar{y} = \dfrac{1}{2A}\int_{-4}^4 (16-x^2)^2\,dx = \dfrac{3}{512}\int_{-4}^4 (256-32x^2+x^4)\,dx = \dfrac{3}{256}\left[256x - \dfrac{32}{3}x^3 + \dfrac{x^5}{5}\right]_0^4$

306

$$= \frac{3}{256} \left(1024 - \frac{2048}{3} + \frac{1024}{5} \right) = 3 \left(4 - \frac{8}{3} + \frac{4}{5} \right) = \frac{32}{5}$$

$$(\overline{x}, \overline{y}) = (0, 6.4)$$

47. $y = 6x - x^2 = -9 + 6x - x^2 + 9 = -(x-3)^2 + 9$

$y - 9 = -(x-3)^2$

Parabola with vertex at $(3, 9)$.

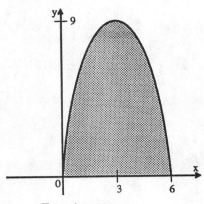

Exercise 47

$$\overline{x} = 3, \text{ by symmetry}$$

$$A = 2 \int_0^3 (6x - x^2)\, dx = 2 \left[3x^2 - \frac{x^3}{3} \right]_0^3 = 2(27 - 9) = 36$$

$$\overline{y} = \frac{1}{2A} \int_0^6 (6x - x^2)^2\, dx = \frac{1}{72} \int_0^6 (36x^2 - 12x^3 + x^4)\, dx = \frac{1}{72} \left[12x^3 - 3x^4 + \frac{x^5}{5} \right]_0^6$$

$$= \frac{1}{72} \left(2592 - 3888 + \frac{7776}{5} \right) = 36 - 54 + \frac{108}{5} = \frac{18}{5}$$

$$(\overline{x}, \overline{y}) = (3, 3.6)$$

49. $y = 6x - x^2, \quad y = 3 - |x - 3|$

$y = 6x - x^2 = -9 + 6x - x^2 + 9 = -(x-3)^2 + 9$

$y - 9 = -(x-3)^2$, a parabola with vertex at $(3, 9)$

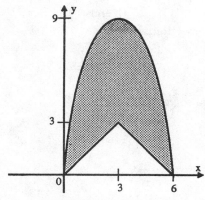

Exercise 49

$$\overline{x} = 3, \text{ by symmetry}$$

307

$$A = 2\int_0^3 (6x - x^2 - x)\, dx = 2\left[\frac{5}{2}x^2 - \frac{x^3}{3}\right]_0^3 = 2\left(\frac{45}{2} - 9\right) = 27$$

$$\overline{y} = \frac{1}{2A}\int_0^6 \left[(6x - x^2)^2 - (3 - |x - 3|)^2\right] dx$$

$$= \frac{1}{54}\left[\int_0^3 (36x^2 - 12x^3 + x^4 - x^2)\, dx + \int_3^6 (36x^2 - 12x^3 + x^4 - 36 + 12x - x^2)\, dx\right]$$

$$= \frac{1}{54}\left[\frac{35}{3}x^3 - 3x^4 + \frac{x^5}{5}\right]_0^3 + \frac{1}{54}\left[\frac{35}{3}x^3 - 3x^4 + \frac{x^5}{5} - 36x + 6x^2\right]$$

$$= \frac{1}{54}\left(315 - 243 + \frac{243}{5}\right) + \frac{1}{54}\left[\frac{35}{3}(216 - 27) - 3(1296 - 81)\right.$$

$$\left. + \frac{1}{5}(7776 - 243) - 36(6 - 3) + 6(36 - 9)\right]$$

$$= \frac{1}{54}\left(\frac{603}{5}\right) + \frac{1}{54}\left(\frac{603}{5}\right) = \frac{67}{15}$$

$$(\overline{x}, \overline{y}) = \left(3, \frac{67}{15}\right)$$

51. $y = |x - 2|$, $y = -\frac{1}{5}x$

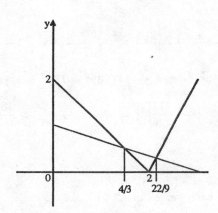

Exercise 51

To obtain the points of intersection, we solve

$$x - 2 = -\frac{x}{5} + \frac{14}{5},$$

which yields $x = 4$. We also solve

$$2 - x = -\frac{x}{5} + \frac{14}{5},$$

which yields $x = -1$.

The area A is

$$A = \int_{-1}^{2}\left(-\frac{x}{5} + \frac{14}{5}\right) - (2 - x)\, dx + \int_{2}^{4}\left(-\frac{x}{5} + \frac{14}{5}\right) - (x - 2)\, dx = 6.$$

Then

$$\overline{x} = \frac{1}{A}\int_{-1}^{2} x\left(\left(-\frac{x}{5}+\frac{14}{5}\right)-(2-x)\right)dx + \frac{1}{A}\int_{2}^{4} x\left(\left(-\frac{x}{5}+\frac{14}{5}\right)-(x-2)\right)dx$$

$$= \frac{1}{6}\left(\frac{18}{5}\right)+\frac{1}{6}\left(\frac{32}{5}\right)=\frac{5}{3},$$

and

$$\overline{y} = \frac{1}{2A}\int_{-1}^{2}\left(-\frac{x}{5}+\frac{14}{5}\right)^2-(2-x)^2\,dx + \frac{1}{2A}\int_{2}^{4}\left(-\frac{x}{5}+\frac{14}{5}\right)^2-(x-2)^2\,dx$$

$$= \frac{1}{12}\left(\frac{324}{25}\right)+\frac{1}{12}\left(\frac{176}{25}\right)=\frac{5}{3}.$$

53. $y = \sqrt{4-x^2}$

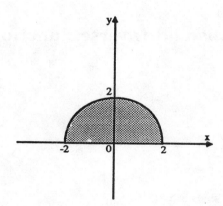

Exercise 53

$$\overline{y} = \frac{1}{2A}\int_{-2}^{2}(4-x^2)\,dx = \frac{2}{4\pi}\left[4x-\frac{x^3}{3}\right]_0^2 = \frac{1}{2\pi}\left(8-\frac{8}{3}\right)=\frac{8}{3\pi}$$

a) about $y = 0$

$$V = 2\pi\overline{y}A = 2\pi\left(\frac{8}{3\pi}\right)(2\pi)=\frac{32\pi}{3}$$

b) about $y = -2$

$$V = 2\pi(\overline{y}+2)A = 2\pi\left(\frac{8}{3\pi}+2\right)(2\pi)=\frac{32}{3}\pi+8\pi^2$$

c) about $y = 6$

$$V = 2\pi(6-\overline{y})A = 2\pi\left(6-\frac{8}{3\pi}\right)(2\pi)=4\pi^2\left(6-\frac{8}{3\pi}\right)=24\pi^2-\frac{32\pi}{3}$$

Chapter 8

Logarithmic and Exponential Functions

8.1 Review of Logarithms and Inverse Functions

1. $\log_{10} 1000 = \log_{10} 10^3 = 3$

3. $\log_8 2 = \log_8 8^{1/3} = 1/3$

5. $\log_5(1/125) = \log_5 5^{-3} = -3$

7. $\log_4 8 = \log_4 2^3 = \log_4 4^{3/2} = 3/2$

9. $\log_{10} 1 = \log_{10} 10^0 = 0$

11. $\log_x 16 = 4 \implies x^4 = 16 \implies x = 2$ since $x > 0$.

13. $\log_{1/2} x = 2 \implies x = (1/2)^2 = 1/4$

15. $\log_3 x = 1/6 \implies x = 3^{1/6}$

17. $\log_x 27 = 3/2 \implies x^{3/2} = 27 = 9^{3/2} \implies x = 9$

19. $\log_{10} x = 0 \implies x = 10^0 = 1$

21. $\log_2 x = -3 \implies x = 2^{-3} = 1/8$

23. $y = f(x) = 3x - 2 \implies x = \frac{1}{3}y + \frac{2}{3}$. Hence the inverse of f is the function $f^{-1} : (-\infty, \infty) \to (-\infty, \infty)$, where $f^{-1}(x) = \frac{1}{3}x + \frac{2}{3}$. The graphs of $y = f(x)$ and $y = f^{-1}(x)$ are shown below.

25. $y = \frac{1}{x+2} \iff x = \frac{1}{y} - 2$. Hence the inverse of the function $f : (-\infty, -2) \cup (-2, \infty) \to (-\infty, 0) \cup (0, \infty)$ is the function $f^{-1} : (-\infty, 0) \cup (0, \infty) \to (-\infty, -2) \cup (-2, \infty)$, where $f^{-1}(x) = (1/x) - 2$.

27. $y = \frac{x}{x+3} \iff x = \frac{3y}{1-y}$. Hence the inverse of the function $f : (-\infty, -3) \cup (-3, \infty) \to (-\infty, 1) \cup (1, \infty)$ is the function $f^{-1} : (-\infty, 1) \cup (1, \infty) \to (-\infty, -3) \cup (-3, \infty)$, where $f^{-1}(x) = \frac{3x}{1-x}$.

29. $y = 4 + \sqrt{x - 1} \iff x = (y - 4)^2 + 1$. Hence the inverse of the function $f : [1, \infty) \to [4, \infty)$ is the function $f^{-1} : [4, \infty) \to [1, \infty)$, where $f^{-1}(x) = (x - 4)^2 + 1$.

310

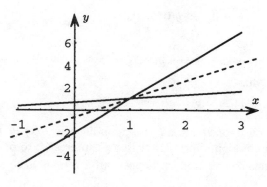

$$y = 3x - 2, \quad y = (1/3)x + 2/3$$

Exercise 23

31. $y = 3 + x^{5/3} \iff x = (y-3)^{3/5}$. Hence the inverse of the function $f : (-\infty, \infty) \to (-\infty, \infty)$ is the function $f^{-1} : (-\infty, \infty) \to (-\infty, \infty)$, where $f^{-1}(x) = (x-3)^{3/5}$.

33. $f'(x) = -6/(7-2x)^2 \implies (f^{-1})'(3/5) = 1/f'(1) = -25/6$

35. $f'(x) = \sec^2 x \implies (f^{-1})'(1) = 1/f'(\pi/4) = \cos^2(\pi/4) = 1/2$

37. Since $f(f(x)) = 1/f(x) = 1/(1/x) = x$, it follows that $f^{-1} = f$.

39. (a) $y = x^2 + 3 \implies x = \pm\sqrt{y-3}$. Since $x \geq 0$, $f^{-1}(x) = \sqrt{x-3}$.

    (b) $y = (x+1)^2 \implies x = -1 \pm \sqrt{y}$. Since $x \leq -1$, $f^{-1}(x) = -1 - \sqrt{x}$.

    (c) $y = (x+2)^2 \implies x = -2 \pm \sqrt{y}$. Since $x \geq -2$, $f^{-1}(x) = -2 + \sqrt{x}$.

    (d) $y = x^2 - 2x - 3 = (x-1)^2 - 4 \implies x = 1 \pm \sqrt{y+4}$. Since $x \leq 1$, $f^{-1}(x) = 1 - \sqrt{x+4}$.

41. (a) Since $f'(x) = \frac{1}{3}(x+5)^{-2/3}$ and $f(3) = 2$, $f'(3) = 1/12$ and hence $(f^{-1})'(2) = 1/f'(3) = 1/(1/12) = 12$.

    (b) $y = \sqrt[3]{x+5} \implies x = y^3 - 5$. Hence $f^{-1}(x) = x^3 - 5$.

    (c) $(f^{-1})'(x) = 3x^2$.

    (d) From part (c), $(f^{-1})'(2) = 3 \cdot 2^2 = 12$.

43. If $g = f^{-1}$, where $f : [-\pi/2, \pi/2] \to [-1, 1]$ is defined by $f(x) = \sin x$ for all $x \in [-\pi/2, \pi/2]$, then $g'(b) = 1/f'(a) = 1/\cos a$, where $b = f(a) = \sin a$. Hence:

    (a) $\sin \pi/4 = \sqrt{2}/2 \implies g'(\sqrt{2}/2) = 1/\cos(\pi/4) = \sqrt{2}$;

    (b) $\sin 0 = 0 \implies g'(0) = 1/\cos 0 = 1$;

    (c) $\sin \pi/3 = \sqrt{3}/2 \implies g'(\sqrt{3}/2) = 1/\cos(\pi/3) = 2$;

    (d) $\sin(-\pi/6) = -1/2 \implies g'(-1/2) = 1/\cos(-\pi/6) = 2\sqrt{3}/3$.

45. (a) $f(x) = x + x^3 = 0 \implies x = 0 \implies g(0) = 0$.

    (b) $f(x) = x + x^3 = 2 \implies x = 1 \implies g(2) = 1$.

    (c) $g'(0) = 1/f'(0) = 1/1 = 1$.

    (d) $g'(2) = 1/f'(1) = 1/4$.

47. (a) $f(x) = -35x - x^5 = 0 \implies x = 0 \implies g(0) = 0.$

    (b) $g'(0) = 1/f'(0) = -1/35.$

    (c) $f(x) = -35x - x^5 = -36 \implies x = 1 \implies g(-36) = 1.$

    (d) $g'(-36) = 1/f'(1) = -1/40.$

    (e) Since $g^{-1} = f$, $g^{-1}(1) = f(1) = -36.$

49. By Exercise 48, the amount on deposit after t compoundings is $P(t) = (1+r)^t P_0$. If there are n compoundings per year, each compounding period has an interest rate of r/n and there are nt such compoundings in t years. Hence the amount on deposit after t years is equal to $\left(1 + \frac{r}{n}\right)^{nt} P_0$.

51. Let $x = \log_b m$ and $y = \log_b n$. Then $b^x = m$ and $b^y = n$, and hence $m/n = b^x/b^y = b^{x-y}$. Now, if z is rational, we showed in Exercise 50 that $\log_b b^z = z$. Hence $\log_b m/n = \log_b b^{x-y} = x - y = \log_b m - \log_b n$.

53. $\log_{10}(I/I_0) = kx \implies I/I_0 = 10^{kx} \implies I = I_0 \cdot 10^{kx}.$

55. It depends on the value of a. To see this, let $n = \log_a x$ and $m = \log_a y$. Then $a^n = x$, $a^m = y$, and hence $x/y = a^{n-m}$. Now, if $0 < x < y$ and $a > 1$, then $a^{n-m} = x/y < 1$ and hence $n - m < 0$; that is, $\log_a x < \log_a y$. But if $0 < a < 1$, then $n - m > 0$; that is, $\log_a x > \log_a y$. To illustrate this last case, consider the numbers 2 and 4. Clearly, $2 < 4$; but $\log_{1/2} 2 = -1 > \log_{1/2} 4 = -2$.

For Exercises 57 and 59 we use parametric plotting programs on the HP-28S or the Casio with the x and y coordinates defined by $x = f(t)$, $y = t$. On the TI-81 use the **Param** mode and set $\mathbf{X_{1T} = f(T)}$ and $\mathbf{Y_{1T} = T}$. With HP-48SX use the **PARA** plot type and enter **'f(t)+i*t'** for the function to be plotted. The range for the variable t should be the same as the domain of f(x). The function $y = f(x)$ can also be drawn by taking $x = t$ and $y = f(t)$.

57. Graph using the range $[-1, 8] \times [-1, 8]$; see the graph below on the left. The inverse relation is a function since $f(x)$ is one-to-one.

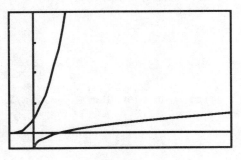
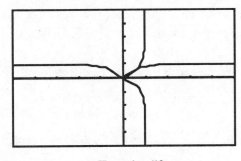

| Exercise 57 | Exercise 59 |

59. Graph using the range $[-10, 10] \times [-10, 10]$; see the graph above on the right. The inverse relation is not a function since $f(x)$ is not one-to-one. Notice that it fails the horizontal line test.

For Exercises 61 and 63 we use parametric plotting programs on the HP-28S or the Casio with the x and y coordinates defined by $x = f(t)$, $y = 1/f'(t)$. On the TI-81 use the **Param** mode and set $\mathbf{X_{1T}=f(T)}$ and$\mathbf{Y_{1T}=1/f'(T)}$. With HP-48SX use the **PARA** plot type and enter **'f(t)+i*INV(f'(t))'** for the function to be plotted. The range for the variable t should be the same as the domain for the function $f(x)$.

61. Graph using the range $[1/3, 1] \times [-9, 1]$; see below on the right. Note that $f(1) = 1$ and $f(3) = 1/3$ so the domain of $f^{-1}(x)$ is $[1/3, 1]$. Now $f^{-1}(x) = 1/x$ and $(f^{-1})'(x) = -1/x^2$. So we plot the pairs defined by $x = 1/t$ and $y = -t^2$. Since $x = 1/t$ we have that $y = -1/x^2 = (f^{-1})'(x)$.

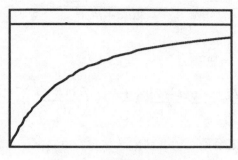

| Exercise 61 | Exercise 63 |

63. Graph using the range $[-1, 1] \times [0, 10]$; see above on the right. The range of $f(x)$ is $[-1, 1]$. This graph is produced by plotting the pair $x = \sin t$, $y = \sec t$.

8.2   The Natural Logarithm Function

1. (a) $\ln x = 0 \implies 2x = e^0 = 1 \implies x = 1/2$.

   (b) $2\ln x = \ln 2x \implies \ln x^2 = \ln 2x \implies x^2 = 2x \implies x = 0, 2$. Since $\ln x$ is only defined for $x > 0$, the only solution is $x = 2$.

   (c) $\ln(2/x) - \ln x = 0 \implies \ln(2/x) = \ln x \implies 2/x = x \implies x = \pm\sqrt{2}$. Since $\ln x$ is only defined for $x > 0$, the only solution is $x = \sqrt{2}$.

   (d) $\ln 3^{x^2} = 0 \implies x^2 \ln 3 = 0 \implies x = 0$ since $\ln 3 \neq 0$.

   (e) $\sqrt{\ln \sqrt{x}} = 1 \implies \ln \sqrt{x} = 1 \implies \sqrt{x} = e^1 = e \implies x = e^2$.

   (f) $3\ln x + x = 2 + \ln x^3 \implies 3\ln x + x = 2 + 3\ln x \implies x = 2$.

   (g) $\int_1^x \frac{3}{t}\, dt = 6 \implies 3\ln x = 6 \implies \ln x = 2 \implies x = e^2$.

   (h) $\int_2^x \frac{1}{t}\, dt = 0 \implies \ln x = \int_1^x \frac{1}{t}\, dt = \int_1^2 \frac{1}{t}\, dt + \int_2^x \frac{1}{t}\, dt = \ln 2 + 0 = \ln 2 \implies x = 2$.

3. True. The graph of $y = \ln x$ shows that every number on the y-axis occurs as the natural logarithm of some number x.

5. $y = x\ln x \implies y' = x(1/x) + (\ln x)(1) = 1 + \ln x$.

7. $f(x) = \ln\sqrt{x^3 - x} = (1/2)\ln(x^3 - x) \implies y' = \frac{1}{2}\frac{1}{x^3 - x}(3x^2 - 1) = \frac{3x^2 - 1}{2(x^3 - x)}$.

9. $y = \sin(\ln x) \implies y' = [\cos(\ln x)](1/x)$.

11. $f(x) = \dfrac{x}{1 + \ln x} \implies f'(x) = \dfrac{(1 + \ln x)(1) - x(1/x)}{(1 + \ln x)^2} = \dfrac{\ln x}{(1 + \ln x)^2}$.

13. $y = (3\ln\sqrt{x})^4 = (81/16)(\ln x)^4 \implies y' = \dfrac{81}{16} 4 (\ln x)^3 \left(\dfrac{1}{x}\right) = \dfrac{81\ln^3 x}{4x}$.

313

15. $f(t) = \dfrac{\ln(a+bt)}{\ln(c+dt)} \implies$

$$f'(t) = \frac{(\ln(c+dt))\left[\frac{1}{a+bt}b\right] - (\ln(a+bt))\left[\frac{1}{c+dt}d\right]}{[\ln(c+dt)]^2} = \frac{\frac{b\ln(c+dt)}{a+bt} - \frac{d\ln(a+bt)}{c+dt}}{[\ln(c+dt)]^2}.$$

17. $f(x) = \ln\cos x \implies f'(x) = \dfrac{1}{\cos x}(-\sin x) = -\dfrac{\sin x}{\cos x} = -\tan x.$

19. $u(t) = \dfrac{t}{1+\ln^2 t} \implies u'(t) = \dfrac{(1+\ln^2 t)(1) - t(2\ln t)(1/t)}{(1+\ln^2 t)^2} = \dfrac{\ln^2 t - 2\ln t + 1}{(\ln^2 t + 1)^2} = \left(\dfrac{\ln t - 1}{\ln^2 t + 1}\right)^2.$

21. $\ln(x+y) + \ln(x-y) = 1 \implies \ln(x^2 - y^2) = 1 \implies \dfrac{2x - 2yy'}{x^2 - y^2} = 0 \implies 2x - 2yy' = 0 \implies y' = x/y.$

23. $x\ln y + y\ln x = x \implies x(1/y)y' + \ln y + y(1/x) + y'\ln x = 1 \implies x^2 y' + xy\ln y + y^2 + xyy'\ln x = xy$
$\implies y'(x^2 + xy\ln x) = xy - xy\ln y - y^2 \implies$

$$y' = \frac{xy(1-\ln y)}{x(x+y\ln x)} = \frac{y(1-\ln y)}{x+y\ln x}.$$

25. $y = \displaystyle\int_x^1 \dfrac{1}{2t}\, dt = -\dfrac{1}{2}\int_1^x \dfrac{1}{t}\, dt = -(1/2)\ln x \implies y' = -1/(2x).$

27. Since $y' = 1 - 1/x = 0 \iff x = 1$, the graph of $y = x - \ln x$ has one critical number, namely $x = 1$. Moreover, $x = 1$ is a local minimum since $y'' = 1/x^2$ and $y''(1) = 1 > 0$. Hence, since $y(1) = 1 - \ln 1 = 1$, the point $(1,1)$ is a local minimum. The graph is shown below.

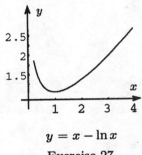

$y = x - \ln x$

Exercise 27

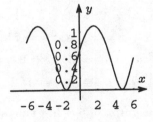

$y = \ln(2 + \sin x)$

Exercise 29

29. $y = \ln(2 + \sin x) \implies y' = (\cos x)/(2 + \sin x)$. Therefore

$$y' = 0 \iff \cos x = 0 \iff x = \ldots, -\pi/2, \pi/2, 3\pi/2, \ldots.$$

Now,

$$y'' = \frac{(2+\sin x)(-\sin x) - (\cos x)(\cos x)}{(2+\sin x)^2} = \frac{-1 - 2\sin x}{(2+\sin x)^2}.$$

Therefore

$$y''(-\pi/2) = 1 > 0 \quad \Longrightarrow \quad x = -\pi/2 \text{ is a relative minimum}$$
$$y''(\pi/2) = -1/3 < 0 \quad \Longrightarrow \quad x = \pi/2 \text{ is a relative maximum}$$
$$y''(3\pi/2) = 1 > 0 \quad \Longrightarrow \quad x = 3\pi/2 \text{ is a relative minimum}$$

Thus, on the interval $[0, 2\pi]$ the graph of $y = \ln(2 + \sin x)$ has two relative extrema: a relative maximum at $(\pi/2, \ln 3)$ and a relative minimum at $(3\pi/2, 0)$. Since $\sin x$ is periodic with period 2π, the graph of $y = \ln(2 + \sin x)$ is also periodic with period 2π.

31. Let $u = x + 3$. Then $du = dx$ and hence

$$\int \frac{dx}{x+3} = \int \frac{du}{u} = \ln|u| + C = \ln|x+3| + C.$$

33. Let $u = 1 - x$. Then $du = -dx$ and hence

$$\int \frac{dx}{1-x} = -\int \frac{du}{u} = -\ln|u| + C = -\ln|1-x| + C.$$

35. Let $u = x^2 - 2x$. Then $du = (2x - 2)dx = 2(x-1)dx$ and hence

$$\int \frac{x-1}{x^2-2x} \, dx = \frac{1}{2} \int \frac{du}{u} = \frac{1}{2} \ln|u| + C = \frac{1}{2} \ln|x^2 - 2x| + C.$$

37. Let $u = x^3 + 9x$. Then $du = (3x^2 + 9)dx = 3(x^2 + 3)dx$ and hence

$$\int \frac{x^2+3}{x^3+9x} \, dx = \frac{1}{3} \int \frac{du}{u} = \frac{1}{3} \ln|u| + C = \frac{1}{3} \ln|x^3 + 9x| + C.$$

39. Let $u = 4 + 2\cos t$. Then $du = -2\sin t \, dt$ and hence

$$\int \frac{\sin t \, dt}{4 + 2\cos t} = -\frac{1}{2} \int \frac{du}{u} = -\frac{1}{2} \ln|u| + C = -\frac{1}{2} \ln(4 + 2\cos t) + C.$$

We need not use the absolute value of $4 + 2\cos t$ in the antiderivative since $4 + 2\cos t$ is always positive.

41. Let $u = \ln x$. Then $du = \dfrac{1}{x} \, dx$ and hence

$$\int \frac{\ln^2 x}{x} \, dx = \int u^2 \, du = \frac{1}{3}u^3 + C = \frac{1}{3} \ln^3 x + C.$$

43. Let $u = 1 - \sqrt{x}$. Then $du = -\dfrac{1}{2\sqrt{x}} \, dx$ and hence

$$\int \frac{1}{\sqrt{x}(1 - \sqrt{x})} \, dx = -2 \int \frac{du}{u} = -2\ln|u| + C = -2\ln|1 - \sqrt{x}| + C.$$

45. Since $\dfrac{x^4 + 3x^2 + x + 1}{x + 1} = x^3 - x^2 + 4x - 3 + \dfrac{4}{x + 1}$, it follows that

$$\int \frac{x^4 + 3x^2 + x + 1}{x + 1}\, dx = \int \left(x^3 - x^2 + 4x - 3 + \frac{4}{x + 1}\right) dx$$
$$= \frac{1}{4}x^4 - \frac{1}{3}x^3 + 2x^2 - 3x + 4\ln|x + 1| + C.$$

47. Let $u = \ln x$. Then $du = \dfrac{1}{x}\, dx$ and hence

$$\int_e^{e^2} \frac{1}{x \ln x}\, dx = \int_1^2 \frac{1}{u}\, du = \ln u\big|_1^2 = \ln 2 - \ln 1 = \ln 2.$$

49. Let $u = \ln(x^2) = 2\ln x$. Then $du = \dfrac{2}{x}\, dx$ and hence

$$\int_e^{e^2} \frac{1}{x \ln(x^2)}\, dx = \frac{1}{2}\int_2^4 \frac{1}{u}\, du = \frac{1}{2}\ln u\Big|_2^4 = \frac{1}{2}(\ln 4 - \ln 2) = \frac{1}{2}\ln\frac{4}{2} = \frac{1}{2}\ln 2.$$

51. $\displaystyle\int_1^2 \left(\frac{1}{1 + x} - \frac{1}{2 + x}\right) dx = (\ln(1 + x) - \ln(2 + x))\big|_1^2 = (\ln 3 - \ln 4) - (\ln 2 - \ln 3) = \ln(9/8).$

53. Let $u = \cos x$. Then $du = \sin x\, dx$ and hence

$$\int_0^{\pi/3} \tan x\, dx = \int_0^{\pi/3} \frac{\sin x}{\cos x}\, dx = -\int_1^{1/2} \frac{1}{u}\, du = -\ln u\big|_1^{1/2} = -\left(\ln\frac{1}{2} - \ln 1\right) = \ln 2.$$

55. If $y = x^3 \ln^2 x$, then $y' = (x^3)\, 2\, (\ln x)(1/x) + (\ln^2 x)(3x^2)$ and hence $y'(1) = 2\ln 1 + (\ln 1)(3) = 0$. Thus, the slope of the tangent line at $x = 1$ is 0. Since $y(1) = 1\ln 1 = 0$, an equation of the tangent line is $y - 0 = 0(x - 1)$, or $y = 0$.

57. We first find the intersection points of the two graphs. Since $x + y = 6$, $y = 6 - x$. Hence $xy = 8 \implies x(6 - x) = 8 \implies x^2 - 6x + 8 = 0 \implies (x - 4)(x - 2) = 0 \implies x = 2, 4$. Thus, the area bounded by the two graphs is equal to

$$\int_2^4 \left[(6 - x) - \frac{8}{x}\right] dx = \left(6x - \frac{1}{2}x^2 - 8\ln x\right)\Big|_2^4 = 6 - 8\ln 2.$$

59. True. To see this, let x be a positive real number. Then the slope of the tangent line to the graph of $y = \ln x$ is $1/x$, while the slope of the tangent line to the graph of $y = \ln ax$ is $(1/ax)(a) = 1/x$. Thus, for a given value of x, the two graphs have the same slope.

61. Let dV be the volume of a differential cross-section disk of thickness dx. Then

$$dV = \pi r^2\, dx = \pi y^2\, dx = \pi\, \frac{\ln^2 x}{x}\, dx$$

and hence the volume of revolution is

$$V = \pi \int_1^2 \frac{\ln^2 x}{x}\, dx = \frac{\pi}{3}\ln^3 x\big|_3^5 = \frac{\pi}{3}(\ln 2)^3 \qquad \text{(by Exercise 41)}.$$

316

63. By definition, the average value of $\ln x$ on the interval $[1, e]$ is

$$\frac{1}{e-1}\int_1^e \ln x\, dx = \frac{1}{e-1}(x\ln x - x)\Big|_1^e = \frac{1}{e-1}[(e\ln e - e) - (\ln 1 - 1)] = \frac{1}{e-1}. \qquad \text{(by Exercise 62)}$$

65. $dy/dx = 1/x \implies y = \displaystyle\int \frac{1}{x}\, dx = \ln x + C_1 = \ln(C_2 x)$, $x > 0$.

67. $\cos x\, dy = \sin x\, dx \implies dy = \tan x\, dx \implies y = \int \tan x\, dx = \ln|\sec x| + C$. (See Exercise 53)

69. Since $(2x+1)dy = y^2 dx$,

$$\frac{1}{y^2}\, dy = \frac{1}{2x+1}\, dx \implies -\frac{1}{y} = \int \frac{1}{2x+1}\, dx = \frac{1}{2}\ln|2x+1| + C \implies y = \frac{-2}{\ln|2x+1| + C}.$$

71. (a) Regarding P as a function of T and differentiating both sides of the equation

$$\ln P = \frac{-2000}{T} + 5.5$$

with respect to T, we obtain

$$\frac{P'}{P} = \frac{2000}{T^2} \implies \frac{dP}{dT} = P' = \frac{2000}{T^2}\, P.$$

   (b) Regarding T as a function of P and differentiating both sides of the equation

$$\ln P = \frac{-2000}{T} + 5.5$$

with respect to P, we obtain

$$\frac{1}{P} = \frac{2000}{T^2}\frac{dT}{dP} \implies \frac{dT}{dP} = \frac{1}{2000}\frac{T^2}{P}.$$

73. Let $f(x) = \ln x^r$, where r is a rational number. Then $f'(x) = \frac{1}{x^r}(rx^{r-1}) = \frac{r}{x} = (r\ln x)'$. Hence $f(x) = r\ln x + C$ for some constant C. Now, $f(x) = \ln x^r \implies f(1) = \ln 1 = 0$, while $f(x) = r\ln x + C \implies f(1) = r\ln 1 + C = C$. Hence $C = 0$ and therefore $\ln x^r = r\ln x$. Letting $x = a$, it follows that $\ln a^r = r\ln a$.

75. Subdividing the interval $[1, 5]$ into 8 equal subintervals of length $1/2$, we obtain the subintervals $[1, 1.5]$, $[1.5, 2], \ldots, [4.5, 5]$, whose midpoints are 1.25, 1.75, 2.25, 2.75, 3.25, 3.75, 4.25, and 4.75. Hence, using the Midpoint Rule with $f(x) = 1/x$,

$$\ln 5 = \int_1^5 \frac{1}{x}\, dx \approx 0.5[f(1.25) + f(1.75) + \cdots + f(4.75)] = 0.5\left(\frac{1}{1.25} + \frac{1}{1.75} + \cdots + \frac{1}{4.75}\right) \approx 1.5998.$$

77. Since the function $f(x) = 1/x$ is decreasing on the interval $[1, 5]$, the actual area under the curve is less than area of the trapezoid on each subinterval. Hence

$$\int_1^5 \frac{1}{x}\, dx < \text{Trapezoidal Rule approximation} \implies \ln 5 < 1.6290.$$

317

Now, it was shown in Theorem 10, Chapter 6, that if the function $f(x)$ is concave up on the interval $[a, b]$, then the Midpoint Rule value is smaller than $\int_a^b f(x)\, dx$, which is smaller than the Trapezoidal Rule value:

$$\text{Midpoint Rule} < \int_a^b f(x)\, dx < \text{Trapezoidal Rule}.$$

Hence, since $f(x) = 1/x$ is concave up on the interval $[1, 5]$, it follows from Exercises 75, 76 that $1.5998 < \ln 5 < 1.6290$.

79. Graph using the range $[0, 10] \times [-2, 1]$; see below on the left. $f'(x) = (1 - \ln x)/x^2$. Domain: $(0, +\infty)$. Range: $(-\infty, e^{-1})$. Absolute maximum: (e, e^{-1}).

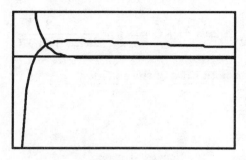

Exercise 79

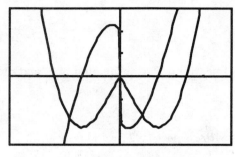

Exercise 81

81. Graph using the range $[-4, 4] \times [-1.5, 1.5]$; see above on right. In this case,

$$f'(x) = \begin{cases} 2x \ln(1 - x) + 1 - x & \text{if } x < 0 \\ 2x \ln(1 + x) - 1 - x & \text{if } x > 0. \end{cases}$$

Domain: $(-\infty, +\infty)$. Range: $[-1.12024037, +\infty)$. Absolute minima: $(\pm 1.37301614, -1.12024037)$.

8.3 The Natural Exponential Function

1. (a) $e^{\ln 2} = 2$

   (b) $e^{-\ln 4} = e^{\ln 1/4} = 1/4$

   (c) $e^{(\ln x - \ln y)} = e^{\ln x}/e^{\ln y} = x/y$

   (d) $\ln e^{-x^2} = -x^2$

   (e) $\ln x e^{\sqrt{x}} - \ln x = \ln x + \ln e^{\sqrt{x}} - \ln x = \sqrt{x}$

   (f) $e^{x \ln 2} = e^{\ln 2^x} = 2^x$

   (g) $e^{\ln(1/x)} = 1/x$

   (h) $e^{4 \ln x} = e^{\ln x^4} = x^4$

   (i) $\ln x e^{x^2} = \ln x + \ln e^{x^2} = \ln x + x^2$

   (j) $e^{x - \ln x} = e^x/e^{\ln x} = e^x/x$

3. $e^{x-y} = x + 3 \implies \ln e^{x-y} = \ln(x + 3) \implies x - y = \ln(x + 3) \implies y = x - \ln(x + 3)$.

5. $y = e^{3x} \implies y' = e^{3x}(3) = 3e^{3x}$.

7. $f(t) = e^{\sqrt{t}} \implies f'(t) = e^{\sqrt{t}}(1/2)t^{-1/2} = e^{\sqrt{t}}/(2\sqrt{t})$.

9. $y = e^{x^2 - x} \implies y' = (2x - 1)e^{x^2 - x}$.

11. $f(x) = e^x \sin x \implies y' = e^x \cos x + e^x \sin x = e^x(\cos x + \sin x)$.

13. $f(x) = \ln \dfrac{e^x + 1}{x + 1} = \ln(e^x + 1) - \ln(x + 1) \implies f'(x) = \dfrac{e^x}{e^x + 1} - \dfrac{1}{x + 1}$.

15. $f(x) = (2 - e^{x^2})^3 \implies f'(x) = 3(2 - e^{x^2})^2(-2xe^{x^2}) = -6xe^{x^2}(2 - e^{x^2})^2$.

17. $y = (1/2)\,(e^x + e^{-x}) \implies y' = (1/2)(e^x - e^{-x})$.

19. $y = e^{\sqrt{x}} \ln \sqrt{x} = (1/2)\,e^{\sqrt{x}} \ln x \implies$

$$y' = \frac{1}{2}e^{\sqrt{x}}\frac{1}{x} + \frac{1}{2}e^{\sqrt{x}}\frac{1}{2\sqrt{x}}\ln x = \frac{e^{\sqrt{x}}(2 + \sqrt{x}\ln x)}{4x}.$$

21. $e^{xy} = x \implies e^{xy}(xy' + y) = 1 \implies xy' + y = e^{-xy} \implies y' = \dfrac{e^{-xy} - y}{x} = \dfrac{1/x - y}{x} = \dfrac{1 - xy}{x^2}$.

23. $\ln(x + 2y) = e^y \implies \dfrac{1}{x + 2y}\,(1 + 2y') = e^y y' \implies y'(e^y(x + 2y) - 2) = 1 \implies y' = \dfrac{1}{e^y(x + 2y) - 2}$.

25. True; if $y = e^{kx}$ for some constant k, then $y' = ke^{kx} = ky$ and hence y' is proportional to y.

27. True; e^x is never zero.

29. Let $u = -x$. Then $du = -dx$ and hence

$$\int e^{-x}\,dx = -\int e^u\,du = -e^u + C = -e^{-x} + C.$$

31. Let $u = x^2 + 3$. Then $du = 2x\,dx$ and hence

$$\int xe^{x^2 + 3}\,dx = \frac{1}{2}\int e^u\,du = \frac{1}{2}e^u + C = \frac{1}{2}e^{x^2 + 3} + C.$$

33. Let $u = 1 + e^{2x}$. Then $du = 2e^{2x}\,dx$ and hence

$$\int e^{2x}(1 + e^{2x})^3\,dx = \frac{1}{2}\int u^3\,du = \left(\frac{1}{2}\right)\left(\frac{1}{4}\right)u^4 + C = \frac{1}{8}(1 + e^{2x})^4 + C.$$

35. Let $u = 1/x$. Then $du = -(1/x^2)dx$ and hence

$$\int \frac{e^{1/x}}{x^2}\,dx = -\int e^u\,du = -e^u + C = -e^{1/x} + C.$$

37. Let $u = 1 + e^x$. Then $du = e^x\,dx$ and hence

$$\int \frac{e^x}{1 + e^x}\,dx = \int \frac{1}{u}\,du = \ln|u| + C = \ln(1 + e^x) + C.$$

Note that the absolute value in the antiderivative is not needed, *i.e.* $\ln|1 + e^x|$, since $1 + e^x$ is always positive.

39. Let $u = 3 + e^x$. Then $du = e^x dx$ and hence

$$\int \frac{e^x}{(3+e^x)^2} \, dx = \int \frac{1}{u^2} \, du = -\frac{1}{u} + C = -\frac{1}{3+e^x} + C.$$

41. Let $u = 1 + e^{\sqrt{x}}$. Then $du = e^{\sqrt{x}}/(2\sqrt{x}) \, dx$ and hence

$$\int \frac{(1+e^{\sqrt{x}})e^{\sqrt{x}}}{\sqrt{x}} \, dx = 2 \int u \, du = u^2 + C = (1 + e^{\sqrt{x}})^2 + C.$$

43. Referring to Exercise 29, we find that

$$\int_1^{\ln 4} e^{-x} \, dx = -e^{-x}\Big|_1^{\ln 4} = -e^{-\ln 4} + e^{-1} = \frac{1}{e} - \frac{1}{4}.$$

45. Referring to Exercise 38, we find that

$$\int_0^{\pi/2} (\cos x)e^{\sin x} \, dx = e^{\sin x}\Big|_0^{\pi/2} = e^1 - e^0 = e - 1.$$

47. Let $u = \ln x$. Then $du = (1/x)dx$ and hence

$$\int_1^{e^2} \frac{\ln x}{x} \, dx = \int_0^2 u \, du = \frac{1}{2}u^2\Big|_0^2 = 2.$$

49. $\int_0^2 e^x \, dx = e^x\big|_0^2 = e^2 - 1.$

51. Let $u = \sin^2 x$. Then $du = 2\sin x \cos x \, dx$ and hence

$$\int_0^{\pi/3} \sin x \cos x \, e^{\sin^2 x} \, dx = \frac{1}{2}\int_0^{3/4} e^u \, du = \frac{1}{2}e^u\Big|_0^{3/4} = \frac{1}{2}(e^{3/4} - 1).$$

53. $y = x^2 e^{1-x^2} \implies y' = x^2 e^{1-x^2}(-2x) + e^{1-x^2}(2x) = 2xe^{1-x^2}(1-x^2) = 0 \implies x = 0, \pm 1$. Thus, the graph of $y = x^2 e^{1-x^2}$ has three critical points: $(0,0)$, $(1,1)$, and $(-1,1)$. Now,

$$y'' = 2xe^{1-x^2}(-2x) + (1-x^2)[2xe^{1-x^2}(-2x) + 2e^{1-x^2}] = e^{1-x^2}(4x^4 - 10x^2 + 2).$$

Thus, $y''(0) = 2e > 0 \implies (0,0)$ is a relative minimum, while $y''(\pm 1) = -4 < 0 \implies (1,1)$ and $(-1,1)$ are relative maximums, as shown above.

55. We find that

$$f'(x) = xe^{2x}(2) + e^{2x} = e^{2x}(2x + 1)$$

and

$$f''(x) = e^{2x}(2) + (2x+1)e^{2x}(2) = e^{2x}(4x + 4).$$

(a) Since $f'(x) > 0 \iff x > -1/2$, $f(x)$ is increasing on the interval $[-1/2, +\infty)$.

(b) Since $f'(x) < 0 \iff x < -1/2$, $f(x)$ is decreasing on the interval $(-\infty, -1/2]$.

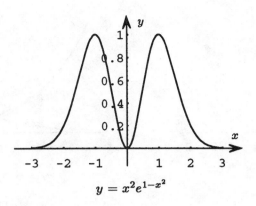

$$y = x^2 e^{1-x^2}$$

Exercise 53

(c) $f'(x) = 0 \iff x = -1/2$. Hence $f(x)$ has a relative extremum at the point $(-1/2, -1/2\, e^{-1})$. Since $f''(-1/2) = e^{-1}(2) > 0$, this is a relative minimum.

(d) Since $f''(x) > 0 \iff x > -1$, the graph is concave up on the interval $[-1, +\infty)$.

(e) Since $f''(x) < 0 \iff x < -1$, the graph is concave down on the interval $(-\infty, -1]$.

(f) Since the concavity changes only at the point $x = -1$, $(-1, -e^{-2})$ is the only inflection point. The graph is shown below.

57. $y = (1/2)(e^x + e^{-x}) \implies y' = (1/2)(e^x - e^{-x})$. Hence the length of graph from $(0,1)$ to $(\ln 2, 5/4)$ is equal to

$$\int_0^{\ln 2} \sqrt{1 + (y')^2}\, dx = \int_0^{\ln 2} \sqrt{1 + \frac{1}{4}(e^x - e^{-x})^2}\, dx = \frac{1}{2}\int_0^{\ln 2} \sqrt{4 + e^{2x} - 2 + e^{-2x}}\, dx$$

$$= \frac{1}{2}\int_0^{\ln 2} \sqrt{e^{2x} + 2 + e^{-2x}}\, dx = \frac{1}{2}\int_0^{\ln 2} (e^x + e^{-x})\, dx = \frac{1}{2}(e^x - e^{-x})\Big|_0^{\ln 2} = \frac{1}{2}(2 - \frac{1}{2}) - \frac{1}{2}(1 - 1) = \frac{3}{4}.$$

59. $\frac{dy}{dx} = 2xy \implies \frac{dy}{y} = 2x\,dx \implies \ln|y| = x^2 + C \implies y = Ce^{x^2}$, where C is any constant.

61. Since $y' = xe^x + e^x = e^x(x+1) = 0 \iff x = -1$, the graph of $y = xe^x$ has only one point at which the tangent line is horizontal, namely $(-1, -1/e)$. Hence $P = (-1, -1/e)$.

63. First observe in the figure on the following page that the region is symmetric with respect to the y-axis. Hence the area of the region is

$$2\int_0^1 (e^x - e^{-x})\, dx = 2(e^x + e^{-x})\big|_0^1 = 2(e + e^{-1}) - 2(1 + 1) = 2(e - 2) + 2/e.$$

65. $y = Ae^{kt} \implies y' = Ake^{kt} = ky$ and $y(0) = Ae^0 = A$.

(a) $y' = y \implies k = 1$. Since $y(0) = 1 = A$, $y = e^t$.

(b) $y' = \pi y \implies k = \pi$. Since $y(0) = -2 = A$, $y = -2e^{\pi t}$.

(c) $y' = -3y \implies k = -3$. Since $y(0) = 2 = A$, $y = 2e^{-3t}$.

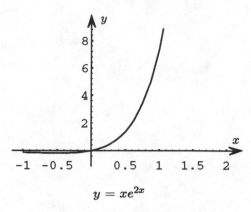

$$y = xe^{2x}$$

Exercise 55

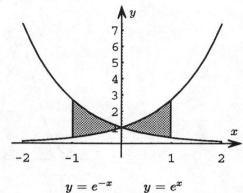

$$y = e^{-x} \qquad y = e^x$$

Exercise 63

67. $f(-x) = (e^{-x} + e^x)/2 = f(x) \implies f(x)$ is an even function.
$g(-x) = (e^{-x} - e^x)/2 = -g(x) \implies g(x)$ is an odd function.

69. $I = I_0 e^{-kl} \implies dI/dl = I_0 e^{-kl}(-k) = -kI$. Hence, if $dI/dl = -kI = 4I$, then $k = -4$.

71. Since $-b_1(t - c_1) < 0$ and $-b_2(t - c_2) < 0$ as $t \to \infty$, $e^{-b_1(t-c_1)} \to 0$ and $e^{-b_2(t-c_2)} \to 0$. Therefore

$$\lim_{t \to \infty} y(t) = a \lim_{t \to \infty} \frac{1}{1 + e^{-b_1(t-c_1)}} + (f - a) \lim_{t \to \infty} \frac{1}{1 + e^{-b_2(t-c_2)}} = a(1) + (f - a)(1) = f.$$

73. Let $f(x) = e^{2x} - 8x + 1$. The graph of $y = f(x)$, drawn by computer and shown below on the right, indicates that $f(x)$ has two zeros, one between 0 and 0.5, the other between 0.5 and 1. Using $r_0 = 0.4$ and $r_0 = 0.9$ as initial values in Newton's Method and setting

$$r_{n+1} = r_n - \frac{f(r_n)}{f'(r_n)},$$

the table below on the left summarizes the successive values of r_n and shows that the roots, to within four decimal places, are 0.4073 and 0.9333.

$r_{n+1} = r_n - f(r_n)/f'(r_n)$	
0.4	0.9
0.407197	0.936678
0.407263	0.933355
0.407263	0.933326
0.407263	0.933326
0.407263	0.933326

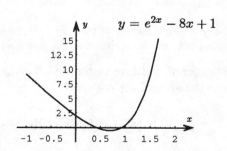

$$y = e^{2x} - 8x + 1$$

75. The values of the polynomials are most easily calculated using a spreadsheet. For example, let the first column contain the values of x. In the second column, cell B1, enter the formula EXP(A1), which means to raise e to the number in cell A1; in the third column, cell C1, enter the formula A1 + 1, which adds one to the number in cell A1 and gives $P_1(x)$; the fourth column the formula

```
C1 + (A1)^2/FACT(2),
```

322

which gives the value of the polynomial $P_2(x)$ when x equals the value in cell A1. Similarly, in the fifth and sixth columns enter the formulas

```
D1 + (A1)^3/FACT(3)
E1 + (A1)^4/FACT(4)
```

which give $P_3(x)$ and $P_4(x)$. Now select cells B1 through D6 and do a FILL DOWN. The table below is the result of these calculations.

x	e^x	$P_1(x)$	$P_2(x)$	$P_3(x)$	$P_4(x)$
1	2.71828183	2	2.5	2.66666667	2.70833333
0.5	1.64872127	1.5	1.625	1.64583333	1.6484375
2	7.3890561	3	5	6.33333333	7
−2	0.13533528	−1	1	−0.3333333	0.33333333

77. Let $y_1 = e^{x_1}$ and $y_2 = e^{x_2}$. Then $x_1 = \ln y_1$, $x_2 = \ln y_2$ and hence $x_1 - x_2 = \ln y_1 - \ln y_2 = \ln(y_1/y_2)$. Therefore

$$e^{x_1-x_2} = e^{\ln(y_1/y_2)} = \frac{y_1}{y_2} = \frac{e^{x_1}}{e^{x_2}}.$$

8.4 Exponentials and Logs to other Bases

1. (a) $2x = e^{\ln 2x}$

   (b) $\pi^3 = e^{\ln \pi^3} = e^{3\ln \pi}$

   (c) $7^{\ln x} = e^{\ln 7^{\ln x}} = e^{(\ln x)(\ln 7)}$

   (d) $4^{\sqrt{2}} = e^{\ln 4^{\sqrt{2}}} = e^{\sqrt{2}\ln 4}$

   (e) $3^{\sin x} = e^{\ln 3^{\sin x}} = e^{(\sin x)\ln 3}$

   (f) $2^x 4^{1-x} = 2^x 2^{2-2x} = 2^{2-x} = e^{\ln 2^{2-x}} = e^{(2-x)\ln 2}$

3. For any real number x,

$$10^x = e^{\ln 10^x} = e^{x \ln 10}$$

and

$$e^x = 10^{\log_{10} e^x} = 10^{x \log_{10} e} = 10^{x/\ln 10}.$$

Note that $\log_{10} e = \ln e / \ln 10 = 1/\ln 10$.

5. $y = 3^x \implies y' = 3^x \ln 3$.

7. $f(x) = \log_{10}(2x - 1) = \frac{\ln(2x-1)}{\ln 10} \implies f'(x) = \frac{2}{(\ln 10)(2x-1)}$.

9. $y = \log_{10}(\ln x) = \frac{\ln(\ln x)}{\ln 10} \implies y' = \frac{1}{x(\ln 10)(\ln x)}$.

11. $g(t) = t^2 \log_2 t = t^2 \frac{\ln t}{\ln 2} \implies$

$$g'(t) = \frac{1}{\ln 2}\left(t^2\left(\frac{1}{t}\right) + (\ln t)(2t)\right) = \frac{t}{\ln 2}(1 + 2\ln t).$$

13. $f(x) = \pi^x + x^\pi \implies f'(x) = \pi^x \ln \pi + \pi x^{\pi - 1}$.

15. $y = x^x = e^{x \ln x} \implies y' = e^{x \ln x}(x\frac{1}{x} + \ln x) = x^x(1 + \ln x)$.

17. $y = x^{\sqrt{x}} = e^{\sqrt{x} \ln x} \implies$

$$y' = e^{\sqrt{x} \ln x}\left(\sqrt{x}\,\frac{1}{x} + (\ln x)\frac{1}{2\sqrt{x}}\right) = \frac{x^{\sqrt{x}}}{\sqrt{x}}\left(1 + \frac{1}{2}\ln x\right).$$

19. $y = (\cos x)^{\sin x} = e^{(\sin x) \ln \cos x} \implies$

$$y' = e^{(\sin x) \ln \cos x}\left(\sin x \frac{1}{\cos x}(-\sin x) + (\ln \cos x)(\cos x)\right) = (\cos x)^{\sin x}\left(\frac{-\sin^2 x}{\cos x} + (\ln \cos x)(\cos x)\right).$$

21. $\int 5^x\,dx = \int e^{x \ln 5}\,dx = \frac{1}{\ln 5}\,5^x + C$.

23. Let $u = \sqrt{x}$. Then $du = \frac{1}{2\sqrt{x}}dx$ and hence

$$\int \frac{\pi^{\sqrt{x}}}{\sqrt{x}}\,dx = 2\int \pi^u\,du = \frac{2}{\ln \pi}\,\pi^u + C = \frac{2}{\ln \pi}\,\pi^{\sqrt{x}} + C.$$

25. Since $a^{2 \ln x} = (e^{\ln a})^{2 \ln x} = x^{2 \ln a}$,

$$\int a^{2 \ln x}\,dx = \int x^{2 \ln a}\,dx = \frac{1}{1 + 2 \ln a}\,x^{1 + 2 \ln a} + C.$$

27. Let $u = x^2$. Then $du = 2x\,dx$ and hence

$$\int_0^2 x 3^{x^2}\,dx = \frac{1}{2}\int_0^4 3^u\,du = \frac{1}{2 \ln 3}\,3^u\Big|_0^4 = \frac{1}{2 \ln 3}(3^4 - 1) = \frac{40}{\ln 3}.$$

29. Let $u = e^{2x}$. Then $du = 2e^{2x}\,dx$ and hence

$$\int_0^{\ln 2} 3^{e^{2x}} e^{2x}\,dx = \frac{1}{2}\int_1^4 3^u\,du = \frac{1}{2}\frac{1}{\ln 3}3^u\Big|_1^4 = \frac{39}{\ln 3}.$$

31. True.

33. For all real numbers x and y, $a^x/a^y = e^{x \ln a}/e^{y \ln a} = e^{(x-y) \ln a} = a^{x-y}$.

35. $y = a^x \implies y' = a^x \ln a$. Hence $y' - 2y = 0 \implies a^x \ln a - 2a^x = a^x(\ln a - 2) = 0 \implies \ln a = 2 \implies a = e^2$.

37. Let dV be the volume of a differential cross-section disk. Then

$$dV = \pi r^2 dx = \pi y^2 dx = \pi 3^{2x}\,dx.$$

Hence the volume of revolution is equal to

$$V = \int_0^1 \pi 3^{2x}\,dx = \pi \int_0^1 e^{2x \ln 3}\,dx$$

$$= \frac{\pi}{2 \ln 3}3^{2x}\Big|_0^1 = \frac{\pi}{2 \ln 3}(3^2 - 3^0) = \frac{4\pi}{\ln 3}.$$

39. $y = \log_{10} x = (\ln x)/(\ln 10) \implies y' = 1/(x \ln 10) \implies y'(1) = 1/\ln 10$. Hence the equation of the tangent line to the graph at $x = 1$ is $y = (1/\ln 10)(x - 1)$.

41. $V(t) = (5000)2^{\sqrt{t}} = 5000e^{\sqrt{t} \ln 2} \implies V'(t) = 5000e^{\sqrt{t} \ln 2}\left(\frac{1}{2\sqrt{t}} \ln 2\right) = (5000)2^{\sqrt{t}}\left(\frac{\ln 2}{2\sqrt{t}}\right)$. Therefore

$$V'(4) = (5000)(2^2)\left(\frac{\ln 2}{4}\right) \approx 3465.74.$$

Hence after 4 years the value of the painting is increasing at the rate of $3,465.74 per year.

43. $y = \sqrt[3]{\dfrac{x+2}{x+3}}$

$$\implies \ln y = \frac{1}{3}(\ln(x+2) - \ln(x+3))$$
$$\implies \frac{1}{y}y' = \frac{1}{3}\left(\frac{1}{x+2} - \frac{1}{x+3}\right)$$
$$\implies y' = \frac{1}{3}y\left(\frac{1}{x+2} - \frac{1}{x+3}\right).$$

45. $y = \dfrac{(x^2+2)^3(x-1)^5}{x\sqrt{x+1}\sqrt{x+2}}$

$$\implies \ln y = 3\ln(x^2+2) + 5\ln(x-1) - \ln x - \frac{1}{2}\ln(x+1) - \frac{1}{2}\ln(x+2)$$
$$\implies \frac{1}{y}y' = \frac{3(2x)}{x^2+2} + \frac{5}{x-1} - \frac{1}{x} - \frac{1}{2}\frac{1}{x+1} - \frac{1}{2}\frac{1}{x+3}$$
$$\implies y' = y\left(\frac{6x}{x^2+2} + \frac{5}{x-1} - \frac{1}{x} - \frac{1}{2(x+1)} - \frac{1}{2(x+2)}\right).$$

47. $y = (x^2+1)^x$

$$\implies \ln y = x\ln(x^2+1)$$
$$\implies \frac{1}{y}y' = x\frac{2x}{x^2+1} + \ln(x^2+1)$$
$$\implies y' = y\left(\frac{2x^2}{x^2+1} + \ln(x^2+1)\right).$$

49. (a) Since $x \to \infty \iff h \to 0^+$,

$$y = \lim_{x\to\infty}\left(1 + \frac{r}{x}\right)^x = \lim_{h\to 0^+}(1+rh)^{1/h}.$$

(b) $f(x) = \ln(1+rx) \implies$

$$\begin{aligned}
f'(0) &= \lim_{h\to 0^+}\frac{f(h) - f(0)}{h}\\
&= \lim_{h\to 0^+}\frac{\ln(1+rh)}{h}\\
&= \lim_{h\to 0^+}\ln(1+rh)^{1/h}\\
&= \ln\lim_{h\to 0^+}(1+rh)^{1/h} \quad \text{since ln is continuous}\\
&= \ln y.
\end{aligned}$$

(c) Since $f'(x) = r/(1+rx)$, $f'(0) = r$. Hence, by part (b), $\ln y = f'(0) = r$. Therefore $y = e^r$; *i.e.*,

$$\lim_{x \to \infty} \left(1 + \frac{r}{x}\right)^x = e^r.$$

51. (a) The intersection points are summarized in the table below on the left:

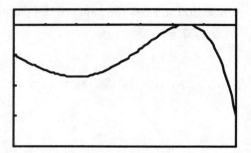

b	Intersection Points
1.5	1.5, 7.408765
2	2,4
2.5	2.5, 2.970287
3	2.478053, 3
4	2, 4
5	1.764922, 5

Notice that for each value of b there are 2 points of intersection. Also notice that the points get close between 2.5 and 3 and as b gets larger one of the points moves towards zero.

(b) Find b so that $f(x) = g(x)$ and $f'(x) = g'(x)$, i.e.,

$$x^b - b^x = 0, \text{ and } bx^{x-1} - b^x \ln b = 0.$$

Note that in part a the points are closer if b is taken between 2.5 and 3. A good value between 2.5 and 3 to check is $x = e$. If x is set equal to e we have $x^e - e^x = 0$ if and only if $x = e$ and $ex^{e-1} - e^x = x^e - e^x = 0$. So if $b = e$ then $f(x) = g(x)$ and $f'(x) = g'(x)$. Below is the graph of $F(x) = x^e - e^x$ using the range $[0, 3.5] \times [-4, .5]$.

8.5 Exponential Growth and Decay

1. $y = e^{2t}$.

3. $y = y(0)e^{5t} \implies y(1) = y(0)e^5 = 1 \implies y(0) = e^{-5}$. Hence $y = e^{-5}e^{5t} = e^{5t-5}$.

5. $y = e^{kt} \implies y(1) = e^k = e^{-2} \implies k = -2$. Hence $y = e^{-2t}$.

7. False. To see this, let y_0 be the amount of isotope present at some time $t = 0$. Then $y = y_0 e^{kt}$ and hence, if $t_{1/2}$ is the half-life of the isotope,

$$y(t_{1/2}) = y_0 e^{kt_{1/2}} = \frac{1}{2}y_0.$$

Therefore $e^{kt_{1/2}} = 1/2$. The solution of this equation for the half-life is $t_{1/2} = -(1/k)\ln 2$, which is independent of y_0. Thus the half-life is independent of the amount present.

9. (a) $v' = (-c/m)v \implies v(t) = v_0 e^{-(c/m)t}$.

(b) $v(t) = 10e^{-(40/10)t} = 10e^{-4t}$.

(c) yes; $\lim_{t \to \infty} v(t) = \lim_{t \to \infty} 10e^{-4t} = 0$.

11. Let $y(t)$ stand for the number of bacteria after t hours. Then $y(t) = 50e^{kt}$. Since $y(12) = 400$,

$$50e^{12k} = 400 \implies 12k = \ln 8 \implies k = \frac{1}{12}\ln 8.$$

Hence $y(t) = 50e^{(t/12)\ln 8}$.

(a) $y(t) = 2(50) \implies 50e^{(t/12)\ln 8} = 2(50) \implies (t/12)\ln 8 = \ln 2 \implies t = 12\frac{\ln 2}{\ln 8} = 4$ hours.

(b) $y(16) = 50e^{(16/12)\ln 8} = 800$ bacteria.

13. Let $R(t)$ be the present value of the revenue received from selling the land t years from now. Then

$$R(t) = V(t)e^{-0.1t} = 10000e^{\sqrt{t}\ln 1.2 - 0.1t}.$$

To maximize this, we set $R'(t) = 0$ and solve for t:

$$R'(t) = 10000e^{\sqrt{t}\ln 1.2 - 0.1t}\left(\frac{\ln 1.2}{2\sqrt{t}} - 0.1\right) = 0$$

$$\implies \frac{\ln 1.2}{2\sqrt{t}} - 0.1 = 0 \implies \sqrt{t} = \frac{\ln 1.2}{2(0.1)} \approx 0.9116 \implies t \approx 0.831 \text{ years} \approx 10 \text{ months.}$$

The original price of the land, \$10,000, does not affect this calculation.

15. Let $y(t)$ be the amount of the substance present after t years. Then $y(t) = y_0 e^{kt}$. If 30% of the substance is excreted after 12 hours, then 70% remains and hence $y(12) = 0.7y_0$. Therefore

$$y_0 e^{12k} = 0.7y_0 \implies k = \frac{1}{12}\ln 0.7.$$

Hence $y(t) = y_0 e^{(t/12)\ln 0.7}$. To find the half-life $t_{1/2}$ of the substance, set $y(t_{1/2}) = \frac{1}{2}y_0$ and solve for $t_{1/2}$:

$$y_0 e^{(t_{1/2}/12)\ln 0.7} = \frac{1}{2}y_0 \implies \frac{t_{1/2}}{12}\ln 0.7 = \ln\frac{1}{2} \implies t_{1/2} = 12\frac{\ln 0.5}{\ln 0.7} \approx 23.32 \text{ hours.}$$

17. Let $y(t)$ be the amount of the iodine isotope present in the body after t days. Then $y(t) = y_0 e^{kt}$. If the half-life is 8 days, then $y(8) = (1/2)y_0$ and hence

$$y_0 e^{8k} = \frac{1}{2}y_0 \implies 8k = \ln 0.5 \implies k = \frac{\ln 0.5}{8}.$$

Therefore $y(t) = y_0 e^{(t/8)\ln 0.5}$. Thus, if $y_0 = 50$, the amount remaining after 3 weeks (21 days) is $y(21) = 50e^{(21/8)\ln 0.5} \approx 8.11$ micrograms.

19. $T(t) = T_e + (T_0 - T_e)e^{-kt} \implies dT/dt = (T_0 - T_e)e^{-kt}(-k) = -k(T - T_e)$ and $T(0) = T_e + (T_0 - T_e)e^0 = T_0$. As $t \to \infty$,

$$\lim_{t\to\infty} T(t) = \lim_{t\to\infty}\left[T_e + (T_0 - T_e)e^{-kt}\right] = T_e.$$

21. $T_e = 20$, $T_0 = 100$, $t_1 = 10$, $T(10) = 50 \implies$

$$k = -\frac{1}{10}\ln\frac{50 - 20}{100 - 20} = -\frac{1}{10}\ln\frac{3}{8} \implies T(t) = 20 + 80e^{(t/10)(\ln 3/8)}.$$

To find the time to cool to $30°C$, set $T(t) = 30$ and solve for t:

$$20 + 80e^{(t/10)(\ln 3/8)} = 30 \implies \frac{t}{10}\ln\frac{3}{8} = \ln\frac{1}{8} \implies t = 10\frac{\ln 1/8}{\ln 3/8} \approx 21.2 \text{ minutes.}$$

23. $T_e = 0$, $T_0 = 100$, $T(5) = 50 \implies k = -(1/5)\ln 0.5 \implies T(t) = 100e^{(t/5)\ln 0.5}$. To find the time to cool to $30°C$, set $T(t) = 30$ and solve for t:

$$100e^{(t/5)\ln 0.5} = 30 \implies \frac{t}{5}\ln 0.5 = \ln\frac{3}{10} \implies t = 5\frac{\ln 0.3}{\ln 0.5} \approx 8.68 \text{ minutes.}$$

After 1 hour, the temperature will be $T(60) = 100e^{(60/5)\ln 0.5} \approx 0.024°C$.

25. Let $R(t)$ be the present value of the revenue received from selling the wine t years from now. Then

$$R(t) = V(t)e^{-0.125t} = 100(1.5)^{\sqrt{t}}e^{-0.125t} = 100e^{\sqrt{t}\ln 1.5 - 0.125t}.$$

To maximize revenue, set $R'(t) = 0$ and solve for t:

$$R'(t) = 100e^{\sqrt{t}\ln 1.5 - 0.125t}\left(\frac{\ln 1.5}{2\sqrt{t}} - 0.125\right) = 0$$

$$\implies \frac{\ln 1.5}{2\sqrt{t}} - 0.125 = 0 \implies \sqrt{t} = \frac{\ln 1.5}{2(0.125)} \approx 1.622 \implies t \approx 2.63 \text{ years.}$$

27. $P(1) = P_0 e^{r(1)} = 0.08P_0 \implies r = \ln 1.08 \approx .077$, or 7.7% compounded continuously.

29. (a) $P(1) = P_0\left(1 + \frac{0.1}{2}\right)^2 = P_0(1 + r) \implies r = 0.1025$, or 10.25%.

   (b) $P(1) = P_0\left(1 + \frac{0.1}{4}\right)^4 = P_0(1 + r) \implies r = 0.1038$, or 10.38%.

   (c) $P(1) = P_0\left(1 + \frac{0.1}{365}\right)^{365} = P_0(1 + r) \implies r = 0.105156$, or 10.5156%.

   (d) $P(1) = P_0 e^{0.1} = P_0(1 + r) \implies r = e^{0.1} - 1 = 0.10517$ or 10.517%.

31. $P(1) = P_0 e^r = P_0(1 + i) \implies i = e^r - 1$.

33. False. The equation $P'(t) = 0.05P(t)$ means that at each moment the growth rate is 5%, which is false. The 5% growth rate refers to the effective yearly growth rate. To find the continuous growth rate r which yields an annual 5% growth, set $P(t) = P_0 e^{rt}$. Then $P(1) = P_0 e^r = (1 + 0.05)P_0 \implies r = \ln 1.05 \approx 0.0488$. Hence $P(t) = P_0 e^{rt\ln 1.05} = P_0(1.05)^r$ and therefore the correct equation is $P'(t) = P_0 e^{r\ln 1.05}(\ln 1.05) = (\ln 1.05)P(t)$. In particular, a continuous growth rate of 4.88% is equivalent to a yearly growth rate of 5%.

35. (a) If $y = f(t)$ is a differentiable solution of the equation $y' = ky$, then $f'(t) = kf(t)$ for all t.

   (b) $g'(t) = f(t)(-ke^{-kt}) + e^{-kt}f'(t) = -kye^{-kt} + (ky)e^{-kt} = 0$.

   (c) Since $g'(t) = 0$ for all t, $g(t) = C$ for some constant C. Hence $f(t) = Ce^{kt}$.

37. (a) Graph of the pairs $(\ln a, \ln A)$ using the range $[-1, 2] \times [0, 4]$.

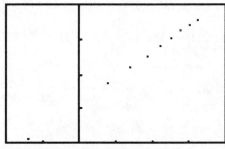

The graph of the pairs $(\ln a, \ln A)$ appears to be a straight line, therefore $\ln A = m\ln a + b$. Exponentiating gives $A = e^{m\ln a + b} = e^b e^{\ln a^m} = ca^m$ or $A = ca^r$ where $c = e^b$ and $m = r$.

(b)   Slope: $m = \frac{\ln 35.1241 - \ln 1.1107}{\ln 5 - \ln 0.5} = 1.500001$

Intercept: $b = \ln 1.1107 - 1.500001 \ln 0.5 = 1.44711911$.

Hence $A = e^b a^{1.500001}$ or $A = 3.1415 a^{1.5}$.

39.   (a) If $N(t) = N_0 e^{kt}$ then $N(0) = 75.995 = N_0 e^{k(0)}$. So $N_0 = 75.995$ and so $N(t) = 75.995 e^{kt}$. Since $N(8) = 226.505$, then choose k so that $226.505 = 75.995 e^{8k}$. So $e^{8k} = 2.980525$ and thus $k = \frac{1}{8} \ln(2.980525) = 0.1365124$. Therefore

$$N(t) = 75.995 e^{0.1365124t}.$$

Graph this function and the values in the table using the range $[0, 10] \times [75, 350]$.

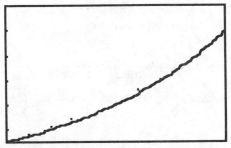

For the year 1990 this model predicts the population to be $N(9) = 259.663$ and for the year 2000, $N(10) = 297.612$. These appear to be fairly resonable estimates.

(b) If $N(0) = 5.308$ then $N_0 = 5.308$ and so $N(t) = 5.308 e^{kt}$. At $t = 11$ (the year 1900) we choose k so than $75.995 = 5.308 e^{11k}$. Thus $k = \frac{1}{11} \ln(75.995/5.308) = 0.2419502$. Therefore $N(t) = 5.308 e^{0.2419502t}$. The graph of this function along with the values in the table using the range $[11, 21] \times [75, 35]$ are shown below on the left.

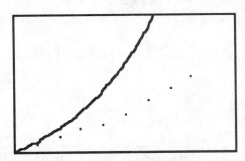 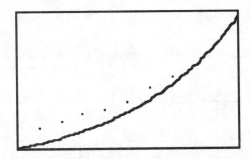

It appears that this model is worse than the one in part a. If instead we choose k so that at $t = 19$ (the year 1980) $5.308 e^{19k} = 226.505$, then

$$k = \frac{1}{19} \ln(226.505/5.308) = 0.1975554.$$

The graph of $N(t) = 5.308 e^{0.1975554t}$ using the range $[11, 21] \times [45, 350]$ is shown above on the right. This model doesn't appear to be a very good predictor either, except for the later part of this century.

329

Review Exercises – Chapter 8

1. (a) $8^{2/3} = 4$

    (b) $36^{-5/2} = 6^{-5} = 1.286 \times 10^{-4}$

    (c) $16^{5/4} = 32$

    (d) $4^{-3/4} = 2^{-3/2} = 1/(2\sqrt{2})$

3. (a) $x = y + 2 \implies y = f^{-1}(x) = x - 2$

    (b) $x = y^2 \implies y = g^{-1}(x) = \sqrt{x}$

    (c) $x = 2y + 1 \implies y = (1/2)x - 1/2$

    (d) $x = \sqrt{y} \implies y = x^2,\ x \geq 0$

5. $y = x^2 \ln(x - a) \implies y' = x^2 \dfrac{1}{x - a} + 2x \ln(x - a)$

7. $f(x) = \ln(\ln^2 x) = 2\ln(\ln x) \implies f'(x) = 2\dfrac{1}{\ln x}\dfrac{1}{x} = \dfrac{2}{x \ln x}$

9. $y = e^{x^2} \tan 2x \implies y' = e^{x^2}(\sec 2x)(2) + (\tan 2x)e^{x^2}(2x) = 2e^{x^2}[\sec^2 2x + x \tan 2x]$

11. $f(t) = \dfrac{te^t}{1 + e^t} \implies$

$$f'(t) = \frac{(1 + e^t)(te^t + e^t) - te^t(e^t)}{(1 + e^t)^2} = \frac{te^t + te^{2t} + e^t + e^{2t} - te^{2t}}{(1 + e^t)^2} = \frac{e^t(t + 1 + e^t)}{(1 + e^t)^2}$$

13. $y = \dfrac{\ln(t - a)}{\ln(t^2 + b)} \implies$

$$y' = \frac{[\ln(t^2 + b)]\left(\frac{1}{t-a}\right) - [\ln(t - a)]\left(\frac{1}{t^2+b}\right)(2t)}{[\ln(t^2 + b)]^2}$$

15. $f(t) = e^{\sqrt{t} - \ln t} \implies f'(t) = e^{\sqrt{t} - \ln t}\left(\dfrac{1}{2\sqrt{t}} - \dfrac{1}{t}\right) = e^{\sqrt{t} - \ln t}\left(\dfrac{\sqrt{t} - 2}{2t}\right) = \left(\dfrac{\sqrt{t} - 2}{2t^2}\right)e^{\sqrt{t}}$

17. $y = \dfrac{1}{3x} + \dfrac{1}{4}\ln\left(\dfrac{1 - 2x}{\sqrt{x}}\right) = \dfrac{1}{3}x^{-1} + \dfrac{1}{4}\ln(1 - 2x) - \dfrac{1}{8}\ln x$

$$\implies \quad y' = -\frac{1}{3}x^{-2} + \frac{1}{4}\frac{1}{1 - 2x}(-2) - \frac{1}{8x} = -\frac{1}{3x^2} - \frac{1}{2(1 - 2x)} - \frac{1}{8x} = \frac{-6x^2 + 13x - 8}{24x^2(1 - 2x)}$$

19. $x \ln y^2 + y \ln x = 1 \implies 2x \ln y + y \ln x = 1$

$$\implies \quad (2x)\left(\frac{1}{y}\,y'\right) + (\ln y)(2) + y\left(\frac{1}{x}\right) + (\ln x)y' = 0$$

$$\implies \quad y'\left(\frac{2x}{y} + \ln x\right) = -2\ln y - \frac{y}{x}$$

$$\implies \quad y' = \frac{-2\ln y - \frac{y}{x}}{\frac{2x}{y} + \ln x} = \frac{-2xy \ln y - y^2}{2x^2 + xy \ln x}$$

21. $x^2 + e^{xy} - y^2 = 2 \implies 2x + e^{xy}(xy' + y) - 2yy' = 0 \implies y'(xe^{xy} - 2y) = -2x - ye^{xy} \implies y' = \dfrac{2x + ye^{xy}}{2y - xe^{xy}}$

23. $e^{xy} = \sqrt{xy}$

$$\implies e^{xy}(xy' + y) = \sqrt{x}\,\frac{1}{2\sqrt{y}}\,y' + \sqrt{y}\,\frac{1}{2\sqrt{x}}$$

$$\implies y'\left[xe^{xy} - \frac{\sqrt{x}}{2\sqrt{y}}\right] = \frac{\sqrt{y}}{2\sqrt{x}} - ye^{xy}$$

$$\implies y' = \frac{\frac{\sqrt{y}}{2\sqrt{x}} - ye^{xy}}{xe^{xy} - \frac{\sqrt{x}}{2\sqrt{y}}} = \frac{y - 2y\sqrt{x}\sqrt{y}e^{xy}}{2x\sqrt{x}\sqrt{y}e^{xy} - x} = \frac{y(1 - 2e^{2xy})}{x(2e^{2xy} - 1)} = -\frac{y}{x}$$

25. $y = \ln\sqrt{e^{2x} + \sin\sqrt{x}} = (1/2)\ln(e^{2x} + \sin\sqrt{x}) \implies$

$$y' = \frac{1}{2}\,\frac{1}{e^{2x} + \sin\sqrt{x}}\left(2e^{2x} + \frac{\cos\sqrt{x}}{2\sqrt{x}}\right)$$

27. Let $u = 1 - x^2$. Then $du = -2x\,dx$ and hence

$$\int \frac{x\,dx}{1 - x^2} = -\frac{1}{2}\int \frac{du}{u} = -\frac{1}{2}\ln|u| + C = -\frac{1}{2}\ln|1 - x^2| + C.$$

29. Let $u = 4x + 2x^2$. Then $du = (4 + 4x)dx = 4(1 + x)dx$ and hence

$$\int \frac{x + 1}{4x + 2x^2}\,dx = \frac{1}{4}\int \frac{du}{u} = \frac{1}{4}\ln|u| + C = \frac{1}{4}\ln|4x + 2x^2| + C.$$

31. $\int \sqrt{e^x}\,dx = \int e^{\frac{1}{2}x}\,dx = 2e^{\frac{1}{2}x} + C = 2\sqrt{e^x} + C.$

33. Let $u = \ln x$. Then $du = (1/x)dx$ and hence

$$\int_e^{e^2} \frac{1}{x\sqrt{\ln x}}\,dx = \int_1^2 u^{-1/2}\,du = 2\sqrt{u}\Big|_1^2 = 2\sqrt{2} - 2.$$

35. $\displaystyle\int_0^1 \frac{x^3 - 1}{x + 1}\,dx = \int_0^1 \left(x^2 - x + 1 - \frac{2}{x + 1}\right)dx = \left[\frac{1}{3}x^3 - \frac{1}{2}x^2 + x - 2\ln(x + 1)\right]\Big|_0^1 = \frac{5}{6} - 2\ln 2.$

37. Let $u = x^2$. Then $du = 2x\,dx$ and hence

$$\int_0^2 x(e^{x^2} + 1)\,dx = \frac{1}{2}\int_0^4 (e^u + 1)\,du = \frac{1}{2}(e^u + u)\Big|_0^4 = \frac{1}{2}e^4 + \frac{3}{2}.$$

39. $\int e^x(1 - e^{2x})^3\,dx = \int e^x(1 - 3e^{2x} + 3e^{4x} - e^{6x})\,dx = \int (e^x - 3e^{3x} + 3e^{5x} - e^{7x})\,dx = e^x - e^{3x} + \frac{3}{5}e^{5x} - \frac{1}{7}e^{7x} + C.$

41. Let $u = x^3 + x^2 - 7$. Then $du = (3x^2 + 2x)dx$ and hence

$$\int \frac{2x + 3x^2}{x^3 + x^2 - 7}\,dx = \int \frac{du}{u} = \ln|u| + C = \ln|x^3 + x^2 - 7| + C.$$

43. Let $u = x^2 + 1$. Then $du = 2x\,dx$ and hence

$$\int_{-3}^{-2} \frac{x}{x^2+1}\,dx = \frac{1}{2}\int_{10}^{5} \frac{du}{u} = \frac{1}{2}\ln u\Big|_{10}^{5} = \frac{1}{2}\ln 5 - \frac{1}{2}\ln 10 = \frac{1}{2}\ln\frac{1}{2} = -\frac{\ln 2}{2}.$$

45. $y = x^{\sqrt{x}} \implies \ln y = \sqrt{x}\ln x \implies \frac{1}{y}y' = \sqrt{x}\left(\frac{1}{x}\right) + (\ln x)\left(\frac{1}{2\sqrt{x}}\right) = \frac{1}{\sqrt{x}} + \frac{\ln x}{2\sqrt{x}} \implies y' = y\left(\frac{1}{\sqrt{x}} + \frac{\ln x}{2\sqrt{x}}\right).$

47. $y = x^{\sin^2 x}$

$$\implies \ln y = (\sin^2 x)(\ln x)$$

$$\implies \frac{1}{y}y' = (\sin^2 x)\left(\frac{1}{x}\right) + (\ln x)2\sin x\cos x$$

$$\implies y' = y\left[\frac{\sin^2 x}{x} + 2(\ln x)(\sin x)(\cos x)\right].$$

49. $y = x^2\ln x \implies y' = x^2\left(\frac{1}{x}\right) + (\ln x)(2x) = x(1 + 2\ln x)$. Hence $y' = 0 \iff x = 0$ or $\ln x = -1/2 \implies x = e^{-1/2}$. Now, $x = 0$ is not in the domain of the function and hence is not a critical number. Since $y'' = x\left(\frac{2}{x}\right) + (1 + 2\ln x) = 3 + 2\ln x$, $y''(e^{-1/2}) = 3 + 2\ln e^{-1/2} = 2 > 0$ and hence $x = e^{-1/2}$ is a local minimum. Thus $y = x^2\ln x$ has only one relative extremum, a local minimum at $(e^{-1/2}, -\frac{1}{2}e^{-1})$. The graph is shown below on the left.

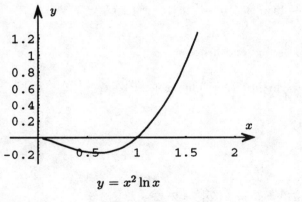

$y = x^2\ln x$

Exercise 49

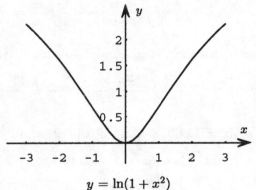

$y = \ln(1 + x^2)$

Exercise 51

51. $y = \ln(1 + x^2) \implies y' = \frac{2x}{1+x^2}$ and $y'' = \frac{2-2x^2}{(1+x^2)^2}$. Hence $y'' = 0 \iff x = \pm 1$. Since $y'' > 0$ when $-1 < x < 1$, the graph is concave up on the interval $[-1, 1]$ and concave down otherwise.

53. The area of T is $A(x) = \frac{1}{2}x\ln x$. Thus the rate of change of the area is equal to

$$\frac{dA}{dt} = \left[\left(\frac{1}{2}x\right)\frac{1}{x} + (\ln x)\frac{1}{2}\right]\frac{dx}{dt} = \frac{1}{2}(1 + \ln x)(4) = 2(1 + \ln x).$$

Hence, when $x = 5$ the area is increasing at a rate of $2(1 + \ln 5)$ square units/sec.

55. Since $f(-x) = -f(x)$, the graph is symmetric about the origin and hence the area is equal to

$$2\int_0^2 \underbrace{\frac{x}{x^2+1}}_{u=x^2+1}\,dx = \int_1^5 \frac{du}{u} = \ln u\big|_1^5 = \ln 5.$$

332

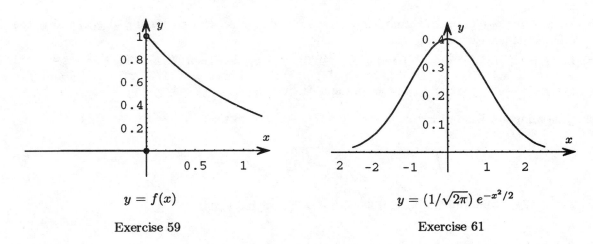

$$y = f(x)$$

Exercise 59

$$y = (1/\sqrt{2\pi})\, e^{-x^2/2}$$

Exercise 61

57. Let dV be the volume of a differential cross-section disk of thickness dx. Then

$$dV = \pi r^2 dx = \pi y^2 dx = \pi \frac{2}{x-1}\, dx$$

and hence the volume of revolution is

$$V = \int_3^5 \frac{2\pi}{x-1}\, dx = 2\pi \ln(x-1)\big|_3^5 = 2\pi \ln 2.$$

59. For $x > 0$, the graph is the decreasing exponential function $y = e^{-x}$, while for $x < 0$ it is identically zero, as illustrated below.

61. We find that

$$f(x) = \frac{1}{\sqrt{2\pi}} e^{-x^2/2}$$

$$f'(x) = \frac{1}{\sqrt{2\pi}} e^{-x^2/2}(-x)$$

$$f''(x) = \frac{-x}{\sqrt{2\pi}} e^{-x^2/2}(-x) + e^{-x^2/2}\left(\frac{-1}{\sqrt{2\pi}}\right) = \frac{e^{-x^2/2}}{\sqrt{2\pi}}(x^2 - 1).$$

Hence $f'(x) = 0 \iff x = 0$. Since $f''(0) = -1/\sqrt{2\pi} < 0$, $x = 0$ is a relative maximum. Moreover, $f''(x) = 0 \iff x = \pm 1$. Since the sign of f'' changes at both ± 1, $x = \pm 1$ are inflection points.

63. $y = x^2 \ln x \implies y' = x^2\left(\frac{1}{x}\right) + 2x \ln x = x + 2x \ln x \implies y'(e) = 3e$. Hence the slope of the line perpendicular to the graph of $y = x^2 \ln x$ at (e, e^2) is $-\frac{1}{3e}$ and therefore its equation is $y - e^2 = -\frac{1}{3e}(x - e)$.

65. $y = x \ln x^2 = 2x \ln x \implies y' = 2x\left(\frac{1}{x}\right) + 2 \ln x = 2 + 2 \ln x \implies y'(e) = 2 + 2 \ln e = 4$. Hence

$$y \approx y(e) + y'(e)(x - e) = e \ln e^2 + 4(x - e) = 4x - 2e \quad \text{for } x \approx e.$$

If $x = 2.7$, $y(2.7) \approx 4(2.7) - 2e \approx 5.363436$. (*Calculator value:* $y(2.7) = 5.36355957$.)

67. (a) $f'(x) = (1/2)(e^x - e^{-x}) = g(x)$, $g'(x) = (1/2)(e^x + e^{-x}) = f(x)$.

(b) Since $f''(x) = g'(x) = f(x)$ and $g''(x) = f'(x) = g(x)$, both $y = f(x)$ and $y = g(x)$ satisfy the differential equation $y'' - y = 0$.

(c) Let $f(x) = (1/2)(e^{kx} + e^{-kx})$ and $g(x) = (1/2)(e^{kx} - e^{-kx})$. Then $f''(x) = (1/2)k^2(e^{kx} + e^{-kx}) = k^2 f(x)$ and $g''(x) = (1/2)k^2(e^{kx} - e^{-kx}) = k^2 g(x)$. Hence $y = (1/2)(e^{kx} + e^{-kx})$ and $y = (1/2)(e^{kx} - e^{-kx})$ are two solutions of the differential equation $y'' - k^2 y = 0$.

69. Using the Midpoint Rule with $f(x) = e^{-x^2}$ and $n = 20$, we find that

$$\int_0^5 e^{-x^2}\, dx \approx \frac{5}{20} \sum_{i=0}^{19} f\left(\frac{5}{2n} + \frac{5i}{n}\right) = 0.886227.$$

71. Set $f(x) = 1/x$. Using the Midpoint Rule with $n = 10$, we find that:

(a) $\ln 4 = \int_1^4 \frac{1}{x}\, dx \approx \frac{3}{10} \sum_{i=1}^{10} f\left(1 + \left(i + \frac{1}{2}\right)\frac{3}{10}\right) = 1.38284$

(b) $\ln 5 = \int_1^5 \frac{1}{x}\, dx \approx \frac{4}{10} \sum_{i=1}^{10} f\left(1 + \left(i + \frac{1}{2}\right)\frac{4}{10}\right) = 1.60321$

(c) $\ln \frac{1}{2} = \int_1^{1/2} \frac{1}{x}\, dx = -\int_{1/2}^1 \frac{1}{x}\, dx \approx -\frac{1}{20} \sum_{i=1}^{10} f\left(1 + \left(i + \frac{1}{2}\right)\frac{1}{20}\right) = -0.692835$

73. $\frac{dy}{dx} = 2y \implies \frac{dy}{y} = 2dx \implies \ln|y| = 2x + x \implies y = Ce^{2x}$. Since $y(0) = C = 1$, $y = e^{2x}$.

75. $2y' + 4y = 0 \implies 2\frac{dy}{dx} = -4y \implies \frac{dy}{y} = -2dx \implies \ln|y| = -2x + C \implies y = Ce^{-2x}$. Since $y(0) = C = \pi$, $y = \pi e^{-2x}$.

77. $4y' - 4y + 4 = 0 \implies \frac{dy}{dx} = y - 1 \implies \frac{dy}{y-1} = dx \implies \ln|y-1| = x + C \implies y - 1 = Ce^x \implies y = 1 + Ce^x$. Since $y(0) = 1 + C = 1$, $C = 0$ and hence $y = 1$.

79. The average value is equal to

$$\frac{1}{\sqrt{\ln 2}} \int_0^{\sqrt{\ln 2}} x e^{x^2 + 1}\, dx.$$

Let $u = x^2 + 1$. Then $du = 2x\,dx$ and hence

$$\frac{1}{\sqrt{\ln 2}} \int_0^{\sqrt{\ln 2}} x e^{x^2 + 1}\, dx = \frac{1}{\sqrt{\ln 2}} \frac{1}{2} \int_1^{1 + \ln 2} e^u\, du = \frac{1}{2\sqrt{\ln 2}} e^u \Big|_1^{1 + \ln 2} = \frac{1}{2\sqrt{\ln 2}}(e^{1 + \ln 2} - e) = \frac{e}{2\sqrt{\ln 2}}.$$

81. The curves $y = 1/x$ and $y = x^2$ intersect when $1/x = x^2 \implies x^3 = 1 \implies x = 1$. Now, on the interval $[1, 3]$, $x^2 > \frac{1}{x}$. Hence the area A of the region is

$$A = \int_1^3 \left(x^2 - \frac{1}{x}\right) dx = \left(\frac{1}{3}x^3 - \ln x\right)\Big|_1^3 = \frac{26}{3} - \ln 3 \approx 7.568.$$

Therefore the centroid $(\overline{x}, \overline{y})$ is given by

$$\overline{x} = \frac{1}{A} \int_1^3 x\left(x^2 - \frac{1}{x}\right) dx = \frac{1}{A} \int_1^3 (x^3 - 1)\, dx = \frac{1}{A}\left(\frac{1}{4}x^4 - x\right)\Big|_1^3 = \frac{18}{A} \approx 2.378$$

and

$$\overline{y} = \frac{1}{2A} \int_1^3 \left[(x^2)^2 - \left(\frac{1}{x}\right)^2 \right] dx = \frac{1}{2A} \int_1^3 (x^4 - x^{-2}) \, dx = \frac{1}{2A} \left(\frac{1}{5}x^5 + \frac{1}{x} \right) \Big|_1^3 = \frac{358}{15A} \approx 3.1536.$$

83. Let $P(t)$ be the population t years after 1980. Then $P(t) = P_0 e^{kt}$, where $P_0 = 203$. Since $P(10) = 227$,

$$203e^{10k} = 227 \quad \Longrightarrow \quad 10k = \ln\frac{227}{203} \quad \Longrightarrow \quad k = \frac{1}{10}\ln\frac{227}{203}.$$

Hence $P(t) = 203e^{(t/10)\ln(227/203)}$. Thus, the population in 2000 will be

$$P(20) = 203e^{\frac{20}{10}\ln\frac{227}{203}} \approx 203(1.25) = 254 \text{ million people,}$$

and in the year 2010 will be

$$P(30) = 203e^{\frac{30}{10}\ln\frac{227}{203}} \approx 203(1.4) = 284 \text{ million people.}$$

85. Let $y(t)$ be the number of fruit flies after t days. Then $y(t) = 100e^{kt}$. Since $y(10) = 500$,

$$100e^{10k} = 500 \quad \Longrightarrow \quad 10k = \ln 5 \quad \Longrightarrow \quad k = \frac{1}{10}\ln 5.$$

Hence $y(t) = 100e^{(t/10)\ln 5}$. Thus, after 4 days the number of fruit flies is equal to

$$y(4) = 100e^{(4/10)\ln 5} \approx 190.$$

87. Let $y(t)$ be the amount of the substance after t hours. Then $y(t) = 100e^{kt}$. Since $y(6) = 40$,

$$100e^{6k} = 40 \quad \Longrightarrow \quad 6k = \ln\frac{4}{10} \quad \Longrightarrow \quad k = \frac{1}{6}\ln\frac{2}{5}.$$

Hence $y(t) = 100e^{(t/6)\ln(2/5)}$. If $t_{1/2}$ is the half-life of the substance, then

$$y(t_{1/2}) = 100e^{(t_{1/2}/6)\ln(2/5)} = 50 \quad \Longrightarrow \quad \frac{t_{1/2}}{6}\ln\frac{2}{5} = \ln\frac{1}{2} \quad \Longrightarrow \quad t_{1/2} = 6\,\frac{\ln 0.5}{\ln 0.4} \approx 4.54 \text{ hours.}$$

Chapter 9

Trigonometric and Inverse Trigonometric Functions

9.1   Integrals of the Trigonometric Functions

1. $\displaystyle\int \cos 3x\ dx = \frac{1}{3}\sin 3x + C$

3. $\displaystyle\int_0^{\pi/8} \sec 2x \tan 2x\ dx = \frac{1}{2}\sec 2x \Big|_0^{\pi/8} = \frac{1}{2}\sqrt{2} - \frac{1}{2}$

5. $\displaystyle\int (\tan^2 x + 1)\ dx = \int \sec^2 x\ dx = \tan x + C$

7. Let $u = x^2$. Then $du = 2x\,dx$ and hence
$$\int \frac{x}{\cos x^2}\ dx = \frac{1}{2}\int \sec u\ du = \frac{1}{2}\ln|\sec u + \tan u| + C = \frac{1}{2}\ln|\sec x^2 + \tan x^2| + C.$$

9. Let $u = 3x^2 - 1$. Then $du = 6x\,dx$ and hence
$$\int x \tan(3x^2 - 1)\ dx = \frac{1}{6}\int \tan u\ du = \frac{1}{6}\ln|\sec u| + C = \frac{1}{6}\ln|\sec(3x^2 - 1)| + C.$$

11. Let $u = \tan x$. Then $du = \sec^2 x\,dx$ and hence
$$\int_{\pi/4}^{\pi/3} \tan^3 x \sec^2 x\ dx = \int_1^{\sqrt{3}} u^3\ du = \frac{1}{4}u^4 \Big|_1^{\sqrt{3}} = 2.$$

13. Let $u = \tan x$. Then $du = \sec^2 x\,dx$ and hence
$$\int_{\pi/6}^{\pi/4} \sec^2 x\, e^{\tan x}\ dx = \int_{1/\sqrt{3}}^1 e^u\ du = e^u \big|_{1/\sqrt{3}}^1 = e - e^{1/\sqrt{3}}.$$

15. Let $u = x + \tan x$. Then $du = (1 + \sec^2 x)dx$ and hence
$$\int \frac{\sec^2 x + 1}{x + \tan x}\ dx = \int \frac{du}{u} = \ln|u| + C = \ln|x + \tan x| + C.$$

17. Let $u = 1 - x^2$. Then $du = -2x dx$ and hence

$$\int x \cot^2(1 - x^2) \, dx = -\frac{1}{2} \int \cot^2 u \, du = -\frac{1}{2} \int (\csc^2 u - 1) \, du$$

$$= -\frac{1}{2}(-\cot u - u) + C = \frac{1}{2} \cot(1 - x^2) + \frac{1}{2}(1 - x^2) + C.$$

19. $\int (\sec x + \tan x)^2 \, dx = \int (\sec^2 x + 2 \sec x \tan x + \underbrace{\tan^2 x}_{\sec^2 x - 1}) \, dx = \int (2 \sec^2 x + 2 \sec x \tan x - 1) \, dx =$
$2 \tan x + 2 \sec x - x + C$

21. Let $u = x^2 - 2x$. Then $du = (2x - 2)dx = 2(x - 1)dx$ and hence

$$\int (1 - x) \sec(x^2 - 2x) \, dx = -\frac{1}{2} \int \sec u \, du$$

$$= -\frac{1}{2} \ln|\sec u + \tan u| + C = -\frac{1}{2} \ln|\sec(x^2 - 2x) + \tan(x^2 - 2x)| + C.$$

23. Let $u = \sin x$. Then $du = \cos x dx$ and hence

$$\int \sqrt{1 - \sin^2 x} e^{\sin x} \, dx = \int |\cos x| e^{\sin x} \, dx = \begin{cases} \int e^u \, du = e^u + C = e^{\sin x} + C, & \cos x \geq 0 \\ -\int e^u \, du = -e^u + C = -e^{\sin x} + C, & \cos x < 0. \end{cases}$$

25. Let $u = \sqrt{x}$. Then $du = \frac{1}{2\sqrt{x}} \, dx$ and hence

$$\int \frac{dx}{\sqrt{x}(1 + \cos \sqrt{x})} = 2 \int \frac{du}{1 + \cos u} \cdot \frac{1 - \cos u}{1 - \cos u} = 2 \int \frac{1 - \cos u}{\sin^2 u} \, du$$

$$= 2 \int (\csc^2 u - \cot u \csc u) \, du = 2(-\cot u + \csc u) + C = 2(-\cot \sqrt{x} + \csc \sqrt{x}) + C.$$

27. Let $u = 1 + \sin 4x$. Then $du = 4 \cos 4x dx$ and hence

$$\int \frac{\cos 4x}{\sqrt{1 + \sin 4x}} \, dx = \frac{1}{4} \int u^{-1/2} \, du = \frac{1}{2} \sqrt{u} + C = \frac{1}{2} \sqrt{1 + \sin 4x} + C.$$

29. Let $u = \tan \theta + 1$. Then $du = \sec^2 \theta \, d\theta$ and hence $\int \frac{\sec^2 \theta \, d\theta}{\sqrt{\tan \theta + 1}} = \int \frac{du}{\sqrt{u}} = 2\sqrt{u} + C = 2\sqrt{\tan \theta + 1} + C.$

31. Area $= \int_0^{\sqrt{\pi/3}} \underbrace{x \sec x^2 \, dx}_{u = x^2} = \frac{1}{2} \int_0^{\pi/3} \sec u \, du = \frac{1}{2} \ln|\sec u + \tan u| \big|_0^{\pi/3} = \frac{1}{2} \ln(2 + \sqrt{3})$

33. Average value on $[\pi/6, \pi/3] =$

$$\frac{1}{\pi/3 - \pi/6} \int_{\pi/6}^{\pi/3} \csc x \, dx = \frac{6}{\pi} \ln|\csc x - \cot x| \Big|_{\pi/6}^{\pi/3} = \frac{6}{\pi} \ln\left|\frac{2}{\sqrt{3}} - \frac{1}{\sqrt{3}}\right| - \frac{6}{\pi} \ln|2 - \sqrt{3}| = -\frac{6}{\pi} \ln(2\sqrt{3} - 3)$$

35. Let $u = \tan x$. Then $\int \tan x \sec^2 x \, dx = \int u \, du = \frac{1}{2} u^2 + C = \frac{1}{2} \tan^2 x + C.$ Alternatively,

$$\int \tan x \sec^2 x \, dx = \int \frac{\sin x}{\cos^3 x} \, dx = -\int \frac{1}{u^3} \, du = \frac{1}{2} \frac{1}{u^2} + C = \frac{1}{2 \cos^2 x} + C = \frac{1}{2} \sec^2 x + C.$$

37. Since $y' = \dfrac{1}{\sin x}(\cos x) = \cot x$,

$$\text{length} = \int_{\pi/6}^{\pi/3} \sqrt{1 + \cot^2 x}\, dx = \int_{\pi/6}^{\pi/3} \csc x\, dx$$

$$= \ln|\csc x - \cot x|\Big|_{\pi/6}^{\pi/3} = \ln\left|\frac{2}{\sqrt{3}} - \frac{1}{\sqrt{3}}\right| - \ln(2 - \sqrt{3}) = -\ln(2\sqrt{3} - 3).$$

39. Let dV be the volume of a differential cross-sectional disk of thickness dx. Then $dV = \pi y^2 dx = \pi \sec^2 x \tan^2 x\, dx$. Since the solid of revolution is symmetric about the y-axis, the total volume of revolution is

$$V = 2 \int_0^{\pi/4} \pi \underbrace{\sec^2 x \tan^2 x}_{u=\tan x}\, dx = 2\pi \int_0^1 u^2\, du = \frac{2\pi}{3} u^3 \Big|_0^1 = \frac{2\pi}{3}.$$

41. The line $y = -2$ intersects the graph of $y = \sec x$ when $\sec x = -2 \Longrightarrow \cos x = -1/2 \Longrightarrow x = 2\pi/3$, $4\pi/3$. Since this region is symmetric about the line $x = \pi$, the area of the region is equal to

$$2 \int_{\pi}^{4\pi/3} (\sec x + 2)\, dx = (2\ln|\sec x + \tan x| + 4x)|_{\pi}^{4\pi/3} = 2\ln(2 - \sqrt{3}) + \frac{4\pi}{3} = \frac{4\pi}{3} - 2\ln(2 + \sqrt{3}).$$

43. Let dV be the differential volume of a shell of thickness dx revolved about the y-axis. Then $dV = 2\pi r h\, dx = 2\pi xy\, dx = 2\pi x \tan 2x^2\, dx$ and hence the volume of revolution is

$$V = 2\pi \int_0^{\sqrt{\pi/8}} \underbrace{x \tan 2x^2}_{u=2x^2}\, dx = \frac{\pi}{2} \int_0^{\pi/4} \tan u\, du = \frac{\pi}{2} \ln|\sec u|\Big|_0^{\pi/4} = \frac{\pi}{2} \ln\sqrt{2}.$$

9.2   Integrals Involving Products of Trigonometric Functions

1. $\displaystyle \int \sin^2 2x\, dx = \frac{1}{2} \int (1 - \cos 4x)\, dx = \frac{1}{2}\left(x - \frac{1}{4}\sin 4x\right) + C$

3. Let $u = \cos x$. Then $du = -\sin x\, dx$ and hence

$$\int_{\pi/4}^{\pi/2} \sin x \cos^2 x\, dx = -\int_{\sqrt{2}/2}^0 u^2\, du = -\frac{1}{3} u^3 \Big|_{\sqrt{2}/2}^0 = \frac{\sqrt{2}}{12}.$$

5. Let $u = \cos x$. Then $du = -\sin x\, dx$ and hence

$$\int \sin^5 x\, dx = \int \sin^4 x (\sin x\, dx) = \int (1 - \cos^2 x)^2 (\sin x\, dx) = -\int (1 - u^2)^2\, du$$

$$= -\int (1 - 2u^2 + u^4)\, du = -\left(u - \frac{2}{3} u^3 + \frac{1}{5} u^5\right) + C = -\cos x + \frac{2}{3}\cos^3 x - \frac{1}{5}\cos^5 x + C.$$

7. Let $u = \sin x$. Then $du = \cos x\, dx$ and hence

$$\int_0^{\pi/2} \sin^2 x \cos^3 x\, dx = \int_0^{\pi/2} \sin^2 x (1 - \sin^2 x)\cos x\, dx = \int_0^1 u^2(1 - u^2)\, du = \left(\frac{1}{3} u^3 - \frac{1}{5} u^5\right)\Big|_0^1 = \frac{2}{15}.$$

9. Let $u = \sec x$. Then $du = \sec x \tan x \, dx$ and hence

$$\int \sec^3 x \tan^3 x \, dx = \int \sec^2 x \underbrace{(\sec^2 x - 1)}_{\tan^2 x}(\sec x \tan x \, dx)$$

$$= \int u^2(u^2 - 1) \, du = \frac{1}{5} u^5 - \frac{1}{3} u^3 + C = \frac{1}{5} \sec^5 x - \frac{1}{3} \sec^3 x + C.$$

11. $\displaystyle\int_0^\pi \sin^6 x \, dx = \int_0^\pi (\sin^2 x)^3 \, dx = \int_0^\pi \left[\frac{1}{2}(1 - \cos 2x)\right]^3 dx$

$$= \frac{1}{8} \int_0^\pi (1 - 3\cos 2x + 3\cos^2 2x - \cos^3 2x) \, dx$$

$$= \frac{1}{8} \int_0^\pi \left(1 - 3\cos 2x + \frac{3}{2}(1 + \cos 4x) - (\cos 2x)(1 - \sin^2 2x)\right) dx$$

$$= \frac{1}{8} \int_0^\pi \left(\frac{5}{2} - 3\cos 2x + \frac{3}{2}\cos 4x\right) dx - \frac{1}{8} \int_0^\pi \underbrace{(1 - \sin^2 2x)\cos 2x}_{u = \sin 2x} \, dx$$

$$= \frac{1}{8} \left[\frac{5}{2} x - \frac{3}{2}\sin 2x + \frac{3}{8}\sin 4x\right]\Big|_0^\pi - \frac{1}{8} \cdot \frac{1}{2} \int_0^0 (1 - u^2) \, du$$

$$= \frac{1}{8} \left(\frac{5}{2} \pi\right)$$

$$= \frac{5\pi}{16}.$$

13. Let $u = \cot x$. Then $du = -\csc^2 x \, dx$ and hence

$$\int \cot^3 x \, dx = \int \cot x \underbrace{(\csc^2 x - 1)}_{\cot^2 x} dx = \int \underbrace{\cot x \csc^2 x}_{u = \cot x} \, dx - \int \cot x \, dx$$

$$= -\int u \, du - \ln|\sin x| + C = -\frac{1}{2} \cot^2 x - \ln|\sin x| + C.$$

15. $\displaystyle\int_0^{\pi/4} \sin x \sin 3x \, dx = \frac{1}{2} \int_0^{\pi/4} (\cos 2x - \cos 4x) \, dx = \frac{1}{2}\left(\frac{1}{2}\sin 2x - \frac{1}{4}\sin 4x\right)\Big|_0^{\pi/4} = \frac{1}{4}$

17. $\displaystyle\int_{\pi/2}^\pi (\sin x + \cos x)^2 \, dx = \int_{\pi/2}^\pi (\sin^2 x + 2\sin x \cos x + \cos^2 x) \, dx = \int_{\pi/2}^\pi (1 + \sin 2x) \, dx$

$$= \left(x - \frac{1}{2}\cos 2x\right)\Big|_{\pi/2}^\pi = \frac{\pi}{2} - 1$$

19. Let $u = \tan x$. Then $du = \sec^2 x \, dx$ and hence

$$\int \tan^4 x \sec^4 x \, dx = \int \tan^4 x \underbrace{(1 + \tan^2 x)}_{\sec^2 x} \sec^2 x \, dx$$

$$= \int u^4(1 + u^2) \, du = \frac{1}{5} u^5 + \frac{1}{7} u^7 + C = \frac{1}{5} \tan^5 x + \frac{1}{7} \tan^7 x + C.$$

21. Let $u = \sec x$. Then $du = \sec x \tan x \, dx$ and hence

$$\int \tan^5 x \sec^3 x \, dx = \int \underbrace{(\sec^2 x - 1)^2}_{\tan^2 x} \sec^2 x (\sec x \tan x) \, dx = \int (u^2 - 1)^2 u^2 \, du$$

$$= \int (u^6 - 2u^4 + u^2) \, du = \frac{1}{7} u^7 - \frac{2}{5} u^5 + \frac{1}{3} u^3 + C = \frac{1}{7} \sec^7 x - \frac{2}{5} \sec^5 x + \frac{1}{3} \sec^3 x + C.$$

23. $\displaystyle \int \sin x \sin 2x \, dx = \frac{1}{2} \int (\cos x - \cos 3x) \, dx = \frac{1}{2} \sin x - \frac{1}{6} \sin 3x + C$

25. Let $u = \sin x$. Then $du = \cos x \, dx$ and hence

$$\int \frac{\cos^3 x}{\sqrt{\sin x}} \, dx = \int \frac{\overbrace{1 - \sin^2 x}^{\cos^2 x}}{\sqrt{\sin x}} \cos x \, dx = \int \frac{1 - u^2}{\sqrt{u}} \, du$$

$$= \int (u^{-1/2} - u^{3/2}) \, du = 2u^{1/2} - \frac{2}{5} u^{5/2} + C = 2\sqrt{\sin x} - \frac{2}{5} \sin^{5/2} x + C.$$

27. Let $u = 1 + \tan x$. Then $du = \sec^2 x \, dx$ and hence

$$\int \frac{\sec^2 x}{(1 + \tan x)^4} \, dx = \int u^{-4} \, du = -\frac{1}{3} \frac{1}{u^3} + C = -\frac{1}{3} \frac{1}{(1 + \tan x)^3} + C.$$

29. $\displaystyle \int \cos 5x \cos 3x \, dx = \frac{1}{2} \int (\cos 8x + \cos 2x) \, dx = \frac{1}{16} \sin 8x + \frac{1}{4} \sin 2x + C$

31. Let $u = \cos \theta$. Then $du = -\sin \theta \, d\theta$ and hence

$$\int \frac{\sin^3 \theta}{\cos \theta} \, d\theta = \int \frac{1 - \cos^2 \theta}{\cos \theta} \sin \theta \, d\theta = -\int \frac{1 - u^2}{u} \, du$$

$$= -\int \left(\frac{1}{u} - u \right) du = -\ln |u| + \frac{1}{2} u^2 + C = -\ln |\cos \theta| + \frac{1}{2} \cos^2 \theta + C.$$

33. Let $u = \tan x$. Then $du = \sec^2 x \, dx$ and hence

$$\int \sec^2 x \sqrt{\tan x} \, dx = \int \sqrt{u} \, du = \frac{2}{3} u^{3/2} + C = \frac{2}{3} \tan^{3/2} x + C.$$

35. Since $\sin ax \cos bx = \frac{1}{2} [\sin(a + b)x + \sin(a - b)x]$,

$$\sin ax \cos bx \cos cx = \frac{1}{2} [\sin(a + b)x \cos cx + \sin(a - b)x \cos cx]$$

$$= \frac{1}{4} [\sin(a + b + c)x + \sin(a + b - c)x + \sin(a - b + c)x - \sin(a - b - c)x].$$

Hence

$$\int \sin ax \cos bx \cos cx \, dx = \frac{1}{4} \int [\sin(a + b + c)x + \sin(a + b - c)x] \, dx$$

$$+ \frac{1}{4} \int [\sin(a - b + c)x + \sin(a - b - c)x] \, dx$$

$$= -\frac{\cos(a + b + c)x}{4(a + b + c)} - \frac{\cos(a + b - c)x}{4(a + b - c)}$$

$$-\frac{\cos(a - b + c)x}{4(a - b + c)} - \frac{\cos(a - b - c)x}{4(a - b - c)} + C.$$

340

37. $\dfrac{1}{\pi}\displaystyle\int_{-\pi}^{\pi}\cos n\theta\,d\theta = \dfrac{1}{n\pi}\sin n\theta\Big|_{-\pi}^{\pi} = 0$

39. Let $u = \dfrac{n\pi x}{L}$. Then $du = \dfrac{n\pi}{L}\,dx$ and hence

$$\int_{-L}^{L}\cos^2\left(\frac{n\pi x}{L}\right)dx = \frac{L}{n\pi}\int_{-n\pi}^{n\pi}\cos^2 u\,du$$

$$= \frac{L}{n\pi}\int_{-n\pi}^{n\pi}\frac{1}{2}(1+\cos 2u)\,du = \frac{L}{2n\pi}\left[u+\frac{1}{2}\sin 2u\right]\Big|_{-n\pi}^{n\pi} = L.$$

41. $\displaystyle\int_0^{2\pi}\sin nx\cos mx\,dx = \frac{1}{2}\int_0^{2\pi}[\sin(n+m)x+\sin(n-m)x]\,dx$

$$= \frac{1}{2}\left[-\frac{1}{n+m}\cos(n+m)x - \frac{1}{n-m}\cos(n-m)x\right]\Big|_0^{2\pi} = 0$$

43. Since $\tan^n x = \tan^{n-2}x\tan^2 x = \tan^{n-2}x(\sec^2 x - 1)$, it follows that

$$\int\tan^n x\,dx = \int\underbrace{\tan^{n-2}x\sec^2 x}_{u=\tan x}\,dx - \int\tan^{n-2}x\,dx$$

$$= \int u^{n-2}\,du - \int\tan^{n-2}x\,dx = \frac{\tan^{n-1}x}{n-1} - \int\tan^{n-2}x\,dx.$$

45. Adding the two identities, we find that

$$\sin(\theta_1+\theta_2)+\sin(\theta_1-\theta_2) = 2\sin\theta_1\cos\theta_2 \implies \sin\theta_1\cos\theta_2 = \frac{1}{2}[\sin(\theta_1+\theta_2)+\sin(\theta_1-\theta_2)].$$

47. Adding the two identities in Exercise 46, we find that

$$\cos(\theta_1-\theta_2)+\cos(\theta_1+\theta_2) = 2\cos\theta_1\cos\theta_2 \implies \cos\theta_1\cos\theta_2 = \frac{1}{2}[\cos(\theta_1+\theta_2)+\cos(\theta_1-\theta_2)].$$

49. $\dfrac{1}{\pi}\displaystyle\int_0^{\pi}\sin^2 x\,dx = \frac{1}{2\pi}\int_0^{\pi}(1-\cos 2x)\,dx = \frac{1}{2\pi}\left(x - \frac{1}{2}\sin 2x\right)\Big|_0^{\pi} = 1/2.$

51. Let dV be the volume of a differential cross-section disk of thickness dx. Then $dV = \pi y^2 dx = \pi\tan^4 x\sec^2 x\,dx$ and hence the volume of revolution is

$$V = \int_0^{\pi/4}\pi\underbrace{\tan^4 x\sec^2 x}_{u=\tan x}\,dx = \pi\int_0^1 u^4\,du = \frac{\pi}{5}u^5\Big|_0^1 = \frac{\pi}{5}.$$

53. Area $= \displaystyle\int_0^{\pi/4}\sec x\,dx = \ln|\sec x+\tan x|\,\Big|_0^{\pi/4} = \ln(1+\sqrt{2})$

55. $s = \displaystyle\int_0^{\pi/6}v(t)\,dt = \int_0^{\pi/6}\sin 3t\cos 2t\,dt = \frac{1}{2}\int_0^{\pi/6}(\sin 5t+\sin t)\,dt = \left(-\frac{1}{10}\cos 5t - \frac{1}{2}\cos t\right)\Big|_0^{\pi/6} = \frac{1}{5}(3-\sqrt{3})$

9.3   The Inverse Trigonometric Functions

1. $\text{Sin}^{-1}(1/2) = \pi/6$

3. $\text{Tan}^{-1}(0) = 0$

5. $\text{Cos}^{-1}(-1) = \pi$

7. $\text{Sin}^{-1}(1) + \text{Cos}^{-1}(1) = \pi/2$

9. $\tan\!\left(\text{Sin}^{-1}(1/2)\right) = \tan(\pi/2) = \sqrt{3}/3$

11. $\sec\!\left(\text{Tan}^{-1}(\sqrt{3})\right) = 2$

13. $\cos\!\left(2\text{Tan}^{-1}(\sqrt{3})\right) = 2\cos^2\!\left(\text{Tan}^{-1}(\sqrt{3})\right) - 1 = -1/2$

15. $\tan^2\!\left(\pi - \text{Sin}^{-1}(\sqrt{2}/2)\right) = \tan^2\left(\pi - \frac{\pi}{4}\right) = 1$

17. $\text{Cos}^{-1}\!\left(2 - \sqrt{2}\sin(\pi/4)\right) = \text{Cos}^{-1}1 = 0$

19. $\tan(\text{Sec}^{-1}2) = \tan\pi/3 = \sqrt{3}$

21. $y = \tan(\text{Sin}^{-1}x) = x/\sqrt{1 - x^2}$

23. $y = \sin\!\left(2\text{Sin}^{-1}(x)\right) = 2\sin(\text{Sin}^{-1}x)\cos(\text{Sin}^{-1}x) = 2x\sqrt{1 - x^2}$

25. $y = \sin(\text{Sec}^{-1}x) = \sqrt{x^2 - 1}/x$

27. $y = \tan(\text{Cot}^{-1}x) = 1/x$

29. $y = \cos(\text{Csc}^{-1}x) = \sqrt{x^2 - 1}/x$

31. No. $\text{Sec}^{-1}(\cos\pi/4) = \text{Sec}^{-1}(\sqrt{2}/2)$, which is undefined since $\sqrt{2}/2 < 1$.

33. $\tan\theta = x/100 \implies \theta = \text{Tan}^{-1}(x/100)$.

35. $\sin\theta = 1/2 \implies \theta = \text{Sin}^{-1}(1/2) = \pi/6$.

37. $\tan\theta = x/20 \implies \theta = \text{Tan}^{-1}(x/20)$.

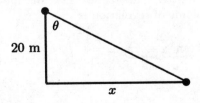

Exercise 37

39. Let θ_1 be the angle between the kite and the horizontal. Then $\sin\theta_1 = 100/(100\sqrt{2}) = \sqrt{2}/2 \implies \theta_1 = \pi/4$. Now, let θ_2 be the new angle between the kite and the horizontal after it has lost half its altitude. Then $\sin\theta_2 = 50/(100\sqrt{2}) = 1/(2\sqrt{2}) \implies \theta_2 = \text{Sin}^{-1}(1/(2\sqrt{2})) \approx 20.705°$. Hence the change in angle is $\theta_1 - \theta_2 = \pi/4 - \text{Sin}^{-1}(1/(2\sqrt{2})) \approx 24.295°$, or 0.428 radians.

41. The graph on the left is $f(x) = \text{Cos}^{-1}(1/x)$ using the range $[1, \pi] \times [0, 2]$. The graph on the right is that of $f(x) = \text{Cos}^{-1}(1/x)$, $g(x) = \sec x$, and $y = x$ using the range $[-1.4, 3.35] \times [-.6, 2.55]$.

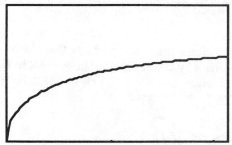 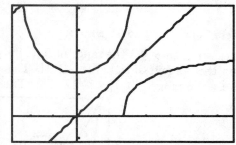

Remark: This range setting was obtained by using the **RANGE** program with **X**=**Y**=1 and **S**=0.05. This is the range setting you would get on the TI-81. The Casio would be similar. For the HP-28S a good range setting is entering **X**=**Y**=1 and **S**=0.075. This will give the range $[-4.1, 6.1] \times [-.125, 2.2]$. For the HP-48SX use **X**=**Y**=1 and **S**=0.05. This will give the range setting $[-2.25, 4.25] \times [-.55, 2.6]$.

From the graph it is seen that if the graph of $f(x)$ is reflected across the line $y = x$, the graph of $g(x)$ is produced for the values of x in $[0, \pi/2]$. To show this set $y = \text{Cos}^{-1}(1/x)$. Then $\cos y = 1/x$. Therefore $\sec y = x$ implying that $y = \text{Sec}^{-1}x$. A similar statement can be made for the functions $f(x) = \text{Sin}^{-1}(1/x)$ and $f(x) = \text{Tan}^{-1}(1/x)$. The function $f(x) = \text{Sin}^{-1}(1/x)$ is the inverse of $\csc x$ and the function $f(x) = \text{Tan}^{-1}(1/x)$ is the inverse of $\cot x$.

43. (a) Since $-1 \le \cos\theta \le 1$ the range of $T_n(x)$ is $[-1, 1]$. Since the domain of $y = \text{Cos}^{-1}x$ is also $[-1, 1]$, then the domain of $T_n(x)$ is $[-1, 1]$. This explains why the range for the graphs in this exercise all use the range $[-1, 1] \times [-1, 1]$. The graph below is that of $T_n(x)$, $n = 1, 2, 3, 4$, and 5.

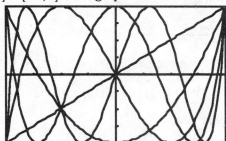

Let $\theta = \text{Cos}^{-1}x$, so $x = \cos\theta$. Then

$$T_2(x) = \cos 2\theta = 2\cos^2\theta - 1 = 2x^2 - 1$$

and

$$
\begin{aligned}
T_3(x) &= \cos 3\theta = \cos(2\theta + \theta) \\
&= \cos 2\theta \cos\theta - \sin 2\theta \sin\theta \\
&= (2\cos^2\theta - 1)\cos\theta - (2\sin\theta\cos\theta)\sin\theta \\
&= 2\cos^3\theta - \cos\theta - 2\sin^2\theta\cos\theta \\
&= 2\cos^3\theta - \cos\theta - 2(1 - \cos^2\theta)\cos\theta \\
&= 2\cos^3\theta - \cos\theta - 2\cos\theta + 2\cos^3\theta \\
&= 4\cos^3\theta - 3\cos\theta \\
&= 4x^3 - 3x.
\end{aligned}
$$

343

Notice that $T_n(x)$ defined in this way cannot be used if $|x| > 1$. For these values of x the following formula is used:

$$T_{n+1}(x) = 2xT_n(x) - T_{n-1}(x),$$

where $T_0(x) = 1$ and $T_1(x) = x$.

(b) Since $-1 \le \sin \theta \le 1$ the range of $U_n(x)$ is $[-1, 1]$. Since the domain of $y = \text{Cos}^{-1}x$ is also $[-1, 1]$, then the domain of $U_n(x)$ is $[-1, 1]$. This explains why the range for the graphs in this exercise all use the range $[-1, 1] \times [-2.5, 2.5]$. The graph below is that of $U_n(x)$, $n = 1, 2, 3, 4$, and 5.

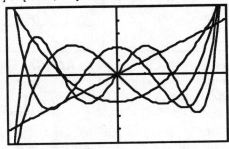

Let $\theta = \text{Cos}^{-1}x$, so $x = \cos \theta$. Then

$$U_1(x) = \frac{\sin 2\theta}{\sin \theta} = \frac{2 \sin \theta \cos \theta}{\sin \theta} = 2 \cos \theta = 2x$$

and

$$U_2(x) = \frac{\sin 3\theta}{\sin \theta} = \frac{\sin(2\theta + \theta)}{\sin \theta} = \frac{\sin 2\theta \cos \theta + \cos 2\theta \sin \theta}{\sin \theta}$$

$$= \frac{2 \sin \theta \cos^2 \theta + (2\cos^2 \theta - 1)\sin \theta}{\sin \theta} = 2\cos^2 \theta + 2\cos^2 \theta - 1 = 4\cos^2 \theta - 1 = 4x^2 - 1.$$

Notice that $U_n(x)$ defined in this way cannot be used if $|x| > 1$. For these values of x the following formula is used:

$$U_{n+1}(x) = 2xU_n(x) - U_{n-1}(x),$$

where $U_0(x) = 1$ and $U_1(x) = 2x$.

9.4 Derivatives of the Inverse Trigonometric Functions

1. $y = \text{Sin}^{-1}3x \implies y' = \dfrac{3}{\sqrt{1 - 9x^2}}$

3. $f(t) = \text{Sin}^{-1}\sqrt{t} \implies f'(t) = \dfrac{1}{\sqrt{1-t}} \dfrac{1}{2\sqrt{t}} = \dfrac{1}{2\sqrt{t}\sqrt{1-t}}$

5. $f(x) = \text{Sin}^{-1}e^{-x} \implies f'(x) = \dfrac{1}{\sqrt{1-(e^{-x})^2}}(-e^{-x}) = \dfrac{-e^{-x}}{\sqrt{1-e^{-2x}}}$

7. $y = \sqrt{\text{Cos}^{-1}x} \implies y' = \dfrac{1}{2}\left(\text{Cos}^{-1}x\right)^{-1/2}\dfrac{-1}{\sqrt{1-x^2}} = \dfrac{-1}{2\sqrt{(\text{Cos}^{-1}x)(1-x^2)}}$

9. $f(x) = \ln \text{Tan}^{-1}x \implies f'(x) = \dfrac{1}{\text{Tan}^{-1}x}\dfrac{1}{1+x^2}$

Exercise Set 9.4

11. $y = \text{Sec}^{-1}(1/x) \implies y' = \dfrac{1}{\frac{1}{x}\sqrt{\left(\frac{1}{x}\right)^2 - 1}}\left(-\dfrac{1}{x^2}\right) = \dfrac{-1}{\sqrt{1 - x^2}}$

13. $f(x) = \dfrac{\text{Tan}^{-1}x}{1 + x^2} \implies f'(x) = \dfrac{(1 + x^2)\frac{1}{1+x^2} - (\text{Tan}^{-1}x)(2x)}{(1 + x^2)^2} = \dfrac{1 - 2x\text{Tan}^{-1}x}{(1 + x^2)^2}$

15. Let $u = x/2$. Then $du = (1/2)dx$ and hence

$$\int \frac{1}{4 + x^2}\,dx = \frac{1}{4}\int \frac{1}{1 + \left(\frac{x}{2}\right)^2}\,dx = \frac{1}{2}\int \frac{1}{1 + u^2}\,du = \frac{1}{2}\text{Tan}^{-1}u + C = \frac{1}{2}\text{Tan}^{-1}\frac{x}{2} + C.$$

17. Let $u = 3x$. Then $du = 3dx$ and hence

$$\int \frac{1}{1 + 9x^2}\,dx = \frac{1}{3}\int \frac{1}{1 + u^2}\,du = \frac{1}{3}\text{Tan}^{-1}u + C = \frac{1}{3}\text{Tan}^{-1}3x + C.$$

19. Let $u = x/4$. Then $du = (1/4)dx$ and hence

$$\int_{8\sqrt{3}/3}^{8} \frac{1}{2x\sqrt{x^2 - 16}}\,dx = \frac{1}{8}\int_{8\sqrt{3}/3}^{8} \frac{(1/4)dx}{\frac{x}{4}\sqrt{\left(\frac{x}{4}\right)^2 - 1}}\,dx = \frac{1}{8}\int_{2\sqrt{3}/2}^{2} \frac{1}{u\sqrt{u^2 - 1}}\,du = \frac{1}{8}\text{Sec}^{-1}u\Big|_{2\sqrt{3}/3}^{2} = \frac{\pi}{48}.$$

21. Let $u = 1 - x^2$. Then $du = -2xdx$ and hence

$$\int \frac{x}{\sqrt{1 - x^2}}\,dx = -\frac{1}{2}\int u^{-1/2}\,du = -\sqrt{u} + C = -\sqrt{1 - x^2} + C.$$

23. Let $u = e^x$. Then $du = e^x dx$ and hence

$$\int \frac{1}{\sqrt{e^{2x} - 1}}\,dx = \int \frac{1}{u\sqrt{u^2 - 1}}\,du = \text{Sec}^{-1}u + C = \text{Sec}^{-1}e^x + C.$$

25. Let $u = e^{\sqrt{x}}$. Then $du = \dfrac{e^{\sqrt{x}}}{2\sqrt{x}}\,dx$ and hence

$$\int \frac{e^{\sqrt{x}}}{\sqrt{x}(1 + e^{2\sqrt{x}})}\,dx = 2\int \frac{du}{1 + u^2} = 2\text{Tan}^{-1}u + C = 2\text{Tan}^{-1}e^{\sqrt{x}} + C.$$

27. Let $u = x^3$. Then $du = 3x^2 dx$ and hence

$$\int \frac{x^2}{1 + x^6}\,dx = \frac{1}{3}\int \frac{1}{1 + u^2}\,du = \frac{1}{3}\text{Tan}^{-1}u + C = \frac{1}{3}\text{Tan}^{-1}x^3 + C.$$

29. Let $u = \text{Tan}^{-1}x$. Then $du = \dfrac{1}{1 + x^2}\,dx$ and hence

$$\int_0^1 \frac{\text{Tan}^{-1}x}{1 + x^2}\,dx = \int_0^{\pi/4} u\,du = \frac{1}{2}u^2\Big|_0^{\pi/4} = \frac{\pi^2}{32}.$$

345

31. Let $u = \text{Cos}^{-1} 2x$. Then $du = \dfrac{-2}{\sqrt{1 - 4x^2}} \, dx$ and hence

$$\int \frac{(\text{Cos}^{-1} 2x)^3}{\sqrt{1 - 4x^2}} \, dx = -\frac{1}{2} \int u^3 \, du = -\frac{1}{8} u^4 + C = -\frac{1}{8} (\text{Cos}^{-1} 2x)^4 + C.$$

33. Let $u = (1/2) \sin x$. Then $du = (1/2) \cos x \, dx$ and hence

$$\int_0^{\pi/2} \frac{\cos x}{\sqrt{4 - \sin^2 x}} \, dx = \frac{1}{2} \int_0^{\pi/2} \frac{\cos x}{\sqrt{1 - \left(\frac{\sin x}{2}\right)^2}} \, dx = \int_0^{1/2} \frac{du}{\sqrt{1 - u^2}} = \left. \text{Sin}^{-1} u \right|_0^{1/2} = \pi/6.$$

35. Let $u = 4x/5$. Then $du = (4/5) dx$ and hence

$$\int_2^3 \frac{dx}{x\sqrt{16x^2 - 25}} = \frac{4}{25} \int_2^3 \frac{dx}{\left(\frac{4x}{5}\right)\sqrt{\left(\frac{4x}{5}\right)^2 - 1}} = \frac{1}{5} \int_{8/5}^{12/5} \frac{du}{u\sqrt{u^2 - 1}} = \frac{1}{5} \left. \text{Sec}^{-1} u \right|_{8/5}^{12/5} \approx 0.05.$$

37. Let $u = \ln x$. Then $du = (1/x) dx$ and hence

$$\int_1^{\sqrt{e}} \frac{1}{x\sqrt{1 - \ln^2 x}} \, dx = \int_0^{1/2} \frac{1}{\sqrt{1 - u^2}} \, du = \left. \text{Sin}^{-1} u \right|_0^{1/2} = \pi/6.$$

39. $\frac{dy}{dx} = \sqrt{1 - y^2} \implies \dfrac{dy}{\sqrt{1 - y^2}} = dx \implies \text{Sin}^{-1} y = x + C \implies y = \sin(x + C)$. Since $y(0) = \sin C = 1$,
$C = \text{Sin}^{-1} 1 = \pi/2$ and hence $y = \sin(x + \pi/2) = \cos x$.

41. Let x be the horizontal distance of the eye from the wall. Then

$$\theta = \text{Tan}^{-1}\left(\frac{4}{x}\right) - \text{Tan}^{-1}\left(\frac{1}{x}\right)$$

$$\implies \quad \frac{d\theta}{dx} = \left[\frac{1}{1 + \left(\frac{4}{x}\right)^2}\left(-\frac{4}{x^2}\right) - \frac{1}{1 + \left(\frac{1}{x}\right)^2}\left(-\frac{1}{x^2}\right) \right] = \frac{12 - 3x^2}{(x^2 + 16)(x^2 + 1)}.$$

It follows that $d\theta/dx = 0 \iff x = \pm 2$. We reject the value $x = -2$. Since $d\theta/dx > 0$ for $x < 2$ and $d\theta/dx < 0$ for $x > 2$, $x = 2$ is a relative maximum for θ. Thus, the maximum visibility occurs when the observer stands 2 meters from the wall.

43. Let x stand for the horizontal distance of the dog from the pole, as illustrated in the figure below, and θ the angle between the pole and the leash. Then $\theta = \text{Tan}^{-1}(x/10)$. Therefore

$$\frac{d\theta}{dt} = \frac{1}{1 + \left(\frac{x}{10}\right)^2}\left(\frac{1}{10}\right)\frac{dx}{dt}.$$

Now, when $x = 10$,

$$\left. \frac{d\theta}{dt} \right|_{x=10} = \frac{1}{1 + \left(\frac{10}{10}\right)^2}\left(\frac{1}{10}\right)(4) = 0.2 \text{ radians/s.}$$

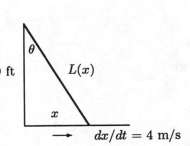

In other words, when the dog is 10 feet from the pole, the angle between the leash and pole is increasing at the rate of 0.2 radians/s. Now, let $L(x)$ stand for the length of the leash when the dog is x feet from the pole. Then $L(x) = \sqrt{x^2 + 100}$. Hence

$$\frac{dL}{dt} = \frac{1}{2\sqrt{x^2 + 100}} \, (2x) \, \frac{dx}{dt}.$$

When $x = 10$,

$$\frac{dL}{dt}\bigg|_{x=10} = \frac{1}{2\sqrt{10^2 + 100}} \, (20)(4) = 2\sqrt{2} \text{ ft/s}.$$

Thus, when the dog is 10 feet from the pole its leash is increasing at the rate of $2\sqrt{2}$ ft/s.

45. $\sin\theta = 1.22\lambda/d \Longrightarrow \theta = \text{Sin}^{-1}(1.22\lambda/d) \Longrightarrow$

$$\theta' = \frac{1}{\sqrt{1 - (1.22\lambda/d)^2}} \left(\frac{-1.22\lambda}{d^2} \right).$$

Since $\theta' < 0$, θ is a decreasing function of d.

47. Area $= \int_0^1 \frac{x^2}{1 + x^6} \, dx$. From Exercise 27 in this section,

$$\int_0^1 \frac{x^2}{1 + x^6} \, dx = \frac{1}{3}\text{Tan}^{-1}x^3 \bigg|_0^1 = \frac{\pi}{12}.$$

49. Let dV stand for the volume of a differential shell of thickness dx. Then $dV = 2\pi rh dx = 2\pi xy dx = (2\pi x/\sqrt{1 - x^4})dx$ and hence the volume of revolution is

$$V = \int_0^{\sqrt{2}/2} \frac{2\pi x}{\sqrt{1 - x^4}} \, dx.$$

Let $u = x^2$. Then $du = 2x dx$ and hence

$$V = \pi \int_0^{1/2} \frac{du}{\sqrt{1 - u^2}} = \pi\text{Sin}^{-1}u \bigg|_0^{1/2} = \frac{\pi^2}{6}.$$

51. Let dV be the differential volume of a cross-section disk of thickness dx. Then $dV = \pi r^2 dx = 4\pi/(1 + 4x^2) \, dx$ and hence the volume of revolution is

$$V = \int_0^{\sqrt{3}} \frac{4\pi}{1 + 4x^2} \, dx.$$

Let $u = 2x$. Then $du = 2dx$ and hence

$$V = 2\pi \int_0^{2\sqrt{3}} \frac{du}{1 + u^2} = 2\,\pi\text{Tan}^{-1}u \bigg|_0^{2\sqrt{3}} = 2\pi\text{Tan}^{-1}2\sqrt{3} \approx 8.10.$$

53. $y = \text{Cos}^{-1}x \Longrightarrow \cos y = x \Longrightarrow -(\sin y)y' = 1 \Longrightarrow y' = -\dfrac{1}{\sin y}$. Since $0 \leq y \leq \pi$, $\sin y > 0$. Therefore $\sin y = +\sqrt{1 - \cos^2 y} = \sqrt{1 - x^2}$ and hence

$$y' = \frac{-1}{\sqrt{1 - x^2}}.$$

55. $y = \text{Sec}^{-1}x \Longrightarrow \sec y = x \Longrightarrow (\sec y \tan y)y' = 1 \Longrightarrow y' = \dfrac{1}{(\sec y)(\tan y)}$. Since $y \in [0, \pi/2) \cup [\pi, 3\pi/2)$, $\tan y > 0$. Therefore $\tan y = +\sqrt{\sec^2 y - 1} = \sqrt{x^2 - 1}$ and hence

$$y' = \frac{1}{x\sqrt{x^2 - 1}}.$$

57. Since $\dfrac{d}{dx}(\text{Cos}^{-1}x) = -1/\sqrt{1-x^2} = \dfrac{d}{dx}(-\text{Sin}^{-1}x)$, the functions $\text{Cos}^{-1}x$ and $-\text{Sin}^{-1}x$ must differ by a constant, say $\text{Cos}^{-1}x + \text{Sin}^{-1}x = C$. Setting $x = 0$, $C = \text{Cos}^{-1}0 + \text{Sin}^{-1}0 = \pi/2$. Hence

$$\text{Cos}^{-1}x = \frac{\pi}{2} - \text{Sin}^{-1}x.$$

59. (a) Since $\theta = \text{Tan}^{-1}(x/50)$, it follows that

$$\frac{d\theta}{dt} = \frac{1/50}{1 + (x/50)^2}\frac{dx}{dt} = \frac{50}{2500 + x^2}\frac{dx}{dt}.$$

If $dx/dt = 1.5$ then $d\theta/dt = 50(1.5)/(2500 + x^2) = 75/(2500 + x^2)$. Graph using the range $[0, 50] \times [0, 0.05]$.

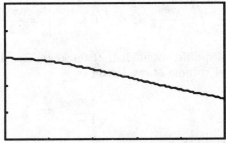

We see from the graph that the maximum of $d\theta/dt$ is when $x = 0$. For $x = 0$, $d\theta/dt = 75/2500 = 0.03$ radians/s.

(b) Since $s = 100\text{Tan}^{-1}(x/50)$,

$$\frac{ds}{dt} = \frac{100(1/50)}{1 + (x/50)^2}\frac{dx}{dt} = \frac{5000}{2500 + x^2}\frac{dx}{dt}.$$

If $dx/dt = 1.5$ then $d\theta/dt = 5000(1.5)/(2500 + x^2) = 7500/(2500 + x^2)$. As in part (a), ds/dt is a maximum when $x = 0$. For $x = 0$, $ds/dt = 3$ feet/s.

9.5 The Hyperbolic Functions

1. $\sinh 0 = (1/2)(e^0 - e^0) = 0$

3. $\cosh(\ln 2) = (1/2)(e^{\ln 2} + e^{-\ln 2}) = 5/4$

5. $\tanh(\ln 2) = \dfrac{e^{\ln 2} - e^{-\ln 2}}{e^{\ln 2} + e^{-\ln 2}} = 3/5$

7. $\coth(\ln 4) = \dfrac{e^{\ln 4} + e^{-\ln 4}}{e^{\ln 4} - e^{-\ln 4}} = 17/15$

9. $y = \sinh^{-1}(1) = \ln(1 + \sqrt{2})$

11. $y = \sinh 2x \implies y' = 2\cosh 2x$

13. $f(x) = \sinh x \tanh x \implies f'(x) = (\sinh x)(\text{sech}^2 x) + (\tanh x)\cosh x = (\sinh x)(1 + \text{sech}^2 x)$

15. $f(x) = \sqrt{\cosh 4x} = (\cosh 4x)^{1/2} \implies f'(x) = \frac{1}{2}(\cosh 4x)^{-1/2}(\sinh 4x)(4) = 2(\sinh 4x)/\sqrt{\cosh 4x}$

17. $y = 1/\cosh x = (\cosh x)^{-1} \implies y' = -(\cosh x)^{-2}(\sinh x) = -(\sinh x)/\cosh^2 x$

19. $f(x) = e^x \text{csch}\, x^2 \implies f'(x) = e^x(-\text{csch}\, x^2 \coth x^2)(2x) + (\text{csch}\, x^2)e^x$

21. $y = \sinh^{-1} 2x \implies y' = \dfrac{1}{\sqrt{(2x)^2 + 1}}\, 2 = \dfrac{2}{\sqrt{4x^2 + 1}}$

23. $f(s) = \tanh^{-1} s^2 \implies f'(s) = \dfrac{1}{1 - (s^2)^2}\,(2s) = \dfrac{2s}{1 - s^4}$

25. $f(x) = \ln \cosh^{-1} \pi x \implies f'(x) = \dfrac{1}{\cosh^{-1} \pi x}\, \dfrac{1}{\sqrt{(\pi x)^2 - 1}}\, \pi$

27. $y = \cosh x \cosh x^2 \implies y' = (\cosh x)(\sinh x^2)(2x) + (\cosh x^2)(\sinh x)$

29. $y = \sinh \ln x^2 \implies y' = (\cosh \ln x^2)\left(\dfrac{1}{x^2}\, 2x\right) = \dfrac{2}{x}\, \cosh \ln x^2 = \dfrac{2}{x}\dfrac{1}{2}\left(x^2 + \dfrac{1}{x^2}\right) = x + \dfrac{1}{x^3}$

31. $y = \tanh \sqrt{1 + x^2} \implies y' = (\text{sech}^2 \sqrt{1 + x^2})\, \dfrac{1}{2}(1 + x^2)^{-1/2}(2x) = \dfrac{x\, \text{sech}^2 \sqrt{1 + x^2}}{\sqrt{1 + x^2}}$

33. $\displaystyle\int_1^2 \dfrac{1}{\sqrt{1 + x^2}}\, dx = \sinh^{-1} x \Big|_1^2 = \sinh^{-1} 2 - \sinh^{-1} 1 = \ln(2 + \sqrt{5}) - \ln(1 + \sqrt{2})$

35. $\displaystyle\int \tanh^2(2x)\, dx = \int (1 - \text{sech}^2 2x)\, dx = x - \dfrac{1}{2}\tanh 2x + C$

37. Let $u = \sinh x$. Then $du = \cosh x\, dx$ and hence

$$\int_0^2 \sinh x \cosh x\, dx = \int_{\sinh 0}^{\sinh 2} u\, du = \dfrac{1}{2} u^2 \Big|_{\sinh 0}^{\sinh 2} = \dfrac{1}{2}\left(\dfrac{e^2 - e^{-2}}{2}\right)^2.$$

39. Let $u = x/2$. Then $du = (1/2)dx$ and hence

$$\int_4^6 \dfrac{dx}{\sqrt{x^2 - 4}} = \dfrac{1}{2}\int_4^6 \dfrac{dx}{\sqrt{\left(\frac{x}{2}\right)^2 - 1}} = \int_2^3 \dfrac{du}{\sqrt{u^2 - 1}} = \cosh^{-1} u \Big|_2^3 = \ln(3 + 2\sqrt{2}) - \ln(2 + \sqrt{3}).$$

41. Let $u = 3x/5$. Then $du = (3/5)dx$ and hence

$$\int_0^1 \dfrac{1}{\sqrt{9x^2 + 25}}\, dx = \dfrac{1}{5}\int_0^1 \dfrac{1}{\sqrt{\left(\frac{3x}{5}\right)^2 + 1}}\, dx = \dfrac{1}{3}\int_0^{3/5} \dfrac{du}{\sqrt{u^2 + 1}} = \dfrac{1}{3}\sinh^{-1} u \Big|_0^{3/5} = \dfrac{1}{3}\ln\left(\dfrac{3}{5} + \dfrac{\sqrt{34}}{5}\right).$$

43. Let $u = \cosh x$. Then $du = \sinh x\, dx$ and hence

$$\int \dfrac{\sinh x}{\cosh x}\, dx = \int \dfrac{du}{u} = \ln|u| + C = \ln|\cosh x| + C = \ln(\cosh x) + C.$$

45. Let $u = \sinh x$. Then $du = \cosh x\, dx$ and hence

$$\int \frac{\cosh x}{\sqrt{\sinh x}}\, dx = \int \frac{du}{\sqrt{u}} = 2\sqrt{u} + C = 2\sqrt{\sinh x} + C.$$

47. Let $u = \tanh x$. Then $du = \operatorname{sech}^2 x\, dx$ and hence

$$\int \tanh^2 x \operatorname{sech}^2 x\, dx = \int u^2\, du = \frac{1}{3} u^3 + C = \frac{1}{3} \tanh^3 x + C.$$

49. False. For example, $\sinh 0 = 0$ but $\sinh 2\pi = \dfrac{e^{2\pi} - e^{-2\pi}}{2} \neq 0$.

51. $\cosh x \pm \sinh x = \dfrac{e^x + e^{-x}}{2} \pm \dfrac{e^x - e^{-x}}{2} = e^{\pm x}$

53. $\cosh 2x = \dfrac{e^{2x} + e^{-2x}}{2} = \dfrac{(e^x)^2 - 2 + (e^{-x})^2 + 2}{2} = 2\left(\dfrac{e^x - e^{-x}}{2}\right)^2 + 1 = 2\sinh^2 x + 1$

55. $\cosh x \cosh y + \sinh x \sinh y$

$$\begin{aligned}
&= \frac{e^x + e^{-x}}{2} \frac{e^y + e^{-y}}{2} + \frac{e^x - e^{-x}}{2} \frac{e^y - e^{-y}}{2} \\
&= \frac{e^{x+y} + e^{y-x} + e^{x-y} + e^{-x-y}}{4} + \frac{e^{x+y} - e^{-x+y} - e^{x-y} + e^{-x-y}}{4} \\
&= \frac{e^{x+y} + e^{-(x+y)}}{2} \\
&= \cosh(x + y)
\end{aligned}$$

57. $g(y) = c_1 \sinh(ky) + c_2 \cosh(ky) \Longrightarrow g'(y) = kc_1 \cosh(ky) + kc_2 \sinh(ky) \Longrightarrow g''(y) = k^2 c_1 \sinh(ky) + k^2 c_2 \cosh(ky) = k^2 g(y) \Longrightarrow g''(y) - k^2 g(y) = 0.$

59. $y = c_1 \sinh k\pi x + c_2 \cosh k\pi x = C_1 e^{k\pi x} + C_2 e^{-k\pi x}$

61. Since $\coth x > \tanh x$,

$$\text{Area} = \int_{\ln 2}^{\ln 4} (\coth x - \tanh x)\, dx = \int_{\ln 2}^{\ln 4} \left(\frac{\cosh x}{\sinh x} - \frac{\sinh x}{\cosh x}\right) dx$$

$$= (\ln |\sinh x| - \ln |\cosh x|)\big|_{\ln 2}^{\ln 4} = \ln |\tanh x|\big|_{\ln 2}^{\ln 4} = \ln \frac{75}{51} = \ln \frac{25}{17}.$$

63. $f(x) = \cosh x - \sinh x = (e^x + e^{-x})/2 - (e^x - e^{-x})/2 = e^{-x}$ has no relative extrema.

65. $f(x) = 5\cosh x - 2\sinh x = (5/2)(e^x + e^{-x}) - (e^x - e^{-x}) = (3/2)e^x + (7/2)e^{-x} \Longrightarrow f'(x) = (3/2)e^x - (7/2)e^{-x} = (1/2)e^{-x}(3e^{2x} - 7) = 0 \Longrightarrow e^{2x} = 7/3 \Longrightarrow 2x = \ln(7/3) \Longrightarrow x = (1/2)\ln(7/3)$. Since $f''(x) = (3/2)e^x + (7/2)e^{-x} > 0$, $x = (1/2)\ln(7/3)$ is a relative minimum for $f(x)$.

67. $f(x) = \cosh 3x - 2\sinh 3x \Longrightarrow f'(x) = 3\sinh 3x - 6\cosh 3x \Longrightarrow f''(x) = 9\cosh 3x - 18\sinh 3x = 9f(x).$

69. $\sinh(-x) = (e^{-x} - e^x)/2 = -(e^x - e^{-x})/2 = -\sinh x \Longrightarrow \sinh x$ is odd
$\cosh(-x) = (e^{-x} + e^x)/2 = \cosh x \Longrightarrow \cosh x$ is even
$\tanh(-x) = \sinh(-x)/\cosh(-x) = -(\sinh x)/(\cosh x) = -\tanh x \Longrightarrow \tanh x$ is odd
$\operatorname{sech}(-x) = 1/\cosh(-x) = 1/\cosh x = \operatorname{sech} x \Longrightarrow \operatorname{sech} x$ is even
$\operatorname{csch}(-x) = 1/\sinh(-x) = -1/\sinh x = -\operatorname{csch} x \Longrightarrow \operatorname{csch} x$ is odd
$\coth(-x) = 1/\tanh(-x) = -1/\tanh x = -\coth x \Longrightarrow \coth x$ is odd

71. $\cosh^2 x - \sinh^2 x = 1 \implies \dfrac{\cosh^2 x}{\sinh^2 x} - 1 = \dfrac{1}{\sinh^2 x} \implies \coth^2 x - 1 = \operatorname{csch}^2 x.$

73. Since $y = \cosh^{-1} x = \ln(x + \sqrt{x^2 - 1}) \geq 0$ when $x \geq 1$, the range of $\cosh^{-1}$ is $[0, \infty)$, which means that the domain of cosh, that is, its principal branch, is the interval $[0, \infty)$.

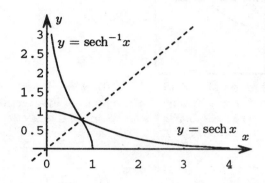

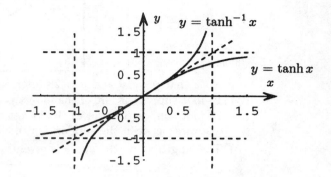

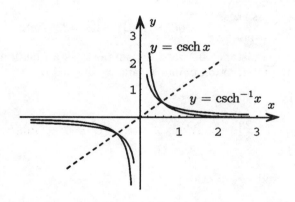

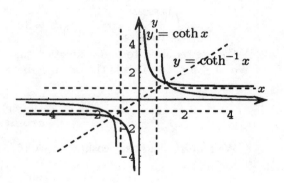

75. The principal branch of sech is the interval $[0, \infty)$; hence,

$$\operatorname{sech} : [0, \infty) \to (0, 1] \implies \operatorname{sech}^{-1} : (0, 1] \to [0, \infty).$$

The graph of $y = \operatorname{sech}^{-1} x$ is shown above on the left. The principal branch of tanh is the interval $(-\infty, \infty)$; hence,

$$\tanh : (-\infty, \infty) \to (-1, 1) \implies \tanh^{-1} : (-1, 1) \to (-\infty, \infty).$$

The graph of $y = \tanh^{-1} x$ is shown above on the right. The principal branch of csch is the interval $(-\infty, 0) \cup (0, \infty)$; hence,

$$\operatorname{csch} : (-\infty, 0) \cup (0, \infty) \to (-\infty, 0) \cup (0, \infty) \implies \operatorname{csch}^{-1} : (-\infty, 0) \cup (0, \infty) \to (-\infty, 0) \cup (0, \infty).$$

The graph of $y = \operatorname{csch}^{-1} x$ is shown above on the left. The principal branch of coth is the interval $(-\infty, 0) \cup (0, \infty)$; hence,

$$\coth : (-\infty, 0) \cup (0, \infty) \to (-\infty, -1) \cup (1, \infty) \implies \coth^{-1} : (-\infty, -1) \cup (1, \infty) \to (-\infty, 0) \cup (0, \infty).$$

The graph of $y = \coth^{-1} x$ is shown above on the right.

77. (a) Graph $y = 10\cosh(x/10)$, $y = 16$ and $y = 28$ using the range $[-20, 20] \times [0, 30]$.

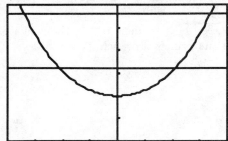

Note that $y = 16$ and $y = 28$ intersect the graph of $y = 10\cosh(x/10)$ twice. If we assume the 16 foot support is left of the lowest point then $x = -10.469679$ is the point of intersection. If the 28 foot support is right of the lowest point the $x = 16.892355$ is the point of intersection. So the distance between the supports is $16.892355 + 10.469679 = 27.362035$ feet.

(b) The length along the curve is given by

$$L = \int_{-10.469679}^{16.892355} \sqrt{1 + \left[10\left(\frac{1}{10}\sinh\left(\frac{x}{10}\right)\right)\right]^2}\, dx = \int_{-10.469679}^{16.892355} \sqrt{1 + \sinh^2\left(\frac{x}{10}\right)}\, dx$$

$$= \int_{-10.469679}^{16.892355} \cosh\left(\frac{x}{10}\right)\, dx = 10\sinh\left(\frac{x}{10}\right)\Big|_{-10.469679}^{16.892355} = 10(3.864339) = 38.64339 \text{ feet.}$$

(c) We are required to find a function that gives the distance between the supports as a function of a, $0 \le a \le 16$. This is the distance between the intersection points x_1 and x_2 defined by

$$a\cosh\left(\frac{x_1}{a}\right) = 16, \quad x_1 < 0$$
$$a\cosh\left(\frac{x_2}{a}\right) = 28, \quad x_2 > 0.$$

We require then that $\cosh(x_1/a) = 16/a$ and $\cosh(x_2/a) = 28/a$. Hence

$$x_1 = a\mathrm{Cosh}^{-1}\left(\frac{16}{a}\right) \quad \text{and} \quad x_2 = a\mathrm{Cosh}^{-1}\left(\frac{28}{a}\right).$$

But the range of $\mathrm{Cosh}^{-1}t$ is positive if $t > 0$ so $x_1 = -a\mathrm{Cosh}^{-1}(16/a)$. Therefore the distance between the supports is

$$D(a) = a\left(\mathrm{Cosh}^{-1}\left(\frac{28}{a}\right) + \mathrm{Cosh}^{-1}\left(\frac{16}{a}\right)\right).$$

To determine the maximum of this function graph **Y=X(Cosh^{-1}(26/X)+Cosh^{-1}(16/X))** using the range $[0, 16] \times [0, 30]$.

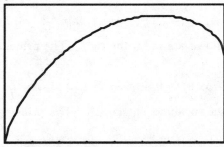

Magnifying the graph at the maximum point we find that if the lowest point on the cable is $a = 11.119420$ feet then the supports will be a maximum of $D(11.119420) = 27.579904$ feet apart.

Review Exercises – Chapter 9

1. $\text{Sin}^{-1}(\sqrt{3}/2) = \pi/3$

3. $\text{Tan}^{-1}\sqrt{3} = \pi/3$

5. $\text{Cos}^{-1}\big(\cos(-\pi/4)\big) = \pi/4$

7. $\cot\big(\text{Sin}^{-1}(\sqrt{2}/2)\big) = 1$

9. $\sinh(\ln 2) = (1/2)(e^{\ln 2} - e^{-\ln 2}) = 3/4.$

11. $y = \text{Sin}^{-1}\sqrt{x} \Longrightarrow y' = \dfrac{1}{\sqrt{1-x}}\,\dfrac{1}{2\sqrt{x}} = \dfrac{1}{2\sqrt{x(1-x)}}$

13. $f(t) = \text{Cot}^{-1}(1-t^2) \Longrightarrow f'(t) = \dfrac{-1}{1+(1-t^2)^2}\,(-2t) = \dfrac{2t}{t^4 - 2t^2 + 2}$

15. $f(x) = x^2 \sinh(1-x) \Longrightarrow$

$$f'(x) = x^2[\cosh(1-x)](-1) + [\sinh(1-x)](2x) = 2x\sinh(1-x) - x^2\cosh(1-x)$$

17. $y = \csc(\cot 6x) \Longrightarrow y' = -\csc(\cot 6x)\cot(\cot 6x)(-\csc^2 6x)(6) = 6\csc(\cot 6x)\cot(\cot 6x)(\csc^2 6x)$

19. $f(x) = \text{Sin}^{-1}[\ln(2x+1)] \Longrightarrow f'(x) = \dfrac{1}{\sqrt{1 - \ln^2(2x+1)}}\,\dfrac{1}{2x+1}\,(2)$

21. $f(x) = \dfrac{\text{Tan}^{-1}x}{1+x^2} \Longrightarrow f'(x) = \dfrac{(1+x^2)\left(\frac{1}{1+x^2}\right) - (\text{Tan}^{-1}x)(2x)}{(1+x^2)^2} = \dfrac{1 - 2x\text{Tan}^{-1}x}{(1+x^2)^2}$

23. $f(x) = \text{Sec}^{-1}\sqrt{x^2+4} \Longrightarrow f'(x) = \dfrac{1}{\sqrt{x^2+4}\sqrt{(x^2+4)-1}}\,\dfrac{1}{2}\,\dfrac{1}{\sqrt{x^2+4}}\,(2x) = \dfrac{x}{(x^2+4)\sqrt{x^2+3}}$

25. $y = 1/(\pi + \tanh x) \Longrightarrow y' = -(\pi + \tanh x)^{-2}(\text{sech}^2 x) = \dfrac{-\text{sech}^2 x}{(\pi + \tanh x)^2}$

27. $y = (\sinh x + \text{Cos}^{-1}2x)^{1/5} \Longrightarrow y' = \dfrac{1}{5}(\sinh x + \text{Cos}^{-1}2x)^{-4/5}\left(\cosh x - \dfrac{2}{\sqrt{1-4x^2}}\right)$

29. $y = \ln(\text{Sin}^{-1}x) \Longrightarrow y' = \dfrac{1}{\text{Sin}^{-1}x}\,\dfrac{1}{\sqrt{1-x^2}}$

31. $f(x) = x\tanh^{-1}(\ln x) \Longrightarrow f'(x) = x\,\dfrac{1}{1 - \ln^2 x}\,\dfrac{1}{x} + \tanh^{-1}(\ln x) = \dfrac{1}{1 - \ln^2 x} + \tanh^{-1}(\ln x)$

33. $y = \ln\sqrt{\tanh^{-1}(x^2)} = \frac{1}{2}\ln\big(\tanh^{-1}(x^2)\big) \Longrightarrow y' = \dfrac{1}{2}\,\dfrac{1}{\tanh^{-1}(x^2)}\,\dfrac{1}{1-x^4}\,(2x) = \dfrac{x}{(1-x^4)\tanh^{-1}(x^2)}$

35. Let $u = 2x$. Then $du = 2dx$ and hence

$$\int \frac{dx}{\sqrt{4x^2-1}} = \frac{1}{2}\int \frac{du}{\sqrt{u^2-1}}$$

$$= \frac{1}{2}\cosh^{-1}u + C = \frac{1}{2}\cosh^{-1}u + C = \frac{1}{2}\cosh^{-1}2x + C = \frac{1}{2}\ln\left|2x + \sqrt{4x^2-1}\right| + C.$$

37. Let $u = \sin 2x$. Then $du = 2\cos 2x dx$ and hence

$$\int \sin^4 2x \cos 2x \, dx = \frac{1}{2} \int u^4 \, du = \frac{1}{10} u^5 + C = \frac{1}{10} \sin^5 2x + C.$$

39. Let $u = \sqrt{x}$. Then $du = \frac{1}{2\sqrt{x}} \, dx$ and hence

$$\int_{\pi^2/16}^{\pi^2/4} \frac{\sin^2 \sqrt{x}}{\sqrt{x}} \, dx = 2 \int_{\pi/4}^{\pi/2} \sin^2 u \, du = \int (1 - \cos 2u) \, du = \left. \left(u - \frac{1}{2} \sin 2u \right) \right|_{\pi/4}^{\pi/2} = \frac{\pi}{4} + \frac{1}{2}.$$

41. Let $u = \frac{x}{\sqrt{2}}$. Then $du = \frac{1}{\sqrt{2}} \, dx$ and hence

$$\int_{-1}^1 \frac{dx}{\sqrt{2 - x^2}} = \frac{1}{\sqrt{2}} \int_{-1}^1 \frac{dx}{\sqrt{1 - \left(\frac{x}{\sqrt{2}}\right)^2}} = \int_{-1/\sqrt{2}}^{1/\sqrt{2}} \frac{du}{\sqrt{1 - u^2}} = \left. \mathrm{Sin}^{-1} u \right|_{-1/\sqrt{2}}^{1/\sqrt{2}} = \frac{\pi}{2}.$$

43. Let $u = \tan x$. Then $du = \sec^2 x dx$ and hence

$$\int_0^{\pi/4} \sqrt{\tan x} \sec^2 x \, dx = \int_0^1 \sqrt{u} \, du = \left. \frac{2}{3} u^{3/2} \right|_0^1 = \frac{2}{3}.$$

45. $\displaystyle \int \frac{1 + \sin^2 2x}{\cos^2 2x} \, dx = \int \frac{2 - \cos^2 2x}{\cos^2 2x} \, dx = 2 \int \sec^2 2x \, dx - \int dx = \tan 2x - x + C.$

47. Let $u = \tan 2x$. Then $du = 2\sec^2 2x dx$ and hence

$$\int_0^{\pi/8} \tan^2(2x) \sec^2(2x) \, dx = \frac{1}{2} \int_0^1 u^2 \, du = \left. \frac{1}{6} u^3 \right|_0^1 = \frac{1}{6}.$$

49. Let $u = x/2$. Then $du = (1/2) dx$ and hence

$$\int \frac{dx}{\sqrt{4 - x^2}} = \frac{1}{2} \int \frac{dx}{\sqrt{1 - \left(\frac{x}{2}\right)^2}} = \int \frac{du}{\sqrt{1 - u^2}} = \mathrm{Sin}^{-1} u + C = \mathrm{Sin}^{-1} \frac{x}{2} + C.$$

51. Let $u = \mathrm{Cos}^{-1} x$. Then $du = \frac{-1}{\sqrt{1 - x^2}} \, dx$ and hence

$$\int \frac{\mathrm{Cos}^{-1} x}{\sqrt{1 - x^2}} \, dx = - \int u \, du = -\frac{1}{2} u^2 + C = -\frac{1}{2} \left(\mathrm{Cos}^{-1} x \right)^2 + C.$$

53. Let $u = e^{\sqrt{x}}$. Then $du = \frac{e^{\sqrt{x}}}{2\sqrt{x}} \, dx$ and hence

$$\int \frac{e^{\sqrt{x}}}{\sqrt{x}(1 + e^{2\sqrt{x}})} \, dx = \int \frac{e^{\sqrt{x}}}{\sqrt{x}(1 + (e^{\sqrt{x}})^2)} \, dx = 2 \int \frac{du}{1 + u^2} = 2 \, \mathrm{Tan}^{-1} u + C = 2 \, \mathrm{Tan}^{-1} e^{\sqrt{x}} + C.$$

55. Let $u = \text{Sin}^{-1}x$. Then $du = dx/\sqrt{1-x^2}$ and hence

$$\int \frac{\sqrt{\text{Sin}^{-1}x}}{\sqrt{1-x^2}}\,dx = \int \sqrt{u}\,du = \frac{2}{3}\,u^{3/2} + C = \frac{2}{3}\,(\text{Sin}^{-1}x)^{3/2} + C.$$

57. Let $u = \text{Sec}^{-1}3x$. Then $du = \dfrac{3}{3x\sqrt{9x^2-1}}\,dx$ and hence

$$\int \frac{\text{Sec}^{-1}3x}{x\sqrt{36x^2-4}}\,dx = \frac{1}{2}\int \frac{\text{Sec}^{-1}3x}{x\sqrt{9x^2-1}}\,dx = \frac{1}{2}\int u\,du = \frac{1}{4}\,u^2 + C = \frac{1}{4}\,(\text{Sec}^{-1}3x)^2 + C.$$

59. Let $u = \sec 2x$. Then $du = 2\sec 2x \tan 2x\,dx$ and hence

$$\int_0^{\pi/9} \sec^3 2x \tan 2x\,dx = \frac{1}{2}\int_1^{\sec 2\pi/9} u^2\,du = \frac{1}{6}\,u^3\bigg|_1^{\sec 2\pi/9} = \frac{1}{6}\,\sec^3\frac{2\pi}{9} - \frac{1}{6}.$$

61. Let $u = \tan x$. Then $du = \sec^2 x\,dx$ and hence

$$\int_0^{\pi/3} \sec^4 x\,dx = \int_0^{\pi/3} (1 + \tan^2 x)\sec^2 x\,dx = \int_0^{\sqrt{3}} (1 + u^2)\,du = \left(u + \frac{1}{3}\right)\bigg|_0^{\sqrt{3}} = 2\sqrt{3}.$$

63. $\displaystyle\int_0^{\pi/3} \cos x \cos 5x\,dx = \frac{1}{2}\int_0^{\pi/3} (\cos 6x + \cos 4x)\,dx = \frac{1}{2}\left(\frac{1}{6}\,\sin 6x + \frac{1}{4}\,\sin 4x\right)\bigg|_0^{\pi/3} = -\frac{\sqrt{3}}{16}$

65. Let $u = \cos x$. Then $du = -\sin x\,dx$ and hence

$$\int \sqrt{\cos x}\,\sin^3 x\,dx = -\int u^{1/2}(1-u^2)\,du = -\left(\frac{2}{3}\,u^{3/2} - \frac{2}{7}\,u^{7/2}\right) + C = -\frac{2}{3}\,\cos^{3/2}x + \frac{2}{7}\,\cos^{7/2}x + C.$$

67. Let $u = \cos x$. Then $du = -\sin x\,dx$ and hence

$$\int \frac{\sin^3 x}{\cos^2 x}\,dx = \int \frac{1 - \cos^2 x}{\cos^2 x}\,(\sin x)\,dx = -\int \frac{1 - u^2}{u^2}\,du$$

$$= -\int (u^{-2} - 1)\,du = u^{-1} + u + C = \sec x + \cos x + C.$$

69. Let $u = 1 + \sinh x$. Then $du = \cosh x\,dx$ and hence

$$\int \frac{\cosh x}{1 + \sinh x}\,dx = \int \frac{du}{u} = \ln|u| + C = \ln|1 + \sinh x| + C.$$

71. Let $u = \cosh x$. Then $du = \sinh x\,dx$ and hence

$$\int \sinh^3 x\,dx = \int (\sinh x)(\underbrace{\cosh^2 - 1}_{\sinh^2 x})\,dx = \int u^2\,du - \cosh x = \frac{1}{3}\,\cosh^3 x - \cosh x + C.$$

73. Area $= \displaystyle\int_0^{\pi} \sin^2 x\,dx = \frac{1}{2}\int_0^{\pi} (1 - \cos 2x)\,dx = \frac{1}{2}\left(x - \frac{1}{2}\,\sin 2x\right)\bigg|_0^{\pi} = \pi/2.$

75. Let dV be the differential volume of a cross-section disk of thickness dx. Then $dV = \pi r^2 dx = \pi y^2 dx$
$= \pi \tan^2 x\, dx$ and hence the volume of revolution is

$$V = \int_0^{\pi/4} \pi \tan^2 x \, dx = \pi \int_0^{\pi/4} (\sec^2 x - 1) \, dx = \pi(\tan x - x)|_0^{\pi/4} = \pi\left(1 - \frac{\pi}{4}\right).$$

77. Let dV be the volume of a differential shell of thickness dx. Then

$$dV = 2\pi r h\, dx = 2\pi x y\, dx = \frac{2\pi x}{1 + x^4}\, dx$$

and hence the volume of revolution is

$$V = 2\pi \int_0^2 \underbrace{\frac{x}{1 + x^4}}_{u = x^2}\, dx = \pi \int_0^4 \frac{du}{1 + u^2} = \pi \mathrm{Tan}^{-1} u \big|_0^4 = \pi \mathrm{Tan}^{-1} 4.$$

79. $y = \cosh^{-1} x^2 y$

$$\implies y' = \frac{1}{\sqrt{(x^2 y)^2 - 1}}\, (x^2 y' + 2xy)$$
$$\implies y'\sqrt{x^4 y^2 - 1} = x^2 y' + 2xy$$
$$\implies y'(\sqrt{x^4 y^2 - 1} - x^2) = 2xy$$
$$\implies y' = \frac{2xy}{\sqrt{x^4 y^2 - 1} - x^2}$$

81. Area $= \displaystyle\int_{-1}^1 \cosh x \, dx = \sinh x \big|_{-1}^1 = 2\sinh 1 = e - 1/e$

83. $\displaystyle\int_0^4 v(t)\, dt = \int_0^4 \sin^2 \pi t \, dt = \frac{1}{2}\int_0^4 (1 - \cos 2\pi t)\, dt = \frac{1}{2}\left(t - \frac{1}{2\pi}\sin 2\pi t\right)\Big|_0^4 = 2$

85. Let x be any real number. Then

$$\cosh x - \sinh x = \frac{e^x + e^{-x}}{2} - \frac{e^x - e^{-x}}{2} = e^{-x} > 0$$

and hence $\cosh x > \sinh x$.

87. Let θ stand for the angle of elevation of the plane when the plane is a horizontal distance of x km from the station. Then $\theta = \mathrm{Tan}^{-1}(3/x)$ and $dx/dt = 400$. Hence

$$\frac{d\theta}{dt} = \frac{1}{1 + \left(\frac{3}{x}\right)^2}\left(-\frac{3}{x^2}\right)\frac{dx}{dt}$$
$$= -\frac{1200}{x^2 + 9}.$$

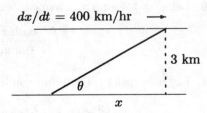

When the plane is a horizontal distance of 4 km from the station, $x = 4$ and hence

$$\frac{d\theta}{dt}\bigg|_{x=4} = -\frac{1200}{25} = -48 \text{ radians/hr.}$$

89. $y = \text{Cot}^{-1}x \implies y' = -1/(1+x^2)$. Now, the slope of the line $x + 5y - 10 = 0$ is $-1/5$. Hence

$$y' = \frac{-1}{1+x^2} = -\frac{1}{5} \implies 1 + x^2 = 5 \implies x = \pm 2.$$

Thus, there are two points on the graph of $y = \text{Cot}^{-1}x$ at which the tangent line is parallel to $x + 5y - 10 = 0$, namely $(2, \text{Cot}^{-1}2)$ and $(-2, \text{Cot}^{-1}(-2))$.

91. For the velocity $v(t)$, we find that

$$v(t) = \int a(t)\, dt = \int \left[-1 - \frac{2t}{(1+t^2)^2} \right] dt = -t + \frac{1}{1+t^2} + C.$$

Since $v(0) = 1 + C = 4$, $C = 3$ and hence

$$v(t) = -t + \frac{1}{1+t^2} + 3.$$

Therefore

$$s(t) = \int v(t)\, dt = \int \left(-t + \frac{1}{1+t^2} + 3 \right) dt = -\frac{1}{2}t^2 + \text{Tan}^{-1}t + 3t + C.$$

Since $s(0) = C = 0$,

$$s(t) = -\frac{1}{2}\, t^2 + \text{Tan}^{-1}t + 3t.$$

Chapter 10

Techniques of Integration

10.1   Integration by Parts

1. Let $u = x$, $dv = e^x dx$. Then $du = dx$, $v = e^x$, and hence

$$\int x e^x \, dx = x e^x - \int e^x \, dx = x e^x - e^x + C.$$

3. Let $u = x$, $dv = \cos x dx$. Then $du = dx$, $v = \sin x$, and hence

$$\int_{\pi/2}^{\pi} x \cos x \, dx = \left[x \sin x - \int \sin x \, dx \right] \Big|_{\pi/2}^{\pi} = (x \sin x + \cos x)|_{\pi/2}^{\pi} = -1 - \frac{\pi}{2}.$$

5. Let $u = \text{Tan}^{-1} x$, $dv = dx$. Then $du = \dfrac{1}{1 + x^2} \, dx$, $v = x$, and hence

$$\int \text{Tan}^{-1} x \, dx = x \text{Tan}^{-1} x - \int \underbrace{\frac{x}{1 + x^2}}_{w = 1 + x^2} \, dx = x \text{Tan}^{-1} x - \frac{1}{2} \int \frac{1}{w} \, dw$$

$$= x \text{Tan}^{-1} x - \frac{1}{2} \ln |w| + C = x \text{Tan}^{-1} x - \frac{1}{2} \ln(1 + x^2) + C.$$

7. Let $u = x$, $dv = \sec^2 \pi x \, dx$. Then $du = dx$, $v = (1/\pi) \tan \pi x$ and hence

$$\int x \sec^2 \pi x \, dx = \frac{x}{\pi} \tan x - \frac{1}{\pi} \int \tan \pi x \, dx = \frac{x}{\pi} \tan \pi x - \frac{1}{\pi^2} \ln |\sec \pi x| + C.$$

9. Let $u = \sin(\ln x)$, $dv = dx$. Then $du = \dfrac{\cos(\ln x)}{x} \, dx$, $v = x$, and hence

$$\int \sin(\ln x) \, dx = x \sin(\ln x) - \underbrace{\int \cos(\ln x) \, dx}.$$

For the last integral, let $u = \cos(\ln x)$, $dv = dx$. Then $du = -\dfrac{\sin(\ln x)}{x} \, dx$, $v = x$, and hence

$$\int \cos(\ln x) \, dx = x \cos(\ln x) + \int \sin(\ln x) \, dx.$$

Therefore

$$\int \sin(\ln x)\, dx = x\sin(\ln x) - \left[x\cos(\ln x) + \int \sin(\ln x)\, dx\right]$$

$$\implies \quad 2\int \sin(\ln x)\, dx = x[\sin(\ln x) - \cos(\ln x)] + C$$

$$\implies \quad \int \sin(\ln x)\, dx = \frac{x}{2}\,[\sin(\ln x) - \cos(\ln x)] + C.$$

11. Let $u = x$, $dv = (x+2)^8 dx$. Then $du = dx$, $v = (1/9)(x+2)^9$, and hence

$$\int_0^1 x(x+2)^8\, dx = \left[\frac{1}{9}\,x(x+2)^9 - \frac{1}{9}\int (x+2)^9\, dx\right]\Bigg|_0^1$$

$$= \left[\frac{1}{9}\,x(x+2)^9 - \frac{1}{90}\,(x+2)^{10}\right]\Bigg|_0^1 = \frac{27761}{18} \approx 1542.28.$$

13. Let $u = (\ln x)^2$, $dv = dx$. Then $du = \dfrac{2\ln x}{x}\, dx$, $v = x$, and hence

$$\int_e^{e^3} (\ln x)^2\, dx = \left[x(\ln x)^2 - 2\underbrace{\int \ln x\, dx}\right]\Bigg|_e^{e^3}.$$

For the last integral, let $u = \ln x$, $dv = dx$. Then $du = (1/x)dx$, $v = x$, and hence

$$\int \ln x\, dx = x\ln x - \int dx = x\ln x - x.$$

Hence

$$\int_e^{e^3} (\ln x)^2\, dx = \left[x(\ln x)^2 - 2x\ln x + 2x\right]\Big|_e^{e^3} = 5e^3 - e.$$

15. Let $u = x$, $dv = e^{2x} dx$. Then $du = dx$, $v = (1/2)e^{2x}$, and hence

$$\int_0^{\ln 2} xe^{2x}\, dx = \left[\frac{1}{2}\,xe^{2x} - \frac{1}{2}\int e^{2x}\, dx\right]\Bigg|_0^{\ln 2} = \left(\frac{1}{2}\,xe^{2x} - \frac{1}{4}\,e^{2x}\right)\Bigg|_0^{\ln 2} = 2\ln 2 - \frac{3}{4}.$$

17. Let $u = x^3$, $dv = e^{2x} dx$. Then $du = 3x^2 dx$, $v = (1/2)e^{2x}$, and hence

$$\int_0^1 x^3 e^{2x}\, dx = \left[\frac{1}{2}\,x^3 e^{2x} - \frac{3}{2}\underbrace{\int x^2 e^{2x}\, dx}\right]\Bigg|_0^1.$$

For the last integral, let $u = x^2$, $dv = e^{2x} dx$. Then $du = 2x\, dx$, $v = (1/2)e^{2x}$, and hence, using Exercise 15,

$$\int x^2 e^{2x}\, dx = \frac{1}{2}\,x^2 e^{2x} - \int xe^{2x}\, dx = \frac{1}{2}\,x^2 e^{2x} - \frac{1}{2}\,xe^{2x} + \frac{1}{4}\,e^{2x}.$$

Therefore

$$\int_0^1 x^3 e^{2x}\, dx = \left[\frac{1}{2}\,x^3 e^{2x} - \frac{3}{2}\left(\frac{1}{2}\,x^2 e^{2x} - \frac{1}{2}\,xe^{2x} + \frac{1}{4}\,e^{2x}\right)\right]\Bigg|_0^1 = \frac{1}{8}\,e^2 + \frac{3}{8}.$$

19. Let $u = 3x^2 - 2x + 1$, $dv = e^{2x}dx$. Then $du = (6x - 2)dx$, $v = (1/2)e^{2x}$, and hence

$$\int_0^1 (3x^2 - 2x + 1)e^{2x}\, dx = \left[\frac{1}{2}(3x^2 - 2x + 1)e^{2x} - \int (3x - 1)e^{2x}\, dx \right]\Bigg|_0^1$$

$$= \left[\frac{1}{2}(3x^2 - 2x + 1)e^{2x} - 3\int xe^{2x}\, dx + \int e^{2x}\, dx \right]\Bigg|_0^1$$

$$= \left[\frac{1}{2}(3x^2 - 2x + 1)e^{2x} - 3\left(\frac{1}{2}xe^{2x} - \frac{1}{4}e^{2x} \right) + \frac{1}{2}e^{2x} \right]\Bigg|_0^1$$

$$= \frac{3}{4}e^2 - \frac{7}{4}.$$

21. Let $u = x$, $dv = \sinh x\,dx$. Then $du = dx$, $v = \cosh x$, and hence

$$\int x\sinh x\, dx = x\cosh x - \int \cosh x\, dx = x\cosh x - \sinh x + C.$$

23. Let $u = e^{2x}$, $dv = \cos x\,dx$. Then $du = 2e^{2x}$, $v = \sin x$, and hence

$$\int e^{2x}\cos x\, dx = e^{2x}\sin x - 2\underbrace{\int e^{2x}\sin x\, dx}.$$

For the last integral, let $u = e^{2x}$, $dv = \sin x\,dx$. Then $du = 2e^{2x}$, $v = -\cos x$, and hence

$$\int e^{2x}\sin x\, dx = -e^{2x}\cos x + 2\int e^{2x}\cos x\, dx.$$

Therefore

$$\int e^{2x}\cos x\, dx = e^{2x}\sin x - 2\left[-e^{2x}\cos x + 2\int e^{2x}\cos x\, dx \right]$$

$$\implies \quad 5\int e^{2x}\cos x\, dx = e^{2x}\sin x + 2e^{2x}\cos x$$

$$\implies \quad \int e^{2x}\cos x\, dx = \frac{e^{2x}}{5}(\sin x + 2\cos x)$$

$$\implies \quad \int_0^\pi e^{2x}\cos x\, dx = -\frac{2e^{2\pi}}{5} - \frac{2}{5}.$$

25. Let $u = (\text{Cos}^{-1}x)^2$, $dv = dx$. Then $du = 2(\text{Cos}^{-1}x)\dfrac{-1}{\sqrt{1 - x^2}}\, dx$, $v = x$, and hence

$$\int (\text{Cos}^{-1}x)^2\, dx = x(\text{Cos}^{-1}x)^2 + 2\underbrace{\int \frac{x\text{Cos}^{-1}x}{\sqrt{1 - x^2}}\, dx}.$$

For the last integral, let $u = \text{Cos}^{-1}x$, $dv = \dfrac{x}{\sqrt{1 - x^2}}\, dx$. Then $du = \dfrac{-1}{\sqrt{1 - x^2}}\, dx$, $v = -\sqrt{1 - x^2}$, and hence

$$\int \frac{x\text{Cos}^{-1}x}{\sqrt{1 - x^2}}\, dx = -\sqrt{1 - x^2}\text{Cos}^{-1}x - \int dx = -\sqrt{1 - x^2}\text{Cos}^{-1}x - x.$$

Therefore

$$\int (\text{Cos}^{-1}x)^2\, dx = x(\text{Cos}^{-1}x)^2 - 2\sqrt{1 - x^2}\,\text{Cos}^{-1}x - 2x + C.$$

27. Let $u = x^2$, $dv = x\sqrt{1+x^2}\,dx$. Then $du = 2x\,dx$, $v = (1/3)(1+x^2)^{3/2}$, and hence

$$\int x^3\sqrt{1+x^2}\,dx = \frac{1}{3}\,x^2(1+x^2)^{3/2} - \frac{2}{3}\int x(1+x^2)^{3/2}\,dx$$

$$= \frac{1}{3}\,x^2(1+x^2)^{3/2} - \frac{2}{15}\,(1+x^2)^{5/2} + C = \frac{1}{15}\,(1+x^2)^{3/2}(3x^2 - 2) + C.$$

29. Let $u = x^2$, $dv = xe^{ax^2}dx$. Then $du = 2x\,dx$, $v = (1/2a)e^{ax^2}$, and hence

$$\int x^3 e^{ax^2}\,dx = \frac{x^2}{2a}\,e^{ax^2} - \frac{1}{a}\int xe^{ax^2}\,dx = \frac{x^2 e^{ax^2}}{2a} - \frac{1}{2a^2}\,e^{ax^2} + C = \frac{e^{ax^2}}{2a^2}\,(ax^2 - 1) + C.$$

31. First observe that

$$\int \sec^3 x \tan^2 x\,dx = \int \sec^3 x(\sec^2 x - 1)\,dx = \int \sec^5\,dx - \int \sec^3 x\,dx.$$

Hence, by Exercise 18 and Example 7,

$$\begin{aligned}
\int \sec^3 x \tan^2 x\,dx &= \frac{1}{4}\,\sec^3 x \tan x + \frac{3}{8}\,\sec x \tan x + \frac{3}{8}\,\ln|\sec x + \tan x| \\
&\quad - \left[\frac{1}{2}\,\sec x \tan x + \frac{1}{2}\,\ln|\sec x + \tan x|\right] + C \\
&= \frac{1}{4}\,\sec^3 x \tan x - \frac{1}{8}\,\sec x \tan x - \frac{1}{8}\,\ln|\sec x + \tan x| + C.
\end{aligned}$$

33. Let $w = \sinh^{-1} x$. Then $x = \sinh w$, $dx = \cosh w\,dw$, and hence

$$\int x \sinh^{-1} x\,dx = \int (\sinh w)w \cosh w\,dw = \frac{1}{2}\int w \sinh 2w\,dw,$$

since $\sinh 2w = 2\sinh w \cosh w$. Now, let $u = w$, $dv = \sinh 2w\,dw$. Then $du = dw$, $v = (1/2)\cosh 2w$, and hence

$$\int w \sinh 2w\,dw = \frac{1}{2}\,w \cosh 2w - \frac{1}{2}\int \cosh 2w\,dw = \frac{1}{2}\,w \cosh 2w - \frac{1}{4}\,\sinh 2w + C.$$

Therefore

$$\int x \sinh^{-1} x\,dx = \frac{1}{4}\,w \cosh 2w - \frac{1}{8}\,\sinh 2w + C$$

$$= \frac{1}{4}\,w(2\sinh^2 w + 1) - \frac{1}{8}\,(2\sinh w \cosh w) + C = \frac{2x^2 + 1}{4}\,\sinh^{-1} x - \frac{1}{4}\,x\sqrt{1+x^2} + C.$$

35. Let $u = 2x^{3/2}$, $dv = \dfrac{e^{-\sqrt{x}}}{\sqrt{x}}\,dx$. Then $du = 3x^{1/2}dx$, $v = -2e^{-\sqrt{x}}$, and hence

$$\begin{aligned}
\int 2xe^{-\sqrt{x}}\,dx &= -4x^{3/2}e^{-\sqrt{x}} + 6\int \sqrt{x}e^{-\sqrt{x}}\,dx \\
&= -4x^{3/2}e^{-\sqrt{x}} + 6\left[-2xe^{-\sqrt{x}} - 4\sqrt{x}e^{-\sqrt{x}} - 4e^{-\sqrt{x}}\right] + C \quad \text{by Exercise 34} \\
&= -4x^{3/2}e^{-\sqrt{x}} - 12xe^{-\sqrt{x}} - 24\sqrt{x}e^{-\sqrt{x}} - 24e^{-\sqrt{x}} + C.
\end{aligned}$$

37. Area $= \int_0^1 \underbrace{x \sin \pi x \, dx}_{u=x, dv=\sin \pi x} = \left(-\frac{x}{\pi} \cos \pi x + \frac{1}{\pi} \int \cos \pi x \, dx \right)\Big|_0^1 = \left(-\frac{x}{\pi} \cos \pi x + \frac{1}{\pi^2} \sin \pi x \right)\Big|_0^1 = \frac{1}{\pi}.$

39. Let dV be the differential volume of a cross-section disk of thickness dx. Then $dV = \pi r^2 dx = \pi y^2 dx = \pi (\ln x)^2 dx$ and hence the volume of revolution is

$$V = \int_1^e \pi (\ln x)^2 \, dx.$$

Let $u = (\ln x)^2$, $dv = dx$. Then $du = \dfrac{2(\ln x)}{x} \, dx$, $v = x$, and hence

$$V = \pi \left[x(\ln x)^2 - 2 \int \ln x \, dx \right]\Big|_1^e = \pi \left[x(\ln x)^2 - 2x \ln x + 2x \right]\Big|_1^e = \pi(e - 2).$$

41. Let dV be the differential volume of a shell of thickness dx revolved about the y-axis. Then $dV = 2\pi r h dx = 2\pi x y dx = 2\pi x \sin x dx$ and hence the volume of revolution is

$$V = 2\pi \int_0^\pi \underbrace{x \sin x \, dx}_{u=x, dv=\sin x \, dx} = 2\pi \left(-x \cos x + \sin x \right)\Big|_0^\pi = 2\pi^2.$$

43. The average value on $[0, 1]$ is equal to $\int_0^1 e^{-x} \sin \pi x \, dx$. Now,

$$
\begin{aligned}
\int \underbrace{e^{-x} \sin \pi x \, dx}_{u=\sin \pi x, dv=e^{-x} \, dx} &= -e^{-x} \sin \pi x + \pi \int \underbrace{e^{-x} \cos \pi x \, dx}_{u=\cos \pi x, dv=e^{-x} \, dx} \\
&= -e^{-x} \sin \pi x + \pi \left[-e^{-x} \cos \pi x - \pi \int e^{-x} \sin \pi x \, dx \right] \\
&= -e^{-x} \sin \pi x - \pi e^{-x} \cos \pi x - \pi^2 \int e^{-x} \sin \pi x \, dx
\end{aligned}
$$

$$\implies (1 + \pi^2) \int e^{-x} \sin \pi x \, dx = -e^{-x} (\sin \pi x + \pi \cos \pi x) + C$$

$$\implies \int e^{-x} \sin \pi x \, dx = -\tfrac{e^{-x}}{1+\pi^2} (\sin \pi x + \pi \cos \pi x) + C.$$

Therefore

$$\text{Average value} = -\frac{e^{-x}}{1+\pi^2} (\sin \pi x + \pi \cos \pi x)\Big|_0^1 = \frac{\pi(1 + e)}{e(1 + \pi^2)}.$$

45. Since $v(t) > 0$ for $0 < t < 1$ and $v(t) < 0$ for $1 < t < 2$, the distance travelled from $t = 0$ to $t = 2$ is

$$s = \int_0^2 |v(t)| \, dt = \int_0^1 t \sin \pi t \, dt + \int_1^2 (-t \sin \pi t) \, dt.$$

Now,

$$\int \underbrace{t \sin \pi t \, dt}_{u=t, dv=\sin \pi t \, dt} = -\frac{t}{\pi} \cos \pi t + \frac{1}{\pi} \int \cos \pi t \, dt = -\frac{t}{\pi} \cos \pi t + \frac{1}{\pi^2} \sin \pi t + C.$$

Therefore

$$s = \left(-\frac{t}{\pi} \cos \pi t + \frac{1}{\pi^2} \sin \pi t \right)\Big|_0^1 + \left(\frac{t}{\pi} \cos \pi t - \frac{1}{\pi^2} \sin \pi t \right)\Big|_1^2 = \frac{4}{\pi} \text{ meters.}$$

47. Work $= \int_1^5 (62.4)\pi x^2 (5-y)\ dy = \int_1^5 (62.4)\pi (\ln y)^2 (5-y)\ dy$. Now,

$$\int \underbrace{(5-y)(\ln y)^2\ dy}_{u=(\ln y)^2, dv=(5-y)\ dy} = (\ln y)^2 \left(5y - \frac{1}{2}y^2\right) - 2 \int \underbrace{(\ln y)\left(5 - \frac{1}{2}y\right)\ dy}_{u=\ln y, dv=(5-\frac{1}{2}y)dy}$$

$$= (\ln y)^2 \left(5y - \frac{1}{2}y^2\right) - 2\left[(\ln y)\left(5y - \frac{1}{4}y^2\right) - \int \left(5 - \frac{1}{4}y\right)\ dy\right]$$

$$= (\ln y)^2 \left(5y - \frac{1}{2}y^2\right) - 2(\ln y)\left(5y - \frac{1}{4}y^2\right) + 2\left(5y - \frac{1}{8}y^2\right) + C.$$

Hence

$$\text{Work} = (62.4)\pi \left[(\ln y)^2 \left(5y - \frac{1}{2}y^2\right) - 2(\ln y)\left(5y - \frac{1}{4}y^2\right) + 2\left(5y - \frac{1}{8}y^2\right)\right]\Big|_1^5$$

$$= (62.4)\pi \left[\frac{25}{2}(\ln 5)^2 - \frac{75}{2}(\ln 5) + 34\right].$$

49. The centroid of the region is the point $(\overline{x}, \overline{y})$, where

$$\overline{x} = \frac{1}{A}\int_1^e x(x - \ln x)\ dx = \frac{1}{A}\int_1^e \underbrace{(x^2 - x\ln x)\ dx}_{u=\ln x, dv=x\ dx} = \frac{1}{A}\left[\frac{1}{3}x^3 - \left(\frac{1}{2}x^2 \ln x - \frac{1}{2}\int x\ dx\right)\right]\Big|_1^e$$

$$= \frac{1}{A}\left[\frac{1}{3}x^3 - \frac{1}{2}x^2 \ln x + \frac{1}{4}x^2\right]\Big|_1^e = \frac{1}{A}\left(\frac{1}{3}e^3 - \frac{1}{4}e^2 - \frac{7}{12}\right),$$

$$\overline{y} = \frac{1}{2A}\int_1^e [x^2 - (\ln x)^2]\ dx = \frac{1}{2A}\left[\frac{1}{3}x^3 - (x(\ln x)^2 - 2x\ln x + 2x)\right]\Big|_1^e = \frac{1}{2A}\left(\frac{1}{3}e^3 - e + \frac{5}{3}\right),$$

and

$$A = \int_1^e (x - \ln x)\ dx = \left[\frac{1}{2}x^2 - (x\ln x - x)\right]\Big|_1^e = \frac{1}{2}e^2 - \frac{3}{2}.$$

51. $\int \sin^2 x\ dx = -\frac{1}{2}\sin x \cos x + \frac{1}{2}\int dx = -\frac{1}{2}\sin x \cos x + \frac{1}{2}x + C.$

53. Let $u = x^n$, $dv = e^x dx$. Then $du = nx^{n-1}dx$, $v = e^x$, and hence

$$\int x^n e^x\ dx = x^n e^x - n\int x^{n-1} e^x\ dx.$$

55. Let $u = x^n$, $dv = \sin ax dx$. Then $du = nx^{n-1}dx$, $v = -\frac{1}{a}\cos ax$, and hence

$$\int x^n \sin ax\ dx = -\frac{1}{a}x^n \cos ax + \frac{n}{a}\int x^{n-1}\cos ax\ dx.$$

57. Let $u = \rho(x)$, $dv = e^x dx$. Then $du = \rho'(x)dx$, $v = e^x$, and hence

$$\int \rho(x)e^x\ dx = \rho(x)e^x - \int \rho'(x)e^x\ dx.$$

Since $\rho(x)$ is a polynomial of degree n, $\rho'(x)$ is a polynomial of degree at most $n-1$. Hence, after at most $n-1$ further applications of this reduction formula one obtains $\int \rho(x)e^x\ dx$ in terms of $\int e^x\ dx$, which is equal to $e^x + C$, and hence gives a complete formula for $\int \rho(x)e^x\ dx$.

10.2 Trigonometric Substitutions

1. Let $x = 2\sin\theta$. Then $dx = 2\cos\theta\,d\theta$ and hence

$$\int \sqrt{4 - x^2}\,dx = \int (2\cos\theta)(2\cos\theta\,d\theta) = 2\int (1 + \cos 2\theta)\,d\theta = 2\theta + \sin 2\theta + C$$

$$= 2\theta + 2\sin\theta\cos\theta + C = 2\text{Sin}^{-1}\frac{x}{2} + 2\left(\frac{x}{2}\right)\frac{\sqrt{4 - x^2}}{2} + C = 2\text{Sin}^{-1}\frac{x}{2} + \frac{x}{2}\sqrt{4 - x^2} + C.$$

3. Let $x = 3\sin\theta$. Then $dx = 3\cos\theta\,d\theta$ and hence

$$\int_0^{3/2} \frac{dx}{\sqrt{9 - x^2}} = \int_0^{\pi/6} \frac{3\cos\theta}{3\cos\theta}\,d\theta = \theta\big|_0^{\pi/6} = \frac{\pi}{6}.$$

5. Let $u = x^2 - 4$. Then $du = 2x\,dx$ and hence

$$\int \frac{x^3}{\sqrt{x^2 - 4}}\,dx = \frac{1}{2}\int \frac{u + 4}{\sqrt{u}}\,du = \frac{1}{2}\int (u^{1/2} + 4u^{-1/2})\,du$$

$$= \frac{1}{3}u^{3/2} + 4u^{1/2} + C = \frac{1}{3}(x^2 - 4)^{3/2} + 4\sqrt{x^2 - 4} + C.$$

7. Let $x = \sin\theta$. Then $dx = \cos\theta\,d\theta$ and hence

$$\int \frac{\sqrt{1 - x^2}}{x}\,dx = \int \frac{\cos\theta}{\sin\theta}\cos\theta\,d\theta = \int \frac{1 - \sin^2\theta}{\sin\theta}\,d\theta = \int (\csc\theta - \sin\theta)\,d\theta$$

$$= \ln|\csc\theta - \cot\theta| + \cos\theta + C = \ln\left|\frac{1}{x} - \frac{\sqrt{1 - x^2}}{x}\right| + \sqrt{1 - x^2} + C.$$

Therefore

$$\int_{1/2}^1 \frac{\sqrt{1 - x^2}}{x}\,dx = \left(\ln\left|\frac{1 - \sqrt{1 - x^2}}{x}\right| + \sqrt{1 - x^2}\right)\Bigg|_{1/2}^1 = -\ln\left(2 - \sqrt{3}\right) - \frac{1}{2}\sqrt{3}.$$

9. Let $x = 2\sec\theta$. Then $dx = 2\sec\theta\tan\theta\,d\theta$ and hence

$$\int \frac{1}{x^3\sqrt{x^2 - 4}}\,dx = \int \frac{1}{8\sec^3\theta\,2\tan\theta}(2\sec\theta\,\tan\theta)\,d\theta = \frac{1}{8}\int \cos^2\theta\,d\theta = \frac{1}{16}\int (1 + \cos 2\theta)\,d\theta$$

$$= \frac{1}{16}\left(\theta + \frac{1}{2}\sin 2\theta\right) + C = \frac{1}{16}(\theta + \sin\theta\cos\theta) + C = \frac{1}{16}\left(\text{Sec}^{-1}(x/2) + \frac{\sqrt{x^2 - 4}}{x}\frac{2}{x}\right) + C.$$

11. Let $x = 4\tan\theta$. Then $dx = 4\sec^2\theta\,d\theta$ and hence

$$\int_0^4 \frac{1}{(16 + x^2)^2}\,dx = \int_0^{\pi/4} \frac{1}{256\sec^4\theta}4\sec^2\theta\,d\theta$$

$$= \frac{1}{64}\int_0^{\pi/4} \cos^2\theta\,d\theta = \frac{1}{128}(\theta + \sin\theta\cos\theta)\Big|_0^{\pi/4} = \frac{1}{128}\left(\frac{\pi}{4} + \frac{1}{2}\right).$$

364

13. Let $x = \sin\theta$. Then $dx = \cos\theta\, d\theta$ and hence

$$\int \frac{x^2}{(1-x^2)^{3/2}} = \int \frac{\sin^2\theta}{\cos^3\theta}\,\cos\theta\, d\theta = \int \tan^2\theta\, d\theta$$

$$= \int (\sec^2\theta - 1)\, d\theta = \tan\theta - \theta + C = \frac{x}{\sqrt{1-x^2}} - \text{Sin}^{-1}x + C.$$

15. Let $u = 1 - x^2$. Then $du = -2x\,dx$ and hence

$$\int_0^1 x^3\sqrt{1-x^2}\, dx = -\frac{1}{2}\int_1^0 (1-u)\sqrt{u}\, du = -\frac{1}{2}\int_1^0 (u^{1/2} - u^{3/2})\, du$$

$$= -\frac{1}{2}\left(\frac{2}{3}\, u^{3/2} - \frac{2}{5}\, u^{5/2}\right)\Big|_1^0 = \frac{2}{15}.$$

17. Let $x = \sqrt{3}\tan\theta$. Then $dx = \sqrt{3}\sec^2\theta\, d\theta$ and hence

$$\int \frac{x^2\, dx}{(x^2+3)^{3/2}} = \int \frac{3\tan^2\theta}{3\sqrt{3}\sec^3\theta}\,\sqrt{3}\sec^2\theta\, d\theta = \int \frac{\sin^2\theta}{\cos\theta}\, d\theta = \int \frac{1-\cos^2\theta}{\cos\theta}\, d\theta = \int (\sec\theta - \cos\theta)\, d\theta$$

$$= \ln|\sec\theta + \tan\theta| - \sin\theta + C = \ln\left|\frac{\sqrt{x^2+3}+x}{\sqrt{3}}\right| - \frac{x}{\sqrt{x^2+3}} + C = \ln(\sqrt{x^2+3}+x) - \frac{x}{\sqrt{x^2+3}} + C.$$

Note that the missing constant, $\ln\sqrt{3}$, is absorbed in the constant C.

19. Let $x = a\tan\theta$. Then $dx = a\sec^2\theta\, d\theta$ and hence

$$\int \frac{x^2}{\sqrt{x^2+a^2}}\, dx = \int \frac{a^2\tan^2\theta}{a\sec\theta}\,a\sec^2\theta\, d\theta = a^2\int \tan^2\theta\sec\theta\, d\theta = \frac{a^2}{2}\,(\sec\theta\tan\theta - \ln|\sec\theta + \tan\theta|) + C$$

$$= \frac{a^2}{2}\left(\frac{\sqrt{x^2+a^2}}{a}\,\frac{x}{a} - \ln\left|\frac{\sqrt{x^2+a^2}}{a} + \frac{x}{a}\right|\right) + C = \frac{x\sqrt{x^2+a^2}}{2} - \frac{a^2}{2}\,\ln(\sqrt{x^2+a^2}+x) + C.$$

21. Let $x = a\sec\theta$. Then $dx = a\sec\theta\tan\theta\, d\theta$ and hence

$$\int \frac{\sqrt{x^2-a^2}}{x^2}\, dx = \int \frac{a\tan\theta}{a^2\sec^2\theta}\,a\sec\theta\tan\theta\, d\theta = \int \frac{\tan^2\theta}{\sec\theta}\, d\theta = \int \frac{\sec^2\theta - 1}{\sec\theta}\, d\theta = \int (\sec\theta - \cos\theta)\, d\theta$$

$$= \ln|\sec\theta + \tan\theta| - \sin\theta + C = \ln\left|\frac{x}{a} + \frac{\sqrt{x^2-a^2}}{a}\right| - \frac{\sqrt{x^2-a^2}}{x} + C = \ln|x + \sqrt{x^2-a^2}| - \frac{\sqrt{x^2-a^2}}{x} + C.$$

Note that the missing constant, $\ln a$, is absorbed into the constant C.

23. Let $u = x^2 + a^2$. Then $du = 2x\,dx$ and hence

$$\int \frac{x\, dx}{\sqrt{(x^2+a^2)^3}} = \frac{1}{2}\int \frac{du}{u^{3/2}} = -u^{-1/2} + C = -\frac{1}{\sqrt{x^2+a^2}} + C.$$

25. Let $x = a\sec\theta$. Then $dx = a\sec\theta\tan\theta\, d\theta$ and hence

$$\int \frac{dx}{x\sqrt{x^2-a^2}} = \int \frac{a\sec\theta\tan\theta\, d\theta}{(a\sec\theta)(a\tan\theta)} = \frac{1}{a}\,\theta + C = \frac{1}{a}\,\text{Sec}^{-1}\frac{x}{a} + C.$$

27. Let $x = a \tan \theta$. Then $dx = a \sec^2 \theta \, d\theta$ and hence

$$\int \frac{\sqrt{x^2 + a^2}}{x^2} \, dx = \int \frac{a \sec \theta}{a^2 \tan^2 \theta} \, a \sec^2 \theta \, d\theta = \int \frac{\sec \theta (1 + \tan^2 \theta)}{\tan^2 \theta} \, d\theta$$

$$= \int \cot \theta \csc \theta \, d\theta + \int \sec \theta \, d\theta = -\csc \theta + \ln|\sec \theta + \tan \theta| + C$$

$$= -\frac{\sqrt{x^2 + a^2}}{x} + \ln\left|\frac{\sqrt{x^2 + a^2}}{a} + \frac{x}{a}\right| + C = -\frac{\sqrt{x^2 + a^2}}{x} + \ln(\sqrt{x^2 + a^2} + x) + C.$$

Note that the missing constant, $\ln a$, is absorbed in the constant C.

29. Let $x = 3 \tan \theta$. Then $dx = 3 \sec^2 \theta \, d\theta$ and hence

$$\int \frac{x^2 + 3}{\sqrt{x^2 + 9}} \, dx = \int \frac{9 \tan^2 \theta + 3}{3 \sec \theta} \, 3 \sec^2 \theta \, d\theta = 9 \int (\sec \theta)(\tan^2 \theta) \, d\theta + 3 \int \sec \theta \, d\theta$$

$$= \frac{9}{2} (\sec \theta \tan \theta - \ln|\sec \theta + \tan \theta|) + 3 \ln|\sec \theta + \tan \theta| + C = \frac{9}{2} \sec \theta \tan \theta - \frac{3}{2} \ln|\sec \theta + \tan \theta| + C$$

$$= \frac{9}{2} \frac{\sqrt{x^2 + 9}}{3} \frac{x}{3} - \frac{3}{2} \ln\left|\frac{\sqrt{x^2 + 9}}{3} + \frac{x}{3}\right| + C = \frac{1}{2} x\sqrt{x^2 + 9} - \frac{3}{2} \ln(\sqrt{x^2 + 9} + x) + C.$$

31. Let $x = 3 \sin \theta$. Then $dx = 3 \cos \theta \, d\theta$ and hence

$$\int \frac{x^2 - 4x + 5}{x\sqrt{9 - x^2}} \, dx = \int \frac{9 \sin^2 \theta - 12 \sin \theta + 5}{(3 \sin \theta)(3 \cos \theta)} (3 \cos \theta \, d\theta) = 3 \int \sin \theta \, d\theta - 4 \int d\theta + \frac{5}{3} \int \csc \theta \, d\theta$$

$$= -3 \cos \theta - 4\theta + \frac{5}{3} \ln|\csc \theta - \cot \theta| + C = -\sqrt{9 - x^2} - 4 \operatorname{Sin}^{-1}\frac{x}{3} + \frac{5}{3} \ln\left|\frac{3}{x} - \frac{\sqrt{9 - x^2}}{x}\right| + C.$$

33. First observe that $y = 5 - \sqrt{x^2 + 9} = 0 \implies x^2 + 9 = 25 \implies x = \pm 4$. Hence the area is equal to

$$2 \int_0^4 \left(5 - \sqrt{x^2 + 9}\right) dx = 10x\big|_0^4 - 2 \int_0^4 \sqrt{x^2 + 9} \, dx = 40 - 2 \int_0^4 \sqrt{x^2 + 9} \, dx.$$

To find the last integral, let $x = 3 \tan \theta$. Then $dx = 3 \sec^2 \, d\theta$ and hence

$$\int \sqrt{x^2 + 9} \, dx = \int (3 \sec \theta)(3 \sec^2 \theta) \, d\theta = 9 \int \sec^3 \theta \, d\theta$$

$$= \frac{9}{2} \sec \theta \tan \theta + \frac{9}{2} \ln|\sec \theta + \tan \theta| + C = \frac{1}{2} x\sqrt{x^2 + 9} + \frac{9}{2} \ln\left|\sqrt{x^2 + 9} + x\right| + C.$$

Therefore

$$\text{Area} = 40 - 2 \left[\frac{1}{2} x\sqrt{x^2 + 9} + \frac{9}{2} \ln\left|\sqrt{x^2 + 9} + x\right|\right]\Bigg|_0^4 = 20 - 9 \ln 3.$$

35. If $0 \le x \le 3$, $\dfrac{1}{\sqrt{x^2 + 16}} \ge -\dfrac{x}{60} + \dfrac{1}{4}$, with equality at $x = 3$. Hence the area between the curves is equal to

$$\int_0^3 \left[\frac{1}{\sqrt{x^2 + 16}} - \left(-\frac{x}{60} + \frac{1}{4}\right)\right] dx.$$

Now, let $x = 4\tan\theta$. Then $dx = 4\sec^2\theta\,d\theta$ and hence

$$\int \frac{1}{\sqrt{x^2+16}}\,dx = \int \frac{1}{4\sec\theta}\,4\sec^2\theta\,d\theta = \int \sec\theta\,d\theta$$

$$= \ln|\sec\theta + \tan\theta| + C = \ln\left|\frac{\sqrt{x^2+16}}{4} + \frac{x}{4}\right| + C = \ln\left|\sqrt{x^2+16} + x\right| + C,$$

the missing constant, $\ln 4$, being absorbed in the constant C. Therefore the area is equal to

$$\left(\ln\left|\sqrt{x^2+16} + x\right| + \frac{x^2}{120} - \frac{1}{4}x\right)\Bigg|_0^3 = \ln 2 - \frac{27}{40}.$$

37. Let dV be the differential volume of a cross-section disk of thickness dx. Then $dV = \pi r^2\,dx = \pi y^2\,dx = \pi x^2\sqrt{16-x^2}\,dx$ and hence, since the region is symmetric about the y-axis, the total volume of revolution is

$$V = 2\pi \int_0^4 x^2\sqrt{16-x^2}\,dx.$$

To evaluate this integral, let $x = 4\sin\theta$. Then $dx = 4\cos\theta\,d\theta$ and hence, as in Exercise 36,

$$V = 2\pi \int_0^{\pi/2} (16\sin^2\theta)(4\cos\theta)(4\cos\theta)\,d\theta$$

$$= 512\pi \int_0^{\pi/2} \sin^2\theta\cos^2\theta\,d\theta = 64\pi\left(\theta - \frac{1}{4}\sin 4\theta\right)\Bigg|_0^{\pi/2} = 32\pi^2.$$

39. Let dV be the differential volume of a cross-section washer of thickness dx. Then $dV = \pi(R^2 - r^2)\,dx = \pi[1/9 - 1/(x^2+9)]\,dx$ and hence the volume of revolution is

$$V = \pi \int_0^3 \left[\frac{1}{9} - \frac{1}{x^2+9}\right]dx = \frac{\pi}{9}x\Big|_0^3 - \frac{\pi}{9}\int_0^3 \underbrace{\frac{1}{(x/3)^2+1}}_{u=x/3}\,dx$$

$$= \frac{\pi}{3} - \frac{\pi}{3}\int_0^1 \frac{1}{1+u^2}\,du = \frac{\pi}{3} - \frac{\pi}{3}\tan^{-1}u\Big|_0^1 = \frac{\pi}{3} - \frac{\pi^2}{12}.$$

41. Let L be the horizontal length of the tank. Then the total volume of the tank is $\pi(5)^2 L$ and hence the volume of water in the tank is $(\text{Area})L$, where Area stands for the area of the circular end when the water is at a depth of 6 meters. Therefore the percentage of the filled capacity is

$$\frac{\text{Vol(water)}}{\text{Vol(tank)}} = \frac{\text{Area }L}{\pi(5)^2 L} = \frac{\text{Area}}{25\pi}.$$

Now,

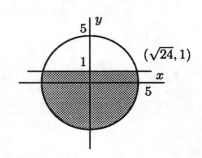

$$\begin{aligned}
\text{Area} &= \pi(5)^2 - 2\int_0^{\sqrt{24}} (\sqrt{25-x^2} - 1)\,dx \\
&= 25\pi - 2\left[\frac{1}{2}\left(x\sqrt{25-x^2} + 25\sin^{-1}\frac{x}{5}\right) - x\right]\Bigg|_0^{\sqrt{24}} \\
&= 25\pi + \sqrt{24} - 25\sin^{-1}\frac{\sqrt{24}}{5} \approx 49.20279.
\end{aligned}$$

Hence the percentage of filled capacity is $49.20279/25\pi \approx 0.6265$, or 62.65%.

43. $y = x^2 \implies y' = 2x \implies$

$$\text{Length} = \int_0^{1/2} \sqrt{1 + 4x^2}\, dx.$$

To evaluate this integral, let $x = (1/2)\tan\theta$. Then $dx = (1/2)\sec^2\theta\, d\theta$ and hence

$$\text{Length} = \int_0^{\pi/4} \sec\theta \left(\frac{1}{2}\sec^2\theta \right) d\theta = \frac{1}{2} \int \sec^3\theta\, d\theta$$

$$= \frac{1}{4}\left[\sec\theta\tan\theta + \ln|\sec\theta + \tan\theta|\right]\Big|_0^{\pi/4} = \frac{1}{4}\sqrt{2} + \frac{1}{4}\ln(\sqrt{2}+1).$$

10.3   Integrals Involving Quadratic Expressions

1. $x^2 - 4x + 4 = (x-2)^2$. Let $u = x - 2$. Then $du = dx$ and hence

$$\int \frac{dx}{x^2 - 4x + 4} = \int \frac{du}{u^2} = -\frac{1}{u} + C = -\frac{1}{x-2} + C.$$

3. $x^2 + 6x + 13 = (x+3)^2 + 4$. Let $u = x + 3$. Then $du = dx$ and hence

$$\int \frac{dx}{\sqrt{x^2 + 6x + 13}} = \int \frac{du}{\sqrt{u^2 + 4}}.$$

Now, let $u = 2\tan\theta$. Then $du = 2\sec^2\theta\, d\theta$ and hence

$$\int \frac{du}{\sqrt{u^2 + 4}} = \int \frac{2\sec^2\theta\, d\theta}{2\sec\theta} = \int \sec\theta\, d\theta = \ln|\sec\theta + \tan\theta| + C = \ln\left|\sqrt{u^2 + 4} + u\right| + C$$

$$\implies \int \frac{dx}{\sqrt{x^2 + 6x + 13}} = \ln(\sqrt{x^2 + 6x + 13} + x + 3) + C.$$

5. $x^2 - 6x = (x-3)^2 - 9$. Let $u = x - 3$. Then $du = dx$ and hence

$$\int \frac{dx}{\sqrt{x^2 - 6x}} = \int \frac{du}{\sqrt{u^2 - 9}}.$$

Now, let $u = 3\sec\theta$. Then $du = 3\sec\theta\tan\theta\, d\theta$ and hence

$$\int \frac{du}{\sqrt{u^2 - 9}} = \int \frac{3\sec\theta\tan\theta\, d\theta}{3\tan\theta} = \int \sec\theta = \ln|\sec\theta + \tan\theta| + C = \ln\left|u + \sqrt{u^2 - 9}\right| + C$$

$$\implies \int \frac{dx}{\sqrt{x^2 - 6x}} = \ln\left|x - 3 + \sqrt{x^2 - 6x}\right| + C.$$

7. $6x - x^2 = 9 - (x-3)^2$. Let $u = x - 3$. Then $du = dx$ and hence

$$\int \frac{x\, dx}{\sqrt{6x - x^2}} = \int \frac{u + 3}{\sqrt{9 - u^2}}\, du = \int \frac{u}{\sqrt{9 - u^2}}\, du + 3\int \frac{1}{\sqrt{9 - u^2}}\, du$$

$$= -\sqrt{9 - u^2} + 3\text{Sin}^{-1}\frac{u}{3} + C = -\sqrt{6x - x^2} + 3\text{Sin}^{-1}\frac{x-3}{3} + C.$$

9. $x^2 + 6x + 18 = (x+3)^2 + 9$. Let $u = x + 3$. Then $du = dx$ and hence

$$\int_0^1 \frac{2x+1}{\sqrt{18+6x+x^2}} \, dx = \int_3^4 \frac{2u-5}{\sqrt{u^2+9}} \, du = 2\int_3^4 \frac{u}{\sqrt{u^2+9}} \, du - 5\int \frac{1}{\sqrt{u^2+9}} \, du$$

$$= \left(2\sqrt{u^2+9} - 5\ln\left|\sqrt{u^2+9}+u\right|\right)\Big|_3^4 = 10 - 6\sqrt{2} - 5\ln 9 + 5\ln(3\sqrt{2}+3).$$

11. $x^2 - 8x - 7 = (x-4)^2 - 23$. Let $u = x - 4$. Then $du = dx$ and hence

$$\int_{-3}^{-2} \frac{x-4}{x^2-8x-7} \, dx = \int_{-7}^{-6} \frac{u}{u^2-23} \, du = \frac{1}{2}\ln\left|u^2-23\right|\Big|_{-7}^{-6} = -\frac{1}{2}\ln 2 = -\ln\sqrt{2}.$$

13. $x^2 + 2x + 2 = (x+1)^2 + 1$. Let $u = x + 1$. Then $du = dx$ and hence

$$\int_0^1 \frac{x}{(x^2+2x+2)^2} \, dx = \int_1^2 \frac{u-1}{(u^2+1)^2} \, du = \int_1^2 \frac{u}{(u^2+1)^2} \, du - \int_1^2 \frac{1}{(u^2+1)^2} \, du$$

$$= \frac{1}{2}\frac{-1}{u^2+1}\Big|_1^2 - \int_1^2 \frac{1}{(u^2+1)^2} \, du = \frac{3}{20} - \int_1^2 \frac{1}{(u^2+1)^2} \, du.$$

Now, let $u = \tan\theta$. Then $du = \sec^2\theta$ and hence

$$\int \frac{1}{(u^2+1)^2} \, du = \int \frac{\sec^2\theta \, d\theta}{\sec^4\theta} = \int \cos^2\theta \, d\theta = \frac{1}{2}\int(1+\cos 2\theta) \, d\theta = \frac{1}{2}\left(\theta + \frac{1}{2}\sin 2\theta\right) + C$$

$$= \frac{1}{2}(\theta + \sin\theta\cos\theta) + C = \frac{1}{2}\left(\text{Tan}^{-1}u + \frac{u}{\sqrt{u^2+1}}\frac{1}{\sqrt{u^2+1}}\right) + C = \frac{1}{2}\left(\text{Tan}^{-1}u + \frac{u}{u^2+1}\right) + C$$

$$\implies \int_1^2 \frac{1}{(u^2+1)^2} \, du = \frac{1}{2}\left(\text{Tan}^{-1}2 - \frac{1}{10} - \frac{\pi}{4}\right)$$

$$\implies \int_0^1 \frac{x}{(x^2+2x+2)^2} \, dx = \frac{3}{20} - \frac{1}{2}\left(\text{Tan}^{-1}2 - \frac{1}{10} - \frac{\pi}{4}\right) = \frac{1}{5} - \frac{1}{2}\text{Tan}^{-1}2 + \frac{\pi}{8}.$$

15. $x^2 - 18x + 88 = (x-9)^2 + 7$. Let $u = x - 9$. Then $du = dx$ and hence

$$\int \frac{dx}{(9-x)(88-18x+x^2)^{3/2}} = -\int \frac{du}{u(u^2+7)^{3/2}}.$$

Now, let $u = \sqrt{7}\tan\theta$. Then $du = \sqrt{7}\sec^2\theta \, d\theta$ and hence

$$\int \frac{du}{u(u^2+7)^{3/2}} = \int \frac{\sqrt{7}\sec^2\theta \, d\theta}{(\sqrt{7}\tan\theta)7\sqrt{7}\sec^3\theta} = \frac{1}{7\sqrt{7}}\int \frac{\cos^2\theta}{\sin\theta} \, d\theta = \frac{1}{7\sqrt{7}}\int(\csc\theta - \sin\theta) \, d\theta$$

$$= \frac{1}{7\sqrt{7}}(\ln|\csc\theta - \cot\theta| + \cos\theta) + C = \frac{1}{7\sqrt{7}}\left(\ln\left|\frac{\sqrt{u^2+7}}{u} - \frac{\sqrt{7}}{u}\right| + \frac{\sqrt{7}}{\sqrt{u^2+7}}\right) + C$$

$$\implies \int \frac{dx}{(9-x)(88-18x+x^2)^{3/2}} = \frac{-1}{7\sqrt{7}}\left(\ln\left|\frac{\sqrt{x^2-18x+88}-\sqrt{7}}{x-9}\right| + \sqrt{\frac{7}{88-18x+x^2}}\right) + C.$$

17. $4x^2 + 20x + 29 = (2x + 5)^2 + 4$. Let $u = 2x + 5$. Then $du = 2dx$ and hence

$$\int \frac{4x^2 + 20x + 25}{\sqrt{4x^2 + 20x + 29}} \, dx = \frac{1}{2} \int \frac{u^2}{\sqrt{u^2 + 4}} \, du.$$

Now, let $u = 2 \tan \theta$. Then $du = 2 \sec^2 \theta \, d\theta$ and hence

$$\int \frac{u^2}{\sqrt{u^2 + 4}} \, du = \int \frac{4 \tan^2 \theta}{2 \sec \theta} 2 \sec^2 \theta \, d\theta = 4 \int \sec \theta \tan^2 \theta \, d\theta$$

$$= 2(\sec \theta \tan \theta - \ln |\sec \theta + \tan \theta|) + C = 2 \left(\frac{\sqrt{u^2 + 4}}{2} \frac{u}{2} - \ln \left| \sqrt{u^2 + 4} + u \right| \right) + C$$

$$\implies \int \frac{4x^2 + 20x + 25}{\sqrt{4x^2 + 20x + 29}} \, dx = \frac{(2x + 5)\sqrt{4x^2 + 20x + 29}}{4} - \ln \left| \sqrt{4x^2 + 20x + 29} + 2x + 5 \right| + C.$$

19. $x^2 - 4x + 8 = (x - 2)^2 + 4$. Let $u = x - 2$. Then

$$\text{Area} = \int_2^4 \frac{1}{x^2 - 4x + 8} \, dx = \int_0^2 \frac{1}{u^2 + 4} \, du = \frac{1}{2} \text{Tan}^{-1} \frac{u}{2} \Big|_0^2 = \frac{\pi}{8}.$$

21. Let dV be the differential volume of a cross-section disk of thickness dx. Then $dV = \pi r^2 dx = \pi y^2 dx$ $= \pi(8x - x^2)^{-1/2} dx$ and hence the volume of revolution is

$$V = \pi \int_2^4 \frac{1}{\sqrt{8x - x^2}} \, dx = \pi \int_2^4 \frac{1}{\sqrt{16 - (4 - x)^2}} \, dx = -\pi \int_2^0 \frac{1}{\sqrt{16 - u^2}} \, du.$$

Now, let $u = 4 \sin \theta$. Then $du = 4 \cos \theta \, d\theta$ and hence

$$V = -\pi \int_{\pi/6}^0 \frac{1}{4 \cos \theta} 4 \cos \theta \, d\theta = -\pi \theta \Big|_{\pi/6}^0 = \frac{\pi^2}{6}.$$

10.4   The Method of Partial Fractions

1. $\frac{1}{x(x + 1)} = \frac{A}{x} + \frac{B}{x + 1}$. Then $x = 0 \implies 1 = A$, $x = -1 \implies -1 = B$, and hence

$$\int \frac{1}{x(x + 1)} \, dx = \int \left(\frac{1}{x} - \frac{1}{x + 1} \right) dx = \ln |x| - \ln |x + 1| + C.$$

3. $\frac{2x}{(x - 3)(x - 1)} = \frac{A}{x - 3} + \frac{B}{x - 1}$. Then $x = 3 \implies 3 = A$, $x = 1 \implies -1 = B$, and hence

$$\int_4^6 \frac{2x}{(x - 3)(x - 1)} \, dx = \int_4^6 \left(\frac{3}{x - 3} + \frac{-1}{x - 1} \right) dx = (3 \ln |x - 3| - \ln |x - 1|) \Big|_4^6 = 4 \ln 3 - \ln 5.$$

5. $\frac{2x + 4}{1 - x^2} = \frac{-2x - 4}{(x - 1)(x + 1)} = \frac{A}{x - 1} + \frac{B}{x + 1}$. Then $x = 1 \implies -3 = A$, $x = -1 \implies 1 = B$, and hence

$$\int \frac{2x + 4}{1 - x^2} \, dx = \int \left(\frac{-3}{x - 1} + \frac{1}{x + 1} \right) dx = -3 \ln |x - 1| + \ln |x + 1| + C.$$

7. $\dfrac{1}{1-x^2} = \dfrac{-1}{(x-1)(x+1)} = \dfrac{A}{x-1} + \dfrac{B}{x+1}$. Then $x = 1 \implies -1/2 = A$, $x = -1 \implies 1/2 = B$, and hence

$$\int \frac{1}{1-x^2}\, dx = \int \left(\frac{-1/2}{x-1} + \frac{1/2}{x+1} \right)\, dx = -\frac{1}{2}\ln|x-1| + \frac{1}{2}\ln|x+1| + C.$$

9. $\dfrac{1}{x+x^3} = \dfrac{1}{x(1+x^2)} = \dfrac{A}{x} + \dfrac{Bx+C}{1+x^2} \implies 1 = A(1+x^2) + x(Bx+C) = A + x(C) + x^2(A+B).$

Equating coefficients of x, we find that

$$
\begin{aligned}
A &= 1 \quad \text{coefficients of constant terms}\\
C &= 0 \quad \text{coefficients of } x\\
A+B &= 0 \quad \text{coefficients of } x^2
\end{aligned}
$$

Therefore $A = 1$, $B = -1$, $C = 0$ and hence

$$\int_e^{e^2} \frac{1}{x+x^3}\, dx = \int_e^{e^2} \left(\frac{1}{x} + \frac{-x}{1+x^2} \right)\, dx = \left[\ln|x| - \frac{1}{2}\ln(1+x^2) \right]\Bigg|_e^{e^2} = 1 - \frac{1}{2}\ln(1+e^4) + \frac{1}{2}\ln(1+e^2).$$

11. $\dfrac{1}{(x+1)(x^2+1)} = \dfrac{A}{x+1} + \dfrac{Bx+C}{x^2+1}$

$$\implies \quad 1 = A(x^2+1) + (Bx+C)(x+1) = (A+C) + x(B+C) + x^2(A+B).$$

Equating coefficients of like powers of x, we find that

$$
\begin{aligned}
A+C &= 1 \quad \text{coefficients of constant terms}\\
B+C &= 0 \quad \text{coefficients of } x\\
A+B &= 0 \quad \text{coefficients of } x^2
\end{aligned}
$$

It follows that $A = C = 1/2$, $B = -1/2$, and hence

$$\int \frac{1}{(x+1)(x^2+1)}\, dx = \int \left(\frac{\frac{1}{2}}{x+1} + \frac{-\frac{1}{2}x + \frac{1}{2}}{x^2+1} \right)\, dx$$

$$= \frac{1}{2}\int \frac{1}{x+1}\, dx - \frac{1}{2}\int \frac{x}{x^2+1}\, dx + \frac{1}{2}\int \frac{1}{x^2+1}\, dx = \frac{1}{2}\ln|x+1| - \frac{1}{4}\ln(x^2+1) + \frac{1}{2}\mathrm{Tan}^{-1}x + C.$$

13. $\dfrac{3x^2 - 2x + 2}{(x-2)(x^2+1)} = \dfrac{A}{x-2} + \dfrac{Bx+C}{x^2+1}$

$$\implies \quad 3x^2 - 2x + 2 = A(x^2+1) + (x-2)(Bx+C) = (A-2C) + x(-2B+C) + x^2(A+B)$$

$$\implies \quad A - 2C = 2,\ -2B+C = -2,\ A+B = 3$$

$$\implies \quad A = 2,\ B = 1,\ C = 0.$$

Therefore

$$\int_3^4 \frac{3x^2 - 2x + 2}{(x-2)(x^2+1)}\, dx = \int_3^4 \left(\frac{2}{x-2} + \frac{x}{x^2+1} \right)\, dx$$

$$= \left(2\ln|x-2| + \frac{1}{2}\ln(x^2+1) \right)\Bigg|_3^4 = 2\ln 2 + \frac{1}{2}\ln 17 - \frac{1}{2}\ln 10.$$

15. $\dfrac{4x^2 + x + 1}{(x+2)(x^2+1)} = \dfrac{A}{x+2} + \dfrac{Bx + C}{x^2 + 1}$

$\implies 4x^2 + x + 1 = A(x^2 + 1) + (x+2)(Bx + C) = (A + 2C) + x(2B + C) + x^2(A + B)$

$\implies A + 2C = 1,\ 2B + C = 1,\ A + B = 4$

$\implies A = 3,\ B = 1,\ C = -1.$

Therefore

$$\int \frac{4x^2 + x + 1}{(x+2)(x^2+1)}\, dx = \int \left(\frac{3}{x+2} + \frac{x-1}{x^2+1} \right) dx$$

$$= 3 \int \frac{1}{x+2}\, dx + \int \frac{x}{x^2+1}\, dx - \int \frac{1}{x^2+1}\, dx = 3 \ln|x+2| + \frac{1}{2} \ln(x^2+1) - \text{Tan}^{-1}x + C.$$

17. $\dfrac{x^2 + 5x + 1}{x(x+1)^2} = \dfrac{A}{x} + \dfrac{B}{x+1} + \dfrac{C}{(x+1)^2}$

$\implies x^2 + 5x + 1 = A(x+1)^2 + Bx(x+1) + Cx = A + x(2A + B + C) + x^2(A + B)$

$\implies A = 1,\ 2A + B + C = 5,\ A + B = 1$

$\implies A = 1,\ B = 0,\ C = 3.$

Therefore

$$\int_e^{e^3} \frac{x^2 + 5x + 1}{x(x+1)^2}\, dx = \int_e^{e^3} \left(\frac{1}{x} + \frac{3}{(x+1)^2} \right) dx = \left(\ln|x| - \frac{3}{x+1} \right)\Bigg|_e^{e^3} = 2 + \frac{3}{e+1} - \frac{3}{e^3 + 1}.$$

19. $\dfrac{2x^2 + 3x + 2}{x^3 + 2x^2 + x} = \dfrac{2x^2 + 3x + 2}{x(x+1)^2} = \dfrac{A}{x} + \dfrac{B}{x+1} + \dfrac{C}{(x+1)^2}$

$\implies 2x^2 + 3x + 2 = A((x+1)^2 + Bx(x+1) + Cx = A + x(2A + B + C) + x^2(A + B)$

$\implies A = 2,\ 2A + B + C = 3,\ A + B = 2$

$\implies A = 2,\ B = 0,\ C = -1$

Therefore

$$\int \frac{2x^2 + 3x + 2}{x^3 + 2x^2 + x}\, dx = \int \left(\frac{2}{x} - \frac{1}{(x+1)^2} \right) dx = 2 \ln|x| + \frac{1}{x+1} + C.$$

21. $\dfrac{1}{x^3 + 3x^2 + 2x} = \dfrac{1}{x(x+1)(x+2)} = \dfrac{A}{x} + \dfrac{B}{x+1} + \dfrac{C}{x+2}.$ Then

$$\begin{array}{rcccl} x = 0 & \implies & 1/2 & = & A \\ x = -1 & \implies & -1 & = & B \\ x = -2 & \implies & 1/2 & = & C. \end{array}$$

and therefore

$$\int_1^2 \frac{1}{x^3 + 3x^2 + 2x}\, dx = \int_1^2 \left(\frac{1/2}{x} - \frac{1}{x+1} + \frac{1/2}{x+2} \right) dx$$

$$= \left(\frac{1}{2} \ln|x| - \ln|x+1| + \frac{1}{2} \ln|x+2| \right)\Bigg|_1^2 = \frac{3}{2} \ln 2 - \frac{3}{2} \ln 3 + \ln 2.$$

23. $\dfrac{5x^3 - 9x^2 + 3x - 3}{x(x - 3)(x^2 + 1)} = \dfrac{A}{x} + \dfrac{B}{x - 3} + \dfrac{Cx + D}{x^2 + 1}$

$\implies \quad 5x^3 - 9x^2 + 3x - 3 = A(x - 3)(x^2 + 1) + Bx(x^2 + 1) + (Cx + D)x(x - 3)$

$\qquad\qquad = (-3A) + x(A + B - 3D) + x^2(-3A - 3C + D) + x^3(A + B + C)$

$\implies \quad -3A = -3, \ A + B - 3D = 3, \ -3A - 3C + D = -9, \ A + B + C = 5$

$\implies \quad A = 1, \ B = 2, \ C = 2, \ D = 0.$

Therefore

$$\int \frac{5x^3 - 9x^2 + 3x - 3}{x(x - 3)(x^2 + 1)} \, dx = \int \left(\frac{1}{x} + \frac{2}{x - 3} + \frac{2x}{x^2 + 1} \right) \, dx$$

$$= \ln |x| + 2\ln |x - 3| + \ln(x^2 + 1) + C.$$

25. $\dfrac{4}{x^3 + x} = \dfrac{4}{x(x^2 + 1)} = \dfrac{A}{x} + \dfrac{Bx + C}{x^2 + 1}$

$\implies \quad 4 = A(x^2 + 1) + (Bx + C)x = A + x(C) + x^2(A + B)$

$\implies \quad A = 4, \ C = 0, \ A + B = 0$

$\implies \quad A = 4, \ B = -4, \ C = 0.$

Therefore

$$\int \frac{4}{x^3 + x} \, dx = \int \left(\frac{4}{x} + \frac{-4x}{x^2 + 1} \right) \, dx = 4\ln |x| - 2\ln(x^2 + 1) + C.$$

27. First observe, by trial and error, that

$$\frac{1}{a^2 - x^2} = \frac{1}{2a} \left(\frac{1}{a - x} + \frac{1}{a + x} \right).$$

Therefore

$$\frac{1}{(a^2 - x^2)^2} = \ = \ \frac{1}{4a^2} \left[\frac{1}{(a - x)^2} + \frac{2}{(a - x)(a + x)} + \frac{1}{(a + x)^2} \right]$$

$$= \ \frac{1}{4a^2} \left[\frac{1}{(a - x)^2} + \frac{1}{a} \left(\frac{1}{a - x} + \frac{1}{a + x} \right) + \frac{1}{(a + x)^2} \right]$$

$$= \ \frac{1}{4a^3} \frac{1}{a - x} + \frac{1}{4a^2} \frac{1}{(a - x)^2} + \frac{1}{4a^3} \frac{1}{a + x} + \frac{1}{4a^2} \frac{1}{(a + x)^2}$$

Hence

$$\int \frac{1}{(a^2 - x^2)^2} \, dx = \frac{1}{4a^3} \int \frac{1}{a - x} \, dx + \frac{1}{4a^2} \int \frac{1}{(a - x)^2} \, dx + \frac{1}{4a^3} \int \frac{1}{a + x} \, dx + \frac{1}{4a^2} \int \frac{1}{(a + x)^2} \, dx$$

$$= -\frac{1}{4a^3} \ln |a - x| + \frac{1}{4a^2} \frac{1}{a - x} + \frac{1}{4a^3} \ln |a + x| - \frac{1}{4a^2} \frac{1}{a + x} + C.$$

29. $\dfrac{2x^3 - 4x^2 - 7x - 18}{x^2 - 2x - 8} = 2x + \dfrac{9x - 18}{(x - 4)(x + 2)} = 2x + \dfrac{A}{x - 4} + \dfrac{B}{x + 2}.$ Thus

$$x = 4 \quad \implies \quad 3 = A$$

$$x = -2 \quad \implies \quad 6 = B.$$

Therefore

$$\int \frac{2x^3 - 4x^2 - 7x - 18}{x^2 - 2x - 8} \, dx = \int \left(2x + \frac{3}{x - 4} + \frac{6}{x + 2} \right) \, dx = x^2 + 3\ln |x - 4| + 6\ln |x + 2| + C.$$

31. $\dfrac{5x^2 + 2x + 2}{x^3 - 1} = \dfrac{5x^2 + 2x + 2}{(x-1)(x^2 + x + 1)} = \dfrac{A}{x-1} + \dfrac{Bx + C}{x^2 + x + 1}$

$\implies 5x^2 + 2x + 2 = A(x^2 + x + 1) + (x-1)(Bx + C) = (A - C) + x(A - B + C) + x^2(A + B)$

$\implies A - C = 2,\ A - B + C = 2,\ A + B = 5$

$\implies A = 3,\ B = 2,\ C = 1.$

Therefore

$$\int \dfrac{5x^2 + 2x + 2}{x^3 - 1}\, dx = \int \left(\dfrac{3}{x-1} + \dfrac{2x+1}{x^2 + x + 1} \right)\, dx = 3\ln|x - 1| + \ln(x^2 + x + 1) + C.$$

33. $\dfrac{9x^2 + 26x - 16}{x^3 + 2x^2 - 8x} = \dfrac{9x^2 + 26x - 16}{x(x+4)(x-2)} = \dfrac{A}{x} + \dfrac{B}{x+4} + \dfrac{C}{x-2}.$ Then

$$x = 0 \implies 2 = A$$
$$x = -4 \implies 1 = B$$
$$x = 2 \implies 6 = C.$$

Therefore

$$\int \dfrac{9x^2 + 26x - 16}{x^3 + 2x^2 - 8x}\, dx = \int \left(\dfrac{2}{x} + \dfrac{1}{x+4} + \dfrac{6}{x-2} \right)\, dx = 2\ln|x| + \ln|x + 4| + 6\ln|x - 2| + C.$$

35. Area $= \displaystyle\int_1^2 \dfrac{7x + 3}{x^2 + x}\, dx = \int_1^2 \left(\dfrac{3}{x} + \dfrac{4}{x+1} \right)\, dx = \left[3\ln x + 4\ln(x+1) \right]\Big|_1^2 = 4\ln 3 - \ln 2.$

37. Let dV be the differential volume of a shell of thickness dx revolved about the y-axis. Then

$$dV = 2\pi rh\, dx = 2\pi xy\, dx = \dfrac{2\pi x}{(x+1)(x+3)}\, dx$$

and hence the volume of revolution is

$$V = 2\pi \int_0^2 \dfrac{x}{(x+1)(x+3)}\, dx = 2\pi \int_0^2 \left(\dfrac{-1/2}{x+1} + \dfrac{3/2}{x+3} \right)\, dx$$

$$= 2\pi \left(-\dfrac{1}{2}\ln(x+1) + \dfrac{3}{2}\ln(x+3) \right)\Bigg|_0^2 = 2\pi \left(-2\ln 3 + \dfrac{3}{2}\ln 5 \right).$$

39. Let dV be the differential volume of a cross-section disk of thickness dx. Then

$$dV = \pi r^2 dx = \pi y^2 dx = \pi\, \dfrac{4x^2 + 1}{x^3 + x}\, dx$$

and hence the volume of revolution is

$$V = \pi \int_1^5 \dfrac{4x^2 + 1}{x^3 + x}\, dx = \pi \int_1^5 \left(\dfrac{1}{x} + \dfrac{3x}{x^2 + 1} \right)\, dx$$

$$= \pi \left(\ln x + \dfrac{3}{2}\ln(x^2 + 1) \right)\Bigg|_1^5 = \pi \left(\ln 5 + \dfrac{3}{2}\ln 13 \right).$$

41. $\dfrac{A}{x-a} + \dfrac{B}{(x-a)^2} = \dfrac{A(x-a)+B}{(x-a)^2} = \dfrac{Ax+B-aA}{(x-a)^2}$, which has the required form with $C = A$ and $D = B - aA$.

43. $\dfrac{dP}{dt} = 2P\left(\dfrac{100-P}{100}\right)$

$$\implies \frac{dP}{P(100-P)} = \frac{1}{50}\,dt$$

$$\implies \frac{1}{50}\int dt = \int \frac{dP}{P(100-P)} = \frac{1}{100}\int\left(\frac{1}{P} + \frac{1}{100-P}\right)\,dt$$

$$\implies \frac{1}{50}\,t = \frac{1}{100}(\ln|P| - \ln|100-P|) + C = \frac{1}{100}\,\ln\frac{P}{100-P} + C$$

$$\implies ce^{2t} = \frac{P}{100-P} = \frac{1}{\frac{100}{P}-1}$$

$$\implies P = \frac{100ce^{2t}}{1+ce^{2t}} = \frac{100c}{c+e^{-2t}}.$$

It follows that the carrying capacity is

$$K = \lim_{t\to\infty} P(t) = \lim_{t\to\infty}\frac{100c}{c+e^{-2t}} = 100.$$

45. $\dfrac{dP}{dt} = P(1-P) \implies \dfrac{dP}{P(1-P)} = dt \implies \displaystyle\int\left(\frac{1}{P} + \frac{1}{1-P}\right)dP = \int dt$

$$\implies \ln|P| - \ln|1-P| = t + C \implies \ln\frac{P}{1-P} = t + C \implies \frac{P}{1-P} = ce^t$$

$$\implies P = \frac{c}{c+e^{-t}} \implies \lim_{t\to\infty} P = 1.$$

10.5   Miscellaneous Substitutions

1. Let Let $x = u^2$. Then $dx = 2u\,du$ and hence

$$\int \frac{\sqrt{x}}{1+\sqrt{x}}\,dx = \int \frac{u}{1+u}\,2u\,du = 2\int\left(u - 1 + \frac{1}{1+u}\right)\,du$$

$$= 2\left(\frac{1}{2}\,u^2 - u + \ln|1+u|\right) + C = x - 2\sqrt{x} + 2\ln(1+\sqrt{x}) + C.$$

3. Let $x = u^6$. Then $dx = 6u^5\,du$ and hence

$$\int \frac{dx}{\sqrt{x}+2\sqrt[3]{x}} = \int \frac{6u^5}{u^3+2u^2}\,du = 6\int\left(u^2 - 2u + 4 - \frac{8}{u+2}\right)\,du$$

$$= 6\left(\frac{1}{3}\,u^3 - u^2 + 4u - 8\ln|u+2|\right) + C = 2\sqrt{x} - 6\sqrt[3]{x} + 24\sqrt[6]{x} - 48\ln\left|\sqrt[6]{x}+2\right| + C.$$

5. Let $u = x + 1$. Then $du = dx$ and hence

$$\int_0^7 \frac{x}{\sqrt[3]{x+1}}\,dx = \int_1^8 \frac{u-1}{\sqrt[3]{u}}\,du = \int_1^8 (u^{2/3} - u^{-1/3})\,du = \left(\frac{3}{5}\,u^{5/3} - \frac{3}{2}\,u^{2/3}\right)\Bigg|_1^8 = \frac{141}{10}.$$

7. Let $u = \tan(x/2)$. Then

$$du = \frac{1}{2} \sec^2 \frac{x}{2} \, dx = \frac{1}{2} \left(1 + u^2\right) dx, \; \sin x = \frac{2u}{1 + u^2},$$

and hence

$$\int \frac{dx}{2 + \sin x} = \int \frac{1}{2 + \frac{2u}{1+u^2}} \frac{2 \, du}{1 + u^2} = \int \frac{1}{(u + 1/2)^2 + 3/4} \, du$$

$$= \frac{2}{\sqrt{3}} \mathrm{Tan}^{-1} \frac{u + 1/2}{\sqrt{3}/2} + C = \frac{2}{\sqrt{3}} \mathrm{Tan}^{-1} \left(\frac{2 \tan(x/2) + 1}{\sqrt{3}} \right) + C.$$

9. Let $u = \tan(x/2)$. Then

$$du = \frac{1}{2} \left(1 + u^2\right) dx, \; \cos x = \frac{1 - u^2}{1 + u^2},$$

and hence

$$\int \frac{dx}{1 - \cos x} = \int \frac{1}{1 - \frac{1-u^2}{1+u^2}} \frac{2}{1 + u^2} \, du = \int \frac{1}{u^2} \, du = -\frac{1}{u} + C = -\cot \frac{x}{2} + C.$$

11. Let $u^3 = 1 + x$. Then $3u^2 du = dx$ and hence

$$\int_0^2 \frac{x^2}{\sqrt[3]{1 + x}} \, dx = \int_1^{\sqrt[3]{3}} \frac{(u^3 - 1)^2}{u} 3u^2 \, du = 3 \int_1^{\sqrt[3]{3}} (u^7 - 2u^4 + u) \, du$$

$$= 3 \left(\frac{1}{8} u^8 - \frac{2}{5} u^5 + \frac{1}{2} u^2 \right) \Bigg|_1^{\sqrt[3]{3}} = \frac{51}{40} \sqrt[3]{9} - \frac{27}{40}.$$

13. Let $u = \tan(x/2)$. Then

$$du = \frac{1}{2} \left(1 + u^2\right) dx, \; \sin x = \frac{2u}{1 + u^2}, \; \cos x = \frac{1 - u^2}{1 + u^2},$$

and hence

$$\int \frac{1}{\sin x + \cos x} \, dx = \int \frac{1}{\frac{2u}{1+u^2} + \frac{1-u^2}{1+u^2}} \frac{2}{1 + u^2} \, du = -2 \int \frac{1}{(u - 1)^2 - 2} \, du.$$

Now, let $u - 1 = \sqrt{2} \sec \theta$. Then $du = \sqrt{2} \sec \theta \tan \theta \, d\theta$ and hence

$$\int \frac{1}{(u - 1)^2 - 2} \, du = \int \frac{1}{2 \sec^2 \theta - 2} \sqrt{2} \sec \theta \tan \theta \, d\theta = \frac{1}{\sqrt{2}} \int \csc \theta \, d\theta = \frac{\sqrt{2}}{2} \ln |\csc \theta - \cot \theta| + C$$

$$= \frac{\sqrt{2}}{2} \ln \left| \frac{u - 1}{\sqrt{(u - 1)^2 - 2}} - \frac{\sqrt{2}}{\sqrt{(u - 1)^2 - 2}} \right| + C = \frac{\sqrt{2}}{2} \ln \left| \frac{u - 1 - \sqrt{2}}{\sqrt{(u - 1 - \sqrt{2})(u - 1 + \sqrt{2})}} \right| + C$$

$$= \frac{\sqrt{2}}{2} \ln \left| \sqrt{\frac{u - 1 - \sqrt{2}}{u - 1 + \sqrt{2}}} \right| + C = \frac{\sqrt{2}}{4} \ln \left| \frac{u - 1 - \sqrt{2}}{u - 1 + \sqrt{2}} \right| + C.$$

Hence

$$\int \frac{1}{\sin x + \cos x} \, dx = -2 \frac{\sqrt{2}}{4} \ln \left| \frac{\tan(x/2) - 1 - \sqrt{2}}{\tan(x/2) - 1 + \sqrt{2}} \right| + C = \frac{\sqrt{2}}{2} \ln \left| \frac{\tan(x/2) - 1 + \sqrt{2}}{\tan(x/2) - 1 - \sqrt{2}} \right| + C.$$

15. Let $u^2 = 1 - 2x$. Then $2u\,du = -2\,dx$ and hence

$$\int \frac{x^3}{\sqrt{1-2x}}\,dx = \int \frac{\left(\frac{1-u^2}{2}\right)^3}{u}\,(-u\,du)$$

$$= -\frac{1}{8}\int (1 - 3u^2 + 3u^4 - u^6)\,du = -\frac{1}{8}\left(u - u^3 + \frac{3}{5}u^5 - \frac{1}{7}u^7\right) + C$$

$$= -\frac{1}{8}\left[(1-2x)^{1/2} - (1-2x)^{3/2} + \frac{3}{5}(1-2x)^{5/2} - \frac{1}{7}(1-2x)^{7/2}\right] + C.$$

17. $\displaystyle\int \sqrt{\frac{1+x}{1-x}}\,dx = \int \sqrt{\frac{1+x}{1-x}\frac{1+x}{1+x}}\,dx = \int \frac{1+x}{\sqrt{1-x^2}}\,dx = \mathrm{Sin}^{-1}x - \sqrt{1-x^2} + C.$

19. Let $u^2 = a + bx$. Then $2u\,du = b\,dx$ and hence

$$\int \frac{dx}{x\sqrt{a+bx}} = \int \frac{\frac{2u}{b}\,du}{\frac{u^2-a}{b}\,u} = 2\int \frac{1}{u^2-a}\,du.$$

Now, if $a > 0$, then

$$\int \frac{dx}{x\sqrt{a+bx}} = 2\int \frac{1}{u^2-a}\,du = \frac{1}{\sqrt{a}}\int \left(\frac{1}{u-\sqrt{a}} - \frac{1}{u+\sqrt{a}}\right)du$$

$$= \frac{1}{\sqrt{a}}\left(\ln|u-\sqrt{a}| - \ln|u+\sqrt{a}|\right) + C = \frac{1}{\sqrt{a}}\ln\left|\frac{\sqrt{a+bx}-\sqrt{a}}{\sqrt{a+bx}+\sqrt{a}}\right| + C.$$

On the other hand, if $a < 0$, let $u = \sqrt{-a}\tan\theta$. Then $du = \sqrt{-a}\sec^2\theta\,d\theta$ and hence

$$\int \frac{dx}{x\sqrt{a+bx}} = 2\int \frac{1}{u^2-a}\,du = 2\int \frac{1}{-a\tan^2\theta - a}\sqrt{-a}\sec^2\theta\,d\theta$$

$$= \frac{2}{-a}\sqrt{-a}\int d\theta = \frac{2}{\sqrt{-a}}\theta + C = \frac{2}{\sqrt{-a}}\mathrm{Tan}^{-1}\frac{u}{\sqrt{-a}} = \frac{2}{\sqrt{-a}}\mathrm{Tan}^{-1}\sqrt{\frac{a+bx}{-a}} + C.$$

21. Let $u^2 = 2 + x$. Then $2u\,du = dx$ and hence

$$\int_0^2 \frac{x}{\sqrt{2+x}}\,dx = \int_{\sqrt{2}}^2 \frac{u^2-2}{u}\,2u\,du = 2\left(\frac{1}{3}u^3 - 2u\right)\Big|_{\sqrt{2}}^2 = -\frac{8}{3} + \frac{8}{3}\sqrt{2}.$$

10.6 The Use of Integral Tables

1. Formula 102: Let $u = 5x$. Then $du = 5\,dx$ and hence

$$\int \frac{1}{1+e^{5x}}\,dx = \frac{1}{5}\int \frac{1}{1+e^u}\,du = \frac{1}{5}\ln\left(\frac{e^u}{1+e^u}\right) + C = \frac{1}{5}\ln\left(\frac{e^{5x}}{1+e^{5x}}\right) + C.$$

3. Formula 32:

$$\int_1^2 \frac{dx}{x^2(9+2x)^2} = \left(-\frac{9+4x}{81x(9+2x)} + \frac{4}{729}\ln\left|\frac{9+2x}{x}\right|\right)\Big|_1^2 = \frac{151}{23166} + \frac{4}{729}(\ln 13 - \ln 2 - \ln 11).$$

5. Formula 42:

$$\int \frac{x+3}{2+5x^2}\,dx = \int \frac{x}{2+5x^2}\,dx + 3\int \frac{1}{2+5x^2}\,dx = \frac{1}{10}\ln(2+5x^2) + \frac{3}{\sqrt{10}}\,\text{Tan}^{-1}\left(x\sqrt{\frac{5}{2}}\right) + C.$$

7. Formula 80: Let $u = \pi x$. Then $du = \pi dx$ and hence

$$\int_0^{1/4} \tan^4 \pi x\,dx = \frac{1}{\pi}\int_0^{\pi/4} \tan^4 u\,du = \frac{1}{\pi}\left(\frac{1}{3}\tan^3 u - \tan u + u\right)\Big|_0^{\pi/4} = \frac{1}{4} - \frac{2}{3\pi}.$$

9. Formula 77: $\displaystyle\int \sin 6x \sin 3x\,dx = \frac{\sin 3x}{6} - \frac{\sin 9x}{18} + C.$

11. Formula 31: $\displaystyle\int \frac{dx}{x^2(2-3x)} = -\frac{1}{2x} - \frac{3}{4}\ln\left|\frac{2-3x}{x}\right| + C.$

13. Formula 38: Let $u = 2x$. Then $du = 2dx$ and hence

$$\int \frac{9\,dx}{25-4x^2} = \frac{9}{2}\int \frac{du}{25-u^2} = \frac{9}{20}\ln\left|\frac{5+u}{5-u}\right| + C = \frac{9}{20}\ln\left|\frac{5+2x}{5-2x}\right| + C.$$

15. Formula 79: Let $u = 5x$. Then $du = 5dx$ and hence

$$\int \tan^3 5x\,dx = \frac{1}{5}\int \tan^3 u\,du = \frac{1}{5}\left(\frac{1}{2}\tan^2 u + \ln|\cos u|\right) + C = \frac{1}{10}\tan^2 5x + \frac{1}{5}\ln|\cos 5x| + C.$$

17. Formula 37: $\displaystyle\int \frac{6x^2\,dx}{\sqrt{6+x}} = \frac{4}{5}(288 - 24x + 3x^2)\sqrt{6+x} + C.$

19. Formula 57: $\displaystyle\int \frac{6\,dx}{(14-x^2)^{3/2}} = \frac{3x}{7\sqrt{14-x^2}} + C.$

21. Formula 73: Let $u = 2x$. Then $du = 2dx$ and hence

$$\int_0^{\pi/6} \frac{\pi}{4+4\sin 2x}\,dx = \frac{\pi}{8}\int_0^{\pi/3} \frac{1}{1+\sin u}\,du = -\frac{\pi}{8}\tan\left(\frac{\pi}{4} - \frac{u}{2}\right)\Big|_0^{\pi/3} = \frac{\pi}{8}\left(1 - \tan\frac{\pi}{12}\right).$$

23. Formula 41: $\displaystyle\int \frac{3\,dx}{x\sqrt{x^2+16}} = -\frac{3}{4}\ln\left|\frac{4+\sqrt{x^2+16}}{x}\right| + C.$

25. Formula 20: Let $u = 3x + 2$. Then $du = 3dx$ and hence

$$\int_{-1}^{-2/3} \frac{dx}{9x^2+12x+5} = \frac{1}{3}\int_{-1}^0 \frac{du}{u^2+1} = \frac{1}{3}\text{Tan}^{-1}u\Big|_{-1}^0 = \frac{\pi}{12}.$$

27. Formula 48: Let $u = x + 4$. Then $du = dx$ and hence

$$\int \frac{7\,dx}{(x+4)\sqrt{x^2+8x+20}} = 7\int \frac{du}{u\sqrt{u^2+4}}$$

$$= -\frac{7}{2}\ln\left(\frac{2+\sqrt{u^2+4}}{u}\right) + C = -\frac{7}{2}\ln\left(\frac{2+\sqrt{x^2+8x+20}}{x+4}\right) + C.$$

29. Formulas 59 and 22: $\displaystyle\int \frac{(2x^2-7)\,dx}{\sqrt{16-x^2}} = 2\int \frac{x^2}{\sqrt{16-x^2}}\,dx - 7\int \frac{dx}{\sqrt{16-x^2}}$

$$= 2\left(-\frac{x}{2}\sqrt{16-x^2} + 8\,\text{Sin}^{-1}\frac{x}{4}\right) - 7\,\text{Sin}^{-1}\frac{x}{4} + C = -x\sqrt{16-x^2} + 9\,\text{Sin}^{-1}\frac{x}{4} + C.$$

Review Exercises – Chapter 10

1. $\int x\sqrt{x^2+9}\,dx = \frac{1}{3}\,(x^2+9)^{3/2}+C.$

3. Let $u = x^2$, $dv = \sin 2x\,dx$. Then $du = 2x\,dx$, $v = -\frac{1}{2}\cos 2x$ and hence

$$\int x^2 \sin 2x\,dx = -\frac{1}{2}\,x^2 \cos 2x + \int \underbrace{x\cos 2x\,dx}_{u=x,\,dv=\cos 2x\,dx}$$

$$= -\frac{1}{2}\,x^2\cos 2x + \frac{1}{2}\,x\sin 2x - \frac{1}{2}\int \sin 2x\,dx = -\frac{1}{2}\,x^2\cos 2x + \frac{1}{2}\,x\sin 2x + \frac{1}{4}\,\cos 2x + C.$$

Evaluating this expression, we find that

$$\int_0^\pi x^2 \sin 2x\,dx = -\frac{\pi^2}{2}.$$

5. Let $x = a\tan\theta$. Then $dx = a\sec^2\theta\,d\theta$ and hence

$$\int \sqrt{x^2+a^2}\,dx = \int (a\sec\theta)(a\sec^2\theta)\,d\theta = a^2 \int \sec^3\theta\,d\theta$$

$$= \frac{a^2}{2}(\sec\theta\tan\theta + \ln|\sec\theta+\tan\theta|) + C = \frac{1}{2}\,x\sqrt{x^2+a^2} + \frac{a^2}{2}\,\ln\left|\sqrt{x^2+a^2}+x\right| + C.$$

Note that the missing constant, $\ln a$, has been absorbed in the constant C.

7. $\displaystyle\int_1^3 \frac{2x+3}{x(x+3)}\,dx = \int_1^3 \left(\frac{1}{x} + \frac{1}{x+3}\right)\,dx = (\ln|x| + \ln|x+3|)\big|_1^3 = \ln 3 + \ln 6 - \ln 4.$

9. Let $u = x$, $dv = \cos\pi x\,dx$. Then $du = dx$, $v = \frac{1}{\pi}\sin\pi x$ and hence

$$\int_{-1/2}^{1/2} x\cos\pi x\,dx = \left(\frac{x}{\pi}\,\sin\pi x - \frac{1}{\pi}\int \sin\pi x\,dx\right)\bigg|_{-1/2}^{1/2} = \left(\frac{x}{\pi}\,\sin\pi x + \frac{1}{\pi^2}\,\cos\pi x\right)\bigg|_{-1/2}^{1/2} = 0.$$

11. Let $u = x$, $dv = \cos(2x+1)dx$. Then $du = dx$, $v = \frac{1}{2}\sin(2x+1)$, and hence

$$\int_{-1/2}^{(\pi-1)/2} x\cos(2x+1)\,dx = \left(\frac{1}{2}\,x\sin(2x+1) - \frac{1}{2}\int \sin(2x+1)\,dx\right)\bigg|_{-1/2}^{(\pi-1)/2}$$

$$= \left(\frac{1}{2}\,x\sin(2x+1) + \frac{1}{4}\,\cos(2x+1)\right)\bigg|_{-1/2}^{(\pi-1)/2} = -\frac{1}{2}.$$

13. Let $u = x^2$, $dv = e^{-5x}dx$. Then $du = 2x\,dx$, $v = -\frac{1}{5}e^{-5x}$ and hence

$$\int x^2 e^{-5x}\,dx = -\frac{1}{5}\,x^2 e^{-5x} + \frac{2}{5}\int \underbrace{xe^{-5x}\,dx}_{u=x,\,dv=e^{-5x}dx}$$

379

$$= -\frac{1}{5}\,x^2 e^{-5x} + \frac{2}{5}\left(-\frac{1}{5}\,xe^{-5x} + \frac{1}{5}\int e^{-5x}\,dx\right) = -\frac{1}{5}\,x^2 e^{-5x} - \frac{2}{25}\,xe^{-5x} - \frac{2}{125}\,e^{-5x} + C.$$

Therefore

$$\int_0^1 x^2 e^{-5x}\,dx = \frac{2}{125} - \frac{37}{125}\,e^{-5}.$$

15. Let $u = (\ln x)^2$, $dv = xdx$. Then $du = 2(\ln x)\,\frac{1}{x}\,dx$, $v = \frac{1}{2}x^2$ and hence

$$\int x\ln^2 x\,dx = \frac{1}{2}\,x^2(\ln x)^2 - \int \underbrace{x\ln x\,dx}_{u=\ln x,\ dv=xdx} = \frac{1}{2}\,x^2(\ln x)^2 - \frac{1}{2}\,x^2(\ln x) + \frac{1}{4}\,x^2 + C.$$

17. $\displaystyle\int \frac{dx}{x^2 - 4x + 9} = \int \frac{dx}{(x-2)^2 + 5} = \frac{1}{\sqrt{5}}\,\mathrm{Tan}^{-1}\frac{x-2}{\sqrt{5}} + C.$

19. Let $u^2 = x + 4$. Then $2udu = dx$ and hence

$$\int \frac{\sqrt{x+4}}{x}\,dx = \int \frac{u}{u^2 - 4}\,2u\,du = 2\int\left(1 + \frac{1}{u-2} - \frac{1}{u+2}\right)du$$

$$= 2(u + \ln|u-2| - \ln|u+2|) + C = 2\left(\sqrt{x+4} + \ln\left|\sqrt{x+4} - 2\right| - \ln\left|\sqrt{x+4} + 2\right|\right) + C.$$

21. Let $u = x + 2$. Then $du = dx$ and hence

$$\int_0^2 x\sqrt{x+2}\,dx = \int_2^4 (u-2)\sqrt{u}\,du = \int_2^4 (u^{3/2} - 2u^{1/2})\,du = \left(\frac{2}{5}\,u^{5/2} - \frac{4}{3}\,u^{3/2}\right)\Bigg|_2^4 = \frac{32}{15} + \frac{16}{15}\,\sqrt{2}.$$

23. Let $x = a\sin\theta$. Then $dx = a\cos\theta\,d\theta$ and hence

$$\int \frac{dx}{\sqrt{a^2 - x^2}} = \int \frac{a\cos\theta\,d\theta}{a\cos\theta} = \theta + C = \mathrm{Sin}^{-1}\frac{x}{a} + C.$$

25. $\displaystyle\int \frac{3x^2 + x + 1}{x^2(x+1)}\,dx = \int\left(\frac{1}{x^2} + \frac{3}{x+1}\right)dx = -\frac{1}{x} + 3\ln|x+1| + C.$

27. $\displaystyle\int_0^{\pi/4} \tan^2 x\,dx = \int_0^{\pi/4}(\sec^2 x - 1)\,dx = (\tan x - x)\big|_0^{\pi/4} = 1 - \frac{\pi}{4}.$

29. Let $x = a\sin\theta$. Then $dx = a\cos\theta\,d\theta$ and hence

$$\int \sqrt{a^2 - x^2}\,dx = \int (a\cos\theta)(a\cos\theta\,d\theta) = \frac{a^2}{2}\int(1 + \cos 2\theta)\,d\theta$$

$$= \frac{a^2}{2}\left(\theta + \frac{1}{2}\,\sin 2\theta\right) + C = \frac{a^2}{2}\,(\theta + \sin\theta\cos\theta) + C = \frac{a^2}{2}\,\mathrm{Sin}^{-1}\frac{x}{a} + \frac{x\sqrt{a^2 - x^2}}{2} + C.$$

31. $\displaystyle\int_0^1 \frac{4x^2 + x + 2}{(x-2)(x^2 + 1)}\,dx = \int_0^1\left(\frac{4}{x-2} + \frac{1}{x^2 + 1}\right)dx = \left(4\ln|x-2| + \mathrm{Tan}^{-1}x\right)\big|_0^1 = \frac{\pi}{4} - 4\ln 2.$

33. $\displaystyle\int \frac{x+1}{\sqrt[3]{x+1}}\,dx = \int (x+1)^{2/3}\,dx = \frac{3}{5}\,(x+1)^{5/3} + C.$

35. Let $x = a \sec \theta$. Then $dx = a \sec \theta \tan \theta \, d\theta$ and hence

$$\int \sqrt{x^2 - a^2} \, dx = \int (a \tan \theta)(a \sec \theta \tan \theta) \, d\theta = a^2 \int (\sec^3 \theta - \sec \theta) \, d\theta$$

$$= a^2 \left[\frac{1}{2} \left(\sec \theta \tan \theta + \ln |\sec \theta + \tan \theta| \right) - \ln |\sec \theta + \tan \theta| \right] + C$$

$$= \frac{1}{2} x \sqrt{x^2 - a^2} - \frac{a^2}{2} \ln \left| x + \sqrt{x^2 - a^2} \right| + C,$$

where the missing constant, $\ln a$, is absorbed in the constant C.

37. $\displaystyle \int \frac{x^2 + 2x + 3}{(x+2)(x^2+x+1)} \, dx = \int \left(\frac{1}{x+2} + \frac{1}{(x+1/2)^2 + 3/4} \right) dx = \ln|x+2| + \frac{2}{\sqrt{3}} \operatorname{Tan}^{-1} \frac{2x+1}{\sqrt{3}} + C.$

39. Let $x = 2 \sec \theta$. Then $dx = 2 \sec \theta \tan \theta \, d\theta$ and hence

$$
\begin{aligned}
\int (x^2 - 4)^{3/2} \, dx &= \int (8 \tan^3 \theta)(2 \sec \theta \tan \theta \, d\theta) \\
&= 16 \int \tan^4 \theta \sec \theta \, d\theta = 16 \int (\sec^5 \theta - 2 \sec^3 \theta + \sec \theta) \, d\theta \\
&= 16 \left[\frac{1}{4} \tan \theta \sec^3 \theta + \frac{3}{4} \int \sec^3 \theta \, d\theta \right] - 32 \int \sec^3 \theta \, d\theta + 16 \int \sec \, d\theta \\
&= 4 \tan \theta \sec^3 \theta - 20 \int \sec^3 \, d\theta + 16 \int \sec \theta \, d\theta \\
&= 4 \tan \theta \sec^3 \theta - 20 \left[\frac{1}{2} \tan \theta \sec \theta + \frac{1}{2} \int \sec \theta \, d\theta \right] + 16 \int \sec \theta \, d\theta \\
&= 4 \tan \theta \sec^3 \theta - 10 \tan \theta \sec \theta + 6 \ln |\sec \theta + \tan \theta| + C \\
&= \frac{1}{4} x^3 \sqrt{x^2 - 4} - \frac{5}{2} x \sqrt{x^2 - 4} + 6 \ln \left| x + \sqrt{x^2 - 4} \right| + C.
\end{aligned}
$$

41. Since $1 + x^3 = (1+x)(1 - x + x^2)$, we find that

$$
\begin{aligned}
\int \frac{3x^3 + x + 2}{x(1+x^3)} \, dx &= \int \frac{2}{x} \, dx + \frac{2}{3} \int \frac{1}{1+x} \, dx + \frac{1}{3} \int \underbrace{\frac{x+1}{x^2 - x + 1}}_{u = x - 1/2} \, dx \\
&= 2 \int \frac{1}{x} \, dx + \frac{2}{3} \int \frac{1}{1+x} \, dx + \frac{1}{3} \int \frac{u + 3/2}{u^2 + 3/4} \, du \\
&= 2 \ln|x| + \frac{2}{3} \ln|x+1| + \frac{1}{6} \ln \left| u^2 + \frac{3}{4} \right| + \frac{1}{\sqrt{3}} \operatorname{Tan}^{-1} \frac{u}{\sqrt{3/4}} + C \\
&= 2 \ln|x| + \frac{2}{3} \ln|x+1| + \frac{1}{6} \ln(x^2 - x + 1) + \frac{1}{\sqrt{3}} \operatorname{Tan}^{-1} \frac{2x-1}{\sqrt{3}} + C.
\end{aligned}
$$

43. Let $u = \csc x$, $dv = \csc^2 x \, dx$. Then $du = -\csc x \cot x \, dx$, $v = -\cot x$ and hence

$$\int \csc^3 x \, dx = \int \csc x \csc^2 x \, dx = -\csc x \cot x - \int \csc x \cot^2 x \, dx$$

$$= -\csc x \cot x - \int \csc x (\csc^2 x - 1) \, dx = -\csc x \cot x - \int \csc^3 x \, dx + \ln|\csc x - \cot x| + C$$

$$\implies \int \csc^3 x \, dx = -\frac{1}{2} \csc x \cot x + \frac{1}{2} \ln|\csc x - \cot x| + C.$$

381

45. Let $x = 3\tan\theta$. Then $dx = 3\sec^2\theta\,d\theta$ and hence

$$\int \frac{dx}{x\sqrt{9+x^2}} = \int \frac{3\sec^2\theta\,d\theta}{3\tan\theta\,3\sec\theta} = \frac{1}{3}\int \csc\theta\,d\theta = \frac{1}{3}\ln|\csc\theta - \cot\theta| + C = \frac{1}{3}\ln\left|\frac{\sqrt{9+x^2}-3}{x}\right| + C.$$

47. $\displaystyle \int \frac{3x^2 - x + 10}{(x^2+2)(4-x)}\,dx = \int \left(\frac{3}{4-x} + \frac{1}{x^2+2}\right)\,dx = -3\ln|4-x| + \frac{1}{\sqrt{2}}\mathrm{Tan}^{-1}\frac{x}{\sqrt{2}} + C.$

49. Let $u = \tan(x/2)$. Then

$$\cos x = \frac{1-u^2}{1+u^2}, \quad du = \frac{1}{2}\sec^2\frac{x}{2}\,dx = \frac{1+u^2}{2}\,dx$$

and hence

$$\int \frac{\cos x}{2 - \cos x}\,dx = \int \left(-1 + \frac{2}{2-\cos x}\right)\,dx = -x + 2\int \frac{1}{2 - \frac{1-u^2}{1+u^2}}\frac{2}{1+u^2}\,du$$

$$= -x + 4\int \frac{1}{1+3u^2}\,du = -x + \frac{4}{\sqrt{3}}\mathrm{Tan}^{-1}u\sqrt{3} + C = -x + \frac{4}{\sqrt{3}}\mathrm{Tan}^{-1}\left(\sqrt{3}\,\tan\frac{x}{2}\right) + C.$$

51. Let $u = e^x$. Then $du = e^x\,dx$ and hence

$$\int_{\ln 2}^{\ln 3} \frac{dx}{e^x - e^{-x}} = \int_{\ln 2}^{\ln 3} \frac{e^x}{(e^x-1)(e^x+1)}\,dx = \frac{1}{2}\int_2^3 \left(\frac{1}{u-1} - \frac{1}{u+1}\right)\,du$$

$$= \frac{1}{2}\left(\ln|u-1| - \ln|u+1|\right)\Big|_2^3 = \frac{1}{2}\left(\ln 2 - \ln 4 + \ln 3\right) = \ln\sqrt{3/2}.$$

53. Let $u = a^2 + x^2$. Then $du = 2x\,dx$ and hence

$$\int \frac{x}{a^2+x^2}\,dx = \frac{1}{2}\int \frac{du}{u} = \frac{1}{2}\ln|u| + C = \frac{1}{2}\ln(a^2+x^2) + C.$$

55. $\displaystyle \int_2^3 \frac{5x^2 - 2x + 12}{x^3 - x^2 + 4x - 4}\,dx = \int_2^3 \left(\frac{3}{x-1} + \frac{2x}{x^2+4}\right)\,dx = \left(3\ln|x-1| + \ln(x^2+4)\right)\Big|_2^3 = \ln 13.$

57. $\displaystyle \int \frac{1}{1+e^x}\,dx = \int \frac{e^{-x}}{e^{-x}+1}\,dx = -\ln(e^{-x}+1) + C.$

59. Let $u = x + 4$. Then $du = dx$ and hence

$$\int \frac{3x+1}{\sqrt{x^2+8x}}\,dx = \int \frac{3u-11}{\sqrt{u^2-16}}\,du = 3\sqrt{u^2-16} - 11\ln\left|u + \sqrt{u^2-16}\right| + C$$

$$= 3\sqrt{x^2+8x} - 11\ln\left|x+4 + \sqrt{x^2+8x}\right| + C.$$

61. Let $u = 1 + x^2$. Then $du = 2x\,dx$ and hence

$$\int \frac{x^5\,dx}{\sqrt{1+x^2}} = \frac{1}{2}\int \frac{(u-1)^2}{\sqrt{u}}\,du = \frac{1}{2}\int (u^{3/2} - 2u^{1/2} + u^{-1/2})\,du$$

$$= \frac{1}{2}\left(\frac{2}{5}u^{5/2} - \frac{4}{3}u^{3/2} + 2u^{1/2}\right) + C = \frac{1}{5}(1+x^2)^{5/2} - \frac{2}{3}(1+x^2)^{3/2} + (1+x^2)^{1/2} + C.$$

382

63. Let $u = x^2 + 2$. Then $du = 2x\,dx$ and hence

$$\int \frac{x^5}{(x^2+2)^2}\,dx = \frac{1}{2}\int \frac{(u-2)^2}{u^2}\,du = \frac{1}{2}\int \left(1 - \frac{4}{u} + \frac{4}{u^2}\right)\,du$$

$$= \frac{1}{2}\left(u - 4\ln|u| - \frac{4}{u}\right) + C = \frac{1}{2}\left(x^2 - 4\ln(x^2+2) - \frac{4}{x^2+2}\right) + C.$$

65. Let $2x = 5\sec\theta$. Then $2dx = 5\sec\theta\tan\theta\,d\theta$ and hence

$$\int \frac{1}{\sqrt{4x^2-25}}\,dx = \int \frac{1}{5\tan\theta}\,\frac{5}{2}\,\sec\theta\tan\theta\,d\theta = \frac{1}{2}\int \sec\theta\,d\theta$$

$$= \frac{1}{2}\,\ln|\sec\theta + \tan\theta| + C = \frac{1}{2}\,\ln\left|2x + \sqrt{4x^2-25}\right| + C.$$

67. Let $u = (\ln x)^2$, $dv = dx$. Then $du = 2(\ln x)/x$, $v = x$ and hence

$$\int_1^e \ln^2 x\,dx = \left(x(\ln x)^2 - 2\int \ln x\,dx\right)\Big|_1^e = [x(\ln x)^2 - 2(x\ln x - x)]\Big|_1^e = e - 2.$$

69. $\displaystyle \int \frac{3\,dx}{x^2-5} = \frac{3}{2\sqrt{5}}\,\ln\left|\frac{x-\sqrt{5}}{x+\sqrt{5}}\right| + C = \frac{3}{2\sqrt{5}}\,\ln\left(\frac{x-\sqrt{5}}{x+\sqrt{5}}\right) + C$ since $x > \sqrt{5}$. (Formula 39)

71. Let $u^2 = 9 - \pi x$. Then $2u\,du = -\pi\,dx$ and hence

$$\int \frac{dx}{x\sqrt{9-\pi x}} = -\frac{1}{\pi}\int \frac{2u\,du}{\frac{9-u^2}{\pi}\,u} = -2\int \frac{1}{9-u^2}\,du = -\frac{1}{3}\int \left(\frac{1}{3-u} + \frac{1}{3+u}\right)\,du$$

$$= -\frac{1}{3}\left(-\ln|3-u| + \ln|3+u|\right) + C = -\frac{1}{3}\left(-\ln\left|3 - \sqrt{9-\pi x}\right| + \ln\left|3 + \sqrt{9-\pi x}\right|\right) + C.$$

73. $\displaystyle \int \frac{5}{x^2(7-2x)}\,dx = \frac{10}{49}\int \frac{1}{x}\,dx + \frac{5}{7}\int \frac{1}{x^2}\,dx + \frac{20}{49}\int \frac{1}{7-2x}\,dx$

$$= \frac{10}{49}\,\ln|x| - \frac{5}{7}\,\frac{1}{x} - \frac{10}{49}\,\ln|7-2x| + C$$

75. Let $u = 9 + 2x$. Then $du = 2dx$ and hence

$$\int \frac{3x\,dx}{\sqrt{9+2x}} = \frac{3}{2}\int \frac{\frac{1}{2}(u-9)}{\sqrt{u}}\,du = \frac{3}{4}\int \left(u^{1/2} - 9u^{-1/2}\right)\,du$$

$$= \frac{3}{4}\left(\frac{2}{3}\,u^{3/2} - 18u^{1/2}\right) + C = \frac{1}{2}\,(9+2x)^{3/2} - \frac{27}{2}\,(9+2x)^{1/2} + C.$$

77. $\displaystyle \int \frac{dx}{x^2\sqrt{8-x^2}}\,dx = -\frac{\sqrt{8-x^2}}{8x} + C.$ (Formula 60)

79. Let $u = \tan(x/2)$. Then

$$\cos x = \frac{1-u^2}{1+u^2},\ du = \frac{1}{2}\,\sec^2\frac{x}{2}\,dx = \frac{1+u^2}{2}\,dx$$

and hence

$$\int \frac{dx}{9+4\cos x} = \int \frac{\frac{2}{1+u^2}\,du}{9 + 4\frac{1-u^2}{1+u^2}} = \int \frac{2\,du}{13+5u^2} = \frac{2}{\sqrt{65}}\,\text{Tan}^{-1}\left(\sqrt{\frac{5}{13}}\,\tan\frac{x}{2}\right) + C.$$

81. Let $u = \ln x$, $dv = x^3\, dx$. Then $du = \dfrac{1}{x}\, dx$, $v = \dfrac{1}{4}\, x^4$ and hence

$$\int x^3 \ln x\, dx = \frac{1}{4}\, x^4 \ln x - \int x^3\, dx = \frac{1}{4}\, x^4 \ln x - \frac{1}{16}\, x^4 + C.$$

83. The graphs intersect when $xe^{-x} = x/e \implies x(e^{-x+1} - 1) = 0 \implies x = 0$, $x = 1$. Hence

$$\text{Area} \;=\; \int_0^1 \left(\underbrace{xe^{-x}}_{u=x,\, dv=e^{-x}dx} - \frac{x}{e} \right) dx = \left(-xe^{-x} - e^{-x} - \frac{x^2}{2e} \right)\Big|_0^1 = 1 - \frac{5}{2e}.$$

85. Let dV be the differential volume of a cross-section disk of thickness dx. Then

$$dV = \pi r^2\, dx = \pi y^2\, dx = \pi \frac{x+5}{x^2(x-5)}\, dx$$

and hence the volume of revolution is

$$V = \pi \int_6^7 \frac{x+5}{x^2(x-5)}\, dx = \pi \int_6^7 \left(\frac{-2/5}{x} - \frac{1}{x^2} + \frac{2/5}{x-5} \right) dx$$

$$= \pi \left(-\frac{2}{5}\, \ln x + \frac{1}{x} + \frac{2}{5}\, \ln(x-5) \right)\Big|_6^7 = \pi \left(-\frac{2}{5}\, \ln 7 + \frac{2}{5}\, \ln 2 + \frac{2}{5}\, \ln 6 - \frac{1}{42} \right) = \pi \left(-\frac{1}{42} + \frac{2}{5}\, \ln \frac{12}{7} \right).$$

87. $y = -\sqrt{1-x^2} \implies y' = x(1-x^2)^{-1/2} \implies$

$$\text{Arc length} \;=\; \int_0^{1/2} \sqrt{1 + \frac{x^2}{1-x^2}}\, dx = \int_0^{1/2} \frac{1}{\sqrt{1-x^2}}\, dx = \operatorname{Sin}^{-1}x\big|_0^{1/2} = \frac{\pi}{6}.$$

89. The center of mass is at the point

$$\bar{x} = \frac{\int_0^4 \frac{4}{(x+1)(5-x)}\, x(3)\, dx}{\int_0^4 \frac{4}{(x+1)(5-x)}\, (3)\, dx} = \frac{\int_0^4 \left(\frac{-1/6}{x+1} + \frac{5/6}{5-x} \right) dx}{\int_0^4 \left(\frac{1/6}{x+1} + \frac{1/6}{5-x} \right) dx} = \frac{\left(-\ln(x+1) - 5\ln(5-x) \right)\big|_0^4}{\left(\ln(x+1) - \ln(5-x) \right)\big|_0^4} = 2.$$

Chapter 11

l'Hôpital's Rule and Improper Integrals

11.1   Indeterminate Forms: l'Hôpital's Rule

1. $\displaystyle\lim_{x\to 1}\frac{1-x}{e^x-e}=\lim_{x\to 1}\frac{-1}{e^x}=-\frac{1}{e}$

3. $\displaystyle\lim_{x\to 0}\frac{\sin x^2}{x}=\lim_{x\to 0}\frac{2x\cos x^2}{1}=0$

5. $\displaystyle\lim_{x\to 0}\frac{1-\cos x}{x^2}=\lim_{x\to 0}\frac{\sin x}{2x}=\lim_{x\to 0}\frac{\cos x}{2}=1/2$

7. $\displaystyle\lim_{x\to 1^-}\frac{\sqrt{1-x^2}}{x-1}=-\lim_{x\to 1^-}\sqrt{\frac{1+x}{1-x}}=-\infty$

9. $\displaystyle\lim_{\theta\to 0}\frac{\tan\theta-\theta}{\theta-\sin\theta}=\lim_{\theta\to 0}\frac{\sec^2\theta-1}{1-\cos\theta}=\lim_{\theta\to 0}\frac{1-\cos^2\theta}{\cos^2\theta(1-\cos\theta)}=\lim_{\theta\to 0}\frac{1+\cos\theta}{\cos^2\theta}=2$

11. $\displaystyle\lim_{x\to 1}\frac{\ln x}{x-1}=\lim_{x\to 1}\frac{1/x}{1}=1$

13. $\displaystyle\lim_{x\to 0^+}\frac{1+\cos\sqrt{x}}{\sin x}=+\infty$

15. $\displaystyle\lim_{x\to 0^+}\frac{x^2}{x-\sin x}=\lim_{x\to 0^+}\frac{2x}{1-\cos x}=\lim_{x\to 0^+}\frac{2}{\sin x}=+\infty$

17. $\displaystyle\lim_{x\to\infty}\frac{\sqrt[3]{x^3+3}}{x^2}=\lim_{x\to\infty}\sqrt[3]{\frac{1}{x^3}+\frac{3}{x^6}}=0$

19. $\displaystyle\lim_{x\to 1}\frac{e^{x^2}-e^x}{x^2-1}=\lim_{x\to 1}\frac{2xe^{x^2}-e^x}{2x}=\frac{e}{2}$

21. $\displaystyle\lim_{x\to 0}\frac{\sin x}{\sqrt[3]{x}}=\lim_{x\to 0}\frac{\cos x}{\frac{1}{3}x^{-2/3}}=3\lim_{x\to 0}x^{2/3}\cos x=0$

23. $\lim\limits_{x \to 0} \dfrac{e^{x^2} - 1}{x^2} = \lim\limits_{x \to 0} \dfrac{2xe^{x^2}}{2x} = 1$

25. $\lim\limits_{x \to \infty} \dfrac{x - \sin x}{x \sin x} = \lim\limits_{x \to \infty} \dfrac{1 - \frac{\sin x}{x}}{\sin x}$ does not exist

27. $\lim\limits_{x \to \infty} \dfrac{9 - x^3}{xe^{\pi x}} = \lim\limits_{x \to \infty} \dfrac{-3x^2}{e^{\pi x}(\pi x + 1)} = \lim\limits_{x \to \infty} \dfrac{-6x}{e^{\pi x}(\pi^2 x + \pi + 1)} = \lim\limits_{x \to \infty} \dfrac{-6}{e^{\pi x}(\pi^2) + \pi e^{\pi x}(\pi^2 x + \pi + 1)} = 0$

29. $\lim\limits_{x \to 0} \dfrac{\operatorname{Sin}^{-1} x}{x} = \lim\limits_{x \to 0} \dfrac{\frac{1}{\sqrt{1 - x^2}}}{1} = 1$

31. $\lim\limits_{x \to \pi/2} \dfrac{\tan(x/2) - 1}{x - \pi/2} = \lim\limits_{x \to \pi/2} \dfrac{\frac{1}{2} \sec^2(x/2)}{1} = 1$

33. $\lim\limits_{x \to \infty} \dfrac{\sqrt{x} - 3x^2}{x(6 - x)} = \lim\limits_{x \to \infty} \dfrac{1/(x\sqrt{x}) - 3}{(6/x) - 1} = 3$

35. $\lim\limits_{x \to \infty} \dfrac{\sqrt{x} - \sqrt{a}}{\sqrt{x} + \sqrt{a}} = \lim\limits_{x \to \infty} \dfrac{1 - \sqrt{a/x}}{1 + \sqrt{a/x}} = 1$

37. $\lim\limits_{x \to \infty} \dfrac{3x^2 - 4}{4x^2 + 3} = \lim\limits_{x \to \infty} \dfrac{3 - 4/x^2}{4 + 3/x^2} = 3/4$

39. $\lim\limits_{x \to \infty} \dfrac{\sin x}{x^2 + \pi} = 0$ since $0 \le \left| \dfrac{\sin x}{x^2 + \pi} \right| \le \dfrac{1}{x^2}$ and $\lim\limits_{x \to \infty} \dfrac{1}{x^2} = 0$

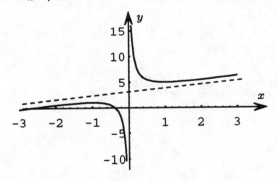

$y = (x^2 + 3x + 1)/x$

Exercise 41

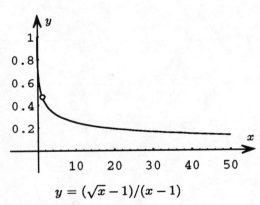

$y = (\sqrt{x} - 1)/(x - 1)$

Exercise 43

41. $f(x) = (x^2 + 3x + 1)/x = x + 3 + 1/x \implies f'(x) = 1 - 1/x^2$ and $f''(x) = 2/x^3$. It follows that the graph of $y = f(x)$ has a relative minimum at $(1, 5)$, a relative maximum at $(-1, 1)$, and is concave down for $x < 0$ and concave up for $x > 0$. Since $\lim_{x \to 0^+} f(x) = +\infty$ and $\lim_{x \to 0^-} f(x) = -\infty$, the line $x = 0$ is a vertical asymptote. Since $\lim_{x \to \pm\infty} f(x) = \pm\infty$, the graph has no horizontal asymptote. However,

$$\lim_{x \to \pm\infty} (f(x) - x - 3) = \lim_{x \to \pm\infty} \dfrac{1}{x} = 0,$$

which means that the graph is asymptotic to the line $y = x + 3$.

43. To obtain the graph of $y = f(x)$, first observe that if $x \neq 1$,

$$f(x) = \frac{\sqrt{x} - 1}{x - 1} = \frac{\sqrt{x} - 1}{(\sqrt{x} - 1)(\sqrt{x} + 1)} = \frac{1}{\sqrt{x} + 1}.$$

Hence the graph lies above the y-axis and is decreasing. For the behavior of $f(x)$ at $x = 1$ we find, using l'Hôpital's Rule, that

$$\lim_{x \to 1} \frac{\sqrt{x} - 1}{x - 1} = \lim_{x \to 1} \frac{\frac{1}{2}\sqrt{x}}{1} = \frac{1}{2},$$

while as $x \to \infty$ we find that

$$\lim_{x \to \infty} \frac{\sqrt{x} - 1}{x - 1} = \lim_{x \to \infty} \frac{\frac{1}{2}\sqrt{x}}{1} = 0.$$

Hence the line $y = 0$ is a horizontal asymptote.

45. If the hypotheses of l'Hôpital's Rule are satisfied, then

$$\lim_{x \to a^-} \frac{f(x)}{g(x)} = \lim_{x \to a^-} \frac{f'(x)}{g'(x)} = \lim_{x \to a^+} \frac{f'(x)}{g'(x)} = \lim_{x \to a^+} \frac{f(x)}{g(x)}.$$

Therefore

$$\lim_{x \to a} \frac{f(x)}{g(x)} = \lim_{x \to a} \frac{f'(x)}{g'(x)}.$$

47. (a) Graph $(e^x - 1)/(x^3 - 2x)$ using the range $[-1, 1] \times [-2, 0]$.

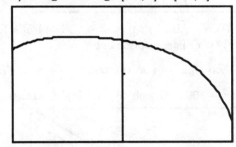

By magnifying the graph at the y-intercept we find that $\lim_{x \to 0}\big(f(x)/g(x)\big) = -1/2$. After magnifying the graph six times with factors of 10 it becomes oscillatory.

(b) Graph $e^x/(3x^2 - 2)$ using the range $[-1, 1] \times [-1, 2]$.

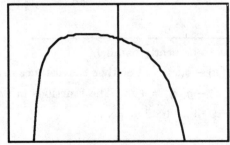

By magnifying the graph at the y-intercept we find that $\lim_{x \to 0}\big(f'(x)/g'(x)\big) = -1/2$.

(c) The method in part (b) is more accurate since the graph doesn't become oscillatory as we magnify it.

49. Graph $f(x) = x \sin \pi x$ and $g(x) = 2x(x-1)$ using the range $[0,2] \times [-2,1]$.

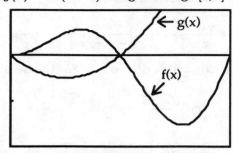

From the graph we see that $f'(1) \neq 0$ and $g'(1) \neq 0$. L'Hôpital's Rule implies therefore that both $f(x)/g(x)$ and $g(x)/f(x)$ are continuous at $x = 1$.

51. Graph $f(x) = \ln(2x+1)$ and $g(x) = x \arcsin x$ using the range $[-1,1] \times [-2,2]$.

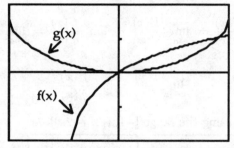

From the graph we see that $f'(0) \neq 0$, but $g'(0) = 0$ so L'Hôpital's Rule implies that $f(x)/g(x)$ is discontinuous at $x = 0$ while $g(x)/f(x)$ is continuous at $x = 0$.

53. (a) If $\lim\limits_{x \to +\infty} \left(f'(x)/g'(x) \right) = 1$ then as $x \to +\infty$ the tangent lines to $f(x)$ and $g(x)$ become parallel.

  (b) Let $f(x) = x^2$ and $g(x) = x^2 - 100x$. Graph these using the range $[500, 1050] \times [25000, 1100000]$.

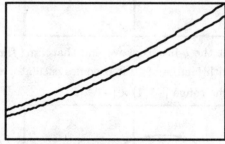

  Note that the graphs appear to be nearly parallel.

  (c) If $f(x) > g(x)$ and $\lim\limits_{x \to +\infty} \left(f(x) - g(x) \right) = +\infty$ then the distance between the graphs must increase as $x \to +\infty$, *i.e.*, $\lim\limits_{x \to +\infty} |f(x) - g(x)| = +\infty$. The functions in part (b) also have this property since $f(x) - g(x) = 100x$ and $\lim\limits_{x \to +\infty} 100x = +\infty$.

11.2 Other Indeterminate Forms

1. $\lim\limits_{x \to 0^+} \left(\dfrac{1}{x} - \dfrac{1}{x^2} \right) = \lim\limits_{x \to 0^+} \dfrac{x(x-1)}{x^2} = \lim\limits_{x \to 0^+} \left(1 - \dfrac{1}{x} \right) = -\infty$

3. $\lim\limits_{x\to 0} x\cot x = \lim\limits_{x\to 0}\dfrac{x\cos x}{\sin x} = \lim\limits_{x\to 0}\dfrac{x(-\sin x)+\cos x}{\cos x} = 1$

5. $\lim\limits_{x\to 0^+} x\ln x = \lim\limits_{x\to 0^+}\dfrac{\ln x}{1/x} = \lim\limits_{x\to 0^+}\dfrac{1/x}{-1/x^2} = \lim\limits_{x\to 0^+}(-x) = 0$

7. $\lim\limits_{x\to 1^+}\left(\dfrac{1}{x-1} - \dfrac{x}{\sqrt{x-1}}\right) = \lim\limits_{x\to 1^+}\left(\dfrac{1-x\sqrt{x-1}}{x-1}\right) = +\infty$

9. $\lim\limits_{x\to 0^+}(1/x)\mathrm{Tan}^{-1}x = \lim\limits_{x\to 0^+}\dfrac{1/(1+x^2)}{1} = 1$

11. $\lim\limits_{x\to 0^+} x\ln x = 0$ by Exercise 5 $\Longrightarrow \lim\limits_{x\to 0^+}x^x = \lim\limits_{x\to 0^+}e^{x\ln x} = e^0 = 1$

13. $\lim\limits_{x\to\infty}(1+x^2)e^{-2x} = \lim\limits_{x\to\infty}\dfrac{1+x^2}{e^{2x}} = \lim\limits_{x\to\infty}\dfrac{2x}{2e^{2x}} = \lim\limits_{x\to\infty}\dfrac{1}{2e^{2x}} = 0$

15. $\lim\limits_{x\to 0^+}(\tan x)\ln(\tan x) = \lim\limits_{x\to 0^+}\dfrac{\ln(\tan x)}{\cot x} = \lim\limits_{x\to 0^+}\dfrac{\sec^2 x/\tan x}{-\csc^2 x} = \lim\limits_{x\to 0^+}\dfrac{\sin x}{\cos x} = 0$

17. $\lim\limits_{x\to\infty}\left(1+\dfrac{2}{x}\right)^{3x} = e^6$ since

$$\lim\limits_{x\to\infty}3x\ln\left(1+\dfrac{2}{x}\right) = 3\lim\limits_{x\to\infty}\dfrac{\ln(1+2/x)}{1/x} = 3\lim\limits_{x\to\infty}\dfrac{\frac{1}{1+2/x}(-2/x^2)}{-1/x^2} = 3\lim\limits_{x\to\infty}\dfrac{2}{1+2/x} = 6.$$

19. $\lim\limits_{x\to 0^+}\cot x\,\mathrm{Tan}^{-1}x = \lim\limits_{x\to 0^+}\dfrac{\mathrm{Tan}^{-1}x}{\tan x} = \lim\limits_{x\to 0^+}\dfrac{1/(1+x^2)}{\sec^2 x} = 1$

21. $\lim\limits_{x\to 0}(2-e^x)^{1/x} = e^{-1}$ since

$$\lim\limits_{x\to 0}\dfrac{1}{x}\ln(2-e^x) = \lim\limits_{x\to 0}\dfrac{\frac{1}{2-e^x}(-e^x)}{1} = -1.$$

23. $\lim\limits_{x\to\infty}\left[\ln\sqrt{4x+2} - \ln\sqrt{x+3}\right] = \lim\limits_{x\to\infty}\ln\sqrt{\dfrac{4+2/x}{1+3/x}} = \ln 2$

25. $\lim\limits_{x\to 0}(e^x+x^2)^{1/x^2} = \infty$ since

$$\lim\limits_{x\to 0}\dfrac{1}{x^2}\ln(e^x+x^2) = \lim\limits_{x\to 0}\dfrac{\frac{1}{e^x+x^2}(e^x+2x)}{2x} = \infty.$$

27. $\lim\limits_{x\to\infty}\left[\ln\sqrt{x^6+3x^2} - \ln(2x^3)\right] = \ln\lim\limits_{x\to\infty}\sqrt{\dfrac{1+3/x^4}{4}} = -\ln 2$

29. $\lim\limits_{x\to\infty}\left(\dfrac{x}{x+1}\right)^{x+1} = \lim\limits_{x\to\infty}(1+1/x)^{-(x+1)} = e^{-1}$ since

$$-\lim\limits_{x\to\infty}\dfrac{\ln(1+1/x)}{(x+1)^{-1}} = -\lim\limits_{x\to\infty}\dfrac{\frac{1}{1+1/x}(-1/x^2)}{-(x+1)^{-2}} = -\lim\limits_{x\to\infty}(1+1/x) = -1.$$

389

31. $f(x) = x^2 e^x \implies f'(x) = xe^x(x + 2)$, $f''(x) = e^x(x^2 + 4x + 2)$. Since $f' = 0 \iff x = 0, -2$, and $f''(0) > 0$ while $f''(-2) < 0$, the graph of $y = f(x)$ has a relative minimum at $(0, 0)$ and a relative maximum at $(-2, 4/e^2)$. Moreover,

$$\lim_{x \to -\infty} x^2 e^x = \lim_{x \to -\infty} \frac{x^2}{e^{-x}} = \lim_{x \to -\infty} \frac{2x}{-e^{-x}} = \lim_{x \to -\infty} \frac{2}{e^{-x}} = \lim_{x \to -\infty} 2e^x = 0$$

and $\lim_{x \to \infty} x^2 e^x = \infty$. Hence the graph is asymptotic to the x-axis as $x \to -\infty$, but increases without bound as $x \to \infty$.

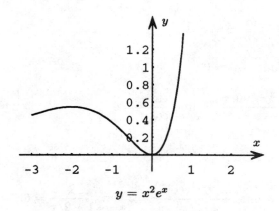

$y = x^2 e^x$

Exercise 31

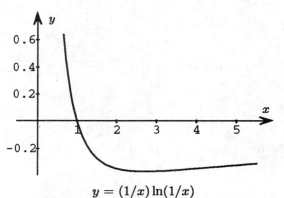

$y = (1/x)\ln(1/x)$

Exercise 33

33. $f(x) = \frac{1}{x}\ln\frac{1}{x} = -\frac{\ln x}{x} \implies f'(x) = -\frac{1 - \ln x}{x^2}$ and $f''(x) = \frac{3 - 2\ln x}{x^3}$. Since $f' = 0 \iff 1 = \ln x \iff x = e$ and $f''(e) > 0$, the graph of $y = f(x)$ has a relative minimum at $(e, -1/e)$. Moreover,

$$\lim_{x \to 0^+} \left(-\frac{\ln x}{x}\right) = +\infty \quad \text{and} \quad \lim_{x \to \infty} \left(-\frac{\ln x}{x}\right) = -\lim_{x \to \infty} \frac{1/x}{1} = 0$$

and hence the graph is asymptotic to the x-axis as x increases without bound.

35. $\lim_{x \to \infty} x^r \ln x = \lim_{x \to \infty} \frac{\ln x}{x^{-r}} = \lim_{x \to \infty} \frac{1/x}{-rx^{-r-1}} = -\frac{1}{r}\lim_{x \to \infty} x^r = 0$ if $r < 0$.

37. (a) Let $y = \left(1 + \frac{1}{t}\right)^{tx}$. Then

$$\lim_{t \to +\infty} \ln y = \lim_{t \to +\infty} \ln\left(1 + \frac{1}{t}\right)^{tx} = \lim_{t \to +\infty}\left[tx \ln\left(1 + \frac{1}{t}\right)\right] = \lim_{t \to +\infty} \frac{x \ln(1 + 1/t)}{1/t}.$$

Now, from Example 5 in the text, with $r = 1$, we know that

$$\lim_{t \to +\infty} \frac{x \ln(1 + 1/t)}{1/t} = x \lim_{t \to +\infty} \frac{\ln(1 + 1/t)}{1/t} = x(1) = x.$$

Hence $\lim_{t \to +\infty} \ln y = x$ and therefore

$$\lim_{t \to +\infty}\left(1 + \frac{1}{t}\right)^{tx} = e^x.$$

390

In particular, if we let $s = 1/t$, then $s \to 0^+$ as $t \to +\infty$ and therefore

$$\lim_{s \to 0^+} (1+s)^{x/s} = \lim_{t \to +\infty} \left(1 + \frac{1}{t}\right)^{tx} = e^x.$$

(b) The following table gives the values of $h(x) = 1 - (1 + 1/t)^{tx}$ for $x = -1, -1/2, 0, 1/2, 1$.

x	$t = 10$	$t = 100$	$t = 1000$	$t = 10000$
-1	-0.480153	-0.0049793	-0.0004998	-0.0000500
$-1/2$	-0.0237262	-0.0024865	-0.0002499	-0.0000250
0	0	0	0	0
$1/2$	0.0237263	0.0024804	0.0002498	0.0000250
1	0.0458155	0.004956	0.0004995	0.0000500

(c) The following table gives the values of $g(x) = 1 - (1 + s)^{x/s}$ for $x = 0.1, 0.01, 0.001, 0.0001$.

x	$s = 0.1$	$s = 0.01$	$s = 0.001$	$s = 0.0001$
-1	-0.480153	-0.0049793	-0.0004998	-0.0000500
$-1/2$	-0.0237262	-0.0024865	-0.0002499	-0.0000250
0	0	0	0	0
$1/2$	0.0237263	0.0024804	0.0002498	0.0000250
1	0.0458155	0.004956	0.0004995	0.0000500

11.3 Improper Integrals

1. $\displaystyle\int_3^\infty \frac{1}{x}\, dx = \lim_{t \to \infty} \ln x \Big|_3^t = \lim_{t \to \infty} (\ln t - \ln 3)$ diverges

3. $\displaystyle\int_1^\infty \frac{x}{\sqrt{1+x^2}}\, dx = \lim_{t \to \infty} \sqrt{1+x^2}\, \Big|_1^t = \lim_{t \to \infty} \left(\sqrt{1+t^2} - \sqrt{2}\right)$ diverges

5. $\displaystyle\int_0^\infty e^{-x}\, dx = \lim_{t \to \infty} (-e^{-x}) \Big|_0^t = \lim_{t \to \infty} \left(\frac{-1}{e^t} + 1\right) = 1$

7. $\displaystyle\int_0^\infty e^{-x} \sin x\, dx = \lim_{t \to \infty} \frac{1}{2} e^{-x}(-\sin x - \cos x) \Big|_0^t = \lim_{t \to \infty} \left[\frac{1}{2} e^{-t}(-\sin t - \cos t) - \frac{1}{2}\right] = \frac{1}{2}$

9. $\displaystyle\int_{-\infty}^2 e^{2x}\, dx = \lim_{t \to -\infty} \frac{1}{2} e^{2x} \Big|_t^2 = \lim_{t \to -\infty} \frac{1}{2}(e^4 - e^{2t}) = \frac{1}{2} e^4$

11. $\displaystyle\int_0^\infty x e^{-x}\, dx = \lim_{t \to \infty} e^{-x}(-x - 1) \Big|_0^t = \lim_{t \to \infty} \left(\frac{-t}{e^t} - \frac{1}{e^t} + 1\right) = \lim_{t \to \infty} \frac{-1}{-e^{-t}} - \lim_{t \to \infty} e^{-t} + 1 = 1$

13. $\displaystyle\int_0^1 \frac{1}{x}\, dx = \lim_{t \to 0^+} \ln x \Big|_t^1 = \lim_{t \to 0^+} (-\ln t)$ diverges

15. $\displaystyle\int_0^3 \frac{1}{\sqrt{9 - x^2}}\, dx = \lim_{t \to 3^-} \mathrm{Sin}^{-1} \frac{x}{3} \Big|_0^t = \lim_{t \to 3^-} \left(\mathrm{Sin}^{-1} \frac{t}{3}\right) = \frac{\pi}{2}$

17. $\displaystyle\int_0^1 x \ln x \, dx = \lim_{t\to 0^+} \frac{x^2}{4}(2\ln x - 1)\Big|_t^1 = \lim_{t\to 0^+}\left[-\frac{1}{4} - \frac{t^2}{4}(2\ln t - 1)\right] = -\frac{1}{4} - \frac{1}{2}\lim_{t\to 0^+}\frac{\ln t}{t^{-2}}$

$\displaystyle = -\frac{1}{4} - \frac{1}{2}\lim_{t\to 0^+}\frac{1/t}{-2t^{-3}} = -\frac{1}{4} - \frac{1}{4}\lim_{t\to 0^+} t^2 = -\frac{1}{4}$

19. $\displaystyle\int_0^e x^2 \ln x \, dx = \lim_{t\to 0^+}\frac{x^3}{9}(3\ln x - 1)\Big|_t^e = \lim_{t\to 0^+}\left[\frac{2e^3}{9} - \frac{t^3}{3}\ln t + \frac{t^3}{9}\right]$

$\displaystyle = \frac{2e^3}{9} - \frac{1}{3}\lim_{t\to 0^+}\frac{\ln t}{t^{-3}} = \frac{2e^3}{9} - \frac{1}{3}\lim_{t\to 0^+} -\frac{1}{3}\lim\frac{1/t}{-3t^{-4}} = \frac{2e^3}{9} + \frac{1}{9}\lim_{t\to 0^+} t^3 = \frac{2e^3}{9}$

21. $\displaystyle\int_{-1}^1 \frac{1}{x}\, dx = \lim_{t\to 0^-}\ln|x|\,\Big|_{-1}^t + \lim_{t\to 0^+}\ln|x|\,\Big|_t^1 = \lim_{t\to 0^-}\ln|t| + \lim_{t\to 0^+}(-\ln t)$ diverges

23. $\displaystyle\int_0^\infty e^{-2x}\cos 2x \, dx = \lim_{t\to\infty}\frac{e^{-2x}}{8}(-2\cos 2x + 2\sin 2x)\Big|_0^t = \lim_{t\to\infty}\left(\frac{-2\cos 2t + 2\sin 2t}{8e^{2t}} + \frac{1}{4}\right) = \frac{1}{4}$

25. $\displaystyle\int_0^\infty e^{-2\sqrt{x}}\sqrt{x}\, dx = \lim_{t\to\infty}\left(-e^{-2\sqrt{x}}\right)\Big|_0^t = \lim_{t\to\infty}\left(-e^{-2\sqrt{t}} + 1\right) = 1$

27. This is a proper integral:

$$\int_0^1 \underbrace{x\ln(1+x)\,dx}_{u=1+x} = \int_1^2 (u-1)\ln u\, du = \left[\frac{u^2}{4}(2\ln u - 1) - u(\ln u - 1)\right]\Big|_1^2 = \frac{1}{4}.$$

29. $\displaystyle\int_0^\infty \sin x \, dx = \lim_{t\to\infty}(-\cos x)\Big|_0^t = \lim_{t\to\infty}(1 - \cos t)$, which does not exist. Hence the improper integral diverges.

31. $\displaystyle\int_0^1 x^p\, dx$ converges if and only if $p > -1$:

$p > -1 \implies \displaystyle\int_0^1 x^p\, dx = \lim_{t\to 0}\frac{x^{1+p}}{1+p}\Big|_t^1 = \lim_{t\to 0}\left(\frac{1 - t^{1+p}}{1+p}\right) = 1/(1+p)$ since $1 + p > 0$

$p = -1 \implies \displaystyle\int_0^1 \frac{1}{x}\, dx = \lim_{t\to 0}\ln|x|\,\Big|_t^1 = \lim_{t\to 0}(-\ln|t|)$ diverges

$p < -1 \implies \displaystyle\int_0^1 x^p\, dx = \lim_{t\to 0}\frac{x^{1+p}}{1+p}\Big|_t^1 = \lim_{t\to 0}\left(\frac{1}{1+p} - \frac{t^{1+p}}{1+p}\right)$ diverges since $1 + p < 0$

33. Suppose $\int_a^\infty g(x)\, dx$ converges to some number A. Since $f(x) \le g(x)$, the area under the graph of $y = f(x)$ from $x = a$ to $x = t$ is bounded above by the area under the graph of $y = g(x)$. Hence

$$\int_a^\infty f(x)\, dx = \lim_{t\to\infty}\int_a^t f(x)\, dx \le \lim_{t\to\infty}\int_a^t g(x)\, dx = \int_a^\infty g(x)\, dx = A$$

and therefore $\int_a^\infty f(x)\, dx$ converges. It follows that if $\int_a^\infty f(x)\, dx$ diverges, so does $\int_a^\infty g(x)\, dx$.

35. diverges; $\displaystyle\int_1^\infty \frac{e^x}{\sqrt{1+x^2}}\, dx > \int_1^\infty \frac{x}{\sqrt{1+x^2}}\, dx = \lim_{t\to\infty}\left(\sqrt{1+t^2} - \sqrt{2}\right)$, which diverges.

37. converges; $\displaystyle\int_1^\infty \frac{|\sin x|}{x^2}\,dx < \int_1^\infty \frac{1}{x^2}\,dx$, which converges ($p$-test, $p = 2 > 1$)

39. diverges; $\displaystyle\int_1^\infty \sqrt{e^x + \sin x}\,dx > \int_1^\infty \sqrt{e-1}\,dx$, which diverges

41. $\displaystyle\int_{-\infty}^\infty x^3\,dx = \lim_{t\to-\infty} \frac{1}{4}x^4\Big|_t^0 + \lim_{t\to\infty} \frac{1}{4}x^4\Big|_0^t$, which diverges, while

$$\lim_{t\to\infty}\int_{-t}^t x^3\,dx = \lim_{t\to\infty} \frac{1}{4}x^4\Big|_{-t}^t = 0.$$

43. (a) $y = 1/x \Longrightarrow y' = -1/x^2 \Longrightarrow$

$$\text{Surface Area} = \int_1^\infty 2\pi\frac{1}{x}\sqrt{1 + \frac{1}{x^4}}\,dx > 2\pi \int_1^\infty \frac{1}{x}\,dx,$$

which diverges. Hence the surface area is infinite.

(b) The volume of revolution is finite.

45. Area $= \displaystyle\int_1^\infty \frac{1}{x^4}\,dx = \lim_{t\to\infty}\left(-\frac{1}{3}\frac{1}{x^3}\right)\Big|_1^t = \lim_{t\to\infty}\left(-\frac{1}{3t^3} + \frac{1}{3}\right) = \frac{1}{3}$

47. Area $= \displaystyle-\int_0^1 \ln x\,dx = -\lim_{t\to 0^+}(x\ln x - x)\Big|_t^1 = -\lim_{t\to)^+}(-1 - t\ln t + t) = 1 + \lim_{t\to 0^+}\frac{\ln t}{1/t}$

$= 1 + \displaystyle\lim_{t\to 0^+}\frac{1/t}{-1/t^2} = 1$

49. Let dV be the differential volume of a shell of thickness dx revolved about the y-axis. Then $dV = 2\pi r h\,dx = 2\pi x(e^{-x} + \frac{1}{x^3})\,dx$ and hence the volume of revolution is

$$V = 2\pi \int_1^\infty \left(xe^{-x} + \frac{1}{x^2}\right)dx = 2\pi \lim_{t\to\infty}\left[e^{-x}(-x-1) - \frac{1}{x}\right]\Big|_1^t$$

$$= 2\pi \lim_{t\to\infty}\left(\frac{-t-1}{e^t} - \frac{1}{t} + \frac{2}{e} + 1\right) = 2\pi\left(\frac{2}{e} + 1\right).$$

51. The centroid is at the point $(\bar{x}, \bar{y})$, where

$$\bar{x} = \frac{\displaystyle\int_1^\infty x\left(\frac{1}{x^3}\right)dx}{\displaystyle\int_1^\infty x^3\,dx} = \frac{\displaystyle\lim_{t\to\infty}\left(-\frac{1}{x}\right)\Big|_1^t}{\displaystyle\lim_{t\to\infty}\left(-\frac{1}{2x^2}\right)\Big|_1^t} = \frac{\displaystyle\lim_{t\to\infty}\left(-\frac{1}{t} + 1\right)}{\displaystyle\lim_{t\to\infty}\left(-\frac{1}{2t^2} + \frac{1}{2}\right)} = 2$$

and

$$\bar{y} = \frac{\displaystyle\int_1^\infty \frac{1}{x^6}\,dx}{2(1/2)} = \lim_{t\to\infty}\left(-\frac{x^{-5}}{5}\right)\Big|_1^t = \lim_{t\to\infty}\left(-\frac{1}{5t^5} + \frac{1}{5}\right) = \frac{1}{5}.$$

53. (a) $\displaystyle\lim_{x\to\infty}[f(x) - g(x)] = \lim_{x\to\infty}\left(\frac{x^2-1}{x} - x\right) = \lim_{x\to\infty}\left(-\frac{1}{x}\right) = 0$

(b) Area $= \int_1^\infty \left(x - \frac{x^2 - 1}{x} \right) dx = \int_1^\infty \frac{1}{x} dx$, which diverges.

55. (a) $\int_0^5 e^{-0.08t} dt = -\frac{25}{2} e^{-0.08t} \Big|_0^5 \approx \$4.12$ million

(b) $\int_0^\infty e^{-0.08t} dt = -\frac{25}{2} \lim_{t \to \infty} \left(e^{-0.08t} - 1 \right) \approx \$12.5$ million

57. If $n = 1$, $\int_0^1 \ln x \, dx = \lim_{t \to 0^+} (x \ln x - x)|_t^1 = \lim_{t \to 0^+} (-1 - t \ln t - t) = -1$. Now, if $n > 1$, assume that $\int_0^1 (\ln x)^{n-1} dx = (-1)^{n-1}(n-1)!$. Then

$$\int_0^1 \underbrace{(\ln x)^n \, dx}_{u=(\ln x)^n, dv=dx} = x(\ln x)^n|_0^1 - n \int_0^1 (\ln x)^{n-1} dx = -n(-1)^{n-1}(n-1)! = (-1)^n n!.$$

59. If $y = \sqrt{r^2 - x^2}$, then the surface area obtained by revolving the semi-circle about the x-axis, which is the surface area of the sphere, is equal to

$$2 \int_0^r 2\pi y \sqrt{1 + (y')^2} \, dx = 4\pi \lim_{t \to r^-} \int_0^t \sqrt{r^2 - x^2} \sqrt{1 + \frac{x^2}{r^2 - x^2}} \, dx = 4\pi \lim_{t \to r^-} \int_0^t r \, dx = 4\pi r^2.$$

For Exercises 61-67, S_{20} denotes the value of Simpson's Rule using 20 points.

61. Let $t = 1/x$. Then

$$\int_2^{+\infty} \frac{dx}{4 + x^2} = \int_{1/2}^0 \frac{-1/t^2}{4 + 1/t^2} \, dt = \int_0^{1/2} \frac{1/t^2}{4 + 1/t^2} \, dt.$$

Since

$$\lim_{t \to 0^+} \frac{1/t^2}{4 + 1/t^2} = \lim_{t \to 0^+} \frac{1}{4t^2 + 1} = 1,$$

we apply Simpson's Rule with $n = 20$ to the integral $\int_0^{1/2} f(t) dt$, where

$$f(t) = \begin{cases} \frac{1/t^2}{4 + 1/t^2}, & t > 0 \\ 1, & t = 0, \end{cases}$$

to obtain $S_{20} = 0.3926991$ and $|S_{20} - \pi/8| = 7.9 \times 10^{-11}$.

63. We first write

$$\int_0^{+\infty} \frac{x \, dx}{(1 + x^2)^{3/2}} = \int_0^1 \frac{x \, dx}{(1 + x^2)^{3/2}} + \int_1^{+\infty} \frac{x \, dx}{(1 + x^2)^{3/2}}.$$

Now apply the transformation $x = 1/t$ to the second integral to obtain

$$\int_1^{+\infty} \frac{x \, dx}{(1 + x^2)^{3/2}} = \int_0^1 \frac{(1/t)(-dt/t^2)}{(1 + 1/t^2)^{3/2}} = \int_0^1 \frac{dt}{(1 + t^2)^{3/2}}.$$

Therefore

$$\int_0^{+\infty} \frac{x \, dx}{(1 + x^2)^{3/2}} = \int_0^1 \frac{x \, dx}{(1 + x^2)^{3/2}} + \int_0^1 \frac{dt}{(1 + t^2)^{3/2}}.$$

Applying Simpson's Rule to both integrals using $n = 20$ gives 0.2928936 for the first integral and 0.7071068 for the second integral. Hence $S_{20} = 0.2928936 + 0.7071068 = 1.0000004$.

65. Let $x = t^2$. Then

$$\int_0^{1/2} \frac{\cos \pi x}{\sqrt{x}} \, dx = \int_0^{\sqrt{1/2}} \frac{\cos \pi t^2}{t} \, 2t dt = 2 \int_0^{\sqrt{1/2}} \cos \pi t^2 dt.$$

Applying Simpson's Rule to the last integral we get $S_{20} = 1.1029373$ and $|S_{20} - EV| = 0.00000152$.

67. Let $x = t^2$. Then

$$\int_0^1 \frac{e^{-x^2/2} \sin 3x}{\sqrt{x}} \, dx = \int_0^1 \frac{e^{-t^4/2} \sin 3t^2}{t} \, 2t \, dt = 2 \int_0^1 e^{-t^4/2} \sin 3t^2 dt.$$

Clearly,

$$\lim_{t \to 0} e^{-t^4/2} \sin 3t^2 = (1)(0) = 0.$$

Applying Simpson's Rule to the last integral we get $S_{20} = 0.910287634$ and $|S_{20} - EV| = 0.000012066$.

Review Exercises – Chapter 11

1. $\displaystyle \lim_{x \to 0} \frac{\sin 3x}{\sin 4x} = \lim_{x \to 0} \frac{3 \cos 3x}{4 \cos 4x} = \frac{3}{4}$

3. $\displaystyle \lim_{x \to 0^+} \frac{\tan x}{x^2} = \lim_{x \to 0^+} \frac{\sec^2 x}{2x} = +\infty$

5. $\displaystyle \lim_{x \to \infty} \frac{\sqrt{x^2 + 2x + 1}}{x} = \lim_{x \to \infty} \frac{x + 1}{x} = \lim_{x \to \infty} \left(1 + \frac{1}{x} \right) = 1$

7. $\displaystyle \lim_{x \to 0} \frac{x - \sin x}{\tan x} = \lim_{x \to 0} \frac{1 - \cos x}{\sec^2 x} = 0$

9. $\displaystyle \lim_{x \to 0^+} \frac{\sin \sqrt{x}}{\sqrt{x}} = \lim_{x \to 0^+} \frac{(\cos \sqrt{x}) \frac{1}{2\sqrt{x}}}{\frac{1}{2\sqrt{x}}} = \lim_{x \to 0^+} \cos \sqrt{x} = 1$

11. $\displaystyle \lim_{x \to \infty} \frac{\ln \sqrt{x}}{\sqrt{x}} = \lim_{x \to \infty} \frac{1/(2x)}{1/(2\sqrt{x})} = \lim_{x \to \infty} \frac{1}{\sqrt{x}} = 0$

13. $\displaystyle \lim_{x \to \infty} \frac{x^3 + e^x}{xe^{2x}} = \text{(four applications of l'Hôpital)} = \lim_{x \to \infty} \frac{1}{16e^x(x + 2)} = 0$

15. $\displaystyle \lim_{x \to \infty} x^{1/x} = e^0 = 1$ since $\displaystyle \lim_{x \to \infty} \frac{\ln x}{x} = \lim_{x \to \infty} \frac{1}{x} = 0$

17. $\displaystyle \lim_{x \to 0^+} x^{\sin 2x} = e^0 = 1$ since

$$\lim_{x \to 0^+} (\sin 2x) \ln x = \lim_{x \to 0^+} \frac{\ln x}{\csc 2x} = \lim_{x \to 0^+} \frac{1/x}{-\csc 2x \cot 2x}$$

$$= -\lim_{x \to 0^+} \frac{\sin^2 2x}{x \cos 2x} = \lim_{x \to 0^+} \frac{2 \sin 2x \cos 2x}{-2x \sin 2x + \cos 2x} = 0.$$

19. $\displaystyle \int_1^\infty xe^{-x} \, dx = \lim_{t \to \infty} e^{-x}(-x - 1) \Big|_1^t = \lim_{t \to \infty} \left(\frac{-t - 1}{e^t} + \frac{2}{e} \right) = \frac{2}{e}$

21. $\displaystyle\int_0^2 \frac{2x+1}{x^2+x-6}\,dx = \lim_{t\to 2^-} \ln|x^2+x-6|\Big|_0^t = \lim_{t\to 2^-}\left[\ln|t^2+t-6|-\ln 6\right]$, which diverges.

23. $\displaystyle\int_{-\infty}^{\infty} xe^{-x^2}\,dx = \lim_{t\to-\infty}\left(-\frac{1}{2}e^{-x^2}\right)\Big|_t^0 + \lim_{t\to\infty}\left(-\frac{1}{2}e^{-x^2}\right)\Big|_0^t$

$\displaystyle = \lim_{t\to-\infty}\left(-\frac{1}{2}+\frac{1}{2}e^{-t^2}\right) + \lim_{t\to\infty}\left(-\frac{1}{2}e^{-t^2}+\frac{1}{2}\right) = 0$

25. $\displaystyle\int_0^{\pi/2}\tan x\,dx = \lim_{t\to\pi/2^-}\ln|\sec x|\Big|_0^t = \lim_{t\to\pi/2^-}\ln|\sec t|$, which diverges.

27. $\displaystyle\int_0^4 \frac{1}{\sqrt{16-x^2}}\,dx = \lim_{t\to 4^-}\operatorname{Sin}^{-1}\frac{x}{4}\Big|_0^t = \lim_{t\to 4^-}\operatorname{Sin}^{-1}\frac{t}{4} = \frac{\pi}{2}$

29. $y = \frac{3x+1}{x-2} \implies y' = \frac{-7}{(x-2)^2} < 0$. Hence the graph of $y = \frac{3x+1}{x-2}$, shown on the following page, is decreasing and has no relative extrema. Since $\lim_{x\to 2^-} y = -\infty$ and $\lim_{x\to 2^+} y = \infty$, the graph has the line $x = 2$ as a vertical asymptote. Since $\lim_{x\to\pm\infty} y = 3$, the graph has $y = 3$ as a horizontal asymptote.

31. $y = \frac{\ln x}{x} \implies y' = \frac{1-\ln x}{x^2} = 0 \iff x = e$. It follows that the point $(e, 1/e)$ is a relative maximum. Since $\lim_{x\to 0^+} y = -\infty$, the graph, shown on the following page, has the line $x = 0$ as a vertical asymptote, and since $\lim_{x\to\infty} y = \lim_{x\to\infty}(1/x) = 0$, the line $y = 0$ is a horizontal asymptote.

33. Area $= \displaystyle\int_1^\infty \left(\frac{x^5+3x+1}{x^3} - x^2\right)dx = \int_1^\infty\left(\frac{3}{x^2}+\frac{1}{x^3}\right)dx = \lim_{t\to\infty}\left(-\frac{3}{x}-\frac{1}{2x^2}\right)\Big|_1^t = \frac{7}{2}$

35. Let dV be the differential volume of a cross-section disk of thickness dx. Then

$$dV = \pi r^2 dx = \pi y^2 dx = \frac{16\pi}{(x^2+1)^2}\,dx$$

and hence the volume of revolution is

$$V = 16\pi\int_0^\infty \frac{1}{(x^2+1)^2}\,dx = 16\pi\lim_{t\to\infty}\left[\frac{1}{2}\operatorname{Tan}^{-1}x + \frac{x}{2(x^2+1)}\right]\Big|_0^t = 4\pi^2.$$

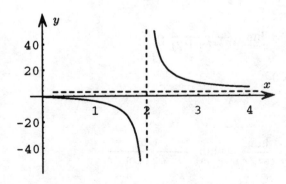

$y = (3x+1)/(x-2)$

Exercise 29

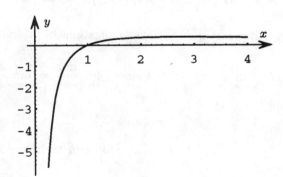

$y = (\ln x)/x$

Exercise 31

Chapter 12

The Theory of Infinite Series

12.1   Infinite Sequences

1. $\dfrac{1}{3}, \dfrac{2}{5}, \dfrac{3}{7}, \dfrac{4}{9}$; converges; $\displaystyle\lim_{n\to\infty} \dfrac{n}{2n+1} = \lim_{n\to\infty} \dfrac{1}{2+1/n} = \dfrac{1}{2}$

3. $-1, \dfrac{-1}{3}, \dfrac{-1}{11}, 0$; converges; $\displaystyle\lim_{n\to\infty} \dfrac{n-4}{n^2+2} = \lim_{n\to\infty} \dfrac{1/n - 4/n^2}{1 + 2/n^2} = 0$

5. $e, e^{1/2}, e^{1/3}, e^{1/4}$; converges; $\displaystyle\lim_{n\to\infty} e^{1/n} = e^0 = 1$

7. $\sqrt{5}, \sqrt{5}, \sqrt{5}, \sqrt{5}$; converges; $\displaystyle\lim_{n\to\infty} \sqrt{5} = \sqrt{5}$

9. $1, 0, -1, 0$; diverges; $\displaystyle\lim_{n\to\infty} \sin\dfrac{n\pi}{2}$ does not exist

11. $10, \dfrac{40}{1+\sqrt{2}}, \dfrac{60}{1+\sqrt{3}}, \dfrac{80}{3}$; diverges; $\displaystyle\lim_{n\to\infty} \dfrac{20n}{1+\sqrt{n}} = \lim_{n\to\infty} \dfrac{20}{1/n + 1/\sqrt{n}} = \infty$

13. $\dfrac{2}{3}, \dfrac{3+\sqrt{2}}{4}, \dfrac{3-\sqrt{3}}{5}, \dfrac{5}{6}$; converges; $\displaystyle\lim_{n\to\infty} \dfrac{3+(-1)^n\sqrt{n}}{n+2} = \lim_{n\to\infty} \dfrac{3/n + (-1)^n/\sqrt{n}}{1 + 2/n} = 0$

15. $\sqrt{2}, \sqrt{\dfrac{3}{2}}, \sqrt{\dfrac{4}{3}}, \sqrt{\dfrac{5}{4}}$; converges; $\displaystyle\lim_{n\to\infty} \sqrt{1+\dfrac{1}{n}} = 1$

17. $1, \cos\dfrac{1}{4}, \cos\dfrac{2}{9}, \cos\dfrac{3}{16}$; converges; $\displaystyle\lim_{n\to\infty} \cos\left(\dfrac{n-1}{n^2}\right) = \lim_{n\to\infty} \cos 0 = 1$

19. $\dfrac{3}{2}, \dfrac{2^{3/2}+2}{2(2^{3/2})}, \dfrac{3^{3/2}+2}{2(3^{3/2})}, \dfrac{5}{8}$; converges; $\displaystyle\lim_{n\to\infty} \dfrac{n^{3/2}+2}{2n^{3/2}} = \lim_{n\to\infty} \dfrac{1+2/n^{3/2}}{2} = \dfrac{1}{2}$

21. $\dfrac{1}{2}, \dfrac{1}{6}, \dfrac{1}{12}, \dfrac{1}{20}$; converges; $\displaystyle\lim_{n\to\infty} \left(\dfrac{1}{n} - \dfrac{1}{n+1}\right) = \lim_{n\to\infty} \dfrac{1}{n(n+1)} = 0$

23. $\sqrt{2}-1, \sqrt{3}-\sqrt{2}, \sqrt{4}-\sqrt{3}, \sqrt{5}-\sqrt{4}$; converges; $\displaystyle\lim_{n\to\infty} (\sqrt{n+1} - \sqrt{n}) = \lim_{n\to\infty} \dfrac{1}{\sqrt{n+1}+\sqrt{n}} = 0$

25. $\sqrt{3}, \dfrac{3}{2}, \dfrac{\sqrt{19}}{3}, \dfrac{\sqrt{33}}{4}$; converges; $\displaystyle\lim_{n\to\infty} \dfrac{\sqrt{2n^2+1}}{n} = \lim_{n\to\infty} \sqrt{2+\dfrac{1}{n^2}} = \sqrt{2}$

27. $\text{Tan}^{-1}2$, $\text{Tan}^{-1}\dfrac{3}{2}$, $\text{Tan}^{-1}\dfrac{4}{3}$, $\text{Tan}^{-1}\dfrac{5}{4}$; converges; $\displaystyle\lim_{n\to\infty}\text{Tan}^{-1}\dfrac{n+1}{n}=\lim_{n\to\infty}\text{Tan}^{-1}\left(1+\dfrac{1}{n}\right)=\dfrac{\pi}{4}$.

29. -1, $\dfrac{1}{2}$, $-\dfrac{1}{3}$, $\dfrac{1}{4}$; converges; $\displaystyle\lim_{n\to\infty}\dfrac{\cos n\pi}{n}=\lim_{n\to\infty}\dfrac{(-1)^n}{n}=0$

31. $\ln\dfrac{3}{2}$, $\ln\dfrac{5}{2}$, $\ln\dfrac{7}{2}$, $\ln\dfrac{9}{2}$; diverges; $\displaystyle\lim_{n\to\infty}\ln\dfrac{2n+1}{2}=\infty$

33. 0, $\left(\dfrac{1}{2}\right)^2$, $\left(\dfrac{2}{3}\right)^3$, $\left(\dfrac{3}{4}\right)^4$; converges; $\displaystyle\lim_{n\to\infty}\left(1-\dfrac{1}{n}\right)^n=e^{-1}$ since

$$\lim_{n\to\infty}n\ln\left(1-\dfrac{1}{n}\right)=\lim_{n\to\infty}\dfrac{\ln(1-1/n)}{1/n}=\lim_{n\to\infty}\dfrac{\dfrac{1}{1-1/n}(1/n^2)}{-1/n^2}=-1.$$

35. Let $\varepsilon>0$. Then

$$\left|\dfrac{1}{2n+1}-0\right|=\dfrac{1}{2n+1}<\varepsilon\iff 2n+1>\dfrac{1}{\varepsilon}\iff n>\dfrac{1}{2}\left(\dfrac{1}{\varepsilon}-1\right).$$

Hence, if we choose any integer $N>\dfrac{1}{2}\left(\dfrac{1}{\varepsilon}-1\right)$ and set $L=0$, it follows that

$$n>N>\dfrac{1}{2}\left(\dfrac{1}{\varepsilon}-1\right)\implies\left|\dfrac{1}{2n+1}-L\right|=\dfrac{1}{2n+1}<\varepsilon,$$

thus proving that $\displaystyle\lim_{n\to\infty}\dfrac{1}{2n+1}=0$.

37. Let $\varepsilon>0$. Then

$$\left|\dfrac{3n-1}{n+1}-3\right|=\left|\dfrac{4}{n+1}\right|<\varepsilon\iff n>\dfrac{4}{\varepsilon}-1.$$

Hence, if we choose any integer $N>\dfrac{4}{\varepsilon}-1$, it follows that

$$n>N>\dfrac{4}{\varepsilon}-1\implies\left|\dfrac{3n-1}{n+1}-3\right|<\varepsilon,$$

thus proving that $\displaystyle\lim_{n\to\infty}\dfrac{3n-1}{n+1}=3$.

39. Since the function $f(x)=e^x$ is continuous for all x and since

$$\lim_{n\to\infty}\ln a^{1/n}=\lim_{n\to\infty}\dfrac{\ln a}{n}=0,$$

it follows that

$$\lim_{n\to\infty}\sqrt[n]{a}=\lim_{n\to\infty}e^{\frac{1}{n}\ln a}=e^0=1.$$

41. Suppose to the contrary that $\displaystyle\lim_{n\to\infty}a_n=L$ and $\displaystyle\lim_{n\to\infty}a_n=M$ for some numbers L, M. If $L\neq M$, there is some $\varepsilon>0$ such that $0<\varepsilon<|L-M|$. But then there are integers N_L and N_M such that

$$n>N_L\implies|a_n-L|<\dfrac{1}{2}\varepsilon$$

and

$$n > N_M \implies |a_n - M| < \frac{1}{2}\,\varepsilon.$$

Now, let n be any integer larger than both N_L and N_M. Then

$$|L - M| = |(L - a_n) - (M - a_n)| \leq |L - a_n| + |M - a_n| < \frac{1}{2}\,\varepsilon + \frac{1}{2}\,\varepsilon = \varepsilon,$$

a contradiction. Therefore $L = M$.

43. $\{(-1)^n\}$ is bounded but does not converge.

45. Suppose $\lim\limits_{n \to \infty} a_n = L$ and let $\varepsilon > 0$. Then there is some number N such that

$$n > N \implies |a_n - L| < \varepsilon$$

and hence $\lim\limits_{n \to \infty} |a_n - L| = 0$. Conversely, if $\lim\limits_{n \to \infty} |a_n - L| = 0$ and $\varepsilon > 0$, there is some N such that

$$n > N \implies |a_n - L| < \varepsilon$$

and hence $\lim\limits_{n \to \infty} a_n = L$.

47. Let $\varepsilon > 0$. Then there are numbers N_L and N_M such that

$$n > N_L \implies |a_n - L| < \frac{\varepsilon}{2} \quad \text{and} \quad n > N_M \implies |b_n - M| < \frac{\varepsilon}{2}.$$

Now, let N be any number larger than both N_L and N_M. Then

$$n > N \implies |(a_n + b_n) - (L + M)| = |(a_n - L) + (b_n - M)| \leq |a_n - L| + |b_n - M| < \frac{\varepsilon}{2} + \frac{\varepsilon}{2} = \varepsilon$$

and hence $\lim\limits_{n \to \infty} (a_n + b_n) = L + M$.

49. First observe that for any integer $n > 0$,

$$|a_n b_n - LM| = |a_n(b_n - M) + M(a_n - L)| \leq |a_n|\,|b_n - M| + |M|\,|a_n - L|.$$

Now, let $\varepsilon > 0$. Since $\{a_n\}$ converges, it is bounded (Exercise 42) and hence there is a number B such that $|a_n| < B + 1$ for all $n > 0$. Moreover, there are integers N_1 and N_2 such that

$$n > N_1 \implies |a_n - L| < \frac{\varepsilon}{2(|M| + 1)}$$

and

$$n > N_2 \implies |b_n - M| < \frac{\varepsilon}{2(B + 1)}.$$

Let N be any integer larger than both N_1 and N_2. Then

$$n > N \implies |a_n b_n - LM| < (B + 1)\,\frac{\varepsilon}{2(B + 1)} + |M|\,\frac{\varepsilon}{2(|M| + 1)} = \frac{\varepsilon}{2} + \frac{|M|}{|M| + 1}\,\frac{\varepsilon}{2} < \frac{\varepsilon}{2} + \frac{\varepsilon}{2} = \varepsilon,$$

and therefore $\lim\limits_{n \to \infty} a_n b_n = LM$.

51. Let $\varepsilon > 0$. Since f is continuous at L, there is some $\delta > 0$ such that

$$0 < |x - L| < \delta \quad \Longrightarrow \quad |f(x) - f(L)| < \varepsilon.$$

Now, $\lim_{n \to \infty} a_n = L$ and hence there is an integer N such that

$$n > N \quad \Longrightarrow \quad |a_n - L| < \delta.$$

Therefore

$$n > N \quad \Longrightarrow \quad |f(a_n) - f(L)| < \varepsilon,$$

and hence $\lim_{n \to \infty} f(a_n) = f(L)$.

12.2   More on Infinite Sequences

1. $\displaystyle \lim_{n \to \infty} n \sin\left(\frac{2}{n}\right) = \lim_{x \to \infty} \frac{\sin(2/x)}{1/x} = \lim_{x \to \infty} \frac{[\cos(2/x)](-2/x^2)}{-1/x^2} = \lim_{x \to \infty} 2\cos\frac{2}{x} = 2$

3. $\displaystyle \lim_{n \to \infty} \sqrt[n]{4n} = \lim_{x \to \infty} (4x)^{1/x} = \lim_{x \to \infty} e^{(1/x)\ln 4x} = e^0 = 1$ since $\displaystyle \lim_{x \to \infty} \frac{\ln 4x}{x} = \lim_{x \to \infty} \frac{(1/4x)(4)}{1} = 0$

5. $\displaystyle \lim_{n \to \infty} (n+1)e^{-n} = \lim_{x \to \infty} \frac{x+1}{e^x} = \lim_{x \to \infty} \frac{1}{e^x} = 0$

7. $\displaystyle \lim_{n \to \infty} \frac{2^{4n+1}}{9^{n/2}} = \lim_{n \to \infty} 2\left(\frac{16}{3}\right)^n = \infty$; diverges

9. $\displaystyle \lim_{n \to \infty} n^{3/n} = \lim_{x \to \infty} e^{(3/x)\ln x} = e^0 = 1$ since $\displaystyle \lim_{x \to \infty} \frac{3\ln x}{x} = \lim_{x \to \infty} \frac{3/x}{1} = 0$

11. $\displaystyle \lim_{n \to \infty} \left(1 - \frac{3}{n}\right)^n = \lim_{x \to \infty} e^{x\ln(1-3/x)} = e^{-3}$ since

$$\lim_{x \to \infty} \frac{\ln(1 - 3/x)}{1/x} = \lim_{x \to \infty} \frac{\dfrac{1}{1-3/x}(3/x^2)}{-1/x^2} = -3.$$

13. $\displaystyle \lim_{n \to \infty} \frac{7^{2n}}{n!} = \lim_{n \to \infty} \frac{49^n}{n!} = 0$

15. $\displaystyle \lim_{n \to \infty} \frac{n^2 \ln n}{2^n} = \lim_{x \to \infty} \frac{x^2 \ln x}{2^x} = \lim_{x \to \infty} \frac{(2x)\ln x + x}{2^x \ln 2} = \lim_{x \to \infty} \frac{2x(1/x) + 2\ln x + 1}{2^x (\ln 2)^2} = \lim_{x \to \infty} \frac{2/x}{2^x (\ln 2)^3} = 0$

17. $\displaystyle \lim_{n \to \infty} \frac{\cos^2 n\pi}{4^{n+3}} = \lim_{n \to \infty} \left(\frac{1}{4}\right)^{n+3} = 0$

19. $\displaystyle \lim_{n \to \infty} \left(\frac{n+3}{n}\right)^n = \lim_{x \to \infty} e^{x\ln(1+3/x)} = e^3$ since

$$\lim_{x \to \infty} x \ln\left(1 + \frac{3}{x}\right) = \lim_{x \to \infty} \frac{\ln(1 + 3/x)}{1/x} = \lim_{x \to \infty} \frac{\dfrac{1}{1+3/x}(-3/x^2)}{-1/x^2} = 3$$

21. $\lim\limits_{n\to\infty}\dfrac{n^n}{n!}=\infty$ since $\lim\limits_{n\to\infty}\dfrac{n^n}{n!}=\lim\limits_{n\to\infty}\dfrac{n\cdot n\cdots n}{n(n-1)\cdots 1}>\lim\limits_{n\to\infty}\dfrac{n}{1}=\infty$; diverges

23. $\left\{\dfrac{3}{n^2}\right\}$ converges since it is a decreasing sequence bounded below by 0.

25. $\left\{\dfrac{e^{-n}}{3^n}\right\}=\{(1/3e)^n\}$ converges since it is a decreasing sequence bounded below by 0.

27. If $|x|>1$, then $|x|=1+M$ for some number $M>0$. Hence if n is a positive integer, $|x^n|=(1+M)^n$ $=1+nM+\cdots+M^n>1+nM$. Since $\lim\limits_{n\to\infty}(1+nM)=\infty$, it follows that $\lim\limits_{n\to\infty}x^n$ does not exist.

29. $a_0=1$, $a_n=2a_{n-1}\implies\{a_n\}=\{1,2,4,\ldots\}$. To show that $a_n=2^n$ for all positive integers n, we use induction on n. Clearly, $a_n=2^n$ for $n=0,1$. Now let $n>1$ and assume, as induction hypothesis, that $a_n=2^n$. Then $a_{n+1}=2a_n=2(2^n)=2^{n+1}$. Hence $a_n=2^n$ for all n by induction.

31. $a_0=-5$, $a_n=a_{n-1}+2\implies\{a_n\}=\{-5,-3,-1,\ldots\}$. We claim that $a_n=-5+2n$ for all positive integers n. To show this, we use induction on n. Clearly, $a_n=-5+2n$ for $n=0,1$. Now let $n>1$ and assume as induction hypothesis that $a_n=-5+2n$. Then $a_{n+1}=(-5+2n)+2=-5+2(n+1)$. Hence $a_n=-5+2n$ for all positive integers n by induction.

33. $\{n\}$ increases but does not converge.

35. Suppose $\lim\limits_{n\to\infty}|a_n|=0$. Then, for any number M, there is some $N>0$ such that $n>N\implies\big||a_n|\big|>M$. Hence, since $\big||a_n|\big|=|a_n|$, $n>N\implies|a_n|>M$ and therefore $\lim\limits_{n\to\infty}a_n=0$.

37. The point of intersection is $x=1.7321$. Table:

n	1	2	3	4	5	6	7	8	9	10
x_n	2	1.75	1.7321	1.7321	1.7321	1.7321	1.7321	1.7321	1.7321	1.7321

39. The point of intersection is $x=11/15=0.7333$. Table:

n	1	2	3	4	5	6	7	8	9	10
x_n	0.7031	0.7828	0.6376	0.8665	0.4339	0.9211	0.2725	0.7433	0.7154	0.7634

It appears that the sequence $\{x_n\}$ does not converge.

41. Computing the terms of these sequences to the 15th term produces

$$a_{15}=0.3183098863,\qquad b_{15}=0.3183098861,$$

and

$$\frac{1}{a_{15}}=3.141592652,\qquad \frac{1}{b_{15}}=3.141592654.$$

It appears that

$$\lim_{n\to+\infty}\frac{1}{a_{15}}=\lim_{n\to+\infty}\frac{1}{a_{15}}=\pi.$$

12.3 Infinite Series

1. $\displaystyle\sum_{k=1}^{\infty} \frac{\cos \pi k}{2^k} = \frac{-1}{2} + \frac{1}{2^2} + \frac{-1}{2^3} + \frac{1}{2^4} + \cdots$

3. $\displaystyle\sum_{k=1}^{\infty} \frac{2^k + 1}{3^k + 2} = \frac{3}{5} + \frac{5}{11} + \frac{9}{29} + \frac{17}{83} + \cdots$

5. $\displaystyle\sum_{k=1}^{\infty} \ln\left(\frac{k}{k+1}\right) = \ln\left(\frac{1}{2}\right) + \ln\left(\frac{2}{3}\right) + \ln\left(\frac{3}{4}\right) + \ln\left(\frac{4}{5}\right) + \cdots$

7. The series converges since it is geometric with ratio 1/7:

$$\sum_{k=0}^{\infty} \frac{1}{7^k} = \sum_{k=0}^{\infty} \left(\frac{1}{7}\right)^k = \frac{1}{1 - \frac{1}{7}} = \frac{7}{6}.$$

9. The series diverges since it is geometric with ratio 4.

11. The series converges since it is geometric with ratio 4/27:

$$\sum_{k=0}^{\infty} \frac{2^{2k}}{3^{3k}} = \sum_{k=0}^{\infty} \left(\frac{4}{27}\right)^k = \frac{1}{1 - \frac{4}{27}} = \frac{27}{23}.$$

13. The series is geometric with ratio $\frac{1}{2+x}$. Since $|x| < 1 \implies \frac{1}{3} < \frac{1}{2+x} < 1$, the series converges:

$$\sum_{k=0}^{\infty} \frac{1}{(2+x)^k} = \frac{1}{1 - \dfrac{1}{2+x}} = \frac{2+x}{1+x}.$$

15. The series converges since the sequence of partial sums converges:

$$S_k = \left(\frac{1}{3} - \frac{1}{2}\right) + \left(\frac{1}{4} - \frac{1}{3}\right) + \cdots + \left(\frac{1}{k+1} - \frac{1}{k}\right) + \left(\frac{1}{k+2} - \frac{1}{k+1}\right) = -\frac{1}{2} - \frac{1}{k} + \frac{1}{k+2}$$

$$\implies \sum_{k=1}^{\infty} \left[\frac{1}{k+2} - \frac{1}{k+1}\right] = \lim_{k \to \infty} S_k = -\frac{1}{2}$$

17. The series diverges since the kth term does not approach zero:

$$\lim_{k \to \infty} \cos \pi k = \lim_{k \to \infty} (-1)^k \quad \text{does not exist.}$$

19. The series converges since the sequence of partial sums converges:

$$\frac{1}{k^2 + 5k + 6} = \frac{1}{k+2} - \frac{1}{k+3}$$

$$\implies S_k = \left(\frac{1}{3} - \frac{1}{4}\right) + \left(\frac{1}{4} - \frac{1}{5}\right) + \left(\frac{1}{5} - \frac{1}{6}\right) + \cdots + \left(\frac{1}{k+2} - \frac{1}{k+3}\right) = \frac{1}{3} - \frac{1}{k+3}$$

$$\implies \sum_{k=1}^{\infty} \frac{1}{k^2 + 5k + 6} = \lim_{k \to \infty} S_k = \frac{1}{3}$$

21. The series converges since the sequence of partial sums converges:

$$\frac{2}{4k^2 + 8k + 3} = \frac{1}{2k+1} - \frac{1}{2k+3}$$

$$\implies S_k = \left(\frac{1}{3} - \frac{1}{5}\right) + \left(\frac{1}{5} - \frac{1}{7}\right) + \cdots + \left(\frac{1}{2k+1} - \frac{1}{2k+3}\right) = \frac{1}{3} - \frac{1}{2k+3}$$

$$\implies \sum_{k=1}^{\infty} \frac{2}{4k^2 + 8k + 3} = \lim_{k\to\infty} S_k = \frac{1}{3}.$$

23. The series diverges since the sequence of partial sums does not have a limit:

$$\ln\left(\frac{k}{k+1}\right) = \ln k - \ln(k+1)$$

$$\implies S_k = (\ln 1 - \ln 2) + (\ln 2 - \ln 3) + \cdots + [\ln k - \ln(k+1)] = \ln 1 - \ln(k+1)$$

$$\implies \sum_{k=1}^{\infty} \ln\left(\frac{k}{k+1}\right) = \lim_{k\to\infty} S_k = -\infty.$$

25. The series converges since it is geometric with ratio $\sqrt{2}/3$:

$$\sum_{k=0}^{\infty} \frac{2^{k/2}}{3^k} = \sum_{k=0}^{\infty} \left(\frac{\sqrt{2}}{3}\right)^k = \frac{1}{1 - \frac{\sqrt{2}}{3}} = \frac{3}{3 - \sqrt{2}}.$$

27. The series diverges since the kth term does not aproach zero:

$$\lim_{k\to\infty} \frac{e^k}{k^2} = \lim_{x\to\infty} \frac{e^x}{x^2} = \lim_{x\to\infty} \frac{e^x}{2x} = \lim_{x\to\infty} \frac{e^x}{2} = \infty.$$

29. (a) $0.333\overline{3} = \dfrac{3}{10} + \dfrac{3}{10^2} + \dfrac{3}{10^3} + \cdots = \sum_{k=1}^{\infty} \dfrac{3}{10^k}$

   (b) $0.333\overline{3} = 3\sum_{k=1}^{\infty} \left(\dfrac{1}{10}\right)^k = 3\left(\dfrac{1}{1 - 1/10} - 1\right) = \dfrac{1}{3}$

31. (a) $0.9292\overline{92} = \dfrac{92}{10^2} + \dfrac{92}{10^4} + \dfrac{92}{10^6} + \cdots = \sum_{k=1}^{\infty} \dfrac{92}{10^{2k}}$

   (b) $0.9292\overline{92} = 92\sum_{k=1}^{\infty} \left(\dfrac{1}{100}\right)^k = 92\left(\dfrac{1}{1 - 1/100} - 1\right) = \dfrac{92}{99}$

33. (a) $0.412412\overline{412} = \sum_{k=1}^{\infty} \dfrac{412}{10^{3k}}$

   (b) $0.412412\overline{412} = 412\sum_{k=1}^{\infty} \left(\dfrac{1}{1000}\right)^k = 412\left(\dfrac{1}{1 - 1/1000} - 1\right) = \dfrac{412}{999}$

35. $\sum_{k=0}^{\infty} (-1)^k x^k = \sum_{k=0}^{\infty} (-x)^k = \dfrac{1}{1 - (-x)} = \dfrac{1}{1 + x}$ if $|x| < 1$.

37. $\displaystyle\sum_{k=0}^{\infty}\frac{x^k}{y^k}=\sum_{k=0}^{\infty}\left(\frac{x}{y}\right)^k=\frac{1}{1-x/y}=\frac{y}{y-x}$ if $|x|<|y|$ since $|x|<|y|\iff|x/y|<1$.

39. Total distance $=h+2\left[\dfrac{2}{3}\,h+\left(\dfrac{2}{3}\right)^2 h+\cdots\right]=h+2\left(\dfrac{2}{3}\,h\right)\displaystyle\sum_{k=0}^{\infty}\left(\dfrac{2}{3}\right)^k=h+\dfrac{4}{3}\,h\,\dfrac{1}{1-2/3}=5h.$

41. No; for example, the series $\sum_{k=0}^{\infty}(1)$ diverges but $\sum_{k=0}^{\infty}0(1)=\sum_{k=0}^{\infty}0=0$ converges. Note, however, that if we assume $c\neq0$, then the series $\sum a_k$ does in fact converge if $\sum ca_k$ converges. For suppose that $\sum ca_k$ converges to some number s. Let S_n and T_n stand for the nth partial sums of $\sum ca_k$ and $\sum a_k$, respectively. Then $S_n=\sum_{k=1}^{n}ca_k=c\sum_{k=1}^{n}a_k=cT_n$ and hence

$$s=\lim_{n\to\infty}S_n=\lim_{n\to\infty}cT_n=c\lim_{n\to\infty}T_n\implies\lim_{n\to\infty}T_n=s/c.$$

Therefore $\sum a_k$ converges to s/c.

43. If $\sum ca_k$ converges and $c\neq0$, then, by Theorem 6(ii), $\frac{1}{c}\sum ca_k$ converges with sum $\sum a_k$, a contradiction. Hence $\sum ca_k$ diverges if $c\neq0$.

45. $S_n=a_1+\cdots+a_n=(a_1+\cdots+a_{n-1})+a_n=S_{n-1}+a_n.$

The table below lists the partial sums for Exercises 47 and 49.

n	$S_n=\displaystyle\sum_{k=0}^{n}\left(\dfrac{2}{3}\right)^k$	$S_n=\displaystyle\sum_{k=5}^{n}(-2)\left(\dfrac{5}{7}\right)^k$
5	2.73663	-0.3718688642
10	2.96531694	-1.128682959
20	2.999398543	-1.295565042
50	2.999999997	-1.301540778
100	3.0	-1.301541025

47. $\displaystyle\sum_{k=0}^{\infty}\left(\frac{2}{3}\right)^k=\frac{1}{1-2/3}=3$

49. $\displaystyle\sum_{k=5}^{\infty}(-2)\left(\frac{5}{7}\right)^k=(-2)\left(\frac{5}{7}\right)^5\sum_{k=0}^{\infty}\left(\frac{5}{7}\right)^k=(-2)\left(\frac{5}{7}\right)^5\frac{1}{1-5/7}=-\frac{3125}{2401}$

51. Since $l_k=l_{k-1}/2$ and $l_1=1$, $l_k=1/2^{k-1}$ for all k. Now, let h_k stand for the altitude of triangle T_k. Then

$$\tan60°=\frac{h_k}{\frac{1}{2}\,l_k}\implies h_k=\frac{1}{2}\,l_k\tan60°=\frac{1}{2}\left(\frac{1}{2^{k-1}}\right)\sqrt{3}=\frac{\sqrt{3}}{2^k}.$$

Hence

$$a_k=\text{Area}\,(T_k)=\frac{1}{2}\,l_k h_k=\frac{1}{2}\left(\frac{1}{2^{k-1}}\right)\left(\frac{\sqrt{3}}{2^k}\right)=\frac{\sqrt{3}}{2^{2k}}$$

and therefore

$$\sum_{k=1}^{\infty}a_k=\sum_{k=1}^{\infty}\frac{\sqrt{3}}{2^{2k}}=\sqrt{3}\sum_{k=1}^{\infty}\left(\frac{1}{4}\right)^k=\sqrt{3}\left(\frac{1}{1-1/4}-1\right)=\frac{\sqrt{3}}{3}.$$

53. Table:

n	5	10	30	50	100
S_n	0.469726563	0.595562935	0.571505107	0.571428814	0.571428571

$\lim\limits_{n \to +\infty} S_n = 4/7.$

55. Table:

n	5	10	30	50	100
S_n	8.8125	11.71484375	11.99999809	12	12

$\lim\limits_{n \to +\infty} S_n = 12.$

57. Table:

k	1	2	3	4	5
a_k	3.5	3.33113883	3.328997488	3.328997143	3.328997143
b_k	3.16227766	3.326856145	3.328996799	3.328997143	3.328997143
d_k	2.25	0.285140947	0.0000045853	1×10^{-11}	1×10^{-11}

From Exercise 40 in Section 12.2, $K(5,2) \approx \dfrac{\pi}{2a_5} = 0.4718527109.$ Therefore

$$
\begin{aligned}
P &= 4(0.4718527109)[25 - d_0/2 - d_1 - 2d_2 - 4d_3 - 8d_4 - 16d_5] \\
&= 1.887410844[25 - 21/2 - d_1 - 2d_2 - 4d_3 - 8d_4 - 16d_5] \\
&= 1.887410844[25 - 12.80704653] \\
&= 23.0131126.
\end{aligned}
$$

The following table gives the results of applying Simpson's Rule to the integral

$$
\int_0^{\pi/2} 4\sqrt{15\sin^2 x + 4\cos^2 x}\, dx
$$

for $n = 4, 6, 8, 10, 12, 14,$ and 16.

n	S_n
4	23.03054483
6	23.01482492
8	23.01331338
10	23.01313866
12	23.01311621
14	23.01311312
16	23.0131126

Thus, using Simpson's Rule with $n = 16$ gives the same accuracy as P.

12.4   The Integral Test

1. The series diverges; for if $f(x) = 1/(2x + 1)$, then $f(x)$ is decreasing and

$$
\int_1^\infty \frac{1}{2x + 1}\, dx = \frac{1}{2} \lim_{t \to \infty} \ln(2x + 1)\Big|_1^t = \frac{1}{2} \lim_{t \to \infty} \left(\ln(2t + 1) - \ln 3\right) = \infty.
$$

3. The series diverges; for if $f(x) = \dfrac{x+1}{2x^2 + 4x + 3}$, then $f(x)$ is decreasing on the interval $[1, \infty)$ since

$$f'(x) = \frac{(2x^2 + 4x + 3)(1) - (x+1)(4x+4)}{(2x^2 + 4x + 3)^2} = \frac{-2x^2 - 4x - 1}{(2x^2 + 4x + 3)^2} = \frac{1 - 2(x+1)^2}{(2x^2 + 4x + 3)^2} < 0$$

for $x > 1$, and

$$\int_1^\infty \frac{x+1}{2x^2 + 4x + 3}\, dx = \frac{1}{4} \lim_{t \to \infty} \ln(2x^2 + 4x + 3)\Big|_1^t = \frac{1}{4} \lim_{t \to \infty} \ln[(2t^2 + 4t + 3) - 9] = \infty.$$

5. The series diverges; for if $f(x) = \dfrac{1}{x \ln x}$, then $f(x)$ is decreasing on the interval $[2, \infty)$ and

$$\int_2^\infty \frac{1}{x \ln x}\, dx = \lim_{t \to \infty} \ln(\ln x)\Big|_2^t = \lim_{t \to \infty} [\ln(\ln t) - \ln(\ln 2)] = \infty.$$

Remark: Note that we begin the series at $k = 2$ since it is not defined at $k = 1$.

7. The series converges; for if $f(x) = \dfrac{1}{1 + x^2}$, then $f(x)$ is decreasing on the interval $[1, \infty)$ and

$$\int_1^\infty \frac{1}{1 + x^2}\, dx = \lim_{t \to \infty} \mathrm{Tan}^{-1} x\Big|_1^t = \lim_{t \to \infty} \left(\mathrm{Tan}^{-1} t - \frac{\pi}{4} \right) = \frac{\pi}{2} - \frac{\pi}{4} = \frac{\pi}{4}.$$

9. The series diverges since it is a p-series with $p = 1/2$:

$$\sum \frac{1}{\sqrt{k+1}} = \sum_{k=2}^\infty \frac{1}{k^{1/2}}.$$

Remark: The fact that the series starts at $k = 2$ instead of $k = 1$ does not effect its divergence.

11. The series converges; for if $f(x) = x^2 2^{-x^3}$, then $f(x)$ is decreasing on the interval $[1, \infty)$ since

$$f'(x) = x^2(-3x^2 2^{-x^3} \ln 2) + 2x 2^{-x^3} = x 2^{-x^3}(2 - 3x^3 \ln 2) < 0$$

for $x > 1$, and

$$\int_1^\infty x^2 2^{-x^3}\, dx = -\frac{1}{3 \ln 2} \lim_{t \to \infty} 2^{-x^3}\Big|_1^t = -\frac{1}{3 \ln 2} \lim_{t \to \infty} (2^{-t^3} - 2^{-1}) = \frac{1}{6 \ln 2}.$$

13. The series diverges since the kth term does not approach zero:

$$\lim_{k \to \infty} \frac{k}{\sqrt{1 + k^2}} = \lim_{k \to \infty} \frac{1}{\sqrt{\frac{1}{k^2} + 1}} = 1.$$

Remark: Note that the Integral Test does not apply. The function $f(x) = \dfrac{x}{\sqrt{1 + x^2}}$ is not decreasing since

$$f'(x) = \frac{\sqrt{1 + x^2} - x \dfrac{1}{2} \dfrac{1}{\sqrt{1 + x^2}} 2x}{1 + x^2} = \frac{1}{(1 + x^2)^{3/2}} > 0$$

for all x.

15. The series diverges since it is a p-series with $p = 1/2$.

17. The series converges; for if $f(x) = \dfrac{x}{(1+x^2)^3}$, then $f(x)$ is decreasing on the interval $[1, \infty)$ since

$$f'(x) = \frac{(1+x^2)^3 - x[3(1+x^2)^2 2x]}{(1+x^2)^6} = \frac{1 - 5x^2}{(1+x^2)^4} < 0$$

for $x > 1$, and

$$\int_1^\infty \frac{x}{(1+x^2)^3}\, dx = -\frac{1}{4}\lim_{t\to\infty}\frac{1}{(1+x^2)^2}\bigg|_1^t = -\frac{1}{4}\lim_{t\to\infty}\left[\frac{1}{(1+t^2)^2} - \frac{1}{4}\right] = \frac{1}{16}.$$

19. The series diverges; for if $f(x) = \dfrac{\ln x}{x}$, then $f(x)$ is decreasing on the interval $[3, \infty)$ since

$$f'(x) = \frac{x(1/x) - \ln x}{x^2} = \frac{1 - \ln x}{x^2} < 0$$

for $x > 3$, and

$$\int_3^\infty \frac{\ln x}{x}\, dx = \frac{1}{2}\lim_{t\to\infty}(\ln x)^2\bigg|_3^t = \frac{1}{2}\lim_{t\to\infty}[(\ln t)^2 - (\ln 3)^2] = \infty.$$

21. The series diverges; for if $f(x) = \dfrac{x^2}{x^3 + 1}$, then $f(x)$ is decreasing on the interval $[2, \infty)$ since

$$f'(x) = \frac{(x^3+1)2x - x^2(3x^2)}{(x^3+1)^2} = \frac{x(2 - x^3)}{(x^3+1)^2} < 0$$

for $x > 2$, and

$$\int_2^\infty \frac{x^2}{x^3+1}\, dx = \frac{1}{3}\lim_{t\to\infty}\ln(x^3+1)\bigg|_2^t = \frac{1}{3}\lim_{t\to\infty}[\ln(t^3+1) - \ln 9] = \infty.$$

23. The series converges; it is the same as the series in Exercise 7 since $\cos^2 \pi k = 1$ for all k.

25. The series converges; for if $f(x) = \dfrac{e^{-2\sqrt{x}}}{\sqrt{x}}$, then $f(x)$ is decreasing on the interval $[1, \infty)$ since

$$f'(x) = \frac{\sqrt{x}e^{-2\sqrt{x}}\left(-\frac{1}{\sqrt{x}}\right) - e^{-2\sqrt{x}}\left(\frac{1}{2\sqrt{x}}\right)}{x} = -e^{-2\sqrt{x}}\left(\frac{2\sqrt{x}+1}{2x\sqrt{x}}\right) < 0$$

for $x > 1$, and

$$\int_1^\infty \frac{e^{-2\sqrt{x}}}{\sqrt{x}}\, dx = -\lim_{t\to\infty}e^{-2\sqrt{x}}\bigg|_1^t = -\lim_{t\to\infty}(e^{-2\sqrt{t}} - e^{-2}) = \frac{1}{e^2}.$$

27. Let $R_{10} = \displaystyle\sum_{11}^\infty 1/k^2$ be the error. By Exercise 26, $\displaystyle\int_{11}^\infty \frac{1}{x^2}\, dx \le R_{10} \le \int_{10}^\infty \frac{1}{x^2}\, dx$. Now,

$$\int_{11}^\infty \frac{1}{x^2}\, dx = \lim_{t\to\infty}\left(-\frac{1}{x}\right)\bigg|_{11}^t = \lim_{t\to\infty}\left(-\frac{1}{t} + \frac{1}{11}\right) = \frac{1}{11}$$

$$\int_{10}^\infty \frac{1}{x^2}\, dx = \lim_{t\to\infty}\left(-\frac{1}{x}\right)\bigg|_{10}^t = \lim_{t\to\infty}\left(-\frac{1}{t} + \frac{1}{10}\right) = \frac{1}{10}.$$

Hence $\dfrac{1}{11} \le R_{10} \le \dfrac{1}{10}$.

29. In this case

$$\int_n^\infty xe^{-x^2} \, dx = \lim_{t\to\infty} \left(-\frac{1}{2}e^{-x^2}\right)\Big|_n^t = -\frac{1}{2}\lim_{t\to\infty}(e^{-t^2} - e^{-n^2}) = \frac{1}{2}e^{-n^2}.$$

Therefore

$$R_n \le \frac{1}{2}e^{-n^2} < 0.02 \iff e^{n^2} > 25 \iff n^2 > \ln 25 \iff n > 2.$$

31. Using the formula in Exercise 30(c), with $n+1$ replaced with n, it follows that $S_{2^n} \ge 1 + \dfrac{n}{2}$.

33. Since

$$\lim_{t\to+\infty}\int_n^t 2^{-x^2} s \, dx = \lim_{t\to+\infty}\left(-\frac{2^{-x^2}}{2\ln 2}\right)\Big|_n^t = \lim_{t\to+\infty}\frac{-1}{2\ln 2}\left(2^{-t^2} - 2^{-n^2}\right) = \frac{2^{-n^2}}{2\ln 2},$$

we choose n so that $2^{-n^2}/2\ln 2 \le 0.001$ or $2^{n^2} \ge 1000/2\ln 2 = 721.3475$. Note that $2^{3^2} = 512$ and $2^{4^2} = 65536$. Taking $n = 4$ we have $S_4 = 0.6309204102$ and $|S_8 - S| = 0.000000149$, where $S = 0.6309205593$.

35. Since

$$\lim_{t\to+\infty}\int_n^t (2x+1)^{-2} x \, dx = \lim_{t\to+\infty}\frac{-1}{2(2x+1)}\Big|_n^t = -\frac{1}{2}\lim_{t\to+\infty}\left[\frac{1}{2t+1} - \frac{1}{2n+1}\right] = \frac{1}{2(2n+1)},$$

we choose n so that $1/(2(2n+1)) \le 0.001$ or $2(2n+1) \ge 1000$. Then $2n+1 \ge 500$ and therefore $n \ge 249.5$ Taking $n = 250$ we have $S_{250} = 0.2327045355$ and $|S_{250} - (\pi^2/8 - 1)| = 0.000996$.

12.5   The Comparison Tests

1. Converges: $\sum \dfrac{1}{1+k^2} < \sum \dfrac{1}{k^2}$, which converges since it is a p-series, $p = 2$.

3. Diverges: $\sum \dfrac{1}{k-4} > \sum \dfrac{1}{k}$, which diverges since it is the tail of a harmonic series.

5. Converges: $\sum e^{-3k} = \sum \left(\dfrac{1}{e^3}\right)^k$ is geometric with ratio $e^{-3} < 1$.

7. Diverges: $\sum \dfrac{3}{\sqrt{k}+2} > 3\sum \dfrac{1}{k+2}$, which diverges since it is the tail of a harmonic series.

9. Converges: $\sum \dfrac{\sqrt{k}}{1+k^3} < \sum \dfrac{1}{k^2}$, which converges since it is a p-series, $p = 2$.

11. Diverges. Choose $\sum b_k = \sum \dfrac{1}{k}$. Then

$$\rho = \lim_{k\to\infty}\frac{\dfrac{k+3}{2k^2+1}}{\dfrac{1}{k}} = \lim_{k\to\infty}\frac{k^2+3k}{2k^2+1} = \frac{1}{2} > 0.$$

Therefore $\sum \dfrac{k+3}{2k^2+1}$ diverges since $\sum b_k$ diverges.

13. Converges. Choose $\sum b_k = \sum \dfrac{1}{k^2}$. Then

$$\rho = \lim_{k \to \infty} \frac{\dfrac{k + \sqrt{k}}{k + k^3}}{\dfrac{1}{k^2}} = \lim_{k \to \infty} \frac{k^3 + k^2 \sqrt{k}}{k^3 + k} = 1 > 0.$$

Therefore $\sum \dfrac{k + \sqrt{k}}{k + k^3}$ converges since $\sum b_k$ is a p-series, $p = 2$, and hence converges.

15. Converges since $\sin(\pi k) = 0$ for all k.

17. Diverges. Use the Limit Comparison Test with $\sum b_k = \sum \dfrac{1}{k}$:

$$\lim_{k \to \infty} \frac{\dfrac{k(k + 1)}{(k + 2)(k^2 + 1)}}{\dfrac{1}{k}} = \lim_{k \to \infty} \frac{k^2(k + 1)}{(k + 2)(k^2 + 1)} = 1 > 0.$$

Therefore $\sum \dfrac{k(k + 1)}{(k + 2)(k^2 + 1)}$ diverges since $\sum b_k$ diverges.

19. Converges. Use the Comparison Test:

$$\sum \frac{2^k}{3^k + 1} < \sum \left(\frac{2}{3}\right)^k,$$

which converges since it is geometric with ratio $2/3 < 1$.

21. Diverges. Use the Comparison Test:

$$k > 4 \quad \Longrightarrow \quad 4k(k + 1) < 5k^2 \quad \Longrightarrow \quad \sum \frac{1}{\sqrt{4k(k + 1)}} > \frac{1}{\sqrt{5}} \sum \frac{1}{k},$$

which is the tail of a harmonic series and hence diverges.

23. Diverges. Use the Limit Comparison Test with $\sum b_k = \sum \dfrac{1}{k^{2/3}}$:

$$\lim_{k \to \infty} \frac{\dfrac{1}{\sqrt[3]{k^2 + 2k}}}{\dfrac{1}{\sqrt[3]{k^2}}} = \lim_{k \to \infty} \sqrt[3]{\frac{k^2}{k^2 + 2k}} = 1 > 0.$$

Therefore $\sum \dfrac{1}{\sqrt[3]{k^2 + 2k}}$ diverges since $\sum b_k$ is a p-series, $p = 2/3$, and hence diverges.

25. Converges. Use the Comparison Test:

$$\sum \frac{\mathrm{Tan}^{-1} k}{k^2} < \frac{\pi}{2} \sum \frac{1}{k^2},$$

which is a p-series, $p = 2$, and hence converges.

27. Diverges. The kth term does not approach zero:

$$\lim_{k\to\infty} \frac{\ln k}{1 + \ln k} = \lim_{k\to\infty} \frac{1}{1 + 1/\ln k} = 1.$$

29. Diverges. The kth term does not approach zero:

$$\lim_{k\to\infty} \frac{3^k}{k+7} = \lim_{x\to\infty} \frac{3^x}{x+7} = \lim_{x\to\infty} \frac{3^x \ln 3}{1} = \infty.$$

31. Use the Limit Comparison Test with $\sum b_k = \sum \frac{1}{k\sqrt{k}}$:

$$\lim_{k\to\infty} \frac{\dfrac{\sqrt{k}}{\cos(2k-6)+k^2}}{\dfrac{1}{k\sqrt{k}}} = \lim_{k\to\infty} \frac{k^2}{\cos(2k-6)+k^2} = 1 > 0.$$

Therefore $\sum \dfrac{\sqrt{k}}{\cos(2k-6)+k^2}$ converges since $\sum b_k$ is a p-series, $p = 3/2$, and hence converges.

33. Let S_n and T_n stand for the partial sums of $\sum a_k$ and $\sum b_k$, respectively, and suppose that $\sum b_k$ diverges. Since $\lim_{k\to\infty} a_k/b_k = \infty$, it follows that $\lim_{k\to\infty} b_k/a_k = 0$. Thus, if $\sum a_k$ converges, then, by Exercise 32, $\sum b_k$ converges, a contradiction. Hence $\sum a_k$ diverges.

12.6 The Ratio and Root Tests

1. Converges since, by the Ratio Test,

$$\lim_{k\to\infty} \frac{\dfrac{1}{(k+2)!}}{\dfrac{1}{(k+1)!}} = \lim_{k\to\infty} \frac{(k+1)!}{(k+2)!} = \lim_{k\to\infty} \frac{1}{k+2} = 0 < 1.$$

3. Diverges since, by the Ratio Test,

$$\lim_{k\to\infty} \frac{\dfrac{2^{k+1}}{k+3}}{\dfrac{2^k}{k+2}} = \lim_{k\to\infty} 2\left(\frac{k+2}{k+3}\right) = 2 > 1.$$

Alternatively, the series diverges since the limit of the kth term is not zero: $\lim_{k\to\infty} 2^k/(k+2) \neq 0$.

5. Converges since, by the Ratio Test,

$$\lim_{k\to\infty} \frac{(k+1)^{10}e^{-(k+1)}}{k^{10}e^{-k}} = \lim_{k\to\infty} \left(\frac{k+1}{k}\right)^{10} \frac{1}{e} = \frac{1}{e} < 1.$$

7. Converges since, by the Ratio Test,

$$\lim_{k \to \infty} \frac{\dfrac{\ln(k+1)}{e^{k+1}}}{\dfrac{\ln k}{e^k}} = \lim_{k \to \infty} \frac{1}{e} \frac{\ln(k+1)}{\ln k} = \frac{1}{e} \lim_{k \to \infty} \frac{1/(k+1)}{1/k} = \frac{1}{e} \lim_{k \to \infty} \frac{k}{k+1} = \frac{1}{e} < 1.$$

9. Converges since, by the Comparison Test,

$$\sum \frac{k+2}{1+k^3} < \sum \frac{k+2}{k^3} = \sum \frac{1}{k^2} + 2 \sum \frac{1}{k^3},$$

which is the sum of two convergent p-series.
Remark: In this case the Ratio Test fails since $\rho = \lim_{k \to \infty} a_{k+1}/a_k = 1$.

11. Converges since, by the Root Test,

$$\lim_{k \to \infty} \left[\frac{1}{(\ln k)^k} \right]^{1/k} = \lim_{k \to \infty} \frac{1}{\ln k} = 0 < 1.$$

13. Diverges since the kth term does not approach zero: $\lim\limits_{k \to \infty} \left(1 + \dfrac{2}{k} \right)^k = e^2$ since

$$\lim_{k \to \infty} k \ln \left(1 + \frac{2}{k} \right) = \lim_{x \to \infty} \frac{\ln(1 + 2/x)}{1/x} = \lim_{x \to \infty} \frac{\dfrac{1}{1+2/x} (-2/x^2)}{-1/x^2} = \lim_{x \to \infty} \frac{2}{1 + 2/x} = 2.$$

15. Diverges since, by the Ratio Test,

$$\lim_{k \to \infty} \frac{\dfrac{3^{k+1}}{(k+1)^3 2^{k+3}}}{\dfrac{3^k}{k^3 2^{k+2}}} = 3 \lim_{k \to \infty} \frac{k^3 \, 2^{k+2}}{(k+1)^3 \, 2^{k+3}} = \frac{3}{2} \lim_{k \to \infty} \left(\frac{k}{k+1} \right)^3 = \frac{3}{2} > 1.$$

17. Converges since, by the Ratio Test,

$$\lim_{k \to \infty} \frac{\dfrac{[(k+1)!]^2}{(3k+3)!}}{\dfrac{(k!)^2}{(3k)!}} = \lim_{k \to \infty} \frac{[(k+1)!]^2}{(k!)^2} \frac{(3k)!}{(3k+3)!} = \lim_{k \to \infty} \frac{(k+1)^2}{(3k+3)(3k+2)(3k+1)} = 0 < 1.$$

19. Converges since, by the Ratio Test,

$$\lim_{k \to \infty} \frac{\dfrac{(k+3)!}{4!\,(k+1)!\,2^{k+1}}}{\dfrac{(k+2)!}{4!\,k!\,2^k}} = \lim_{k \to \infty} \frac{(k+3)!}{(k+2)!} \frac{k!\,2^k}{(k+1)!\,2^{k+1}} = \frac{1}{2} \lim_{k \to \infty} \frac{k+3}{k+1} = \frac{1}{2} < 1.$$

21. Converges. First note that $\sum \dfrac{k^2 \sin^2(k\pi/2)}{2k!} = \sum \dfrac{(2k+1)^2}{2(2k+1)!}$ since only the odd terms are nonzero. This series converges since, by the Ratio Test,

$$\lim_{k\to\infty} \frac{\dfrac{(2k+3)^2}{2(2k+3)!}}{\dfrac{(2k+1)^2}{2(2k+1)!}} = \lim_{k\to\infty} \left(\frac{2k+3}{2k+1}\right)^2 \frac{(2k+1)!}{(2k+3)!} = \lim_{k\to\infty} \left(\frac{2k+3}{2k+1}\right)^2 \frac{1}{(2k+3)(2k+2)} = 0 < 1.$$

23. Converges since, by the Root Test,

$$\lim_{k\to\infty} \left[\left(\frac{\sqrt{k}}{3+k^2}\right)^{2k}\right]^{1/k} = \lim_{k\to\infty} \left(\frac{\sqrt{k}}{3+k^2}\right)^2 = \lim_{k\to\infty} \left(\frac{1/(k\sqrt{k})}{1+3/k^2}\right)^2 = 0 < 1.$$

25. Diverges since, by the Integral Test,

$$\int_2^\infty \frac{1}{x\sqrt{x}}\, dx = 2 \lim_{t\to\infty} \sqrt{\ln x}\,\Big|_2^t = 2 \lim_{t\to\infty} (\sqrt{\ln t} - \sqrt{\ln 2}) = \infty.$$

27. By the Ratio Test, the series converges if

$$\rho = \lim_{k\to\infty} \frac{\dfrac{x^{2k+2}}{2k+2}}{\dfrac{x^{2k}}{2k}} = \lim_{k\to\infty} x^2 \frac{2k}{2k+2} = x^2 < 1.$$

If $x = 1$, the series is harmonic and hence diverges. Therefore the only positive numbers x for which it converges are $0 < x < 1$.

29. By the Ratio Test, the series converges if

$$\rho = \lim_{k\to\infty} \frac{\dfrac{x^{k+1}}{(k+1)!}}{\dfrac{x^k}{k!}} = \lim_{k\to\infty} x \frac{k!}{(k+1)!} = \lim_{k\to\infty} \frac{x}{k+1} = 0 < 1.$$

Therefore the series converges for all positive numbers x.

31. Let γ be a real number such that $\rho > \gamma > 1$. Then, since $\rho = \lim_{k\to\infty} \sqrt[k]{a_k} > 1$, there is some integer N such that

$$k \geq N \implies \sqrt[k]{a_k} > \gamma \implies a_k > \gamma^k.$$

Hence $\sum_{k=N}^\infty a_k > \sum \gamma^k$. But $\sum \gamma^k$ is geometric with ratio $\gamma > 1$ and hence diverges. Therefore $\sum_{k=N}^\infty a_k$ diverges by the Comparison Test and hence $\sum a_k$ diverges.

33. Let γ be a real number such that $\rho > \gamma > 1$. Since $\rho = \lim_{k\to\infty} \sqrt[k]{a_k} > \gamma > 1$, there is some integer N such that $a_k > \gamma^k$ when $k \geq N$. Therefore $\lim_{k\to\infty} a_k > \lim_{k\to\infty} \gamma^k = \infty$ and hence $\lim_{k\to\infty} a_k = \infty$.

12.7   Absolute and Conditional Convergence

1. Converges absolutely since

$$\sum_{k=1}^{\infty} \left| \frac{(-1)^k}{k^3} \right| = \sum_{k=1}^{\infty} \frac{1}{k^3}$$

is a p-series, $p = 3$, and hence converges.

3. Converges absolutely since, by the Ratio Test,

$$\lim_{k\to\infty} \frac{\dfrac{(k+1)^2}{(k+3)!}}{\dfrac{k^2}{(k+2)!}} = \lim_{k\to\infty} \left(\frac{k+1}{k} \right)^2 \frac{(k+2)!}{(k+3)!} = \lim_{k\to\infty} \left(1 + \frac{1}{k} \right)^2 \frac{1}{k+3} = 0 < 1.$$

5. Converges absolutely since, by the Ratio Test,

$$\lim_{k\to\infty} \frac{\dfrac{(k+1)!}{(2k+3)!}}{\dfrac{k!}{(2k+1)!}} = \lim_{k\to\infty} \frac{(k+1)!}{k!} \frac{(2k+1)!}{(2k+3)!} = \lim_{k\to\infty} \frac{k+1}{(2k+3)(2k+2)} = 0 < \infty.$$

7. Converges conditionally since it is an alternating series whose kth terms are decreasing and approach zero, but

$$\sum_{k=1}^{\infty} \left| \frac{(-1)^k}{1 + \sqrt{k}} \right| = \sum_{k=1}^{\infty} \frac{1}{1 + \sqrt{k}} > \sum_{k=1}^{\infty} \frac{1}{k+1} = \infty.$$

9. Converges conditionally since

$$\sum_{k=1}^{\infty} \frac{\sin(\pi k/2)}{k} = \sum_{k=1}^{\infty} \frac{(-1)^{k-1}}{2k-1},$$

which is an alternating series whose kth terms are decreasing and approach zero, but

$$\sum_{k=1}^{\infty} \left| \frac{(-1)^{k-1}}{2k-1} \right| = \sum_{k=1}^{\infty} \frac{1}{2k-1} > \frac{1}{2} \sum_{k=1}^{\infty} \frac{1}{k} = \infty.$$

11. Converges absolutely since, by the Ratio Test,

$$\lim_{k\to\infty} \frac{\dfrac{(k+1)^3}{2^{k+3}}}{\dfrac{k^3}{2^{k+2}}} = \lim_{k\to\infty} \frac{2^{k+2}}{2^{k+3}} \left(\frac{k+1}{k} \right)^3 = \frac{1}{2} < 1.$$

13. Diverges since the kth term does not approach zero:

$$\lim_{k\to\infty} \frac{k^2}{\ln k} = \lim_{x\to\infty} \frac{2x}{1/x} = \lim_{x\to\infty} 2x^2 = \infty.$$

15. Converges conditionally; for, since

$$\frac{d}{dx}\left(\frac{\sqrt{x}}{x+1}\right) = \frac{(x+1)\frac{1}{2\sqrt{x}} - \sqrt{x}}{(x+1)^2} = \frac{1-x}{2(x+1)^2\sqrt{x}} < 0$$

for $x > 1$, the kth terms are decreasing, and since

$$\lim_{k\to\infty}\frac{\sqrt{k}}{k+1} = \lim_{x\to\infty}\frac{1/2\sqrt{x}}{1} = 0,$$

the terms approach zero. Hence the series converges. But

$$\sum_{k=1}^{\infty}\left|\frac{(-1)^k\sqrt{k}}{k+1}\right| = \sum_{k=1}^{\infty}\frac{\sqrt{k}}{k+1} > \sum_{k=1}^{\infty}\frac{\sqrt{k}}{2k} = \frac{1}{2}\sum_{k=1}^{\infty}\frac{1}{\sqrt{k}} = \infty.$$

Thus the series converges conditionally.

17. Converges absolutely since, by the Integral Test,

$$\int_2^{\infty}\frac{1}{x\ln^2 x}\,dx = \lim_{t\to\infty}\left(-\frac{1}{\ln x}\right)\Big|_2^{\infty} = \lim_{t\to\infty}\left(\frac{1}{\ln 2} - \frac{1}{\ln t}\right) = \frac{1}{\ln 2}.$$

19. Converges absolutely since

$$\sum_{k=1}^{\infty}\left|\frac{(-1)^k}{1+k^2}\right| = \sum_{k=1}^{\infty}\frac{1}{1+k^2} < \sum_{k=1}^{\infty}\frac{1}{k^2},$$

which is a p-series, $p = 2$, and hence converges.

21. Diverges since the kth terms do not approach zero:

$$\lim_{k\to\infty}\frac{k}{\ln\sqrt{k}} = 2\lim_{k\to\infty}\frac{k}{\ln k} = 2\lim_{x\to\infty}\frac{1}{1/x} = \infty.$$

23. Converges absolutely since

$$\sum_{k=1}^{\infty}\left|\frac{(-1)^k\sinh k}{3e^{2k}}\right| = \sum_{k=1}^{\infty}\frac{e^k - e^{-k}}{6e^{2k}} = \sum_{k=1}^{\infty}\frac{e^{2k} - 1}{6e^{3k}} < \sum_{k=1}^{\infty}\frac{e^{2k}}{6e^{3k}} = \frac{1}{6}\sum_{k=1}^{\infty}\left(\frac{1}{e}\right)^k,$$

which is geometric with ratio $\frac{1}{e} < 1$ and hence converges.

25. Converges conditionally since

$$\sum_{k=1}^{\infty}\frac{\cos\pi k}{k+2} = \sum_{k=1}^{\infty}\frac{(-1)^k}{k+2},$$

which is the tail of the alternating harmonic series.

27. Converges absolutely. To show this, first observe that

$$\sum_{k=1}^{\infty}\left|\frac{k\sin(k+1/2)\pi}{1+e^{\sqrt{k}}}\right| = \sum_{k=1}^{\infty}\frac{k}{1+e^{\sqrt{k}}}.$$

Now,

$$\lim_{x\to\infty}\frac{(\ln x)^2}{x} = \lim_{x\to\infty}\frac{2(\ln x)(1/x)}{1} = \lim_{x\to\infty}\frac{2\ln x}{x} = \lim_{x\to\infty}\frac{2/x}{1} = 0.$$

414

Hence there is an integer N such that

$$\frac{(\ln k)^2}{k} < \frac{1}{9} \quad \text{for all integers } k > N$$
$$\implies \quad 9(\ln k)^2 < k$$
$$\implies \quad 3\ln k < \sqrt{k}$$
$$\implies \quad k^3 < e^{\sqrt{k}}$$
$$\implies \quad \frac{k}{e^{\sqrt{k}}} < \frac{1}{k^2} \quad \text{for all integers } k > N.$$

Therefore

$$\sum_{k=1}^{\infty} \frac{k}{1+e^{\sqrt{k}}} < \sum_{k=1}^{N} \frac{k}{1+e^{\sqrt{k}}} + \sum_{k=N}^{\infty} \frac{k}{e^{\sqrt{k}}} < \sum_{k=1}^{N} \frac{k}{1+e^{\sqrt{k}}} + \sum_{k=N}^{\infty} \frac{1}{k^2},$$

which is a *p*-series, $p = 2$, and hence converges.

29. Converges absolutely; for,

$$\sum_{k=1}^{\infty} \frac{\sinh k - \cosh k}{k} = \sum_{k=1}^{\infty} \frac{e^k - e^{-k}}{2} - \frac{e^k + e^{-k}}{2} above1ptk = -\sum_{k=1}^{\infty} \frac{1}{ke^k},$$

which converges by the Ratio Test since

$$\lim_{k\to\infty} \frac{\dfrac{1}{(k+1)e^{k+1}}}{\dfrac{1}{ke^k}} = \lim_{k\to\infty} \frac{k}{k+1}\frac{e^k}{e^{k+1}} = \lim_{k\to\infty} \frac{1}{1+1/k}\frac{1}{e} = \frac{1}{e} < 1.$$

31. Converges absolutely for all real x since, by the Ratio Test,

$$\lim_{k\to\infty} \left| \frac{\dfrac{x^{k+1}}{(k+1)!}}{\dfrac{x^k}{k!}} \right| = \lim_{k\to\infty} |x|\frac{1}{k+1} = 0.$$

33. Converges absolutely for all real x since

$$\lim_{k\to\infty} \left| \frac{\dfrac{x^{2k+3}}{(2k+3)!}}{\dfrac{x^{2k+1}}{(2k+1)!}} \right| = \lim_{k\to\infty} x^2\frac{(2k+1)!}{(2k+3)!} = \lim_{k\to\infty} \frac{x^2}{(2k+3)(2k+2)} = 0.$$

35. $n \geq 98$ since

$$|R_n| < a_{n+1} \leq 0.01 \quad \Longleftrightarrow \quad \frac{1}{n+2} \leq 0.01 \quad \Longleftrightarrow \quad n \geq 98.$$

37. Let $\sum |a_k| = S$ and $\sum |b_k| = T$. Then

$$\sum |a_k + b_k| \leq \sum |a_k| + \sum |b_k| = S + T,$$

and hence $\sum |a_k + b_k|$ converges absolutely.

415

39. (a) Suppose $0 \leq \rho < 1$. Then, by the Ratio Test, the series $\sum |a_k|$ converges. Hence $\sum a_k$ converges absolutely.

   (b) If $\rho > 1$, then there is some integer N such that $|a_{k+1}| > |a_k|$ for $k > N$. Therefore $\lim_{k \to \infty} |a_k| \neq 0$ and hence the series $\sum a_k$ diverges.

   (c) The series may or may not converge if $\rho = 1$. For example, $\rho = 1$ for both of the series $\sum \frac{1}{k}$ and $\sum \frac{(-1)^k}{k}$, although the first one diverges, the second one converges conditionally.

41. Graph using the range $[-6, 6] \times [-1.5, 1.5]$.

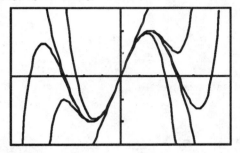

   This looks like the graph of $f(x) = \sin x$.

43. Graph using the range $[-1, 1] \times [-3, 3]$.

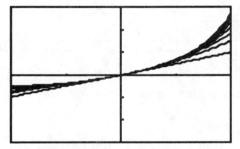

   This looks like the graph of $f(x) = -\ln(1 - x)$.

45. The $(n+1)$st term of the seris is $a_{n+1} = (n+1)/3^{n+2}$. Hence we choose n so that $(n+1)/3^{n+2} \leq 0.001$ or $3^{n+2}/(n+2) \geq 1000$. Taking $n = 7$ gives $3^9/9 = 2187$, $S_7 = 0.1871665905$, and $|S_7 - 3/16| = 0.0003341$.

47. The $(n+1)$st term of the series is $a_{n+1} = 1/(4(n+1)^2 - 1)$. Hence we choose n so that $1/(4(n+1)^2 - 1) \leq 0.001$ or $4(n+1)^2 - 1 \geq 1000$ or $(n+1)^2 \geq 250.25$. Taking $n = 15$ gives $(n+1)^2 = 256$, $S_{15} = 0. - 0.2859173808$, and $|S_9 - (2 - \pi)/4| = 0.0005192$.

Review Exercises – Chapter 12

1. Diverges; $\lim\limits_{n \to \infty} \sin \dfrac{n\pi}{2}$ does not exist.

3. Diverges since

$$\lim_{x \to \infty} \frac{\sqrt{x}}{\ln(\sqrt{x} + 1)} = \lim_{x \to \infty} \frac{\dfrac{1}{2\sqrt{x}}}{\dfrac{1}{\sqrt{x} + 1}\left(\dfrac{1}{2\sqrt{x}}\right)} = \lim_{x \to \infty} (\sqrt{x} + 1) = \infty.$$

5. Converges; $\lim\limits_{n\to\infty}\dfrac{3^n}{n!}=0$. For, by the Ratio Test, the series $\sum\dfrac{3^n}{n!}$ converges since

$$\lim_{n\to\infty}\frac{\dfrac{3^{n+1}}{(n+1)!}}{\dfrac{3^n}{n!}}=\lim_{n\to\infty}3\,\frac{n!}{(n+1)!}=3\lim_{n\to\infty}\frac{1}{n+1}=0.$$

Hence the nth term of the series approaches zero.

7. Diverges; for, if n is even,

$$\lim_{n\to\infty}\frac{(-1)^{n+1}n^3}{n(1-n+n^2)}=\lim_{n\to\infty}\frac{-1}{1-1/n+1/n^2}=-1,$$

but if n is odd,

$$\lim_{n\to\infty}\frac{(-1)^{n+1}n^3}{n(1-n+n^2)}=\lim_{n\to\infty}\frac{1}{1-1/n+1/n^2}=1.$$

9. Diverges; $\lim\limits_{n\to\infty}e^{2\ln n}=\lim\limits_{n\to\infty}n^2=\infty$

11. $1+\dfrac{1}{6}+\dfrac{1}{36}+\dfrac{1}{216}+\cdots=\sum\limits_{k=0}^{\infty}\left(\dfrac{1}{6}\right)^k=\dfrac{1}{1-1/6}=\dfrac{6}{5}$

13. $\dfrac{2}{3}+\dfrac{4}{9}+\dfrac{8}{27}+\cdots=\sum\limits_{k=1}^{\infty}\left(\dfrac{2}{3}\right)^k=\dfrac{1}{1-2/3}-1=2$

15. $0.37\overline{37}\cdots=\dfrac{37}{10^2}+\dfrac{37}{10^4}+\cdots=\dfrac{37}{10^2}\sum\limits_{k=0}^{\infty}\left(\dfrac{1}{10^2}\right)^k=\dfrac{37}{100}\dfrac{1}{1-1/100}=\dfrac{37}{99}$

17. $\sum\limits_{k=3}^{\infty}\left(\dfrac{1}{4}\right)^k=\dfrac{1}{1-1/4}-\left(1+\dfrac{1}{4}+\dfrac{1}{16}\right)=\dfrac{1}{48}$

19. $\sum\limits_{k=2}^{\infty}\dfrac{20}{3^{k+1}}=20\sum\limits_{k=3}^{\infty}\left(\dfrac{1}{3}\right)^k=20\left(\dfrac{1}{1-1/3}-1-\dfrac{1}{3}-\dfrac{1}{9}\right)=\dfrac{10}{9}$

21. Converges since

$$\frac{1}{k(k+1)}=\frac{1}{k}-\frac{1}{k+1}$$
$$\implies\ S_k=\left(1-\frac{1}{2}\right)+\left(\frac{1}{2}-\frac{1}{3}\right)+\cdots+\left(\frac{1}{k}-\frac{1}{k+1}\right)=1-\frac{1}{k+1}$$
$$\implies\ \sum_{k=1}^{\infty}\frac{1}{k(k+1)}=\lim_{k\to\infty}S_k=1.$$

23. Diverges since, by the Ratio Test,

$$\lim_{k\to\infty}\frac{\dfrac{2^{k+1}}{(k+1)^5}}{\dfrac{2^k}{k^5}}=\lim_{k\to\infty}2\left(\frac{k}{k+1}\right)^5=2>1.$$

25. Converges since, by the Ratio Test,

$$\lim_{k \to \infty} \frac{\dfrac{(k+1)^3}{(k+1)!}}{\dfrac{k^3}{k!}} = \lim_{k \to \infty} \left(\frac{k+1}{k}\right)^3 \frac{k!}{(k+1)!} = \lim_{k \to \infty} \left(1 + \frac{1}{k}\right)^3 \frac{1}{k+1} = 0 < 1.$$

27. Converges since it is an alternating series whose kth terms are decreasing and approach zero:

$$\sum_{k=2}^{\infty} \frac{\cos \pi k}{\sqrt{k}} = \sum_{k=2}^{\infty} \frac{(-1)^k}{\sqrt{k}}.$$

29. Converges; for, by the Limit Comparison Test, $\displaystyle\sum_{k=3}^{\infty} \frac{1}{k^{3/2}}$ is a convergent p-series, $p = 3/2$, and

$$\lim_{k \to \infty} \frac{\dfrac{\ln k}{k^2}}{\dfrac{1}{k^{3/2}}} = \lim_{x \to \infty} \frac{\ln x}{\sqrt{x}} = \lim_{x \to \infty} \frac{1/x}{(1/2)x^{-1/2}} = \lim_{x \to \infty} \frac{2}{\sqrt{x}} = 0. \qquad \text{(see Exercise 32, Sec. 12.5)}$$

31. Converges since, by the Comparison Test,

$$\sum_{k=1}^{\infty} \frac{1}{9 + k^2} < \sum_{k=1}^{\infty} \frac{1}{k^2},$$

which is a p-series, $p = 2$, and hence converges.

33. Diverges since

$$\sum_{k=2}^{\infty} \frac{2}{k \ln(k+1)} > 2 \sum_{k=2}^{\infty} \frac{1}{(k+1)\ln(k+1)},$$

which, by the Integral Test, diverges since

$$\int_{2}^{\infty} \frac{1}{(x+1)\ln(x+1)} \, dx = \lim_{t \to \infty} \ln \ln(x+1) \Big|_{2}^{\infty} = \lim_{t \to \infty} [\ln(t+1) - \ln 3] = \infty.$$

35. Converges since, by the Ratio Test,

$$\lim_{k \to \infty} \frac{\dfrac{(k+1)^4 \, 2^{k+1}}{(k+3)!}}{\dfrac{k^4 \, 2^k}{(k+2)!}} = \lim_{k \to \infty} 2\left(\frac{k+1}{k}\right)^4 \frac{(k+2)!}{(k+3)!} = \lim_{k \to \infty} 2(1 + 1/k)^4 \frac{1}{k+3} = 0.$$

37. Diverges since

$$\sum_{k=0}^{\infty} \frac{1}{\sqrt{4 + k^2}} > \sum_{k=0}^{\infty} \frac{1}{k+2},$$

which is harmonic and hence diverges.

418

39. Converges by the alternating series test; for, since

$$\frac{d}{dx}\left(\frac{\sqrt{x}}{2x+1}\right) = \frac{(2x+1)\dfrac{1}{2\sqrt{x}} - \sqrt{x}(2)}{(2x+1)^2} = \frac{1-2x}{2\sqrt{x}(2x+1)^2} < 0,$$

the terms are decreasing, while

$$\lim_{x\to\infty}\frac{\sqrt{x}}{2x+1} = \lim_{x\to\infty}\frac{1/(2\sqrt{x})}{2} = 0$$

and hence the terms approach zero.

41. Converges since, by the Root Test,

$$0 \le \lim_{k\to\infty}\left[\left(\frac{k!}{k^k}\right)^k\right]^{1/k} = \lim_{k\to\infty}\frac{k!}{k^k} = \lim_{k\to\infty}\frac{k-1}{k}\cdots\frac{2}{k}\frac{1}{k} \le \lim_{k\to\infty}\frac{1}{k} = 0$$

$$\implies \lim_{k\to\infty}\left[\left(\frac{k!}{k^k}\right)^k\right]^{1/k} = 0.$$

43. Converges since, by the Ratio Test,

$$\lim_{k\to\infty}\frac{2(k+1)e^{-(k+1)}}{2ke^{-k}} = \lim_{k\to\infty}\frac{k+1}{k}\frac{1}{e} = \frac{1}{e} < 1.$$

45. Converges: $\displaystyle\sum_{k=0}^{\infty}(-1)^k\sin\pi k = \sum_{k=0}^{\infty} 0 = 0.$

47. Diverges since, by the Ratio Test,

$$\lim_{k\to\infty}\frac{\dfrac{(k+1)!}{e^{2k+2}}}{\dfrac{k!}{e^{2k}}} = \lim_{k\to\infty}\frac{(k+1)!}{k!}\frac{1}{e^2} = \lim_{k\to\infty}\frac{k+1}{e^2} = \infty.$$

49. Converges by the alternating series test since the kth terms are decreasing and approach zero.

51. Converges since, by the Ratio Test,

$$\lim_{k\to\infty}\frac{\dfrac{2^{2k+1}}{(2k+1)!}}{\dfrac{2^{2k-1}}{(2k-1)!}} = \lim_{k\to\infty} 4\frac{(2k-1)!}{(2k+1)!} = \lim_{k\to\infty}\frac{4}{(2k+1)(2k)} = 0.$$

53. Converges by the alternating series test since the kth terms are decreasing and approach zero.

55. Converges; for, by the Ratio Test,

$$\lim_{k\to\infty}\frac{(k+1)\left(\dfrac{\pi}{k+1}\right)^{k+1}}{k\left(\dfrac{\pi}{k}\right)^k} = \lim_{k\to\infty}\pi\frac{k+1}{k}\frac{k^k}{(k+1)^k} = \pi\lim_{k\to\infty}\left(\frac{k}{k+1}\right)^{k-1} = \frac{\pi}{e} > 1$$

since

$$\lim_{x\to\infty}(x-1)\ln\frac{x}{x+1} = \lim_{x\to\infty}\frac{\ln x - \ln(x+1)}{\frac{1}{x-1}} = \lim_{x\to\infty}\frac{\frac{1}{x}-\frac{1}{x+1}}{\frac{-1}{(x-1)^2}} = \lim_{x\to\infty}\frac{-(x-1)^2}{x(x+1)} = -1.$$

Remark: The Root Test may also be used to show convergence.

57. Converges since, by the Root Test,

$$\lim_{k\to\infty}\left[\left(\frac{k}{2k+1}\right)^k\right]^{1/k} = \lim_{k\to\infty}\frac{k}{2k+1} = \frac{1}{2} < 1.$$

59. Converges conditionally since the kth terms are decreasing and approach zero, but

$$\sum_{k=0}^{\infty}\left|\frac{(-1)^{k+1}}{3k+2}\right| = \sum_{k=0}^{\infty}\frac{1}{3k+2} > \frac{1}{3}\sum_{k=0}^{\infty}\frac{1}{k+1},$$

which is harmonic and hence diverges.

61. Diverges since the kth terms do not approach zero:

$$\lim_{k\to\infty}\frac{k^k}{k!} = \lim_{k\to\infty}\left(\frac{k}{k-1}\cdots\frac{k}{2}\frac{k}{1}\right) > 1.$$

63. Converges absolutely since

$$\sum_{k=1}^{\infty}\left|\frac{(-1)^{k+1}}{k(k+1)}\right| = \sum_{k=1}^{\infty}\frac{1}{k(k+1)} = \lim_{t\to\infty}\sum_{k=1}^{t}\left(\frac{1}{k}-\frac{1}{k+1}\right) = \lim_{t\to\infty}\left(1-\frac{1}{t+1}\right) = 1.$$

65. Diverges since, by the Ratio Test,

$$\lim_{k\to\infty}\frac{\dfrac{(k+1)!}{2^{k+1}}}{\dfrac{k!}{2^k}} = \lim_{k\to\infty}\frac{1}{2}\frac{(k+1)!}{k!} = \lim_{k\to\infty}\frac{1}{2}(k+1) = \infty.$$

67. Converges absolutely since, by the Ratio Test,

$$\lim_{k\to\infty}\frac{\dfrac{(k+1)^2}{(k+2)!}}{\dfrac{k^2}{(k+1)!}} = \lim_{k\to\infty}\left(\frac{k+1}{k}\right)^2\frac{(k+1)!}{(k+2)!} = \lim_{k\to\infty}\left(1+\frac{1}{k}\right)^2\frac{1}{k+2} = 0.$$

69. $|R_4| < |a_5| = \dfrac{2}{26} = 0.0769.$

71. $|R_4| < |a_5| = \dfrac{1}{11} = 0.0909.$

420

73. $\{a_k\} = \{c^k\}$ converges $\Longleftrightarrow -1 < c \leq 1$.

75. $\sum b_k$ diverges; for, if both $\sum a_k$ and $\sum b_k$ converge, then $\sum(a_k + b_k)$ converges and is equal to $\sum a_k + \sum b_k$.

77. (a) After the first fall, the ball has travelled 10 meters. After 2 falls, it travelled

$$10 + 2\left[10\left(\frac{5}{6}\right)\right] \text{ meters.}$$

Continuing in this manner, after 10 falls it has travelled

$$10 + 20\left(\frac{5}{6}\right) + \cdots + 20\left(\frac{5}{6}\right)^9 = 10 + 20\left[\frac{1 - (5/6)^{10}}{1 - 5/6} - 1\right] \approx 90.62 \text{ meters.}$$

(b) The total distance travelled when the ball comes to rest is equal to

$$10 + 20\sum_{k=1}^{\infty}\left(\frac{5}{6}\right)^k = 10 + 20\left(\frac{1}{1 - 5/6} - 1\right) = 110 \text{ meters.}$$

79. (a) $\sum a_k^2 \leq \sum(a_k + b_k)^2 = \sum(a_k^2 + 2a_kb_k + b_k^2) \Longrightarrow \sum a_k^2$ converges
    (b) $\sum b_k^2 \leq \sum(a_k + b_k)^2 = \sum(a_k^2 + 2a_kb_k + b_k^2) \Longrightarrow \sum b_k^2$ converges
    (a) $\sum a_kb_k \leq \sum(a_k + b_k)^2 = \sum(a_k^2 + 2a_kb_k + b_k^2) \Longrightarrow \sum a_kb_k$ converges

Chapter 13

Taylor Polynomials and Power Series

13.1 The Approximation Problem and Taylor Polynomials

1. Since

$$f(x) = e^{-x}, \qquad f(0) = 1$$
$$f'(x) = -e^{-x}, \qquad f'(0) = -1$$
$$f''(x) = e^{-x}, \qquad f''(0) = 1$$
$$f'''(x) = -e^{-x}, \qquad f'''(0) = -1$$
$$f^{(4)}(x) = e^{-x}, \qquad f^{(4)}(0) = 1,$$

it follows that

$$P_4(x) = 1 - x + \frac{x^2}{2!} - \frac{x^3}{3!} + \frac{x^4}{4!}.$$

3. Since

$$f(x) = \cos x, \qquad f(\pi/4) = \sqrt{2}/2$$
$$f'(x) = -\sin x, \qquad f'(\pi/4) = -\sqrt{2}/2$$
$$f''(x) = -\cos x, \qquad f''(\pi/4) = -\sqrt{2}/2$$
$$f'''(x) = \sin x, \qquad f'''(\pi/4) = \sqrt{2}/2$$
$$f^{(4)}(x) = \cos x, \qquad f^{(4)}(\pi/4) = \sqrt{2}/2$$
$$f^{(5)}(x) = -\sin x, \qquad f^{(5)}(\pi/4) = -\sqrt{2}/2$$
$$f^{(6)}(x) = -\cos x, \qquad f^{(6)}(\pi/4) = -\sqrt{2}/2,$$

it follows that

$$P_6(x) = \frac{\sqrt{2}}{2} - \frac{\sqrt{2}}{2}\left(x - \frac{\pi}{4}\right) - \frac{\sqrt{2}}{2}\frac{1}{2!}\left(x - \frac{\pi}{4}\right)^2 + \frac{\sqrt{2}}{2}\frac{1}{3!}\left(x - \frac{\pi}{4}\right)^3$$

$$+ \frac{\sqrt{2}}{2}\frac{1}{4!}\left(x - \frac{\pi}{4}\right)^4 - \frac{\sqrt{2}}{2}\frac{1}{5!}\left(x - \frac{\pi}{4}\right)^5 - \frac{\sqrt{2}}{2}\frac{1}{6!}\left(x - \frac{\pi}{4}\right)^6.$$

5. Since

$$f(x) = \ln(1+x), \qquad f(0) = 0$$
$$f'(x) = (1+x)^{-1}, \qquad f'(0) = 1$$
$$f''(x) = -(1+x)^{-2}, \qquad f''(0) = -1$$
$$f'''(x) = 2(1+x)^{-3}, \qquad f'''(0) = 2$$
$$f^{(4)}(x) = -6(1+x)^{-4}, \qquad f^{(4)}(0) = -6,$$

it follows that

$$P_4(x) = x - \frac{1}{2!}x^2 + \frac{2}{3!}x^3 - \frac{6}{4!}x^4 = x - \frac{x^2}{2} + \frac{x^3}{3} - \frac{x^4}{4}.$$

7. Since

$$f(x) = \tan x, \qquad f(0) = 0$$
$$f'(x) = \sec^2 x, \qquad f'(0) = 1$$
$$f''(x) = 2\sec^2 x \tan x, \qquad f''(0) = 0$$
$$f'''(x) = 2\sec^4 x + 4\sec^2 x \tan x, \qquad f'''(0) = 2,$$

it follows that

$$P_3(x) = x + \frac{2}{3!}x^3 = x + \frac{x^3}{3}.$$

9. Since

$$f(x) = \sin(-\pi x), \qquad f(0) = 0$$
$$f'(x) = -\pi \cos(-\pi x), \qquad f'(0) = -\pi$$
$$f''(x) = -\pi^2 \sin(-\pi x), \qquad f''(0) = 0$$
$$f'''(x) = \pi^3 \cos(-\pi x), \qquad f'''(0) = \pi^3$$
$$f^{(4)}(x) = \pi^4 \sin(-\pi x), \qquad f^{(4)}(0) = 0$$
$$f^{(5)}(x) = -\pi^5 \cos(-\pi x), \qquad f^{(5)}(0) = -\pi^5,$$

it follows that

$$P_5(x) = -\pi x + \frac{\pi^3}{3!}x^3 - \frac{\pi^5}{5!}x^5.$$

11. Since

$$f(x) = e^{x^2}, \qquad f(0) = 1$$
$$f'(x) = 2xe^{x^2}, \qquad f'(0) = 0$$
$$f''(x) = e^{x^2}(4x^2 + 2), \qquad f''(0) = 2$$
$$f'''(x) = e^{x^2}(8x^3 + 12x), \qquad f'''(0) = 0,$$

it follows that

$$P_3(x) = 1 + \frac{2}{2!}x^2 = 1 + x^2.$$

13. Since

$$f(x) = \sqrt{1-x}, \qquad f(0) = 1$$
$$f'(x) = (-1/2)(1-x)^{-1/2}, \qquad f'(0) = -1/2$$
$$f''(x) = (-1/4)(1-x)^{-3/2}, \qquad f''(0) = -1/4$$
$$f'''(x) = (-3/8)(1-x)^{-5/2}, \qquad f'''(0) = -3/8$$

it follows that

$$P_3(x) = 1 - \frac{1}{2}\,x - \frac{1}{4}\,\frac{1}{2!}\,x^2 - \frac{3}{8}\,\frac{1}{3!}\,x^3 = 1 - \frac{x}{2} - \frac{x^2}{8} - \frac{x^3}{16}.$$

15. Since

$$\begin{aligned}
f(x) &= \sec x, & f(\pi/4) &= \sqrt{2} \\
f'(x) &= \sec x \tan x, & f'(\pi/4) &= \sqrt{2} \\
f''(x) &= \sec^3 x + \sec x \tan^2 x, & f''(\pi/4) &= 3\sqrt{2} \\
f'''(x) &= 5\sec^3 x \tan x + \sec x \tan^3 x, & f'''(\pi/4) &= 11\sqrt{2},
\end{aligned}$$

it follows that

$$P_4(x) = \sqrt{2} + \sqrt{2}\left(x - \frac{\pi}{4}\right) + \frac{3\sqrt{2}}{2!}\left(x - \frac{\pi}{4}\right)^2 + \frac{11\sqrt{2}}{3!}\left(x - \frac{\pi}{4}\right)^3.$$

17. Since

$$\begin{aligned}
f(x) &= \operatorname{Tan}^{-1}x, & f(0) &= 0 \\
f'(x) &= (1 + x^2)^{-1}, & f'(0) &= 1 \\
f''(x) &= -(1 + x^2)^{-2}(2x), & f''(0) &= 0 \\
f'''(x) &= [-(1 + x^2)^{-2}](2) + (2x)[2(1 + x^2)^{-3}(2x)], & f'''(0) &= -2,
\end{aligned}$$

it follows that

$$P_3(x) = x - \frac{2}{3!}\,x^3 = x - \frac{x^3}{3}.$$

19. Since

$$\begin{aligned}
f(x) &= \frac{2x}{2 + x^2}, & f(1) &= 2/3 \\[2mm]
f'(x) &= \frac{4 - 2x^2}{(2 + x^2)^2}, & f'(1) &= 2/9 \\[2mm]
f''(x) &= \frac{4x^3 - 24x}{(2 + x^2)^3}, & f''(1) &= -20/27,
\end{aligned}$$

it follows that

$$P_2(x) = \frac{2}{3} + \frac{2}{9}\,(x - 1) - \frac{20}{27}\,\frac{1}{2!}\,(x - 1)^2 = \frac{2}{3} + \frac{2}{9}\,(x - 1) - \frac{10}{27}\,(x - 1)^2.$$

21. Since

$$\begin{aligned}
f(x) &= 3x^3 + 2x + 1, & f(-1) &= -4 \\
f'(x) &= 9x^2 + 2, & f'(-1) &= 11 \\
f''(x) &= 18x, & f''(-1) &= -18 \\
f'''(x) &= 18, & f'''(-1) &= 18,
\end{aligned}$$

it follows that

$$P_3(x) = -4 + 11(x + 1) - \frac{18}{2!}\,(x + 1)^2 + \frac{18}{3!}\,(x + 1)^3 = -4 + 11(x + 1) - 9(x + 1)^2 + 3(x + 1)^3.$$

23. Let $P_n(f; x)$ stand for the Taylor polynomial of degree n of the function f. Then $P_n'(f; x) = P_{n-1}(f'; x)$ since

$$
\begin{aligned}
P_n'(f; x) &= \left[f(a) + f'(a)(x - a) + \frac{f''(a)}{2!}(x - a)^2 + \cdots + \frac{f^{(n)}(a)}{n!}(x - a)^n \right]' \\
&= f'(a) + f''(a)(x - a) + \cdots + \frac{f^{(n)}(a)}{(n - 1)!}(x - a)^{n-1} \\
&= P_{n-1}(f'; x)
\end{aligned}
$$

25. Recall that if f is even, then f' is odd. Moreover, if f is an odd function, then $f'(0) = 0$. Hence, since f is even, f', f''', $\ldots$, are odd and therefore $f'(0) = 0$, $f'''(0) = 0$, $\ldots$. Thus the Taylor polynomial for f contains only even powers of x.

27. Here

$$
P_5(x) = 1 + 2x + 2x^2 + \frac{4x^3}{3} + \frac{x^4}{6} + \frac{4x^5}{15}.
$$

The table below gives the values of $E_n = \max\{\,|e^{2x} - P_n(x)|\mid -1/2 \leq x \leq 1/2\,\}$.

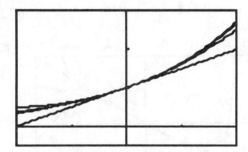

n	E_n
1	0.71828
2	0.218282
3	0.051615
4	0.041198
5	0.032865

The graphs of $f(x) = e^{2x}$ and $P_1(x)$, $P_2(x)$, $P_3(x)$, $P_4(x)$, and $P_5(x)$ are shown using the range $[-.5, .5] \times [-.5, 3]$.

29. Here

$$
\begin{aligned}
P_5(x) &= \sqrt{2} + \frac{x - 1}{2\sqrt{2}} - \frac{(x - 1)^2}{16\sqrt{2}} + \frac{(x - 1)^3}{64\sqrt{2}} - \frac{5(x - 1)^4}{1024\sqrt{2}} + \frac{7(x - 1)^5}{4096\sqrt{2}} \\
&= 1.4142135624 + 0.3535533906(x - 1) - 0.441941738(x - 1)^2 + 0.0110485435(x - 1)^3 \\
&\quad - 0.0034526698(x - 1)^4 + 0.0012084344(x - 1)^5.
\end{aligned}
$$

The table below gives the values of $E_n = \max\{\,|\sqrt{x + 1} - P_n(x)|\mid -1/2 \leq x \leq 2\,\}$.

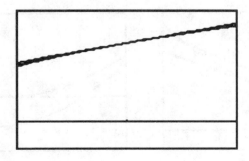

n	E_n
1	0.060660
2	0.016466
3	0.005417
4	0.001965
5	0.000756

The graphs of $f(x) = \sqrt{x+1}$ and $P_1(x)$, $P_2(x)$, $P_3(x)$, $P_4(x)$, and $P_5(x)$ are shown using the range $[0,2] \times [-.5, 2]$.

31. (a) Graph $f(x)$ and $P_{2n}(x)$ using the range $[-2,2] \times [0, 1.25]$.

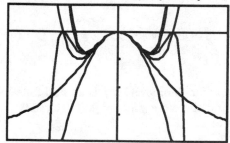

The polynomials $P_{2n}(x)$ do not approximate $f(x)$ very well on the intervals $[-2,-1]$ and $[1,2]$. The reason is that the series $\sum_{k=0}^{\infty}(-1)^k x^{2k}$ is a geometric series with common ratio $r = -x^2$. Therefore, for the series to converge we must have $\left|-x^2\right| = x^2 < 1$, *i.e.*, $-1 < x < 1$.

(b) Now,

$$Q_{2n+1}(x) = \int_0^x P_{2n}(t)\, dt = \sum_{k=0}^{n}(-1)^k \frac{x^{2k+1}}{2k+1}.$$

Graph $F(x) = \mathrm{Tan}^{-1} x$ and $Q_{2n+1}(x)$, $n = 0, 1, 2, 3, 4$, and 5 using the range $[-2,2] \times [-2,2]$.

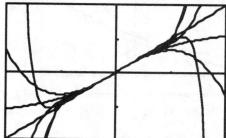

It appears that the following conjecture is true: If $P_n(x)$ is the Taylor polynomial for the function $f(x)$ then the polynomials $Q_{n+1}(x) = \int_0^x P_n(t)\, dt$ are the Taylor polynomials for $F(x) = \int_0^x f(t)\, dt$.

33. The graph below on the left is that of $f(x) = e^x$, $P_1(x) = 1 + x$, and $q_1(x)$ using the range $[-1,1] \times [-.5, 3]$. The graph on the right is that of $e^x - P_1(x)$ and $e^x - q_1(x)$ using the range $[-1,1] \times [-0.5, 0.75]$. Note that

$$\max\left\{\, |e^x - P_1(x)| \,\big|\, -1 \le x \le 1 \,\right\} = 0.7183 \quad \text{and} \quad \max\left\{\, |e^x - q_1(x)| \,\big|\, -1 \le x \le 1 \,\right\} = 0.2788.$$

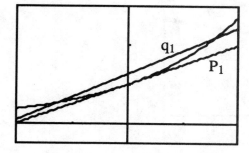

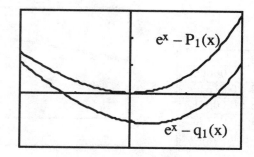

426

The graph below on the left is $f(x) = e^x$, $P_2(x) = 1 + x + x^2/2$, and $q_2(x)$ using the range $[-1, 1] \times [-.5, 3]$. The graph on the right is that of $e^x - P_2(x)$ and $e^x - q_2(x)$ using the range $[-1, 1] \times [-0.25, 0.25]$. Note that

$$\max\{\, |e^x - P_2(x)| \,\big|\, -1 \le x \le 1 \,\} = 0.2183 \quad \text{and} \quad \max\{\, |e^x - q_2(x)| \,\big|\, -1 \le x \le 1 \,\} = 0.0550.$$

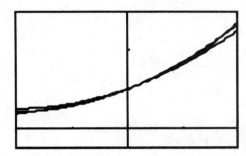
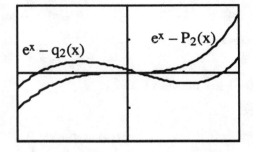

The graph below is that of $e^x - P_3(x)$, $P_3(x) = 1 + x + x^2/2 + x^3/6$, and $e^x - q_3(x)$ using the range $[-1, 1] \times [-0.05, 0.05]$. Note that

$$\max\{\, |e^x - P_3(x)| \,\big|\, -1 \le x \le 1 \,\} = 0.0516 \quad \text{and} \quad \max\{\, |e^x - q_3(x)| \,\big|\, -1 \le x \le 1 \,\} = 0.0055.$$

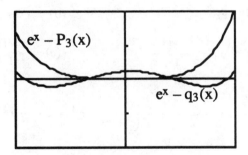

13.2 Taylor's Theorem

1. Since

$$
\begin{aligned}
f(x) &= e^{-x}, & f(0) &= 1 \\
f'(x) &= -e^{-x}, & f'(0) &= -1 \\
f''(x) &= e^{-x}, & f''(0) &= 1 \\
f'''(x) &= -e^{-x}, & f'''(0) &= -1 \\
f^{(4)}(x) &= e^{-x}, & f^{(4)}(c) &= e^{-c}
\end{aligned}
$$

it follows that

$$e^{-x} = 1 - x + \frac{x^2}{2!} - \frac{x^3}{3!} + \frac{e^{-c}}{4!}\, x^4,$$

where c lies between 0 and x.

3. Since

$$f(x) = \cos x, \qquad f(0) = 1$$
$$f'(x) = -\sin x, \qquad f'(0) = 0$$
$$f''(x) = -\cos x, \qquad f''(0) = -1$$
$$f'''(x) = \sin x, \qquad f'''(0) = 0$$
$$f^{(4)}(x) = \cos x, \qquad f^{(4)}(0) = 1$$
$$f^{(5)}(x) = -\sin x, \qquad f^{(5)}(c) = \sin c,$$

it follows that

$$\cos x = 1 - \frac{x^2}{2!} + \frac{x^4}{4!} - \frac{\sin c}{5!} x^5,$$

where c lies between 0 and x.

5. Since

$$f(x) = \text{Tan}^{-1}(x), \qquad f(0) = 0$$

$$f'(x) = \frac{1}{1+x^2}, \qquad f'(0) = 1$$

$$f''(x) = \frac{-2x}{(1+x^2)^2}, \qquad f''(0) = 0$$

$$f'''(x) = \frac{6x^2 - 2}{(1+x^2)^3}, \qquad f'''(0) = -2$$

$$f^{(4)}(x) = \frac{24x(1-x^2)}{(1+x^2)^4}, \qquad f^{(4)}(c) = \frac{24c(1-c^2)}{(1+c^2)^4},$$

it follows that

$$\text{Tan}^{-1}(x) = x - \frac{2}{3!} x^3 + \frac{24c(1-c^2)}{4!\,(1+c^2)^4} x^4 = x - \frac{1}{3} x^3 + \frac{c(1-c^2)}{(1+c^2)^4} x^4,$$

where c lies between 0 and x.

7. Since

$$f(x) = \frac{1}{1+x^2}, \qquad f(1) = 1$$

$$f'(x) = \frac{-2x}{(1+x^2)^2}, \qquad f'(1) = -\tfrac{1}{2}$$

$$f''(x) = \frac{6x^2 - 2}{(1+x^2)^3}, \qquad f''(1) = \tfrac{1}{2}$$

$$f'''(x) = \frac{24x(1-x^2)}{(1+x^2)^4}, \qquad f'''(c) = \frac{24c(1-c^2)}{(1+c^2)^4},$$

it follows that

$$\frac{1}{1+x^2} = \frac{1}{2} - \frac{1}{2}(x-1) + \frac{1}{2}\frac{1}{2!}(x-1)^2 + \frac{24c(1-c^2)}{(1+c^2)^4}\frac{1}{3!}(x-1)^3,$$

where c lies between 1 and x.

9. Since

$$f(x) = \sec x, \qquad\qquad f(\pi/4) = \sqrt{2}$$
$$f'(x) = \sec x \tan x, \qquad\qquad f'(\pi/4) = \sqrt{2}$$
$$f''(x) = \sec^3 x + \sec x \tan^2 x, \qquad f''(\pi/4) = 3\sqrt{2}$$
$$f'''(x) = 5\sec^3 x \tan x + \sec x \tan^3 x, \qquad f'''(c) = 5\sec^3 c \tan c + \sec c \tan^3 c,$$

it follows that

$$\sec x = \sqrt{2} + \sqrt{2}\left(x - \frac{\pi}{4}\right) + \frac{3\sqrt{2}}{2!}\left(x - \frac{\pi}{4}\right)^2 + \frac{5\sec^3 c \tan c + \sec c \tan^3 c}{3!}\left(x - \frac{\pi}{4}\right)^3,$$

where c lies between $\pi/4$ and x.

11. Since

$$f(x) = \sinh x, \qquad f(0) = 0$$
$$f'(x) = \cosh x, \qquad f'(0) = 1$$
$$f''(x) = \sinh x, \qquad f''(0) = 0$$
$$f'''(x) = \cosh x, \qquad f'''(0) = 1$$
$$f^{(4)}(x) = \sinh x, \qquad f^{(4)}(0) = 0$$
$$f^{(5)}(x) = \cosh x, \qquad f^{(5)}(c) = \cosh c,$$

it follows that

$$\sinh x = x + \frac{x^3}{3!} + \frac{\cosh c}{5!}\, x^5,$$

where c lies between 0 and x.

13. Since

$$f(x) = \cosh x, \qquad f(\ln 2) = 5/4$$
$$f'(x) = \sinh x, \qquad f'(\ln 2) = 3/4$$
$$f''(x) = \cosh x, \qquad f''(\ln 2) = 5/4$$
$$f'''(x) = \sinh x, \qquad f'''(\ln 2) = 3/4$$
$$f^{(4)}(x) = \cosh x, \qquad f^{(4)}(c) = \cosh c$$

it follows that

$$\cosh x = \frac{5}{4} + \frac{3}{4}(x - \ln 2) + \frac{5}{4}\frac{1}{2!}(x - \ln 2)^2 + \frac{3}{4}\frac{1}{3!}(x - \ln 2)^3 + \frac{\cosh c}{4!}(x - \ln 2)^4,$$

where c lies between $\ln 2$ and x.

15. If f has degree n, then $f^{(n+1)}(x) = 0$. Therefore $R_n(x) = 0$ and hence, by Theorem 1, $f = P_n$.

17. Let r be a root of a function $f(x)$, x_n the nth approximation to r using Newton's Method, and $\varepsilon_n = r - x_n$ the error term. Using Taylor's Theorem on an open interval around the point x_n, it follows that

$$f(x) = f(x_n) + f'(x_n)(x - x_n) + \frac{f''(c)}{2}(x - x_n)^2,$$

where c lies between x and x_n. Therefore

$$0 = f(r) = f(x_n) + f'(x_n)(r - x_n) + \frac{f''(c)}{2}(r - x_n)^2,$$

and hence, solving for r, we find that

$$r = x_n - \frac{f(x_n)}{f'(x_n)} - \frac{f''(c)}{2f'(x_n)}(r - x_n)^2 = \underbrace{x_n - \frac{f(x_n)}{f'(x_n)}}_{x_{n+1}} - \underbrace{\frac{f''(c)}{2f'(x_n)}\varepsilon_n^2}_{\varepsilon_{n+1}}.$$

It follows that

$$|\varepsilon_{n+1}| = \left|\frac{f''(c)}{2f'(x_n)}\right|\varepsilon_n^2.$$

Thus, if the quotient $f''(c)/2f'(x_n)$ is bounded, say

$$\left|\frac{f''(c)}{2f'(x_n)}\right| \le M,$$

then

$$|\varepsilon_{n+1}| \le M\varepsilon_n^2,$$

and hence, if the initial approximation x_1 is chosen close to r, the error sequence approaches zero.

19. (a) Since

$$\begin{aligned}
f(x) &= (1-x)^{-1}, & f(0) &= 1 \\
f'(x) &= (1-x)^{-2}, & f'(0) &= 1 \\
f''(x) &= 2(1-x)^{-3}, & f''(0) &= 2 \\
f'''(x) &= 3!(1-x)^{-4}, & f'''(0) &= 3!,
\end{aligned}$$

and, in general, $f^{(n)}(0) = n!$, it follows that

$$P_n(x) = 1 + x + x^2 + \cdots + x^n.$$

(b) We find that

$$\left(1 + x + x^2 + \cdots + x^n + \frac{x^{n+1}}{1-x}\right)(1-x) = (1 + x + x^2 + \cdots + x^n)(1-x) + x^{n+1}$$

$$= (1 + x + x^2 + \cdots + x^n) - (x + x^2 + x^3 + \cdots + x^{n+1}) + x^{n+1} = 1.$$

Therefore

$$\frac{1}{1-x} = 1 + x + x^2 + \cdots + +x^n + \frac{x^{n+1}}{1-x}.$$

(c) $R_n(x) = f(x) - P_n(x) = \frac{1}{1-x} - P_n(x) = \frac{x^{n+1}}{1-x}.$ This does not contradict Taylor's Theorem; it is just another way to represent the remainder term.

21. From Exercise 20, with x replaced by x^2, it follows that

$$\frac{1}{1+x^2} = 1 - x^2 + x^4 - \cdots + (-1)^n x^{2n} + \frac{(-1)^{n+1}x^{2n+2}}{1+x^2}.$$

Therefore the odd derivatives of $f(x) = 1/(1+x^2)$ are equal to 0 when $x = 0$ and hence the $2n$th Taylor polynomial for $f(x)$ at $a = 0$ is

$$P_{2n}(x) = 1 - x^2 + x^4 - x^6 + \cdots,$$

where only the even power terms of degree $\leq 2n$ are used. To show that this polynomial is in fact the Taylor polynomial of degree $2n$ for $f(x)$ at $a = 0$, let

$$Z_n(x) = (-1)^{n+1} \frac{x^{2n+2}}{1 + x^2}.$$

Then $f(x) = P_{2n}(x) + Z_n(x)$. Now, we claim that $Z_n^{(k)}(0) = 0$ for $k = 0,1, \ldots, 2n$. For clearly,

$$Z_n'(x) = (-1)^{n+1} \frac{(1 + x^2)(2n + 2)x^{2n+1} - x^{2n+2}(2x)}{(1 + x^2)^2} \implies Z_n'(0) = 0.$$

In general,

$$Z_n^{(k)}(x) = (-1)^{n+1} x^{2n+2-k} \frac{g(x)}{(1 + x^2)^{2^k}}$$

for some polynomial $g(x)$, and hence for $k \leq 2n$,

$$\begin{aligned}
Z_n^{(k+1)}(x) &= (-1)^{n+1} \left[x^{2n+2-k} \frac{(1 + x^2)^{2^k} g'(x) - g(x)2^k(1 + x^2)^{2^k-1}(2x)}{(1 + x^2)^{2^{k+1}}} \right.\\
&\quad \left. + (2n + 2 - k)x^{2n-k+1} \frac{g(x)}{(1 + x^2)^{2^k}} \right] \\
\implies Z_n^{(k+1)}(0) &= 0.
\end{aligned}$$

Hence $Z_n'(0) = \cdots = Z_n^{(2n)}(0) = 0$ and therefore

$$f^{(k)}(0) = P_{2n}^{(k)}(0) \qquad \text{for } k = 0,1, \ldots, 2n.$$

It now follows that $P_{2n}(x)$ is the $2n$th Taylor polynomial for $f(x)$ at $a = 0$.

23. Let $f(x) = (x + a)^n$. Then f is a polynomial of degree n and hence $f = P_n$ by Exercise 15. Now,

$$\begin{aligned}
f(x) &= (x + a)^n, & f(0) &= a^n \\
f'(x) &= n(x + a)^{n-1}, & f'(0) &= na^{n-1} \\
f''(x) &= n(n - 1)(x + a)^{n-2}, & f''(0) &= n(n - 1)a^{n-2},
\end{aligned}$$

and, in general, $f^{(r)}(0) = n(n - 1) \cdots (n - r + 1)a^{n-r}$. Hence

$$\begin{aligned}
P_n(x) &= a^n + na^{n-1}x + \frac{n(n - 1)}{2} a^{n-2}x^2 \\
&\quad + \cdots + \frac{n(n - 1) \cdots (n - r + 1)}{r!} a^{n-r}x^r + \cdots + \frac{n(n - 1) \cdots 2 \cdot 1}{n!} x^n.
\end{aligned}$$

Since the binomial coefficient

$$\binom{n}{r} = \frac{n!}{r!(n - r)!} = \frac{n(n - 1) \cdots (n - r + 1)(n - r)!}{r!(n - r)!} = \frac{n(n - 1) \cdots (n - r + 1)}{r!},$$

it follows that

$$f(x) = (x + a)^n = P_n(x) = a^n + na^{n-1}x + \cdots + \binom{n}{r}a^{n-r}x^r + \cdots + nax^{n-1} + x^n.$$

25. (a) Graph of $f(x) = e^x$ and $r(x)$ using the range $[-1, 1] \times [0, 3]$. There is no visible difference between the graphs.

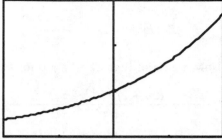

(b) Graph of $|e^x - r(x)|$ using the range $[-1, 1] \times [0, 0.0005]$.

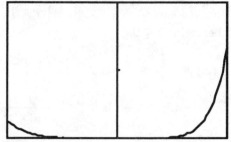

From the graph we see that

$$\max\{|e^x - r(x)| \mid -1 \le x \le 1\} = |e - r(1)| = 0.000333.$$

If we take the Taylor polynomial of degree 6 then

$$\max\{|e^x - P_6(x)| \mid -1 \le x \le 1\} = |e - P_6(1)| = 0.000263.$$

13.3 Applications of Taylor's Theorem

1. In Exercise 5, Section 13.1, we showed that the Taylor polynomial of degree 3 for $f(x) = \ln(x+1)$ at $a = 0$ is

$$P_3(x) = x - \frac{x^2}{2} + \frac{x^3}{3}.$$

Hence, setting $x = 0.5$, we find that

$$\ln 1.5 \approx 0.5 - \frac{(0.5)^2}{2} + \frac{(0.5)^3}{3} = 0.416\overline{6}.$$

Since $f^{(4)}(x) = \dfrac{-6}{(1+x)^4}$, and since $(1+c)^{-4} < 1$ for $0 < c < 0.5$, the maximum error in this approximation is

$$|R_3(0.5)| = \left| \frac{-6}{4!(1+c)^4} (0.5)^4 \right| = \frac{0.015625}{(1+c)^4} < 0.015625.$$

3. Since

$$\begin{array}{ll} f(x) = \sin x, & f(\pi/2) = 1 \\ f'(x) = \cos x, & f'(\pi/2) = 0 \\ f''(x) = -\sin x, & f''(\pi/2) = -1 \\ f'''(x) = -\cos x, & f'''(\pi/2) = 0, \end{array}$$

432

it follows that the Taylor polynomial of degree 3 for $\sin x$ at $a = \pi/2$ is

$$P_3(x) = 1 - \frac{1}{2}\left(x - \frac{\pi}{2}\right)^2.$$

Hence, since $80° = 1.3963$ radians,

$$\sin 80° \approx P_3(1.3963) = 1 - \frac{1}{2}\left(1.3963 - \frac{\pi}{2}\right)^2 = 0.9848.$$

Since $f^{(4)}(x) = \sin x$ and $|\sin c| < 1$ for $1.3963 < c < \pi/2$, the maximum error in this approximation is

$$|R_3(1.3963)| = \left|\frac{\sin c}{4!}\left(1.3963 - \frac{\pi}{2}\right)^4\right| < 0.0000386.$$

5. It follows from Example 1, Section 13.1, that the Taylor polynomial of degree 3 for e^x at $a = 0$ is

$$P_3(x) = 1 + x + \frac{x^2}{2!} + \frac{x^3}{3!}.$$

Hence

$$e^{0.2} \approx P_3(0.2) = 1 + 0.2 + \frac{(0.2)^2}{2!} + \frac{(0.2)^3}{3!} = 1.22133.$$

Since $f^{(4)}(x) = e^x$ and $e^c < 2$ for $0 < c < 0.2$, the maximum error in this approximation is

$$|R_3(0.2)| < \left|\frac{e^c}{4!}(0.2)^4\right| < 0.00013.$$

7. Since

$$\begin{aligned}
f(x) &= \sqrt{x}, & f(9) &= 3 \\
f'(x) &= (1/2)x^{-1/2}, & f'(9) &= 1/6 \\
f''(x) &= (-1/4)x^{-3/2}, & f''(9) &= -1/108,
\end{aligned}$$

the Taylor polynomial of degree 2 for $\sqrt{x}$ at $a = 9$ is

$$P_2(x) = 3 + \frac{1}{6}(x - 9) - \frac{1}{108}\frac{1}{2!}(x - 9)^2.$$

Hence

$$\sqrt{9.2} \approx P_2(9.2) = 3 + \frac{1}{6}(9.2 - 9) - \frac{1}{216}(9.2 - 9)^2 = 3.033148.$$

Since $f'''(x) = (3/8)x^{-5/2}$ and $c^{-5/2} < 9^{-5/2} = 1/243$ for $9 < c < 9.2$, the maximum error in this approximation is

$$|R_2(9.2)| = \left|\frac{3}{8}c^{-5/2}\frac{1}{3!}(9.2 - 9)^3\right| < 0.0000021.$$

9. Since

$$\begin{aligned}
f(x) &= x^{1/3}, & f(8) &= 2 \\
f'(x) &= (1/3)x^{-2/3}, & f'(8) &= 1/12 \\
f''(x) &= (-2/9)x^{-5/3}, & f''(8) &= -1/144,
\end{aligned}$$

the Taylor polynomial of degree 2 for $\sqrt[3]{x}$ at $a = 8$ is

$$P_2(x) = 2 + \frac{1}{12}(x - 8) - \frac{1}{144}\frac{1}{2!}(x - 8)^2.$$

Hence

$$\sqrt[3]{10} \approx P_2(10) = 2 + \frac{1}{12}(10-8) - \frac{1}{288}(10-8)^2 = 2.15278.$$

Since $f'''(x) = (10/27)x^{-8/3}$ and $c^{-8/3} < 8^{-8/3} = 1/256$ for $8 < c < 10$, the maximum error in this approximation is

$$|R_2(10)| = \left| \frac{10}{27} c^{-8/3} \frac{1}{3!}(10-8)^3 \right| < 0.0019.$$

11. If $|x| < 0.05$, the maximum error in the approximation is

$$|R_2(x)| = \left| \frac{\sin c}{3!} x^3 \right| < \left| \frac{x^3}{3!} \right| < \frac{(0.05)^3}{3!} = 2.08 \times 10^{-5}.$$

13. If $|x - \pi/3| < 0.05$, the maximum error in the approximation is

$$|R_1(x)| = \left| \frac{-\cos c}{2!} \left(x - \frac{\pi}{3} \right)^2 \right| < \frac{(x-\pi/3)^2}{2} < \frac{(0.05)^2}{2} = 1.25 \times 10^{-3}.$$

15. If $|x| < 0.025$, the maximum error in the approximation is

$$|R_1(x)| = \left| \frac{-2}{9(1+c)^{5/3}} \frac{1}{2!} x^2 \right| < \left| \frac{1}{9(1+c)^{5/3}} \right| (0.025)^2.$$

Since c lies between 0 and x, $(1+c)^{5/3} > (1-0.025)^{5/3}$ and therefore

$$|R_1(x)| < \left| \frac{1}{9(1-0.025)^{5/3}} \right| (0.025)^2 = 7.24 \times 10^{-5}.$$

17. If $0 < x < 0.02$, the maximum error in the approximation is

$$|R_1(x)| = \left| -\frac{1}{4}(1+c)^{-3/2} \frac{1}{2!} x^2 \right| < \left| \frac{1}{8(1+c)^{3/2}} \right| (0.02)^2.$$

Since $1 + c > 1$, $(1+c)^{-3/2} < 1$ and hence

$$|R_1(x)| < \frac{1}{8}(0.02)^2 = 5 \times 10^{-5}.$$

19. The error in approximating $\ln 1.3$ by $P_n(0.3)$ is

$$R_n(0.3) = \frac{f^{(n+1)}(c)}{(n+1)!}(0.3)^{n+1}.$$

Since $f^{(n+1)}(x) = (-1)^n n!(x+1)^{-(n+1)}$ and $c + 1 > 1$, it follows that

$$|R_n(0.3)| = \left| \frac{(-1)^n n!(c+1)^{-(n+1)}}{(n+1)!}(0.3)^{n+1} \right| < \frac{(0.3)^{n+1}}{n+1}.$$

n	1	2	3	4	5	6	7
$\dfrac{(0.3)^{n+1}}{n+1}$	0.045	9×10^{-3}	2.02×10^{-3}	4.86×10^{-4}	1.21×10^{-4}	3.12×10^{-5}	8.2×10^{-6}

The table lists the upper bounds on $R_n(0.3)$ and shows that n must be at least 6 in order to obtain $\ln 0.3$ to four decimal place accuracy.

21. Since $\cos 42° = \cos 0.733$, the error in approximating $\cos 42°$ by $P_n(0.733)$ is

$$|R_n(0.733)| = \left| \frac{f^{(n+1)}(c)}{(n+1)!} \left(0.733 - \frac{\pi}{4} \right)^{n+1} \right| = \left| \frac{f^{(n+1)}(c)}{(n+1)!} (0.052)^{n+1} \right|.$$

Now, $f^{(n+1)}(x)$ is either $\pm \sin x$ or $\pm \cos x$. Therefore

$$|R_n(0.733)| < \frac{(0.052)^{n+1}}{(n+1)!}.$$

n	1	2	3
$\dfrac{(0.052)^{n+1}}{(n+1)!}$	1.35×10^{-3}	2.34×10^{-5}	3.05×10^{-7}

The table lists the upper bounds on $R_n(0.733)$ and shows that n must be at least 2 in order to obtain $\cos 42°$ to four decimal place accuracy.

23. The error in approximating $\sqrt[3]{9.2}$ by $P_{9.2}$ is

$$|R_n(9.2)| = \left| \frac{f^{(n+1)}(c)}{(n+1)!} (9.2 - 8)^{n+1} \right| = \left| \frac{f^{(n+1)}(c)}{(n+1)!} (1.2)^{n+1} \right|.$$

Now, $c > 8$ and hence

$$f''(x) = -\frac{2}{9} x^{-5/3} \implies R_1(9.2) = \left| -\frac{2}{9} c^{-5/3} \frac{1}{2!} (1.2)^2 \right| < \frac{2}{9} \frac{1}{8^{5/3}} \frac{1}{2!} (1.2)^2 = 5 \times 10^{-3}$$

$$f'''(x) = \frac{10}{27} x^{-8/3} \implies R_2(9.2) = \left| \frac{10}{27} c^{-8/3} \frac{1}{3!} (1.2)^3 \right| < \frac{10}{27} \frac{1}{8^{8/3}} \frac{1}{3!} (1.2)^3 = 4.17 \times 10^{-4}$$

$$f^{(4)}(x) = -\frac{80}{81} x^{-11/3} \implies R_3(9.2) = \left| -\frac{80}{81} c^{-11/3} \frac{1}{4!} (1.2)^4 \right| < \frac{80}{81} \frac{1}{8^{11/3}} \frac{1}{4!} (1.2)^4 = 4.17 \times 10^{-5}$$

$$f^{(5)}(x) = \frac{880}{243} x^{-14/3} \implies R_4(9.2) = \left| \frac{880}{243} c^{-14/3} \frac{1}{5!} (1.2)^5 \right| < \frac{880}{243} \frac{1}{8^{14/3}} \frac{1}{5!} (1.2)^5 = 4.58 \times 10^{-6}.$$

Hence n must be at least 3 in order to obtain $\sqrt[3]{9.2}$ to four decimal places of accuracy.

25. Using the error approximation for e^x obtained in Exercise 22, with $x = 1$, we find that

$$|R_n(1)| = \left| \frac{f^{(n+1)}(c)}{(n+1)!} 1^{n+1} \right| < \frac{e^c}{(n+1)!}.$$

Now, since $c < 1$, $e^c < e < 3$. Therefore

$$|R_n(1)| < \frac{3}{(n+1)!}.$$

n	4	5	6	7	8
$\dfrac{3}{(n+1)!}$	0.025	4.17×10^{-3}	5.95×10^{-4}	7.44×10^{-5}	2.20×10^{-5}

Hence n must be at least 8 to obtain four decimal place accuracy in the approximation. Now, the Taylor polynomial of degree 8 for $f(x) = e^x$ at $a = 0$ is

$$P_8(x) = 1 + x + \frac{1}{2!} x^2 + \frac{1}{3!} x^3 + \cdots + \frac{1}{8!} x^8.$$

Hence

$$e \approx P_8(1) = 1 + 1 + \frac{1}{2!} + \frac{1}{3!} + \cdots + \frac{1}{8!} = 2.7183,$$

accurate to four decimal places.

27. First recall that the Taylor $P_1(x)$ for $\cos x$ at $a = 0$ is $P_1(x) = 1$. Hence, since $6° = 0.105$ radians, the error in approximating $\cos x$ by 1 when $|x| < 6°$ is

$$|R_1(x)| = \left| \frac{f''(c)}{2!} x^2 \right| < \left| \frac{1}{2} x^2 \right| < \frac{1}{2} (0.105)^2 = 0.0055.$$

(*Calculator value:* $\cos 6° = 0.994522$, which differs from the approximation 1 by 0.0054781.)

29. Suppose that $x = c$ is a relative extremum for a function $f(x)$ and that $f''(c)$ exists. Then

$$f(x) \approx P_2(x) = f(c) + f'(c)(x - c) + \frac{f''(c)}{2!} (x - c)^2.$$

If $f''(c) > 0$, then

$$\frac{f''(c)}{2!} (x - c)^2 > 0$$

and hence $f(x) > f(c) + f'(c)(x - c)$ for $x \approx c$, which shows that the graph of $y = f(x)$ lies above the tangent line $y = f(c) + f'(c)(x - c)$ at $x = c$. Therefore $x = c$ is a relative minimum. If $f''(c) < 0$, then

$$\frac{f''(c)}{2!} (x - c)^2 < 0$$

and hence the graph of $y = f(x)$ lies below the tangent line, in which case $x = c$ is a relative maximum.

31. At each step we need to find the vertex of the parabola defined by

$$P_2(x) = f(a) + f'(a)(x - a) + \frac{f''(a)}{2} (x - a)^2.$$

This is done by solving $P_2''(x) = 0$.

(a) Graph $f(x) = e^{-x} + \cos 2x$ using the range $[0, 5] \times [-1, 1]$.

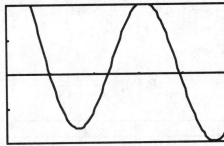

From the graph we see that $f''(3) < 0$ so choose $a_0 = 3$.

At $a_0 = 3$ we find that:

$$\begin{aligned}
P_2(x) &= 1.0099574 + 0.5090439(x - 3) - 1.8954470(x - 3)^2 \\
&= -17.5761978 + 11.8817262x - 1.8954470x^2.
\end{aligned}$$

$$\text{Vertex at } a_1 = 3.1342807.$$

At $a_1 = 3.1342807$ we find that:

$$\begin{aligned}
P_2(x) &= 1.0434241 + 0.0142843(x - a_1) - 1.9780206(x - a_1)^2 \\
&= -18.3433167 + 12.3850594x - 1.9780206x^2.
\end{aligned}$$

$$\text{Vertex at } a_2 = 3.1306699.$$

At $a_2 = 3.1306699$ we find that:

$$\begin{aligned}
P_2(x) &= 1.0434499 + 0.0000012(x - a_2) - 1.9780206(x - a_2)^2 \\
&= -18.3433167 + 12.3850594x - 1.9776785x^2.
\end{aligned}$$

$$\text{Vertex at } a_3 = 3.1306697.$$

At $a_3 = 3.1306697$, $f'(a_3) = -1.9008 \times 10^{-7}$ and $f''(a_3) = -3.9553570$. So a_3 is a good estimate of the maximum of $f(x) = e^{-x} + \cos 2x$.

(b) Graph $f(x) = xe^{-x^2}$ using the range $[0, 2] \times [0, 1]$.

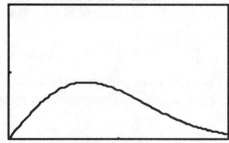

From the graph we see that $f''(1) < 0$ so choose $a_0 = 1$.
At $a_0 = 1$ we find that:

$$\begin{aligned}
P_2(x) &= 0.3678794 - 0.3678794(x - 1) - 0.3678794(x - 1)^2 \\
&= 0.3678794 + 0.3678794x - 0.3678794x^2.
\end{aligned}$$

$$\text{Vertex at } a_1 = 0.5.$$

At $a_1 = 0.5$ we find that:

$$\begin{aligned}
P_2(x) &= 0.3894004 - 0.3894004(x - a_1) - 0.9735010(x - a_1)^2 \\
&= -0.0486750 + 1.3629014x - 0.9735010x^2.
\end{aligned}$$

$$\text{Vertex at } a_2 = 0.7.$$

At $a_2 = 0.7$ we find that:

$$\begin{aligned}
P_2(x) &= 0.4288385 - 0.0122525(x - a_2) - 0.8662537(x - a_2)^2 \\
&= -0.0042026 + 1.2250077x - 0.8662537x^2.
\end{aligned}$$

$$\text{Vertex at } a_3 = 0.7070721.$$

At $a_3 = 0.7070721$, $f'(a_3) = 5.9496 \times 10^{-5}$ and $f''(a_3) = -1.7156119$. So a_3 is a good estimate of the maximum of $f(x) = xe^{-x^2}$.

(c) Graph $f(x) = \ln\left(\dfrac{2x+1}{x^2+1}\right)$ using the range $[0, 0.5] \times [0, 0.5]$.

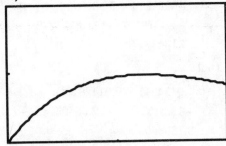

From the graph we see that $f''(1/2) < 0$, so choose $a_0 = 1/2$.
At $a_0 = 0.5$ we find that:

$$\begin{aligned}
P_2(x) &= 0.4700036 + 0.2(x - 0.5) - 0.98(x - 0.5)^2 \\
&= 0.1250036 + 0.1.18x - 0.98x^2.
\end{aligned}$$

$$\text{Vertex at } a_1 = 0.6020408.$$

At $a_1 = 0.0.6020408$ we find that:

$$\begin{aligned}
P_2(x) &= 0.4810241 + 0.0236474(x - a_1) - 0.7551484(x - a_1)^2 \\
&= 0.1930815 + 0.9329077x - 0.7551484x^2.
\end{aligned}$$

$$\text{Vertex at } a_2 = 0.6176983.$$

At $a_2 = 0.6176983$ we find that:

$$\begin{aligned}
P_2(x) &= 0.4810241 + 0.0004861(x - a_2) - 0.7242589(x - a_2)^2 \\
&= 0.2045697 + 0.8952330x - 0.7242589x^2.
\end{aligned}$$

$$\text{Vertex at } a_3 = 0.6180338.$$

At $a_3 = 0.6180338$, $f'(a_3) = 2.7316 \times 10^{-7}$ and $f''(a_3) = -1.4472143$. So a_3 is a good estimate of the maximum of $f(x) = \ln\left(\dfrac{2x+1}{x^2+1}\right)$.

13.4 Power Series

1. $[-1, 1)$; for,

$$\rho = \lim_{k \to \infty} \frac{\left|\dfrac{x^{k+1}}{k+3}\right|}{\left|\dfrac{x^k}{k+2}\right|} = \lim_{k \to \infty} |x| \frac{k+2}{k+3} = |x| < 1$$

and

$$x = 1 \quad \Longrightarrow \quad \sum \frac{x^k}{k+2} = \sum \frac{1}{k+2},$$

which is harmonic and diverges, while

$$x = -1 \quad \Longrightarrow \quad \sum \frac{x^k}{k+2} = \sum \frac{(-1)^k}{k+2},$$

which is alternating harmonic and hence converges.

3. $(-\infty, \infty)$; for,

$$\rho = \lim_{k \to \infty} \frac{\left|\dfrac{(-1)^{k+2}x^{k+1}}{(k+1)!}\right|}{\left|\dfrac{(-1)^k x^k}{k!}\right|} = \lim_{k \to \infty} |x| \frac{k!}{(k+1)!} = \lim_{k \to \infty} \frac{|x|}{k+1} = 0,$$

and hence the series converges for all x.

5. $(-\infty, \infty)$; for,

$$\rho = \lim_{k \to \infty} \frac{\left|\dfrac{[(k+1)^2+1]x^{k+1}}{(k+1)!}\right|}{\left|\dfrac{(k^2+1)x^k}{k!}\right|} = \lim_{k \to \infty} |x| \frac{k^2+2k+2}{k^2+1} \frac{k!}{(k+1)!} = \lim_{k \to \infty} |x| \frac{1+2/k+2/k^2}{1+1/k^2} \frac{1}{k+1} = 0,$$

and hence the series converges for all x.

7. $[-1, 1)$; for,

$$\rho = \lim_{k \to \infty} \frac{\left|\dfrac{x^{k+1}}{\ln(k+1)}\right|}{\left|\dfrac{x^k}{\ln k}\right|} = \lim_{k \to \infty} |x| \frac{\ln k}{\ln(k+1)} = \lim_{k \to \infty} |x| \frac{1/k}{1/(k+1)} = |x| < 1,$$

and

$$x = 1 \quad \Longrightarrow \quad \sum \frac{x^k}{\ln k} = \sum \frac{1}{\ln k} > \sum \frac{1}{k},$$

which diverges, while

$$x = -1 \quad \Longrightarrow \quad \sum \frac{x^k}{\ln k} = \sum \frac{(-1)^k}{\ln k},$$

which converges since it is an alternating series whose terms are decreasing and approach zero.

9. $(-1, 1]$; for, since $\cos \pi k = (-1)^k$,

$$\rho = \lim_{k \to \infty} \frac{\left|\dfrac{(-1)^{k+1}x^{k+1}}{2+k}\right|}{\left|\dfrac{(-1)^k x^k}{1+k}\right|} = \lim_{k \to \infty} |x| \frac{1+k}{2+k} = |x| < 1,$$

and

$$x = -1 \quad \Longrightarrow \quad \sum \frac{(-1)^k x^k}{1+k} = \sum \frac{1}{1+k},$$

which is harmonic and diverges, while

$$x = 1 \quad \Longrightarrow \quad \sum \frac{(-1)^k x^k}{1+k} = \sum \frac{(-1)^k}{1+k},$$

which is alternating harmonic and hence converges.

11. $(-1, 1)$; for,

$$\rho = \lim_{k \to \infty} \frac{\left| \dfrac{(2k+3)!}{(2k+2)!} \, x^{2k+2} \right|}{\left| \dfrac{(2k+1)!}{(2k)!} \, x^{2k} \right|} = \lim_{k \to \infty} \frac{(2k+3)!}{(2k+1)!} \, \frac{(2k)!}{(2k+2)!} \, x^2 = \lim_{k \to \infty} \frac{2k+3}{2k+1} \, x^2 = x^2 < 1,$$

and

$$x = \pm 1 \quad \Longrightarrow \quad \sum \frac{(2k+1)!}{(2k)!} \, x^{2k} = \sum (2k+1),$$

which diverges.

13. $[-1, 1]$; for

$$\rho = \lim_{k \to \infty} \frac{\left| \dfrac{(-1)^k}{(k+1)(k+2)} \, x^{k+1} \right|}{\left| \dfrac{(-1)^k}{k(k+1)} \, x^k \right|} = \lim_{k \to \infty} \frac{k}{k+1} \, \frac{k+1}{k+2} \, |x| = |x| < 1,$$

and

$$x = 1 \quad \Longrightarrow \quad \sum \frac{(-1)^k}{k(k+1)} \, x^k = \sum \frac{(-1)^k}{k(k+1)},$$

which converges since it is an alternating series whose terms are decreasing and approach zero, while

$$x = -1 \quad \Longrightarrow \quad \sum \frac{(-1)^k}{k(k+1)} \, x^k = \sum \frac{1}{k(k+1)} < \sum \frac{1}{k^2},$$

which converges since it is a p-series, $p = 2 > 1$.

15. $(-\infty, \infty)$; for,

$$\rho = \lim_{k \to \infty} \frac{\left| \dfrac{(-1)^{k+1}(x-3)^{k+1}}{(k+1)!} \right|}{\left| \dfrac{(-1)^k(x-3)^k}{k!} \right|} = \lim_{k \to \infty} |x-3| \, \frac{k!}{(k+1)!} = \lim_{k \to \infty} \frac{|x-3|}{k+1} = 0,$$

and hence the series converges for all x.

17. $x = 1$; for

$$\rho = \lim_{k \to \infty} \left| \frac{(k+1)!(x-1)^{k+1}}{k!(x-1)^k} \right| = \lim_{k \to \infty} (k+1)|x-1| = \infty.$$

Hence the series converges only when $x = 1$.

19. $(1/3, 1]$; for

$$\rho = \lim_{k \to \infty} \frac{\left| \dfrac{(-1)^{k+1}(3x-2)^{k+1}}{\ln(k+1)} \right|}{\left| \dfrac{(-1)^k(3x-2)^k}{\ln k} \right|} = \lim_{k \to \infty} |3x-2| \, \frac{\ln k}{\ln(k+1)}$$

$$= |3x-2| \lim_{k \to \infty} \frac{1/k}{1/(k+1)} = \lim_{k \to \infty} |3x-2| \, \frac{k+1}{k} = |3x-2| < 1 \quad \Longrightarrow \quad \frac{1}{3} < x < 1,$$

and

$$x = 1/3 \quad \Longrightarrow \quad \sum \frac{(-1)^k}{\ln k} (3x-2)^k = \sum \frac{1}{\ln k} > \sum \frac{1}{k},$$

which diverges, while

$$x = 1 \quad \Longrightarrow \quad \sum \frac{(-1)^k}{\ln k} (3x-2)^k = \sum \frac{(-1)^k}{\ln k},$$

which converges since it is an alternating series whose terms are decreasing and approach zero.

21. $(-1, 1)$; for,

$$\rho = \lim_{k\to\infty} \frac{\left|\dfrac{x^{2k+3}}{k+3}\right|}{\left|\dfrac{x^{2k+1}}{k+2}\right|} = \lim_{k\to\infty} x^2 \frac{k+2}{k+3} = x^2 < 1 \quad \Longrightarrow \quad -1 < x < 1,$$

and

$$x = -1 \quad \Longrightarrow \quad \sum \frac{x^{2k+1}}{k+2} = \sum \frac{-1}{k+2},$$

which is harmonic and hence diverges, while

$$x = 1 \quad \Longrightarrow \quad \sum \frac{x^{2k+1}}{k+2} = \sum \frac{1}{k+2},$$

which is harmonic and diverges.

23. $[2, 4]$; for,

$$\rho = \lim_{k\to\infty} \frac{\left|\dfrac{(x-3)^{k+1}}{(k+1)(k+2)}\right|}{\left|\dfrac{(x-3)^k}{k(k+1)}\right|} = \lim_{k\to\infty} |x-3| \frac{k}{k+2} = |x-3| < 1 \quad \Longrightarrow \quad 2 < x < 4,$$

and

$$x = 2 \quad \Longrightarrow \quad \sum \frac{(x-3)^k}{k(k+1)} = \sum \frac{(-1)^k}{k(k+1)},$$

which converges since it is an alternating series whose terms are decreasing and approach zero, while

$$x = 4 \quad \Longrightarrow \quad \sum \frac{(x-3)^k}{k(k+1)} = \sum \frac{1}{k(k+1)} < \sum \frac{1}{k^2},$$

which converges since it is a p-series, $p = 2 > 1$.

25. $[-1/2, 1/2]$; for,

$$\rho = \lim_{k\to\infty} \frac{\left|\dfrac{(2x)^{k+1}}{\sqrt{(k+1)^3 + 2}}\right|}{\left|\dfrac{(2x)^k}{\sqrt{k^3 + 2}}\right|} = \lim_{k\to\infty} |2x| \sqrt{\frac{k^3 + 2}{(k+1)^3 + 2}} = |2x| < 1 \quad \Longrightarrow \quad -\frac{1}{2} < x < \frac{1}{2},$$

and

$$x = -1/2 \quad \Longrightarrow \quad \sum \frac{(2x)^k}{\sqrt{k^3 + 2}} = \sum \frac{(-1)^k}{\sqrt{k^3 + 2}},$$

which converges since it is an alternating series whose terms are decreasing and approach zero, while

$$x = 1/2 \quad \Longrightarrow \quad \sum \frac{(2x)^k}{\sqrt{k^3+2}} = \sum \frac{1}{\sqrt{k^3+2}} < \sum \frac{1}{k^{3/2}},$$

which converges since it is a *p*-series, $p = 3/2 > 1$.

27. $(2-e, 2+e)$; for,

$$\rho = \lim_{k\to\infty} \frac{\left|\dfrac{(k+1)(x-2)^{k+1}}{e^{k+1}}\right|}{\left|\dfrac{k(x-2)^k}{e^k}\right|} = \lim_{k\to\infty} |x-2| \frac{1}{e} \frac{k+1}{k} = \frac{1}{e}|x-2| < 1 \quad \Longrightarrow \quad 2-e < x < 2+e,$$

and

$$x = 2 - e \quad \Longrightarrow \quad \sum \frac{k(x-2)^k}{e^k} = \sum (-1)^k,$$

which diverges since the terms do not approach zero, while

$$x = 2 + e \quad \Longrightarrow \quad \sum \frac{k(x-2)^k}{e^k} = \sum k,$$

which diverges.

29. $(-3/7, 1/7)$; for,

$$\rho = \lim_{k\to\infty} \frac{\left|\dfrac{(k+1)(7x+1)^{k+1}}{2^{k+1}}\right|}{\left|\dfrac{k(7x+1)^k}{2^k}\right|} = \lim_{k\to\infty} |7x+1| \frac{1}{2} \frac{k+1}{k} = \frac{1}{2}|7x+1| < 1 \quad \Longrightarrow \quad -\frac{3}{7} < x < \frac{1}{7},$$

and

$$x = -3/7 \quad \Longrightarrow \quad \sum \frac{k(7x+1)^k}{2^k} = \sum (-1)^k,$$

which diverges since the terms do not approach zero, while

$$x = 1/7 \quad \Longrightarrow \quad \sum \frac{k(7x+1)^k}{2^k} = \sum k,$$

which diverges.

31. $(-1, 1)$; for,

$$\rho = \lim_{k\to\infty} \left|\frac{(k+1)x^{k+1}}{kx^k}\right| = \lim_{k\to\infty} |x| \frac{k+1}{k} = |x| < 1 \quad \Longrightarrow \quad -1 < x < 1,$$

and

$$x = -1 \quad \Longrightarrow \quad \sum kx^k = \sum (-1)^k k,$$

which diverges since the terms do not approach zero, while

$$x = 1 \quad \Longrightarrow \quad \sum kx^k = \sum k,$$

which diverges.

33. $[-2, 4)$; for,

$$\rho = \lim_{k \to \infty} \frac{\left| \dfrac{(x-1)^{k+1}}{3^{k+1}\sqrt{k+2}} \right|}{\left| \dfrac{(x-1)^k}{3^k\sqrt{k+1}} \right|} = \lim_{k \to \infty} |x-1| \frac{1}{3} \sqrt{\frac{k+1}{k+2}} = \frac{1}{3}|x-1| < 1 \implies -2 < x < 4,$$

and

$$x = -2 \implies \sum \frac{(x-1)^k}{3^k\sqrt{k+1}} = \sum \frac{(-1)^k}{\sqrt{k+1}},$$

which converges since it is an alternating series whose terms decrease and approach zero, while

$$x = 4 \implies \sum \frac{(x-1)^k}{3^k\sqrt{k+1}} = \sum \frac{1}{\sqrt{k+1}},$$

which diverges since it is a p-series, $p = 1/2 < 1$.

35. $[-e, e)$; for,

$$\rho = \lim_{k \to \infty} \frac{\left| \dfrac{(k+5)x^{k+1}}{(k+2)(k+3)e^{k+1}} \right|}{\left| \dfrac{(k+4)x^k}{(k+1)(k+2)e^k} \right|} = \lim_{k \to \infty} |x| \frac{1}{e} \frac{k+5}{k+4} \frac{k+1}{k+3} = \frac{1}{e}|x| < 1 \implies -e < x < e,$$

and

$$x = -e \implies \sum \frac{(k+4)x^k}{(k+1)(k+2)e^k} = \sum \frac{(-1)^k(k+4)}{(k+1)(k+2)},$$

which converges since it is an alternating series whose terms decrease and approach zero, while

$$x = e \implies \sum \frac{(k+4)x^k}{(k+1)(k+2)e^k} = \sum \frac{k+4}{(k+1)(k+2)} > \sum \frac{1}{k+4},$$

which is harmonic and hence diverges.

37. $(-\infty, \infty)$; for,

$$\rho = \lim_{k \to \infty} \frac{\left| \dfrac{(-1)^{k+1}x^{2k+3}}{2(k+2)!} \right|}{\left| \dfrac{(-1)^k x^{2k+1}}{2(k+1)!} \right|} = \lim_{k \to \infty} \frac{x^2}{k+2} = 0$$

and hence the series converges for all x.

39. $[-1, 1]$; for,

$$\rho = \lim_{k \to \infty} \frac{\left| \dfrac{x^{k+1}}{(k+1)(k+2)} \right|}{\left| \dfrac{x^k}{k(k+1)} \right|} = \lim_{k \to \infty} |x| \frac{k}{k+1} \frac{k+1}{k+2} = |x| < 1 \implies -1 < x < 1,$$

and

$$x = -1 \implies \sum \frac{x^k}{k(k+1)} = \sum \frac{(-1)^k}{k(k+1)},$$

443

which converges since it is an alternating series whose terms are decreasing and approach zero, while,

$$x = 1 \quad \Longrightarrow \quad \sum \frac{x^k}{k(k+1)} = \sum \frac{1}{k(k+1)} < \sum \frac{1}{k^2},$$

which converges since it is a p-series, $p = 2 > 1$.

41. If $[-r, r)$ is the interval of convergence, then the series $\sum a_k(-r)^k = \sum (-1)^k a_k r^k$ converges, but $\sum a_k r^k$ diverges. Hence the series is conditionally, but not absolutely, convergent when $x = -r$.

43. If c is the smaller of r_a and r_b, then both $\sum a_k x^k$ and $\sum b_k x^k$ converge when $|x| < c$ and hence

$$\sum (a_k + b_k)x^k = \sum a_k x^k + \sum b_k x^k$$

converges.

45. First observe that f has a removable discontinuity at $x = 1$. Now, if $R = 1/2$, then the interval of convergence is defined by $|x - 1| < 1/2$ or $1/2 < x < 3/2$. Graph of $f(x) = \ln(3 - 2x)/(2(1 - x))$ and $P_3(x)$, $P_6(x)$ and $P_9(x)$ using the range $[0, 2] \times [-2, 4]$.

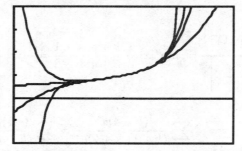

The graph shows that P_3, P_6, and P_9 get closer to $f(x)$ at $x = 1/2$ but not very close at $x = 3/2$. So we conjecture that the series converges at $x = 1/2$ but not at $x = 3/2$. To prove this, note that the series at $x = 1/2$ is

$$\sum_{k=0}^{\infty} \frac{(-1)^k}{k + 1}.$$

This is the alternating harmonic series which is shown to converge by the Alternating Series Test. At $x = 3/2$ the series is

$$\sum_{k=0}^{\infty} \frac{1}{k + 1}.$$

This is the harmonic series which we know diverges.

47. If $R = 1$ then the interval of convergence is defined by $|x| < 1$ or $-1 < x < 1$. Graph of $f(x) = \ln(x + \sqrt{x^2 + 1})$ and $P_3(x)$, $P_6(x)$ and $P_9(x)$ using the range $[-1.5, 1.5] \times [-2, 2]$.

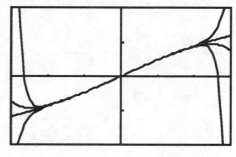

The graph shows that the series converges at both 1 and −1. We conjecture that the series converges at $x = \pm 1$. To prove this, note that because all of the powers of x in the series are odd then at either endpoint the series is the alternating series

$$\pm\left(1 + \sum_{k=1}^{\infty} \frac{(-1)^k (2k)!}{(2^k k!)^2 (2k+1)}\right),$$

where the plus sign is used if $x = 1$ and the negative sign if $x = -1$. Since $(2k)! = [1 \cdot 3 \cdot 5 \cdot \ldots \cdot (2k-1)]2^k k!$ and $2^k k! = 2 \cdot 4 \cdot 6 \cdot \ldots \cdot 2k$, the series becomes (excluding the first term)

$$\sum_{k=1}^{\infty} \frac{(-1)^k [1 \cdot 3 \cdot 5 \cdot \ldots \cdot (2k-1)]}{2^k k! (2k+1)} = \sum_{k=1}^{\infty} \frac{(-1)^k [1 \cdot 3 \cdot 5 \cdot \ldots \cdot (2k-1)]}{2 \cdot 4 \cdot 6 \cdot \ldots \cdot 2k (2k+1)}.$$

Now,

$$\frac{1 \cdot 3 \cdot 5 \cdot \ldots \cdot (2k-1)}{2 \cdot 4 \cdot 6 \cdot \ldots \cdot 2k} = \left(\frac{1}{2}\right)\left(\frac{3}{4}\right)\left(\frac{5}{6}\right) \cdots \left(\frac{2k-1}{2k}\right).$$

Each of the terms of this product is less than 1 so the entire product is less than one. Therefore

$$\frac{1 \cdot 3 \cdot 5 \cdot \ldots \cdot (2k-1)}{2 \cdot 4 \cdot 6 \cdot \ldots \cdot 2k (2k+1)} \le \frac{1}{2k+1}$$

and hence

$$\lim_{k \to +\infty} \frac{1 \cdot 3 \cdot 5 \cdot \ldots \cdot (2k-1)}{2 \cdot 4 \cdot 6 \cdot \ldots \cdot 2k (2k+1)} = 0.$$

To show that the sequence

$$a_k = \frac{1 \cdot 3 \cdot 5 \cdot \ldots \cdot (2k-1)}{2 \cdot 4 \cdot 6 \cdot \ldots \cdot 2k (2k+1)}$$

is decreasing, note that

$$a_{k+1} \le a_k$$

$$\Longleftrightarrow \quad \frac{1 \cdot 3 \cdot 5 \cdot \ldots \cdot (2k-1)(2k+1)}{2 \cdot 4 \cdot 6 \cdot \ldots \cdot 2k(2k+2)(2k+3)} \le \frac{1 \cdot 3 \cdot 5 \cdot \ldots \cdot (2k-1)}{2 \cdot 4 \cdot 6 \cdot \ldots \cdot 2k(2k+1)}$$

$$\Longleftrightarrow \quad (2k+1)^2 \le (2k+2)(2k+3)$$

$$\Longleftrightarrow \quad 4k^2 + 4k + 1 \le 4k^2 + 10k + 6$$

$$\Longleftrightarrow \quad 0 \le 6k + 5.$$

Therefore the Alternating Series Test implies convergence for $x = \pm 1$.

13.5   Differentiation and Integration of Power Series

1. Using (1), with x replaced by $2x$, we obtain

$$\frac{1}{1-2x} = \sum_{k=0}^{\infty} (2x)^k = \sum_{k=0}^{\infty} 2^k x^k = 1 + 2x + 4x^2 + \cdots$$

for $|2x| < 1$. Hence $|x| < 1/2 \Longrightarrow$ radius of convergence $= 1/2$.

3. Using (1), with x replaced by $-x$ and multiplying by x^2, we obtain

$$\frac{x^2}{1-x} = x^2 \sum_{k=0}^{\infty} x^k = \sum_{k=0}^{\infty} x^{k+2} = x^2 + x^3 + \cdots$$

for $|x| < 1$. Hence radius of convergence $= 1$.

5. Using (1), with x replaced by $-4x^2$, we obtain

$$\frac{1}{1+4x^2} = \sum_{k=0}^{\infty} (-4x^2)^k = \sum_{k=0}^{\infty} (-1)^k 4^k x^{2k} = 1 - 4x^2 + 16x^4 - \cdots$$

for $|4x^2| < 1$. Hence $|x| < 1/2 \implies$ radius of convergence $= 1/2$.

7. Using (1), with x replaced by $-x^2$ and multiplying by x, we obtain

$$\frac{x}{1+x^2} = x \sum_{k=0}^{\infty} (-x^2)^k = \sum_{k=0}^{\infty} (-1)^k x^{2k+1} = x - x^3 + x^5 - \cdots$$

for $|x^2| < 1$. Hence $|x| < 1 \implies$ radius of convergence $= 1$.

9. Using (1), with x replaced by x^4, we obtain

$$\frac{1}{1-x^4} = \sum_{k=0}^{\infty} (x^4)^k = \sum_{k=0}^{\infty} x^{4k} = 1 + x^4 + x^8 + \cdots$$

for $|x^4| < 1$. Hence $|x| < 1 \implies$ radius of convergence $= 1$.

11. Since $\dfrac{1}{9-x^2} = \dfrac{1}{9} \dfrac{1}{1-x^2/9}$, we use (1), with x replaced by $x^2/9$ and multiply by $1/9$, to obtain

$$\frac{1}{9-x^2} = \frac{1}{9} \sum_{k=0}^{\infty} \left(\frac{x^2}{9}\right)^k = \sum_{k=0}^{\infty} \frac{x^{2k}}{9^{k+1}} = \frac{1}{9} + \frac{x^2}{81} + \frac{x^4}{729} + \cdots$$

for $|x/3| < 1$. Hence $|x| < 3 \implies$ radius of convergence $= 3$.

13. Since $\dfrac{x-1}{x+1} = 1 - \dfrac{2}{x+1}$, we use (1), with x replaced by $-x$, multiply by 2 and subtract from 1 to obtain

$$\frac{x-1}{x+1} = 1 - 2 \sum_{k=0}^{\infty} (-x)^k = -1 + \sum_{k=1}^{\infty} (-1)^{k+1} 2x^k = -1 + 2x - 2x^2 + \cdots$$

for $|x| < 1$. Hence radius of convergence $= 1$.

15. Since $f(x) = -2 \dfrac{d}{dx}\left(\dfrac{1}{1+x}\right)$,

$$\frac{2}{(1+x)^2} = -2 \frac{d}{dx}\left[\sum_{k=0}^{\infty} (-x)^k\right] = -2 \sum_{k=1}^{\infty} (-1)^k k x^{k-1} = \sum_{k=1}^{\infty} (-1)^{k+1} (2k) x^{k-1} = 2 - 4x + 6x^2 - \cdots$$

for $|x| < 1$. Hence radius of convergence $= 1$.

446

17. Since $f(x) = -\frac{1}{2}\frac{d}{dx}\left(\frac{1}{1+x^2}\right)$,

$$\frac{x}{(1+x^2)^2} = -\frac{1}{2}\frac{d}{dx}\left[\sum_{k=0}^{\infty}(-x^2)^k\right] = -\frac{1}{2}\sum_{k=1}^{\infty}(-1)^k(2k)x^{2k-1} = \sum_{k=1}^{\infty}(-1)^{k+1}kx^{2k-1} = x - 2x^3 + 3x^5 - \cdots$$

for $|x|^2 < 1$. Hence $|x| < 1 \implies$ radius of convergence = 1.

19. Since $f(x) = \frac{d}{dx}\left(\frac{x}{1-x^2}\right)$,

$$\frac{1+x^2}{(1-x^2)^2} = \frac{d}{dx}\left[x\sum_{k=0}^{\infty}(x^2)^k\right] = \frac{d}{dx}\left[\sum_{k=0}^{\infty}x^{2k+1}\right] = \sum_{k=1}^{\infty}(2k+1)x^{2k} = 1 + 3x^2 + 5x^4 + 7x^6 + \cdots$$

for $|x|^2 < 1$. Hence $|x| < 1 \implies$ radius of convergence = 1.

21. Since $f(x) = -\frac{d}{dx}\left(\frac{1}{1+4x^2}\right)$,

$$\frac{8x}{(1+4x^2)^2} = -\frac{d}{dx}\left[\sum_{k=0}^{\infty}(-4x^2)^k\right] = \frac{d}{dx}\left[\sum_{k=0}^{\infty}(-1)^{k+1}4^k x^{2k}\right]$$

$$= \sum_{k=1}^{\infty}(-1)^{k+1}4^k(2k)x^{2k-1} = 8x - 64x^3 + 384x^5 - \cdots$$

for $|4x^2| < 1$. Hence $|x| < 1/2 \implies$ radius of convergence = 1/2.

23. Since $f(x) = \frac{d}{dx}\left(\frac{x-1}{x+1}\right)$, it follows from Exercise 13 that

$$\frac{2}{(x+1)^2} = \frac{d}{dx}\left[-1 + \sum_{k=1}^{\infty}(-1)^{k-1}2x^k\right] = \sum_{k=1}^{\infty}(-1)^{k-1}2kx^{k-1} = 2 - 4x + \cdots$$

for $|x| < 1$. Hence radius of convergence = 1.

25. Since $f(x) = -\int\frac{1}{1-x}\,dx$,

$$\ln(1-x) = -\int\left[\sum_{k=0}^{\infty}x^k\right]dx = -\sum_{k=0}^{\infty}\frac{x^{k+1}}{k+1} + C$$

for $|x| < 1$. Hence radius of convergence = 1. Letting $x = 0$, it follows that $C = \ln 1 = 0$. Therefore

$$\ln(1-x) = -\sum_{k=0}^{\infty}\frac{x^{k+1}}{k+1} = -x - \frac{x^2}{2} - \frac{x^3}{3} - \frac{x^4}{4} - \cdots.$$

27. Since $f(x) = 2\int\frac{1}{1+4x^2}\,dx$, it follows from Exercise 5 that

$$\text{Tan}^{-1}(2x) = 2\int\left[\sum_{k=0}^{\infty}(-1)^k4^k x^{2k}\right]dx = 2\sum_{k=0}^{\infty}\frac{(-1)^k4^k}{2k+1}x^{2k+1} + C$$

for $|x| < 1/2$. Hence radius of convergence $= 1/2$. Letting $x = 0$, we find that $C = \text{Tan}^{-1}(0) = 0$. Therefore

$$\text{Tan}^{-1}(2x) = 2\sum_{k=0}^{\infty} \frac{(-1)^k 4^k}{2k+1} x^{2k+1} = 2x - \frac{8}{3}x^3 + \frac{32}{5}x^5 - \cdots .$$

29. Since $f(x) = \frac{1}{4} \int \frac{1}{1+x/4} \, dx$,

$$\ln(4+x) = \frac{1}{4}\int \left[\sum_{k=0}^{\infty}\left(-\frac{x}{4}\right)^k\right] dx = \int \left[\sum_{k=0}^{\infty} \frac{(-1)^k}{4^{k+1}} x^k\right] dx = \sum_{k=0}^{\infty} \frac{(-1)^k}{4^{k+1}(k+1)} x^{k+1} + C$$

for $|x/4| < 1$. Hence $|x| < 4 \implies$ radius of convergence $= 4$. Letting $x = 0$, it follows that $C = \ln 4$. Therefore

$$\ln(4+x) = \ln 4 + \sum_{k=0}^{\infty} \frac{(-1)^k}{4^{k+1}(k+1)} x^{k+1} = \ln 4 + \frac{x}{4} - \frac{x^2}{32} + \frac{x^3}{192} - \frac{x^4}{1024} + \cdots .$$

31. Since $f(x) = 2\int \frac{x}{1+x^2} \, dx$, it follows from Exercise 7 that

$$\ln(1+x^2) = 2\int \left[\sum_{k=0}^{\infty}(-1)^k x^{2k+1}\right] dx = \sum_{k=0}^{\infty} \frac{(-1)^k(2)}{2k+2} x^{2k+2} + C = \sum_{k=0}^{\infty} \frac{(-1)^k}{k+1} x^{2k+2} + C$$

for $|x| < 1$. Hence radius of convergence $= 1$. Letting $x = 0$, it follows that $C = \ln 1 = 0$. Therefore

$$\ln(1+x^2) = \sum_{k=0}^{\infty} \frac{(-1)^k}{k+1} x^{2k+2} = x^2 - \frac{x^4}{2} + \frac{x^6}{3} - \cdots .$$

33. (a) By the Ratio Test,

$$\rho = \lim_{k\to\infty} \frac{\left|\frac{(-1)^{k+1}x^{2k+2}}{(2k+2)!}\right|}{\left|\frac{(-1)^k x^{2k}}{(2k)!}\right|} = \lim_{k\to\infty} \frac{|x|^2}{(2k+2)(2k+1)} = 0.$$

Hence the series converges absolutely for all x.

(b) By Theorem 4,

$$f'(x) = \sum_{k=1}^{\infty}(-1)^k \frac{(2k)x^{2k-1}}{(2k)!} = \sum_{k=1}^{\infty}(-1)^k \frac{x^{2k-1}}{(2k-1)!}$$

$$\implies f''(x) = \sum_{k=1}^{\infty}(-1)^k \frac{(2k-1)x^{2k-2}}{(2k-1)!} = \sum_{k=1}^{\infty}(-1)^k \frac{x^{2k-2}}{(2k-2)!} = \sum_{k=0}^{\infty}(-1)^{k+1} \frac{x^{2k}}{(2k)!} = -f(x).$$

(c) $f(0) = 1 + 0 + 0 + \cdots = 1$

(d) $f(x) = \cos x$.

35. Using Theorem 4 and the Chain Rule,

$$f(x) = \sum_{k=0}^{\infty} a_k(bx+c)^k \implies f'(x) = \sum_{k=0}^{\infty} ka_k(bx+c)^{k-1}(b) = \sum_{k=0}^{\infty} kba_k(bx+c)^{k-1}$$

for $|bx+c| < r$.

13.6   Taylor and Maclaurin Series

1. Since $e^x = \sum_{k=0}^{\infty} \dfrac{x^k}{k!}$, which converges for all x, it follows that

$$e^{2x} = \sum_{k=0}^{\infty} \frac{(2x)^k}{k!} = \sum_{k=0}^{\infty} \frac{2^k x^k}{k!} = 1 + 2x + 2x^2 + \frac{4x^3}{3} + \cdots$$

for all x.

3. Since

$$\begin{aligned}
f(x) &= \sin x &\implies& f(0) = 0 \\
f'(x) &= \cos x &\implies& f'(0) = 1 \\
f''(x) &= -\sin x &\implies& f''(0) = 0 \\
f'''(x) &= -\cos x &\implies& f'''(0) = -1 \\
f^{(4)}(x) &= \sin x &\implies& f^{(4)}(0) = 0,
\end{aligned}$$

it follows that

$$\sin x = x - \frac{x^3}{3!} + \cdots + \frac{(-1)^k x^{2k+1}}{(2k+1)!} + \cdots = \sum_{k=0}^{\infty} \frac{(-1)^k x^{2k+1}}{(2k+1)!}.$$

Using the Ratio Test, we find that

$$\rho = \lim_{k \to \infty} \frac{\left| \dfrac{(-1)^{k+1} x^{2k+3}}{(2k+3)!} \right|}{\left| \dfrac{(-1)^k x^{2k+1}}{(2k+1)!} \right|} = \lim_{k \to \infty} \frac{|x|^2}{(2k+3)(2k+2)} = 0.$$

Hence the series converges for all x.

5. Since

$$\begin{aligned}
f(x) &= \cos x &\implies& f(\pi/4) = \sqrt{2}/2 \\
f'(x) &= -\sin x &\implies& f'(\pi/4) = -\sqrt{2}/2 \\
f''(x) &= -\cos x &\implies& f''(\pi/4) = -\sqrt{2}/2 \\
f'''(x) &= \sin x &\implies& f'''(\pi/4) = \sqrt{2}/2 \\
f^{(4)}(x) &= \cos x &\implies& f^{(4)}(\pi/4) = \sqrt{2}/2,
\end{aligned}$$

it follows that

$$\begin{aligned}
\cos x &= \frac{\sqrt{2}}{2} - \frac{\sqrt{2}}{2}\left(x - \frac{\pi}{4}\right) - \frac{\sqrt{2}}{2}\frac{1}{2!}\left(x - \frac{\pi}{4}\right)^2 \\
&\quad + \frac{\sqrt{2}}{2}\frac{1}{3!}\left(x - \frac{\pi}{4}\right)^3 + \frac{\sqrt{2}}{2}\frac{1}{4!}\left(x - \frac{\pi}{4}\right)^4 + \cdots \\
&= \sum_{k=0}^{\infty} \left(\frac{\sqrt{2}}{2}\right) \frac{(-1)^{k(k+1)/2}}{k!}\left(x - \frac{\pi}{4}\right)^k
\end{aligned}$$

Now, the kth term of this series has the form

$$\pm \frac{\sqrt{2}}{2} \frac{1}{k!} \left(x - \frac{\pi}{4}\right)^k.$$

Hence, using the Ratio Test,

$$\rho = \lim_{k \to \infty} \frac{\left| \dfrac{\sqrt{2}}{2} \dfrac{(x - \pi/4)^{k+1}}{(k+1)!} \right|}{\left| \dfrac{\sqrt{2}}{2} \dfrac{(x - \pi/4)^k}{k!} \right|} = \lim_{k \to \infty} \left| x - \frac{\pi}{4} \right| \frac{1}{k+1} = 0$$

and therefore the series converges for all x.

7. Since

$$\begin{aligned}
f(x) &= 1 + x^2 &\implies& \quad f(2) = 5 \\
f'(x) &= 2x &\implies& \quad f'(2) = 4 \\
f''(x) &= 2 &\implies& \quad f''(2) = 2
\end{aligned}$$

and $f^{(k)}(x) = 0$ for $k \geq 3$, it follows that

$$1 + x^2 = 5 + 4(x - 2) + (x - 2)^2,$$

which converges for all x.

9. It follows from equation (1), Section 13.5, with x replaced by $-x$, that

$$\frac{1}{1+x} = \sum_{k=0}^{\infty} (-x)^k = \sum_{k=0}^{\infty} (-1)^k x^k = 1 - x + x^2 - x^3 + \cdots$$

for $|x| < 1$.

11. It follows from the Maclaurin series for $\sin x$ in Exercise 3, with x replaced by $2x$, that

$$\sin 2x = \sum_{k=0}^{\infty} \frac{(-1)^k (2x)^{2k+1}}{(2k+1)!} = \sum_{k=0}^{\infty} \frac{(-1)^k 2^{2k+1} x^{2k+1}}{(2k+1)!} = 2x - \frac{4x^3}{3} + \cdots$$

for all x. Hence

$$x \sin 2x = x \left(\sum_{k=0}^{\infty} \frac{(-1)^k 2^{2k+1} x^{2k+1}}{(2k+1)!} \right) = \sum_{k=0}^{\infty} \frac{(-1)^k 2^{2k+1} x^{2k+2}}{(2k+1)!} = 2x^2 - \frac{4x^4}{3} + \cdots$$

for all x.

13. Since $2^x = e^{x \ln 2}$, it follows from the Maclaurin series for e^x in Exercise 1, with x replaced by $x \ln 2$, that

$$2^x = e^{x \ln 2} = \sum_{k=0}^{\infty} \frac{(-x \ln 2)^k}{k!} = \sum_{k=0}^{\infty} \frac{(-1)^k (\ln 2)^k}{k!} x^k = 1 + x \ln 2 + \frac{x^2 (\ln 2)^2}{2!} + \cdots$$

for all x.

15. Using equation (11) for the Binomial Series, with $r = 3/2$, we find that

$$(1+x)^{3/2} = 1 + \sum_{k=1}^{\infty} \frac{\frac{3}{2}\left(\frac{3}{2}-1\right)\cdots\left(\frac{3}{2}-k+1\right)x^k}{k!} = 1 + \frac{3}{2}x + \frac{3}{8}x^2 + \cdots$$

for $|x| < 1$.

17. Since $\sqrt{x} = 2\sqrt{1 + \frac{x-4}{4}}$, it follows from the Binomial Series, with $r = 1/2$ and x replaced by $(x-4)/4$, that

$$\sqrt{x} = 2\left[1 + \frac{1}{2}\left(\frac{x-4}{4}\right) + \frac{\left(\frac{1}{2}\right)\left(-\frac{1}{2}\right)}{2}\left(\frac{x-4}{4}\right)^2 + \cdots\right] = 2 + \frac{x-4}{4} - \frac{(x-4)^2}{64} + \cdots$$

for $|(x-4)/4| < 1$, or $0 < x < 8$.

19. Since $\frac{1}{x} = \frac{1}{2}\frac{1}{1+(x-2)/2}$, it follows that

$$\frac{1}{x} = \frac{1}{2}\sum_{k=0}^{\infty}\left(-\frac{x-2}{2}\right)^k = \sum_{k=0}^{\infty}\frac{(-1)^k(x-2)^k}{2^{k+1}} = \frac{1}{2} - \frac{x-2}{2^2} + \frac{(x-2)^2}{2^3} - \frac{(x-2)^3}{2^4} + \cdots$$

for $|(x-2)/2| < 1$, or $0 < x < 4$.

21. Using the Maclaurin series for e^x in Exercise 1, with x replaced by $-x$, we find that

$$x^2 e^{-x} = x^2\left(\sum_{k=0}^{\infty}\frac{(-x)^k}{k!}\right) = \sum_{k=0}^{\infty}\frac{(-1)^k x^{k+2}}{k!} = x^2 - x^3 + \frac{x^4}{2!} - \frac{x^5}{3!} + \cdots$$

for all x.

23. Using the Maclaurin series for $\ln(1+x)$ in Exercise 24, Section 13.5, we find that

$$x^2 \ln(1+x) = x^2\left(\sum_{k=0}^{\infty}\frac{(-1)^k x^{k+1}}{k+1}\right) = \sum_{k=0}^{\infty}\frac{(-1)^k x^{k+3}}{k+1} = x^3 - \frac{x^4}{2} + \frac{x^5}{3} - \frac{x^6}{4} + \cdots,$$

which converges for $-1 < x \leq 1$.

25. Since $\sin^2 x = 1 - \cos^2 x$, it follows from Exercise 24 that

$$\sin^2 x = 1 - \left(\frac{1}{2} + \frac{1}{2}\sum_{k=0}^{\infty}\frac{(-1)^k 2^{2k} x^{2k}}{(2k)!}\right) = \frac{1}{2}\left[1 - \sum_{k=0}^{\infty}\frac{(-1)^k 2^{2k} x^{2k}}{(2k)!}\right] = x^2 - \frac{1}{3}x^4 + \cdots$$

for all x.

27. Since $\sec^2 = (\tan x)'$, it follows from Exercise 26 that

$$\sec^2 x = \left(x + \frac{2x^3}{3!} + \cdots\right)' = 1 + x^2 + \cdots.$$

29. Using the Maclaurin series for $\cos x$, with x replaced by x^2, it follows that

$$\cos x^2 = \sum_{k=0}^{\infty}\frac{(-1)^k (x^2)^{2k}}{(2k)!} = \sum_{k=0}^{\infty}\frac{(-1)^k x^{4k}}{(2k)!} = 1 - \frac{x^4}{2!} + \cdots.$$

31. Using the Maclaurin series for e^x and the fact that $\cosh x = (e^x + e^{-x})/2$, it follows that

$$x \cosh x = \frac{x}{2} \left(\sum_{k=0}^{\infty} \frac{x^k}{k!} \right) + \frac{x}{2} \left(\sum_{k=0}^{\infty} \frac{(-x)^k}{k!} \right) = \sum_{k=0}^{\infty} \frac{x^{2k+1}}{(2k)!} = x + \frac{x^3}{2!} + \cdots .$$

33. Using the Maclaurin series for e^x, we find that

$$e^2 \approx 1 + 2 + \frac{2^2}{2!} + \cdots + \frac{2^{10}}{10!} = 7.389,$$

with error no more than

$$\frac{e^c}{11!} 2^{11} < \frac{3^2}{11!} 2^{11} = 4.6 \times 10^{-4},$$

where $0 < c < 2$.

35. Integrating the Maclaurin series for $\sin x^2$ in Exercise 28, we obtain

$$\int_0^{\pi/4} \sin x^2 \, dx = \int_0^{\pi/4} \left(x^2 - \frac{x^6}{3!} + \cdots \right) dx$$

$$= \left(\frac{x^3}{3} - \frac{x^7}{42} + \frac{x^{11}}{1320} + \cdots \right) \Big|_0^{\pi/4} \approx \frac{(0.785)^3}{3} - \frac{(0.785)^7}{42} = 0.157,$$

with error no more than

$$\frac{(0.785)^{11}}{1320} = 5.28 \times 10^{-5}.$$

37. Using the Maclaurin series for $\sin x / x$ in Exercise 20, we find that

$$\int_0^1 \frac{\sin x}{x} \, dx = \int_0^1 \left(1 - \frac{x^2}{3!} + \frac{x^4}{5!} - \frac{x^6}{7!} + \cdots \right) dx$$

$$= \left(x - \frac{x^3}{3(3!)} + \frac{x^5}{5(5!)} - \frac{x^7}{7(7!)} + \cdots \right) \Big|_0^1 \approx 1 - \frac{1}{3(3!)} + \frac{1}{5(5!)} = 0.946,$$

with error no more than

$$\frac{1}{7(7!)} = 2.83 \times 10^{-5}.$$

39. Using the Maclaurin series for e^x, with x replaced by $-x^2$, we find that

$$\int_0^1 e^{-x^2} \, dx = \int_0^1 \left(1 - x^2 + \frac{x^4}{2!} - \frac{x^6}{3!} + \cdots \right) dx = \left(x - \frac{x^3}{3} + \frac{x^5}{5(2!)} - \frac{x^7}{7(3!)} + \cdots \right) \Big|_0^1$$

$$\approx 1 - \frac{1}{3} + \frac{1}{5(2!)} - \frac{1}{7(3!)} + \frac{1}{9(4!)} - \frac{1}{11(5!)} = 0.747,$$

with error no more than

$$\frac{1}{13(6!)} = 1.07 \times 10^{-4}.$$

41. Set $r = 1/2$ and replace x by $2x$ to obtain

$$\sqrt{1 + 2x} = 1 + \frac{1}{2}(2x) + \sum_{k=2}^{\infty} \frac{\frac{1}{2}\left(\frac{1}{2} - 1\right) \cdots \left(\frac{1}{2} - k + 1\right)}{k!}(2x)^k$$

$$= 1 + x - \sum_{k=2}^{\infty} \frac{(-1)^k 1 \cdot 3 \cdots (2k - 3)}{k!} x^k$$

$$= 1 + x - \frac{x^2}{2} + \cdots$$

for $|2x| < 1$. Hence $|x| < 1/2 \implies$ radius of convergence $= 1/2$.

43. Since $(9 + 3x)^{3/2} = [9(1 + x/3)]^{3/2} = 27(1 + x/3)^{3/2}$, set $r = 3/2$ and replace x by $x/3$ to obtain

$$(9 + 3x)^{3/2} = 27\left[1 + \frac{3}{2}\left(\frac{x}{3}\right) + \sum_{k=2}^{\infty} \frac{\frac{3}{2}\left(\frac{3}{2} - 1\right) \cdots \left(\frac{3}{2} - k + 1\right)}{k!}\left(\frac{x}{3}\right)^k\right]$$

$$= 27 + \frac{27x}{2} + \sum_{k=2}^{\infty} \frac{\frac{3}{2}\left(\frac{3}{2} - 1\right) \cdots \left(\frac{3}{2} - k + 1\right)}{3^{k-3} k!} x^k$$

$$= 27 + \frac{27x}{2} + \frac{9x^2}{8} + \cdots$$

for $|x/3| < 1$. Hence $|x| < 3 \implies$ radius of convergence $= 3$.

45. (a) By definition,

$$f'(0) = \lim_{x \to 0} \frac{f(x) - f(0)}{x} = \lim_{x \to 0} \frac{e^{-1/x^2}}{x} = \lim_{t \to \infty} \frac{\sqrt{t}}{e^t} = \lim_{t \to \infty} \frac{1}{2\sqrt{t}e^t} = 0,$$

where $t = 1/x^2$. In general,

$$f^{(n+1)}(0) = \lim_{x \to 0} \frac{f^{(n)}(x) - f^{(n)}(0)}{x}.$$

Now, if $x \neq 0$,

$$f'(x) = e^{-1/x^2}\left(2x^{-3}\right)$$

and

$$f''(x) = e^{-1/x^2}\left(-6x^{-4}\right) + e^{-1/x^2}\left(4x^{-6}\right).$$

We claim that $f^{(n)}(x)$ is in fact the sum of terms of the form $cx^{-k}e^{-1/x^2}$ for all $n \geq 1$. To show this, use induction on n. Clearly, this is true for $n = 1$. If $f^{(n)}(x)$ has this form, then

$$f^{(n+1)}(x) = f^{(n)}(x)' = \left(\sum c_i x^{-k_i} e^{-1/x^2}\right)'$$

$$= \sum \left[c_i x^{-k_i} e^{-1/x^2}(2x^{-3}) - k_i c_i x^{-k_i - 1} e^{-1/x^2}\right] = \sum d_i x^{-m_i} e^{-1/x^2}$$

and hence $f^{(n+1)}(x)$ has this form. Thus, by induction, $f^{(n)}(x)$ is the sum of terms of the form $cx^{-k}e^{-1/x^2}$ for all integers $n \geq 1$. Now, $f'(0) = 0$. Assume, as induction hypothesis, that $f^{(n)}(0) = 0$. Then

$$f^{(n+1)}(0) = \lim_{x \to 0} \frac{f^{(n)}(x)}{x} = \lim_{x \to 0} \frac{1}{x} \sum c_i x^{-k_i} e^{-1/x^2} = \sum c_i \lim_{x \to 0} x^{-k_i - 1} e^{-1/x^2}.$$

453

If we let $t = 1/x^2$, then $x = t^{-1/2}$ and hence the typical term in this sum has the form $\lim_{t \to \infty} t^{m/2}/e^t$. Since, by repeated use of l'Hôpital's Rule,

$$\lim_{t \to \infty} \frac{t^{m/2}}{e^t} = 0,$$

it follows that $f^{(n+1)}(0) = 0$. Thus, by induction, $f^{(n)}(0) = 0$ for all integers $n \geq 1$.

(b) Since $f(0) = 0$ and $f^{(n)}(0) = 0$ for all integers $n \geq 1$, the Maclaurin series for $f(x)$ is identically zero:

$$0 + f'(0)x + \frac{f''(0)}{2!} x^2 + \cdots + \frac{f^{(n)}(0)}{n!} x^n + \cdots = 0.$$

47. (a) The Maclaurin series for $(1 - x)^{-1/2}$ is

$$1 + \sum_{k=1}^{\infty} \frac{[1 \cdot 3 \cdot 5 \cdot \ldots \cdot (2k - 1)]x^k}{2^k k!},$$

obtained by using the Binomial Series with x replaced by $-x$ and $r = -1/2$. The kth term of the series is then

$$\frac{(-\frac{1}{2})(-\frac{3}{2})(-\frac{5}{2}) \cdots (-\frac{1}{2} - k + 1)(-x)^k}{k!}$$

$$= \frac{(-1)^k \left(\frac{1}{2}\right)^k [1 \cdot 3 \cdot 5 \cdot \ldots \cdot (2k - 1)](-1)^k x^k}{k!} = \frac{[1 \cdot 3 \cdot 5 \cdot \ldots \cdot (2k - 1)]x^k}{2^k k!}.$$

To obtain the series for $(1 - x^2)^{-1/2}$ replace x with x^2 in the series above. This gives

$$(1 - x^2)^{-1/2} = 1 + \sum_{k=1}^{\infty} \frac{[1 \cdot 3 \cdot 5 \cdot \ldots \cdot (2k - 1)]x^{2k}}{2^k k!}$$

This series converges for $-1 < x < 1$.

(b) Now,

$$\int_0^x \frac{dt}{\sqrt{1 - t^2}} = \operatorname{Sin}^{-1} x$$

and hence

$$\operatorname{Sin}^{-1} x = \int_0^x dt + \sum_{k=1}^{\infty} \frac{[1 \cdot 3 \cdot 5 \cdot \ldots \cdot (2k - 1)]}{2^k k!} \int_0^x t^{2k} \, dt = x + \sum_{k=1}^{\infty} \frac{[1 \cdot 3 \cdot 5 \cdot \ldots \cdot (2k - 1)]x^{2k+1}}{2^k k!(2k + 1)}.$$

Graph of $f(x) = \operatorname{Sin}^{-1} x$ and

$$P_1(x) = x$$

$$P_2(x) = x + \frac{x^3}{4}$$

$$P_3(x) = x + \frac{x^3}{4} + \frac{x^5}{8}$$

$$P_4(x) = x + \frac{x^3}{4} + \frac{x^5}{8} + \frac{5x^7}{112}$$

$$P_5(x) = x + \frac{x^3}{4} + \frac{x^5}{8} + \frac{5x^7}{112} + \frac{35x^9}{1152}.$$

using the range $[-1, 1] \times [-2, 2]$.

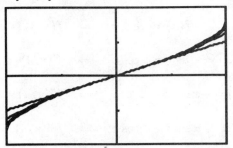

Very little difference between $f(x)$ and $P_1(x)$, $P_2(x)$, $P_3(x)$, $P_4(x)$, and $P_5(x)$ can be seen on the interval $[-0.5, 0.5]$.

(c) Since $(2k)! = 2^k k! [1 \cdot 3 \cdot 5 \cdot \ldots (2k-1)]$ it follows that

$$\frac{(2k)!}{(2^k k!)^2} = \frac{[1 \cdot 3 \cdot 5 \cdot \ldots (2k-1)]}{2^k k!}$$

and therefore

$$\text{Sin}^{-1} x = x + \sum_{k=1}^{\infty} \frac{(2k)!\, x^{2k+1}}{(2^k k!)^2 (2k+1)}.$$

At $x = 1/2$, $\text{Sin}^{-1}(1/2) = \pi/6$ and hence

$$\frac{\pi}{6} = \frac{1}{2} + \sum_{k=1}^{\infty} \frac{(2k)!}{(2^k k!)^2 (2k+1) 2^{2k+1}}.$$

The following table gives the value of the partial sums of this series and the error $|S_k - \pi/6|$.

k	S_n	$|S_k - \pi/6|$
1	0.5208333333	0.00276544
2	0.5231770833	0.00042169
3	0.5235258577	0.00007292
4	0.5235851954	0.00001358
5	0.5235961193	0.00000266
6	0.5235982376	5.380×10^{-7}
7	0.5235986637	1.119×10^{-7}
8	0.5235987519	2.374×10^{-8}
9	0.5235987705	5.120×10^{-9}
10	0.5235987745	1.119×10^{-9}

We have then that $6S_{10} = 3.141592647$.

Review Exercises – Chapter 13

1. $P_5(x) = (2x) - \dfrac{(2x)^3}{3!} + \dfrac{(2x)^5}{5!} = 2x - \dfrac{4x^3}{3} + \dfrac{4x^5}{15}$

3. Since

$$f(x) = x \cos x \implies f(\pi/4) = \frac{\pi\sqrt{2}}{8}$$

$$f'(x) = -x \sin x + \cos x \implies f'(\pi/4) = \frac{\sqrt{2}}{2}\left(1 - \frac{\pi}{4}\right)$$

$$f''(x) = -x \cos x - 2 \sin x \implies f''(\pi/4) = -\frac{\sqrt{2}}{2}\left(2 + \frac{\pi}{4}\right)$$

$$f'''(x) = x \sin x - 3 \cos x \implies f'''(\pi/4) = \frac{\sqrt{2}}{2}\left(\frac{\pi}{4} - 3\right),$$

it follows that

$$P_3(x) = \frac{\pi\sqrt{2}}{8} + \frac{\sqrt{2}}{2}\left(1 - \frac{\pi}{4}\right)\left(x - \frac{\pi}{4}\right) - \frac{\sqrt{2}}{2}\left(2 + \frac{\pi}{4}\right)\frac{1}{2!}\left(x - \frac{\pi}{4}\right)^2 + \frac{\sqrt{2}}{2}\left(\frac{\pi}{4} - 3\right)\frac{1}{3!}\left(x - \frac{\pi}{4}\right)^3.$$

5. $P_3(x) = 1 + x^2$

7. Since

$$f(x) = \text{Tan}^{-1}x \implies f(1) = \pi/4$$

$$f'(x) = (1 + x^2)^{-1} \implies f'(1) = 1/2$$

$$f''(x) = -2x(1 + x^2)^{-2} \implies f''(1) = -1/2$$

$$f'''(x) = (6x^2 - 2)(1 + x^2)^{-3} \implies f'''(1) = 1/2,$$

it follows that

$$P_3(x) = \frac{\pi}{4} + \frac{1}{2}(x - 1) - \frac{1}{2}\frac{1}{2!}(x - 1)^2 + \frac{1}{2}\frac{1}{3!}(x - 1)^3.$$

9. $P_3(x) = e^{x \ln 2} = 1 + x \ln 2 + \frac{(x \ln 2)^2}{2!} + \frac{(x \ln 2)^3}{3!}$

11. Since $34° = 17\pi/90$ radians, the maximum error is no more than

$$\left|R_3\left(\frac{17\pi}{90}\right)\right| = \left|\frac{\sin c}{4!}\left(\frac{17\pi}{90} - \frac{\pi}{6}\right)^4\right| < \frac{|c|}{24}\left(\frac{17\pi}{90} - \frac{\pi}{6}\right)^4 < \frac{17\pi/90}{24}\left(\frac{17\pi}{90} - \frac{\pi}{6}\right)^4 = 5.87 \times 10^{-7},$$

where $\pi/6 < c < 17\pi/90$.

13. Since the Taylor series for $f(x) = \sqrt[5]{x}$ at $a = 32$ is an alternating series, the error is no more than

$$|R_2(35)| = \left|\frac{f'''(32)}{3!}(35 - 32)^3\right| = \frac{36}{125}\frac{1}{2^{14}}\frac{1}{3!}3^3 = 7.91 \times 10^{-5}.$$

15. The error term is

$$|R_3(0.25)| = \frac{e^c}{4!}(0.25)^4,$$

where $0 < c < 0.25$. Since $e^c < e^{0.25} < 2$, it follows that the error is no more than

$$\frac{2}{4!}(0.25)^4 = 3.26 \times 10^{-4}.$$

Remark: The answer in the text, 0.00021, is incorrect.

456

17. We showed in Exercise 19, Section 13.2, that

$$\frac{1}{1-x} = 1 + x + \cdots + x^{n-1} + \frac{x^n}{1-x}.$$

Hence

$$\ln\frac{1}{1-x} = \int_0^x \frac{1}{1-t}\,dt = \int_0^x \left(1 + t + \cdots + t^{n-1}\right)\,dt + \int_0^x \frac{t^n}{1-t}\,dt = t + \frac{t^2}{2} + \cdots + \frac{t^n}{n} + \int_0^x \frac{t^n}{1-t}\,dt.$$

19. Since $f'(x) = (7/2)x^{5/2}$, $f''(x) = (35/4)x^{3/2}$, $f'''(x) = (105/8)x^{1/2}$, $f(0) = f'(0) = f''(0) = f'''(0) = 0$. But for $n \geq 4$, $f^{(n)}(x)$ has the form $cx^{1/2-k}$, where $k = 1, 2, \ldots$, and hence $f^{(n)}(0)$ is undefined. Therefore $P_n(0)$ is undefined for $n \geq 4$.

21. $(2, 4)$; for, by the Ratio Test,

$$\rho = \lim_{k\to\infty}\left|\frac{(k+1)(x-3)^{k+1}}{k(x-3)^k}\right| = \lim_{k\to\infty}|x-3|\,\frac{k+1}{k} = |x-3| < 1 \quad\Longrightarrow\quad 2 < x < 4,$$

and

$$x = 2 \quad\Longrightarrow\quad \sum k(x-3)^k = \sum(-1)^k k,$$

which diverges, while

$$x = 4 \quad\Longrightarrow\quad \sum k(x-3)^k = \sum k,$$

which diverges.

23. $[1, 3)$; for, by the Ratio Test,

$$\rho = \lim_{k\to\infty}\frac{\left|\dfrac{\sqrt{k+1}(x-2)^{k+1}}{k+4}\right|}{\left|\dfrac{\sqrt{k}(x-2)^k}{k+3}\right|} = \lim_{k\to\infty}|x-2|\sqrt{\frac{k+1}{k}}\,\frac{k+3}{k+4} = |x-2| < 1 \quad\Longrightarrow\quad 1 < x < 3,$$

and

$$x = 1 \quad\Longrightarrow\quad \sum\frac{\sqrt{k}(x-2)^k}{k+1} = \sum\frac{(-1)^k\sqrt{k}}{k+1} = \sum\frac{(-1)^k}{\sqrt{k}+1/\sqrt{k}},$$

which converges since it is an alternating series whose terms are decreasing and approach zero, while

$$x = 3 \quad\Longrightarrow\quad \sum\frac{\sqrt{k}(x-2)^k}{k+1} = \sum\frac{\sqrt{k}}{k+1} > \sum\frac{1}{k},$$

which diverges.

25. $(-1, 1]$; for, by the Ratio Test,

$$\rho = \lim_{k\to\infty}\frac{\left|\dfrac{(-1)^{k+1}x^{k+1}}{(k+1)\ln(k+1)}\right|}{\left|\dfrac{(-1)^k x^k}{k\ln k}\right|} = \lim_{k\to\infty}|x|\,\frac{k}{k+1}\,\frac{\ln k}{\ln(k+1)}$$

$$= \lim_{k\to\infty}|x|\,\frac{1/k}{1/(k+1)} = |x| < 1 \quad\Longrightarrow\quad -1 < x < 1,$$

457

and

$$x = -1 \quad \Longrightarrow \quad \sum \frac{(-1)^k x^k}{k \ln k} = \sum \frac{1}{k \ln k},$$

which diverges since, by the Integral Test,

$$\int_2^\infty \frac{1}{x \ln x}\, dx = \lim_{t \to \infty} \int_2^t \frac{1}{x \ln x}\, dx = \lim_{t \to \infty} \ln \ln x \Big|_2^t = \infty,$$

while

$$x = 1 \quad \Longrightarrow \quad \sum \frac{(-1)^k x^k}{k \ln k} = \sum \frac{(-1)^k}{k \ln k},$$

which converges since it is an alternating series whose terms are decreasing and approach zero.

27. Multiplying the series for $1/(1+x)$, with x replaced by $-x^2$, by x, we obtain

$$\frac{x}{1-x^2} = x\left(1 + x^2 + x^4 + \cdots\right) = x + x^3 + x^5 + \cdots$$

for $|x|^2 < 1$. Hence $|x| < 1 \Longrightarrow$ radius of convergence $= 1$.

29. Since $\ln(4 + x) = \ln 4(1 + x/4) = \ln 4 + \ln(1 + x/4)$, we use the series for $\ln(1 + x)$ in Exercise 24, Section 13.5, to obtain

$$\ln(4 + x) = \ln 4 + \frac{x}{4} - \frac{1}{2}\left(\frac{x}{4}\right)^2 + \frac{1}{3}\left(\frac{x}{4}\right)^3 - \cdots$$

for $|x/4| < 1$. Hence $|x| < 4 \Longrightarrow$ radius of convergence $= 4$.

31. Since $\sin x^2 + 2x^2 \cos x^2 = (x \sin x^2)'$, it follows that

$$\sin x^2 + 2x^2 \cos x^2 = \frac{d}{dx}\left[x\left(x^2 - \frac{(x^2)^3}{3!} + \cdots\right)\right] = \frac{d}{dx}\left(x^3 - \frac{x^7}{3!} + \cdots\right) = 3x^2 - \frac{7x^6}{3!} + \cdots$$

for all x.

33. $\displaystyle \sum_{k=0}^\infty a_k x^{k+1} = a_0 x + a_1 x^2 + \cdots = \sum_{k=1}^\infty a_{k-1} x^k$

35. Since

$$\sum n a_n x^{n-1} + c \sum a_n x^n = \sum_{n=0}^\infty \left[(n+1)a_{n+1} + ca_n\right] x^n = 0$$

$$\Longrightarrow \quad (n+1)a_{n+1} + ca_n = 0 \quad \text{for } n = 0, 1, \ldots,$$

it follows that

$$\begin{aligned}
n = 0 &\quad \Longrightarrow \quad a_1 + ca_0 = 0 &\quad \Longrightarrow \quad a_1 = -ca_0 \\
n = 1 &\quad \Longrightarrow \quad 2a_2 + ca_1 &\quad \Longrightarrow \quad a_2 = -\frac{c}{2}a_1 = \frac{c^2}{2}a_0 \\
n = 2 &\quad \Longrightarrow \quad 3a_3 + ca_2 = 0 &\quad \Longrightarrow \quad a_3 = -\frac{c}{3}a_2 = -\frac{c^3}{3!}a_0
\end{aligned}$$

and, in general,

$$a_n = (-1)^n \frac{c^n}{n!} a_0,$$

where a_0 is arbitrary. Hence

$$f(x) = \sum_{k=0}^\infty a_k x^k = \sum_{k=0}^\infty (-1)^k \frac{c^k}{k!} a_0 x^k = a_0 \sum_{k=0}^\infty \frac{(-cx)^k}{k!},$$

which is the function $a_0 e^{-cx}$.

37. Using the series for $1/(1+x)$, with x replaced by x^4, we obtain

$$\frac{1}{1+x^4} = \sum_{k=0}^{\infty}(-x^4)^k = \sum_{k=0}^{\infty}(-1)^k x^{4k} = 1 - x^4 + x^8 - \cdots$$

for $|x| < 1$.

39. Using the Binomial Series, with $r = 1/3$ and x replaced by x^2, we obtain

$$\sqrt[3]{1+x^2} = 1 + \frac{1}{3}x^2 + \sum_{k=2}^{\infty}\frac{\frac{1}{3}\left(\frac{1}{3}-1\right)\cdots\left(\frac{1}{3}-k+1\right)}{k!}x^{2k} = 1 + \frac{x^2}{3} - \frac{x^4}{9} + \cdots.$$

41. Since

$$
\begin{aligned}
f(x) &= \sin\sqrt{x} &&\implies& f(\pi^2/4) &= 1 \\
f'(x) &= \frac{\cos\sqrt{x}}{2\sqrt{x}} &&\implies& f'(\pi^2/4) &= 0 \\
f''(x) &= \frac{-\sqrt{x}\sin\sqrt{x}-\cos\sqrt{x}}{4x\sqrt{x}} &&\implies& f''(\pi^2/4) &= -1/\pi^2 \\
f'''(x) &= \frac{(\sqrt{x}\cos\sqrt{x})(6-2x)+6x\sin\sqrt{x}}{16x^3} &&\implies& f'''(\pi^2/4) &= 6/\pi^4,
\end{aligned}
$$

it follows that the first three terms of the series for $\sin\sqrt{x}$ are

$$1 - \frac{1}{\pi^2}\frac{1}{2!}\left(x - \frac{\pi^2}{4}\right)^2 + \frac{6}{\pi^4}\frac{1}{3!}\left(x - \frac{\pi^2}{4}\right)^3.$$

43. Since

$$\sqrt{x+4} = \sqrt{6\left(1 + \frac{x-2}{6}\right)} = \sqrt{6}\sqrt{1 + \frac{x-2}{6}},$$

we use the Binomial Series, with $r = 1/2$ and x replaced by $(x-2)/6$, to obtain

$$
\begin{aligned}
\sqrt{x+4} &= \sqrt{6}\left[1 + \sum_{k=1}^{\infty}\frac{\left(\frac{1}{2}\right)\left(\frac{1}{2}-1\right)\cdots\left(\frac{1}{2}-k+1\right)}{k!}\left(\frac{x-2}{6}\right)^k\right] \\
&= \sqrt{6} + \frac{\sqrt{6}}{2}\left(\frac{x-2}{6}\right) - \frac{\sqrt{6}}{2^3}\left(\frac{x-2}{6}\right)^2 + \cdots,
\end{aligned}
$$

where $|(x-2)/6| < 1$. If x is either -4 or 8, the series is an alternating series whose terms are decreasing and approach zero. Hence the series converges for $-4 \le x \le 8$.

45. Using the series for e^x, with $x = 1/2$, we obtain

$$\sqrt{e} = e^{1/2} \approx 1 + \frac{1}{2} + \frac{(1/2)^2}{2!} + \frac{(1/2)^3}{3!} + \frac{(1/2)^4}{4!} + \frac{(1/2)^5}{5!} = 1.649,$$

where the error is

$$|R_5(1/2)| = \left|\frac{e^c}{6!}\left(\frac{1}{2}\right)^6\right|.$$

Since $0 < c < 1/2$, $e^c < e^{1/2} < 2$ and hence

$$|R_5(1/2)| < \frac{2}{6!}\left(\frac{1}{2}\right)^6 = 4.34 \times 10^{-5}.$$

Thus, $\sqrt{e} \approx 1.649$ to within three decimal places.

Chapter 14

The Conic Sections

14.1 Parabolas

1. Standard form: $x^2 = 4(1/8)y \implies$ vertex = (0,0); focal length $c = 1/8$; axis is $x = 0$; focus = $(0, 1/8)$

3. Standard form: $x^2 = 4(-1/8)(y - 2) \implies$ vertex = (0,2); focal length $c = -1/8$; axis is $x = 0$; focus = $(0, 2 - 1/8) = (0, 15/8)$.

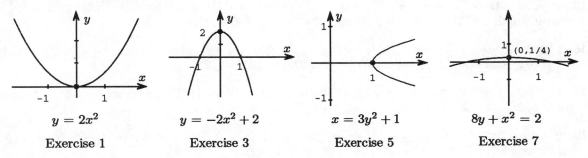

$y = 2x^2$	$y = -2x^2 + 2$	$x = 3y^2 + 1$	$8y + x^2 = 2$
Exercise 1	Exercise 3	Exercise 5	Exercise 7

5. Standard form: $y^2 = 4(1/12)(x - 1) \implies$ vertex = (1,0); focal length $c = 1/12$; axis is $y = 0$; focus = $(1 + 1/12, 0) = (13/12, 0)$.

7. Standard form: $x^2 = 4(-2)(y - 1/4) \implies$ vertex = (0,1/4); focal length $c = -2$; axis is $x = 0$; focus = $(0, 1/4 - 2) = (0, -7/4)$.

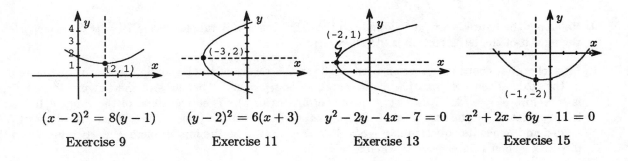

$(x - 2)^2 = 8(y - 1)$	$(y - 2)^2 = 6(x + 3)$	$y^2 - 2y - 4x - 7 = 0$	$x^2 + 2x - 6y - 11 = 0$
Exercise 9	Exercise 11	Exercise 13	Exercise 15

461

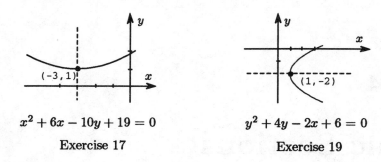

$$x^2 + 6x - 10y + 19 = 0$$

Exercise 17

$$y^2 + 4y - 2x + 6 = 0$$

Exercise 19

9. Standard form: $(x-2)^2 = 4(2)(y-1) \implies$ vertex $= (2,1)$; focal length $c = 2$; axis is $x = 2$; focus $= (2, 1+2) = (2,3)$.

11. Standard form: $(y-2)^2 = 4(3/2)(x+3) \implies$ vertex $= (-3,2)$; focal length $c = 3/2$; axis is $y = 2$; focus $= (-3+3/2, 2) = (-3/2, 2)$.

13. $y^2 - 2y - 4x - 7 = 0 \implies y^2 - 2y + 1 = 4x + 8 \implies (y-1)^2 = 4(1)(x+2)$. Hence vertex $= (-2,1)$; focal length $c = 1$; axis is $y = 1$; focus $= (-2+1, 1) = (-1, 1)$.

15. $x^2 + 2x - 6y - 11 = 0 \implies (x+1)^2 = 6(y+2) = 4(3/2)(y+2)$. Hence vertex $= (-1,-2)$; focal length $c = 3/2$; axis is $x = -1$; focus $= (-1, -2+3/2) = (-1, -1/2)$.

17. $x^2 + 6x - 10y + 19 = 0 \implies (x+3)^2 = 10y - 10 = 4(5/2)(y-1)$. Hence vertex $= (-3,1)$; focal length $c = 5/2$; axis is $x = -3$; focus $= (-3, 1+5/2) = (-3, 7/2)$.

19. $y^2 + 4y - 2x + 6 = 0 \implies (y+2)^2 = 2x - 2 = 4(1/2)(x-1)$. Hence vertex $= (1,-2)$; focal length $c = 1/2$; axis is $y = -2$; focus $= (1+1/2, -2) = (3/2, -2)$.

21. $x^2 = 4(3)y = 12y$

23. $c = 3 + 5 \implies (x-1)^2 = 4(8)(y-3) = 32(y-3)$

25. $c = 4 \implies (y+6)^2 = 4(4)(x+1) = 16(x+1)$

27. Since the axis is $x = 0$, the vertex must be at the point $(0,1)$ and hence $c = 4$. Therefore the equation is $x^2 = 4(4)(y-1) = 16(y-1)$.

29. Referring the figure below, it follows from the definition of a parabola that $|FP| = |PQ| = |2c|$. Hence the length of the latus rectum is $2|FP| = |4c|$.

31. First choose a coordinate system with origin at the vertex of the parabola, as shown in Figure 1.12 in the text. Then the equation of the parabola is $4cy = x^2$. The slope of the tangent line at (x, y) is therefore $y' = x/2c$. Now, let P have coordinates (a, b). Then the slope of the tangent line at P is $a/2c$. Let L' be the line through $F(0, c)$ perpendicular to the tangent line. Then the slope of L' is $-2c/a$ and hence its equation is $y - c = (-2c/a)x$. Let L be the line through P perpendicular to the directrix. Then L intersects L' when $x = a$ and

$$y - c = -\frac{2c}{a}(a) = -2c \implies y = -c.$$

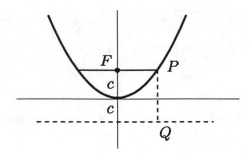

Exercise 29

Thus, L and L' intersect at the point $Q(a, -c)$, which is on the directrix. It then follows from the definition of a parabola that

$$|FP| = |PQ| \quad \Longrightarrow \quad \angle QFP = \angle FQP.$$

Therefore the complements of these angles are equal; $\beta = \alpha$.

33. From the given information, it follows that vertex $= (0, -1)$ and $c = 1$. Hence the equation of the parabola is

$$x^2 = 4(1)(y + 1) \quad \Longrightarrow \quad y = \frac{x^2}{4} - 1.$$

This parabola intersects the x-axis at $(\pm 2, 0)$ and hence, by symmetry, the area of the region is equal to

$$2 \int_0^2 \left(-\frac{x^2}{4} + 1 \right) \, dx = 2 \left(x - \frac{x^3}{12} \right) \Big|_0^2 = \frac{8}{3}.$$

35. From the given information, $c = -2$ and hence the equation of the parabola is

$$(x - 4)^2 = 4(-2)(y - 2) = -8(y - 2) \quad \Longrightarrow \quad y = -\frac{1}{8}(x - 4)^2 + 2.$$

The parabola intersects the x-axis at $(0, 0)$ and $(8, 0)$. Hence the area of the region is equal to

$$\int_0^8 \left[2 - \frac{1}{8}(x - 4)^2 \right] \, dx = \left[2x - \frac{1}{24}(x - 4)^3 \right] \Big|_0^8 = \frac{32}{3}.$$

37. From the given information, $c = 4$ and hence the equation of the parabola is

$$(x - 2)^2 = 4(4)(y + 1) = 16(y + 1) \quad \Longrightarrow \quad y = \frac{1}{16}(x - 2)^2 - 1.$$

Hence the average value on $[0, 5]$ is equal to

$$\frac{1}{5} \int_0^5 \left[\frac{1}{16}(x - 2)^2 - 1 \right] \, dx = \frac{1}{5} \left[\frac{1}{48}(x - 2)^3 - x \right] \Big|_0^5 = -\frac{41}{48}.$$

39. Let $P(x, y)$ be a point on the parabola. Then the distance from the focus $F(0, c)$ to P is equal to

$$|FP| = \sqrt{x^2 + (y - c)^2} = \sqrt{4cy + y^2 - 2cy + c^2} = \sqrt{y^2 + 2cy + c^2} = \sqrt{(y + c)^2} = |y + c|,$$

which is the distance from $P(x, y)$ to the directrix $y = -c$.

14.2 The Ellipse

1. Center $= (0, 0)$; $a = 3$, $b = 1 \implies c = \sqrt{8} \implies$ foci at $(\pm\sqrt{8}, 0)$; length of major axis $= 6$; length of minor axis $= 2$.

3. Standard form: $x^2/16 + y^2/4 = 1 \implies$ Center $= (0, 0)$; $a = 4$, $b = 2 \implies c = \sqrt{12} \implies$ foci at $(\pm\sqrt{12}, 0)$; length of major axis $= 8$; length of minor axis $= 4$.

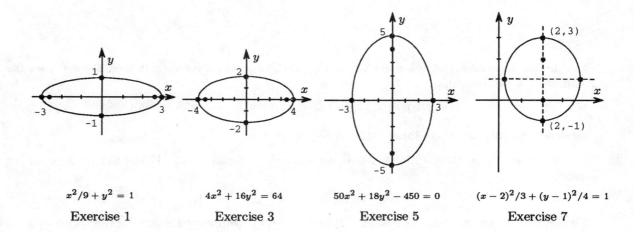

$x^2/9 + y^2 = 1$	$4x^2 + 16y^2 = 64$	$50x^2 + 18y^2 - 450 = 0$	$(x - 2)^2/3 + (y - 1)^2/4 = 1$
Exercise 1	Exercise 3	Exercise 5	Exercise 7

5. Standard form: $x^2/9 + y^2/25 = 1 \implies$ Center $= (0, 0)$; $a = 5$, $b = 3 \implies c = 4 \implies$ foci at $(0, \pm 4)$; length of major axis $= 10$; length of minor axis $= 6$.

7. Center $= (2, 1)$; $a = 2$, $b = \sqrt{3} \implies c = 1 \implies$ foci at $(2, 1 \pm 1) = (2, 2)$, $(2, 0)$; length of major axis $= 4$; length of minor axis $= 2\sqrt{3}$.

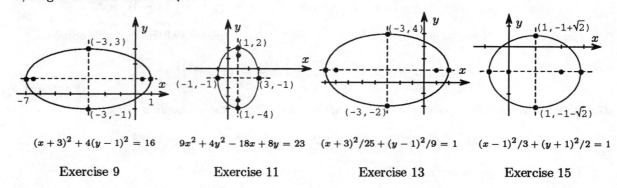

$(x + 3)^2 + 4(y - 1)^2 = 16$	$9x^2 + 4y^2 - 18x + 8y = 23$	$(x + 3)^2/25 + (y - 1)^2/9 = 1$	$(x - 1)^2/3 + (y + 1)^2/2 = 1$
Exercise 9	Exercise 11	Exercise 13	Exercise 15

9. Standard form: $(x + 3)^2/16 + (y - 1)^2/4 = 1 \implies$ Center $= (-3, 1)$; $a = 4$, $b = 2 \implies c = \sqrt{12} \implies$ foci at $(-3 \pm \sqrt{12}, 1)$; length of major axis $= 8$; length of minor axis $= 4$.

11. Standard form: $(x-1)^2/4 + (y+1)^2/9 = 1 \implies$ Center $= (1,-1)$; $a = 3$, $b = 2 \implies c = \sqrt{5} \implies$ foci at $(1, -1 \pm \sqrt{5})$; length of major axis $= 6$; length of minor axis $= 4$.

13. Standard form: $(x+3)^2/25 + (y-1)^2/9 = 1 \implies$ Center $= (-3, 1)$; $a = 5$, $b = 3 \implies c = 4 \implies$ foci at $(-3 \pm 4, 1) = (1, 1)$, $(-7, 1)$; length of major axis $= 10$; length of minor axis $= 6$.

15. Standard form: $(x-1)^2/3 + (y+1)^2/2 = 1 \implies$ Center $= (1, -1)$; $a = \sqrt{3}$, $b = \sqrt{2} \implies c = 1 \implies$ foci at $(1 \pm 1, -1) = (2, -1)$, $(0, -1)$; length of major axis $= 2\sqrt{3}$; length of minor axis $= 2\sqrt{2}$.

17. From the given information, $a = 4$, $b = 3$. Hence the equation is $x^2/16 + y^2/9 = 1$.

19. From the given information, $c = 4$, $b = 3$ and center $= (0, 0)$. Therefore $a = \sqrt{4^2 + 3^2} = 5$ and hence the equation is $x^2/25 + y^2/9 = 1$.

21. From the given information, $c = 3$, $b = 4$ and center $= (4, 2)$. Therefore $a = \sqrt{3^2 + 4^2} = 5$ and hence the equation is $(x-4)^2/16 + (y-2)^2/25 = 1$.

23. From the given information, $c = 3$, $b = 4$ and center $= (-2, 6)$. Therefore $a = \sqrt{3^2 + 4^2} = 5$ and hence the equation is $(x+2)^2/16 + (y-6)^2/25 = 1$.

25. Let dV be the volume of a differential disk of thickness dx. Then

$$dV = \pi r^2 dx = \pi y^2 dx = \pi \left(25 - \frac{25}{9} x^2 \right) dx$$

and hence the volume of revolution is

$$V = \int_0^3 \pi \left(25 - \frac{25}{9} x^2 \right) dx = \pi \left(25x - \frac{25}{27} x^3 \right) \Bigg|_0^3 = 50\pi.$$

27. If $x = c$,

$$y = \sqrt{b^2 \left(1 - \frac{c^2}{a^2} \right)} = \frac{b^2}{a}.$$

Hence the length of the latus rectum is $2b^2/a$.

29. True. If $a = b$, $x^2/a^2 + y^2/b^2 = 1 \iff x^2 + y^2 = a^2$, which is the equation of a circle.

31. (a) Differentiating with respect to x, we find that

$$\frac{2x}{a^2} + \frac{2yy'}{b^2} = 0 \implies y' = -\frac{b^2}{a^2} \frac{x}{y}.$$

(b) The slope of the tangent line to the ellipse at (x_0, y_0) is $-b^2 x_0/a^2 y_0$. Hence the equation of the tangent line is

$$y - y_0 = -\frac{b^2 x_0}{a^2 y_0} (x - x_0) \implies a^2 y y_0 - a^2 y_0^2 + b^2 x_0 x - b^2 x_0^2 = 0.$$

Since $x_0^2/a^2 + y_0^2/b^2 = 1$, $x_0^2 = a^2(1 - y_0^2/b^2)$, it follows that

$$a^2 y y_0 - a^2 y_0^2 + b^2 x_0 x - b^2 a^2 \left(1 - \frac{y_0^2}{b^2} \right) = 0 \implies x(b^2 x_0) + y(a^2 y_0) - a^2 b^2 = 0.$$

33. The foci are at $(2,3)$ and $(-4,3)$. Hence $c = 3$ and center $= (-1,3)$. Now, $2a = 12 \Longrightarrow a = 6$ and $b = \sqrt{6^2 - 3^2} = \sqrt{27}$. Hence the equation is $(x+1)^2/36 + (y-3)^2/27 = 1$.

35. Let dV be the volume of a differential disk of thickness dy. Then

$$dV = \pi r^2 dy = \pi x^2 dy = \pi \left(9 - \frac{9}{16}\, y^2 \right)\, dy$$

and hence by symmetry the volume of revolution is

$$V = 2 \int_0^4 \pi \left(9 - \frac{9}{16}\, y^2 \right)\, dy = 2\pi \left(9y - \frac{3}{16}\, y^3 \right)\Big|_0^4 = 48\pi.$$

37. If the tank is only half full,

$$W = 2218.7 \int_{-9}^0 (9 - y)\sqrt{81 - y^2}\, dy$$

$$= 2218.7 \left(\frac{9}{2} \right) \left[y\sqrt{81 - y^2} + 81\mathrm{Sin}^{-1}\frac{y}{9} \right]\Big|_{-9}^0 - 2218.7 \left(-\frac{1}{2} \right) \left(\frac{2}{3} \right) (81 - y^2)^{3/2}\Big|_{-9}^0 = 1{,}809{,}470 \text{ ft lbs.}$$

39. Referring to Figure 2.2 in the text, and using the fact that $y^2 = b^2(1 - x^2/a^2)$, we find that

$$
\begin{aligned}
|F_1 P| + |F_2 P| &= \sqrt{(x+c)^2 + y^2} + \sqrt{(x-c)^2 + y^2} \\[2mm]
&= \sqrt{x^2 + 2cx + c^2 + b^2 \left(1 - \frac{x^2}{a^2} \right)} + \sqrt{x^2 - 2cx + c^2 + b^2 \left(1 - \frac{x^2}{a^2} \right)} \\[2mm]
&= \sqrt{\left(1 - \frac{b^2}{a^2} \right) x^2 + 2cx + c^2 + b^2} + \sqrt{\left(1 - \frac{b^2}{a^2} \right) x^2 - 2cx + c^2 + b^2} \\[2mm]
&= \sqrt{\frac{c^2}{a^2}\, x^2 + 2cx + a^2} + \sqrt{\frac{c^2}{a^2}\, x^2 - 2cx + a^2} \qquad \text{since } c^2 = a^2 - b^2 \\[2mm]
&= \frac{1}{a}\sqrt{(cx + a^2)^2} + \frac{1}{a}\sqrt{(cx - a^2)^2} \\[2mm]
&= \frac{cx + a^2}{a} + \frac{a^2 - cx}{a} \qquad \text{since } cx \le a^2 \\[2mm]
&= 2a.
\end{aligned}
$$

14.3   The Hyperbola

1. Center $= (0,0)$; asymptotes: $y = \pm(3/2)x$.

3. Center $= (0,0)$; asymptotes: $y = \pm(3/\sqrt{5})x$.

5. Standard form: $x^2/4 - y^2/16 = 1 \Longrightarrow$ Center $= (0,0)$; asymptotes: $y = \pm 2x$.

7. Standard form: $y^2/7 - x^2/4 = 1 \implies$ Center $= (0,0)$; asymptotes: $y = \pm(\sqrt{7}/2)x$.

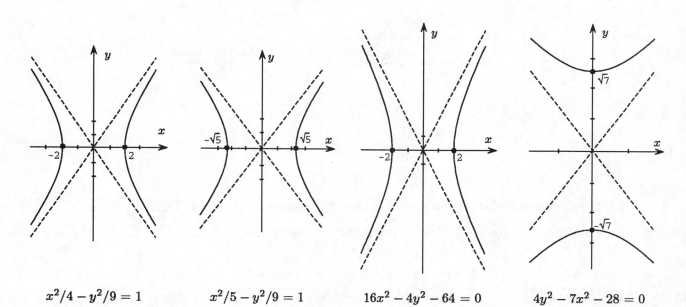

$x^2/4 - y^2/9 = 1$

Exercise 1

$x^2/5 - y^2/9 = 1$

Exercise 3

$16x^2 - 4y^2 - 64 = 0$

Exercise 5

$4y^2 - 7x^2 - 28 = 0$

Exercise 7

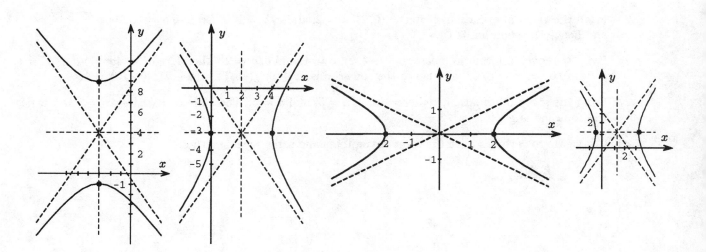

$2(y-4)^2/51 - 3(x+3)^2/34 = 1$

Exercise 9

$(x-2)^2/4 - (y+3)^2/9 = 1$

Exercise 11

$x^2 - 4y^2 = 4$

Exercise 13

$2x^2 - y^2 - 4\sqrt{2}x + 2\sqrt{2}y = 6$

Exercise 15

9. Standard form: $(y-4)^2/(51/2) - (x+3)^2/(34/3) = 1 \implies$ Center $= (-3, 4)$; asymptotes: $y = \pm(3/2)(x+3) + 4$.

11. Standard form: $(x-2)^2/4 - (y+3)^2/9 = 1 \implies$ Center $= (2, -3)$; asymptotes: $y = \pm(3/2)(x-2) - 3$.

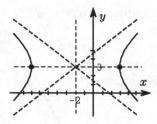

$(x+2)^2/25 - (y-3)^2/16 = 1$

Exercise 17

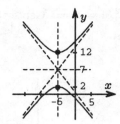

$(y-7)^2/9 - 16(x+5)^2/225 = 1$

Exercise 19

13. Standard form: $x^2/4 - y^2 = 1 \implies$ Center $= (0,0)$; asymptotes: $y = \pm(1/2)x$.

15. Standard form: $(x - \sqrt{2})^2/4 - (y - \sqrt{2})^2/8 = 1 \implies$ Center$= (\sqrt{2}, \sqrt{2})$; asymptotes: $y = \pm\sqrt{2}(x - \sqrt{2}) + \sqrt{2}$.

17. Standard form: $(x+2)^2/25 - (y-3)^2/16 = 1 \implies$ Center $= (-2, 3)$; asymptotes: $y = 3 \pm (4/5)(x+2)$.

19. Standard form: $(y-7)^2/25 - (x+5)^2/(225/16) = 1 \implies$ Center $= (-5, 7)$; asymptotes: $y = 7 \pm (4/3)(x+5)$.

21. From the given information, center $= (0,0)$ and $b = \sqrt{2^2 - 1^2} = \sqrt{3}$. Hence the equation is $x^2 - y^2/3 = 1$.

23. From the given information, center $= (0,2)$, $a = 2$, and $a/b = 2$. Therefore $b = 1$, $c = \sqrt{2^2 - 1^2} = \sqrt{3}$ and hence the equation is $(y - 2)^2/4 - x^2 = 1$.

25. From the given information, center $= (-1, 2)$, $c = 4$, and $b/a = 2$. Therefore $b^2 = 4a^2 = 4^2 - a^2 \implies a = 4/\sqrt{5} \implies b = 8/\sqrt{5}$ and hence the equation is $(x+1)^2/(16/5) - (y-2)^2/(64/5) = 1$.

27. From the given information, center $= (0,0)$, $c = 5$, and $a = 4$. Therefore $b = \sqrt{5^2 - 4^2} = 3$ and hence the equation is $y^2/16 - x^2/9 = 1$.

29. Conjugate hyperbolas have the same asymptotes and same focal length c.

31. $y - y_0 = \dfrac{b^2 x_0}{a^2 y_0}(x - x_0)$.

33. If $x = c$, then

$$\frac{x^2}{a^2} - \frac{y^2}{b^2} = 1 \quad \implies \quad y = b\sqrt{\frac{c^2}{a^2} - 1} = \frac{b}{a}\sqrt{c^2 - a^2} = \frac{b^2}{a}$$

since $c^2 - a^2 = b^2$. Hence the length of the latus rectum is $2b^2/a$.

35. Referring to Figure 3.2 in the text and using the fact that $y^2 = b^2(x^2/a^2 - 1)$, we find that

$$|F_1 P| - |F_2 P| = \sqrt{(x+c)^2 + y^2} - \sqrt{(x-c)^2 + y^2}$$

$$= \sqrt{x^2 + 2cx + c^2 + b^2\left(\frac{x^2}{a^2} - 1\right)} - \sqrt{x^2 - 2cx + c^2 + b^2\left(\frac{x^2}{a^2} - 1\right)}$$

$$= \sqrt{\left(1+\frac{b^2}{a^2}\right)x^2 + 2cx + c^2 - b^2} - \sqrt{\left(1+\frac{b^2}{a^2}\right)x^2 - 2cx + c^2 - b^2}$$

$$= \sqrt{\frac{c^2}{a^2}x^2 + 2cx + a^2} - \sqrt{\frac{c^2}{a^2}x^2 - 2cx + a^2} \qquad \text{since } a^2 = c^2 - b^2$$

$$= \frac{1}{a}\sqrt{(cx+a^2)^2} - \frac{1}{a}\sqrt{(cx-a^2)^2}$$

$$= \frac{cx+a^2}{a} - \frac{cx-a^2}{a} \qquad \text{since } a^2 \le cx$$

$$= 2a.$$

14.4 Rotation of Axes

1. $\cot 2\theta = (3-1)/2 = 1 \implies 2\theta = \pi/4 \implies \theta = \pi/8$.

3. $\cot 2\theta = (1-1)/2 = 0 \implies 2\theta = \pi/2 \implies \theta = \pi/4$.

5. $\cot 2\theta = (5-2)/2 = 3/4 \implies 2\theta = \text{Tan}^{-1}(4/3) \implies \theta = (1/2)\text{Tan}^{-1}(4/3)$.

7. $\cot 2\theta = (5-5)/6 = 0 \implies 2\theta = \pi/2 \implies \theta = \pi/4$. Therefore

$$x = \frac{\sqrt{2}}{2}(x'-y'), \quad y = \frac{\sqrt{2}}{2}(x'+y')$$

and hence the equation of the conic becomes

$$
\begin{aligned}
5x^2 + 6xy + 5y^2 - 32 &= 5\left(\frac{\sqrt{2}}{2}\right)^2 (x'-y')^2 + 6\left(\frac{\sqrt{2}}{2}\right)\left(\frac{\sqrt{2}}{2}\right)(x'-y')(x'+y') \\
&\quad + 5\left(\frac{\sqrt{2}}{2}\right)^2 (x'+y')^2 - 32 \\
&= \frac{5}{2}\left(x'^2 - 2x'y' + y'^2\right) + 3\left(x'^2 - y'^2\right) + \frac{5}{2}\left(x'^2 + 2x'y' + y'^2\right) - 32 \\
&= 8x'^2 + 2y'^2 - 32 \\
&= 0 \\
\implies \quad &\frac{x'^2}{4} + \frac{y'^2}{16} = 1.
\end{aligned}
$$

This is the equation of an ellipse in the x', y' coordinate system, center $(0,0)$, $a = 4$, $b = 2$.

9. $\cot 2\theta = (3-1)/2\sqrt{3} \implies 2\theta = \pi/3 \implies \theta = \pi/6$. Therefore

$$x = \frac{1}{2}(\sqrt{3}x' - y'), \quad y = \frac{1}{2}(x' + \sqrt{3}y')$$

and hence the equation of the conic becomes

$$3x^2 + 2\sqrt{3}xy + y^2 + 2x - 2\sqrt{3}y + 16 = 3\left(\frac{1}{2}\right)^2 \left(\sqrt{3}x' - y'\right)^2$$

$$+2\sqrt{3}\left(\frac{1}{2}\right)\left(\frac{1}{2}\right)\left(\sqrt{3}x'-y'\right)\left(x'+\sqrt{3}y'\right)$$

$$+\left(\frac{1}{2}\right)^2\left(x'+\sqrt{3}y'\right)^2+2\left(\frac{1}{2}\right)\left(\sqrt{3}x'-y'\right)$$

$$-2\sqrt{3}\left(\frac{1}{2}\right)\left(x'+\sqrt{3}y'\right)+16$$

$$=\;\frac{3}{4}\left(3x'^2-2\sqrt{3}x'y'+y'^2\right)+\frac{\sqrt{3}}{2}\left(\sqrt{3}x'^2+2x'y'-\sqrt{3}y'^2\right)$$

$$+\frac{1}{4}\left(x'^2+2\sqrt{3}x'y'+3y'^2\right)+\left(\sqrt{3}x'-y'\right)$$

$$-\sqrt{3}(x'+\sqrt{3}y')+16$$

$$=\;x'^2-y'+4$$

$$=\;0$$

$$\Longrightarrow\qquad y'=x'^2+4.$$

This is the equation of a parabola in the x', y' coordinate system, vertex at $(0,4)$.

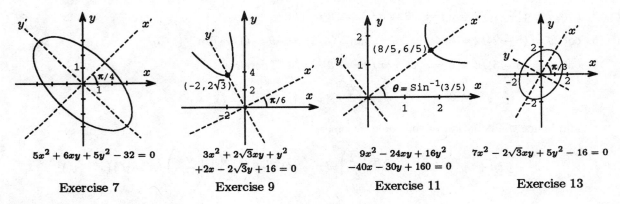

$5x^2+6xy+5y^2-32=0$	$3x^2+2\sqrt{3}xy+y^2$ $+2x-2\sqrt{3}y+16=0$	$9x^2-24xy+16y^2$ $-40x-30y+160=0$	$7x^2-2\sqrt{3}xy+5y^2-16=0$
Exercise 7	Exercise 9	Exercise 11	Exercise 13

11. $\cot 2\theta=(9-16)/(-24)=7/24$. Therefore

$$\tan 2\theta=\frac{24}{7}=\frac{2\tan\theta}{1-\tan^2\theta}$$

$$\Longrightarrow\quad 14\tan\theta=24-24\tan^2$$

$$\Longrightarrow\quad 12\tan^2\theta+7\tan\theta-12=0$$

$$\Longrightarrow\quad \tan\theta=\frac{3}{4}\quad\text{(we reject the negative root since θ is acute)}$$

$$\Longrightarrow\quad \sin\theta=\frac{3}{5},\cos\theta=\frac{4}{5}.$$

Therefore

$$x=\frac{1}{5}\left(4x'-3y'\right),\quad y=\frac{1}{5}\left(3x'+4y'\right)$$

and hence the equation of the conic becomes

$$9x^2-24xy+16y^2-40x-30y+100\;=\;9\left(\frac{1}{5}\right)^2(4x'-3y')^2-24\left(\frac{1}{5}\right)\left(\frac{1}{5}\right)(4x'-3y')(3x'+4y')$$

470

$$+16\left(\frac{1}{5}\right)^2(3x'+4y')^2 - 40\left(\frac{1}{5}\right)(4x'-3y')$$

$$-30\left(\frac{1}{5}\right)(3x'+4y')+100$$

$$= \quad 25y'^2 - 50x' + 100$$

$$= \quad 0$$

$$\implies \quad y'^2 = 2x' - 4 = 2(x'-2).$$

This is the equation of a parabola in the x', y' coordinate system, vertex at $(2,0)$.

13. $\cot 2\theta = (7-5)/(-2\sqrt{3}) = -1/\sqrt{3} \implies 2\theta = 2\pi/3 \implies \theta = \pi/3$. Therefore

$$x = \frac{1}{2}\left(x' - \sqrt{3}\,y'\right), \quad y = \frac{1}{2}\left(\sqrt{3}\,x' + y'\right)$$

and hence the equation of the conic becomes

$$7x^2 - 2\sqrt{3}xy + 5y^2 - 16 \quad = \quad \left(\frac{1}{2}\right)^2(x'-\sqrt{3}\,y')^2 - 2\sqrt{3}\left(\frac{1}{2}\right)\left(\frac{1}{2}\right)(x'-\sqrt{3}\,y')(\sqrt{3}\,x'+y')$$

$$+5\left(\frac{1}{2}\right)^2(\sqrt{3}\,x'+y')^2 - 16$$

$$= \quad 4x'^2 + 8y'^2 - 16$$

$$= \quad 0$$

$$\implies \quad \frac{x'^2}{4} + \frac{y'^2}{2} = 1.$$

This is the equation of an ellipse in the x', y' coordinate system, center at $(0,0)$, $a=2$, $b=\sqrt{2}$.

15. $\cot 2\theta = (1-1)/(2) = 0 \implies 2\theta = \pi/2 \implies \theta = \pi/4$. Therefore

$$x = \frac{\sqrt{2}}{2}\left(x' - y'\right), \quad y = \frac{\sqrt{2}}{2}\left(x' + y'\right)$$

and hence the equation of the conic becomes

$$x^2 + 2\sqrt{3}xy + y^2 + \sqrt{2}\,x - \sqrt{2}\,y$$

$$= \quad (x+y)^2 + \sqrt{2}(x-y)$$

$$= \quad (\sqrt{2}\,x')^2 + \sqrt{2}(-\sqrt{2}\,y')$$

$$= \quad 2x'^2 - 2y'$$

$$= \quad 0$$

$$\implies \quad y' = x'^2.$$

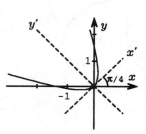

This is the equation of a parabola in the x', y' coordinate system, vertex at $(0,0)$.

17. In the new coordinate system, the endpoints are $(x', y') = (0, \pm 3)$. Since $x = (1/2)(\sqrt{3}\, x' - y')$ and $y = (1/2)(x' + \sqrt{3}\, y')$, the original coordinates are

$$(x', y') = (0, 3) \quad \Longrightarrow \quad x = -3/2, y = 3\sqrt{3}/2 \quad \Longrightarrow \quad (x, y) = (-3/2, 3\sqrt{3}/2)$$
$$(x', y') = (0, -3) \quad \Longrightarrow \quad x = 3/2, y = -3\sqrt{3}/2 \quad \Longrightarrow \quad (x, y) = (3/2, -3\sqrt{3}/2)$$

19. In the new coordinate system, the vertices are $(x', y') = (\pm 3, 0)$. Since $x = (1/2)x' - (\sqrt{3}/2)y'$ and $y = (\sqrt{3}/2)x' + (1/2)y'$, the original coordinates are

$$(x', y') = (3, 0) \quad \Longrightarrow \quad x = 3/2, y = 3\sqrt{3}/2 \quad \Longrightarrow \quad (x, y) = (3/2, 3\sqrt{3}/2)$$
$$(x', y') = (-3, 0) \quad \Longrightarrow \quad x = -3/2, y = -3\sqrt{3}/2 \quad \Longrightarrow \quad (x, y) = (-3/2, -3\sqrt{3}/2).$$

Review Exercises – Chapter 14

1. $(x - 3)^2 + (y + 4)^2 = 36$

3. $(x - 2)^2 + (y - 3)^2 = (2 - 2)^2 + (-1 - 3)^2 = 16$

5. From the given information, $c = 2$. Hence the equation is $(y - 4)^2 = 4(2)x = 8x$.

7. Let $y = ax^2 + bx + c$ be the general form for the equation of the parabola. Substituting in the points $(0, 5)$, $(2, 5)$, and $(3, 14)$, we find that:

$$c = 5$$
$$4a + 2b + c = 5$$
$$9a + 3b + c = 14$$

Therefore $a = 3$, $b = -6$, $c = 5$ and hence $y = 3x^2 - 6x + 5$.

9. From the given information, $c = 2$ and $a = 3$. Therefore $b = \sqrt{3^2 - 2^2} = \sqrt{5}$ and hence the equation is

$$\frac{(x + 1)^2}{9} + \frac{(y - 1)^2}{5} = 1.$$

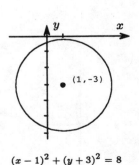

$(x - 1)^2 + (y + 3)^2 = 8$

Exercise 17

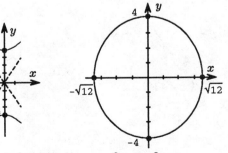

$y^2/16 - x^2/7 = 1$

Exercise 19

$x^2/12 + y^2/16 = 1$

Exercise 21

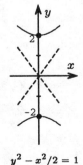

$y^2 - x^2/2 = 1$

Exercise 23

11. Since $(-4)^2 + 4(-3)^2 = 52 > 4$, the origin lies outside the ellipse.

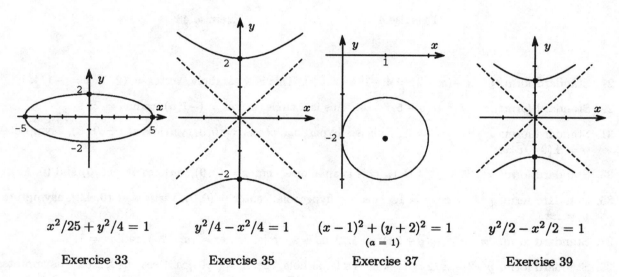

$x^2/9 - y^2/4 = 1$

Exercise 25

$y^2 = -(x - 2)$

Exercise 27

$(x + 7)^2 + (y - 5)^2 = 4$

Exercise 29

$x^2/8 - y^2/9 = 1$

Exercise 31

$x^2/25 + y^2/4 = 1$

Exercise 33

$y^2/4 - x^2/4 = 1$

Exercise 35

$(x - 1)^2 + (y + 2)^2 = 1$
$(a = 1)$

Exercise 37

$y^2/2 - x^2/2 = 1$

Exercise 39

13. $x = y + 2 \implies y = 4 - x^2 = 4 - (y + 2)^2 = 4 - y^2 - 4y - 4 \implies y(y + 5) = 0 \implies y = 0, -5.$ Hence the intersection points are $(2, 0)$ and $(-3, -5)$.

15. $6 - x^2 = x^2 - 2 \implies 2x^2 - 8 = 0 \implies x = \pm 2.$ Hence the intersection points are $(2, 2)$ and $(-2, 2)$.

17. Standard form: $(x - 1)^2 + (y + 3)^2 = 8.$ This is a circle, center $= (1, -3)$, radius $= \sqrt{8}$.

19. Standard form: $y^2/16 - x^2/7 = 1.$ This is a hyperbola, center $= (0, 0$, vertices at $(0, \pm 4)$, asymptotes $y = \pm(4/\sqrt{7})x$.

21. Standard form: $x^2/12 + y^2/16 = 1.$ This is an ellipse, center $= (0, 0)$, vertices at $(\pm\sqrt{12}, 0)$ and $(0, \pm 4)$.

23. Standard form: $y^2/4 - x^2/2 = 1.$ This is a hyperbola, center $= (0, 0)$, vertices at $(0, \pm 2)$, asymptotes $y = \pm\sqrt{2}\, x$.

25. Standard form: $x^2/9 - y^2/4 = 1.$ This is a hyperbola, center $= (0, 0)$, vertices at $(\pm 3, 0)$, asymptotes $y = \pm(2/3)x$.

473

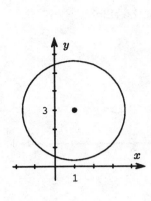

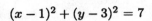

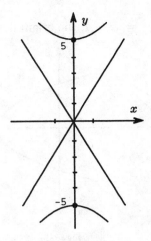

$$(x-1)^2 + (y-3)^2 = 7$$

Exercise 41

$$y^2/25 - x^2/9 = 1$$

Exercise 43

27. Standard form: $y^2 = -(x-2) = 4(-1/4)(x-2)$. This is a parabola, vertex $= (2,0)$, $c = -1/4$.

29. Standard form: $(x+7)^2 + (y-5)^2 = 4$. This is a circle, center $= (-7,5)$, radius $= 2$.

31. Standard form: $x^2/8 - y^2/9 = 1$. This is a hyperbola, center $= (0,0)$, vertices at $(\pm\sqrt{8}, 0)$, asymptotes $y = \pm(3/\sqrt{8})x$.

33. Standard form: $x^2/25 + y^2/4 = 1$. This is an ellipse, center $= (0,0)$, vertices at $(\pm 5, 0)$ and $(0, \pm 2)$.

35. Standard form: $y^2/4 - x^2/4 = 1$. This is a hyperbola, center $= (0,0)$, vertices at $(0, \pm 2)$, asymptotes $y = \pm x$.

37. Standard form: $(x-a)^2 + (y+2a)^2 = 1$. This is a circle, center $= (a, -2a)$, radius $= 1$.

39. Standard form: $y^2/2 - x^2/2 = 1$. This is a hyperbola, center $= (0,0)$, vertices at $(0, \pm\sqrt{2})$, asymptotes $y = \pm x$.

41. Standard form: $(x-1)^2 + (y-3)^2 = 7$. This is a circle, center $= (1,3)$, radius $= \sqrt{7}$.

43. Standard form: $y^2/25 - x^2/9 = 1$. This is a hyperbola, center $= (0,0)$, vertices at $(0, \pm 5)$, asymptotes $y = \pm(5/3)x$.

Chapter 15

Polar Coordinates and Parametric Equations

15.1   The Polar Coordinate System

1. $x = 1\cos(\pi/2) = 0$ and $y = 1\sin(\pi/2) = 1$, so the point is $(0, 1)$.

3. $x = 0\cos\pi = 0$ and $y = 0\sin\pi = 0$, so the point is $(0, 0)$.

5. $x = \sqrt{2}\cos(-\pi/4) = 1$ and $y = \sqrt{2}\sin(-\pi/4) = -1$, so the point is $(1, -1)$.

7. $x = -3\cos\pi = 3$ and $y = -3\sin\pi = 0$, so the point is $(3, 0)$.

9. $r^2 = 1^2 + 1^2 = 2$, and thus $r = \pm\sqrt{2}$. $\tan\theta = 1/1 = 1$ and $0 < \theta < \pi$ implies $\theta = \pi/4$. Since the point is in the first quadrant, the polar coordinates are $(\sqrt{2}, \pi/4)$.

11. $r^2 = (-3)^2 + 0^2 = 9$, and thus $r = \pm3$. $\tan\theta = 0/(-3) = 0$ and $-\pi/2 < \theta < \pi/2$ implies $\theta = 0$. Since the point is on the negative portion of the x-axis, the polar coordinates are $(-3, 0)$.

13. $r^2 = 1^2 + (-\sqrt{3})^2 = 4$, and thus $r = \pm2$. $\tan\theta = -\sqrt{3}/1 = -\sqrt{3}$ and $0 < \theta < \pi$ implies $\theta = 2\pi/3$. Since the point is in the fourth quadrant, the polar coordinates are $(-2, 2\pi/3)$.

15. $r^2 = (-2)^2 + 2^2 = 8$, and thus $r = \pm\sqrt{8} = \pm2\sqrt{2}$. $\tan\theta = 2/(-2) = -1$ and $3\pi < \theta < 4\pi$ implies $\theta = 15\pi/4$. Since the point is in the second quadrant, the polar coordinates are $(-2\sqrt{2}, 15\pi/4)$.

17. Symmetric about the x-axis, since if (r, θ) is on the graph, then $4\cos(\pi - \theta) = -4\cos\theta = -r$ implies that $(-r, \pi - \theta)$ is on the graph.

19. Symmetric about the y-axis, since if (r, θ) is on the graph, then $1 + \sin(\pi - \theta) = 1 + \sin\theta = r$ implies that $(r, \pi - \theta)$ is on the graph.

21. Symmetric about the x-axis, y-axis and origin, since if (r, θ) is on the graph, then $r^2 = \cos 2\theta = \cos 2(-\theta)$ implies that $(r, -\theta)$ is on the graph, $(-r)^2 = r^2 = \cos 2\theta = \cos 2(-\theta)$ implies that $(-r, -\theta)$ is on the graph, and $(-r)^2 = r^2 = \cos 2\theta$ implies that $(-r, \theta)$ is on the graph.

23. Symmetric about the x-axis, y-axis and origin, since if (r, θ) is on the graph, then $r = 2$ and thus $(r, -\theta)$, $(r, \pi - \theta)$ and $(r, \pi + \theta)$ are all on the graph. (Note that the graph is a circle of radius 2 centered at the origin.)

25. $(r\cos\theta)^2 + (r\sin\theta)^2 = 4$
$r^2(\cos^2\theta + \sin^2\theta) = 4$
$r^2 = 4$
$r = 2$
(Note $(-2, \theta)$ corresponds to $(2, -\theta)$.)

27. $4(r\cos\theta)^2 + (r\sin\theta)^2 = 4$
$r^2(4\cos^2\theta + \sin^2\theta) + 4$
$r^2 = 4/(4\cos^2\theta + \sin^2\theta)$, or
$r^2 = 4/(4 - 3\sin^2\theta)$, or
$r^2 = 4/(1 + 3\cos^2\theta)$
(using $\cos^2\theta + \sin^2\theta = 1$).

29. $(r\cos\theta)^2 + 2(r\sin\theta)^2 = 1$
$r^2(\cos^2\theta + 2\sin^2\theta) = 1$
$r^2 = 1/(\cos^2\theta + 2\sin^2\theta)$, or
$r^2 = 1/(1 + \sin^2\theta)$, or
$r^2 = 1/(2 - \cos^2\theta)$

31. $(r\cos\theta)^2 - (r\sin\theta)^2 = 1$
$r^2(\cos^2\theta - \sin^2\theta) = 1$
$r^2(\cos 2\theta) = 1$
$r^2 = \sec 2\theta$

33. $r\cos\theta = 6$
$r = 6\sec\theta$

35. $(r\cos\theta)^2 + (r\sin\theta)^2 + 2(r\sin\theta) = 0$
$r^2(\cos^2\theta + \sin^2\theta) + 2r\sin\theta = 0$
$r^2 = -2r\sin\theta$
$r = -2\sin\theta$

37. $r\sin\theta = (r\cos\theta)/2$
$r\tan\theta = r/2$
$\tan\theta = 1/2$
$\theta = \text{Tan}^{-1}(1/2) + n\pi$

39. $r = 4\sin\theta$
$r^2 = 4r\sin\theta$
$x^2 + y^2 = 4y$
$x^2 + y^2 - 4y = 0$

41. $r = \tan\theta$
$r\cos\theta = \sin\theta$
$r^2\cos\theta = r\sin\theta$
$r^4\cos^2\theta = r^2\sin^2\theta$
$(x^2 + y^2)x^2 = y^2$
$x^4 + x^2y^2 - y^2 = 0$

43. $r = 4\sec\theta$
$r\cos\theta = 4$
$x = 4$

45. $r = 1/(1 - \cos\theta)$
$r - r\cos\theta = 1$
$r = 1 + r\cos\theta$
$r^2 = (1 + r\cos\theta)^2$
$r^2 = 1 + 2r\cos\theta + r^2\cos^2\theta$
$x^2 + y^2 = 1 + 2x + x^2$
$y^2 = 2x + 1$

47. $r^2 = 4\sec\theta$
$r^2\cos\theta = 4$
$r^4\cos^2\theta = 16$
$(x^2 + y^2)x^2 = 16$
$x^4 + x^2y^2 = 16$

49.

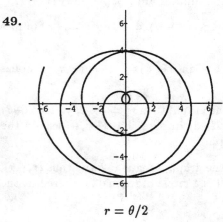

$r = \theta/2$

51.

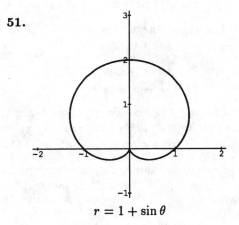

$r = 1 + \sin\theta$

53.

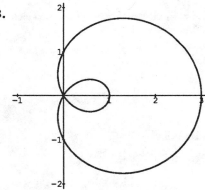

$$r = 1 + 2\cos\theta$$

55.

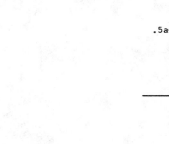

$$r = e^\theta,\ \theta \geq 0$$

57.

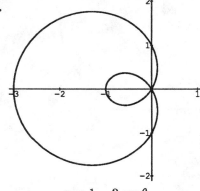

$$r = 1 - 2\cos\theta$$

59. Change to an equation in rectangular coordinates.
$r = 2\sin\theta - 2\cos\theta$
$r^2 = 2r\sin\theta - 2r\cos\theta$
$x^2 + y^2 = 2y - 2x$
$x^2 + 2x + 1 + y^2 - 2y + 1 = 2$
$(x + 1)^2 + (y - 1)^2 = 2$
Circle with center $(-1, 1)$ and radius $\sqrt{2}$.

61. $r = a\cos\theta$
$r^2 = ar\cos\theta$
$x^2 + y^2 = ax$
$x^2 - ax + (a^2/4) + y^2 = a^2/4$
$(x - a/2)^2 + y^2 = (a/2)^2$
This is the equation of a circle.

$$r = a\cos\theta,\ a > 0$$

63. $y = ax$
$r\sin\theta = ar\cos\theta$
Divide by $r\cos\theta$.
$\tan\theta = a$
Dividing by r does not eliminate a solution at $r = 0$ since $(0, \mathrm{Tan}^{-1} a)$ is a solution of the resulting equation. Dividing by $\cos\theta$ is also legal since no θ for which $\cos\theta = 0$ can be a solution to the equation.

65. This is the graph of a circle with equation $(x - 3/2)^2 + y^2 = (3/2)^2$, which has a radius of $3/2$. The area of the circle is therefore $\pi r^2 = 9\pi/4$.

67. In rectangular coordinates, these are the lines $y = 0$, $y = x$, and $x = 4$ respectively. The vertices are therefore $(0, 0)$, $(4, 0)$, and $(4, 4)$.

15.2 Graphing Techniques for Polar Equations

1.

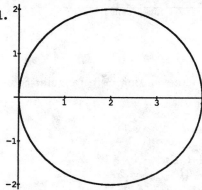

$r = 4\cos\theta$

3.

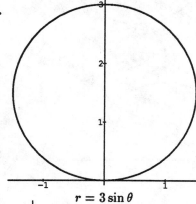

$r = 3\sin\theta$

5.

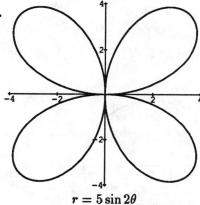

$r = 5\sin 2\theta$

7.

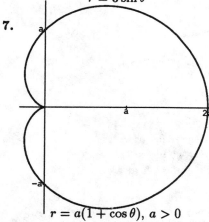

$r = a(1 + \cos\theta)$, $a > 0$

9.

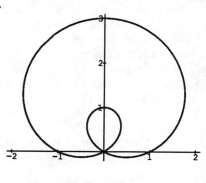

$r = 1 + 2\sin\theta$

11.

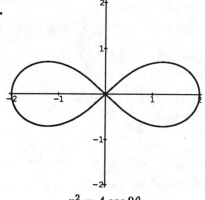

$r^2 = 4\cos 2\theta$

478

13.

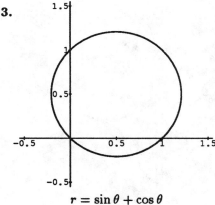

$$r = \sin\theta + \cos\theta$$

15.

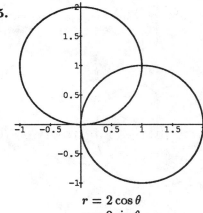

$$r = 2\cos\theta$$
$$r = 2\sin\theta$$

The graphs intersect at the origin $(0, \theta)$ since $r = 0$ satisfies both equations. For the other point use $2\cos\theta = 2\sin\theta$
$$1 = \tan\theta$$
$$\theta = \pi/4$$
$$r = 2\cos(\pi/4) = \sqrt{2}$$
$$(\sqrt{2}, \pi/4)$$

17.

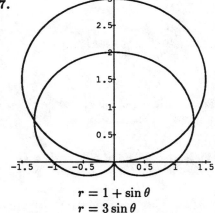

$$r = 1 + \sin\theta$$
$$r = 3\sin\theta$$

The graphs intersect at the origin $(0, \theta)$, since $r = 0$ satisfies the first equation at $\theta = -\pi/2$ and the second equation at $\theta = 0$. For the other points use $1 + \sin\theta = 3\sin\theta$
$$1 = 2\sin\theta$$
$$\sin\theta = 1/2$$
$$\theta = \pi/6 \text{ or } 5\pi/6$$
In either case $r = 3/2$
$(3/2, \pi/6)$ and $(3/2, 5\pi/6)$

19.

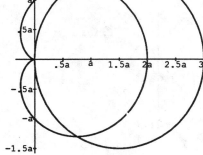

$$r = a(1 + \cos\theta)$$
$$r = 3a\cos\theta$$

The graphs intersect at the origin, since $r = 0$ satisfies the first equation at $\theta = \pi$ and the second equation at $\theta = \pi/2$. For the other points use $a(1 + \cos\theta) = 3a\cos\theta$
$$1 + \cos\theta = 3\cos\theta$$
$$1 = 2\cos\theta$$
$$\cos\theta = 1/2$$
$$\theta = \pm\pi/3$$
In either case, $r = 3a/2$
$(3a/2, \pi/3)$ or $(3a/2, -\pi/3)$

21.

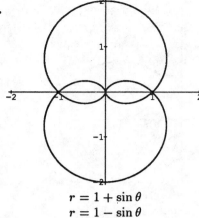

$$r = 1 + \sin\theta$$
$$r = 1 - \sin\theta$$

The graphs intersect at the origin, since $r = 0$ satisfies the first equation at $\theta = -\pi/2$ and the second equation at $\theta = \pi/2$. For the other points use $1 + \sin\theta = 1 - \sin\theta$

$2\sin\theta = 0$
$\sin\theta = 0$
$\theta = 0$ or π
In either case $r = 1$
$(1, 0)$ and $(1, \pi)$

23.

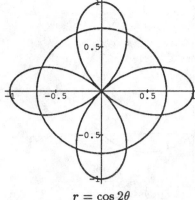

$$r = \cos 2\theta$$
$$r = \sqrt{2}/2$$

When the r-coordinates are equal or opposite in sign, one has:

$\cos 2\theta = \sqrt{2}/2$ $\cos 2\theta = -\sqrt{2}/2$
$2\theta = \pm\pi/4 + 2n\pi$ $2\theta = \pm 3\pi/4 + 2n\pi$
$\theta = \pm\pi/8 + n\pi$ $\theta = \pm 3\pi/8 + n\pi$

The eight points are therefore
$(\sqrt{2}/2, \pi/8), (\sqrt{2}/2, 3\pi/8), (\sqrt{2}/2, 5\pi/8), \ldots,$
$(\sqrt{2}/2, 15\pi/8)$.

25.

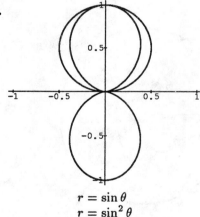

$$r = \sin\theta$$
$$r = \sin^2\theta$$

$\sin^2\theta = \sin\theta$
$\sin^2\theta - \sin\theta = 0$
$\sin\theta(\sin\theta - 1) = 0$
$\sin\theta = 0$ or $\sin\theta = 1$
$\theta = 0$ or $\theta = \pi/2$
$r = 0$ or $r = 1$ respectively
$(0, 0)$ and $(1, \pi/2)$

27. $r = a/(1 \pm \cos\theta)$
$r \pm r\cos\theta = a$
$r = a \mp r\cos\theta$
$r^2 = (a \mp r\cos\theta)^2$
$r^2 = a^2 \mp 2ar\cos\theta + (r\cos\theta)^2$
$x^2 + y^2 = a^2 \mp 2ax + x^2$
$\mp 2ax = y^2 - a^2$
$x = \mp(y^2 - a^2)/2a$
This is a parabola which opens to the right or left, depending on the sign of a.

$r = a/(1 \pm \sin\theta)$
This is the same as the above, with $\cos\theta$ replaced by $\sin\theta$, and therefore with x and y exchanged. The rectangular equation is $y = \mp(x^2 - a^2)/2a$. This is a parabola which opens upward or downward, depending on the sign of a.

29. $y = 1/8 - 2x^2 = -2(x^2 - 1/16) = -\frac{1}{2(1/4)}(x^2 - (1/4)^2)$
$a = 1/4$
$r = 1/(4 + 4\sin\theta)$

480

31. The coefficient of y^2 is $3/4 = (1 - b^2)/a^2 b^2$. The coefficient of the x term is $9/16 = (1 - b^2)^2/a^2 b^2 = (3/4)(1 - b^2)$, so $3/4 = (1 - b^2)$ and $b = 1/2$. The number $2/3$ subtracted from x must be $ab^2/(1 - b^2)$, and replacing b by $1/2$ we get $2/3 = a/3$ and so $a = 2$. Using these values for a and b in the equation derived in Exercise 28 produces the equation in this exercise. Therefore, the equation in polar coordinates is $r = \frac{1}{1 - (1/2)\cos\theta}$.

33.

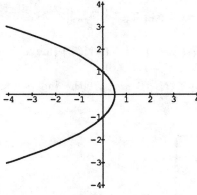

$$r = 1/(1 + \cos\theta)$$

Parabola, $a = 1$
$x = -(1/2)(y^2 - 1)$

35.

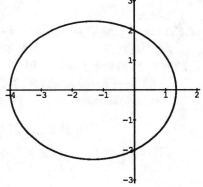

$$r = \frac{2}{1 + (1/2)\cos\theta}$$

Ellipse, $b = 1/2$, $a = 4$
$(9/64)(x + (4/3))^2 + (3/16)y^2 = 1$

37.

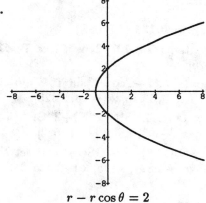

$$r - r\cos\theta = 2$$

$r = 2/(1 - \cos\theta)$
Parabola, $a = 2$
$x = (1/4)(y^2 - 4)$

For all of these exercises, except where indicated, the calculator should be in the radian mode. The range for these graphs was set using the **RANGE** programs of Appendix V.

39. Using **RANGE**, **X=0**, **Y=0**, and **S=0.1**.

Casio: Use **PTPLOT**, **A=0**, **B=2π**, **N=101**. Program location **0** contains **(1+tan T)cos T→ X**: **(1+tan T)sin T→ Y**. Range: $[-4.7, 4.7] \times [-3.1, 3.1]$.

TI-81: Use **PARAM** mode and **X₁ᴛ=(1+tan T)cos T**; **Y₁ᴛ=(1+tan T)sin T**, **Tmin=0**, **Tmax=6.283185307**(2π), **Tstep=0.063821853**, (2π/100). Range: $[-4.8, 4.7] \times [-3.2, 3.1]$.

HP-28S: Use **POLPLT** and on the stack before execution should be: Level 4: **'1+TAN(T)'**; Level 3: **0**; Level 2: **6.28318530718**; Level 1: **0.06283185307**. Range: $[-6.8, 6.8] \times [-1.5, 1.6]$.

HP-48SX: Use the **POLAR** plot type. The equation should be **'1+TAN(θ)'** and set the independent variable to be '**θ**' and its range as 0, 6.283185307. Range $[-6.5, 6.5] \times [-3.1, 3.2]$.

Graph:

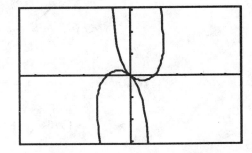

41. Using **RANGE**, **X=0**, **Y=0**, and **S=0.075** (Casio, TI-81, HP-48SX), **S=0.125** (HP-28S).

Casio: Use **PTPLOT**, **A=0**, **B=2π**, **N=101**. Program location **0** contains **(sin 2T+2cos T)cos T→ X**: **(sin 2T+2cos T)sin T→ Y**. Range: $[-3.525, 3.525] \times [-2.325, 2.325]$.

TI-81: Use **PARAM** mode and **X₁ᴛ=(sin 2T+2cos T)cos T**; **Y₁ᴛ=(sin 2T+2cos T)sin T**, **Tmin=0**, **Tmax=6.283185307**(2π), **Tstep=0.063821853**, (2π/100). Range: $[-3.6, 3.525] \times [-2.4, 2.325]$.

HP-28S: Use **POLPLT** and on the stack before execution should be: Level 4: **'SIN(2*T)+2*COS(T)'**; Level 3: **0**; Level 2: **6.28318530718**; Level 1: **0.06283185307**. Range: $[-8.5, 8.5] \times [-1.875, 2]$.

HP-48SX: Use the **POLAR** plot type. The equation should be **'SIN(2*θ)+2*COS(θ)'** and set the independent variable to be '**θ**' and its range as 0, 6.283185307. Range $[-4.875, 4.875] \times [-2.325, 2.4]$.

Graph:

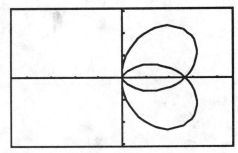

43. In all cases take $0 \le \theta \le 2\pi$. Range settings: **RANGE** program with **X=Y=0**, **S=0.1** (Casio, TI-81, HP-48SX), **S=0.12** (HP-28S). Casio: $[-4.7, 4.7] \times [-3.1, 3.1]$, TI-81: $[-4.8, 4.7] \times [-3.2, 3.1]$, HP-28S: $[-13.6, 13.6] \times [-3, 3.2]$, HP-48SX: $[-6.5, 6.5] \times [-3.1, 3.2]$. Below are the graphs of $r = 2 + 2\cos\theta$, $r = 1 - 2\cos\theta$, $r = 2 - \cos\theta$, and $r = 2 + \frac{7}{4}\cos\theta$.

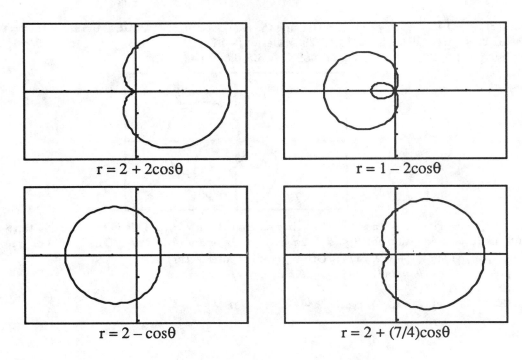

$$r = 2 + 2\cos\theta \qquad r = 1 - 2\cos\theta$$

$$r = 2 - \cos\theta \qquad r = 2 + (7/4)\cos\theta$$

45. Graph $r = \theta/2$, $0 \le \theta \le 35$ using the range settings: **RANGE** program with **X=Y=0**, **S=0.5** (Casio, TI-81, HP-48SX), **S=1** (HP-28S). Casio: $[-23.5, 23.5] \times [-15.5, 15.5]$, TI-81: $[-24, 23.5] \times [-16, 15.5]$, HP-28S: $[-68, 68] \times [-15, 16]$, HP-48SX: $[-32.5, 32.5] \times [-15.5, 16]$.

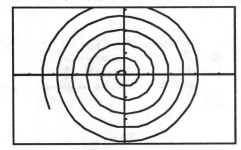

47. Graph $r = 5/\theta$, $0.01 \le \theta \le 20$ using the range settings: **RANGE** program with **X=Y=0**, **S=0.15** (Casio, TI-81, HP-48SX), **S=0.3** (HP-28S). Casio: $[-7.05, 7.05] \times [-4.65, 4.65]$, TI-81: $[-7.2, 7.05] \times [-4.8, 4.65]$, HP-28S: $[-20.4, 20.4] \times [-4.5, 4.8]$, HP-48SX: $[-9.75, 9.75] \times [-4.65, 4.8]$.

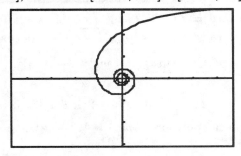

49. Graph of $r_1 = 1 + 2\cos\theta$, $r_2 = 1 - 2\sin\theta$ using the range settings: **RANGE** program with **X=Y=0, S=0.1** (Casio, TI-81, HP-48SX), **S=0.2** (HP-28S). Casio: $[-4.7, 4.7] \times [-3.1, 3.1]$, TI-81: $[-4.8, 4.7] \times [-3.2, 3.1]$, HP-28S: $[-13.6, 13.6] \times [-3, 3.2]$, HP-48SX: $[-6.5, 6.5] \times [-3.1, 3.2]$.

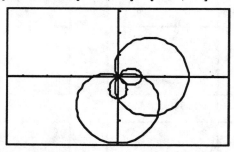

Setting $r_1 = r_2$ gives $1 + 2\cos\theta = 1 - 2\sin\theta$ so $\cos\theta + \sin\theta = 0$. This is the same as $\tan\theta = -1$. The solutions to this equation for $0 \leq \theta \leq 2\pi$ are $\theta = 3\pi/4$ and $\theta = 7\pi/4$. The points of intersection are $(1 - \sqrt{2}, 3\pi/4)$ and $(1 + \sqrt{2}, 7\pi/4)$. Other points of intersection are:

pole: r_1; $\theta = 2\pi/3, 4\pi/3$, r_2; $\theta = \pi/6, 5\pi/6$

(0,1): r_1; $(\pi, -1)$, r_2; $(2\pi, 1)$

$(3\pi/2, 1)$: r_1; $(3\pi/2, 1)$, r_2; $(\pi/2, -1)$.

51. Graph of $r_1 = 1 + \sin\theta$ and $r_2 = \sin 2\theta$ using the range settings: **RANGE** program with **X=Y=0, S=0.05** (Casio, TI-81, HP-48SX), **S=0.** (HP-28S). Casio: $[-2.35, 2.35] \times [-1.55, 1.55]$, TI-81: $[-2.4, 2.35] \times [-1.6, 1.55]$, HP-28S: $[-6.8, 6.8] \times [-1.5, 1.6]$, HP-48SX: $[-3.25, 3.25] \times [-1.55, 1.6]$.

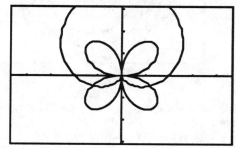

Setting $r_1 = r_2$ gives $1 + \sin\theta = \sin 2\theta$. Using a double angle formula for the sine functions gives $1 + \sin\theta = 2\sin\theta\cos\theta$. Replace now $\cos\theta$ with $\sqrt{1 - \sin^2\theta}$ and squaring both sides gives the equation

$$(1 + \sin\theta)^2 = 4\sin^2\theta(1 - \sin\theta)^2.$$

Simplifying gives

$$4\sin^4\theta - 3\sin^2\theta + 2\sin\theta + 1 = 0.$$

If we set $x = \sin\theta$ and graph the polynomial $p(x) = 4x^4 - 3x^2 + 2x + 1$ using the range $[-2, 2] \times [-5, 5]$ gives

484

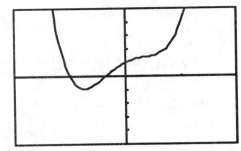

We see that this polynomial has two real roots; $x_1 = -1$ and $x_2 = -0.3478103848$. Therefore $\sin \theta = -0.3478103848$, or $\theta = \text{Sin}^{-1}(-0.3478103848) = -0.3552346611$. From the graph of $r_1 = 1 + \sin \theta$ and $r_2 = \sin 2\theta$ we see that one point of intersection is in the third quadrant so this point is $(0.6521896152\pi + 0.3552346611) = (0.6521896152, 3.496827315)$. The other point of intersection is in the fourth quadrant so for r_1 this point is $(0.6521896152, 2\pi - 0.3552346611) = (0.6521896152, 5.927950646)$ and for r_2 this point is $(-0.6521896152, \pi - 0.3552346611) = (0.6521896152, -2.786357993)$. The other point of intersection is the pole: r_1, $(0, 3\pi/2)$ and r_2, $(0, 0)$, $(0, \pi/2)$, $(0, \pi)$, $(0, 3\pi/2)$.

15.3   Calculating Area in Polar Coordinates

1. $\int_{-\pi/2}^{\pi/2} (1/2)(1 + \sin \theta)^2 \, d\theta$

$= \int_{-\pi/2}^{\pi/2} (1/2)(1 + 2 \sin \theta + \sin^2 \theta) \, d\theta$

$= \int_{-\pi/2}^{\pi/2} (1/2) \, d\theta + \int_{-\pi/2}^{\pi/2} \sin \theta \, d\theta + \int_{-\pi/2}^{\pi/2} (1/2)(1/2)(1 - \cos(2\theta)) \, d\theta$

$= (1/2)\theta - \cos \theta + (1/4)\theta - (1/8) \sin(2\theta)|_{-\pi/2}^{\pi/2}$

$= (\pi/4 - 0 + \pi/8 - 0) - (-\pi/4 - 0 - \pi/8 - 0)$

$= 3\pi/4$

3. $\int_0^\pi (1/2)(2 \sin \theta)^2 \, d\theta$

$= 2 \int_0^\pi \sin^2 \theta \, d\theta$

$= 2 \int_0^\pi (1/2)(1 - \cos 2\theta) \, d\theta$

$= \theta - (1/2) \sin 2\theta|_0^\pi$

$= (\pi - 0) - (0 - 0)$

$= \pi$

5. $\int_0^\pi (1/2)(2 + \sin \theta)^2 \, d\theta$

$= \int_0^\pi (1/2)(4 + 4 \sin \theta + \sin^2 \theta) \, d\theta$

$= \int_0^\pi 2 \, d\theta + 2 \int_0^\pi \sin \theta \, d\theta + (1/2) \int_0^\pi (1/2)(1 - \cos 2\theta) \, d\theta$

$= 2\theta - 2 \cos \theta + (1/4)\theta - (1/8) \sin 2\theta|_0^\pi$

$= (2\pi - 2(-1) + (1/4)\pi - 0) - (0 - 2 + 0 - 0)$

$= 4 + 9\pi/4$

7. $\int_{\pi/4}^{3\pi/4}(1/2)(4+\sin\theta)^2\,d\theta$

$= \int_{\pi/4}^{3\pi/4}(1/2)(16+8\sin\theta+\sin^2\theta)\,d\theta$

$= \int_{\pi/4}^{3\pi/4}8\,d\theta + 4\int_{\pi/4}^{3\pi/4}\sin\theta\,d\theta + (1/2)\int_{\pi/4}^{3\pi/4}(1/2)(1-\cos 2\theta)\,d\theta$

$= 8\theta - 4\cos\theta + (1/4)\theta - (1/8)\sin 2\theta\Big|_{\pi/4}^{3\pi/4}$

$= (6\pi - 4(-\sqrt{2}/2) + 3\pi/16 - (1/8)(-1)) - (2\pi - 4(\sqrt{2}/2) + \pi/6 - (1/8))$

$= 4\pi + 4\sqrt{2} + \pi/8 + 1/4$

$= 33\pi/8 + 4\sqrt{2} + 1/4$

9. $\int_{-\pi/2}^{\pi/2}(1/2)(2\cos\theta)^2\,d\theta$

$= 2\int_{-\pi/2}^{\pi/2}\cos^2\theta\,d\theta$

$= 2\int_{-\pi/2}^{\pi/2}(1/2)(1+\cos 2\theta)\,d\theta$

$= \theta + (1/2)\sin 2\theta\Big|_{-\pi/2}^{\pi/2}$

$= (\pi/2 + 0) - (-\pi/2 + 0)$

$= \pi$

11. $\int_0^{2\pi}(1/2)(1+\sin\theta)^2\,d\theta$

$= \int_0^{2\pi}(1/2)(1+2\sin\theta+\sin^2\theta)\,d\theta$

$= \int_0^{2\pi}(1/2)\,d\theta + \int_0^{2\pi}\sin\theta\,d\theta + (1/2)\int_0^{2\pi}(1/2)(1-\cos 2\theta)\,d\theta$

$= (1/2)\theta - \cos\theta + (1/4)\theta - (1/8)\sin 2\theta\Big|_0^{2\pi}$

$= (\pi - 1 + \pi/2 - 0) - (0 - 1 + 0 - 0)$

$= 3\pi/2$

13. Compute the area of one leaf and multiply by 4.

$4\int_0^{\pi/2}(1/2)(4\sin 2\theta)^2\,d\theta$

$= 32\int_0^{\pi/2}\sin^2(2\theta)\,d\theta$

$= 32\int_0^{\pi/2}(1/2)(1-\cos 2\theta)\,d\theta$

$= 16(\theta - (1/2)\sin 2\theta)\Big|_0^{\pi/2}$

$= 16(\pi/2 - 0) - 16(0 - 0)$

$= 8\pi$

15. Compute the area of one leaf and multiply by 4.

$4\int_{-\pi/4}^{\pi/4}(1/2)(\cos 2\theta)^2\,d\theta$

$= 2\int_{-\pi/4}^{\pi/4}\cos^2(2\theta)\,d\theta$

$= 2\int_{-\pi/4}^{\pi/4}(1/2)(1+\cos 4\theta)\,d\theta$

$= \theta + (1/4)\sin 4\theta\Big|_{-\pi/4}^{\pi/4}$

$= (\pi/4 + 0) - (-\pi/4 + 0)$

$= \pi/2$

17. Compute the area of one leaf and multiply by 8.

$8\int_0^{\pi/4}(1/2)(a\sin 4\theta)^2\,d\theta$

$= 4a^2\int_0^{\pi/4}\sin^2(4\theta)\,d\theta$

$= 4a^2\int_0^{\pi/4}(1/2)(1-\cos 8\theta)\,d\theta$

$= 2a^2(\theta - (1/8)\sin 8\theta)\Big|_0^{\pi/4}$

$= 2a^2(\pi/4 - 0) - 2a^2(0 - 0)$

$= a^2\pi/2$

19. Compute the area for $r = +\sqrt{\cos\theta}$, then multiply by 2.

$2\int_{-\pi/2}^{\pi/2}(1/2)(\sqrt{\cos\theta})^2\,d\theta$

$= \int_{-\pi/2}^{\pi/2}\cos\theta\,d\theta$

$= \sin\theta\Big|_{-\pi/2}^{\pi/2}$

$= 1 - (-1)$

$= 2$

21. Compute the area for each, and then subtract.

$\int_0^{2\pi}(1/2)(1-\cos\theta)^2\,d\theta - \int_{\pi/2}^{3\pi/2}(1/2)(-2\cos\theta)^2\,d\theta$

$= \int_0^{2\pi}(1/2)(1-2\cos\theta+\cos^2\theta)\,d\theta - 2\int_{\pi/2}^{3\pi/2}\cos^2\theta\,d\theta$

$= \int_0^{2\pi}(1/2)\,d\theta - \int_0^{2\pi}\cos\theta\,d\theta + (1/2)\int_0^{2\pi}(1/2)(1+\cos2\theta)\,d\theta - 2\int_{\pi/2}^{3\pi/2}(1/2)(1+\cos2\theta)\,d\theta$

$= ((1/2)\theta - \sin\theta + (1/4)\theta + (1/8)\sin2\theta)|_0^{2\pi} - (\theta + (1/2)\sin2\theta)|_{\pi/2}^{3\pi/2}$

$= [(\pi - 0 + \pi/2 + 0) - (0 - 0 + 0 + 0)] - [(3\pi/2 + 0) - (\pi/2 + 0)]$

$= 3\pi/2 - \pi$

$= \pi/2$

23. Compute the area of the half of the upper region which lies to the right of the y-axis, and then multiply by 4.

$4\int_0^{\pi/2}(1/2)a^2(1-\cos\theta)^2\,d\theta$

$= 2a^2\int_0^{\pi/2}(1-2\cos\theta+\cos^2\theta)\,d\theta$

$= 2a^2\int_0^{\pi/2}d\theta - 4a^2\int_0^{\pi/2}\cos\theta\,d\theta + 2a^2\int_0^{\pi/2}(1/2)(1+\cos2\theta)\,d\theta$

$= 2a^2\theta - 4a^2\sin\theta + a^2\theta + (a^2/2)\sin2\theta|_0^{\pi/2}$

$= (a^2\pi - 4a^2 + a^2\pi/2 + 0) - (0 - 0 + 0 + 0)$

$= 3a^2\pi/2 - 4a^2$ or $a^2(3\pi/2 - 4)$

25. The circles are symmetric across the line $y = x$ or $\theta = \pi/4$. Compute the area between the graph of $\theta = \pi/4$ and the circle $r = 2\sin\theta$ and then multiply by 2.

$2\int_0^{\pi/4}(1/2)(2\sin\theta)^2\,d\theta$

$= 4\int_0^{\pi/4}\sin^2\theta\,d\theta$

$= 4\int_0^{\pi/4}(1/2)(1-\cos2\theta)\,d\theta$

$= 2\theta - \sin2\theta|_0^{\pi/4}$

$= (\pi/2 - 1) - (0 - 0)$

$= \pi/2 - 1$

27. The graphs intersect at $\theta = \pi/6, 5\pi/6, 7\pi/6$ and $11\pi/6$. Compute the area by first computing the area of the half of the right region which is above the x-axis, and then multiplying by 4. The area of this region should be computed by splitting it into a region bounded by the circle on the interval $[0, \pi/6]$, and a region bounded by the lemniscate on the interval $[\pi/6, \pi/4]$.

$4\left[\int_0^{\pi/6}(1/2)(\sqrt{2}/2)^2\,d\theta + \int_{\pi/6}^{\pi/4}(1/2)(\sqrt{\cos2\theta})^2\,d\theta\right]$

$= 2\int_0^{\pi/6}(1/2)\,d\theta + 2\int_{\pi/6}^{\pi/4}\cos2\theta\,d\theta$

$= \theta|_0^{\pi/6} + \sin2\theta|_{\pi/6}^{\pi/4}$

$= (\pi/6 - 0) + (1 - \sqrt{3}/2)$

$= 1 + \pi/6 - \sqrt{3}/2$

29. Compute the area of each, then subtract.

$\int_0^{2\pi}(1/2)(3+\sin\theta)^2\,d\theta - \int_0^{\pi}(1/2)(4\sin\theta)^2\,d\theta$

$= \int_0^{2\pi}(1/2)(9+6\sin\theta+\sin^2\theta)\,d\theta - 8\int_0^{\pi}\sin^2\theta\,d\theta$

$= \int_0^{2\pi}(9/2)\,d\theta + 3\int_0^{2\pi}\sin\theta\,d\theta + (1/2)\int_0^{2\pi}(1/2)(1-\cos2\theta)\,d\theta - 8\int_0^{\pi}(1/2)(1-\cos2\theta)\,d\theta$

$= ((9/2)\theta - 3\cos\theta + (1/4)\theta - (1/8)\sin2\theta)|_0^{2\pi} - (4\theta - 2\sin2\theta)|_0^{\pi}$

$= [(9\pi - 3 + \pi/2 - 0) - (0 - 3 + 0 - 0)] - [(4\pi - 0) - (0 - 0)]$

$= 11\pi/2$

31. Solve for r to get $r = 2/(1 - \cos\theta)$. By Exercise 27 in the previous section, the graph of this equation is a parabola which opens to the right, whose equation in rectangular coordinates is $x = y^2/4 - 1$. We want the area of the region bounded by this parabola to the left of the y-axis ($\theta = \pi/2$). The y-intercepts are $y = \pm 2$.

$$\int_{-2}^{2} |y^2/4 - 1| \, dy$$
$$= \int_{-2}^{2} (1 - y^2/4) \, dy$$
$$= y - y^3/12 \Big|_{-2}^{2}$$
$$= (2 - 2/3) - (-2 + 2/3)$$
$$= 8/3$$

33. The difference between the values of the integrals is:

$$\int_a^b (1/2)[f(\theta) - g(\theta)]^2 \, d\theta - \int_a^b (1/2)[f(\theta)^2 - g(\theta)^2] \, d\theta$$
$$= (1/2) \int_a^b \left([f(\theta) - g(\theta)]^2 - [f(\theta)^2 - g(\theta)^2]\right) d\theta$$
$$= (1/2) \int_a^b [f(\theta)^2 - 2f(\theta)g(\theta) + g(\theta)^2 - f(\theta)^2 + g(\theta)^2] \, d\theta$$
$$= (1/2) \int_a^b [2g(\theta)^2 - 2f(\theta)g(\theta)] \, d\theta$$
$$= \int_a^b [g(\theta)^2 - f(\theta)g(\theta)] \, d\theta.$$

The value of the integral will not in general be 0, and therefore the values of the two integrals with which we started will not in general be the same.

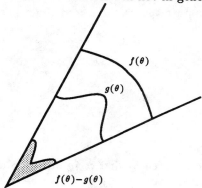

35. Rewrite the equation $r^4 + b^4 - 2r^2 b^2 \cos 2\theta = k^4$ as $r^4 - (2b^2 \cos 2\theta)r^2 + (b^4 - k^4) = 0$ and using the quadratic formula gives

$$r^2 = \frac{1}{2}\left(2b^2 \cos 2\theta \pm \sqrt{4b^4 \cos^2 2\theta - 4(b^4 - k^4)}\right)$$
$$= b^2 \cos 2\theta \pm \sqrt{b^4(\cos^2 2\theta - 1) + k^4}$$
$$= b^2 \cos 2\theta \pm \sqrt{k^4 - b^4 \sin^2 2\theta}.$$

Therefore

$$r = \pm\sqrt{b^2 \cos 2\theta \pm \sqrt{k^4 - b^4 \sin^2 2\theta}}.$$

If $b = 2$ and $k = 17/8 = 2.125$ then by using symmetry we only need to graph the equation

$$r = \sqrt{4 \cos 2\theta + \sqrt{20.39086914 - 16 \sin^2 2\theta}}.$$

using the range settings: **RANGE** program with **X=Y=0**, **S=0.07** (Casio, TI-81, HP-48SX), **S=0.14** (HP-28S). Casio: $[-3.29, 3.29] \times [-2.17, 2.17]$, TI-81: $[-3.36, 3.29] \times [-2.24, 2.17]$, HP-28S: $[-9.52, 9.52] \times [-2.1, 2.24]$, HP-48SX: $[-4.55, 4.55] \times [-2.17, 2.24]$. Since the expressions under the radicals are all positive nothing special is needed for the Casio or the HP-28S.

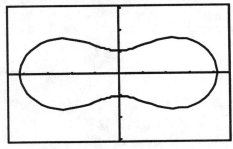

The area bounded by this curve is given by

$$A = 4 \int_0^{\pi/2} 4\cos 2\theta + \sqrt{20.39086914 - 16\sin^2 2\theta} \, d\theta.$$

Using Simpson's Rule with $n = 20$ gives $A \approx 21.4712$.

15.4 Parametric Equations in the Plane

1.

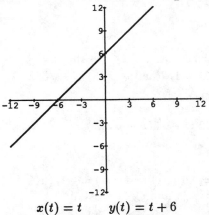

$$x(t) = t \qquad y(t) = t + 6$$
$$y = x + 6$$

3.

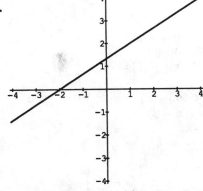

$$x(t) = 1 + 3t \quad y(t) = 2t + 2$$
$$t = (x-1)/3$$
$$y = 2(x-1)/3 + 2$$
$$y = 2x/3 + 4/3$$

5.

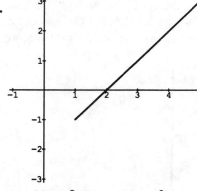

$$x(t) = t^2 + 1 \quad y(t) = t^2 - 1$$
$$t^2 = x - 1$$
$$y = (x - 1) - 1$$
$$y = x - 2, \; x \geq 1$$

7.

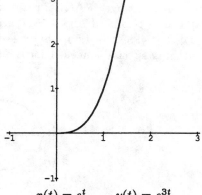

$$x(t) = e^t \qquad y(t) = e^{3t}$$
$$y = (e^t)^3$$
$$y = x^3, \; x > 0$$

9.

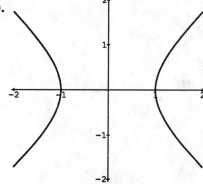

$$x(t) = \sec t \qquad y(t) = \tan t$$
$$y^2 + 1 = \tan^2 t + 1 = \sec^2 t = x^2$$
$$x^2 - y^2 = 1$$

11.

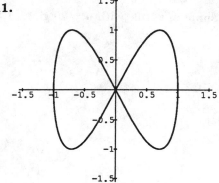

$$x(t) = \sin t \qquad y(t) = \sin 2t$$
$$y = 2 \sin t \cos t$$
$$y^2 = 4 \sin^2 t \cos^2 t = 4 \sin^2 t (1 - \sin^2 t)$$
$$y^2 = 4x^2(1 - x^2)$$
$$y^2 = 4x^2 - 4x^4$$

13. $x'(t) = 1$ and $y'(t) = 2t$
$x'(2) = 1$ and $y'(2) = 4$
$m = 4$
$x(2) = 2$ and $y(2) = 5$
$y - 5 = 4(x - 2)$
$y = 4x - 3$

15. $x'(t) = \cos t$ and $y'(t) = -\sin(t)$
$x'(\pi/3) = 1/2$ and $y'(\pi/3) = -\sqrt{3}/2$
$m = -\sqrt{3}$
$x(\pi/3) = \sqrt{3}/2$ and $y(\pi/3) = 1/2$
$y - 1/2 = -\sqrt{3}(x - \sqrt{3}/2)$
$y = -\sqrt{3}x + 2$

17. $x'(t) = -1/t^2$ and $y'(t) = 6t$
$x'(2) = -1/4$ and $y'(2) = 12$
$m = -48$
$x(2) = 1/2$ and $y(2) = 5$
$y - 5 = -48(x - 1/2)$
$y = -48x + 29$

19. $x'(t) = 1/(2\sqrt{t})$ and $y'(t) = -1/(2\sqrt{t})$
$x'(4) = 1/4$ and $y'(4) = -1/4$
$m = -1$
$x(4) = 3$ and $y(4) = -1$
$y + 1 = -1(x - 3)$
$y = -x + 2$

21. $(x, y) = (8, 4)$ at $t = 2$
$x'(t) = 3t^2$ and $y'(t) = 2t$
$x'(2) = 12$ and $y'(2) = 4$
$m = 1/3$
$y - 4 = (1/3)(x - 8)$
$y = x/3 + 4/3$

23. $(x, y) = (e, 0)$ at $t = 1$
$x'(t) = e^t$ and $y'(t) = 2/t$
$x'(1) = e$ and $y'(1) = 2$
$m = 2/e$
$y - 0 = (2/e)(x - e)$
$y = 2x/e - 2$

25. $x(\theta) = \cos^2 \theta$ and $y(\theta) = \cos \theta \sin \theta$
$x'(\theta) = -2 \cos \theta \sin \theta$ and $y'(\theta) = \cos^2 \theta - \sin^2 \theta$
$x'(\pi/4) = -1$ and $y'(\pi/4) = 0$
$m = 0$

27. $x(\theta) = a \sin 2\theta \cos \theta$ and $y(\theta) = a \sin 2\theta \sin \theta$
$x'(\theta) = 2a \cos 2\theta \cos \theta - a \sin 2\theta \sin \theta$ and $y'(\theta) = 2a \cos 2\theta \sin \theta + a \sin 2\theta \cos \theta$
$x'(\pi/6) = a\sqrt{3}/4$ and $y'(\pi/6) = 5a/4$
$m = 5/\sqrt{3}$

29. $x(\theta) = \theta \cos \theta$ and $y(\theta) = \theta \sin \theta$
$x'(\theta) = \cos \theta - \theta \sin \theta$
$y'(\theta) = \sin \theta + \theta \cos \theta$
$x'(\pi/2) = -\pi/2$ and $y'(\pi/2) = 1$
$m = -2/\pi$

31. $x'(t) = -\sin t$ and $y'(t) = \cos t$
$0 = x'(t) = -\sin t$ when $t = n\pi$
$0 = y'(t) = \cos t$ when $t = \pi/2 + n\pi$
a. $t = n\pi$
$(1, 0)$ and $(-1, 0)$
b. $t = \pi/2 + n\pi$
$(0, 1)$ and $(0, -1)$

33. $x'(t) = 2t$ and $y'(t) = 6t - 6$
$0 = x'(t) = 2t$ when $t = 0$
$0 = y'(t) = 6t - 6$ when $t = 1$
a. $t = 0$
$(4, 2)$
b. $t = 1$
$(5, -1)$

35. $x'(t) = 6t$ and $y'(t) = 1 - 2t$
$0 = x'(t) = 6t$ when $t = 0$
$0 = y'(t) = 1 - 2t$ when $t = 1/2$
a. $t = 0$
$(6, 0)$
b. $t = 1/2$
$(27/4, 1/4)$

37. The line C_2 has slope $y'(t)/x'(t) = 4/1 = 4$.
For C_1: $x'(t) = 1$ and $y'(t) = 4t$
$4 = y'(t)/x'(t) = 4t$
$t = 1$
$(1, 5)$

39. $x(\theta) = a \sin 3\theta \cos \theta$
$y(\theta) = a \sin 3\theta \sin \theta$

41. $x(\theta) = (1 + \sin \theta) \cos \theta$ and $y(\theta) = (1 + \sin \theta) \sin \theta$
A vertical tangent occurs when $x'(\theta) = 0$ and $y'(\theta) \neq 0$.
$x'(\theta) = \cos^2 \theta - \sin \theta - \sin^2 \theta$ and $y'(\theta) = \cos \theta + 2 \sin \theta \cos \theta$
$x'(\theta) = (1 - \sin^2 \theta) - \sin \theta - \sin^2 \theta = 0$
$-2 \sin^2 \theta - \sin \theta + 1 = 0$
$(-2 \sin \theta + 1)(\sin \theta + 1) = 0$
$\sin \theta = 1/2$ or $\sin \theta = -1$
$\theta = \pi/6 + 2n\pi,\ 5\pi/6 + 2n\pi$ or $3\pi/2 + 2n\pi$
Now check whether $y'(\theta) \neq 0$.
$y'(3\pi/2 + 2n\pi) = 0$ for all n
Vertical tangents occur only at $\theta = \pi/6 + 2n\pi,\ 5\pi/6 + 2n\pi$
These are the points $(r, \theta) = (3/2, \pi/6)$ and $(3/2, 5\pi/6)$

43. $x(\theta) = (1 + \cos\theta)\cos\theta$ and $y(\theta) = (1 + \cos\theta)\sin\theta$
A vertical tangent occurs when $x'(\theta) = 0$ and $y'(\theta) \neq 0$.
$x'(\theta) = -\sin\theta - 2\sin\theta\cos\theta$ and $y'(\theta) = -\sin^2\theta + \cos\theta + \cos^2\theta$
$x'(\theta) = -\sin\theta - 2\sin\theta\cos\theta = 0$
$-\sin\theta(1 + 2\cos\theta) = 0$
$\sin\theta = 0$ or $\cos\theta = -1/2$
$\theta = n\pi,\ 2\pi/3 + 2n\pi,$ or $4\pi/3 + 2n\pi$
Now check whether $y'(\theta) \neq 0$.
$y'(\pi + 2n\pi) = 0$ for all n
Vertical tangents occur only at $\theta = 2n\pi,\ 2\pi/3 + 2n\pi,\ 4\pi/3 + 2n\pi$
These are the points $(r, \theta) = (2, 0),\ (1/2, 2\pi/3),$ and $(1/2, 4\pi/3)$

45. a. $y_1 = 8 - (x_1 - 4)^2 = -x_1^2 + 8x_1 - 8$
$y_2 = (x_2 - 4) + 6 = x_2 + 2$

b. Set $y_1 = y_2$ to determine intersections.
$-x^2 + 8x - 8 = x + 2$
$x^2 - 7x + 10 = 0$
$(x - 5)(x - 2) = 0$
Paths cross at $(2, 4),\ (5, 7)$
If $t \geq 0$, then paths only cross at $(5, 7)$

c. Each particle would be at $(2, 4)$ at $t = -2$, and at $(5, 7)$ at $t = 1$. Assuming that $t \geq 0$, the particles collide at time $t = 1$, at the point $(5, 7)$.

47. a. $x_1^2 + y_1^2 = 1$
$x_2 = 0$ and $-1 \leq y_2 \leq 1$

b. Substituting $x = 0$ in the first equation gives $y^2 = 1$, and thus $y = \pm 1$. The only points at which the paths cross are $(0, 1)$ and $(0, -1)$.

c. The first particle is at $(0, 1)$ at times $t = \pi/2 + 2n\pi$, and at $(0, -1)$ at times $t = -\pi/2 + 2n\pi$. The second particle is at $(0, 1)$ at times $\pi/4 + n\pi$ and at $(0, -1)$ at times $t = -\pi/4 + n\pi$. The particles do not collide.

49. Proof. Let $\varepsilon = |x'(t_0)| > 0$. Since $x'(t_0) = \lim_{h \to 0}(x(t_0 + h) - x(t_0))/h$, there exists some $\delta > 0$ such that if $0 < |h| < \delta$, then $|x'(t_0) - (x(t_0 + h) - x(t_0))/h| < \varepsilon = |x'(t_0)|$. By h sufficiently small we mean that $0 < |h| < \delta$. In this case, the fact that the distance between $(x(t_0 + h) - x(t_0))/h$ and $x'(t_0)$ is less than $|x'(t_0)|$ implies that $(x(t_0 + h) - x(t_0))/h \neq 0$, and therefore that the numerator $x(t_0 + h) - x(t_0) \neq 0$. Thus $x(t_0 + h) \neq x(t_0)$.

51.

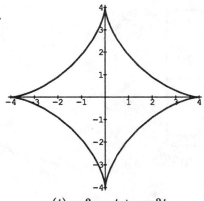

$$x(t) = 3\cos t + \cos 3t$$
$$y(t) = 3\sin t - \sin 3t$$
$$0 \le t \le 2\pi$$

53. a. $y^2 = t^2(t^2 - 1)^2 = (x + 1)x^2 = x^3 + x^2$

b. Graph is in textbook, Figure 4.1.

55. Graph $x(t) = 1 + \sqrt{t}$, $y(t) = 2 - \sqrt{t}$, $0 \le t \le 36$ using the range settings: **RANGE** program with **X=3**, **Y=0**, **S=0.1** (Casio, TI-81, HP-48SX), **S=0.2** (HP-28S). Casio: $[-1.7, 7.7] \times [-3.1, 3.1]$, TI-81: $[-1.8, 7.7] \times [-3.2, 3.1]$, HP-28S: $[-10.6, 16.6] \times [-3, 3.2]$, HP-48SX: $[-3.5, 9.5] \times [-3.1, 3.2]$.

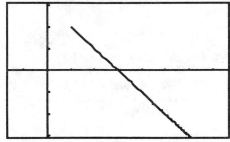

This is the graph of a line with slope $m = -1$ and x-intercept $(3,0)$. To show this note that $\sqrt{t} = x - 1$ so $y = 2 - (x - 1)$ or $y = -x + 3$.

57. Graph $x(t) = \sin t$, $y(t) = \csc t$, $0 \le t \le 2\pi$ using the range settings: **RANGE** program with **X=Y=0**, **S=0.1** (Casio, TI-81, HP-48SX), **S=0.2** (HP-28S). Casio: $[-4.7, 4.7] \times [-3.1, 3.1]$, TI-81: $[-4.8, 4.7] \times [-3.2, 3.1]$, HP-28S: $[-13.6, 13.6] \times [-3, 3.2]$, HP-48SX: $[-6.5, 6.5] \times [-3.1, 3.2]$.

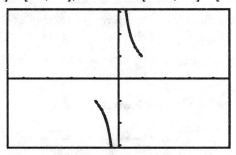

This looks like part of the graph of $y = 1/x$. To show this note that $y = \csc t = 1/\sin t$. So $\sin t = 1/y$. Therefore $x = 1/y$ or $y = 1/x$.

59. Graph $x = 2t/(1 + t^2)$, $y = (1 - t^2)/(1 + t^2)$, $-10 \le t \le 10$ using the range settings: **RANGE** program with **X=Y=0**, **S=0.05** (Casio, TI-81, HP-48SX), **S=0.1** (HP-28S). Casio: $[-2.35, 2.35] \times [-1.55, 1.55]$, TI-81: $[-2.4, 2.35] \times [-1.6, 1.55]$, HP-28S: $[-6.8, 6.8] \times [-1.5, 1.6]$, HP-48SX: $[-3.25, 3.25] \times [-1.55, 1.6]$.

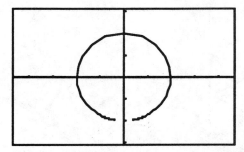

This is the graph of a circle centered at the origin with radius 1. To show this note that

$$x^2 + y^2 = \frac{4t^2}{(1+t^2)^2} + \frac{(1-t^2)^2}{(1+t^2)^2}$$

$$= \frac{4t^2 + 1 - 2t^2 + t^4}{(1+t^2)^2}$$

$$= \frac{1 + 2t^2 + t^4}{(1+t^2)^2}$$

$$= \frac{(1+t^2)^2}{(1+t^2)^2} = 1.$$

61. Graph $x = 5\cot t$, $y = 5\sin^2 t$, $0 \le t \le \pi$ using the range settings: **RANGE** program with **X=0, Y=3, S=0.15** (Casio, TI-81, HP-48SX), **S=0.3** (HP-28S). Casio: $[-7.05, 7.05] \times [-1.65, 7.65]$, TI-81: $[-7.2, 7.05] \times [-1.8, 7.65]$, HP-28S: $[-20.4, 20.4] \times [-1.5, 7.8]$, HP-48SX: $[-9.75, 9.75] \times [-1.65, 7.8]$.

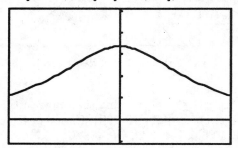

This looks like the graph of a function of the form $y = a/(b + x^2)$. To show this recall that $1 + \cot^2 t = \csc^2 t$ so $5 + 5\cot^2 t = 5\csc^2 t$. Now $y = 5\sin^2 t$ so $5/y = \csc^2 t$. This gives $25/y = 5\csc^2 t$ and therefore, $5 + 5\cot^2 t = 25/y$ or

$$y = \frac{25}{5 + t\cot^2 t}.$$

But, $x^2 = 25\cot^2 t$ so $x^2/5 = 5\cot^2 t$. We have then the

$$y = \frac{25}{5 + \frac{x^2}{5}} = \frac{125}{25 + x^2}.$$

63. Graph $x = 3.5\cos t + 0.5\cos 7t$, $y = 3.5\sin t - 0.5\sin 7t$, $0 \le t \le 2\pi$ using the range settings: **RANGE** program with **X=Y=0, S=0.125** (Casio, TI-81, HP-48SX), **S=0.25** (HP-28S). Casio: $[5.875, 5.875] \times [-3.875, 3.875]$, TI-81: $[-6, 5.875] \times [-4, 3.875]$, HP-28S: $[-17, 17] \times [-3.75, 4]$, HP-48SX: $[-8.125, 8.125] \times [-3.875, 4]$.

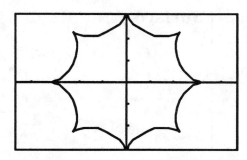

65. Graph $x = -3t(4 - t^6)$, $y = -3t^2(4 - t^6)$, $-2 \le t \le 2$ using the range settings: **RANGE** program with **X=0, Y=3, S=0.5** (Casio, TI-81, HP-48SX), **S=1** (HP-28S). Casio: $[-23.5, 23.5] \times [-12.5, 18.5]$, TI-81: $[-24, 23.5] \times [-13, 18.5]$, HP-28S: $[-68, 68] \times [-12, 19]$, HP-48SX: $[-32.5, 32.5] \times [-12.5, 19]$.

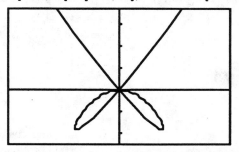

For Exercises 66-71, the range settings are: Casio: $[1, 95] \times [1, 63]$, TI-81: $[1, 96] \times [1, 64]$, HP-28S: $[1, 272] \times [1, 63]$, HP-48SX: $[1, 131] \times [1, 64]$.

67. $\{(50, 60), (20, 30), (60, 60), (50, 40)\}$

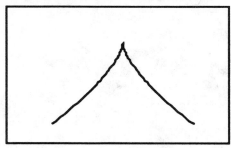

69. $\{(50, 60), (20, 30), (60, 60), (50, 40)\}$
$\{(50, 35), (80, 60), (80, 20), (50, 35)\}$

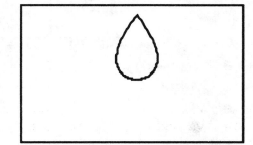

71. $\{(20, 20), (25, 30), (23, 40), (20, 44)\}$
$\{(22, 40), (30, 50), (40, 40), (40, 20)\}$
$\{(35, 40), (45, 50), (50, 40), (55, 20)\}$

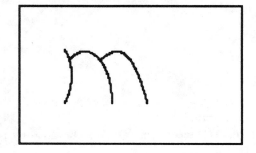

15.5   Arc Length and Surface Area Revisited

1. $x'(t) = 1$ and $y'(t) = (3/2)t^{1/2}$
$L = \int_0^{4/9} \sqrt{1 + (9/4)t}\, dt \quad (u = 1 + (9/4)t)$
$\quad = (4/9)\int_1^2 \sqrt{u}\, du$
$\quad = (4/9)(2/3)u^{3/2}\big|_1^2$
$\quad = (8/27)(2^{3/2} - 1)$
$\quad = (16\sqrt{2} - 8)/27$

3. $x'(t) = 6t$ and $y'(t) = 6t^2$
$L = \int_0^{\sqrt{2}} \sqrt{36t^2 + 36t^4}\, dt$
$\quad = 6\int_0^{\sqrt{2}} t\sqrt{1 + t^2}\, dt \quad (u = 1 + t^2)$
$\quad = 3\int_1^3 \sqrt{u}\, du$
$\quad = 3(2/3)u^{3/2}\big|_1^3$
$\quad = 2(3^{3/2} - 1)$
$\quad = 6\sqrt{3} - 2$

5. $x'(t) = 6(2t + 3)^{1/2}$ and $y'(t) = 6(t + 1)$
$L = \int_0^2 \sqrt{36(2t + 3) + 36(t + 1)^2}\, dt$
$\quad = 6\int_0^2 \sqrt{2t + 3 + t^2 + 2t + 1}\, dt$
$\quad = 6\int_0^2 \sqrt{t^2 + 4t + 4}\, dt$
$\quad = 6\int_0^2 \sqrt{(t + 2)^2}\, dt$
$\quad = 6\int_0^2 (t + 2)\, dt$
$\quad = 3t^2 + 12t\big|_0^2$
$\quad = (12 + 24) - (0 + 0)$
$\quad = 36$

7. $x'(t) = -3\cos^2 t \sin t$ and $y'(t) = 3\sin^2 t \cos t$
$L = \int_0^\pi \sqrt{9\cos^4 t \sin^2 t + 9\sin^4 t \cos^2 t}\, dt$
$\quad = 3\int_0^\pi \sqrt{\cos^2 t \sin^2 t(\cos^2 t + \sin^2 t)}\, dt$
$\quad = 3\int_0^\pi \sqrt{\cos^2 t \sin^2 t}\, dt$
$\quad = 3\int_0^{\pi/2} \cos t \sin t\, dt + 3\int_{\pi/2}^\pi (-\cos t)\sin t\, dt$
The integral was split due to the fact that
$\sqrt{\cos^2 t} = |\cos t|$, $\cos t$ is positive on $[0, \pi/2]$,
and $\cos t$ is negative on $[\pi/2, \pi]$.
For each, substitute $u = \sin t$.
$L = 3\int_0^1 u\, du - 3\int_1^0 u\, du$
$\quad = 3u^2/2\big|_0^1 - 3u^2/2\big|_1^0$
$\quad = (3/2 - 0) - (0 - 3/2)$
$\quad = 3$

9. $x'(t) = e^t(\cos t - \sin t)$ and $y'(t) = e^t(\sin t + \cos t)$
$L = \int_0^\pi \sqrt{e^{2t}(\cos t - \sin t)^2 + e^{2t}(\sin t + \cos t)^2}\, dt$
$\quad = \int_0^\pi \sqrt{e^{2t}(\cos^2 t - 2\cos t \sin t + \sin^2 t + \sin^2 t + 2\cos t \sin t + \cos^2 t)}\, dt$
$\quad = \int_0^\pi e^t \sqrt{2}\, dt$
$\quad = e^t \sqrt{2}\big|_0^\pi$
$\quad = e^\pi \sqrt{2} - \sqrt{2}$
$\quad = (e^\pi - 1)\sqrt{2}$

11. $x'(t) = 3t^2 - 6t$ and $y'(t) = 6t$
$L = \int_0^1 \sqrt{(3t^2 - 6t)^2 + 36t^2}\, dt$
$\quad = \int_0^1 \sqrt{9t^4 - 36t^3 + 72t^2}\, dt$
$\quad = 3\int_0^1 t\sqrt{t^2 - 4t + 8}\, dt$
$\quad = 3\int_0^1 t\sqrt{(t - 2)^2 + 4}\, dt \quad (t - 2 = 2\tan\theta)$
$\quad = 3\int_{t=0}^{t=1} (2\tan\theta + 2)\sqrt{4\tan^2\theta + 4}(2\sec^2\theta)\, d\theta$
$\quad = 24\int_{t=0}^{t=1} (\tan\theta + 1)\sec^3\theta\, d\theta$
$\quad = 24\int_{t=0}^{t=1} \tan\theta \sec^3\theta\, d\theta + 24\int_{t=0}^{t=1} \sec^3\theta\, d\theta$
For the first integral, let $u = \sec^3\theta$ and $du = 3\sec^3\theta \tan\theta$. The second integral can be done by parts, or its antiderivative can be looked up in the table of integrals.

$$L = 24 \int_{t=0}^{t=1} (1/3)\, du + 24 \left((1/2) \sec\theta \tan\theta + (1/2) \ln|\sec\theta + \tan\theta|\right)\Big|_{t=0}^{t=1}$$
$$= 8u + 12 \sec\theta \tan\theta + 12 \ln|\sec\theta + \tan\theta|\Big|_{t=0}^{t=1}$$
$$= 8\sec^3\theta + 12\sec\theta \tan\theta + 12\ln|\sec\theta + \tan\theta|\Big|_{t=0}^{t=1}$$
$$= 8(t^2 - 4t + 8)\sqrt{t^2 - 4t + 8}/8 + 12\sqrt{t^2 - 4t + 8}(t-2)/4 + 12\ln|\sqrt{t^2 - 4t + 8}/2 + (t-2)/2|\Big|_0^1$$
$$= (5\sqrt{5} - 3\sqrt{5} + 12\ln((\sqrt{5}-1)/2)) - (16\sqrt{2} - 12\sqrt{2} + 12\ln(\sqrt{2}-1))$$
$$= 2\sqrt{5} - 4\sqrt{2} + 12\ln((\sqrt{5}-1)(\sqrt{2}+1)/2)$$

13. $x'(t) = (2t+1)^{1/2}$ and $y'(t) = 1$
$$L = \int_{3/2}^{15/2} \sqrt{(2t+1) + 1}\, dt$$
$$= \int_{3/2}^{15/2} \sqrt{2t+2}\, dt \quad (u = 2t+2)$$
$$= (1/2) \int_5^{17} \sqrt{u}\, du$$
$$= (1/2)(2/3) u^{3/2}\Big|_5^{17}$$
$$= (1/3)17\sqrt{17} - (1/3)5\sqrt{5}$$
$$= (17\sqrt{17} - 5\sqrt{5})/3$$

15. $x'(t) = 2\sqrt{2}t$ and $y'(t) = 2t^2 - 1$
$$L = \int_0^{\sqrt{2}} \sqrt{8t^2 + (2t^2 - 1)^2}\, dt$$
$$= \int_0^{\sqrt{2}} \sqrt{4t^4 + 4t^2 + 1}\, dt$$
$$= \int_0^{\sqrt{2}} \sqrt{(2t^2 + 1)^2}\, dt$$
$$= \int_0^{\sqrt{2}} (2t^2 + 1)\, dt$$
$$= (2t^3/3 + t)\Big|_0^{\sqrt{2}}$$
$$= (4\sqrt{2}/3 + \sqrt{2}) - (0 + 0)$$
$$= 7\sqrt{2}/3$$

17. $x'(t) = 2t$ and $y'(t) = 4$
$$S = \int_0^2 2\pi(4t)\sqrt{4t^2 + 16}\, dt \quad (u = 4t^2 + 16)$$
$$= \pi \int_{16}^{32} \sqrt{u}\, du$$
$$= \pi(2/3) u^{3/2}\Big|_{16}^{32}$$
$$= 256\pi\sqrt{2}/3 - 128\pi/3$$

19. $x'(t) = 1$ and $y'(t) = 1/2\sqrt{t}$
$$S = \int_1^2 2\pi\sqrt{t}\sqrt{1 + 1/(4t)}\, dt$$
$$= 2\pi \int_1^2 \sqrt{t(1 + 1/(4t))}\, dt$$
$$= 2\pi \int_1^2 \sqrt{t + 1/4}\, dt \quad (u = t + 1/4)$$
$$= 2\pi \int_{5/4}^{9/4} \sqrt{u}\, du$$
$$= 2\pi(2/3) u^{3/2}\Big|_{5/4}^{9/4}$$
$$= (4\pi/3)(27/8 - 5\sqrt{5}/8)$$
$$= (27 - 5\sqrt{5})\pi/6$$

21. $x'(t) = 6t$ and $y'(t) = 6t^2$
$$S = \int_1^2 2\pi(2t^3)\sqrt{36t^2 + 36t^4}\, dt$$
$$= 24\pi \int_1^2 t^4\sqrt{1 + t^2}\, dt \quad (t = \tan\theta)$$
$$= 24\pi \int_{t=1}^{t=2} \tan^4\theta\sqrt{1 + \tan^2\theta} \sec^2\theta\, d\theta$$
$$= 24\pi \int_{t=1}^{t=2} \tan^4\theta \sec^3\theta\, d\theta$$
$$= 24\pi \int_{t=1}^{t=2} (\sec^2\theta - 1)^2 \sec^3\theta\, d\theta$$
$$= 24\pi \int_{t=1}^{t=2} (\sec^7\theta - 2\sec^5\theta + \sec^3\theta)\, d\theta$$
$$= 24\pi \int_{t=1}^{t=2} \sec^7\theta\, d\theta - 48\pi \int_{t=1}^{t=2} \sec^5\theta\, d\theta + 24\pi \int_{t=1}^{t=2} \sec^3\theta\, d\theta$$
$$= 24\pi \left((1/6)\tan\theta \sec^5\theta\Big|_{t=1}^{t=2} + (5/6)\int_{t=1}^{t=2} \sec^5\theta\, d\theta\right) - 48\pi \int_{t=1}^{t=2} \sec^5\theta\, d\theta + 24\pi \int_{t=1}^{t=2} \sec^3\theta\, d\theta$$
$$= 4\pi \tan\theta \sec^5\theta\Big|_{t=1}^{t=2} - 28\pi \int_{t=1}^{t=2} \sec^5\theta\, d\theta + 24\pi \int_{t=1}^{t=2} \sec^3\theta\, d\theta$$
$$= 4\pi \tan\theta \sec^5\theta\Big|_{t=1}^{t=2} - 28\pi \left((1/4)\tan\theta \sec^3\theta\Big|_{t=1}^{t=2} + (3/4)\int_{t=1}^{t=2} \sec^3\theta\, d\theta\right) + 24\pi \int_{t=1}^{t=2} \sec^3\theta\, d\theta$$
$$= 4\pi \tan\theta \sec^5\theta - 7\pi \tan\theta \sec^3\theta\Big|_{t=1}^{t=2} + 3\pi \int_{t=1}^{t=2} \sec^3\theta\, d\theta$$
$$= 4\pi \tan\theta \sec^5\theta - 7\pi \tan\theta \sec^3\theta + 3\pi((1/2)\tan\theta \sec\theta + (1/2)\ln|\sec\theta + \tan\theta|)\Big|_{t=1}^{t=2}$$
$$= 4\pi t(t^2 + 1)^2\sqrt{t^2 + 1} - 7\pi t(t^2 + 1)\sqrt{t^2 + 1} + (3\pi/2)t\sqrt{t^2 + 1} + (3\pi/2)\ln|\sqrt{t^2 + 1} + t|\Big|_1^2$$
$$= (200\pi\sqrt{5} - 70\pi\sqrt{5} + 3\pi\sqrt{5} + (3\pi/2)\ln(2 + \sqrt{5})) - (16\pi\sqrt{2} - 14\pi\sqrt{2} + 3\pi\sqrt{2}/2 + (3\pi/2)\ln(\sqrt{2} + 1))$$
$$= 133\pi\sqrt{5} - 7\pi\sqrt{2}/2 + (3\pi/2)\ln((2 + \sqrt{5})(\sqrt{2} - 1))$$

23. $f(\theta) = \cos^2(\theta/2)$ and $f'(\theta) = -\cos(\theta/2)\sin(\theta/2)$

$L = \int_0^{2\pi} \sqrt{\cos^4(\theta/2) + \cos^2(\theta/2)\sin^2(\theta/2)}\, d\theta$

$= \int_0^{2\pi} \sqrt{\cos^2(\theta/2)\left(\cos^2(\theta/2) + \sin^2(\theta/2)\right)}\, d\theta$

$= \int_0^{2\pi} \sqrt{\cos^2(\theta/2)}\, d\theta$

The integral must be split at π since $\cos(\theta/2)$ is positive on $[0, \pi]$ and negative on $[\pi, 2\pi]$.

$L = \int_0^{\pi} \cos(\theta/2)\, d\theta + \int_{\pi}^{2\pi} (-\cos(\theta/2))\, d\theta$

$= 2\sin(\theta/2)|_0^{\pi} - 2\sin(\theta/2)|_{\pi}^{2\pi}$

$= (2 - 0) - (0 - 2)$

$= 4$

25. $f(\theta) = e^{2\theta}$ and $f'(\theta) = 2e^{2\theta}$

$L = \int_0^{\pi} \sqrt{e^{4\theta} + 4e^{4\theta}}\, d\theta$

$= \int_0^{\pi} e^{2\theta}\sqrt{5}\, d\theta$

$= (\sqrt{5}/2)e^{2\theta}|_0^{\pi}$

$= (\sqrt{5}/2)e^{2\pi} - (\sqrt{5}/2)$

$= (e^{2\pi} - 1)\sqrt{5}/2$

27. $x'(t) = -2\sin t$ and $y'(t) = 2\cos t$

$L = \int_0^{2\pi} \sqrt{4\sin^2 t + 4\cos^2 t}\, dt$

$= \int_0^{2\pi} 2\, dt$

$= 2t|_0^{2\pi}$

$= 4\pi$

29. $f(\theta) = 3\cos\theta$ and $f'(\theta) = -3\sin\theta$

$L = \int_0^{\pi} \sqrt{9\cos^2\theta + 9\sin^2\theta}\, d\theta$

$= \int_0^{\pi} 3\, d\theta$

$= 3\theta|_0^{\pi}$

$= 3\pi$

31. $f(\theta) = e^{\theta}$ and $f'(\theta) = e^{\theta}$

$L = \int_0^{2\pi} \sqrt{(e^{\theta})^2 + (e^{\theta})^2}\, d\theta$

$= \int_0^{2\pi} \sqrt{2(e^{\theta})^2}\, d\theta$

$= \sqrt{2}\int_0^{2\pi} e^{\theta}\, d\theta$

$= \sqrt{2}e^{\theta}|_0^{2\pi}$

$= \sqrt{2}e^{2\pi} - \sqrt{2}$

$= (e^{2\pi} - 1)\sqrt{2}$

33. $x(t) = e^t \cos t$ and $y(t) = e^t \sin t$ for $0 \le t \le \pi$

$x'(t) = e^t(\cos t - \sin t)$ and $y'(t) = e^t(\cos t + \sin t)$

$S = \int_0^{\pi} 2\pi e^t \sin t \sqrt{e^{2t}(\cos t - \sin t)^2 + e^{2t}(\cos t + \sin t)^2}\, dt$

$= 2\pi \int_0^{\pi} e^{2t} \sin t \sqrt{\cos^2 t - 2\cos t \sin t + \sin^2 t + \cos^2 t + 2\cos t \sin t + \sin^2 t}\, dt$

$= 2\pi \int_0^{\pi} e^{2t} \sin t \sqrt{2}\, dt$

$= 2\pi\sqrt{2} \int_0^{\pi} e^{2t} \sin t\, dt$

Integrate $\int e^{2t} \sin t\, dt$ by parts.

$u = e^{2t}$ and $dv = \sin t\, dt$

$du = 2e^{2t}\, dt$ and $v = -\cos t$

$\int e^{2t} \sin t\, dt = -e^{2t} \cos t + 2\int e^{2t} \cos t\, dt$

Integrate by parts again.

$u = e^{2t}$ and $dv = \cos t\, dt$

$du = 2e^{2t}\, dt$ and $v = \sin t$

$\int e^{2t} \sin t\, dt = -e^{2t} \cos t + 2e^{2t} \sin t - 4\int e^{2t} \sin t\, dt$

$\int e^{2t} \sin t\, dt = (2/5)e^{2t} \sin t - (1/5)e^{2t} \cos t + C$

$S = 2\pi\sqrt{2}\left[(2/5)e^{2t} \sin t - (1/5)e^{2t} \cos t\right]|_0^{\pi}$

$= 2\pi\sqrt{2}e^{2\pi}/5 - 2\pi\sqrt{2}(-1/5)$

$= 2\pi\sqrt{2}(e^{2\pi} + 1)/5$

35. All that has been done is to exchange the variables x and y. The formula for surface area will be:

$$S = \int_a^b 2\pi x(t)\sqrt{[x'(t)]^2 + [y'(t)]^2}\, dt$$

for $a \le t \le b$

37. $x'(t) = 1$ and $y'(t) = 1/(2\sqrt{t})$

$$S = \int_1^2 2\pi t \sqrt{1 + 1/(4t)}\, dt$$
$$= 2\pi \int_1^2 \sqrt{t^2 + t/4}\, dt$$
$$= 2\pi \int_1^2 \sqrt{(t + 1/8)^2 - 1/64}\, dt \quad (t + 1/8 = (1/8)\sec\theta)$$
$$= 2\pi \int_{t=1}^{t=2} \sqrt{(1/64)(\sec^2\theta - 1)}(1/8)\tan\theta\sec\theta\, d\theta$$
$$= (\pi/32)\int_{t=1}^{t=2} \sqrt{\tan^2\theta}\,\tan\theta\sec\theta\, d\theta$$
$$= (\pi/32)\int_{t=1}^{t=2} \tan^2\theta\sec\theta\, d\theta$$
$$= (\pi/32)\int_{t=1}^{t=2} \sec^3\theta\, d\theta - (\pi/32)\int_{t=1}^{t=2} \sec\theta\, d\theta$$
$$= (\pi/32)(1/2)\tan\theta\sec\theta + (\pi/32)(1/2)\ln|\tan\theta + \sec\theta| - (\pi/32)\ln|\tan\theta + \sec\theta|\big|_{t=1}^{t=2}$$
$$= (\pi/64)(8t + 1)\sqrt{64t^2 + 16t} - (\pi/64)\ln|8t + 1 + \sqrt{64t^2 + 16t}|\big|_1^2$$
$$= \left(51\pi\sqrt{2}/16 - (\pi/64)\ln(17 + 12\sqrt{2})\right) - \left(9\pi\sqrt{5}/16 - (\pi/64)\ln(9 + 4\sqrt{5})\right)$$
$$= 51\pi\sqrt{2}/16 - 9\pi\sqrt{5}/16 + (\pi/64)\ln\left((9 + 4\sqrt{5})(17 - 12\sqrt{2})\right)$$

39. Assume that this is not true; that is, that C is a differentiable curve, $x'(t) \ne 0$, and that there are two distinct values $t = a$ and $t = b$ ($a < b$) such that $x(a) = x(b)$. By the Mean Value Theorem, there exists some number c in the interval (a, b) such that $x'(c) = (x(b) - x(a))/(b - a) = 0$. This contradicts the hypothesis. Therefore, the assumption is false, and the claim is true.

Review Exercises — Chapter 15

1.

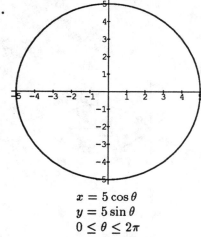

$x = 5\cos\theta$
$y = 5\sin\theta$
$0 \le \theta \le 2\pi$

3.

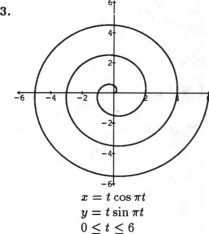

$x = t\cos\pi t$
$y = t\sin\pi t$
$0 \le t \le 6$

5.

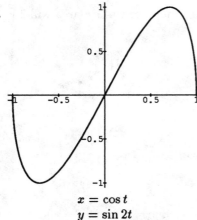

$$x = \cos t$$
$$y = \sin 2t$$
$$0 \le t \le \pi$$

7. $x^2 + y^2 = t^2 \cos^2 \pi t + t^2 \sin^2 \pi t = t^2$
$t = \sqrt{x^2 + y^2}$ since t is nonnegative
$y = \sqrt{x^2 + y^2}\, \sin(\pi\sqrt{x^2 + y^2})$
and $x^2 + y^2 \le 36$

9. $y = 2 \sin t \cos t = 2x \sin t$
$\sin t = +\sqrt{1 - \cos^2 t}$ since $0 \le t \le \pi$
$\sin t = \sqrt{1 - x^2}$
$y = 2x\sqrt{1 - x^2}$

11. $(2x)^2 + (3y)^2 = 6^2$
$2x = 6 \cos t$ and $3y = 6 \sin t$
$x = 3 \cos t$
$y = 2 \sin t$

13. $x = 2t^2 + 4t + 5$
$y = t$

15.

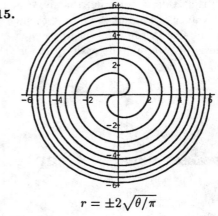

$$r = \pm 2\sqrt{\theta/\pi}$$

17. $t = y/3$
$x = (y/3)^2 - 1$
$x = y^2/9 - 1$
This is the equation of a parabola.

19.

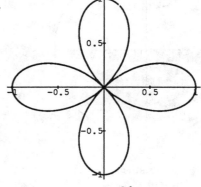

$$r = \cos 2\theta$$

21.

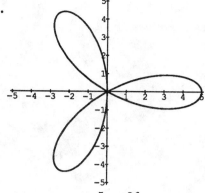

$$r = 5\cos 3\theta$$

23.

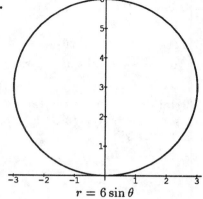

$$r = 6\sin\theta$$

27. The graph is the y-axis.

25.

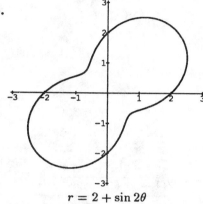

$$r = 2 + \sin 2\theta$$

29. $(x - a)^2 + y^2 = a^2$
$x^2 + y^2 = 2ax$
$r^2 = 2ar\cos\theta$
$r = 2a\cos\theta$

31. $x^2 + (y - a)^2 = a^2$
$x^2 + y^2 = 2ay$
$r^2 = 2ar\sin\theta$
$r = 2a\sin\theta$

33. $A = \int_0^{2\pi}(1/2)4(1 + \sin 2\theta)^2\,d\theta$
$= 2\int_0^{2\pi}(1 + 2\sin 2\theta + \sin^2 2\theta)\,d\theta$
$= 2\int_0^{2\pi}(1 + 2\sin 2\theta + (1 - \cos 4\theta)/2)\,d\theta$
$= 2(\theta - \cos 2\theta + \theta/2 - (1/8)\sin 4\theta)\big|_0^{2\pi}$
$= 2(2\pi - 1 + \pi - 0) - 2(0 - 1 + 0 - 0)$
$= 6\pi$

35. $A = \int_0^{2\pi}(1/2)a^2(1 + \sin\theta)^2\,d\theta$
$= (a^2/2)\int_0^{2\pi}(1 + 2\sin\theta + \sin^2\theta)\,d\theta$
$= (a^2/2)\int_0^{2\pi}(1 + 2\sin\theta + (1 - \cos 2\theta)/2)\,d\theta$
$= (a^2/2)(\theta - 2\cos\theta + \theta/2 - (1/4)\sin 2\theta)\big|_0^{2\pi}$
$= (a^2/2)(2\pi - 2 + \pi - 0) - (a^2/2)(0 - 2 + 0 - 0)$
$= 3a^2\pi/2$

37. $A = 3\int_{-\pi/6}^{\pi/6}(1/2)36\cos^2 3\theta\,d\theta$
$= 54\int_{-\pi/6}^{\pi/6}(1 + \cos 6\theta)/2\,d\theta$
$= 54(\theta/2 + (1/12)\sin 6\theta)\big|_{-\pi/6}^{\pi/6}$
$= 54(\pi/12 + 0) - 54(-\pi/12 + 0)$
$= 9\pi$

39. $x'(t) = 1$ and $y'(t) = (3/2)(1 + t)^{1/2}$
$L = \int_0^1\sqrt{1 + (9/4)(1 + t)}\,dt$
$= \int_0^1\sqrt{(13 + 9t)/4}\,dt \quad (u = (13 + 9t)/4)$
$= (4/9)\int_{13/4}^{11/2}\sqrt{u}\,du$
$= (4/9)(2/3)u^{3/2}\big|_{13/4}^{11/2}$
$= (8/27)(11/2)\sqrt{11/2} - (8/27)(13/4)\sqrt{13/4}$
$= 22\sqrt{22}/27 - 13\sqrt{13}/27$
$= (22\sqrt{22} - 13\sqrt{13})/27$

501

41. $x'(t) = 4t$ and $y'(t) = 4$

$L = \int_0^1 \sqrt{16t^2 + 16} \, dt$

$= 4 \int_0^1 \sqrt{t^2 + 1} \, dt \quad (t = \tan\theta)$

$= 4 \int_0^{\pi/4} \sqrt{\tan^2\theta + 1} \sec^2\theta \, d\theta$

$= 4 \int_0^{\pi/4} \sec^3\theta \, d\theta$

$= 4(1/2)\tan\theta \sec\theta + 4(1/2)\ln|\tan\theta + \sec\theta|\big|_0^{\pi/4}$

$= (2\sqrt{2} + 2\ln(1 + \sqrt{2})) - (0 + 2\ln(1))$

$= 2\sqrt{2} + \ln(1 + \sqrt{2})^2$

$= 2\sqrt{2} + \ln(3 + 2\sqrt{2})$

43. $x'(t) = \sin t$ and $y'(t) = \cos t$

$L = \int_0^\pi \sqrt{\sin^2 t + \cos^2 t} \, dt$

$= \int_0^\pi 1 \, dt$

$= t\big|_0^\pi$

$= \pi$

45. $f(\theta) = \cos\theta$ and $f'(\theta) = -\sin\theta$

$L = \int_{-\pi/4}^{\pi/4} \sqrt{\cos^2\theta + \sin^2\theta} \, d\theta$

$= \int_{-\pi/4}^{\pi/4} 1 \, d\theta$

$= \theta\big|_{-\pi/4}^{\pi/4}$

$= \pi/4 - (-\pi/4)$

$= \pi/2$

47. $f(\theta) = 1 + \sin\theta$ and $f'(\theta) = \cos\theta$

$L = \int_0^{\pi/2} \sqrt{(1 + \sin\theta)^2 + \cos^2\theta} \, d\theta$

$= \int_0^{\pi/2} \sqrt{1 + 2\sin\theta + \sin^2\theta + \cos^2\theta} \, d\theta$

$= \int_0^{\pi/2} \sqrt{2 + 2\sin\theta} \, d\theta$

$= \int_0^{\pi/2} \sqrt{4(1 + \sin\theta)/2} \, d\theta \quad (\theta = \pi/2 - u)$

$= -\int_{\pi/2}^0 2\sqrt{(1 + \sin(\pi/2 - u))/2} \, du$

$= 2\int_0^{\pi/2} \sqrt{(1 + \cos u)/2} \, du$

$= 2\int_0^{\pi/2} \sqrt{\cos^2(u/2)} \, du$

$= 2\int_0^{\pi/2} \cos(u/2) \, du$

$= 4\sin(u/2)\big|_0^{\pi/2}$

$= 2\sqrt{2}$

49. The graphs intersect at the origin, and when $\sin\theta = \cos\theta$, which occurs at $\theta = \pi/4$.

$A = \int_0^{\pi/4} (1/2)\sin^2\theta \, d\theta + \int_{\pi/4}^{\pi/2} (1/2)\cos^2\theta \, d\theta$

$= \int_0^{\pi/4} (1/4)(1 - \cos 2\theta) \, d\theta + \int_{\pi/4}^{\pi/2} (1/4)(1 + \cos 2\theta) \, d\theta$

$= \theta/4 - (1/8)\sin 2\theta\big|_0^{\pi/4} + \theta/4 + (1/8)\sin 2\theta\big|_{\pi/4}^{\pi/2}$

$= (\pi/16 - 1/8) - (0 - 0) + (\pi/8 + 0) - (\pi/16 + 1/8)$

$= \pi/8 - 1/4$

$= (\pi - 2)/8$

51. The graphs intersect when:
$6\cos\theta = 2(1 + \cos\theta)$
$6\cos\theta = 2 + 2\cos\theta$
$4\cos\theta = 2$
$\cos\theta = 1/2$
$\theta = \pm\pi/3$

$A = \int_{-\pi/3}^{\pi/3}(1/2)(36\cos^2\theta - 4(1+\cos\theta)^2)\,d\theta$
$= \int_{-\pi/3}^{\pi/3}(18\cos^2\theta - 2(1 + 2\cos\theta + \cos^2\theta))\,d\theta$
$= \int_{-\pi/3}^{\pi/3}(16\cos^2\theta - 4\cos\theta - 2)\,d\theta$
$= \int_{-\pi/3}^{\pi/3}(8(1 + \cos 2\theta) - 4\cos\theta - 2)\,d\theta$
$= 8\theta + 4\sin 2\theta - 4\sin\theta - 2\theta\big|_{-\pi/3}^{\pi/3}$
$= (8\pi/3 + 2\sqrt{3} - 2\sqrt{3} - 2\pi/3) - (-8\pi/3 - 2\sqrt{3} + 2\sqrt{3} + 2\pi/3)$
$= 4\pi$

53. The graphs intersect when:
$4\sin\theta = 2$
$\sin\theta = 1/2$
$\theta = \pi/6$ or $5\pi/6$

$A = \int_0^{\pi/6}(1/2)16\sin^2\theta\,d\theta + \int_{\pi/6}^{5\pi/6}(1/2)4\,d\theta + \int_{5\pi/6}^{\pi}(1/2)16\sin^2\theta\,d\theta$
$= \int_0^{\pi/6}4(1 - \cos 2\theta)\,d\theta + \int_{\pi/6}^{5\pi/6}2\,d\theta + \int_{5\pi/6}^{\pi}4(1 - \cos 2\theta)\,d\theta$
$= 4\theta - 2\sin 2\theta\big|_0^{\pi/6} + 2\theta\big|_{\pi/6}^{5\pi/6} + 4\theta - 2\sin 2\theta\big|_{5\pi/6}^{\pi}$
$= (2\pi/3 - \sqrt{3}) - (0 - 0) + (5\pi/3 - \pi/3) + (4\pi - 0) - (10\pi/3 + \sqrt{3})$
$= 8\pi/3 - 2\sqrt{3}$

55. The point $(1, -11)$ corresponds to $t = -2$.
$x'(t) = 2t + 2$ and $y'(t) = 4$
$x'(-2) = -2$ and $y'(-2) = 4$
$m = 4/(-2) = -2$
$y + 11 = -2(x - 1)$
$y = -2x - 9$

57. The path of particle p has equation $y = 2x$, and the path of particle q has equation $x = -y^2$. The paths intersect when $y/2 = -y^2$.
$y^2 + y/2 = 0$
$y(y + 1/2) = 0$
$y = 0$ and $y = -1/2$
The points where the paths of the particles cross are $(0,0)$ and $(-1/4, -1/2)$. Particle p is at $(0,0)$ at time $t = 0$ and at $(-1/4, -1/2)$ at time $t = 1/4$. Particle q is at $(0,0)$ at time $t = 0$ and at $(-1/4, -1/2)$ at time $t = 1/2$. The particles collide immediately at time $t = 0$ at the point $(0,0)$.
As $t \to \infty$, both particles move off to infinity at the lower left of the graph (both $x \to -\infty$ and $y \to -\infty$). The particles also separate, and the distance between them grows larger without bound.

Chapter 16

Vectors, Lines, and Planes

16.1   Vectors in the Plane and in Space

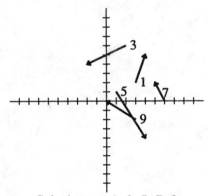

Solutions to 1, 3, 5, 7, 9

11. $\mathbf{v} = \langle -1 - 2, 2 - 5 \rangle = \langle -3, -3 \rangle$

13. $\mathbf{v} = 2\langle -1 - 2, 2 - 5 \rangle - \langle -1 - 2, 2 - (-6) \rangle = \langle -6, -6 \rangle - \langle -3, 8 \rangle = \langle -3, -14 \rangle$

15. $\mathbf{v} = 2\langle -1 - 2, 2 - (-6) \rangle - 2\langle 2 - 2, 5 - (-6) \rangle = \langle -6, 16 \rangle - \langle 0, 22 \rangle = \langle -6, -6 \rangle$

17. $\langle 3, 1 \rangle + 2\langle -2, 4 \rangle = \langle 3, 1 \rangle + \langle -4, 8 \rangle = \langle -1, 9 \rangle$

19. $2\langle 3, 1 \rangle + 2\langle -2, 4 \rangle - 2\langle -4, -2 \rangle = \langle 6, 2 \rangle + \langle -4, 8 \rangle - \langle -8, -4 \rangle = \langle 10, 14 \rangle$

21. $\langle -2, 4 \rangle - \langle 3, 1 \rangle - \langle -4, -2 \rangle = \langle -1, 5 \rangle$

23. $(3\mathbf{i} - \mathbf{j}) + 2(2\mathbf{i} + 6\mathbf{j}) = 3\mathbf{i} - \mathbf{j} + 4\mathbf{i} + 12\mathbf{j} = 7\mathbf{i} + 11\mathbf{j}$

25. $(3\mathbf{i} - \mathbf{j}) + (2\mathbf{i} + 6\mathbf{j}) + (-\mathbf{i} + \mathbf{j}) = 4\mathbf{i} + 6\mathbf{j}$

27. $3(3\mathbf{i} - \mathbf{j}) + 4(2\mathbf{i} + 6\mathbf{j}) - 2(-\mathbf{i} + \mathbf{j}) = 9\mathbf{i} - 3\mathbf{j} + 8\mathbf{i} + 24\mathbf{j} + 2\mathbf{i} - 2\mathbf{j} = 19\mathbf{i} + 19\mathbf{j}$

29. $3(3\mathbf{i} - \mathbf{j}) + 3(2\mathbf{i} + 6\mathbf{j}) + 3(-\mathbf{i} + \mathbf{j}) = 9\mathbf{i} - 3\mathbf{j} + 6\mathbf{i} + 18\mathbf{j} - 3\mathbf{i} + 3\mathbf{j} = 12\mathbf{i} + 18\mathbf{j}$

31.

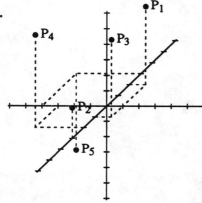

33. $\langle 2, -1, 5 \rangle + 6 \langle -3, 5, 0 \rangle = \langle 2, -1, 5 \rangle + \langle -18, 30, 0 \rangle = \langle -16, 29, 5 \rangle$

35. $3 \langle 2, -1, 5 \rangle - 2 \langle -3, 5, 0 \rangle = \langle 6, -3, 15 \rangle - \langle -6, 10, 0 \rangle = \langle 12, -13, 15 \rangle$

37. $\mathbf{u} + 4\mathbf{v} = \langle 2, -1, 5 \rangle + 4 \langle -3, 5, 0 \rangle = \langle 2, -1, 5 \rangle + \langle -12, 20, 0 \rangle = \langle -10, 19, 5 \rangle$
$|\mathbf{u} + 4\mathbf{v}| = \sqrt{(-10)^2 + 19^2 + 5^2} = \sqrt{486} = 9\sqrt{6}$

39. $\sqrt{(-3)^2 + 4^2} = \sqrt{25} = 5$

41. $|\langle 1, 0 \rangle| = \sqrt{1^2 + 0^2} = \sqrt{1} = 1$

43. $\sqrt{3^2 + 4^2} = \sqrt{25} = 5$

45. $4(\mathbf{i} - 3\mathbf{j}) = 4\mathbf{i} - 12\mathbf{j}$
$|4(\mathbf{i} - 3\mathbf{j})| = \sqrt{4^2 + (-12)^2} = \sqrt{160} = 4\sqrt{10}$

47. Let $D = (x, y, z)$
$\overrightarrow{AB} = \langle -2 - 3, 1 - 1, 5 - 6 \rangle = \langle -5, 0, -1 \rangle$
$\overrightarrow{CD} = \langle x - 6, y + 2, z - 2 \rangle$
$\langle -5, 0, 1 \rangle = \langle x - 6, y + 2, z - 2 \rangle$
$x - 6 = -5$ implies $x = 1$
$y + 2 = 0$ implies $y = -2$
$z - 2 = -1$ implies $z = 1$
$D = (1, -2, 1)$

49. a. $\overrightarrow{PQ} = \langle 2, -1, 3 \rangle$
$|\overrightarrow{PQ}| = \sqrt{2^2 + (-1)^2 + 3^2} = \sqrt{14}$

b. $\overrightarrow{QP} = \langle -2, 1, -3 \rangle$
$|\overrightarrow{QP}| = \sqrt{(-2)^2 + 1^2 + (-3)^2} = \sqrt{14}$

c. $\overrightarrow{PR} = \langle 1, 0, 5 \rangle$
$|\overrightarrow{PR}| = \sqrt{1^2 + 0^2 + 5^2} = \sqrt{26}$

d. $\overrightarrow{QR} = \langle -1, 1, 2 \rangle$
$|\overrightarrow{QR}| = \sqrt{(-1)^2 + 1^2 + 2^2} = \sqrt{6}$

e. $\overrightarrow{PQ} + \overrightarrow{QR} = \overrightarrow{PR}$
$|\overrightarrow{PR}| = \sqrt{26}$

f. $|-2\overrightarrow{PR}| = |-2||\overrightarrow{PR}| = 2\sqrt{26}$

g. $\overrightarrow{PQ} + 2\overrightarrow{QR} = \langle 2, -1, 3 \rangle + \langle -2, 2, 4 \rangle$
$= \langle 0, 1, 7 \rangle$
$|\overrightarrow{PQ} + 2\overrightarrow{QR}| = \sqrt{0^2 + 1^2 + 7^2}$
$= \sqrt{50}$
$= 5\sqrt{2}$

51. $-\mathbf{i} = \langle -1, 0, 0 \rangle$

53. $\langle \cos 120°, \sin 120° \rangle = \langle -1/2, \sqrt{3}/2 \rangle$

55. $|\mathbf{w}| = \sqrt{1^2 + 4^2 + 3^2} = \sqrt{26}$
$\mathbf{w}/|\mathbf{w}| = (\sqrt{26}/26)\mathbf{i} + (2\sqrt{26}/13)\mathbf{j} + (3\sqrt{26}/26)\mathbf{k}$

57. $(3\mathbf{i} - 2\mathbf{j} + 2\mathbf{k}) - 3(5\mathbf{i} - \mathbf{j} + 3\mathbf{k})$
$= 3\mathbf{i} - 2\mathbf{j} + 2\mathbf{k} - 15\mathbf{i} + 3\mathbf{j} - 9\mathbf{k}$
$= -12\mathbf{i} + \mathbf{j} - 7\mathbf{k}$

59. $|\mathbf{v}| = \sqrt{3^2 + (-2)^2 + 2^2} = \sqrt{17}$
$|\mathbf{w}| = \sqrt{5^2 + (-1)^2 + 3^2} = \sqrt{35}$
$(\sqrt{17}/17)(3\mathbf{i} - 2\mathbf{j} + 2\mathbf{k}) + (\sqrt{35}/35)(5\mathbf{i} - \mathbf{j} + 3\mathbf{k})$
$= \left(\frac{3\sqrt{17}}{17} + \frac{\sqrt{35}}{7}\right)\mathbf{i} + \left(\frac{-2\sqrt{17}}{17} - \frac{\sqrt{35}}{35}\right)\mathbf{j} + \left(\frac{2\sqrt{17}}{17} + \frac{3\sqrt{35}}{35}\right)\mathbf{k}$
$= \frac{21\sqrt{17} + 17\sqrt{35}}{119}\mathbf{i} - \frac{70\sqrt{17} + 17\sqrt{35}}{595}\mathbf{j} + \frac{70\sqrt{17} + 51\sqrt{35}}{595}\mathbf{k}$

61. For $f(x) = x^3$, one has $f'(x) = 3x^2$ and $f'(1) = 3$. Thus the tangent line at the point $(1, 1)$ has slope 3, and one such vector pointing along this line is $\mathbf{v} = \mathbf{i} + 3\mathbf{j} = \langle 1, 3 \rangle$. $|\mathbf{v}| = \sqrt{1^2 + 3^2} = \sqrt{10}$. The two unit vectors parallel to $\mathbf{v}$ are

$$\pm\frac{\mathbf{v}}{|\mathbf{v}|} = \pm\frac{\sqrt{10}}{10}\langle 1, 3 \rangle = \pm\left\langle \frac{\sqrt{10}}{10}, \frac{3\sqrt{10}}{10} \right\rangle.$$

The desired vectors with initial point at $(1, 1)$ are $\overrightarrow{PQ}$ and $\overrightarrow{PR}$, where

$$P = (1, 1), \, Q = \left(\frac{10 + \sqrt{10}}{10}, \frac{10 + 3\sqrt{10}}{10} \right), \text{ and } R = \left(\frac{10 - \sqrt{10}}{10}, \frac{10 - 3\sqrt{10}}{10} \right).$$

63. Suppose $D = (x, y)$

a. $\overrightarrow{AD} = \overrightarrow{BC}$
$\overrightarrow{AD} = \langle x - 1, y - 1 \rangle$ and $\overrightarrow{BC} = \langle 1, 2 \rangle$
$x - 1 = 1$ implies that $x = 2$ and $y - 1 = 2$ implies that $y = 3$
Therefore $D = (2, 3)$

b. $\overrightarrow{BD} = \pm\overrightarrow{AC}$
$\overrightarrow{BD} = \langle x - 5, y - 1 \rangle$ and $\overrightarrow{AC} = \langle 5, 2 \rangle$
$\overrightarrow{BD} = \overrightarrow{AC}$ would imply that $x = 10$ and $y = 3$, so $D = (10, 3)$.
$\overrightarrow{BD} = -\overrightarrow{AC}$ would imply that $x = 0$ and $y = -1$, so $D = (0, -1)$.
D is either $(10, 3)$ or $(0, -1)$.

65. By choice of D and E, we have $\overrightarrow{BD} = .5\overrightarrow{BA}$, and $\overrightarrow{BE} = .5\overrightarrow{BC}$. Thus
$\overrightarrow{DE} = \overrightarrow{DB} + \overrightarrow{BE} = -\overrightarrow{BD} + \overrightarrow{BE} = -.5\overrightarrow{BA} + .5\overrightarrow{BC} = .5(-\overrightarrow{BA} + \overrightarrow{BC}) = .5(\overrightarrow{AB} + \overrightarrow{BC}) = .5\overrightarrow{AC}$.

67. The proofs below are for vectors in space. For vectors in the plane, simply drop the third component throughout the proofs. Let $\mathbf{u} = \langle u_1, u_2, u_3 \rangle$, $\mathbf{v} = \langle v_1, v_2, v_3 \rangle$, and $\mathbf{w} = \langle w_1, w_2, w_3 \rangle$.

a. Proof that $(\mathbf{u} + \mathbf{v}) + \mathbf{w} = \mathbf{u} + (\mathbf{v} + \mathbf{w})$:
$(\mathbf{u} + \mathbf{v}) + \mathbf{w} = \langle u_1 + v_1, u_2 + v_2, u_3 + v_3 \rangle + \langle w_1, w_2, w_3 \rangle$
$= \langle (u_1 + v_1) + w_1, (u_2 + v_2) + w_2, (u_3 + v_3) + w_3 \rangle.$
By the associative law of addition for real numbers, this is equal to
$\langle u_1 + (v_1 + w_1), u_2 + (v_2 + w_2), u_3 + (v_3 + w_3) \rangle$
$= \langle u_1, u_2, u_3 \rangle + \langle v_1 + w_1, v_2 + w_2, v_3 + w_3 \rangle$
$= \mathbf{u} + (\mathbf{v} + \mathbf{w}).$

b. Proof that $\mathbf{0} + \mathbf{v} = \mathbf{v}$:

$\mathbf{0} + \mathbf{v} = \langle 0, 0, 0 \rangle + \langle v_1, v_2, v_3 \rangle = \langle 0 + v_1, 0 + v_2, 0 + v_3 \rangle.$

Since 0 is an additive identity for the real numbers, that is, $0 + a = a$ for any real number a, this is equal to $\langle v_1, v_2, v_3 \rangle = \mathbf{v}$.

c. Proof that $\mathbf{v} + (-\mathbf{v}) = \mathbf{0}$:

$\mathbf{v} + (-\mathbf{v}) = \langle v_1, v_2, v_3 \rangle + \langle -v_1, -v_2, -v_3 \rangle = \langle v_1 + (-v_1), v_2 + (-v_2), v_3 + (-v_3) \rangle.$

By the property of additive inverses for the real numbers, that is, that $a + (-a) = 0$ for any real number a, this is equal to $\langle 0, 0, 0 \rangle = \mathbf{0}$.

d. Proof that $1\mathbf{v} = \mathbf{v}$:

$1\mathbf{v} = 1\langle v_1, v_2, v_3 \rangle = \langle 1v_1, 1v_2, 1v_3 \rangle.$

Since 1 is a multiplicative identity for the real numbers, that is, $1a = a$ for any real number a, this is equal to $\langle v_1, v_2, v_3 \rangle = \mathbf{v}$.

e. Proof that $a(\mathbf{v} + \mathbf{w}) = a\mathbf{v} + a\mathbf{w}$:

$a(\mathbf{v} + \mathbf{w}) = a\langle v_1 + w_1, v_2 + w_2, v_3 + w_3 \rangle = \langle a(v_1 + w_1), a(v_2 + w_2), a(v_3 + w_3) \rangle.$

By the distributive law of multiplication over addition for real numbers, that is, $a(b + c) = ab + ac$ for any real numbers a, b and c, this is equal to:

$\langle av_1 + aw_1, av_2 + aw_2, av_3 + aw_3 \rangle = \langle av_1, av_2, av_3 \rangle + \langle aw_1, aw_2, aw_3 \rangle = a\mathbf{v} + a\mathbf{w}.$

f. Proof that $(a + b)\mathbf{v} = a\mathbf{v} + b\mathbf{v}$:

$(a + b)\mathbf{v} = \langle (a + b)v_1, (a + b)v_2, (a + b)v_3 \rangle.$

By the distributive law of multiplication over addition for real numbers, this is equal to:

$\langle av_1 + bv_1, av_2 + bv_2, av_3 + bv_3 \rangle = \langle av_1, av_2, av_3 \rangle + \langle bv_1, bv_2, bv_3 \rangle = a\mathbf{v} + l\,\mathbf{v}.$

g. Proof that $a(b\mathbf{v}) = (ab)\mathbf{v}$:

$a(b\mathbf{v}) = a\langle bv_1, bv_2, bv_3 \rangle = \langle a(bv_1), a(bv_2), a(bv_3) \rangle.$

By the associative law of multiplication for real numbers, this is equal to:

$\langle (ab)v_1, (ab)v_2, (ab)v_3 \rangle = (ab)\mathbf{v}$

69. Let $R = (x, y, z)$ be the point in question. Then $\overrightarrow{PR} = (1/3)\overrightarrow{PQ} = (1/3)\langle 3, -9, -12 \rangle = \langle 1, -3, -4 \rangle$. Since $\overrightarrow{PR} = \langle x + 4, y - 6, z - 1 \rangle$, one obtains the equations $x + 4 = 1$, $y - 6 = -3$, and $z - 1 = -4$. Thus $x = -3$, $y = 3$, and $z = -3$. Therefore $R = (-3, 3, -3)$.

71. Suppose that $\mathbf{w} = a\mathbf{i} + b\mathbf{j}$, $\mathbf{v}_1 = x_1\mathbf{i} + y_1\mathbf{j}$, $\mathbf{v}_2 = x_2\mathbf{i} + y_2\mathbf{j}$, that neither of $\mathbf{v}_1$ nor $\mathbf{v}_2$ is $\mathbf{0}$, and that $\mathbf{v}_1$ and $\mathbf{v}_2$ are not parallel. The goal is to find constants c_1 and c_2, expressed in terms of a, b, x_1, y_1, x_2 and y_2, such that $\mathbf{w} = c_1\mathbf{v}_1 + c_2\mathbf{v}_2$. Rewrite this last equation as

$$a\mathbf{i} + b\mathbf{j} = c_1(x_1\mathbf{i} + y_1\mathbf{j}) + c_2(x_2\mathbf{i} + y_2\mathbf{j}) = (c_1x_1 + c_2x_2)\mathbf{i} + (c_1y_1 + c_2y_2)\mathbf{j}.$$

From this equation one obtains the pair of simultaneous equations

$$c_1x_1 + c_2x_2 = a$$
$$c_1y_1 + c_2y_2 = b$$

which one wishes to solve for c_1 and c_2.

First solve for c_1 by eliminating c_2. Multiply the first equation on both sides by y_2 and the second equation by $-x_2$, and then add the two equations, obtaining $c_1x_1y_2 - c_1x_2y_1 = ay_2 - bx_2$. Solve by factoring out c_1 on the left side and dividing by the other factor. Proceed in a symmetric manner for c_2, multiplying the first equation by $-y_1$ and the second equation by x_1. The solutions obtained are

$$c_1 = \frac{ay_2 - bx_2}{x_1y_2 - x_2y_1} \quad \text{and} \quad c_2 = \frac{bx_1 - ay_1}{x_1y_2 - x_2y_1}.$$

507

Note, however, that these solution are undefined if the quantity $x_1y_2 - x_2y_1$ is zero, which occurs if $x_1y_2 = x_2y_1$. If both sides are nonzero, then $x_1/x_2 = y_1/y_2$ is some number r, in which case

$$\mathbf{v}_1 = x_1\mathbf{i} + y_1\mathbf{j} = rx_2\mathbf{i} + ry_2\mathbf{j} = r(x_2\mathbf{i} + y_2\mathbf{j}) = r\mathbf{v}_2$$

would imply that v_1 and v_2 are parallel, which contradicts our hypotheses. The other possibility is that both x_1y_2 and x_2y_1 are zero. For the first product, if $x_1 = 0$ then $y_1 \neq 0$ since $\mathbf{v}_1 \neq \mathbf{0}$, and so one must have $x_2 = 0$. But this implies that $\mathbf{v}_1 = y_1\mathbf{j}$ and $\mathbf{v}_2 = y_2\mathbf{j}$ must be parallel, again contradicting our hypotheses. The same sort of contradiction will arise if we suppose that $y_2 = 0$ (that $\mathbf{v}_1 = x_1\mathbf{i}$ and $\mathbf{v}_2 = x_2\mathbf{i}$ are parallel). This exhausts all possibilities. Therefore, given our hypotheses, the quantity $x_1y_2 - x_2y_1$ must be nonzero, and therefore our solutions for c_1 and c_2 will exist.

73. **a.** $|\mathbf{v}| = \sqrt{2^2 + (-4)^2 + 4^2} = \sqrt{36} = 6$. The vectors in question are $\pm(2/|\mathbf{v}|)\mathbf{v} = \pm(1/3)(2\mathbf{i} - 4\mathbf{j} + 4\mathbf{k})$. One vector is $(2/3)\mathbf{i} - (4/3)\mathbf{j} + (4/3)\mathbf{k}$ and the other is $-(2/3)\mathbf{i} + (4/3)\mathbf{j} - (4/3)\mathbf{k}$.

   b. One must have $\mathbf{w} = c\mathbf{v}$ for some real number c. $a\mathbf{i} + 9\mathbf{j} - 12\mathbf{k} = \mathbf{w} = c\mathbf{v} = c\mathbf{i} - 3c\mathbf{j} + 4c\mathbf{k}$. Thus $c = -3$, and $a = -3$.

   c. $|\mathbf{v}| = \sqrt{1^2 + (-2)^2 + 3^2} = \sqrt{14}$.
   The vector is $-(5\sqrt{14}/14)\mathbf{v} = -(5\sqrt{14}/14)\mathbf{i} + (5\sqrt{14}/7)\mathbf{j} - (15\sqrt{14}/14)\mathbf{k}$.

   d. There must be a real number c such that $a\mathbf{i} + b\mathbf{j} - 2\mathbf{k} = c(3\mathbf{i} - \mathbf{j} + 4\mathbf{k}) = 3c\mathbf{i} - c\mathbf{j} + 4c\mathbf{k}$. Since $4c = -2$, one must have $c = -1/2$ and thus $a = -3/2$ and $b = 1/2$.

75. Suppose that $\mathbf{w} = x\mathbf{i} + y\mathbf{j} + z\mathbf{k}$. In order to have $\mathbf{w} = c_1\mathbf{v}_1 + c_2\mathbf{v}_2 + c_3\mathbf{v}_3$, one must have $x\mathbf{i} + y\mathbf{j} + z\mathbf{k} = c_1(\mathbf{i} + \mathbf{j}) + c_2(\mathbf{j} + \mathbf{k}) + c_3(\mathbf{i} + 2\mathbf{j} + 3\mathbf{k}) = (c_1 + c_3)\mathbf{i} + (c_1 + c_2 + 2c_3)\mathbf{j} + (c_2 + 3c_3)\mathbf{k}$. In this manner, one obtains the system of equations $x = c_1 + c_3$, $y = c_1 + c_2 + 2c_3$, and $z = c_2 + 3c_3$. In order to solve for c_1, c_2 and c_3, solve for c_1 in the first equation, $c_1 = x - c_3$, and substitute this into the second equation: $y = x - c_3 + c_2 + 2c_3$. The second and third equations then become the pair of simultaneous equations $c_2 + c_3 = y - x$ and $c_2 + 3c_3 = z$. Subtracting the first of this pair from the second gives $2c_3 = z - (y - x)$ and $c_3 = (x - y + z)/2$. Substituting this value for c_3 in the first equation of the pair gives $c_2 + (x - y + z)/2 = y - x$ and solving for c_2 yields $c_2 = -(3x - 3y + z)/2$. Substituting the value obtained for c_3 into the original first equation gives $c_1 + (x - y + z)/2 = x$, and solving for c_1 yields $c_1 = (x + y - z)/2$. Therefore, for any vector $\mathbf{w} = x\mathbf{i} + y\mathbf{j} + z\mathbf{k}$, one may express $\mathbf{w}$ as $c_1\mathbf{v}_1 + c_2\mathbf{v}_2 + c_3\mathbf{v}_3$ by choosing $c_1 = (x + y - z)/2$, $c_2 = -(3x - 3y + z)/2$ and $c_3 = (x - y + z)/2$.

77. Ignoring all downstream motion, her velocity perpendicular to the banks is 30 meters per minute, and she will therefore cover 150 meters in 5 minutes.

79. Let $\mathbf{i}$ be a unit vector (length 1 kilometer) due east, and let $\mathbf{j}$ be a unit vector due north. Let $\mathbf{0}$ be the position of the airplane when our observation of it begins. The velocity of the airplane is $30\mathbf{i} + 200\mathbf{j}$ in kilometers per hour. After t hours, the position of the airplane will be $30t\mathbf{i} + 200t\mathbf{j}$.

   a. At $t = 2$, the position is $60\mathbf{i} + 400\mathbf{j}$, or 400 km north and 60 km east of the starting point.

   b. One has a right triangle with legs of length 60 and 400. The distance traveled is $\sqrt{60^2 + 400^2} = \sqrt{163600} = 20\sqrt{409}$ or approximately 404.5 km.

   c. Let the velocity be $x\mathbf{i} + y\mathbf{j}$. Then the actual velocity of the plane is $(x + 30)\mathbf{i} + y\mathbf{j}$. If movement is to be due north, then one must have $x + 30 = 0$ or $x = -30$. Assuming that its speed is still 200 km/hr, one obtains a right triangle with hypotenuse of length 200 and sides of length $|x| = 30$ and y. If θ is the angle with respect to due north, then θ is the angle between the sides of length y and 200, and thus $\sin\theta = 30/200 = 0.15$. The heading desired is therefore $\operatorname{Sin}^{-1}(0.15)$, or approximately 0.15 radians (8.6°) west of due north.

81. Let w be the weight of the block. $\sin 45° = 50/w$ implies that $w = 50/\sin 45° = 50/(1/\sqrt{2}) = 50\sqrt{2}$. The weight of the block is $50\sqrt{2}$ or approximately 70.7 pounds.

16.2   Lines in Space

1. $\langle 1, 2, 3 \rangle + t \langle 1, 1, -1 \rangle = \langle 1 + t, 2 + t, 3 - t \rangle$
$x(t) = 1 + t$, $y(t) = 2 + t$, $z(t) = 3 - t$

3. $\langle 3, -1, 5 \rangle + t \langle 3, -1, 5 \rangle = \langle 3 + 3t, -1 - t, 5 + 5t \rangle$
$x(t) = 3 + 3t$, $y(t) = -1 - t$, $z(t) = 5 + 5t$

5. Direction vector $\overrightarrow{PQ} = \langle 1, 3, 2 \rangle$
Line $\langle -4, 2, 1 \rangle + t \langle 1, 3, 2 \rangle = \langle -4 + t, 2 + 3t, 1 + 2t \rangle$
$x(t) = -4 + t$, $y(t) = 2 + 3t$, $z(t) = 1 + 2t$

7. Direction vector $\langle \pi, 2, \sqrt{2} \rangle$
Line $\langle 0, 0, 0 \rangle + t \langle \pi, 2, \sqrt{2} \rangle = \langle \pi t, 2t, \sqrt{2} t \rangle$
$x(t) = \pi t$, $y(t) = 2t$, $z(t) = \sqrt{2} t$

9. Solve for t.
$$x - 7 = \frac{y + 6}{4} = \frac{3 - z}{2}$$

11. Solve for t.
$$x = \frac{y + 6}{8} = \frac{z - 4}{4}$$

13. $x(t) = 3 + 3t$, $y(t) = -1 - t$, $z(t) = 5 + 5t$
$$\frac{x - 3}{3} = -y - 1 = \frac{z - 5}{5}$$

15. Set t equal to each expression, then solve for x, y and z.
$x(t) = 3t$, $y(t) = 2t$, $z(t) = 5t$

17. $x(t) = -4 + 4t$, $y(t) = 2 - 2t$, $z(t) = -3 + 3t$

19. The line is $\langle t + 7, 4t - 6, 3 - 2t \rangle = \langle 7, -6, 3 \rangle + t \langle 1, 4, -2 \rangle$.
The desired vector is $\langle 1, 4, -2 \rangle$.

21. The parametric equations are $x(t) = -4 + 4t$, $y(t) = 2 - 2t$, $z(t) = -3 + 3t$.
The line is $\langle -4 + 4t, 2 - 2t, -3 + 3t \rangle = \langle -4, 2, -3 \rangle + t \langle 4, -2, 3 \rangle$.
The desired vector is $\langle 4, -2, 3 \rangle$.

23. $\mathbf{r}_1(t) = \langle 2 + 5t, 1 + t, 2 + 3t \rangle$
$\mathbf{r}_2(t) = \langle -4 + 3t, 7 - 3t, 10 - 4t \rangle$
$2 + 5t_1 = -4 + 3t_2$
$1 + t_1 = 7 - 3t_2$
$2 + 3t_1 = 10 - 4t_2$
Adding the first two equations gives $3 + 6t_1 = 3$, which implies that $t_1 = 0$. Substituting this value into the first equation yields $t_2 = 2$. $t_1 = 0$ and $t_2 = 2$ are solutions to all three equations. The point of intersection is the terminal point of $\mathbf{r}_1(0)$, $(2, 1, 2)$.

25. $1 + t_1 = 2 + 2t_2$
$2 - 2t_1 = 5 - 9t_2$
$t_1 + 5 = 2 + 6t_2$
Subtracting the first equation from the third gives $4 = 4t_2$, which implies that $t_2 = 1$. Substituting this value into the first equation yields $t_1 = 3$. $t_1 = 3$ and $t_2 = 1$ are solutions to all three equations. The point of intersection has coordinates $x(3) = 4$, $y(3) = -4$, $z(3) = 8$, and is therefore the point $(4, -4, 8)$.

27. The parametric equations for the first line are $x(t) = 3 - t$, $y(t) = -2 + t$, $z(t) = -3 + t$, and for the second line are $x(t) = -1 + t$, $y(t) = t$, $z(t) = -1 + 2t$.
$3 - t_1 = -1 + t_2$
$-2 + t_1 = t_2$
$-3 + t_1 = -1 + 2t_2$
Substituting the value for t_2 from the second equation into the first equation gives $3 - t_1 = -3 + t_1$, which implies that $t_1 = 3$. Substituting this value into the second equation yields $t_2 = 1$. However, $t_1 = 3$ and $t_2 = 1$ are not solutions to the third equation. Therefore these two lines do not intersect.

29. Direction vector $\langle 1, 8, 4 \rangle$
Line $\langle 1, 6, -5 \rangle + t \langle 1, 8, 4 \rangle = \langle 1 + t, 6 + 8t, -5 + 4t \rangle$
$x(t) = 1 + t$, $y(t) = 6 + 8t$, $z(t) = -5 + 4t$

31. Direction vector $\langle 3, 2, 5 \rangle$
Line $\langle 1, 2, 1 \rangle + t \langle 3, 2, 5 \rangle = \langle 1 + 3t, 2 + 2t, 1 + 5t \rangle$
$x(t) = 1 + 3t$, $y(t) = 2 + 2t$, $z(t) = 1 + 5t$
Solve for t.
$$\frac{x - 1}{3} = \frac{y - 2}{2} = \frac{z - 1}{5}$$

33. Direction vector $\langle 1, 4, -2 \rangle$
$\mathbf{r}(t) = \langle 1, 2, 4 \rangle + t \langle 1, 4, -2 \rangle$ or
$\mathbf{r}(t) = (\mathbf{i} + 2\mathbf{j} + 4\mathbf{k}) + t(\mathbf{i} + 4\mathbf{j} - 2\mathbf{k})$

35. The direction vectors are $l_1 = \langle 4, -2, 2 \rangle$, $l_2 = \langle 1, 1, -1 \rangle$, $l_3 = \langle 1, 1, 1 \rangle$, $l_4 = \langle 2, -1, 1 \rangle$ and $l_5 = \langle 2, 2, 2 \rangle$. Two lines are parallel if the direction vector for one is a multiple of the direction vector for the other. From this we see that l_1 and l_4 are parallel, and l_3 and l_5 are parallel. Two parallel lines are equal if and only if a vector from a point on one line to a point on the other line is parallel to their direction vectors.

For l_1 and l_4, choose point $P = (3, 1, 2)$ on l_1 and point $Q = (1, 2, 1)$ on l_4. $\overrightarrow{PQ} = \langle -2, 1, -1 \rangle$ is parallel to their direction vectors, and therefore the lines l_1 and l_4 are equal.

For l_3 and l_5, choose point $P = (1, 2, -1)$ on l_3 and point $Q = (3, 4, 2)$ on l_5. $\overrightarrow{PQ} = \langle 2, 2, 3 \rangle$, which is not parallel to the direction vectors. Therefore the lines l_3 and l_5 are parallel but not equal.

16.3 The Dot Product

1. $\mathbf{v} \cdot \mathbf{w} = (1)(3) + (2)(-1) = 3 - 2 = 1$

3. $\mathbf{v} \cdot \mathbf{w} = (-3)(6) + (5)(-2) = -18 - 10 = -28$

5. $\mathbf{v} \cdot \mathbf{w} = (2)(1) + (1)(-2) + (-1)(3) = 2 - 2 - 3 = -3$

7. $\mathbf{v} \cdot \mathbf{w} = (1)(0) + (4)(1) + (-1)(2) = 0 + 4 - 2 = 2$

9. $\mathbf{v} \cdot \mathbf{w} = (1)(-4) + (-3)(1) = -7$
$|\mathbf{v}| = \sqrt{1^2 + (-3)^2} = \sqrt{10}$
$|\mathbf{w}| = \sqrt{(-4)^2 + 1^2} = \sqrt{17}$
$\cos\theta = (\mathbf{v} \cdot \mathbf{w})/(|\mathbf{v}||\mathbf{w}|) = -7/\sqrt{170}$
$\cos\theta = -7\sqrt{170}/170$

11. $\mathbf{v} \cdot \mathbf{w} = (2)(1) + (-1)(-1) + (2)(1) = 5$
$|\mathbf{v}| = \sqrt{2^2 + (-1)^2 + 2^2} = \sqrt{9} = 3$
$|\mathbf{w}| = \sqrt{1^2 + (-1)^2 + 1^2} = \sqrt{3}$
$\cos\theta = (\mathbf{v} \cdot \mathbf{w})/(|\mathbf{v}||\mathbf{w}|) = 5/(3\sqrt{3})$
$\cos\theta = 5\sqrt{3}/9$

13. a. $\mathbf{u} \cdot \mathbf{v} = (-2)(3) + (4)(5) = 14$
 b. $\mathbf{u} \cdot \mathbf{w} = (-2)(6) + (4)(-4) = -28$
 c. $\mathbf{u} \cdot (\mathbf{v} + \mathbf{w}) = \mathbf{u} \cdot (9\mathbf{i} + \mathbf{j}) = (-2)(9) + (4)(1) = -14$
 d. $\mathbf{u} \cdot (\mathbf{v} - \mathbf{w}) = \mathbf{u} \cdot (-3\mathbf{i} + 9\mathbf{j}) = (-2)(-3) + (4)(9) = 42$
 e. $(\mathbf{u} + \mathbf{v}) \cdot (\mathbf{v} - \mathbf{w}) = (\mathbf{i} + 9\mathbf{j}) \cdot (-3\mathbf{i} + 9\mathbf{j}) = (1)(-3) + (9)(9) = 78$
 f. $\mathbf{u} \cdot (2\mathbf{v} + 3\mathbf{w}) = \mathbf{u} \cdot ((6\mathbf{i} + 10\mathbf{j}) + (18\mathbf{i} - 12\mathbf{j})) = \mathbf{u} \cdot (24\mathbf{i} - 2\mathbf{j}) = (-2)(24) + (4)(-2) = -56$
 g. $\text{comp}_\mathbf{v}\, \mathbf{u} = (\mathbf{v} \cdot \mathbf{u})/|\mathbf{v}| = ((3)(-2) + (5)(4))/\sqrt{3^2 + 5^2} = 14/\sqrt{34} = 14\sqrt{34}/34 = 7\sqrt{34}/17$
 h. $\text{comp}_\mathbf{w}\, \mathbf{v} = (\mathbf{w} \cdot \mathbf{v})/|\mathbf{w}| = ((6)(3) + (-4)(5))/\sqrt{6^2 + (-4)^2} = -2/\sqrt{52} = -2/(2\sqrt{13}) = -\sqrt{13}/13$

i. $\mathbf{proj_u\,v} = [(\mathbf{u\cdot v})/|\mathbf{u}|^2]\,\mathbf{u} = [((-2)(3)+(4)(5))/((-2)^2+4^2)]\,\mathbf{u} = [14/20]\mathbf{u} = [7/10]\mathbf{u}$
$= -(7/5)\mathbf{i}+(14/5)\mathbf{j}$

j. $\mathbf{proj_{\perp w}\,u} = \mathbf{u} - \mathbf{proj_w\,u}$
$\mathbf{proj_w\,u} = [(\mathbf{w\cdot u})/|\mathbf{w}|^2]\,\mathbf{w} = [((6)(-2)+(-4)(4))/(6^2+(-4)^2)]\,\mathbf{w} = [-28/52]\mathbf{w} = -7/13\mathbf{w} =$
$-(42/13)\mathbf{i}+(28/13)\mathbf{j}$
$\mathbf{proj_{\perp w}\,u} = (-2\mathbf{i}+4\mathbf{j}) - (-(42/13)\mathbf{i}+(28/13)\mathbf{j}) = (16/13)\mathbf{i}+(24/13)\mathbf{j}$

15. a. $0 = \mathbf{v\cdot w} = 3a+4$. Thus $a = -4/3$.

   b. $10 = \mathbf{v\cdot w} = 3a+4$. Thus $a = 2$.

   c. This will occur precisely if $\mathbf{v}$ is parallel to $\mathbf{w}$, if there is a real number c such that $\mathbf{v}=c\mathbf{w}$.
   $a\mathbf{i}+\mathbf{j} = c(3\mathbf{i}+4\mathbf{j}) = 3c\mathbf{i}+4c\mathbf{j}$
   $4c = 1$ implies that $c = 1/4$
   $a = 3c = 3/4$

   d. This will occur precisely if $\mathbf{v}$ is orthogonal to $\mathbf{w}$, if $\mathbf{v\cdot w}=0$. This is the requirement for part (a).
   Thus $a = -4/3$.

17. $\mathbf{v\cdot(u+2w)} = \mathbf{v}\cdot(\langle 2,-1,5\rangle + \langle 6,6,2\rangle) = \mathbf{v}\cdot\langle 8,5,7\rangle = (-3)(8)+(5)(5)+(0)(7) = -24+25+0 = 1$

19. $\mathbf{proj_w\,v} = [(\mathbf{w\cdot v})/|\mathbf{w}|^2]\,\mathbf{w} = [6/(3^2+3^2+1^2)]\,\mathbf{w} = [6/19]\mathbf{w} = \langle 18/19,18/19,6/19\rangle$

21. $\mathbf{comp_w\,v} = (\mathbf{w\cdot v})/|\mathbf{w}| = 6/\sqrt{3^2+3^2+1^2} = 6/\sqrt{19} = 6\sqrt{19}/19$

23. $\mathbf{proj_u\,u} = [(\mathbf{u\cdot u})/|\mathbf{u}|^2]\,\mathbf{u} = [|\mathbf{u}|^2/|\mathbf{u}|^2]\,\mathbf{u} = \mathbf{u} = \langle 2,-1,5\rangle$

25. $|\mathbf{v\cdot w}+\mathbf{w\cdot u}| = |(3)(5)+(-2)(-1)+(2)(3)+(5)(1)+(-1)(2)+(3)(-1)| = |23+0| = |23| = 23$

27. $\mathbf{proj_w\,v} = [(\mathbf{w\cdot v})/|\mathbf{w}|^2]\,\mathbf{w} = [23/(5^2+(-1)^2+3^2)]\,\mathbf{w} = [23/35]\mathbf{w} = (23/7)\mathbf{i}-(23/35)\mathbf{j}+(69/35)\mathbf{k}$

29. $\mathbf{comp_u\,v} = (\mathbf{u\cdot v})/|\mathbf{u}| = ((1)(3)+(2)(-2)+(-1)(2))/\sqrt{1^2+2^2+(-1)^2} = -3/\sqrt{6} = -3\sqrt{6}/6 = -\sqrt{6}/2$

31. a. $\mathbf{v\cdot w} = 10-4 = 6 \neq 0$: not orthogonal

   b. $\mathbf{v\cdot w} = 30-30 = 0$: orthogonal

   c. $\mathbf{v\cdot w} = 6-6 = 0$: orthogonal

   d. $\mathbf{v\cdot w} = -1-1 = -2 \neq 0$: not orthogonal

33. Let $\mathbf{u} = a\mathbf{i}+b\mathbf{j}$.
   $0 = \mathbf{u\cdot v} = 2a+b$
   One solution is $a=1$, $b=-2$.
   $\mathbf{u} = \mathbf{i}-2\mathbf{j}$ is orthogonal to $\mathbf{v}$.
   Unit vectors orthogonal to $\mathbf{v}$ are $\pm\mathbf{u}/|\mathbf{u}|$.
   $|\mathbf{u}| = \sqrt{1^2+(-2)^2} = \sqrt{5}$
   $\pm\mathbf{u}/|\mathbf{u}| = (\pm\sqrt{5}/5)\mathbf{i}+(\mp 2\sqrt{5}/5)\mathbf{j}$

35. The points form the vertices of a right triangle if some pair of the vectors $\overrightarrow{PQ}, \overrightarrow{QR}$ and $\overrightarrow{RP}$ are orthogonal.

   a. $\overrightarrow{PQ} = \langle 3,-2,2\rangle$, $\overrightarrow{QR} = \langle 2,4,1\rangle$, $\overrightarrow{RP} = \langle -5,-2,-3\rangle$
   $\overrightarrow{PQ}\cdot\overrightarrow{QR} = 6-8+2 = 0$
   This is a right triangle.

b. $\overrightarrow{PQ} = \langle 1, 1, -4 \rangle$, $\overrightarrow{QR} = \langle 0, 1, 4 \rangle$, $\overrightarrow{RP} = \langle -1, -2, 0 \rangle$

$\overrightarrow{PQ} \cdot \overrightarrow{QR} = 0 + 1 - 16 \neq 0$

$\overrightarrow{QR} \cdot \overrightarrow{RP} = 0 - 2 + 0 \neq 0$

$\overrightarrow{RP} \cdot \overrightarrow{PQ} = -1 - 2 + 0 \neq 0$

This is not a right triangle.

c. $\overrightarrow{PQ} = \langle 4, -4, 0 \rangle$, $\overrightarrow{QR} = \langle 1, 1, 0 \rangle$, $\overrightarrow{RP} = \langle -5, 3, 0 \rangle$

$\overrightarrow{PQ} \cdot \overrightarrow{QR} = 4 - 4 + 0 = 0$

This is a right triangle.

37. $x + 2y - z = d$
$d = 1 + 2(2) - (-3) = 8$
$x + 2y - z = 8$

39. $2x - 2y + 3z = d$
$d = 2(0) - 2(-2) + 3(5) = 19$
$2x - 2y + 3z = 19$

41. $x + 2y + z = d$
$d = 0 + 2(-1) + 3 = 1$
$x + 2y + z = 1$

43. $\mathbf{n} = \langle 1, 2, 1 \rangle$
Unit vectors will be $\pm \mathbf{n}/|\mathbf{n}|$
$|\mathbf{n}| = \sqrt{1^2 + 2^2 + 1^2} = \sqrt{6}$
$\pm \mathbf{n}/|\mathbf{n}| = \pm(\sqrt{6}/6)\langle 1, 2, 1 \rangle$
$= \pm\langle \sqrt{6}/6, \sqrt{6}/3, \sqrt{6}/6 \rangle$

45. $\mathbf{i} = \mathbf{proj_v}\,\mathbf{i} + \mathbf{proj_{\perp v}}\,\mathbf{i}$
$\mathbf{proj_v}\,\mathbf{i} = [(\mathbf{v} \cdot \mathbf{i})/|\mathbf{v}|^2]\,\mathbf{v}$
$= [2/5]\mathbf{v}$
$= (4/5)\mathbf{i} - (2/5)\mathbf{j}$
$\mathbf{proj_{\perp v}}\,\mathbf{i} = \mathbf{i} - \mathbf{proj_v}\,\mathbf{i}$
$= \mathbf{i} - ((4/5)\mathbf{i} - (2/5)\mathbf{j})$
$= (1/5)\mathbf{i} + (2/5)\mathbf{j}$
$\mathbf{i} = (0.8\mathbf{i} - 0.4\mathbf{j}) + (0.2\mathbf{i} + 0.4\mathbf{j})$

47. The line has parametric equations $x = t$, $y = 1 - t$, and thus its vector equation is $\langle t, 1 - t \rangle = \langle 0, 1 \rangle + t\langle 1, -1 \rangle$. Its direction vector is therefore $\mathbf{v} = \langle 1, -1 \rangle$, and one point on the line is $Q = (0, 1)$. Let P be the point $(3, 6)$. The distance from P to the line is $|\mathbf{proj_{\perp v}}\,\overrightarrow{PQ}|$. Note that $\overrightarrow{PQ} = \langle -3, -5 \rangle$.

$\mathbf{proj_v}\,\overrightarrow{PQ} = \left[(\mathbf{v} \cdot \overrightarrow{PQ})/|\mathbf{v}|^2\right]\mathbf{v} = [2/2]\mathbf{v} = \langle 1, -1 \rangle$

$\mathbf{proj_{\perp v}}\,\overrightarrow{PQ} = \overrightarrow{PQ} - \mathbf{proj_v}\,\overrightarrow{PQ} = \langle -3, -5 \rangle - \langle 1, -1 \rangle = \langle -4, -4 \rangle$

$|\mathbf{proj_{\perp v}}\,\overrightarrow{PQ}| = \sqrt{(-4)^2 + (-4)^2} = \sqrt{32} = 4\sqrt{2}$

The distance from the point to the line is $4\sqrt{2}$.

49. The line has parametric equations $x = 2 + 3t$, $y = 1 - 6t$, $z = -2 + 2t$, and thus has vector equation $\langle 2 + 3t, 1 - 6t, -2 + 2t \rangle = \langle 2, 1, -2 \rangle + t\langle 3, -6, 2 \rangle$. Its direction vector $\langle 3, -6, 2 \rangle$ is a normal vector for the plane. The plane therefore has equation
$3x - 6y + 2z = d$
$d = 3(3) - 6(-2) + 2(5) = 31$
$3x - 6y + 2z = 31$

51. Normal vectors for the planes are $\mathbf{n_1} = \langle 1, 1, 1 \rangle$ and $\mathbf{n_2} = \langle 3, -1, 2 \rangle$. The angle θ between them has cosine $\cos\theta = (\mathbf{n_1} \cdot \mathbf{n_2})/(|\mathbf{n_1}||\mathbf{n_2}|) = 4/(\sqrt{3}\sqrt{14}) = 4\sqrt{42}/42 = 2\sqrt{42}/21$
$\theta = \text{Cos}^{-1}(2\sqrt{42}/21) \approx 0.9056$ (or $51.89°$)

53. The normal vector for the plane is $\mathbf{n} = \langle 6, -1, 1 \rangle$ and one point on the plane is $Q = (0, 0, 2)$. The distance from P to the plane is $|\mathbf{proj_n}\,\overrightarrow{PQ}|$. Note that $\overrightarrow{PQ} = \langle -3, 6, -3 \rangle$.

$$\mathbf{proj_n}\,\overrightarrow{PQ} = \left[(\mathbf{n} \cdot \overrightarrow{PQ})/|\mathbf{n}|^2 \right] \mathbf{n} = [-27/38]\mathbf{n}$$

$$|\mathbf{proj_n}\,\overrightarrow{PQ}| = |-27/38||\mathbf{n}| = (27/38)\sqrt{38} = 27\sqrt{38}/38$$

The distance from P to the plane is $27\sqrt{38}/38$.

55. A direction vector for the line is the normal vector $\langle 2, 3, -7 \rangle$ for the plane. The vector equation for the line is $\langle 2, 4, -3 \rangle + t \langle 2, 3, -7 \rangle = \langle 2 + 2t, 4 + 3t, -3 - 7t \rangle$. Therefore, the parametric equations for the line are $x(t) = 2 + 2t$, $y(t) = 4 + 3t$, $z(t) = -3 - 7t$

57. Let $\mathbf{i}$ be a unit vector (1 foot) pointing along the ground in the direction which the sled moves, and let $\mathbf{j}$ be a unit vector pointing straight up from the ground. The distance vector is $\mathbf{d} = 100\mathbf{j}$. If $\mathbf{F}$ is the force vector, then $|\mathbf{F}| = 20$ pounds, and the angle θ between $\mathbf{F}$ and $\mathbf{j}$ is $45°$. This is also the angle between $\mathbf{F}$ and $\mathbf{d}$. The work done is $W = \mathbf{F} \cdot \mathbf{d} = |\mathbf{F}||\mathbf{d}|\cos\theta = (20)(100)\cos 45° = 2000\sqrt{2}/2 = 1000\sqrt{2}$ foot-pounds.

59. a. $|\mathbf{v}| = \sqrt{3^2 + (-1)^2 + 2^2} = \sqrt{14}$

$\cos\alpha = 3\sqrt{14}/14$, $\cos\beta = -\sqrt{14}/14$, $\cos\gamma = 2\sqrt{14}/14 = \sqrt{14}/7$

b. $|\mathbf{v}| = \sqrt{6^2 + (-2)^2 + 1^2} = \sqrt{41}$

$\cos\alpha = 6\sqrt{41}/41$, $\cos\beta = -2\sqrt{41}/41$, $\cos\gamma = \sqrt{41}/41$

c. $\overrightarrow{PQ} = \langle -1, 2, -4 \rangle$

$|\overrightarrow{PQ}| = \sqrt{(-1)^2 + 2^2 + (-4)^2} = \sqrt{21}$

$\cos\alpha = -\sqrt{21}/21$, $\cos\beta = 2\sqrt{21}/21$, $\cos\gamma = -4\sqrt{21}/21$

61. $\mathbf{u} = 1\,[(\cos 45°)\mathbf{i} + (\cos 60°)\mathbf{j} + (\cos\gamma)\mathbf{k}] = (\sqrt{2}/2)\mathbf{i} + (1/2)\mathbf{j} + (\cos\gamma)\mathbf{k}$

$1 = |\mathbf{u}| = \sqrt{(\sqrt{2}/2)^2 + (1/2)^2 + \cos^2\gamma} = \sqrt{3/4 + \cos^2\gamma}$

$1 = 1^2 = 3/4 + \cos^2\gamma$

$\cos^2\gamma = 1 - 3/4 = 1/4$

$\cos\gamma = \pm 1/2$

$\mathbf{u} = (\sqrt{2}/2)\mathbf{i} + (1/2)\mathbf{j} + (1/2)\mathbf{k}$ or $\mathbf{u} = (\sqrt{2}/2)\mathbf{i} + (1/2)\mathbf{j} - (1/2)\mathbf{k}$

63. $(\mathbf{v} + \mathbf{w}) \cdot (\mathbf{v} - \mathbf{w}) = \mathbf{v} \cdot \mathbf{v} - \mathbf{v} \cdot \mathbf{w} + \mathbf{w} \cdot \mathbf{v} - \mathbf{w}.\mathbf{w} = |\mathbf{v}|^2 - |\mathbf{w}|^2$. This is equal to zero if $|\mathbf{v}| = |\mathbf{w}|$, and therefore $\mathbf{v} + \mathbf{w}$ and $\mathbf{v} - \mathbf{w}$ would be orthogonal. Given any rhombus $ABCD$, one has $|\overrightarrow{AB}| = |\overrightarrow{AD}| \neq 0$, and therefore $\overrightarrow{AB} + \overrightarrow{AD} = \overrightarrow{AC}$ and $\overrightarrow{AB} - \overrightarrow{AD} = \overrightarrow{AB} + \overrightarrow{DA} = \overrightarrow{DB}$ are orthogonal. Thus the diagonals AC and DB are perpendicular.

65. If $|\mathbf{v}||\mathbf{w}| = \mathbf{v} \cdot \mathbf{w} = |\mathbf{v}||\mathbf{w}|\cos\theta$, and $\mathbf{v} \neq \mathbf{0}$ and $\mathbf{w} \neq \mathbf{0}$, then $\cos\theta = 1$, so $\theta = 0$. Therefore $\mathbf{v} \cdot \mathbf{w} = |\mathbf{v}||\mathbf{w}|$ only if $\mathbf{v}$ and $\mathbf{w}$ point in the same direction. (Saying that $\mathbf{v}$ and $\mathbf{w}$ are parallel is insufficient, since this includes the case where $\theta = \pi$, for which $\mathbf{v} \cdot \mathbf{w} = -|\mathbf{v}||\mathbf{w}|$.)

67. Given that $\mathbf{v}_1$, $\mathbf{v}_2$ and $\mathbf{v}_3$ are mutually orthogonal vectors, so that the dot product of any pair of them is zero. If $c_1\mathbf{v}_1 + c_2\mathbf{v}_2 + c_3\mathbf{v}_3 = \mathbf{0}$, then

$$0 = |\mathbf{0}|^2 = |c_1\mathbf{v}_1 + c_2\mathbf{v}_2 + c_3\mathbf{v}_3|^2$$
$$= (c_1\mathbf{v}_1 + c_2\mathbf{v}_2 + c_3\mathbf{v}_3) \cdot (c_1\mathbf{v}_1 + c_2\mathbf{v}_2 + c_3\mathbf{v}_3)$$
$$= c_1^2\mathbf{v}_1 \cdot \mathbf{v}_1 + c_1c_2\mathbf{v}_1 \cdot \mathbf{v}_2 + c_1c_3\mathbf{v}_1 \cdot \mathbf{v}_3 + c_2c_1\mathbf{v}_2 \cdot \mathbf{v}_1 + c_2^2\mathbf{v}_2 \cdot \mathbf{v}_2 + c_2c_3\mathbf{v}_2 \cdot \mathbf{v}_3 +$$
$$c_3c_1\mathbf{v}_3 \cdot \mathbf{v}_1 + c_3c_2\mathbf{v}_3 \cdot \mathbf{v}_2 + c_3^2\mathbf{v}_3 \cdot \mathbf{v}_3$$
$$= c_1^2|\mathbf{v}_1|^2 + c_2^2|\mathbf{v}_2|^2 + c_3^2|\mathbf{v}_3|^2.$$

A sum of nonnegative values can only be zero if each term is zero. Since $|\mathbf{v}_1|$, $|\mathbf{v}_2|$ and $|\mathbf{v}_3|$ are all nonzero, one must have $c_1 = c_2 = c_3 = 0$.

69. Let $P = (a_1, b_1)$ and $Q = (a_2, b_2)$, and let $X = (x, y)$ be any point on the circle. Note that $\overrightarrow{XP} = \langle a_1 - x, b_1 - y \rangle$ and $\overrightarrow{XQ} = \langle a_2 - x, b_2 - y \rangle$. By the previous exercise:

$$0 = \overrightarrow{XP} \cdot \overrightarrow{XQ}$$
$$= (a_1 - x)(a_2 - y) + (b_2 - x)(b_2 - y)$$
$$= a_1a_2 - (a_1 + a_2)x + x^2 + b_1b_2 - (b_1 + b_2)y + y^2$$
$$= x^2 + y^2 - (a_1 + a_2)x - (b_1 + b_2)y + (a_1a_2 + b_1b_2).$$

71. Two planes are parallel if their normal vectors are parallel. The normal vectors of the five planes are $\mathbf{n}_1 = \langle 1, 1, -2 \rangle$, $\mathbf{n}_2 = \langle 1, -2, 1 \rangle$, $\mathbf{n}_3 = \langle -1, 4, 1 \rangle$, $\mathbf{n}_4 = \langle 2, -4, 2 \rangle$ and $\mathbf{n}_5 = \langle 3, 3, -6 \rangle$. The vectors which are parallel are $\mathbf{n}_1$ and $\mathbf{n}_5$, and $\mathbf{n}_2$ and $\mathbf{n}_4$. Therefore the planes σ_1 and σ_5 are parallel, and the planes σ_2 and σ_4 are parallel.

A pair of parallel planes are equal if they have at least one point in common, and if there is at least one point on one which is not on the other, then they are not equal. The point $(4, 0, 0)$ is on plane σ_1, but is not on σ_5, and therefore these planes are not equal. The point $(4, 0, 0)$ is also on both planes σ_2 and σ_4, and therefore these planes are equal.

16.4 The Cross Product

1. $\mathbf{v} \times \mathbf{w} = \det \begin{bmatrix} \mathbf{i} & \mathbf{j} & \mathbf{k} \\ 2 & -1 & 2 \\ 3 & 4 & -1 \end{bmatrix}$
$= (1 - 8)\mathbf{i} + (6 - (-2))\mathbf{j} + (8 + 3)\mathbf{k}$
$= -7\mathbf{i} + 8\mathbf{j} + 11\mathbf{k}$

3. $\mathbf{w} \times \mathbf{u} = \det \begin{bmatrix} \mathbf{i} & \mathbf{j} & \mathbf{k} \\ 3 & 4 & -1 \\ 1 & 1 & 1 \end{bmatrix}$
$= (4 - (-1))\mathbf{i} + (-1 - 3)\mathbf{j} + (3 - 4)\mathbf{k}$
$= 5\mathbf{i} - 4\mathbf{j} - \mathbf{k}$

5. $\mathbf{u} \cdot (\mathbf{v} \times \mathbf{w}) = \mathbf{u} \cdot (-7\mathbf{i} + 8\mathbf{j} + 11\mathbf{k})$
$= -7 + 8 + 11 = 12$

7. $\mathbf{u} \times \mathbf{v} = \det \begin{bmatrix} \mathbf{i} & \mathbf{j} & \mathbf{k} \\ 2 & -1 & 4 \\ 1 & 0 & -3 \end{bmatrix}$
$= (3 - 0)\mathbf{i} + (4 - (-6))\mathbf{j} + (0 - (-1))\mathbf{k}$
$= 3\mathbf{i} + 10\mathbf{j} + \mathbf{k}$

9. $\mathbf{u} \times \mathbf{w} = \det \begin{bmatrix} \mathbf{i} & \mathbf{j} & \mathbf{k} \\ 2 & -1 & 4 \\ 2 & 3 & 4 \end{bmatrix}$

$= (-4 - 12)\mathbf{i} + (8 - 8)\mathbf{j} + (6 - (-2))\mathbf{k}$

$= -16\mathbf{i} + 8\mathbf{k}$

11. $\mathbf{v} \times \mathbf{w} = \det \begin{bmatrix} \mathbf{i} & \mathbf{j} & \mathbf{k} \\ 1 & 0 & -3 \\ 2 & 3 & 4 \end{bmatrix}$

$= (0 - (-9))\mathbf{i} + (-6 - 4)\mathbf{j} + (3 - 0)\mathbf{k}$

$= 9\mathbf{i} - 10\mathbf{j} + 3\mathbf{k}$

$\mathbf{u} \times (\mathbf{v} \times \mathbf{w}) = \mathbf{u} \times (9\mathbf{i} - 10\mathbf{j} + 3\mathbf{k})$

$= \det \begin{bmatrix} \mathbf{i} & \mathbf{j} & \mathbf{k} \\ 2 & -1 & 4 \\ 9 & -10 & 3 \end{bmatrix}$

$= (-3 - (-40))\mathbf{i} + (36 - 6)\mathbf{j} + (-20 - (-9))\mathbf{k}$

$= 37\mathbf{i} + 30\mathbf{j} - 11\mathbf{k}$

13. $\mathbf{v} \times \mathbf{w} = 9\mathbf{i} - 10\mathbf{j} + 3\mathbf{k}$ (see 8 or 11)

$\mathbf{u} \cdot (\mathbf{v} \times \mathbf{w}) = \mathbf{u} \cdot (9\mathbf{i} - 10\mathbf{j} + 3\mathbf{k})$

$= 18 + 10 + 12$

$= 40$

15. $\mathbf{v} \times \mathbf{w} = \det \begin{bmatrix} \mathbf{i} & \mathbf{j} & \mathbf{k} \\ 2 & 1 & 1 \\ 3 & -1 & -2 \end{bmatrix}$

$= (-2 - (-1))\mathbf{i} + (3 - (-4))\mathbf{j} + (-2 - 3)\mathbf{k}$

$= -\mathbf{i} + 7\mathbf{j} - 5\mathbf{k}$

Area $= |\mathbf{v} \times \mathbf{w}|$

$= \sqrt{(-1)^2 + 7^2 + (-5)^2}$

$= \sqrt{75}$

$= 5\sqrt{3}$

17. Area $= \left| \det \begin{bmatrix} 2 & -1 \\ 4 & 2 \end{bmatrix} \right|$

$= |4 - (-4)|$

$= 8$

19. $\overrightarrow{AB} = \langle -1, -1, 1 \rangle$

$\overrightarrow{AC} = \langle -2, 3, -3 \rangle$

$\overrightarrow{AB} \times \overrightarrow{AC} = \det \begin{bmatrix} \mathbf{i} & \mathbf{j} & \mathbf{k} \\ -1 & -1 & 1 \\ -2 & 3 & -3 \end{bmatrix}$

$= (3 - 3)\mathbf{i} + (-2 - 3)\mathbf{j} + (-3 - 2)\mathbf{k}$

$= -5\mathbf{j} - 5\mathbf{k}$

Area $= (1/2)|-5\mathbf{j} - 5\mathbf{k}|$

$= (1/2)\sqrt{(-5)^2 + (-5)^2}$

$= 5\sqrt{2}/2$

21. $\overrightarrow{AB} = \langle -4, -1, -3 \rangle$

$\overrightarrow{AC} = \langle 1, -1, -3 \rangle$

$\overrightarrow{AB} \times \overrightarrow{AC} = \det \begin{bmatrix} \mathbf{i} & \mathbf{j} & \mathbf{k} \\ -4 & -1 & -3 \\ 1 & -1 & -3 \end{bmatrix}$

$= (3 - 3)\mathbf{i} + (-3 - 12)\mathbf{j} + (4 - (-1))\mathbf{k}$

$= -15\mathbf{j} + 5\mathbf{k}$

Area $= (1/2)|-15\mathbf{j} + 5\mathbf{k}|$

$= (1/2)\sqrt{(-15)^2 + 5^2}$

$= 5\sqrt{10}/2$

23. Two of the sides correspond to vectors $\langle -3 - 1, -4 - 5 \rangle = \langle -4, -9 \rangle$ and $\langle 4 - 1, 6 - 5 \rangle = \langle 3, 1 \rangle$.

Area $= (1/2) \left| \det \begin{bmatrix} -4 & -9 \\ 3 & 1 \end{bmatrix} \right| = (1/2)(-4 - (-27)) = 23/2$

25. Since $\mathbf{v}$ and $\mathbf{w}$ both lie entirely within the plane containing $\mathbf{i}$ and $\mathbf{j}$, and $\mathbf{k}$ and $-\mathbf{k}$ are unit vectors orthogonal to this plane, both $\mathbf{k}$ and $-\mathbf{k}$ are unit vectors orthogonal to both $\mathbf{v}$ and $\mathbf{w}$.

27. An orthogonal vector is $\mathbf{v} \times \mathbf{w} = \det \begin{bmatrix} \mathbf{i} & \mathbf{j} & \mathbf{k} \\ 1 & 1 & 0 \\ 2 & -1 & 3 \end{bmatrix} = (3-0)\mathbf{i} + (0-3)\mathbf{j} + (-1-2)\mathbf{k} = 3\mathbf{i} - 3\mathbf{j} - 3\mathbf{k}.$

Obtain a unit vector which is also orthogonal to $\mathbf{v}$ and $\mathbf{w}$ by dividing this vector by its magnitude.

$|\mathbf{v} \times \mathbf{w}| = \sqrt{3^2 + (-3)^2 + (-3)^2} = 3\sqrt{3}$

The unit vectors are $(\sqrt{3}/3)\mathbf{i} - (\sqrt{3}/3)\mathbf{j} - (\sqrt{3}/3)\mathbf{k}$ and $-(\sqrt{3}/3)\mathbf{i} + (\sqrt{3}/3)\mathbf{j} + (\sqrt{3}/3)\mathbf{k}$.

29. A direction vector for the line is:

$(\mathbf{i} + \mathbf{j}) \times (\mathbf{i} + \mathbf{j} + 2\mathbf{k}) = \det \begin{bmatrix} \mathbf{i} & \mathbf{j} & \mathbf{k} \\ 1 & 1 & 0 \\ 1 & 1 & 2 \end{bmatrix} = (2-0)\mathbf{i} + (0-2)\mathbf{j} + (1-1)\mathbf{k} = 2\mathbf{i} - 2\mathbf{j}.$

The vector equation for the line is $(\mathbf{i} + 2\mathbf{j} + 2\mathbf{k}) + t(2\mathbf{i} - 2\mathbf{j}) = (1 + 2t)\mathbf{i} + (2 - 2t)\mathbf{j} + 2\mathbf{k}.$

The parametric equations for the line are: $x = 1 + 2t$, $y = 2 - 2t$, $z = 2$.

Solve for t to obtain the symmetric equations.

$\dfrac{x-1}{2} = \dfrac{2-y}{2}$ and $z = 2$

31. A direction vector for the line is $\mathbf{b} = \mathbf{i} + 2\mathbf{j} + \mathbf{k}$, and a point on the line is $Q = (3, -1, -1)$. The distance from the point P to the line is $|\overrightarrow{QP} \times \mathbf{b}|/|\mathbf{b}|$.

$\overrightarrow{QP} = \langle -1, 0, 5 \rangle$

$\overrightarrow{QP} \times \mathbf{b} = \det \begin{bmatrix} \mathbf{i} & \mathbf{j} & \mathbf{k} \\ -1 & 0 & 5 \\ 1 & 2 & 1 \end{bmatrix} = (0 - 10)\mathbf{i} + (5 - (-1))\mathbf{j} + (2 - 0)\mathbf{k} = -10\mathbf{i} + 6\mathbf{j} + 2\mathbf{k}$

$|\overrightarrow{QP} \times \mathbf{b}| = \sqrt{(-10)^2 + 6^2 + 2^2} = \sqrt{140} = 2\sqrt{35}$

$|\mathbf{b}| = \sqrt{1^2 + 2^2 + 1^2} = \sqrt{6}$

The distance from the point to the line is $2\sqrt{35}/\sqrt{6} = 2\sqrt{210}/6 = \sqrt{210}/3$.

33. Parametric equations for the line are $x = 1 + 4t$, $y = -3 - 2t$, $z = -1 + 5t$. The vector equation for the line is thus $\langle 1 + 4t, -3 - 2t, -1 + 5t \rangle = \langle 1, -3, -1 \rangle + t\langle 4, -2, 5 \rangle$. A direction vector for the line is $\mathbf{b} = \langle 4, -2, 5 \rangle$, and a point on the line is $Q = (1, -3, -1)$. The distance from the point P to the line is $|\overrightarrow{QP} \times \mathbf{b}|/|\mathbf{b}|$.

$\overrightarrow{QP} = \langle -1, 1, 2 \rangle$

$\overrightarrow{QP} \times \mathbf{b} = \det \begin{bmatrix} \mathbf{i} & \mathbf{j} & \mathbf{k} \\ -1 & 1 & 2 \\ 4 & -2 & 5 \end{bmatrix} = (5 - (-4))\mathbf{i} + (8 - (-5))\mathbf{j} + (2 - 4)\mathbf{k} = 9\mathbf{i} + 13\mathbf{j} - 2\mathbf{k}$

$|\overrightarrow{QP} \times \mathbf{b}| = \sqrt{9^2 + 13^2 + (-2)^2} = \sqrt{254}$

$|\mathbf{b}| = \sqrt{4^2 + (-2)^2 + 5^2} = \sqrt{45} = 3\sqrt{5}$

The distance from the point to the line is $\sqrt{254}/(3\sqrt{5}) = \sqrt{1270}/3$.

35. Line l_1 has vector equation $\langle 1 + t, 5t, 1 - t \rangle = \langle 1, 0, 1 \rangle + t\langle 1, 5, -1 \rangle$, and therefore has direction vector $\mathbf{b}_1 = \langle 1, 5, -1 \rangle$ and contains point $P = (1, 0, 1)$. Line l_2 has vector equation $\langle 2 + t, 2 - 3t, 1 + 5t \rangle = \langle 2, 2, 1 \rangle + t\langle 1, -3, 5 \rangle$, and therefore has direction vector $\mathbf{b}_2 = \langle 1, -3, 5 \rangle$ and contains point $Q = (2, 2, 1)$. The vector orthogonal to both lines is:

$\mathbf{n} = \mathbf{b}_1 \times \mathbf{b}_2 = \det \begin{bmatrix} \mathbf{i} & \mathbf{j} & \mathbf{k} \\ 1 & 5 & -1 \\ 1 & -3 & 5 \end{bmatrix} = (25 - 3)\mathbf{i} + (-1 - 5)\mathbf{j} + (-3 - 5)\mathbf{k} = 22\mathbf{i} - 6\mathbf{j} - 8\mathbf{k} = \langle 22, -6, -8 \rangle$

The distance d between the lines is

$|\operatorname{comp}_{\mathbf{n}} \overrightarrow{PQ}| = |\operatorname{comp}_{\mathbf{n}} \langle 1, 2, 0 \rangle|$

$= |\mathbf{n} \cdot \langle 1, 2, 0 \rangle| / |\mathbf{n}|$

$= |(22)(1) + (-6)(2) + (-8)(0)| / \sqrt{22^2 + (-6)^2 + (-8)^2}$

$= |10| / \sqrt{584}$

$= 10/(2\sqrt{146})$

$= 5\sqrt{146}/146$

37. $\overrightarrow{PQ} = \langle 3, 2, 5 \rangle$ and $\overrightarrow{PR} = \langle 0, -5, 2 \rangle$

A normal vector for the plane is:

$\overrightarrow{PQ} \times \overrightarrow{PR} = \det \begin{bmatrix} \mathbf{i} & \mathbf{j} & \mathbf{k} \\ 3 & 2 & 5 \\ 0 & -5 & 2 \end{bmatrix} = (4 - (-25))\mathbf{i} + (0 - 6)\mathbf{j} + (-15 - 0)\mathbf{k} = 29\mathbf{i} - 6\mathbf{j} - 15\mathbf{k}.$

The equation of the plane is of the form $29x - 6y - 15z = d$. Since point P is on the plane, $d = 29(-2) - 6(3) - 15(-4) = -16$. Therefore the equation of the plane is $29x - 6y - 15z = -16$.

39. $\overrightarrow{PQ} = \langle 1, 1, -2 \rangle$ and $\overrightarrow{PR} = \langle 4, 4, -3 \rangle$

A normal vector for the plane is:

$\overrightarrow{PQ} \times \overrightarrow{PR} = \det \begin{bmatrix} \mathbf{i} & \mathbf{j} & \mathbf{k} \\ 1 & 1 & -2 \\ 4 & 4 & -3 \end{bmatrix} = (-3 - (-8))\mathbf{i} + (-8 - (-3))\mathbf{j} + (4 - 4)\mathbf{k} = 5\mathbf{i} - 5\mathbf{j}.$

The equation of the plane is of the form $5x - 5y = d$. Since point P is on the plane, $d = 5(-1) - 5(1) = -10$. Therefore the equation of the plane is $5x - 5y = -10$, or $x - y = -2$.

41. Let $P = (-2, 1, 3)$, $Q = (5, 1, 5)$ and $R = (0, 0, 2)$.

A vector normal to the plane containing these points is:

$\overrightarrow{PQ} \times \overrightarrow{PR} = \langle 7, 0, 2 \rangle \times \langle 2, -1, -1 \rangle = \det \begin{bmatrix} \mathbf{i} & \mathbf{j} & \mathbf{k} \\ 7 & 0 & 2 \\ 2 & -1 & -1 \end{bmatrix} = (0 - (-2))\mathbf{i} + (4 - (-7))\mathbf{j} + (-7 - 0)\mathbf{k} = 2\mathbf{i} + 11\mathbf{j} - 7\mathbf{k}$

43. Since the line lies in both planes, it is perpendicular to both of the normal vectors for the planes. The normal vectors for the planes are $\mathbf{n}_1 = \langle 1, 1, -2 \rangle$ and $\mathbf{n}_2 = \langle 2, -4, 1 \rangle$. A direction vector for the line is:

$\mathbf{n}_1 \times \mathbf{n}_2 = \det \begin{bmatrix} \mathbf{i} & \mathbf{j} & \mathbf{k} \\ 1 & 1 & -2 \\ 2 & -4 & 1 \end{bmatrix} = (1 - 8)\mathbf{i} + (-4 - 1)\mathbf{j} + (-4 - 2)\mathbf{k} = -7\mathbf{i} - 5\mathbf{j} - 6\mathbf{k}$

A point on the line is a point (x, y, z) which satisfies both of the equations $x + y - 2z = 6$ and $2x - 4y + z = 3$. Solve for z in the second equation, $z = 3 - 2x + 4y$, and substitute into the first equation to obtain

$x + y - 2(3 - 2x + 4y) = 6$

$x + y - 6 + 4x - 8y = 6$

$5x - 7y = 12$

One possible solution is $x = 1$ and $y = -1$, for which we obtain $z = 3 - 2(1) + 4(-1) = -3$. Therefore the point $(1, -1, -3)$ is on the line, and a vector equation for the line is $(\mathbf{i} - \mathbf{j} - 3\mathbf{k}) + t(-7\mathbf{i} - 5\mathbf{j} - 6\mathbf{k}) = (1 - 7t)\mathbf{i} + (-1 - 5t)\mathbf{j} + (-3 - 6t)\mathbf{k}$. Parametric equations for the line are

$x(t) = 1 - 7t$

$y(t) = -1 - 5t$

$z(t) = -3 - 6t$

45. In each part, let A, B, C and D be the four points, in that order. The points are coplanar if and only if a vector orthogonal to both $\overrightarrow{AB}$ and $\overrightarrow{AC}$ is also orthogonal to $\overrightarrow{AD}$. Use $\overrightarrow{AB} \times \overrightarrow{AC}$ as a vector orthogonal to both $\overrightarrow{AB}$ and $\overrightarrow{AC}$, then compute the dot product of the result with $\overrightarrow{AD}$ to test orthogonality. If the result is zero, then the four points are coplanar.

a. $\overrightarrow{AB} = \langle 1, -1, 5 \rangle$, $\overrightarrow{AC} = \langle -1, 0, 1 \rangle$ and $\overrightarrow{AD} = \langle 0, 2, -12 \rangle$

$$\mathbf{n} = \overrightarrow{AB} \times \overrightarrow{AC} = \det \begin{bmatrix} \mathbf{i} & \mathbf{j} & \mathbf{k} \\ 1 & -1 & 5 \\ -1 & 0 & 1 \end{bmatrix} = (-1 - 0)\mathbf{i} + (-5 - 1)\mathbf{j} + (0 - 1)\mathbf{k} = \langle -1, -6, -1 \rangle$$

$\mathbf{n} \cdot \overrightarrow{AD} = 0 - 12 + 12 = 0$
The four points are coplanar.

b. $\overrightarrow{AB} = \langle 2, 1, 9 \rangle$, $\overrightarrow{AC} = \langle 3, 6, 2 \rangle$ and $\overrightarrow{AD} = \langle 0, 11, -4 \rangle$

$$\mathbf{n} = \overrightarrow{AB} \times \overrightarrow{AC} = \det \begin{bmatrix} \mathbf{i} & \mathbf{j} & \mathbf{k} \\ 2 & 1 & 9 \\ 3 & 6 & 2 \end{bmatrix} = (2 - 54)\mathbf{i} + (27 - 4)\mathbf{j} + (12 - 3)\mathbf{k} = \langle -52, 23, 9 \rangle$$

$\mathbf{n} \cdot \overrightarrow{AD} = 0 + 253 - 36 \neq 0$
The four points are not coplanar.

c. $\overrightarrow{AB} = \langle 2, -3, -4 \rangle$, $\overrightarrow{AC} = \langle 0, -2, -1 \rangle$ and $\overrightarrow{AD} = \langle 0, 2, 1 \rangle$

$$\mathbf{n} = \overrightarrow{AB} \times \overrightarrow{AC} = \det \begin{bmatrix} \mathbf{i} & \mathbf{j} & \mathbf{k} \\ 2 & -3 & -4 \\ 0 & -2 & -1 \end{bmatrix} = (3 - 8)\mathbf{i} + (0 - (-2))\mathbf{j} + (-4 - 0)\mathbf{k} = \langle -5, 2, -4 \rangle$$

$\mathbf{n} \cdot \overrightarrow{AD} = 0 + 4 - 4 = 0$
The four points are coplanar.

47. Vector equations for the two lines are $t \langle 5, 1, 1 \rangle$ and $t \langle 1, 1, 1 \rangle$, and therefore direction vectors for the two lines are $\mathbf{b_1} = \langle 5, 1, 1 \rangle$ and $\mathbf{b_2} = \langle 1, 1, 1 \rangle$. Note that these two lines do in fact lie in a common plane since the origin $(0, 0, 0)$ is on both lines. A normal vector for the plane is:

$$\mathbf{b_1} \times \mathbf{b_2} = \det \begin{bmatrix} \mathbf{i} & \mathbf{j} & \mathbf{k} \\ 5 & 1 & 1 \\ 1 & 1 & 1 \end{bmatrix} = (1 - 1)\mathbf{i} + (1 - 5)\mathbf{j} + (5 - 1)\mathbf{k} = -4\mathbf{j} + 4\mathbf{k}.$$

An equation for the plane is of the form $-4y + 4z = d$. Using the fact that the origin is on the plane, one sees that $d = 0$. Therefore the equation of the plane is $-4y + 4z = 0$, or $-y + z = 0$, or $y - z = 0$.

49. If $\mathbf{v}$ and $\mathbf{w}$ are orthogonal, then the angle θ between them is $\pi/2$.
Therefore $|\mathbf{v} \times \mathbf{w}| = |\mathbf{v}||\mathbf{w}| \sin(\pi/2) = |\mathbf{v}||\mathbf{w}|$.

51. $V = |\mathbf{u} \cdot (\mathbf{v} \times \mathbf{w})| = \left| \mathbf{u} \cdot \det \begin{bmatrix} \mathbf{i} & \mathbf{j} & \mathbf{k} \\ -1 & 3 & -2 \\ 4 & -1 & 3 \end{bmatrix} \right|$
$= |\mathbf{u} \cdot ((9 - 2)\mathbf{i} + (-8 - (-3))\mathbf{j} + (1 - 12)\mathbf{k})|$
$= |\mathbf{u} \cdot (7\mathbf{i} - 5\mathbf{j} - 11\mathbf{k})|$
$= |14 - 5 + 44|$
$= 53$

53. $\mathbf{u} \times (\mathbf{v} \times \mathbf{w}) = \mathbf{u} \times ((v_2 w_3 - v_3 w_2)\mathbf{i} + (v_3 w_1 - v_1 w_3)\mathbf{j} + (v_1 w_2 - v_2 w_1)\mathbf{k})$

$= (u_2(v_1 w_2 - v_2 w_1) - u_3(v_3 w_1 - v_1 w_3))\mathbf{i} + (u_3(v_2 w_3 - v_3 w_2) - u_1(v_1 w_2 - v_2 w_1))\mathbf{j} +$
$\quad (u_1(v_3 w_1 - v_1 w_3) - u_2(v_2 w_3 - v_3 w_2))\mathbf{k}$

$= (u_2 v_1 w_2 - u_2 v_2 w_1 - u_3 v_3 w_1 + u_3 v_1 w_3)\mathbf{i} + (u_3 v_2 w_3 - u_3 v_3 w_2 - u_1 v_1 w_2 + u_1 v_2 w_1)\mathbf{j} +$
$\quad (u_1 v_3 w_1 - u_1 v_1 w_3 - u_2 v_2 w_3 + u_2 v_3 w_2)\mathbf{k}$

$= ((u_2 v_1 w_2 + u_3 v_1 w_3)\mathbf{i} + (u_3 v_2 w_3 + u_1 v_2 w_1)\mathbf{j} + (u_1 v_3 w_1 + u_2 v_3 w_2)\mathbf{k}) -$
$\quad ((u_2 v_2 w_1 + u_3 v_3 w_1)\mathbf{i} + (u_3 v_3 w_2 + u_1 v_1 w_2)\mathbf{j} + (u_1 v_1 w_3 + u_2 v_2 w_3)\mathbf{k})$

$= ((u_2 w_2 + u_3 w_3)v_1 \mathbf{i} + (u_1 w_1 + u_3 w_3)v_2 \mathbf{j} + (u_1 w_1 + u_2 w_2)v_3 \mathbf{k}) -$
$\quad ((u_2 v_2 + u_3 v_3)w_1 \mathbf{i} + (u_1 v_2 + u_3 v_3)w_2 \mathbf{j} + (u_1 v_1 + u_2 v_2)w_3 \mathbf{k})$

$= ((u_2 w_2 + u_3 w_3)v_1 \mathbf{i} + (u_1 w_1 + u_3 w_3)v_2 \mathbf{j} + (u_1 w_1 + u_2 w_2)v_3 \mathbf{k}) + (u_1 v_1 w_1 \mathbf{i} + u_2 v_2 w_2 \mathbf{j} + u_3 v_3 w_3 \mathbf{k}) -$
$\quad (u_1 v_1 w_1 \mathbf{i} + u_2 v_2 w_2 \mathbf{j} + u_3 v_3 w_3 \mathbf{k}) - ((u_2 v_2 + u_3 v_3)w_1 \mathbf{i} + (u_1 v_2 + u_3 v_3)w_2 \mathbf{j} + (u_1 v_1 + u_2 v_2)w_3 \mathbf{k})$

$= ((u_1 w_1 + u_2 w_2 + u_3 w_3)v_1 \mathbf{i} + (u_1 w_1 + u_2 w_2 + u_3 w_3)v_2 \mathbf{j} + (u_1 w_1 + u_2 w_2 + u_3 w_3)v_3 \mathbf{k}) -$
$\quad ((u_1 v_1 + u_2 v_2 + u_3 v_3)w_1 \mathbf{i} + (u_1 v_2 + u_2 v_2 + u_3 v_3)w_2 \mathbf{j} + (u_1 v_1 + u_2 v_2 + u_3 v_3)w_3 \mathbf{k})$

$= (u_1 w_1 + u_2 w_2 + u_3 w_3)\mathbf{v} - (u_1 v_1 + u_2 v_2 + u_3 v_3)\mathbf{w}$

$= (\mathbf{u} \cdot \mathbf{w})\mathbf{v} - (\mathbf{u} \cdot \mathbf{v})\mathbf{w}$

55. $\mathbf{u} \cdot (\mathbf{v} \times \mathbf{w}) = \mathbf{u} \cdot (-\mathbf{w} \times \mathbf{v})$

$= -\mathbf{u} \cdot (\mathbf{w} \times \mathbf{v})$

$= -\mathbf{w} \cdot (\mathbf{v} \times \mathbf{u})$ by exercise 54

$= \mathbf{w} \cdot (-\mathbf{v} \times \mathbf{u})$

$= \mathbf{w} \cdot (\mathbf{u} \times \mathbf{v})$

57. First rearrange the equation as $\mathbf{u} + \mathbf{v} = -\mathbf{w}$ in order to prove that $\mathbf{v} \times \mathbf{w} = \mathbf{w} \times \mathbf{u}$.
$\mathbf{v} \times \mathbf{w} - \mathbf{w} \times \mathbf{u} = \mathbf{v} \times \mathbf{w} + \mathbf{u} \times \mathbf{w} = (\mathbf{v} + \mathbf{u}) \times \mathbf{w} = (-\mathbf{w} \times \mathbf{w}) = -\mathbf{w} \times \mathbf{w} = -\mathbf{0} = \mathbf{0}$
Therefore $\mathbf{v} \times \mathbf{w} = \mathbf{w} \times \mathbf{u}$. Likewise, one can use $\mathbf{u} + \mathbf{w} = -\mathbf{v}$ in order to prove that $\mathbf{u} \times \mathbf{v} = \mathbf{v} \times \mathbf{w}$, and from these two equations it follows that all three cross products are equal.

One possible geometric interpretation is that, given any parallelogram, if one takes any side of the parallelogram together with either diagonal of the parallelogram, then these form two sides of a second parallelogram which will have the same area as the first.

Review Exercises — Chapter 16

1. $\sqrt{(1 - 2)^2 + (2 - (-5))^2 + (-4 - 2)^2} = \sqrt{1 + 49 + 36} = \sqrt{86}$

3. a. $\mathbf{u} \cdot \mathbf{v} = (2)(1) + (1)(1) + (-1)(1) = 2$

b. $\mathbf{u} \times \mathbf{v} = \det \begin{bmatrix} \mathbf{i} & \mathbf{j} & \mathbf{k} \\ 2 & 1 & -1 \\ 1 & 1 & 1 \end{bmatrix} = (1 - (-1))\mathbf{i} + (-1 - 2)\mathbf{j} + (2 - 1)\mathbf{k} = 2\mathbf{i} - 3\mathbf{j} + \mathbf{k}$

c. $\mathbf{v} \times \mathbf{u} = -\mathbf{u} \times \mathbf{v} = -2\mathbf{i} + 3\mathbf{j} - \mathbf{k}$

d. $\text{comp}_{\mathbf{w}} \mathbf{u} = (\mathbf{w} \cdot \mathbf{u})/|\mathbf{w}| = (0 + 1 + 1)/\sqrt{0^2 + 1^2 + (-1)^2} = 2/\sqrt{2} = \sqrt{2}$

e. $\text{proj}_{\mathbf{w}} \mathbf{u} = [(\mathbf{w} \cdot \mathbf{u})/|\mathbf{w}|^2]\,\mathbf{w} = [(0 + 1 + 1)/(0^2 + 1^2 + (-1)^2)]\,\mathbf{w} = 1\mathbf{w} = \mathbf{w}$, or $\mathbf{j} - \mathbf{k}$.

f. $\mathbf{u} \cdot (\mathbf{v} \times \mathbf{w}) = \mathbf{u} \cdot \det \begin{bmatrix} \mathbf{i} & \mathbf{j} & \mathbf{k} \\ 1 & 1 & 1 \\ 0 & 1 & -1 \end{bmatrix}$

$= \mathbf{u} \cdot ((-1 - 1)\mathbf{i} + (0 - (-1))\mathbf{j} + (1 - 0)\mathbf{k})$

$= \mathbf{u} \cdot (-2\mathbf{i} + \mathbf{j} + \mathbf{k})$

$= -4 + 1 - 1 = -4$

5. $\cos\theta = (\mathbf{v} \cdot \mathbf{w})/(|\mathbf{v}||\mathbf{w}|) = (3 - 9 + 4)/(\sqrt{1 + 9 + 4}\sqrt{9 + 9 + 4}) = -2/\sqrt{308} = -2/(2\sqrt{77}) = -\sqrt{77}/77$

7. l_1 has vector equation $\langle 2t, 3+t, 5t \rangle = \langle 0, 3, 0 \rangle + t \langle 2, 1, 5 \rangle$ and thus has direction vector $\mathbf{b}_1 = \langle 2, 1, 5 \rangle$. l_2 has vector equation $\langle t+5, 6-4t, 3t+4 \rangle = \langle 5, 6, 4 \rangle + t \langle 1, -4, 3 \rangle$ and thus has direction vector $\mathbf{b}_2 = \langle 1, -4, 3 \rangle$. A vector perpendicular to both lines is:

$$\mathbf{b}_1 \times \mathbf{b}_2 = \det \begin{bmatrix} \mathbf{i} & \mathbf{j} & \mathbf{k} \\ 2 & 1 & 5 \\ 1 & -4 & 3 \end{bmatrix} = (3 - (-20))\mathbf{i} + (5-6)\mathbf{j} + (-8-1)\mathbf{k} = 23\mathbf{i} - \mathbf{j} - 9\mathbf{k}$$

Use this as a direction vector for the desired line, obtaining the vector equation:

$\langle 2, 1, -3 \rangle + t \langle 23, -1, -9 \rangle = \langle 2+23t, 1-t, -3-9t \rangle$.

Therefore the line has parametric equations $x(t) = 2 + 23t$, $y(t) = 1 - t$, $z(t) = -3 - 9t$

9. Let $P = (-1, 2, 1)$, $Q = (2, 5, 3)$ and $R = (-4, 0, 2)$. Two vectors within the plane are $\overrightarrow{PQ} = \langle 3, 3, 2 \rangle$ and $\overrightarrow{PR} = \langle -3, -2, 1 \rangle$. A normal vector for the plane is:

$$\overrightarrow{PQ} \times \overrightarrow{PR} = \det \begin{bmatrix} \mathbf{i} & \mathbf{j} & \mathbf{k} \\ 3 & 3 & 2 \\ -3 & -2 & 1 \end{bmatrix} = (3 - (-4))\mathbf{i} + (-6-3)\mathbf{j} + (-6-(-9))\mathbf{k} = 7\mathbf{i} - 9\mathbf{j} + 3\mathbf{k}.$$

Thus an equation for the plane is of the form $7x - 9y + 3z = d$. Using the fact that P is in the plane, one sees that $d = 7(-1) - 9(2) + 3(1) = -22$. Therefore, an equation for the plane is $7x - 9y + 3z = -22$.

11. The normal $\langle 8, 1, -2 \rangle$ can be read directly from the equation.

13. Since the line lies in both planes, it is orthogonal to both normal vectors. Normal vectors for the planes are $\mathbf{n}_1 = \langle 2, 3, -1 \rangle$ and $\mathbf{n}_2 = \langle 1, -3, 5 \rangle$. A vector orthogonal to both of these is:

$$\mathbf{n}_1 \times \mathbf{n}_2 = \det \begin{bmatrix} \mathbf{i} & \mathbf{j} & \mathbf{k} \\ 2 & 3 & -1 \\ 1 & -3 & 5 \end{bmatrix} = (15-3)\mathbf{i} + (-1-10)\mathbf{j} + (-6-3)\mathbf{k} = \langle 12, -11, -9 \rangle$$

This is a direction vector for the line of intersection. Next we need to find a point (x, y, z) on the line, and thus on both planes. Its coordinates must satisfy both of the equations $2x + 3y - z = 4$ and $x - 3y + 5z = 2$. Solving the first equation for z gives $z = 2x + 3y - 4$, and substituting this value for z in the second equation gives:

$x - 3y + 5(2x + 3y - 4) = 2$
$x - 3y + 10x + 15y - 20 = 2$
$11x + 12y = 22$

One possible solution is $x = 2$ and $y = 0$, for which $z = 2(2) + 3(0) - 4 = 0$. Thus $(2, 0, 0)$ is a point on both planes. The vector equation for the line is therefore $\langle 2, 0, 0 \rangle + t \langle 12, -11, -9 \rangle$, or $2\mathbf{i} + t(12\mathbf{i} - 11\mathbf{j} - 9\mathbf{k})$.

15. $(\mathbf{u} + \mathbf{v}) \times (\mathbf{u} - \mathbf{v}) = \mathbf{u} \times \mathbf{u} - \mathbf{u} \times \mathbf{v} + \mathbf{v} \times \mathbf{u} - \mathbf{v} \times \mathbf{v}$
$= 0 - (-\mathbf{v} \times \mathbf{u}) + \mathbf{v} \times \mathbf{u} - 0$
$= 2\mathbf{v} \times \mathbf{u}$

17. If $P = (1, 2, -1)$ and $Q = (2, 3, 1)$, then a direction vector for the line is $\overrightarrow{PQ} = \langle 1, 1, 2 \rangle$, and a vector equation for the line is therefore $\langle 1, 2, -1 \rangle + t \langle 1, 1, 2 \rangle = \langle 1+t, 2+t, -1+2t \rangle$. Parametric equations for the line are $x = 1 + t$, $y = 2 + t$, $z = -1 + 2t$. Solving for t, one obtains the symmetric equations

$$x - 1 = y - 2 = \frac{z+1}{2}$$

19. If $P = (1, 1, 1)$, $Q = (2, 1, 4)$ and $R = (3, 2, 3)$, then the vectors $\overrightarrow{PQ} = \langle 1, 0, 3 \rangle$ and $\overrightarrow{PR} = \langle 2, 1, 2 \rangle$ lie in the plane. A normal vector for the plane is:

$$\overrightarrow{PQ} \times \overrightarrow{PR} = \det \begin{bmatrix} \mathbf{i} & \mathbf{j} & \mathbf{k} \\ 1 & 0 & 3 \\ 2 & 1 & 2 \end{bmatrix} = (0 - 3)\mathbf{i} + (6 - 2)\mathbf{j} + (1 - 0)\mathbf{k} = -3\mathbf{i} + 4\mathbf{j} + \mathbf{k}$$

An equation for the plane is of the form $-3x + 4y + z = d$. Using the fact that P is in the plane, one sees that $d = -3(1) + 4(1) + 1 = 2$. Therefore, an equation for the plane is $-3x + 4y + z = 2$.

21. The lines have vector equations $\langle 2t, t, 1 - t \rangle = \langle 0, 0, 1 \rangle + t \langle 2, 1, -1 \rangle$ and $\langle t, t, (t + 2)/2 \rangle = \langle 0, 0, 1 \rangle + t \langle 1, 1, 1/2 \rangle$, and thus have direction vectors $\mathbf{b}_1 = \langle 2, 1, -1 \rangle$ and $\mathbf{b}_2 = \langle 1, 1, 1/2 \rangle$. Since any multiple of a direction vector for a line is also parallel to the line, it will be more convenient to use the vector $\mathbf{b}_2' = 2 \langle 1, 1, 1/2 \rangle = \langle 2, 2, 1 \rangle$. Note that the lines intersect at the point $(0, 0, 1)$, and therefore do in fact lie in a common plane. A normal vector for this plane is:

$$\mathbf{b}_1 \times \mathbf{b}_2' = \det \begin{bmatrix} \mathbf{i} & \mathbf{j} & \mathbf{k} \\ 2 & 1 & -1 \\ 2 & 2 & 1 \end{bmatrix} = (1 - (-2))\mathbf{i} + (-2 - 2)\mathbf{j} + (4 - 2)\mathbf{k} = 3\mathbf{i} - 4\mathbf{j} + 2\mathbf{k}.$$

An equation for the plane is $3x - 4y + 2z = d$. Using the fact that $(0, 0, 1)$ is in the plane, one sees that $d = 3(0) - 4(0) + 2(1) = 2$. Therefore, an equation for the plane is $3x - 4y + 2z = 2$.

23. A direction vector for the line is $\mathbf{b} = \langle 1, -2, 5 \rangle$, a point on the line is $Q = (3, 4, 5)$, and the distance from the point $P = (-4, 2, 5)$ to the line is $|\overrightarrow{QP} \times \mathbf{b}|/|\mathbf{b}|$.

$$\overrightarrow{QP} = \langle -7, -2, 0 \rangle$$

$$\overrightarrow{QP} \times \mathbf{b} = \det \begin{bmatrix} \mathbf{i} & \mathbf{j} & \mathbf{k} \\ -7 & -2 & 0 \\ 1 & -2 & 5 \end{bmatrix} = (-10 - 0)\mathbf{i} + (0 - (-35))\mathbf{j} + (14 - (-2))\mathbf{k} = -10\mathbf{i} + 35\mathbf{j} + 16\mathbf{k}$$

$$|\overrightarrow{QP} \times \mathbf{b}| = \sqrt{(-10)^2 + 35^2 + 16^2} = \sqrt{1581}$$
$$|\mathbf{b}| = \sqrt{1^2 + (-2)^2 + 5^2} = \sqrt{30}$$

The distance from point P to the line is $\sqrt{1581}/\sqrt{30} = \sqrt{47430}/30 = 3\sqrt{5270}/30 = \sqrt{5270}/10$

25. $|\mathbf{v}| = \sqrt{1^2 + (-3)^2 + 4^2} = \sqrt{26}$

The unit vector in the same direction as $\mathbf{v}$ is
$\mathbf{v}/|\mathbf{v}| = (\sqrt{26}/26)\mathbf{i} - (3\sqrt{26}/26)\mathbf{j} + (2\sqrt{26}/13)\mathbf{k}$

27. Solve for t.
$$\frac{x - 1}{3} = \frac{y - 2}{7} = 3 - z$$

29. a. $\text{comp}_\mathbf{w} \mathbf{v} = (\mathbf{w} \cdot \mathbf{v})/|\mathbf{w}| = (2 + 9 - 1)/\sqrt{1^2 + 3^2 + 1^2} = 10/\sqrt{11} = 10\sqrt{11}/11$

  b. $\text{proj}_\mathbf{v} \mathbf{w} = [(\mathbf{v} \cdot \mathbf{w})/|\mathbf{v}|^2] \mathbf{v}$
  $\qquad = [(2 + 9 - 1)/(2^2 + 3^2 + (-1)^2)] \mathbf{v}$
  $\qquad = [10/14]\mathbf{v}$
  $\qquad = (10/7)\mathbf{i} + (15/7)\mathbf{j} - (5/7)\mathbf{k}$

31. A normal vector for the plane is $\mathbf{n} = \langle 1, 1, 1 \rangle$ and a point on the plane is the origin. If $P = (1, 2, 2)$, then the distance from P to the plane is:

$$|\text{comp}_\mathbf{n} \overrightarrow{PO}| = |\mathbf{n} \cdot \overrightarrow{PO}|/|\mathbf{n}| = |\mathbf{n} \cdot \langle -1, -2, -2 \rangle|/\sqrt{1^2 + 1^2 + 1^2} = |-1 - 2 - 2|/\sqrt{3} = 5\sqrt{3}/3$$

33. The normal vectors for the planes are $\mathbf{n}_1 = \langle 1, 1, 1 \rangle$ and $\mathbf{n}_2 = \langle 2, 1, -1 \rangle$. The direction vector for the line of intersection of the two planes is orthogonal to both normal vectors, so one may use:

$$\mathbf{b} = \mathbf{n}_1 \times \mathbf{n}_2 = \det \begin{bmatrix} \mathbf{i} & \mathbf{j} & \mathbf{k} \\ 1 & 1 & 1 \\ 2 & 1 & -1 \end{bmatrix} = (-1-1)\mathbf{i} + (2 - (-1))\mathbf{j} + (1-2)\mathbf{k} = \langle -2, 3, -1 \rangle$$

In order to determine a point on the line, solve for z in the second equation, $z = 2x + y - 1$, and substitute this value for z in the first equation, obtaining:

$x + y + 2x + y - 1 = 4$

$3x + 2y = 5$.

One possible solution is $x = 1$ and $y = 1$, for which one will have $z = 2(1) + 1 = 1 = 2$. Thus $(1, 1, 2)$ is a point on the line. The vector equation for the line is $\langle 1, 1, 2 \rangle + t \langle -2, 3, -1 \rangle = \langle 1 - 2t, 1 + 3t, 2 - t \rangle$. Parametric equations for the line are $x = 1 - 2t$, $y = 1 + 3t$, $z = 2 - t$. Solve for t in order to obtain the symmetric equations:

$$\frac{1-x}{2} = \frac{y-1}{3} = 2 - z$$

35. The line has vector equation $\langle 2 + 3t, -1 + 2t, 3 - 2t \rangle = \langle 2, -1, 3 \rangle + t \langle 3, 2, -2 \rangle$, and thus has direction vector $\langle 3, 2, -2 \rangle$, which will be a normal vector for the plane. Thus an equation for the plane is of the form $3x + 2y - 2z = d$. Using the fact that the point $(3, -2, 3)$ is in the plane, one sees that $d = 3(3) + 2(-2) - 2(3) = -1$. Therefore, an equation for the plane is $3x + 2y - 2z = -1$.

37. First find a vector $\mathbf{n}$ orthogonal to both $\mathbf{v}$ and $\mathbf{w}$.

$$\mathbf{n} = \mathbf{v} \times \mathbf{w} = \det \begin{bmatrix} \mathbf{i} & \mathbf{j} & \mathbf{k} \\ 3 & 2 & 1 \\ 1 & 1 & -2 \end{bmatrix} = (-4-1)\mathbf{i} + (1 - (-6))\mathbf{j} + (3-2)\mathbf{k} = -5\mathbf{i} + 7\mathbf{j} + \mathbf{k}$$

The desired vectors are the vectors of length 2 parallel to $\mathbf{n}$, namely $\pm 2\mathbf{n}/|\mathbf{n}|$.

$|\mathbf{n}| = \sqrt{(-5)^2 + 7^2 + 1^2} = \sqrt{75} = 5\sqrt{3}$

$\pm 2\mathbf{n}/|\mathbf{n}| = \pm (2\sqrt{3}/15)\mathbf{n} = \pm ((-2\sqrt{3}/3)\mathbf{i} + (14\sqrt{3}/15)\mathbf{j} + (2\sqrt{3}/15)\mathbf{k})$

39. The planes are perpendicular if and only if their normal vectors $\mathbf{n}_1 = \langle a_1, b_1, c_1 \rangle$ and $\mathbf{n}_2 = \langle a_2, b_2, c_2 \rangle$ are orthogonal, which is true if and only if $0 = \mathbf{n}_1 \cdot \mathbf{n}_2 = a_1 a_2 + b_1 b_2 + c_1 c_2$.

Chapter 17

Curves and Surfaces

17.1   Vector-Valued Functions; Curves in Space

1.

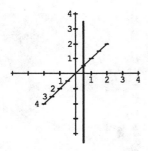

$$\mathbf{f}(t) = \mathbf{i} + \mathbf{j} + t\mathbf{k}$$

3.

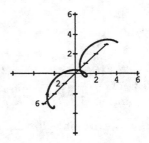

$$\mathbf{f}(t) = t\mathbf{i} + \cos t\,\mathbf{j} + \sin t\,\mathbf{k}$$

5.

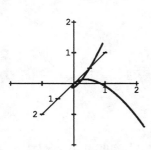

$$\mathbf{f}(t) = t\mathbf{i} + t^2\mathbf{j} + t^3\mathbf{k}$$

7.

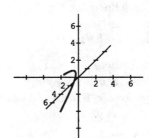

$$\mathbf{f}(t) = e^t\mathbf{i} + t\mathbf{j} + 2t\mathbf{k}$$

9. $\sqrt{t}$ is undefined for $t < 0$. The domain is $[0, \infty)$.

11. $1/\sqrt{1-t}$ is undefined for $1 - t \le 0$, or $1 \le t$. The domain is $(-\infty, 1)$.

523

13. $\sec t$ is undefined for $t = \pi/2 + n\pi$.
The domain is $\{t | t \neq \pi/2 + n\pi\}$.

15. $\ln t$ is undefined for $t \leq 0$.
$\sqrt{9 - t^2}$ is undefined for $9 - t^2 < 0$ or $9 < t^2$.
This occurs if $t < -3$ or $t > 3$.
The domain is $(0, 3]$.

17. $\mathbf{f}(t) + \mathbf{g}(t) = (t + \sin t)\mathbf{i} + (3t + \cos t)\mathbf{j} + (t^2 + 1)\mathbf{k}$

19. $3\mathbf{f}(t) + 2\mathbf{g}(t) = (3t + 2\sin t)\mathbf{i} + (9t + 2\cos t)\mathbf{j} + (3t^2 + 2)\mathbf{k}$

21. $\mathbf{f}(t) \cdot \mathbf{g}(t) = (t)(\sin t) + (3t)(\cos t) + (t^2)(1) = t\sin t + 3t\cos t + t^2$

23. $\mathbf{f}(t) + 3\mathbf{g}(t) = (1 + t + 3\sqrt{t})\mathbf{i} + (1 - t^2 + 3(1 - t))\mathbf{j} + (t + 3e^t)\mathbf{k} = (1 + t + 3\sqrt{t})\mathbf{i} + (4 - 3t - t^2)\mathbf{j} + (t + 3e^t)\mathbf{k}$

25. $h(t)[\mathbf{f}(t) - \mathbf{g}(t)] = (\sin t)\left[(1 + t - \sqrt{t})\mathbf{i} + (1 - t^2 - (1 - t))\mathbf{j} + (t - e^t)\mathbf{k}\right]$
$= (1 + t - \sqrt{t})\sin t\,\mathbf{i} + (t - t^2)\sin t\,\mathbf{j} + (t - e^t)\sin t\,\mathbf{k}$

27. $\mathbf{f}(t) \cdot \mathbf{g}(t) = (1 + t)\sqrt{t} + (1 - t^2)(1 - t) + te^t = (1 + t)\sqrt{t} + 1 - t - t^2 + t^3 + te^t$

29. $(\sqrt{2}/2)\mathbf{i} + \mathbf{j} + (\sqrt{2}/2)\mathbf{k}$

31. Either use l'Hôpital's Rule, or go back to the original discussion of the derivatives of $\sin x$ and $\cos x$, to obtain $\lim_{t \to 0}(\sin t)/t = 1$ and $\lim_{t \to 0}(1 - \cos t)/t = 0$. Also note that $\lim_{t \to 0} e^{2t} = 1$. Therefore, the limit is $\mathbf{i} + \mathbf{k}$.

33. The limit does not exist, since $\lim_{t \to 0} |t|/t$ does not exist.

35. $\lim_{t \to 3}(9 - t^2)/(3 - t) = \lim_{t \to 3}(3 - t)(3 + t)/(3 - t) = \lim_{t \to 3} 3 + t = 6$
$\lim_{t \to 3}(t^2 + t - 12)/(t - 3) = \lim_{t \to 3}(t + 4)(t - 3)/(t - 3) = \lim_{t \to 3} t + 4 = 7$
$\lim_{t \to 3}(t^3 - 13t + 12)/(t - 3) = \lim_{t \to 3}(t - 3)(t^2 + 3t - 4)/(t - 3) = \lim_{t \to 3} t^2 + 3t - 4 = 14$
Therefore, the limit is $6\mathbf{i} + 7\mathbf{j} + 14\mathbf{k}$.

37. The limit does not exist since $\lim_{t \to \infty}(3t^2 + 1)/(t - 4)$ does not exist as a real number.

39. $\tan t$ is undefined for $t = \pi/2 + n\pi$.
$\mathbf{f}(t)$ is continuous on all intervals of the form $(\pi/2 + n\pi, 3\pi/2 + n\pi)$, where n is an integer.

41. $1/(1 + \sqrt{t})$ cannot have a denominator of zero, so it is only undefined for $t < 0$. It is continuous from the right at $t = 0$.
$1/(1 - \sqrt{t})$ is undefined for $t < 0$ and for $t = 1$. It is continuous from the right at $t = 0$.
$t^{3/2}$ is undefined for $t < 0$, and is continuous from the right at $t = 0$.
$\mathbf{f}(t)$ is continuous on $[0, 1) \cup (1, \infty)$.

43. $\ln(1 + t^2)$ is continuous for all real numbers since $1 + t^2$ is always positive.
$\ln(1 - t^2)$ is undefined for $1 - t^2 \leq 0$ or $1 \leq t^2$. This is true for $t \leq -1$ or $t \geq 1$.
$\sqrt{t}$ is undefined for $t < 0$, and is continuous from the right at $t = 0$
$\mathbf{f}(t)$ is continuous on $[0, 1)$.

45. The only possible discontinuity would be at $t = 1$. $\lim_{t \to 1^-} \mathbf{f}(t) = \mathbf{i} + 2\mathbf{j} + 4\mathbf{k}$ and $\lim_{t \to 1^+} \mathbf{f}(t) = \mathbf{i} + 5\mathbf{j} = \mathbf{f}(1)$. Since these do not match, the limit as $t \to 1$ does not exist, and therefore $\mathbf{f}(t)$ is discontinuous at $t = 1$, although it is continuous from the right at $t = 1$. $\mathbf{f}(t)$ is continuous on the interval $(-\infty, 1)$, and also on the interval $[1, \infty)$.

47. $\sqrt{\sin t}$ is undefined for $\sin t < 0$, which occurs on intervals of the form $((2n - 1)\pi, 2n\pi)$. For any integer n, it is continuous from the left at $(2n - 1)\pi$, and continuous from the right at $2n\pi$.
$\mathbf{f}(t)$ is continuous on intervals of the form $[2n\pi, (2n + 1)\pi]$.

49. One complete revolution would occur between $t = 0$ and $t = 2\pi/10^3$. Over this interval, the tip of the vector has traveled from $\mathbf{E}(0) = \mathbf{i}$ to $\mathbf{E}(2\pi/10^3) = \mathbf{i} + (3 \times 10^8)(2\pi/10^3)\mathbf{k} = \mathbf{i} + (6\pi \times 10^5)\mathbf{k}$. The distance advanced along the z-axis is $6\pi \times 10^5$, or approximately 1.885×10^6.

51. a. Let $\mathbf{f}(t) = x_1(t)\mathbf{i} + y_1(t)\mathbf{j} + z_1(t)\mathbf{k}$ and $\mathbf{g}(t) = x_2(t)\mathbf{i} + y_2(t)\mathbf{j} + z_2(t)\mathbf{k}$. Furthermore, let $\mathbf{L} = L_1\mathbf{i} + L_2\mathbf{j} + L_3\mathbf{k}$ and $\mathbf{M} = M_1\mathbf{i} + M_2\mathbf{j} + M_3\mathbf{k}$. Then

$$\lim_{t \to t_0} x_1(t) = L_1 \quad \text{and} \quad \lim_{t \to t_0} x_2(t) = M_1$$

$$\lim_{t \to t_0} y_1(t) = L_2 \quad \text{and} \quad \lim_{t \to t_0} y_2(t) = M_2$$

$$\lim_{t \to t_0} z_1(t) = L_3 \quad \text{and} \quad \lim_{t \to t_0} z_2(t) = M_3$$

b. $(\mathbf{f} + \mathbf{g})(t) = (x_1 + x_2)(t)\mathbf{i} + (y_1 + y_2)(t)\mathbf{j} + (z_1 + z_2)(t)\mathbf{k}$
Considering each component, we have

$$\lim_{t \to t_0}(x_1 + x_2)(t) = \lim_{t \to t_0} x_1(t) + \lim_{t \to t_0} x_2(t) = L_1 + M_1$$

$$\lim_{t \to t_0}(y_1 + y_2)(t) = \lim_{t \to t_0} y_1(t) + \lim_{t \to t_0} y_2(t) = L_2 + M_2$$

$$\lim_{t \to t_0}(z_1 + z_2)(t) = \lim_{t \to t_0} z_1(t) + \lim_{t \to t_0} z_2(t) = L_3 + M_3$$

Therefore, $\lim_{t \to t_0}(\mathbf{f} + \mathbf{g})(t) = (L_1 + M_1)\mathbf{i} + (L_2 + M_2)\mathbf{j} + (L_3 + M_3)\mathbf{k} = \mathbf{L} + \mathbf{M}$.

53. Given that $\mathbf{f}(t)$ and $\mathbf{g}(t)$ are continuous. Let $\mathbf{f}(t) = x_1(t)\mathbf{i} + y_1(t)\mathbf{j} + z_1(t)\mathbf{k}$ and $\mathbf{g}(t) = x_2(t)\mathbf{i} + y_2(t)\mathbf{j} + z_2(t)\mathbf{k}$. Then $x_1(t)$, $y_1(t)$, $z_1(t)$, $x_2(t)$, $y_2(t)$ and $z_2(t)$ are all continuous.

a. It follows from above that $x_1 x_2 + y_1 y_2 + z_1 z_2$ is continuous. This is $\mathbf{f} \cdot \mathbf{g}$.

b. It follows from above that $[x_1(t)]^2 + [y_1(t)]^2 + [z_1(t)]^2$ is continuous.
$|\mathbf{f}|(t) = \sqrt{[x_1(t)]^2 + [y_1(t)]^2 + [z_1(t)]^2}$, which is continuous since the radicand is continuous and nonnegative.

55. It does not follow in either case that $\mathbf{f}$ or $\mathbf{g}$ are continuous anywhere. Let $\mathbf{f}$ be the function $\mathbf{f}(t) = f(t)\mathbf{i}$, where

$$f(t) = \begin{cases} 1 & \text{if } t \text{ is rational} \\ -1 & \text{if } t \text{ is irrational} \end{cases}$$

and let $\mathbf{g} = -\mathbf{f}$. Then $\mathbf{f}(t)$ and $\mathbf{g}(t)$ are not continuous at any value of t. But $(\mathbf{f} + \mathbf{g})(t) = f(t)\mathbf{i} + (-f(t)\mathbf{i}) = 0$ for all t, so it is continuous, and $(\mathbf{f} \cdot \mathbf{g})(t) = [f(t)\mathbf{i}] \cdot [-f(t)\mathbf{i}] = -[f(t)]^2 = -1$ for all t, so it is continuous.

57.

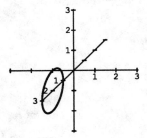

$h = 0, k = 1$

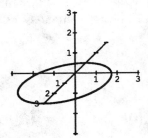

$h = 1, k = 0$

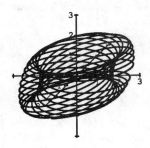

$h = \pi, k = 1, -20 \le t \le 20$

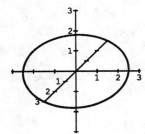

$h = -1, k = 1$

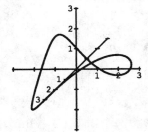

$h = 1, k = 2$

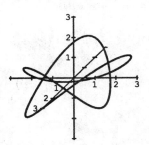

$h = 2, k = 3$

17.2 Derivatives and Integrals of Vector-Valued Functions

1. $z(t) = \sqrt{t}$ is differentiable on $(0, \infty)$, and therefore this is the interval on which $\mathbf{f}(t)$ is differentiable.

3. $x(t) = \ln(3 + t)$ is differentiable on $(-3, \infty)$.
$y(t) = 1/(t + 2)$ is differentiable on $(-\infty, -2) \cup (-2, \infty)$.
$\mathbf{f}(t)$ is differentiable on $(-3, -2) \cup (-2, \infty)$.

5. $\mathbf{f}'(t) = \mathbf{i} + (1/(2\sqrt{t}))\mathbf{j}$

7. $\mathbf{f}'(t) = (1/(2\sqrt{t}))\mathbf{i} - (3/2)t^{-5/2}\mathbf{j} + (2/(2t-1))\mathbf{k}$

9. $\mathbf{f}'(t) = \left[e^t(2t) + e^t(t^2 - 1)\right]\mathbf{i} + (t^2 \cos t + 2t \sin t)\mathbf{j} + (1/2)(t^3 - 2t - 1)^{-1/2}(3t^2 - 2)\mathbf{k}$

$= e^t(t^2 + 2t - 1)\mathbf{i} + (t^2 \cos t + 2t \sin t)\mathbf{j} + \dfrac{3t^2 - 2}{2\sqrt{t^3 - 2t - 1}}\mathbf{k}$

11. $\mathbf{f}'(t) = (1/\sqrt{1 - t^2})\mathbf{i} + (1/2)(1 + t^2)^{-1/2}2t\mathbf{j} - 3t^2 e^{-t^3}\mathbf{k}$

$= \dfrac{1}{\sqrt{1 - t^2}}\mathbf{i} + \dfrac{t}{\sqrt{1 + t^2}}\mathbf{j} - 3t^2 e^{-t^3}\mathbf{k}$

13. $\mathbf{f}'(t) = (1/2)t^{-1/2}\mathbf{i} - (3/2)t^{-5/2}\mathbf{j} + 2(2t - 1)^{-1}\mathbf{k}$
$\mathbf{f}''(t) = (-1/4)t^{-3/2}\mathbf{i} + (15/4)t^{-7/2}\mathbf{j} - 2(2t - 1)^{-2}2\mathbf{k}$
$= -(1/4t^{3/2})\mathbf{i} + (15/4t^{7/2})\mathbf{j} - (4/(2t - 1)^2)\mathbf{k}$

15. $\mathbf{f}'(t) = \omega A e^{\omega t}\mathbf{i} - \omega B e^{-\omega t}\mathbf{j}$
$\mathbf{f}''(t) = \omega^2 A e^{\omega t}\mathbf{i} + \omega^2 B e^{-\omega t}\mathbf{j}$
Note that $\mathbf{f}''(t) = \omega^2 \mathbf{f}(t)$.
Therefore $\mathbf{f}''(t) - \omega^2 \mathbf{f}(t) = 0$.

17. $\mathbf{r}'(t) = 3\mathbf{f}'(t) - 2\mathbf{g}'(t)$
$= 3(\cos t\,\mathbf{i} + 2t\mathbf{k}) - 2(\mathbf{i} - \sin t\,\mathbf{k})$
$= (3\cos t - 2)\mathbf{i} + (6t + 2\sin t)\mathbf{k}$

19. $\mathbf{r}'(t) = \mathbf{f}(t) \times \mathbf{g}'(t) + \mathbf{f}'(t) \times \mathbf{g}(t)$

$= \det \begin{bmatrix} \mathbf{i} & \mathbf{j} & \mathbf{k} \\ \sin t & 1 & t^2 \\ 1 & 0 & -\sin t \end{bmatrix} + \det \begin{bmatrix} \mathbf{i} & \mathbf{j} & \mathbf{k} \\ \cos t & 0 & 2t \\ t & 0 & \cos t \end{bmatrix}$

$= [(-\sin t - 0)\mathbf{i} + (t^2 - (-\sin^2 t))\mathbf{j} + (0 - 1)\mathbf{k}] + [(0 - 0)\mathbf{i} + (2t^2 - \cos^2 t)\mathbf{j} + (0 - 0)\mathbf{k}]$

$= -\sin t\,\mathbf{i} + (3t^2 + \sin^2 t - \cos^2 t)\mathbf{j} - \mathbf{k}$

21. $\mathbf{r}'(t) = \mathbf{f}'(h(t))h'(t)$
$= [\cos(h(t))\mathbf{i} + 2h(t)\mathbf{k}]\,3e^{3t}$
$= 3e^{3t}\cos(e^{3t})\mathbf{i} + 6e^{6t}\mathbf{k}$

23. $\mathbf{r}'(2) = \mathbf{f}'(2) + \mathbf{g}'(2)$
$= 5\mathbf{i} + 14\mathbf{j} + 4\mathbf{k}$

25. $r'(2) = \mathbf{f}(2) \cdot \mathbf{g}'(2) + \mathbf{f}'(2) \cdot \mathbf{g}(2) = [(2)(1) + (8)(2) + (4)(4)] + [(4)(-1) + (12)(9) + (0)(4)] = 138$

27. This occurs at $t_0 = 0$. Note that $\mathbf{f}(0) = \mathbf{k}$, $\mathbf{g}(0) = \mathbf{k}$, $\mathbf{f}'(0) = \mathbf{i} + \mathbf{k}$, and $\mathbf{g}'(0) = \mathbf{i} + \mathbf{j}$.
$(\mathbf{f} \cdot \mathbf{g})'(0) = \mathbf{f}(0) \cdot \mathbf{g}'(0) + \mathbf{f}'(0) \cdot \mathbf{g}(0)$
$= [(0)(1) + (0)(1) + (1)(0)] + [(1)(0) + (0)(0) + (1)(1)]$
$= 1$

29. $\mathbf{f}'(t) = -2\sin 2t\,\mathbf{i} + 2\cos 2t\,\mathbf{j}$
$\mathbf{f}(t) \cdot \mathbf{f}'(t) = (\cos 2t)(-2\sin 2t) + (\sin 2t)(2\cos 2t) + (3)(0)$
$= -2\sin 2t \cos 2t + 2\sin 2t \cos 2t$
$= 0$ for all t.
Therefore $\mathbf{f}(t)$ and $\mathbf{f}'(t)$ are orthogonal for all t.

31. $\int \mathbf{f}(t)\,dt = (2t^{3/2}/3)\mathbf{i} + (e^{2t}/2)\mathbf{j} + \ln|t|\mathbf{k} + \mathbf{C}$

33. $\int \ln t\,dt = t\ln t - t + C$ using integration by parts with $u = \ln t$ and $dv = dt$.
$\int (1/(t\ln t))\,dt = \ln|\ln t| + C$ using substitution with $u = \ln t$.
$\int \mathbf{f}(t)\,dt = (t\ln t - t)\mathbf{i} + \ln|\ln t|\mathbf{j} + \mathbf{C}$

35. $\mathbf{f}(t) = \int \mathbf{f}'(t)\,dt = t\mathbf{i} + (t^3/3)\mathbf{j} + \mathbf{C}$
$3\mathbf{i} + 5\mathbf{j} = \mathbf{f}(0) = \mathbf{C}$
$\mathbf{f}(t) = (t + 3)\mathbf{i} + (t^3/3 + 5)\mathbf{j}$

37. $\mathbf{f}(t) = \int \mathbf{f}'(t)\,dt$
$\int \cos^2 t\,dt = \int (1/2)(1 + \cos 2t)\,dt = t/2 + (1/4)\sin 2t + C$
$\int \sin^2 t\,dt = \int (1/2)(1 - \cos 2t)\,dt = t/2 - (1/4)\sin 2t + C$
$\mathbf{f}(t) = (1/4)(2t + \sin 2t)\mathbf{i} + (1/4)(2t - \sin 2t)\mathbf{j} - e^{-t}\mathbf{k} + \mathbf{C}$
$3\mathbf{i} - 6\mathbf{j} + 2\mathbf{k} = \mathbf{f}(0) = -\mathbf{k} + \mathbf{C}$ so $\mathbf{C} = 3\mathbf{i} - 6\mathbf{j} + 3\mathbf{k}$
$\mathbf{f}(t) = (1/4)(2t + 12 + \sin 2t)\mathbf{i} + (1/4)(2t - 24 - \sin 2t)\mathbf{j} + (3 - e^{-t})\mathbf{k}$

39. $\mathbf{f}'(t) = -4\mathbf{f}(t)$, so considering components, one has $x'(t) = -4x(t)$, $y'(t) = -4y(t)$ and $z'(t) = -4z(t)$. The solutions to these differential equations are $x(t) = c_1 e^{-4t}$, $y(t) = c_2 e^{-4t}$ and $z(t) = c_3 e^{-4t}$. Thus $\mathbf{f}(t) = c_1 e^{-4t}\mathbf{i} + c_2 e^{-4t}\mathbf{j} + c_3 e^{-4t}\mathbf{k}$. Now use the fact that $\mathbf{i} + 4\mathbf{k} = \mathbf{f}(0) = c_1\mathbf{i} + c_2\mathbf{j} + c_3\mathbf{k}$ to determine that $c_1 = 1$, $c_2 = 0$ and $c_3 = 4$. Therefore $\mathbf{f}(t) = e^{-4t}\mathbf{i} + 4e^{-4t}\mathbf{k}$.

41. $\left(\int_a^b t\, dt\right)\mathbf{i} + \left(\int_a^b t^2\, dt\right)\mathbf{j} - \left(\int_a^b t^3\, dt\right)\mathbf{k}$

$= \left(t^2/2\big|_a^b\right)\mathbf{i} + \left(t^3/3\big|_a^b\right)\mathbf{j} - \left(t^4/4\big|_a^b\right)\mathbf{k}$

$= [(b^2 - a^2)/2]\mathbf{i} + [(b^3 - a^3)/3]\mathbf{j} - [(b^4 - a^4)/4]\mathbf{k}$

43. $\sin t\,\mathbf{i} - (1/2)\cos 2t\,\mathbf{j} + (t/2 + (1/4)\sin 2t)\mathbf{k}\big|_0^{\pi/4}$

$= [(\sqrt{2}/2)\mathbf{i} - 0\mathbf{j} + (\pi/8 + 1/4)\mathbf{k}] - [0\mathbf{i} - (1/2)\mathbf{j} + 0\mathbf{k}]$

$= (\sqrt{2}/2)\mathbf{i} + (1/2)\mathbf{j} + (\pi/8 + 1/4)\mathbf{k}$

45. If $\mathbf{f}'(t) = 0$ for all $t \in I$, then $x'(t) = 0$, $y'(t) = 0$ and $z'(t) = 0$ for all $t \in I$, from which it follows that $x(t) = c_1$, $y(t) = c_2$ and $z(t) = c_3$ are constant functions on the interval I. Therefore $\mathbf{f}(t) = c_1\mathbf{i} + c_2\mathbf{j} + c_3\mathbf{k} = \mathbf{C}$ is a constant function on I.

47. False. There exist such constants c on each component, but that does not mean that a single constant c can be found which will apply to all three components. For example, if $\mathbf{f}(t) = t^2\mathbf{i} + t^3\mathbf{j}$, then $\mathbf{f}$ is differentiable on the interval $[0,1]$, $[\mathbf{f}(1) - \mathbf{f}(0)]/(1 - 0) = [\mathbf{i} + \mathbf{j}]/1 = \mathbf{i} + \mathbf{j}$, $\mathbf{f}'(c) = 2c\mathbf{i} + 3c^2\mathbf{j}$ for all $c \in [0,1]$, and one cannot simultaneously have $2c = 1$ and $3c^2 = 1$.

49. a. $(h\mathbf{f})'(t) = \lim_{s\to 0} \frac{1}{s}[(h\mathbf{f})(t+s) - (h\mathbf{f})(t)]$

$= \lim_{s\to 0} \frac{1}{s}[h(t+s)\mathbf{f}(t+s) - h(t)\mathbf{f}(t)]$

$= \lim_{s\to 0} \frac{1}{s}[h(t+s)\mathbf{f}(t+s) - h(t)\mathbf{f}(t+s) + h(t)\mathbf{f}(t+s) - h(t)\mathbf{f}(t)]$

$= \lim_{s\to 0} \left(\frac{1}{s}[h(t+s)\mathbf{f}(t+s) - h(t)\mathbf{f}(t+s)] + \frac{1}{s}[h(t)\mathbf{f}(t+s) - h(t)\mathbf{f}(t)]\right)$

$= \lim_{s\to 0} \frac{1}{s}[h(t+s) - h(t)]\mathbf{f}(t+s) + \lim_{s\to 0} \frac{1}{s}h(t)[\mathbf{f}(t+s) - \mathbf{f}(t)]$

b. $= \left(\lim_{s\to 0} \frac{1}{s}[h(t+s) - h(t)]\right)\left(\lim_{s\to 0}\mathbf{f}(t+s)\right) + h(t)\lim_{s\to 0}\frac{1}{s}[\mathbf{f}(t+s) - \mathbf{f}(t)]$

$= h'(t)\mathbf{f}(t) + h(t)\mathbf{f}'(t)$

17.3   Curves: Tangents and Arc Length

1. $\mathbf{r}'(t) = 3t^2\mathbf{i} + (1/2)t^{-1/2}\mathbf{j}$
$\mathbf{r}'(4) = 48\mathbf{i} + (1/4)\mathbf{j}$

3. $\mathbf{r}'(t) = \tan t \sec t\,\mathbf{i} + \sec^2 t\,\mathbf{j}$
$\mathbf{r}'(\pi/4) = \sqrt{2}\mathbf{i} + 2\mathbf{j}$

5. This point occurs only at $t = -1$.
$\mathbf{r}'(t) = 2t\mathbf{i} + \mathbf{j} + 6\mathbf{k}$
$\mathbf{r}'(-1) = -2\mathbf{i} + \mathbf{j} + 6\mathbf{k}$

7. The point occurs when $t^2 - 6 = -2$, $t^3 - 9t + 4 = -6$ and $t^2 - 4t + 1 = -3$. The only solution for the last equation is $t = 2$, which is also a solution for the other two equations.
$\mathbf{r}'(t) = 2t\mathbf{i} + (3t^2 - 9)\mathbf{j} + (2t - 4)\mathbf{k}$
$\mathbf{r}'(2) = 4\mathbf{i} + 3\mathbf{j}$
$|\mathbf{r}'(2)| = \sqrt{4^2 + 3^2} = 5$
$\mathbf{T} = \mathbf{r}'(2)/|\mathbf{r}'(2)| = (4/5)\mathbf{i} + (3/5)\mathbf{j}$

9. The point occurs when $(t + 4)^{3/2} = 8$, $t + 4 = 8^{2/3} = 4$, or $t = 0$. The is also consistent with the other two components.
$\mathbf{r}'(t) = \cos t\,\mathbf{i} - \sin t\,\mathbf{j} + (3/2)\sqrt{t + 4}\,\mathbf{k}$
$\mathbf{r}'(0) = \mathbf{i} + 3\mathbf{k}$
$|\mathbf{r}'(0)| = \sqrt{1^2 + 3^2} = \sqrt{10}$
$\mathbf{T} = \mathbf{r}'(0)/|\mathbf{r}'(0)| = (\sqrt{10}/10)\mathbf{i} + (3\sqrt{10}/10)\mathbf{k}$

11. $\mathbf{r}'(t) = 2t\mathbf{i} + 3\mathbf{j}$. $\mathbf{r}(t)$ and $\mathbf{r}'(t)$ are orthogonal if and only if their dot product is zero.
$$\begin{aligned} 0 &= \mathbf{r}(t) \cdot \mathbf{r}'(t) \\ &= (t^2 + 2)(2t) + (3t)(3) \\ &= 2t^3 + 13t \\ &= t(2t^2 + 13) \end{aligned}$$
The only solution is $t = 0$.

13. $\mathbf{r}(t)$ and $\mathbf{r}'(t)$ have opposite direction if $\mathbf{r}(t) = c\mathbf{r}'(t)$ for some negative constant c. Thus one has $(t^2 + 2)\mathbf{i} + 3t\mathbf{j} = 2ct\mathbf{i} + 3c\mathbf{j}$. This yields the simultaneous equations $t^2 + 2 = 2ct$ and $3t = 3c$. The second equation implies that $t = c$, and replacing t by c in the first equation, one obtains $t^2 + 2 = 2t^2$, whose solutions are $t = \pm\sqrt{2}$. Note that $c = t = \pm\sqrt{2}$. Since c must be negative, the only solution is $t = -\sqrt{2}$.

15. The direction vector for the line is $\mathbf{r}'(0) = \mathbf{i} + 3\mathbf{k}$, and a point on the line is the terminal point of $\mathbf{r}(0) = \mathbf{j} + \mathbf{k}$. A vector equation for the line is $\mathbf{j} + \mathbf{k} + t(\mathbf{i} + 3\mathbf{k}) = t\mathbf{i} + \mathbf{j} + (1 + 3t)\mathbf{k}$. Parametric equations for the line are $x(t) = t$, $y(t) = 1$ and $z(t) = 1 + 3t$.

17. The radius vector is $\mathbf{r}(t) = \cos t\, \mathbf{i} + \sin t\, \mathbf{j}$. The tangent line is parallel to the tangent vector $\mathbf{r}'(t) = -\sin t\, \mathbf{i} + \cos t\, \mathbf{j}$. Consider $\mathbf{r}(t) \cdot \mathbf{r}'(t) = (\cos t)(-\sin t) + (\sin t)(\cos t) = 0$. Therefore, the two are orthogonal.

19. The direction vector for $\mathbf{L}(t) = 3\mathbf{i} + 4\mathbf{j} + t(-2\mathbf{i} + 8\mathbf{j})$ is $\mathbf{b} = -2\mathbf{i} + 8\mathbf{j}$. The point sought is the point where $\mathbf{r}'(t)$ and $\mathbf{b}$ are parallel; where $\mathbf{r}'(t)$ is a multiple of $\mathbf{b}$.
$\mathbf{r}'(t) = 2t\mathbf{i} + 8\mathbf{j}$
$2t\mathbf{i} + 8\mathbf{j} = c\mathbf{b} = -2c\mathbf{i} + 8c\mathbf{j}$
$2t = -2c$ and $8 = 8c$
$t = -c$ and $1 = c$
Therefore, $t = -1$ and the point is the terminal point of $\mathbf{r}(-1) = 2\mathbf{i} - 8\mathbf{j}$, or $(2, -8)$.

21. The point occurs at $t = 2$.
$\mathbf{r}'(t) = (2/(2t + 1))\mathbf{i} + (2t/(t^2 - 2))\mathbf{j} + (3/t)\mathbf{k}$
$\mathbf{r}'(2) = (2/5)\mathbf{i} + 2\mathbf{j} + (3/2)\mathbf{k}$ is a normal vector for the plane. The plane has equation of the form $(2/5)x + 2y + (3/2)z = d$, and since $(\ln 5, \ln 2, \ln 8)$ is on the plane, $d = (2/5)\ln 5 + 2\ln 2 + (3/2)\ln 8 = \ln\left(5^{2/5}2^2(2^3)^{3/2}\right) = \ln(5^{2/5}2^{13/2})$. Therefore, the equation of the plane is:
$(2/5)x + 2y + (3/2)z = \ln(5^{2/5}2^{13/2})$ or
$4x + 20y + 15z = 10\ln(5^{2/5}2^{13/2})$ or
$4x + 20y + 15z = \ln(5^4 2^{65})$ or
$4x + 20y + 15z = \ln(23058430092136939520000)$

23. The point occurs at $t = 1$.
$\mathbf{r}'(t) = -2\pi\cos 2\pi t\,\mathbf{i} - 4\pi\sin 4\pi t\,\mathbf{j} + \mathbf{k}$
$\mathbf{r}'(1) = -2\pi\mathbf{i} + \mathbf{k}$ is a normal vector for the plane. The plane has equation of the form $-2\pi x + z = d$, and since $(0, 1, 1)$ is on the plane, $d = -2\pi(0) + 1 = 1$. Therefore, the equation of the plane is $-2\pi x + z = 1$.

25. Note that $\mathbf{r}'(t) = 3t^2\mathbf{i} - 2t^2\mathbf{j} - (2t^2/3)\mathbf{k}$ and that the normal vector for the plane is $\mathbf{n} = \mathbf{i} + \mathbf{j} + \mathbf{k}$. The point of intersection occurs when:

$t^3 + (1 - 2t^3/3) - (1 + 2t^3/9) = 3$

$9t^3 + 9 - 6t^3 - 9 - 2t^3 = 27$

$t^3 = 27$

$t = 3$

Therefore, the only point of intersection occurs at $t = 3$. The direction of the curve at this point is $\mathbf{r}'(3) = 27\mathbf{i} - 18\mathbf{j} - 6\mathbf{k}$. The cosine of the angle between the curve and the normal vector is:

$\cos\alpha = (\mathbf{n} \cdot \mathbf{r}'(3))/(|\mathbf{n}||\mathbf{r}'(3)|) = 3/(\sqrt{3}\sqrt{1089}) = \sqrt{3}/33$,

and the sine of this angle is:

$\sin\alpha = \sqrt{1 - \cos^2\alpha} = \sqrt{1 - 3/1089} = \sqrt{1 - 1/363} = \sqrt{362}/(11\sqrt{3}) = \sqrt{1086}/33$.

The cosine of the angle of intersection between the curve and the plane is:

$\cos\theta = \cos(\pi/2 - \alpha) = \sin\alpha = \sqrt{1086}/33$.

27. $\mathbf{r}_1'(t) = 2t\mathbf{i} + \mathbf{j}$ and $\mathbf{r}_2'(t) = -\mathbf{i} - \mathbf{j} + \mathbf{k}$

Points of intersection:

$\mathbf{r}_1(t_1) = \mathbf{r}_2(t_2)$

$t_1^2 = 1 - t_2$ and $t_1 = 1 - t_2$ and $-1 = t_2 - 2$

From the last equation, $t_2 = 1$, and thus $t_1 = 0$. The point of intersection is the terminal point of $\mathbf{r}_1(0) = -\mathbf{k}$, or $(0, 0, -1)$. The directions of the curves at this point are $\mathbf{r}_1'(0) = \mathbf{j}$ and $\mathbf{r}_2'(1) = -\mathbf{i} - \mathbf{j} + \mathbf{k}$, and the cosine of the angle of intersection θ is $\cos\theta = (\mathbf{r}_1'(1) \cdot \mathbf{r}_2'(1))/(|\mathbf{r}_1'(1)||\mathbf{r}_2'(1)|) = -1/(1\sqrt{3}) = -\sqrt{3}/3$.

29. $\mathbf{r}_1'(t) = \cos t\,\mathbf{i} + \mathbf{j} + \cos t\,\mathbf{k}$ and $\mathbf{r}_2'(t) = \mathbf{j} + (2t - \pi)\mathbf{k}$

Points of intersection:

$\mathbf{r}_1(t_1) = \mathbf{r}_2(t_2)$

$\sin t_1 = 1$ and $t_1 = t_2$ and $\sin t_1 = (t_2 - \pi/2)^2 + 1$

From the first equation, substitute 1 for $\sin t_1$ in the last equation in order to obtain $1 = (t_2 - \pi/2)^2 + 1$ or $0 = (t_2 - \pi/2)^2$, which implies that $t_2 = \pi/2$. From the middle equation, one also has $t_1 = \pi/2$, which is consistent with $\sin t_1 = 1$. The point of intersection is the terminal point of $\mathbf{r}_1(\pi/2) = \mathbf{i} + (\pi/2)\mathbf{j} + \mathbf{k}$, or $(1, \pi/2, 1)$. The directions of the curves at this point are $\mathbf{r}_1'(\pi/2) = \mathbf{j}$ and $\mathbf{r}_2'(\pi/2) = \mathbf{j}$. Since the directions are equal, the angle of intersection is zero, and the cosine of the angle is 1.

31. $\mathbf{r}'(t) = -\sin t\,\mathbf{i} + (\sqrt{2}/2)\cos t\,\mathbf{j} + (\sqrt{2}/2)\cos t\,\mathbf{k}$

$|\mathbf{r}'(t)| = 1$

Length $= \int_0^{\pi/2} |\mathbf{r}'(t)|\,dt$

$= \int_0^{\pi/2} dt$

$= t\big|_0^{\pi/2}$

$= \pi/2$

33. $\mathbf{r}'(t) = \mathbf{i} + 2t\mathbf{j} + \mathbf{k}$

$|\mathbf{r}'(t)| = \sqrt{1^2 + 4t^2 + 1^2} = \sqrt{2 + 4t^2}$

Length $= \int_0^1 |\mathbf{r}'(t)|\, dt$

$= \int_0^1 \sqrt{2 + 4t^2}\, dt$

$= \sqrt{2} \int_0^1 \sqrt{1 + 2t^2}\, dt \quad (t\sqrt{2} = \tan\theta)$

$= \sqrt{2} \int_{t=0}^{t=1} \sqrt{1 + \tan^2\theta}\,(1/\sqrt{2})\sec^2\theta\, d\theta$

$= \int_{t=0}^{t=1} \sec^3\theta\, d\theta$

$= (1/2)\tan\theta\sec\theta + (1/2)\ln|\tan\theta + \sec\theta|\big|_{t=0}^{t=1}$

$= (1/2)(t\sqrt{2})\sqrt{1 + 2t^2} + (1/2)\ln|t\sqrt{2} + \sqrt{1 + 2t^2}|\big|_0^1$

$= (\sqrt{2}\sqrt{3}/2 + (1/2)\ln|\sqrt{2} + \sqrt{3}|) - (0 + \ln|1|)$

$= \sqrt{6}/2 + (1/2)\ln(\sqrt{2} + \sqrt{3})$

35. $(0,0,0) = \mathbf{r}(0)$ and $(18,0,9) = \mathbf{r}(3)$

$\mathbf{r}'(t) = 4t\mathbf{i} + t^2\mathbf{k}$

$|\mathbf{r}'(t)| = \sqrt{16t^2 + t^4} = t\sqrt{16 + t^2}$

$(\sqrt{t^2} = |t| = t$ since $0 \le t \le 3)$

Length $= \int_0^3 |\mathbf{r}'(t)|\, dt$

$= \int_0^3 t\sqrt{16 + t^2}\, dt \quad (u = 16 + t^2)$

$= (1/2)\int_{16}^{25} \sqrt{u}\, du$

$= u^{3/2}/3 \big|_{16}^{25}$

$= (125/3) - (64/3)$

$= 61/3$

37. Using $x = t$, one has $y = t^2 + 3$, so the curve is also the graph of:

$\mathbf{r}(t) = t\mathbf{i} + (t^2 + 3)\mathbf{j}$

$\mathbf{r}'(t) = \mathbf{i} + 2t\mathbf{j}$

$\mathbf{r}'(2) = \mathbf{i} + 4\mathbf{j}$

$|\mathbf{r}'(2)| = \sqrt{1^2 + 4^2} = \sqrt{17}$

$\mathbf{T}(2) = \mathbf{r}'(2)/|\mathbf{r}'(2)|$

$= (\sqrt{17}/17)\mathbf{i} + (4\sqrt{17}/17)\mathbf{j}$

39. This point occurs at $t = -1$.

$\mathbf{r}'(t) = 2t\mathbf{i} + \mathbf{j} + 6\mathbf{k}$

$\mathbf{r}'(-1) = -2\mathbf{i} + \mathbf{j} + 6\mathbf{k}$ is a direction vector for the line. Since $(2, -5, -6)$ is a point on the line, the line has vector equation $2\mathbf{i} - 5\mathbf{j} - 6\mathbf{k} + t(-2\mathbf{i} + \mathbf{j} + 6\mathbf{k}) = (2 - 2t)\mathbf{i} + (-5 + t)\mathbf{j} + (-6 + 6t)\mathbf{k}$. The parametric equations are $x = 2 - 2t$, $y = -5 + t$ and $z = -6 + 6t$. Solving for t, one obtains the symmetric equations:

$$\frac{2 - x}{2} = y + 5 = \frac{z + 6}{6}.$$

41. One wants to have $|\mathbf{r}'(t)| = 1$ for $t \in [0, 2\pi/\alpha]$.

$\mathbf{r}'(t) = -\alpha\sin\alpha t\,\mathbf{i} + \alpha\cos\alpha t\,\mathbf{j} + \alpha\mathbf{k} = \alpha(-\sin\alpha t\,\mathbf{i} + \cos\alpha t\,\mathbf{j} + \mathbf{k})$

$|\mathbf{r}'(t)| = \alpha\sqrt{2}$

In order for this to be 1, one must have $\alpha = 1/\sqrt{2} = \sqrt{2}/2$.

43. In both cases, the tangent line will pass through the point $(x(t_0), y(t_0))$, so they will be the same line if they have the same slope. The earlier definition gives the slope of the tangent line as $y'(t_0)/x'(t_0)$. From Definition 5, the tangent line is parameterized by $\mathbf{r}(t_0) + t\mathbf{r}'(t_0) = (x(t_0) + tx'(t_0))\mathbf{i} + (y(t_0) + ty'(t_0))\mathbf{j}$, and therefore the slope is

$$m = \frac{[y(t_0) + ty'(t_0)] - y(t_0)}{[x(t_0) + tx'(t_0)] - x(t_0)} = \frac{ty'(t_0)}{tx'(t_0)} = \frac{y'(t_0)}{x'(t_0)}.$$

Since the lines pass through a common point and have the same slope, they are equal.

531

45. $\mathbf{r}'(t) = -\sqrt{2}\sin t\,\mathbf{i} + \sqrt{2}\cos t\,\mathbf{j} = \sqrt{2}(-\sin t\,\mathbf{i} + \cos t\,\mathbf{j})$

$|\mathbf{r}'(t)| = \sqrt{2}$

$\mathbf{T}(t) = \mathbf{r}'(t)/|\mathbf{r}'(t)| = -\sin t\,\mathbf{i} + \cos t\,\mathbf{j}$

$\mathbf{T}'(t) = -\cos t\,\mathbf{i} - \sin t\,\mathbf{j}$

$|\mathbf{T}'(t)| = 1$

$\mathbf{N}(t) = \mathbf{T}'(t)/|\mathbf{T}'(t)| = -\cos t\,\mathbf{i} - \sin t\,\mathbf{j}$

47. Note that $|\mathbf{r}'(t)| = \sqrt{8} \neq 1$, and therefore this is not a parameterization by arc length. However, the substitution $t = s/\sqrt{8}$ gives the following parameterization of the same curve:

$\mathbf{r}(s) = 2\cos(s/\sqrt{8})\mathbf{i} + 2\sin(s/\sqrt{8})\mathbf{j} + (2s/\sqrt{8})\mathbf{k}$

$\mathbf{r}'(s) = -(2/\sqrt{8})\sin(s/\sqrt{8})\mathbf{i} + (2/\sqrt{8})\cos(s/\sqrt{8})\mathbf{j} + (2/\sqrt{8})\mathbf{k} = (2/\sqrt{8})\left(-\sin(s/\sqrt{8})\mathbf{i} + \cos(s/\sqrt{8})\mathbf{j} + \mathbf{k}\right)$

$|\mathbf{r}'(s)| = (2/\sqrt{8})\sqrt{\sin^2(s/\sqrt{8}) + \cos^2(s/\sqrt{8}) + 1} = (1/\sqrt{2})\sqrt{2} = 1$

Hence this is a parameterization by arc length for the curve. Now the point $(\sqrt{2}, \sqrt{2}, \pi/2)$ occurs at $t = \pi/4$, which corresponds to $s = \pi\sqrt{8}/4 = \pi\sqrt{2}/2$.

$\mathbf{r}''(s) = -(1/4)\cos(s/\sqrt{8})\mathbf{i} - (1/4)\sin(s/\sqrt{8})\mathbf{j}$

$\mathbf{r}''(\pi\sqrt{2}/2) = -(\sqrt{2}/8)\mathbf{i} - (\sqrt{2}/8)\mathbf{j}$

Using Exercise 46, a vector equation for the normal line is:

$\mathbf{R}(\omega) = \sqrt{2}\,\mathbf{i} + \sqrt{2}\,\mathbf{j} + (\pi/2)\mathbf{k} + \omega\left(-(\sqrt{2}/8)\mathbf{i} - (\sqrt{2}/8)\mathbf{j}\right).$

49. $\mathbf{r}'(t) = (-2\sin t - 2\sin 2t)\mathbf{i} + (2\cos t - 2\cos 2t)\mathbf{j}$

$\quad = -2\left[(\sin t + \sin 2t)\mathbf{i} + (\cos 2t - \cos t)\mathbf{j}\right]$

$|\mathbf{r}'(t)| = 2\sqrt{(\sin t + \sin 2t)^2 + (\cos 2t - \cos t)^2}$

Length $= \int_0^{2\pi} |\mathbf{r}'(t)|\,dt$

The result should be approximately 16.

17.4 Velocity and Acceleration

1. $\mathbf{v}(t) = 2\mathbf{j}$

$\mathbf{a}(t) = \mathbf{0}$

3. $\mathbf{v}(t) = -3\sin 3t\,\mathbf{i} + 3\cos 3t\,\mathbf{j}$

$\mathbf{a}(t) = -9\cos 3t\,\mathbf{i} - 9\sin 3t\,\mathbf{j}$

5. $\mathbf{v}(t) = (2/t)\mathbf{i} - 2\sin 2t\,\mathbf{j} - (1/t^2)\mathbf{k}$

$\mathbf{a}(t) = -(2/t^2)\mathbf{i} - 4\cos 2t\,\mathbf{j} + (2/t^3)\mathbf{k}$

7. $\mathbf{v}(t) = e^t(\cos t - \sin t)\mathbf{i} + e^t(\cos t + \sin t)\mathbf{j} - \sin t\,\mathbf{k}$

$\mathbf{a}(t) = -2e^t\sin t\,\mathbf{i} + 2e^t\cos t\,\mathbf{j} - \cos t\,\mathbf{k}$

9. $\mathbf{v}(t) = 2t\mathbf{i} + 2\mathbf{j} + 3t^2\mathbf{k}$

$\mathbf{v}(2) = 4\mathbf{i} + 2\mathbf{j} + 12\mathbf{k}$

The speed at time $t = 2$ is $|\mathbf{v}(2)| = \sqrt{4^2 + 2^2 + 12^2} = \sqrt{164} = 2\sqrt{41}$.

11. $\mathbf{r}(t) = \int \mathbf{v}(t)\,dt = \sin t\,\mathbf{i} - \cos t\,\mathbf{j} + \mathbf{C}$

$3\mathbf{i} + 2\mathbf{j} = \mathbf{r}(0) = -\mathbf{j} + \mathbf{C}$ so $\mathbf{C} = 3\mathbf{i} + 3\mathbf{j}$

$\mathbf{r}(t) = (3 + \sin t)\mathbf{i} + (3 - \cos t)\mathbf{j}$

13. $\mathbf{r}(t) = \int \mathbf{v}(t)\,dt = \mathrm{Tan}^{-1}t\,\mathbf{i} + (e^{t^2}/2)\mathbf{j} + \mathbf{C}$

$-2\mathbf{i} + \mathbf{j} + 4\mathbf{k} = \mathbf{r}(0) = (1/2)\mathbf{j} + \mathbf{C}$

$\mathbf{r}(t) = (-2 + \mathrm{Tan}^{-1}t)\mathbf{i} + ((1 + e^{t^2})/2)\mathbf{j} + 4\mathbf{k}$

15. $\mathbf{v}(t) = \int \mathbf{a}(t)\,dt = 2t\mathbf{i} + t\mathbf{k} + \mathbf{C}$

$\mathbf{0} = \mathbf{v}(0) = \mathbf{C}$

$\mathbf{v}(t) = 2t\mathbf{i} + t\mathbf{k}$

$\mathbf{r}(t) = \int \mathbf{v}(t)\,dt = t^2\mathbf{i} + (t^2/2)\mathbf{k} + \mathbf{C}$

$3\mathbf{i} - \mathbf{j} + 4\mathbf{k} = \mathbf{r}(0) = \mathbf{C}$

$\mathbf{r}(t) = (t^2 + 3)\mathbf{i} - \mathbf{j} + (t^2/2 + 4)\mathbf{k}$

17. $\mathbf{v}(t) = \int \mathbf{a}(t)\,dt = \sin t\,\mathbf{i} - \cos t\,\mathbf{j} + \mathbf{C}$

$\mathbf{k} = \mathbf{v}(0) = -\mathbf{j} + \mathbf{C}$ so $\mathbf{C} = \mathbf{j} + \mathbf{k}$

$\mathbf{v}(t) = \sin t\,\mathbf{i} + (1 - \cos t)\mathbf{j} + \mathbf{k}$

$\mathbf{r}(t) = \int \mathbf{v}(t)\,dt = -\cos t\,\mathbf{i} + (t - \sin t)\mathbf{j} + t\mathbf{k} + \mathbf{C}$

$3\mathbf{i} - \mathbf{j} + 2\mathbf{k} = \mathbf{r}(0) = -\mathbf{i} + \mathbf{C}$ so $\mathbf{C} = 4\mathbf{i} - \mathbf{j} + 2\mathbf{k}$

$\mathbf{r}(t) = (4 - \cos t)\mathbf{i} + (t - 1 - \sin t)\mathbf{j} + (t + 2)\mathbf{k}$

19. Recall from the example that $\mathbf{v}(t) = -\alpha \sin \alpha t \, \mathbf{i} + \alpha \cos \alpha t \, \mathbf{j}$ and $\mathbf{a}(t) = -\alpha^2 \cos \alpha t \, \mathbf{i} - \alpha^2 \sin \alpha t \, \mathbf{j}$. In order to verify that they are orthogonal, show that their dot product is zero.
$$\mathbf{v}(t) \cdot \mathbf{a}(t) = (-\alpha \sin \alpha t)(-\alpha^2 \cos \alpha t) + (\alpha \cos \alpha t)(-\alpha^2 \sin \alpha t)$$
$$= \alpha^3 \sin \alpha t \cos \alpha t - \alpha^3 \sin \alpha t \cos \alpha t$$
$$= 0.$$

21. Note the following:
$$\mathbf{r}(t) = \rho(\cos \alpha t \, \mathbf{i} + \sin \alpha t \, \mathbf{j})$$
$$\mathbf{r}'(t) = -\rho\alpha \sin \alpha t \, \mathbf{i} + \rho\alpha \cos \alpha t \, \mathbf{j} = \rho\alpha(-\sin \alpha t \, \mathbf{i} + \cos \alpha t \, \mathbf{j})$$
The radius is $|\mathbf{r}(t)| = |\rho|$ and the speed is $|\mathbf{r}'(t)| = |\rho\alpha| = |\rho| \, |\alpha|$. Therefore, the speed is $|\alpha|$ times the radius.

23. The initial conditions are $\mathbf{r}(0) = \mathbf{0}$ and $\mathbf{v}(0) = 100\cos(30°)\mathbf{i} + 100\sin(30°)\mathbf{j} = 50\sqrt{3}\mathbf{i} + 50\mathbf{j}$. The acceleration due to the force of gravity is $\mathbf{a}(t) = -9.81\mathbf{j}$. Therefore
$$\mathbf{v}(t) = \int \mathbf{a}(t)\,dt = -9.81t\mathbf{j} + \mathbf{C}$$
$$50\sqrt{3}\mathbf{i} + 50\mathbf{j} = \mathbf{v}(0) = \mathbf{C}$$
$$\mathbf{v}(t) = 50\sqrt{3}\mathbf{i} + (-9.81t + 50)\mathbf{j}$$
$$\mathbf{r}(t) = \int \mathbf{v}(t)\,dt = 50t\sqrt{3}\mathbf{i} + (-4.905t^2 + 50t)\mathbf{j} + \mathbf{C}$$
$$\mathbf{0} = \mathbf{r}(0) = \mathbf{C}$$
$$\mathbf{r}(t) = 50t\sqrt{3}\mathbf{i} + (-4.905t^2 + 50t)\mathbf{j}$$

a. $\mathbf{r}(t) = 50t\sqrt{3}\mathbf{i} + (-4.905t^2 + 50t)\mathbf{j}$

b. This is the maximum of the y component of $\mathbf{r}(t)$, $y(t) = -4.905t^2 + 50t$. The critical numbers are the zeros of $y'(t) = -9.81t + 50$, or $t = 50/9.81 \approx 5.10$. Thus the maximum altitude is reached after approximately 5.10 seconds, and is approximately $y(5.10) \approx 127$ meters.

c. The flight time is the time until it strikes the ground, which occurs when the y component of $\mathbf{r}(t)$ becomes zero after $t = 0$. Solve:
$$-4.905t^2 + 50t = 0$$
$$t(-4.905t + 50) = 0$$
$$t = 0 \text{ or } t = 50/4.905 \approx 10.2$$
The flight time is approximately 10.2 seconds.

d. The distance from launch point to point of impact is the value of the x component of $\mathbf{r}(t)$ at the moment of impact: $x(10.2) = 50(10.2)\sqrt{3} \approx 883$ meters.

e. The speed on impact is the value of $|\mathbf{v}(10.2)|$.
$$\mathbf{v}(10.2) = 50\sqrt{3}\mathbf{i} + ((-9.81)(10.2) + 50)\mathbf{j} \approx 86.6025\mathbf{i} - 50\mathbf{j}$$
$$|\mathbf{v}(8.65)| = \sqrt{86.6025^2 + (-50)^2} \approx 100$$
The speed on impact is 100 meters per second.

25. If $\mathbf{i}$ is the horizontal unit vector in the direction which the first slug was fired, and $\mathbf{j}$ is height, then the $\mathbf{j}$ component of the initial velocity for both slugs is zero, and the $\mathbf{j}$ component of the initial position for both slugs is the same (they are at the same height). Since both slugs have the same acceleration function $\mathbf{a}(t) = -9.81\mathbf{j}$, the $\mathbf{j}$ components of their velocity and position will always be equal. Therefore, they will strike the ground at the same moment, when the $\mathbf{j}$ components of their position functions are zero.

27. a. Recall that $\mathbf{r}(t) = 25t\mathbf{i} + (25t\sqrt{3} - gt^2/2)\mathbf{j}$ and that the moment of impact occurs at time $t = 50\sqrt{3}/g$, where $g = 9.81$.
$$\mathbf{r}'(t) = 25\mathbf{i} + (25\sqrt{3} - gt)\mathbf{j} \text{ and } |\mathbf{r}'(t)| = \sqrt{25^2 + (25\sqrt{3} - gt)^2} = \sqrt{2500 - 50\sqrt{3}gt + g^2t^2}$$
$$\text{Length} = \int_0^{50\sqrt{3}/g} \sqrt{2500 - 50\sqrt{3}gt + g^2t^2}\,dt$$

b. The result should be approximately 304.6.

17.5 Curvature

1. $\mathbf{r}'(s) = (1/2)\mathbf{i} + (\sqrt{3}/2)\mathbf{j}$
$|\mathbf{r}'(s)| = \sqrt{1/4 + 3/4} = 1$
Therefore, it is parameterized by arc length.

   a. $\mathbf{T}(s) = \mathbf{r}'(s) = (1/2)\mathbf{i} + (\sqrt{3}/2)\mathbf{j}$

   b. $\mathbf{r}''(s) = 0$

   c. $\kappa(s) = |0| = 0$

3. $\mathbf{r}'(s) = -(\sqrt{2}/2)\sin s\,\mathbf{i} + (\sqrt{2}/2)\cos s\,\mathbf{j} + (\sqrt{2}/2)\mathbf{k}$
$\phantom{\mathbf{r}'(s)} = (\sqrt{2}/2)[-\sin s\,\mathbf{i} + \cos s\,\mathbf{j} + \mathbf{k}$
$|\mathbf{r}'(s)| = (\sqrt{2}/2)\sqrt{\sin^2 s + \cos^2 s + 1} = (\sqrt{2}/2)\sqrt{2} = 1$
Therefore, it is parameterized by arc length.

   a. $\mathbf{T}(s) = \mathbf{r}'(s) = -(\sqrt{2}/2)\sin s\,\mathbf{i} + (\sqrt{2}/2)\cos s\,\mathbf{j} + (\sqrt{2}/2)\mathbf{k}$

   b. $\mathbf{r}''(s) = -(\sqrt{2}/2)\cos s\,\mathbf{i} - (\sqrt{2}/2)\sin s\,\mathbf{j}$

   c. $\kappa(s) = |\mathbf{r}''(s)| = (\sqrt{2}/2)\sqrt{\cos^2 s + \sin^2 s} = \sqrt{2}/2$

5. $x'(t) = 0$ and $y'(t) = 2t$
$x''(t) = 0$ and $y''(t) = 2$
$\kappa(t) = |x'(t)y''(t) - y'(t)x''(t)|/[x'(t)^2 + y'(t)^2]^{3/2}$
$ = |0 - 0|/[0 + 4t^2]^{3/2}$
$ = 0$

7. $x'(t) = 1$ and $y'(t) = e^t$
$x''(t) = 0$ and $y''(t) = e^t$
$\kappa(t) = |x'(t)y''(t) - y'(t)x''(t)|/[x'(t)^2 + y'(t)^2]^{3/2}$
$ = |e^t - 0|/[1 + (e^t)^2]^{3/2}$
$ = e^t/(1 + e^{2t})^{3/2}$

9. $x'(t) = 2\cos t$ and $y'(t) = -2\sin t$
$x''(t) = -2\sin t$ and $y''(t) = -2\cos t$
$\kappa(t) = |x'(t)y''(t) - y'(t)x''(t)|/[x'(t)^2 + y'(t)^2]^{3/2}$
$ = |-4\cos^2 t - 4\sin^2 t|/[4\cos^2 t + 4\sin^2 t]^{3/2} = 4/4^{3/2}$
$ = 4/8$
$ = 1/2$

11. $f'(x) = 2x$ and $f''(x) = 2$
$\kappa(t) = |f''(x)|/[1 + f'(x)^2]^{3/2}$
$ = 2/(1 + 4x^2)^{3/2}$

13. $f'(x) = 3x^2$ and $f''(x) = 6x$
$\kappa(t) = |f''(x)|/[1 + f'(x)^2]^{3/2}$
$ = 6|x|/(1 + 9x^4)^{3/2}$

15. $f'(x) = -\sin x$ and $f''(x) = -\cos x$
$\kappa(t) = |f''(x)|/[1 + f'(x)^2]^{3/2}$
$ = |\cos x|/[1 + \sin^2 x]^{3/2}$

17. $\mathbf{v}(t) = \mathbf{r}'(t) = \mathbf{j} + 2t\mathbf{k}$
$|\mathbf{v}(t)| = \sqrt{1 + 4t^2}$
$\mathbf{a}(t) = \mathbf{r}''(t) = 2\mathbf{k}$
$\mathbf{v}(t) \times \mathbf{a}(t) = \det\begin{bmatrix} \mathbf{i} & \mathbf{j} & \mathbf{k} \\ 0 & 1 & 2t \\ 0 & 0 & 2 \end{bmatrix} = 2\mathbf{i}$
$|\mathbf{v}(t) \times \mathbf{a}(t)| = 2$
$\kappa(t) = |\mathbf{v}(t) \times \mathbf{a}(t)|/|\mathbf{v}(t)|^3$
$ = 2/(1 + 4t^2)^{3/2}$

19. $\mathbf{v}(t) = \mathbf{r}'(t) = e^t(\cos t - \sin t)\mathbf{i} + e^t(\sin t + \cos t)\mathbf{j} + \mathbf{k}$

$|\mathbf{v}(t)| = \sqrt{e^{2t}(\cos^2 t - 2\cos t \sin t + \sin^2 t) + e^{2t}(\sin^2 t + 2\cos t \sin t + \cos^2 t) + 1}$

$= \sqrt{2e^{2t} + 1}$

$\mathbf{a}(t) = \mathbf{r}''(t) = -2e^t \sin t\, \mathbf{i} + 2e^t \cos t\, \mathbf{j}$

$\mathbf{v}(t) \times \mathbf{a}(t) = \det \begin{bmatrix} \mathbf{i} & \mathbf{j} & \mathbf{k} \\ e^t(\cos t - \sin t) & e^t(\sin t + \cos t) & 1 \\ -2e^t \sin t & 2e^t \cos t & 0 \end{bmatrix}$

$= (0 - 2e^t \cos t)\mathbf{i} + (-2e^t \sin t - 0)\mathbf{j} + [2e^{2t}(\cos^2 t - \cos t \sin t) + 2e^{2t}(\sin^2 t + \cos t \sin t)]\mathbf{k}$

$= -2e^t \cos t\, \mathbf{i} - 2e^t \sin t\, \mathbf{j} + 2e^{2t}\mathbf{k}$

$|\mathbf{v}(t) \times \mathbf{a}(t)| = 2e^t\sqrt{\cos^2 t + \sin^2 t + e^{2t}}$

$= 2e^t\sqrt{1 + e^{2t}}$

$\kappa(t) = |\mathbf{v}(t) \times \mathbf{a}(t)|/|\mathbf{v}(t)|^3$

$= 2e^t\sqrt{1 + e^{2t}}/(2e^{2t} + 1)^{3/2}$

21. One needs to determine the radius of curvature, and the unit normal vector $\mathbf{N}$, which points toward the center of curvature. In order to determine $\mathbf{N}$, use the fact that it is orthogonal to $\mathbf{r}'(t)$ where $\mathbf{r}(t)$ is a parametrization of the graph, that it points towards the concave side of the graph, and that if $a\mathbf{i} + b\mathbf{j}$ is a vector in the plane, then $\pm(b\mathbf{i} - a\mathbf{j})$ are orthogonal vectors.

For $f(x) = \sin x$, one has $f'(x) = \cos x$ and $f''(x) = -\sin x$. Thus $f'(\pi/2) = 0$ and $f''(\pi/2) = -1$. The curvature at the point in question is $\kappa(\pi/2) = |f''(\pi/2)|/[1 + f'(\pi/2)^2]^{3/2} = 1/[1 + 0]^{3/2} = 1$. Thus the radius of curvature is $\rho = 1/\kappa = 1$.

A parameterization of the graph is $\mathbf{r}(t) = t\mathbf{i} + \sin t\, \mathbf{j}$, for which $\mathbf{r}'(t) = \mathbf{i} + \cos t\, \mathbf{j}$ and $\mathbf{r}'(\pi/2) = \mathbf{i}$. Orthogonal vectors are $\pm\mathbf{j}$. Since the graph is concave down at $(\pi/2, 1)$, one wants a unit vector parallel to these whose $\mathbf{j}$ component is negative. Therefore $\mathbf{N} = -\mathbf{j}$. The center of curvature is at $\mathbf{r}(\pi/2) + \rho\mathbf{N} = (\pi/2)\mathbf{i} + \mathbf{j} + (-\mathbf{j}) = (\pi/2)\mathbf{i}$, which is the point $(\pi/2, 0)$.

One could also have obtained $\mathbf{N}$ by using the parametrization $\mathbf{r}(t)$ above and computing:

$\mathbf{r}'(t) = \mathbf{i} + \cos t\, \mathbf{j}$

$|\mathbf{r}'(t)| = \sqrt{1 + \cos^2 t}$

$|\mathbf{r}'(\pi/2)| = \sqrt{1 + 0} = 1$

$\mathbf{T}(t) = \mathbf{r}'(t)/|\mathbf{r}'(t)| = (1 + \cos^2 t)^{-1/2}(\mathbf{i} + \cos t\, \mathbf{j})$

$\mathbf{T}'(t) = \cos t \sin t(1 + \cos^2 t)^{-3/2}(\mathbf{i} + \cos t\, \mathbf{j}) + (1 + \cos^2 t)^{-1/2}(-\sin t\, \mathbf{j})$

$\mathbf{T}'(\pi/2) = -\mathbf{j}$

$\mathbf{N}(\pi/2) = \mathbf{T}'(\pi/2)/|\mathbf{r}'(\pi/2)| = (-\mathbf{j})/1 = -\mathbf{j}$

23. See the solution to exercise 21 for a discussion on finding the center of curvature for the graph of $y = f(x)$.

a. For $f(x) = \sqrt{x}$, one has $f'(x) = 1/(2\sqrt{x})$ and $f''(x) = -1/(4x^{3/2})$. Thus $f'(1) = 1/2$ and $f''(1) = -1/4$. Thus curvature at the point in question is:

$\kappa(1) = |f''(1)|/[1 + f'(1)^2]^{3/2}$

$= (1/4)/[1 + 1/4]^{3/2}$

$= 1/(4(5/4)^{3/2})$

$= 4^{3/2}/(4(5^{3/2}))$

$= 2\sqrt{5}/25$

b. The radius of curvature is $\rho = 1/\kappa = 25/2\sqrt{5} = 5\sqrt{5}/2$. A parameterization of the graph is $\mathbf{r}(t) = t\mathbf{i} + \sqrt{t}\,\mathbf{j}$, for which $\mathbf{r}'(t) = \mathbf{i} + (1/(2\sqrt{t}))\mathbf{j}$ and $\mathbf{r}'(1) = \mathbf{i} + (1/2)\mathbf{j}$. Orthogonal vectors are $\pm((1/2)\mathbf{i} - \mathbf{j})$. Since the graph is concave down at $(1,1)$ ($f''(1)$ is negative), one wants a unit vector parallel to these whose $\mathbf{j}$ component is negative, or a unit vector with the same direction as $(1/2)\mathbf{i} - \mathbf{j}$, the magnitude of which is $\sqrt{1/4 + 1} = \sqrt{5}/2$. Therefore $\mathbf{N} = (1/(\sqrt{5}/2))((1/2)\mathbf{i} - \mathbf{j}) = (2\sqrt{5}/5)((1/2)\mathbf{i} - \mathbf{j}) = (\sqrt{5}/5)(\mathbf{i} - 2\mathbf{j})$. The center of curvature is at $\mathbf{r}(1) + \rho\mathbf{N} = \mathbf{i} + \mathbf{j} + (5\sqrt{5}/2)(\sqrt{5}/5)(\mathbf{i} - 2\mathbf{j}) = \mathbf{i} + \mathbf{j} + (5/2)\mathbf{i} - 5\mathbf{j} = (7/2)\mathbf{i} - 4\mathbf{j}$, which is the point $(7/2, -4)$.

c. The curvature is:
$$\kappa(x) = |f''(x)|/[1 + f'(x)^2]^{3/2} = \frac{1}{4x^{3/2}}\frac{1}{[1 + (1/4x)]^{3/2}} = \frac{1}{4x^{3/2}}\frac{1}{(4x)^{-3/2}(4x + 1)^{3/2}} = 2/(4x + 1)^{3/2}$$
Therefore $\lim_{x \to 0+} \kappa(x) = 2/(0 + 1)^{3/2} = 2$.

25. Let $\mathbf{a}$ be a position vector for a point on the line, and let $\mathbf{b}$ be a unit direction vector for the line. Then $\mathbf{r}(s) = \mathbf{a} + s\mathbf{b}$ is an arc length parametrization of the line since $|\mathbf{r}'(s)| = |\mathbf{b}| = 1$. Since $\mathbf{r}'(s) = \mathbf{b}$ is constant, $\mathbf{r}''(s) = \mathbf{0}$, and the curvature is $\kappa(s) = |\mathbf{r}''(s)| = 0$.

27. $f(\theta) = 1 + \sin\theta$, $f'(\theta) = \cos\theta$ and $f''(\theta) = -\sin\theta$
$$\kappa(\theta) = |(1 + \sin\theta)(-\sin\theta) - 2\cos^2\theta - (1 + \sin\theta)^2|/[\cos^2\theta + (1 + \sin\theta)^2]^{3/2}$$
$$= |-\sin\theta - \sin^2\theta - 2\cos^2\theta - 1 - 2\sin\theta - \sin^2\theta|/[\cos^2\theta + 1 + 2\sin\theta + \sin^2\theta]^{3/2}$$
$$= |-3 - 3\sin\theta|/[2 + 2\sin\theta]^{3/2}$$
$$= 3(1 + \sin\theta)/(2^{3/2}[1 + \sin\theta]^{3/2})$$
$$= 3\sqrt{2}/(4\sqrt{1 + \sin\theta})$$

29. $f(\theta) = 1 - \cos\theta$, $f'(\theta) = \sin\theta$ and $f''(\theta) = \cos\theta$
$$\kappa(\theta) = |(1 - \cos\theta)\cos\theta - 2\sin^2\theta - (1 - \cos\theta)^2|/[\sin^2\theta + (1 - \cos\theta)^2]^{3/2}$$
$$= |\cos\theta - \cos^2\theta - 2\sin^2\theta - 1 + 2\cos\theta - \cos^2\theta|/[\sin^2\theta + 1 - 2\cos\theta + \cos^2\theta]^{3/2}$$
$$= |3\cos\theta - 3|/[2 - 2\cos\theta]^{3/2}$$
$$= 3(1 - \cos\theta)/(2^{3/2}[1 - \cos\theta]^{3/2})$$
$$= 3\sqrt{2}/(4\sqrt{1 - \cos\theta})$$

31. $\mathbf{r}''(s)$ is the rate of change of $\mathbf{r}'(s)$, or the rate of change of the direction of the curve. Clearly, the change in direction is toward the concave side of the curve.

33. A circle of radius ρ with center at the terminal point of $\mathbf{a}$ has arc length parameterization:
$\mathbf{r}(s) = \mathbf{a} + \rho(\cos(s/\rho)\mathbf{i} + \sin(s/\rho)\mathbf{j})$
$\mathbf{r}'(s) = -\sin(s/\rho)\mathbf{i} + \cos(s/\rho)\mathbf{j}$
Note that $|\mathbf{r}'(s)| = \sqrt{\sin^2(s/\rho) + \cos^2(s/\rho)} = 1$, so this is an arc length parameterization.
The curvature is:
$\kappa(s) = |\mathbf{r}''(s)| = |-(1/\rho)\cos(s/\rho)\mathbf{i} - (1/\rho)\sin(s/\rho)\mathbf{j}|$
$= (1/\rho)|\cos(s/\rho)\mathbf{i} + \sin(s/\rho)\mathbf{j}|$
$= (1/\rho)\sqrt{\cos^2(s/\rho) + \sin^2(s/\rho)}$
$= 1/\rho$

536

35. $\mathbf{v}(t) = 2\mathbf{i} + 2t\cos t^2\,\mathbf{j} - 2t\sin t^2\,\mathbf{k}$

$|\mathbf{v}(t)| = \sqrt{4 + 4t^2\cos^2 t^2 + 4t^2\sin^2 t^2}$

$\quad = \sqrt{4 + 4t^2}$

$\quad = 2\sqrt{1 + t^2}$

$\mathbf{a}(t) = (2\cos t^2 - 4t^2\sin t^2)\mathbf{j} - (2\sin t^2 + 4t^2\cos t^2)\mathbf{k}$

$|\mathbf{a}(t)| = \sqrt{(2\cos t^2 - 4t^2\sin t^2)^2 + (2\sin t^2 + 4t^2\cos t^2)^2}$

$\quad = \sqrt{4\cos^2 t^2 - 16t^2\cos t^2\sin t^2 + 16t^4\sin^2 t^2 + 4\sin^2 t^2 + 16t^2\cos t^2\sin t^2 + 16t^4\cos^2 t^2}$

$\quad = \sqrt{4 + 16t^4}$

$\quad = 2\sqrt{1 + 4t^4}$

$a_T = \frac{d}{dt}|\mathbf{v}(t)|$

$\quad = 2(1/2)(1 + t^2)^{-1/2}(2t)$

$\quad = 2t/\sqrt{1 + t^2}$

$a_N = \sqrt{|\mathbf{a}|^2 - a_T^2}$

$\quad = \sqrt{(4 + 16t^4) - (4t^2/(1 + t^2))}$

$\quad = 2\sqrt{1 + 4t^4 - t^2/(1 + t^2)}$

$\quad = 2\sqrt{(1 + 4t^4 + 4t^6)/(1 + t^2)}$

17.6 Surfaces

1. $x^2 + y^2 + z^2 - 4z + 4 = 5 + 4$
$x^2 + y^2 + (z - 2)^2 = 3^2$
center: $(0, 0, 2)$
radius: 3

3. $x^2 - 4x + 4 + y^2 + 2y + 1 + z^2 - 6z + 9 = 2 + 4 + 1 + 9$
$(x - 2)^2 + (y + 1)^2 + (z - 3)^2 = 4^2$
center: $(2, -1, 3)$
radius: 4

5. $x^2 - 6x + 9 + y^2 + 2y + 1 + z^2 + 4z + 4$
$\quad = 11 + 9 + 1 + 4$
$(x - 3)^2 + (y + 1)^2 + (z + 2)^2 = 5^2$
center: $(3, -1, -2)$
radius: 5

7. Sphere with center $(0, 0, 0)$ and radius $\sqrt{10}$.

9. Elliptic cone with vertex $(0, 0, 0)$ and symmetric about the z-axis.

11. $4x^2 + 9y^2 - 36z^2 = -36$
$x^2/9 + y^2/4 - z^2 = -1$
Hyperboloid of two sheets, symmetric about the z-axis.

13. $2x^2 + 8x - 3y^2 + 6y - 6z = -5$
$2(x^2 + 4x) - 3(y^2 - 2y) - 6z = -5$
$2(x^2 + 4x + 4) - 3(y^2 - 2y + 1) - 6z = -5 + 8 - 3$
$2(x + 2)^2 - 3(y - 1)^2 - 6z = 0$
$(x + 2)^2/3 - (y - 1)^2/2 = z$
Hyperbolic paraboloid, translated to move the origin to $(-2, 1, 0)$.

15. $x^2 + 6x + 2y^2 - 4y - 2z^2 = -7$
$x^2 + 6x + 9 + 2(y^2 - 2y + 1) - 2z^2 = -7 + 9 + 2$
$(x + 3)^2 + 2(y - 1)^2 - 2z^2 = 4$
$(x + 3)^2/4 + (y - 1)^2/2 - z^2/2 = 1$
Hyperboloid of one sheet, translated to move the origin to $(-3, 1, 0)$, and symmetric about the line through this point parallel to the z-axis.

17. $6x^2 + 9(y^2 + 4y) - 54z = -36$
$6x^2 + 9(y^2 + 4y + 4) - 54z = -36 + 36$
$6x^2 + 9(y + 2)^2 = 54z$
$x^2/9 + (y + 2)^2/6 = z$
Elliptic paraboloid with vertex $(0, -2, 0)$, and symmetric about the line through this point parallel to the z-axis.

19. $9z^2 + 4y^2 = 36x$
$z^2/4 + y^2/9 = x$
Elliptic paraboloid with vertex $(0, 0, 0)$, and symmetric about the x-axis.

21. $-x^2 + y^2 + z^2 = 1$
Hyperboloid of one sheet, symmetric about the x-axis.

23. $x^2 - 6x + y^2 + 2y + z^2 + 6z = -18$
$x^2 - 6x + 9 + y^2 + 2y + 1 + z^2 + 6z + 9 = -18 + 9 + 1 + 9$
$(x - 3)^2 + (y + 1)^2 + (z + 3)^2 = 1$
Sphere with center $(3, -1, -3)$ and radius 1.

25. $x^2 + 4x + y^2 - 2y - 36z = -175$
$x^2 + 4x + 4 + y^2 - 2y + 1 - 36z = -175 + 4 + 1$
$(x + 2)^2 + (y - 1)^2 = 36z - 170$
$(x + 2)^2/36 + (y - 1)^2/36 = z - 85/18$
Elliptic paraboloid (actually, a circular paraboloid), with vertex $(-2, 1, 85/18)$, and symmetric about the line through this point parallel to the z-axis.

27. Plane parallel to the yz-plane.

29. Plane perpendicular to the xy-plane, whose traces in planes $z = d$ are identical lines.

31. Hyperbolic cylinder, whose traces in planes $z = d$ are identical hyperbolas.

33. Hyperbolic cylinder, whose traces in planes $z = d$ are identical hyperbolas.

35. Parabolic cylinder, whose traces in the planes $x = d$ are identical parabolas.

37. Hyperbolic cylinder, whose traces in planes $y = d$ are identical hyperbolas.

39. Parabolic cylinder, whose traces in the planes $x = d$ are identical parabolas.

41. $z = 3x^2 + 12x + 16$
Parabolic cylinder, whose traces in the planes $y = d$ are identical parabolas.

43. The result will be an elliptic (circular) cone, with vertex at the origin and symmetric about the y-axis. Thus its equation will be of the form $x^2/a^2 + z^2/b^2 = y^2$. In any plane $y = d$, the trace will have equation $x^2/a^2 + z^2/b^2 = d^2$, and will also be a circle of radius $d/2$, implying that its equation is $x^2 + z^2 = d^2/4$, or $d^2 = 4x^2 + 4z^2$. Therefore $1/a^2 = 1/b^2 = 4$, and the equation of the surface is $4x^2 + 4z^2 = y^2$.

45. The trace in any plane $y = d$ will be a circle of radius $\sqrt{d}$, which will therefore have equation $x^2 + z^2 = d$. The equation of the surface must be $x^2 + z^2 = y$, which is an elliptic paraboloid.

47. a. S_1 is the trace of Q_2 in the plane $x = 0$.
S_2 is the trace of Q_2 in the planes $x = 1$ and $x = -1$.
S_3 is the trace of Q_1 in the planes $z = \pm\sqrt{6}/2$, since $x^2 + y^2 = z^2 - 1 = 6/4 - 1 = 1/2$.

b. The hyperboloid of two sheets is Q_1 (the left surface below), and the hyperboloid of one sheet is Q_2 (the right surface below). The trace S_1 is drawn on Q_2 as a hyperbola in the yz-plane. The trace S_2 is drawn on Q_2 as a union of four lines, which form two figures "X". The trace S_3 is drawn on Q_1 as two circles.

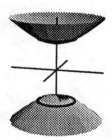 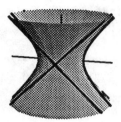

17.7 Cylindrical and Spherical Coordinates

1. a. $z = 0$ and $r = \sqrt{1^2 + 1^2} = \sqrt{2}$

$\tan\theta = 1/1 = 1$ implies $\theta = \pi/4$ or $5\pi/4$. Since $(1, 1)$ is in the first quadrant, $\theta = \pi/4$.
$(\sqrt{2}, \pi/4, 0)$

b. $z = 3$ and $r = \sqrt{(\sqrt{3})^2 + 1^2} = \sqrt{4} = 2$

$\tan\theta = 1/\sqrt{3}$ implies $\theta = \pi/6$ or $7\pi/6$. Since $(\sqrt{3}, 1)$ is in the first quadrant, $\theta = \pi/6$.
$(2, \pi/6, 3)$

c. $z = -2$ and $r = \sqrt{(-1)^2 + 1^2} = \sqrt{2}$

$\tan\theta = 1/(-1) = -1$ implies $\theta = 3\pi/4$ or $7\pi/4$. Since $(-1, 1)$ is in the second quadrant, $\theta = 3\pi/4$.
$(\sqrt{2}, 3\pi/4, -2)$

d. $z = 4$ and $r = \sqrt{(-1)^2 + (\sqrt{3})^2} = \sqrt{4} = 2$

$\tan\theta = \sqrt{3}/(-1) = -\sqrt{3}$ implies $\theta = 2\pi/3$ or $5\pi/3$.
Since $(-1, \sqrt{3})$ is in the second quadrant, $\theta = 2\pi/3$.
$(2, 2\pi/3, 4)$

3. a. $z = -3$

$x = 2\cos(\pi/4) = 2\sqrt{2}/2 = \sqrt{2}$

$y = 2\sin(\pi/4) = 2\sqrt{2}/2 = \sqrt{2}$

$(\sqrt{2}, \sqrt{2}, -3)$

b. $z = 4$

$x = 1\cos(\pi/6) = \sqrt{3}/2$

$y = 1\sin(\pi/6) = 1/2$

$(\sqrt{3}/2, 1/2, 4)$

c. $z = -5$

$x = 4\cos(\pi/3) = 4(1/2) = 2$

$y = 4\sin(\pi/3) = 4(\sqrt{3}/2) = 2\sqrt{3}$

$(2, 2\sqrt{3}, -5)$

d. $z = 2$

$x = 2\cos(4\pi/3) = 2(-1/2) = -1$

$y = 2\sin(4\pi/3) = 2(-\sqrt{3}/2) = -\sqrt{3}$

$(-1, -\sqrt{3}, 2)$

5. a. $\rho = \sqrt{1^2 + 0^2 + 0^2} = 1$

$\tan\theta = y/x = 0$ implies $\theta = 0$ or π. Since x is positive, $\theta = 0$.

$\cos\phi = z/\rho = 0$ implies $\phi = \pi/2$.

$(1, 0, \pi/2)$

b. $\rho = \sqrt{1^2 + 1^2 + (\sqrt{2})^2} = \sqrt{4} = 2$

$\tan\theta = y/x = 1$ implies $\theta = \pi/4$ or $5\pi/4$. Since $(1, 1)$ is in the first quadrant, $\theta = \pi/4$.

$\cos\phi = z/\rho = \sqrt{2}/2$ implies $\phi = \pi/4$.

$(2, \pi/4, \pi/4)$

c. $\rho = \sqrt{1^2 + (-1)^2 + (\sqrt{2})^2} = \sqrt{4} = 2$

$\tan\theta = y/x = -1$ implies $\theta = 3\pi/4$ or $7\pi/4$. Since $(1, -1)$ is in the fourth quadrant, $\theta = 7\pi/4$.

$\cos\phi = z/\rho = \sqrt{2}/2$ implies $\phi = \pi/4$.

$(2, 7\pi/4, \pi/4)$

d. $\rho = \sqrt{1^2 + (\sqrt{3})^2 + 2^2} = \sqrt{8} = 2\sqrt{2}$

$\tan\theta = y/x = \sqrt{3}$ implies $\theta = \pi/3$ or $4\pi/3$. Since $(1, \sqrt{3})$ is in the first quadrant, $\theta = \pi/3$.

$\cos\phi = z/\rho = 2/(2\sqrt{2}) = 1/\sqrt{2}$ implies $\phi = \pi/4$.

$(2\sqrt{2}, \pi/3, \pi/4)$

7. a. $x = 2\sin(\pi/3)\cos(\pi/4) = 2(\sqrt{3}/2)(\sqrt{2}/2) = \sqrt{6}/2$

$y = 2\sin(\pi/3)\sin(\pi/4) = 2(\sqrt{3}/2)(\sqrt{2}/2) = \sqrt{6}/2$

$z = 2\cos(\pi/3) = 2(1/2) = 1$

$(\sqrt{6}/2, \sqrt{6}/2, 1)$

b. $x = 1\sin(\pi)\cos(\pi/2) = 1(0)(0) = 0$

$y = 1\sin(\pi)\sin(\pi/2) = 1(0)(1) = 0$

$z = 1\cos(\pi) = 1(-1) = -1$

$(0, 0, -1)$

c. $x = 2\sin(\pi/4)\cos(3\pi/4) = 2(\sqrt{2}/2)(-\sqrt{2}/2) = -1$
$y = 2\sin(\pi/4)\sin(3\pi/4) = 2(\sqrt{2}/2)(\sqrt{2}/2) = 1$
$z = 2\cos(\pi/4) = 2(\sqrt{2}/2) = \sqrt{2}$
$(-1, 1, \sqrt{2})$

d. $x = 3\sin(2\pi/3)\cos(3\pi/2) = 3(\sqrt{3}/2)(0) = 0$
$y = 3\sin(2\pi/3)\sin(3\pi/2) = 3(\sqrt{3}/2)(-1) = -3\sqrt{3}/2$
$z = 3\cos(2\pi/3) = 3(-1/2) = -3/2$
$(0, -3\sqrt{3}/2, -3/2)$

9. For these conversions, use the following:
$\rho = \sqrt{x^2 + y^2 + z^2} = \sqrt{r^2 + z^2}$
$\theta = \theta$
$\cos\phi = z/\rho$

   a. $\rho = \sqrt{1^2 + 0^2} = 1$
   $\cos\phi = 0/1 = 0$ implies $\phi = \pi/2$.
   $(1, 0, \pi/2)$

   b. $\rho = \sqrt{(\sqrt{2})^2 + (\sqrt{2})^2} = \sqrt{4} = 2$
   $\cos\phi = \sqrt{2}/2$ implies $\phi = \pi/4$.
   $(2, -\pi/4, \pi/4)$

   c. $\rho = \sqrt{2^2 + 2^2} = \sqrt{8} = 2\sqrt{2}$
   $\cos\phi = 2/(2\sqrt{2}) = 1/\sqrt{2}$ implies $\phi = \pi/4$.
   $(2\sqrt{2}, \pi/3, \pi/4)$

   d. $\rho = \sqrt{2^2 + 2^2} = \sqrt{8} = 2\sqrt{2}$
   $\cos\phi = 2/(2\sqrt{2}) = 1/\sqrt{2}$ implies $\phi = \pi/4$.
   $(2\sqrt{2}, \pi/4, \pi/4)$

11. For these conversions, use the following:
$r = \rho\sin\phi$
$\theta = \theta$
$z = \rho\cos\phi$

   a. $r = 2\sin(\pi/2) = 2(1) = 2$
   $z = 2\cos(\pi/2) = 2(0) = 0$
   $(2, \pi/4, 0)$

   b. $r = 2\sin(\pi/4) = 2(\sqrt{2}/2) = \sqrt{2}$
   $z = 2\cos(\pi/4) = 2(\sqrt{2}/2) = \sqrt{2}$
   $(\sqrt{2}, \pi/2, \sqrt{2})$

   c. $r = 3\sin(\pi/2) = 3(1) = 3$
   $z = 3\cos(\pi/2) = 3(0) = 0$
   $(3, 2\pi/3, 0)$

   d. $r = 1\sin(2\pi/3) = \sqrt{3}/2$
   $z = 1\cos(2\pi/3) = -1/2$
   $(\sqrt{3}/2, \pi/2, -1/2)$

13.

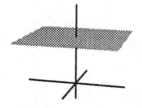

$z = 3$

$z = 3$

15.

$z = 2r$
$z = 2\sqrt{x^2 + y^2}$

17.

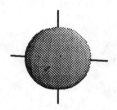

$$r^2 + z^2 = 4$$
$$x^2 + y^2 + z^2 = 4$$

19.

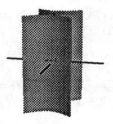

$$\cos^2 \theta - \sin^2 \theta = a^2/r^2$$
$$r^2 \cos^2 \theta - r^2 \sin^2 \theta = a^2$$
$$x^2 - y^2 = a^2$$

21. $r^2 = 9$

23. $r^2 = 9z$

25. $r^2 \cos^2 \theta + z^2 = 4$

27. $r^2 \cos \theta \sin \theta - ar \cos \theta = 4$

29. $\rho^2 \sin^2 \phi \cos^2 \theta + \rho^2 \sin^2 \phi \sin^2 \theta = 9$
$\rho^2 \sin^2 \phi = 9$
$\rho \sin \phi = 3$
Note ρ and $\sin \phi$ are always nonnegative.

31. $\rho^2 \sin^2 \phi \cos^2 \theta + \rho^2 \sin^2 \phi \sin^2 \theta = 9\rho \cos \phi$
$\rho^2 \sin^2 \phi = 9\rho \cos \phi$
$\rho \sin^2 \phi = 9 \cos \phi$

33. $\rho^2 \sin^2 \phi \cos^2 \theta + \rho^2 \cos^2 \phi = 4$

35. $\rho^2 \sin^2 \phi \cos \theta \sin \theta - a\rho \sin \phi \cos \theta = 4$

Review Exercises — Chapter 17

1. The parametric equations for this curve are $x = a \cos t$ and $y = b \sin t$. Note that $x/a = \cos t$ and $y/b = \sin t$. Therefore $x^2/a^2 + y^2/b^2 = 1$.

3. $\sqrt{1 - t^2}$ is only defined for $1 - t^2 \geq 0$, which is only satisfied for $-1 \leq t \leq 1$. $1/t$ is undefined at $t = 0$. Therefore, the implicit domain of $\mathbf{r}(t)$ is $[-1, 0) \cup (0, 1]$.

5.

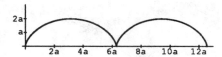

$\mathbf{r}(t) = a(t - \sin t)\mathbf{i} + a(1 - \cos t)\mathbf{j}$
$\mathbf{r}'(t) = a(1 - \cos t)\mathbf{i} + a \sin t\, \mathbf{j}$ for $t \geq 0$.
The tangent vector is zero if $\mathbf{r}'(t) = 0$, which occurs if and only if $a(1 - \cos t) = 0$ and $a \sin t = 0$. The first equation implies that $\cos t = 1$, or $t = 2n\pi$ for n a nonnegative integer. These values of t also satisfy the second equation, and are therefore the numbers desired.

7. Using l'Hôpital's Rule on the first and last terms, one obtains
$\lim_{t \to 0} [(3 \cos 3t/1)\mathbf{i} + \sqrt{2}\,\mathbf{j} + (2 \sec^2 2t/1)\mathbf{k}] = 3\mathbf{i} + \sqrt{2}\,\mathbf{j} + 2\mathbf{k}$

9. $\sec t$, and therefore $\mathbf{r}(t)$, is discontinuous at numbers of the form $\pi/2 + n\pi$. The numbers of this form within the interval $[0, 2\pi]$ are $\pi/2$ and $3\pi/2$.

11. $1/(t^2 - 9)$ is undefined for $t = \pm 3$.
$(t + 2)/(t^2 - 4)$ is undefined for $t = \pm 2$.
Within the interval $[-4, 4]$, $\mathbf{r}(t)$ is discontinuous at $t = -3, -2, 2$ and 3.

13. $\mathbf{r}'(t) = (t + 1)e^t\mathbf{i} + (1 - t)e^{-t}\mathbf{j}$
$\mathbf{r}''(t) = (t + 2)e^t\mathbf{i} + (t - 2)e^{-t}\mathbf{j}$

15. $\mathbf{r}'(t) = (1/t)\mathbf{i} + [e^{\sqrt{t}}/(2\sqrt{t})]\mathbf{j}$
$\mathbf{r}''(t) = -(1/t^2)\mathbf{i} + \dfrac{2\sqrt{t}e^{\sqrt{t}}/(2\sqrt{t}) - e^{\sqrt{t}}/\sqrt{t}}{(2\sqrt{t})^2}\mathbf{j}$
$= -(1/t^2)\mathbf{i} + \dfrac{2\sqrt{t}}{2\sqrt{t}}\dfrac{2\sqrt{t}e^{\sqrt{t}}/(2\sqrt{t}) - e^{\sqrt{t}}/\sqrt{t}}{4t}\mathbf{j}$
$= -(1/t^2)\mathbf{i} + \dfrac{2\sqrt{t}e^{\sqrt{t}} - 2e^{\sqrt{t}}}{8t^{3/2}}\mathbf{j}$
$= -(1/t^2)\mathbf{i} + [(\sqrt{t} - 1)e^{\sqrt{t}}/4t^{3/2}]\mathbf{j}$

17. $(te^t - e^t)\mathbf{i} + (-te^{-t} - e^{-t})\mathbf{j}\big|_0^1$
$= -2e^{-1}\mathbf{j} - (-\mathbf{i} - \mathbf{j})$
$= \mathbf{i} + (1 - 2/e)\mathbf{j}$

19. Considering components, one has $x''(t) = 3^2 x(t)$, $y''(t) = 3^2 y(t)$ and $z''(t) = 3^2 z(t)$.
$\mathbf{r}(t) = (a_1 e^{3t} + a_2 e^{-3t})\mathbf{i} + (b_1 e^{3t} + b_2 e^{-3t})\mathbf{j} + (c_1 e^{3t} + c_2 e^{-3t})\mathbf{j}$
$4\mathbf{i} + 3\mathbf{j} = \mathbf{r}(0) = (a_1 + a_2)\mathbf{i} + (b_1 + b_2)\mathbf{j} + (c_1 + c_2)\mathbf{k}$
Any choice of coefficients is valid which satisfies the equations $a_1 + a_2 = 4$, $b_1 + b_2 = 3$ and $c_1 + c_2 = 0$. In particular, letting $a_2 = b_2 = c_2 = 0$, one obtains the solution $\mathbf{r}(t) = 4e^{3t}\mathbf{i} + 3e^{3t}\mathbf{j}$. More generally, $\mathbf{r}(t) = (4\mathbf{i} + 3\mathbf{j})e^{3t} + \mathbf{c}(e^{3t} - e^{-3t})$ is a solution for any constant vector $\mathbf{c}$.

Since the solution to the differential equation $y'' = 9y$ may also be written as $y = c_1 \cosh 3t + c_2 \sinh 3t$, an equivalent form for the answer above is $\mathbf{r}(t) = (4\mathbf{i} + 3\mathbf{j}) \cosh 3t + \mathbf{c} \sinh 3t$, for any constant vector $\mathbf{c}$.

21. $\mathbf{r}(t) = 5\cos(t/5)\mathbf{i} + 5\sin(t/5)\mathbf{j}$
The components may be exchanged, and the sign of either component may be negative. Other variations are possible.

23. $x^2 - 6x + y^2 + 4y + z^2 - 2z = -10$
$x^2 - 6x + 9 + y^2 + 4y + 4 + z^2 - 2z + 1 = -10 + 9 + 4 + 1$
$(x-3)^2 + (y+2)^2 + (z-1)^2 = 2^2$
This is a sphere with radius 2 and center $(3, -2, 1)$.

25.

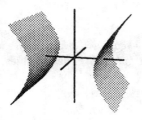

$y^2 = 9 + z^2$
$y^2/3^2 - z^2/3^2 = 1$

27. $x = \rho\sin\phi\cos\theta = 2\sin(\pi/3)\cos(\pi/4) = 2(\sqrt{3}/2)(\sqrt{2}/2) = \sqrt{6}/2$
$y = \rho\sin\phi\sin\theta = 2\sin(\pi/3)\sin(\pi/4) = 2(\sqrt{3}/2)(\sqrt{2}/2) = \sqrt{6}/2$
$z = \rho\cos\phi = 2\cos(\pi/3) = 2(1/2) = 1$
$(\sqrt{6}/2, \sqrt{6}/2, 1)$

29.

$$\phi = \pi/4$$
$\cos\phi = \cos(\pi/4) = \sqrt{2}/2$
$\rho\cos\phi = \rho\sqrt{2}/2$
$(\rho\cos\phi)^2 = \rho^2/2$
$z^2 = (x^2 + y^2 + z^2)/2$
$2z^2 = x^2 + y^2 + z^2$
$z^2 = x^2 + y^2$

31.

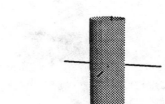

$$\rho\sin\phi = 2\cos\theta$$
$r = 2\cos\theta$
$r^2 = 2r\cos\theta$
$x^2 + y^2 = 2x$

33. $\mathbf{r}'(t) = \mathbf{i} - \sin t\,\mathbf{j} + \cos t\,\mathbf{k}$
$|\mathbf{r}'(t)| = \sqrt{1 + \sin^2 t + \cos^2 t} = \sqrt{2}$
$\mathbf{T}(t) = \mathbf{r}'(t)/|\mathbf{r}'(t)| = (\sqrt{2}/2)\mathbf{r}'(t)$
$\mathbf{T}'(t) = (\sqrt{2}/2)\mathbf{r}''(t) = (\sqrt{2}/2)(-\cos t\,\mathbf{j} - \sin t\,\mathbf{k})$
$|\mathbf{T}'(t)| = (\sqrt{2}/2)\sqrt{\cos^2 t + \sin^2 t} = \sqrt{2}/2$

  a. $\kappa(t) = |\mathbf{T}'(t)|/|\mathbf{r}'(t)| = (\sqrt{2}/2)/\sqrt{2} = 1/2$

  b. $\mathbf{T}(t) = (\sqrt{2}/2)\mathbf{i} - (\sqrt{2}/2)\sin t\,\mathbf{j} + (\sqrt{2}/2)\cos t\,\mathbf{k}$

35. $\mathbf{r}'(t) = 3\mathbf{i} - 2\sin 2t\,\mathbf{j} + 2\cos 2t\,\mathbf{k}$
$|\mathbf{r}'(t)| = \sqrt{9 + 4\sin^2 2t + 4\cos^2 2t} = \sqrt{9 + 4} = \sqrt{13}$
Length $= \int_0^3 |\mathbf{r}'(t)|\,dt = \int_0^3 \sqrt{13}\,dt = \sqrt{13}\,t\big|_0^3 = 3\sqrt{13}$

37. $f(x) = x^3$, $f'(x) = 3x^2$ and $f''(x) = 6x$.
$f'(2) = 12$ and $f''(2) = 12$
$\kappa(2) = |f''(2)|/[1 + f'(2)^2]^{3/2}$
   $= 12/[1 + 144]^{3/2}$
   $= 12\sqrt{145}/145^2$
   $= 12\sqrt{145}/21025$

39. The nonnegative, real-valued function $f(t) = |\mathbf{r}(t)|^2 = \mathbf{r}(t) \cdot \mathbf{r}(t)$ is differentiable on the interval $[a, b]$ and has an absolute extremum at t_0. Therefore $f'(t_0) = 0$. It follows that

$$\frac{d}{dt}(\mathbf{r}(t) \cdot \mathbf{r}(t)) = \mathbf{r}(t) \cdot \mathbf{r}'(t) + \mathbf{r}'(t) \cdot \mathbf{r}(t) = 2\mathbf{r}(t) \cdot \mathbf{r}'(t)$$

has a value of zero at $t = t_0$, and thus that $\mathbf{r}(t_0) \cdot \mathbf{r}'(t_0) = 0$. Therefore, $\mathbf{r}(t_0)$ and $\mathbf{r}'(t_0)$ are orthogonal.

41. $x^2 + y^2/6 - z^2/3 = 1$
Hyperboloid of one sheet, symmetric about the z-axis.

43. Hyperbolic paraboloid.

45. $9x^2 + 18x + 9y^2 - 4z^2 - 16z = 43$
$9(x^2 + 2x) + 9y^2 - 4(z^2 + 4z) = 43$
$9(x^2+2x+1)+9y^2-4(z^2+4z+4) = 43+9-16$
$9(x + 1)^2 + 9y^2 - 4(z + 2)^2 = 36$
$(x + 1)^2/4 + y^2/4 - (z + 2)^2/9 = 1$
Hyperboloid of one sheet, translated to move the origin to $(-1, 0, -2)$, symmetric about the line through this point parallel to the z-axis.

47. $\mathbf{a} = (1/m)\mathbf{F} = \mathbf{0}$

  a. $\mathbf{v} = \int \mathbf{a}\,dt = \int \mathbf{0}\,dt = \mathbf{c}$ is constant.
    $\mathbf{r} = \int \mathbf{v}\,dt = \int \mathbf{c}\,dt = t\mathbf{c} + \mathbf{a}$ is a parameterization of a line.

  b. $|\mathbf{v}(t)| = |\mathbf{c}| = c$ for some constant c
    $a_T = \frac{d}{dt}|\mathbf{v}(t)| = \frac{d}{dt}c = 0$
    $a_N = \sqrt{|\mathbf{a}|^2 + a_T^2} = \sqrt{0 + 0} = 0$

49. See the solution to exercise 21 of section 17.5 for a discussion on finding the center of curvature for the graph of $y = f(x)$. For $f(x) = \cos x$, one has $f'(x) = -\sin x$ and $f''(x) = -\cos x$. Thus $f'(0) = 0$ and $f''(0) = -1$. Thus curvature at the point in question is

$$\kappa(0) = |f''(0)|/[1 + f'(0)^2]^{3/2} = 1/[1]^{3/2} = 1.$$

The radius of curvature is $\rho = 1/\kappa = 1$. A parameterization of the graph is $\mathbf{r}(t) = t\mathbf{i} + \cos t\,\mathbf{j}$, for which $\mathbf{r}'(t) = \mathbf{i} - \sin t\,\mathbf{j}$ and $\mathbf{r}'(0) = \mathbf{i}$. Orthogonal vectors are $\pm\mathbf{j}$. Since the graph is concave down at $(0, 1)$ ($f''(0)$ is negative), one wants a unit vector parallel to these whose $\mathbf{j}$ component is negative, or a unit vector with the same direction as $-\mathbf{j}$. Thus use $\mathbf{N} = -\mathbf{j}$. The center of curvature is at $\mathbf{r}(0) + \rho\mathbf{N} = \mathbf{j} + 1(-\mathbf{j}) = \mathbf{0}$, which is the origin. Since the center is at $(0, 0)$ and the radius is 1, the equation of the circle is $x^2 + y^2 = 1$.

51. $\mathbf{v}(t) = (\cos t - t \sin t)\mathbf{i} + (\sin t + t \cos t)\mathbf{j}$

The speed is:

$$|\mathbf{v}(t)| = \sqrt{(\cos t - t \sin t)^2 + (\sin t + t \cos t)^2}$$
$$= \sqrt{\cos^2 t - 2t \cos t \sin t + t^2 \sin^2 t + \sin^2 t + 2t \cos t \sin t + t^2 \cos^2 t}$$
$$= \sqrt{1 + t^2}$$

53. $\mathbf{r}(0) = \mathbf{0}$

$\mathbf{v}(0) = 20 \cos(30°)\mathbf{i} + 20 \sin(30°)\mathbf{j} = 10\sqrt{3}\,\mathbf{i} + 10\mathbf{j}$

$\mathbf{a}(t) = -9.81\mathbf{j}$

$\mathbf{v}(t) = -9.81t\mathbf{j} + \mathbf{c}$

$10\sqrt{3}\,\mathbf{i} + 10\mathbf{j} = \mathbf{v}(0) = \mathbf{c}$

$\mathbf{v}(t) = 10\sqrt{3}\,\mathbf{i} + (-9.81t + 10)\mathbf{j}$

$\mathbf{r}(t) = 10t\sqrt{3}\,\mathbf{i} + (-4.905t^2 + 10t)\mathbf{j} + \mathbf{c}$

$\mathbf{0} = \mathbf{r}(0) = \mathbf{c}$

$\mathbf{r}(t) = 10t\sqrt{3}\,\mathbf{i} + (-4.905t^2 + 10t)\mathbf{j}$

The horizontal component of $\mathbf{r}(t)$ has a value of 10 when $t = 1/\sqrt{3} = \sqrt{3}/3$. The vertical component of $\mathbf{r}(t)$ at this time is $-4.905(1/3) + 10\sqrt{3}/3 \approx 4.14$. The projectile strikes the wall at a height of approximately 4.14 meters.

55. **a.** $Q_\theta = (3 - \cos(\theta/2), \theta, -\sin(\theta/2))$

b. Within the plane determined by θ, let $\mathbf{j}$ be a unit vector pointing in the direction of the positive z-axis, and let $\mathbf{i}$ be the unit vector from the origin to the point $(1, \theta, 0)$. The center of the circle is on the diameter, and corresponds to the vector $3\mathbf{i}$. The vector with initial point at the center and terminal point at P_θ is $(3 + \cos(\theta/2) - 3)\mathbf{i} + (\sin(\theta/2) - 0)\mathbf{j}$, or just $\cos(\theta/2)\mathbf{i} + \sin(\theta/2)\mathbf{j}$, and this vector is a direction vector for the diameter. A parameterization for the diameter is therefore $\mathbf{r}(t) = 3\mathbf{i} + t(\cos(\theta/2)\mathbf{i} + \sin(\theta/2)\mathbf{j}) = (3 + t \cos(\theta/2))\mathbf{i} + t \sin(\theta/2)\mathbf{j}$. Note that $\mathbf{r}(1) = P_\theta$ and $\mathbf{r}(-1) = Q_\theta$. Thus the diameter is the part of the line for $-1 \le t \le 1$. The $\mathbf{i}$ component is $r(t) = 3 + t \cos(\theta/2)$, and the $\mathbf{j}$ component is $z(t) = t \sin(\theta/2)$.

c. $x = r \cos \theta = (3 + t \cos(\theta/2)) \cos \theta$

$y = r \sin \theta = (3 + t \cos(\theta/2)) \sin \theta$

$z = t \sin(\theta/2)$

Chapter 18

The Differentiation of Functions of Several Variables

18.1 Functions of Several Variables

1. $f(x,y)$ is undefined if $x^2 + y^2 = 0$, which only occurs at $(0,0)$. The domain is $\{(x,y) \mid (x,y) \neq (0,0)\}$.

3. $f(x,y)$ is undefined if $y - x < 0$, which occurs if $y < x$. The domain is $\{(x,y) \mid y \geq x\}$.

5. The domain is $\mathbf{R}^2$.

7. $f(x,y)$ is undefined if $\text{Sin}^{-1}(x^2 + y^2) = 0$, which occurs if $x^2 + y^2 = 0$, which only occurs at $(0,0)$. Also, the domain of the inverse sine function is $[-1, 1]$, and therefore one must also have $x^2 + y^2 \leq 1$.
The domain is $\{(x,y) \mid x^2 + y^2 \leq 1 \text{ and } (x,y) \neq (0,0)\}$.

9. The domain is $\mathbf{R}^3$.

11. $f(x,y,z)$ is undefined if $xyz = 0$, which occurs if $x = 0$, $y = 0$ or $z = 0$.
The domain is $\{(x,y,z) \mid x \neq 0,\ y \neq 0 \text{ and } z \neq 0\}$.

13. $\ln(x^2yz^2)$ is undefined if $x^2yz^2 \leq 0$. This quantity can only equal zero if $x = 0$, $y = 0$ or $z = 0$. It can only be less than zero if $y < 0$, since x^2 and z^2 are nonnegative. Thus $\ln(x^2yz^2)$ is undefined if $x = 0$, $y \leq 0$ or $z = 0$. $f(x,y,z)$ is also undefined if $\ln(x^2yz^2) = 0$, which occurs if $x^2yz^2 = 1$.
The domain is $\{(x,y,z) \mid x \neq 0,\ y > 0,\ z \neq 0 \text{ and } x^2yz^2 \neq 1\}$.

15. a. $h(x,y) = \sqrt{x + y^2}$

    b. The domain of $f(x,y)$ is $\mathbf{R}^2$, and the domain of $g(z)$ is $z \geq 0$.
       Therefore, the domain of h is $\{(x,y) \mid f(x,y) \geq 0\}$, which is $\{(x,y) \mid x + y^2 \geq 0\}$.

17. $S(r,h) = 2\pi r^2 + 2\pi rh$
Each of the top and bottom has an area of πr^2, giving $2\pi r^2$. The side can be imagined as a rectangle of height h and length $2\pi r$ (the circumference), which has been rolled into a cylinder. Multiplying height by length gives the area of the side as $2\pi rh$.

19. The volume of the cone is $\pi r^2 h/3$, and the volume of the hemisphere is $2\pi r^3/3$.
Therefore $V(r,h) = (\pi r^2 h + 2\pi r^3)/3$.

21. $1^2 - 2(3) = -5$

23. $1/\sqrt{3 + (-1)} = \sqrt{2}/2$

25. $(\pi/2)\cos((\pi/2)1) = 0$

27. $\ln \sin((\pi/2)1) = \ln 1 = 0$

29. Definition 1 applies only to functions that are defined in a deleted neighborhood of the point $\mathbf{x}_0$ in question. Since $\sin(xy)/(xy)$ is undefined if $x = 0$ or $y = 0$, we must use a modified version of Definition 1. We consider only those (x, y) such that $x \neq 0$ and $y \neq 0$. This modification is consistent with the notion of continuity that is developed in this section.

We show that the limit is 1 using this modified definition. Let ε be an arbitrary positive number. Since $\lim_{x \to 0} \sin(x)/x = 1$, there exists a number $\delta_1 > 0$ such that if $0 < |x| < \delta_1$, then $|(\sin(x)/x) - 1| < \varepsilon$. Let $\delta = \sqrt{\delta_1}$. Suppose that (x, y) is a point in the domain of $\sin(xy)/xy$, such that $0 < \sqrt{x^2 + y^2} < \delta$. Then $|x| = \sqrt{x^2} < \delta$ and $|y| = \sqrt{y^2} < \delta$. Consequently, $|xy| < \delta^2 = \delta_1$. By the choice of δ_1, this implies that $|(\sin(xy)/xy) - 1| < \varepsilon$. It follows from the definition of limit that $\lim_{(x,y) \to (0,0)} \sin(xy)/xy = 1$.

31. iii and b

The lack of x in the function produces a cylinder.

33. ii and d

Note that $z = x^2 + 3y^2$ is an elliptic paraboloid. Distinguish (ii) from (vi) by the fact that this function increases more rapidly in the y direction. Distinguish (d) from (e) by the fact that one must go farther out on the x-axis to obtain values of the same height, and distinguish (d) from (a) by the fact that as one moves away from the origin, the height is changing more rapidly (since the level curves are closer together).

35. v and a

Distinguish (v) from (iv) by the fact that $f(0, 0) = 1$. Distinguish (a) from (d) and (e) by the fact that as one moves away from the origin, the height is changing less rapidly (since the level curves are farther apart).

37.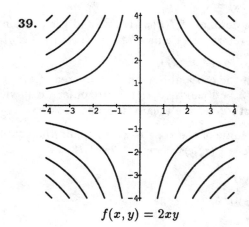

$$f(x, y) = y - x^2$$

39.

$$f(x, y) = 2xy$$

41.

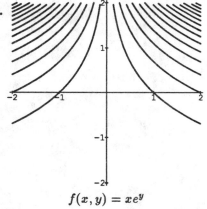

$$f(x, y) = xe^y$$

43.

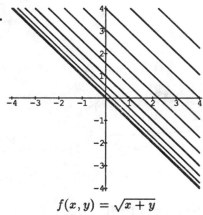

$$f(x, y) = \sqrt{x + y}$$

45.

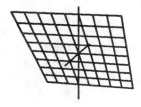

$$f(x, y) = x$$

47.

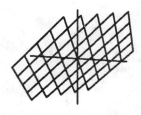

$$f(x, y) = x + y$$

49.

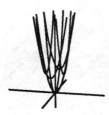 

$$f(x, y) = x^2 + 4y^2$$

51.

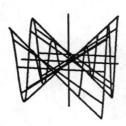 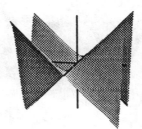

$$f(x, y) = y \sin x$$

53.

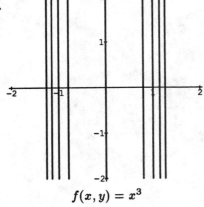  

$$f(x, y) = x^3$$

55.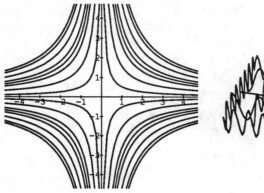

$$f(x, y) = \sin xy$$

57. $\displaystyle\lim_{(x,y)\to(1,0)} \frac{(x-1)y}{(x-1)^2 + y^2}$

The limit does not exist, since along the line $y = 0$ the expression has a constant value of 0, and along the line $y = x - 1$, the expression has a constant value of $1/2$.

59. No. Along the line $y = x$, the values of f approach $1/2$ as $(x, x) \to (0, 0)$.

61. Yes. Show that $\lim_{(x,y)\to(0,0)} f(x, y) = 0$. For any $\varepsilon > 0$, choose $\delta = \varepsilon$. Suppose that (x, y) satisfies $0 < |\langle x, y\rangle - \langle 0, 0\rangle| < \delta$. Show that $|f(x, y)| < \varepsilon$. One has $\sqrt{x^2 + y^2} < \delta$. Consider polar coordinates (r, θ) for (x, y) where $r > 0$. Note that $r = \sqrt{x^2 + y^2} < \delta$. It follows that:
$|f(x, y)| = |x^3/(x^2 + y^2)| = |r^3 \cos^3 \theta / r^2| = |r \cos^3 \theta| \leq r < \delta = \varepsilon$

63. Let $g(x, y) = \sin y$ and $h(x, y) = x/(\sin x \cos y)$. Since $g(x, y)$ is continuous, $\lim_{(x,y)\to(0,0)} g(x, y) = \sin 0 = 0$. By exercise 62, $\lim_{(x,y)\to(0,0)} h(x, y) = 1/1 = 1$. Therefore:

$$\lim_{(x,y)\to(0,0)} f(x, y) = \left[\lim_{(x,y)\to(0,0)} g(x, y)\right]\left[\lim_{(x,y)\to(0,0)} h(x, y)\right] = (0)(1) = 0$$

65. False. This has already been demonstrated in Example 8 with $f(x, y) = -xy/(x^2 + y^2)$ at the point $(0, 0)$.

67. For any $\varepsilon > 0$, let $\delta = \sqrt{\varepsilon}$. If $0 < |\langle x, y\rangle - \langle 0, 0\rangle| < \delta$, then $\sqrt{x^2 + y^2} < \delta = \sqrt{\varepsilon}$, and therefore $x^2 + y^2 < \varepsilon$. It follows that $|x^2 + y^2 - 0| = x^2 + y^2 < \varepsilon$. Therefore $\lim_{(x,y)\to(0,0)}(x^2 + y^2) = 0$.

69. Note that the theorem is trivial if $c = 0$. Therefore, suppose that $c \neq 0$. Let $L = \lim_{\mathbf{x}\to\mathbf{x}_0} f(\mathbf{x})$, which exists by hypothesis. Let ε be an arbitrary positive number. There exists a number $\delta > 0$ such that $|f(\mathbf{x}) - L| < \varepsilon/|c|$ whenever $0 < |\mathbf{x} - \mathbf{x}_0| < \delta$. Suppose that $\mathbf{x}$ is a point such that $0 < |\mathbf{x} - \mathbf{x}_0| < \delta$. Then: $|cf(\mathbf{x}) - cL| = |c||f(\mathbf{x}) - L| < |c|\varepsilon/|c| = \varepsilon$.
Therefore $\lim_{\mathbf{x}\to\mathbf{x}_0} cf(\mathbf{x}) = cL$.

71. Suppose that $f(\mathbf{x})$ and $g(\mathbf{x})$ are continuous at $\mathbf{x}_0$. Then by Theorem 1:
$\lim_{\mathbf{x}\to\mathbf{x}_0}[f(\mathbf{x})g(\mathbf{x})] = [\lim_{\mathbf{x}\to\mathbf{x}_0} f(\mathbf{x})][\lim_{\mathbf{x}\to\mathbf{x}_0} g(\mathbf{x})] = f(\mathbf{x}_0)g(\mathbf{x}_0)$,
and the limit exists. Therefore, $f(\mathbf{x})g(\mathbf{x})$ is continuous at $\mathbf{x}_0$.

18.2 Partial Differentiation

1. $f_x(x, y) = y$
$f_y(x, y) = x$

3. $\partial z/\partial x = \tan y^2$
$\partial z/\partial y = 2xy \sec^2 y^2$

5. $f_x(x, y) = 2xe^{x^2+y^2}$
$f_y(x, y) = 2ye^{x^2+y^2}$

7. $f_r(r, \theta) = \cos\theta$
$f_\theta(r, \theta) = -r\sin\theta$

9. $\partial z/\partial x = (y^2 + 1)/(xy^2 + x - y)$
$\partial z/\partial y = (2xy - 1)/(xy^2 + x - y)$

11. $f_x(x, y) = yx^{y-1}$
$f_y(x, y) = x^y \ln x$

13. $f_r(r, \theta) = 2r\cos\theta$
$f_\theta(r, \theta) = -r^2 \sin\theta$

15. $f_x(x, y, z) = y^3$
$f_y(x, y, z) = 3xy^2 - z^2$
$f_z(x, y, z) = -2yz$

17. $f_x(x, y, z) = z\left(\dfrac{x-y}{x+y}\right)^{z-1} \dfrac{(x+y)1 - (x-y)1}{(x+y)^2} = \dfrac{2yz(x-y)^{z-1}}{(x+y)^{z+1}}$

$f_y(x, y, z) = z\left(\dfrac{x-y}{x+y}\right)^{z-1} \dfrac{(x+y)(-1) - (x-y)1}{(x+y)^2} = \dfrac{-2xz(x-y)^{z-1}}{(x+y)^{z+1}}$

$f_z(x, y, z) = \left(\dfrac{x-y}{x+y}\right)^{z} \ln\left(\dfrac{x-y}{x+y}\right)$

19. $f_r(r, s, t) = \dfrac{[\sqrt{s^2 - 2r + t}/(2\sqrt{r})] - [\sqrt{r}(-1/\sqrt{s^2 - 2r + t})]}{s^2 - 2r + t} s\ln t$

$= \dfrac{(s^2 - 2r + t) + 2r}{2\sqrt{r}(s^2 - 2r + t)^{3/2}} s\ln t$

$= \dfrac{(s^2 + t)s\ln t}{2\sqrt{r}(s^2 - 2r + t)^{3/2}}$

$f_s(r, s, t) = \dfrac{\sqrt{s^2 - 2r + t}\,(1) - s[s/\sqrt{s^2 - 2r + t}]}{s^2 - 2r + t} \sqrt{r}\ln t$

$= \dfrac{(s^2 - 2r + t) - s^2}{(s^2 - 2r + t)^{3/2}} \sqrt{r}\ln t$

$= \dfrac{(t - 2r)\sqrt{r}\ln t}{(s^2 - 2r + t)^{3/2}}$

$f_t(r, s, t) = \dfrac{[\sqrt{s^2 - 2r + t}/t] - [(\ln t)/(2\sqrt{s^2 - 2r + t})]}{s^2 - 2r + t} \sqrt{rs}$

$= \dfrac{2(s^2 - 2r + t) - t\ln t}{2t(s^2 - 2r + t)^{3/2}} \sqrt{rs}$

$= \dfrac{(2s^2 - 4r + 2t - t\ln t)\sqrt{rs}}{2t(s^2 - 2r + t)^{3/2}}$

21. $f_x(x, y) = y^3$
$f_x(2, 5) = 125$

23. $z_x = 1/(2\sqrt{x + y^2})$
$z_x(2, 1) = 1/(2\sqrt{3}) = \sqrt{3}/6$

25. Recall $f_r(r, \theta) = 2r\cos\theta$
and $f_\theta(r, \theta) = -r^2 \sin\theta$.
$f_{rr}(r, \theta) = 2\cos\theta$
$f_{r\theta}(r, \theta) = f_{\theta r} = -2r\sin\theta$
$f_{\theta\theta}(r, \theta) = -r^2 \cos\theta$

27. Recall $w_x = 2x/(x^2 + y^2 + z^2)$, $w_y = 2y/(x^2 + y^2 + z^2)$ and $w_z = 2z/(x^2 + y^2 + z^2)$.
$w_{xx} = [(x^2 + y^2 + z^2)2 - 2x(2x)]/(x^2 + y^2 + z^2)^2 = 2(-x^2 + y^2 + z^2)/(x^2 + y^2 + z^2)^2$
$w_{yy} = 2(x^2 - y^2 + z^2)/(x^2 + y^2 + z^2)^2$
$w_{zz} = 2(x^2 + y^2 - z^2)/(x^2 + y^2 + z^2)^2$
$w_{xx} + w_{yy} + w_{zz} = (2x^2 + 2y^2 + 2z^2)/(x^2 + y^2 + z^2)^2 = 2/(x^2 + y^2 + z^2)$

29. a. $v_2(A_2, A_1, v_1) = A_1 v_1/A_2$

   b. $\partial v_2/\partial A_2 = -A_1 v_1/(A_2)^2$

   c. $\dfrac{\partial v_2}{\partial A_2}(3, 5, 20) = -(5)(20)/(3)^2$

       $= -100/9 \text{ cm}^{-1}\text{s}^{-1}$

   d. $v_2(3, 5, 20) = (5)(20)/3 = 100/3$

     $A_2(v_2, v_1, A_1) = A_1 v_1/v_2$

     $\partial A_2/\partial v_1 = A_1/v_2$

     $(\partial A_2/\partial v_1)(100/3, 20, 5) = 3/20 \text{ cm s}$

31. $\partial y/\partial\theta = \rho\sin\phi\cos\theta$

33. $\partial x/\partial\rho = \sin\phi\cos\theta$

35. $\partial z/\partial\phi = -\rho\sin\phi$

37. a. $\partial f/\partial x = \dfrac{\partial}{\partial x}\int_x^{x+y}\cos t^2\, dt$

     $= \dfrac{\partial}{\partial x}\left(-\int_c^x \cos t^2\, dt + \int_c^{x+y}\cos t^2\, dt\right)$

     $= -\cos x^2 + \cos\big((x+y)^2\big)$

   b. $\partial f/\partial y = \dfrac{\partial}{\partial y}\int_x^{x+y}\cos t^2\, dt$

     $= \cos\big((x+y)^2\big)$

39. $f_x(x, y) = \dfrac{1}{\sqrt{1 - [(x-y)/(x+y)]^2}}\dfrac{(x+y)1 - (x-y)1}{(x+y)^2} = \dfrac{2y}{|x+y|\sqrt{(x+y)^2 - (x-y)^2}}$

    $f_y(x, y) = \dfrac{1}{\sqrt{1 - [(x-y)/(x+y)]^2}}\dfrac{(x+y)(-1) - (x-y)1}{(x+y)^2} = \dfrac{-2x}{|x+y|\sqrt{(x+y)^2 - (x-y)^2}}$

Note that $xf_x(x, y) = -yf_y(x, y)$.

Therefore $xf_x(x, y) + yf_y(x, y) = 0$.

41. a. $\partial V/\partial a = [\sigma/(2\epsilon_0)](a/\sqrt{a^2 + r^2}) = \sigma a/(2\epsilon_0\sqrt{a^2 + r^2})$

   b. $\partial V/\partial r = [\sigma/(2\epsilon_0)][(r/\sqrt{a^2 + r^2}) - 1]$

43. Let $w = f(x, y, z) = e^{x+y+z}$. Note that $\partial w/\partial x = \partial w/\partial y = \partial w/\partial z = e^{x+y+z} = w$. Thus all first order derivatives of f are equal to f. Now suppose that all n^{th} order derivatives of f are equal to f. Any $(n+1)^{\text{st}}$ derivative may be expressed as one of $\partial u/\partial x$, $\partial u/\partial y$ or $\partial u/\partial z$, where u is an nth order derivative. But by supposition, $u = w$, and therefore the $(n+1)^{\text{st}}$ order derivative is w, and hence any $(n+1)^{\text{st}}$ order derivative of f is equal to f. By induction, one may conclude that all partial derivatives of all orders are equal to f.

45. $f_x(x, y) = y\cos xy$

   $f_y(x, y) = x\cos xy$

   $xf_x(x, y) - yf_y(x, y)$

   $= xy\cos xy - yx\cos xy = 0$

47. $f(x, y) = \sin xy$

   For any n:

   $\partial^{2n} f/\partial x^{2n} = (-1)^n y^{2n}\sin xy$

   $\partial^{2n} f/\partial y^{2n} = (-1)^n x^{2n}\sin xy$

   $\partial^{2n+1} f/\partial x^{2n+1} = (-1)^n y^{2n+1}\cos xy$

   $\partial^{2n+1} f/\partial y^{2n+1} = (-1)^n x^{2n+1}\cos xy$

   The differential equation is clearly satisfied.

49. The goal is to obtain $\partial^2 f/\partial x^2$ and $\partial^2 f/\partial y^2$ and then observe that they differ in sign.

   a. $\partial f/\partial x = e^x\sin y$

     $\partial f/\partial y = e^x\cos y$

     $\partial^2 f/\partial x^2 = e^x\sin y$

     $\partial^2 f/\partial y^2 = -e^x\sin y$

b. $\partial f/\partial x = -e^{-x}\cos y$
$\partial f/\partial y = -e^{-x}\sin y$
$\partial^2 f/\partial x^2 = e^{-x}\cos y$
$\partial^2 f/\partial y^2 = -e^{-x}\cos y$

c. $\partial f/\partial x = [1/\sqrt{x^2+y^2}][1/(2\sqrt{x^2+y^2})][2x]$
$\quad = x/(x^2+y^2)$
$\partial f/\partial y = y/(x^2+y^2)$
$\partial^2 f/\partial x^2 = [(x^2+y^2)1 - x(2x)]/(x^2+y^2)^2$
$\quad = (-x^2+y^2)/(x^2+y^2)^2$
$\partial^2 f/\partial y^2 = (x^2-y^2)/(x^2+y^2)^2$

51. Since y is constant, the slope is the partial derivative with respect to x, $\partial z/\partial x = 3x^2$, which has a value of $3(1)^2 = 3$ at the point $(1,2,5)$. The equation of this line is $z - 5 = 3(x-1)$, and it will intersect the xy-plane when $z = 0$:
$-5 = 3(x-1) = 3x - 3$
$x = -2/3$
The point at which the line intersects the xy-plane is therefore $(-2/3, 2, 0)$.

18.3   Tangent Planes

1. $f_x(x,y) = 2x$ and $f_y(x,y) = 2y$
$f_x(1,3) = 2$ and $f_y(1,3) = 6$
$\mathbf{n} = \langle 2, 6, -1 \rangle$
Plane: $2(x-1) + 6(y-3) - (z-10) = 0$
$2x + 6y - z = 10$

3. $z_x = 2x - y - 4$ and $z_y = 2y - x - 2$
$z_x(1,-1) = -1$ and $z_y(1,-1) = -5$
$\mathbf{n} = \langle -1, -5, -1 \rangle$
$-\mathbf{n} = \langle 1, 5, 1 \rangle$
Plane: $(x-1) + 5(y+1) + (z-1) = 0$
$x + 5y + z = -3$

5. $f_x(x,y) = 1/(y+2)$
$f_y(x,y) = -(x-2)/(y+2)^2$
$f_x(4,-1) = 1$ and $f_y(4,-1) = -2$
$\mathbf{n} = \langle 1, -2, -1 \rangle$
Plane: $(x-4) - 2(y+1) - (z-2) = 0$
$x - 2y - z = 4$

7. $f(x,y) = \ln x + \ln y$
$f_x(x,y) = 1/x$ and $f_y(x,y) = 1/y$
$f_x(1,1) = 1$ and $f_y(x,y) = 1$
$\mathbf{n} = \langle 1, 1, -1 \rangle$
Plane: $(x-1) + (y-1) - (z-0) = 0$
$x + y - z = 2$

9. $z_x = [(x^2+y^2)1 - x(2x)]/(x^2+y^2)^2$
$\quad = (-x^2+y^2)/(x^2+y^2)^2$
$z_y = -2y/(x^2+y^2)^2$
$z_x(1,1) = 0$ and $z_y(1,1) = -1/2$
$\mathbf{n} = \langle 0, -1/2, -1 \rangle$
$-2\mathbf{n} = \langle 0, 1, 2 \rangle$
Plane: $(y-1) + 2(z - (1/2)) = 0$
$y + 2z = 2$

11. $z = x \ln y$
$z_x = \ln y$ and $z_y = x/y$
$z_x(1,1) = 0$ and $z_y(1,1) = 1$
$\mathbf{n} = \langle 0, 1, -1 \rangle$
Plane: $(y-1) - (z-0) = 0$
$y - z = 1$

13. $f_x(x, y) = -\sin x \sin y$

$f_y(x, y) = \cos x \cos y$

$f_x(0, \pi/4) = 0$ and $f_y(0, \pi/4) = 1/\sqrt{2}$

$\mathbf{n} = \langle 0, 1/\sqrt{2}, -1 \rangle$

$\sqrt{2}\,\mathbf{n} = \langle 0, 1, -\sqrt{2} \rangle$

Plane: $(y - (\pi/4)) - \sqrt{2}(z - (\sqrt{2}/2)) = 0$

$y - \sqrt{2}\,z = (\pi/4) - 1$

15. $f_x(x, y) = \dfrac{y + x}{y - x}\dfrac{(y + x)(-1) - (y - x)1}{(y + x)^2}$

$= -2y/(y^2 - x^2)$

$f_y(x, y) = \dfrac{y + x}{y - x}\dfrac{(y + x)1 - (y - x)1}{(y + x)^2}$

$= 2x/(y^2 - x^2)$

$f_x(0, e) = -2/e$ and $f_y(0, e) = 0$

$\mathbf{n} = \langle -2/e, 0, -1 \rangle$

$-e\mathbf{n} = \langle 2, 0, e \rangle$

Plane: $2(x - 0) + e(z - 0) = 0$

$2x + ez = 0$

17. Point $(3, -1, 4)$ and direction vector $\langle 4, 4, -1 \rangle$

Vector equation $\mathbf{r}(t) = 3\mathbf{i} - \mathbf{j} + 4\mathbf{k} + t(4\mathbf{i} + 4\mathbf{j} - \mathbf{k})$

19. Point $(1, -2, 2)$ and direction vector $\langle 1, -2, 2 \rangle$

Vector equation $\mathbf{r}(t) = \mathbf{i} - 2\mathbf{j} + 2\mathbf{k} + t(\mathbf{i} - 2\mathbf{j} + 2\mathbf{k})$

21. Point $(1, 1, 0)$ and direction vector $\langle 0, 1, -1 \rangle$

Vector equation $\mathbf{r}(t) = \mathbf{i} + \mathbf{j} + t(\mathbf{j} - \mathbf{k})$

23. Find the point on the graph where the normal vector $\mathbf{n} = (\partial z/\partial x)\mathbf{i} + (\partial z/\partial y)\mathbf{j} - \mathbf{k}$ is parallel to the normal vector $\mathbf{i} - 14\mathbf{j} + \mathbf{k}$ for the given plane. From the $\mathbf{k}$ component, one deduces that $\mathbf{n} = -(\mathbf{i} - 14\mathbf{j} + \mathbf{k})$, and therefore that $\partial z/\partial x = -1$ and $\partial z/\partial y = 14$. The partial derivatives of $z = -x^2 + xy + 2y^2$ are $\partial z/\partial x = -2x + y$ and $\partial z/\partial y = x + 4y$. Thus, one wishes to solve the simultaneous equations $-2x + y = -1$ and $x + 4y = 14$. Solving these simultaneous equations yields $x = 2$ and $y = 3$. For these values of x and y, $z = 20$. Therefore, the desired point on the graph is $(2, 3, 20)$.

25. Recall that the xy-plane is the plane $z = 0$. The partial derivatives of z are $\partial z/\partial x = 2x$ and $\partial z/\partial y = 4y - 4$, and their values at the point $(2, 1, 4)$ are $\partial z/\partial x = 4$ and $\partial z/\partial y = 0$. It follows that the normal vector for the tangent plane is $\langle 4, 0, -1 \rangle$, and that the equation for the tangent plane is $4(x - 2) - (z - 4) = 0$, or $4x - z = 4$. The intersection of this plane with the plane $z = 0$ will have equation $4x = 4$, or $x = 1$. The points on the line are therefore of the form $(1, y, 0)$, and an equation for this line is $\mathbf{r}(t) = \mathbf{i} + t\mathbf{j}$.

27. Show that the normal vectors at all points $(x, \pi/2)$ are parallel. $f_x(x, y) = \sin y$ and $f_y(x, y) = x \cos y$. At points $(x, \pi/2)$ one has $f_x(x, \pi/2) = 1$ and $f_y(x, \pi/2) = 0$. Thus all such points have the same normal vector, $\langle 1, 0, -1 \rangle$.

29. $z_x = 18x$ and $z_y = 8y$

$z_x(1, 2) = 18$ and $z_y(1, 2) = 16$

$\mathbf{n} = \langle 18, 16, -1 \rangle$

Plane: $18(x - 1) + 16(y - 2) - (z - 25) = 0$

$18x + 16y - z = 25$

31. The point is on the graph of one of the two functions $z = \pm\sqrt{r^2 - x^2 - y^2}$. For these functions the partial derivatives are:

$\partial z/\partial x = \mp x/\sqrt{r^2 - x^2 - y^2} = \mp x/|z| = -x/z$ and

$\partial z/\partial y = \mp y/\sqrt{r^2 - x^2 - y^2} = \mp y/|z| = -y/z$.

The normal vector at the point (x_0, y_0, z_0) is $\mathbf{n} = \langle -x_0/z_0, -y_0/z_0, -1 \rangle$, or $-z_0\mathbf{n} = \langle x_0, y_0, z_0 \rangle$. The equation of the tangent plane is $x_0(x - x_0) + y_0(y - y_0) + z_0(z - z_0) = 0$, or $x_0 x + y_0 y + z_0 z = r^2$.

33. The equation of the plane is $f_x(x_0, z_0)(x - x_0) - (y - y_0) + f_z(x_0, z_0)(z - z_0) = 0$.

$\partial y/\partial x = [1/((z/x)^2 + 1)](-z/x^2) = -z/(z^2 + x^2)$

$\partial y/\partial z = [1/((z/x)^2 + 1)](1/x) = x/(z^2 + x^2)$

At $x = z = 1$, $y = \pi/4$, $\partial y/\partial x = -1/2$ and $\partial y/\partial z = 1/2$. The equation of the plane is $(-1/2)(x - 1) - (y - (\pi/4)) + (1/2)(z - 1) = 0$, or $x + 2y - z = \pi/2$.

35. $z_x = 2 - 2x$ and $z_y = 4 - 2y$

$z_x(1,2) = 0$ and $z_y(1,2) = 0$

The normal vector for the surface is $\mathbf{n} = \langle 0, 0, -1 \rangle$ or $-\mathbf{k}$.

$\mathbf{r}'(t) = -\pi \sin(\pi t)\mathbf{i} + \pi \cos(\pi t/2)\mathbf{j} + 5\mathbf{k}$

The tangent vector for the curve is $\mathbf{r}'(1) = 5\mathbf{k}$. The angle α in the range $[0, \pi/2]$ between $\mathbf{r}'(1)$ and vectors parallel to $\mathbf{n}$ is 0, since these two vectors are parallel. Therefore, the angle of intersection is $\theta = \pi/2$.

37. For the first surface:

$z_x = 2 - 2x$ and $z_y = 4 - 2y$

$z_x(1,1) = 0$ and $z_y(1,1) = 2$

The normal vector is $\mathbf{n}_1 = \langle 0, 2, -1 \rangle$.

For the second surface:

$z_x = (x-1)/\sqrt{(x-1)^2 + (2y+1)^2}$

$z_y = (4y+2)/\sqrt{(x-1)^2 + (2y+1)^2}$

$z_x(1,1) = 0$ and $z_y(1,1) = 6/3 = 2$

The normal vector is $\mathbf{n}_2 = \langle 0, 2, -1 \rangle$.

Since the two normal vectors are identical, the angle between them is zero.

39. Since the normal lines of a sphere will look the same no matter where its center is located, one may assume without loss of generality that the center is at the origin. Let the equation of the sphere be $x^2 + y^2 + z^2 = r^2$, where $r > 0$. Recall from the solution to Exercise 31 that for any point (x_0, y_0, z_0) with $z_0 \neq 0$, a normal vector is $\langle x_0, y_0, z_0 \rangle$. If $z_0 = 0$, then one of x_0 or y_0 must be nonzero. Choose one that is nonzero and, depending on the choice and whether it is positive or negative, consider one of the four functions $x = \pm\sqrt{r^2 - y^2 - z^2}$ or $y = \pm\sqrt{r^2 - x^2 - z^2}$. In a manner analogous to Exercise 31, one may use these functions to prove that a normal vector at (x_0, y_0, z_0) is $\langle x_0, y_0, z_0 \rangle$. A vector equation for the normal line is thus $\mathbf{r}(t) = \langle x_0, y_0, z_0 \rangle + t \langle x_0, y_0, z_0 \rangle$, and $\mathbf{r}(-1) = \langle 0, 0, 0 \rangle$. Therefore, the normal line passes through the origin.

18.4 Relative and Absolute Extrema

1. $\partial f/\partial x = 2x$ and $\partial f/\partial y = 2y + 4$

Critical points must satisfy $x = 0$ and $y = -2$. The only critical point is $(0, -2)$. At this point:

$A = \partial^2 f/\partial x^2 = 2$, $B = \partial^2 f/(\partial y \partial x) = 0$ and $C = \partial^2 f/\partial y^2 = 2$

$D = B^2 - AC = -4 < 0$ and $A > 0$

Relative minimum at $(0, -2)$. Note that this is not only a relative minimum, but is also an absolute minimum.

3. $\partial f/\partial x = 2x + 6$ and $\partial f/\partial y = -2y + 4$

Critical points must satisfy $x = -3$ and $y = 2$. The only critical point is $(-3, 2)$. At this point:

$A = \partial^2 f/\partial x^2 = 2$, $B = \partial^2 f/(\partial y \partial x) = 0$ and $C = \partial^2 f/\partial y^2 = -2$

$D = B^2 - AC = 4 > 0$

Saddle point at $(-3, 2)$

5. $\partial f/\partial x = y$ and $\partial f/\partial y = x$

Critical points must satisfy $y = 0$ and $x = 0$. The only critical point is $(0, 0)$. At this point:

$A = \partial^2 f/\partial x^2 = 0$, $B = \partial^2 f/(\partial y \partial x) = 1$ and $C = \partial^2 f/\partial y^2 = 0$

$D = B^2 - AC = 1 > 0$

Saddle point at $(0, 0)$

7. $\partial f/\partial x = 10x - 10$ and $\partial f/\partial y = 2y - 6$
Critical points must satisfy $x = 1$ and $y = 3$. The only critical point is $(1,3)$. At this point:
$A = \partial^2 f/\partial x^2 = 10$, $B = \partial^2 f/(\partial y \partial x) = 0$ and $C = \partial^2 f/\partial y^2 = 2$
$D = B^2 - AC = -20 < 0$ and $A > 0$
Relative minimum at $(1,3)$ Note that this is not only a relative minimum, but is also an absolute minimum.

9. $\partial f/\partial x = 3x^2$ and $\partial f/\partial y = 3y^2$
Critical points must satisfy $x = 0$ and $y = 0$. The only critical point is $(0,0)$. At this point:
$A = \partial^2 f/\partial x^2 = 6x = 0$, $B = \partial^2 f/(\partial y \partial x) = 0$ and $C = \partial^2 f/\partial y^2 = 6y = 0$
$D = B^2 - AC = 0$ fails to provide information. Note that $f(0,0) = 0$. For values of h close to 0, $f(h,0) = h^3$ can be both positive and negative. Therefore, there is neither a relative minimum nor a relative maximum at $(0,0)$, and it is a saddle point.

11. $\partial f/\partial x = 2x - y$ and $\partial f/\partial y = -x$
Critical points must satisfy $2x - y = 0$ and $x = 0$. Substituting $x = 0$ into the first gives $y = 0$. The only critical point is $(0,0)$. At this point:
$A = \partial^2 f/\partial x^2 = 2$, $B = \partial^2 f/(\partial y \partial x) = -1$ and $C = \partial^2 f/\partial y^2 = 0$
$D = B^2 - AC = 1 > 0$
Saddle point at $(0,0)$.

13. $\partial f/\partial x = 3x^2 + 4y$ and $\partial f/\partial y = 3y^2 + 4x$
Critical points must satisfy $3x^2 + 4y = 0$ and $3y^2 + 4x = 0$. Solving for y in the first equation gives $y = -3x^2/4$, and substituting this for y in the second equation gives $(27x^4/16) + 4x = 0$. Multiplying by 16 and factoring gives $x(27x^3 + 64) = 0$. The solutions are $x = 0$, for which $y = 0$, and $x = (-64/27)^{1/3} = -4/3$, for which $y = -4/3$. The critical points are $(0,0)$ and $(-4/3, -4/3)$.
$\partial^2 f/\partial x^2 = 6x$, $\partial^2 f/(\partial y \partial x) = 4$ and $\partial^2 f/\partial y^2 = 6y$
At $(0,0)$: $A = 0$, $B = 4$, $C = 0$ and $D = B^2 - AC = 16 > 0$. Saddle point.
At $(-4/3, -4/3)$: $A = -8 < 0$, $B = 4$, $C = -8$ and $D = B^2 - AC = -48 < 0$. Relative maximum.

15. $\partial f/\partial x = \cos y$ and $\partial f/\partial y = -x \sin y$
Critical points must satisfy $\cos y = 0$ and $-x \sin y = 0$. The first equation implies that y is of the form $(\pi/2) + n\pi$. Note that $\sin y \neq 0$ for any such number, and thus the second equation gives us $x = 0$. The critical points are all points of the form $(0, (\pi/2) + n\pi)$. At any such point:
$A = \partial^2 f/\partial x^2 = 0$, $B = \partial^2 f/(\partial y \partial x) = -\sin y = \pm 1$ and $C = \partial^2 f/\partial y^2 = -x \cos y = 0$
$D = B^2 - AC = 1 > 0$
These are all saddle points.

17. $\partial f/\partial x = \cos(x - y)$ and $\partial f/\partial y = -\cos(x - y)$
Critical points must satisfy $\cos(x - y) = 0$. Thus $x - y$ must be a number of the form $(\pi/2) + n\pi$. The critical points are all points along these lines (one line for each integer n).
$\partial^2 f/\partial x^2 = -\sin(x - y)$, $\partial^2 f/(\partial y \partial x) = \sin(x - y)$ and $\partial^2 f/\partial y^2 = -\sin(x - y)$
At any point (x, y), $D = B^2 - AC = \sin^2(x - y) - \sin^2(x - y) = 0$ and the second derivative test will be inconclusive. For any critical point (x_0, y_0), $f(x_0, y_0) = \sin(x - y) = \pm 1$. Since 1 and -1 are the absolute maximum and minimum values, respectively, of the sine function, it follows that for any point (x, y) in a neighborhood of (x_0, y_0), $f(x, y) = \sin(x - y)$ is either less than or equal to $f(x_0, y_0)$ if this is 1, or greater than or equal to $f(x_0, y_0)$ if this is -1. Therefore, points along lines of the form $x - y = (\pi/2) + 2n\pi$ are relative maxima, and points along lines of the form $x - y = -(\pi/2) + 2n\pi$ are relative minima. Note that both of these are not only relative extrema, but also absolute extrema.

19. $\partial f/\partial x = 2x/(x^2 + y^2 + 1)$ and $\partial f/\partial y = 2y/(x^2 + y^2 + 1)$

$\partial f/\partial x$ and $\partial f/\partial y$ are defined everywhere since $x^2 + y^2 + 1$ is always positive. They are both zero when $x = 0$ and $y = 0$. The only critical point is $(0, 0)$. At this point:

$A = \partial^2 f/\partial x^2 = [(x^2 + y^2 + 1)2 - 2x(2x)]/(x^2 + y^2 + 1)^2 = 2$
$B = \partial^2 f/(\partial y \partial x) = -4xy/(x^2 + y^2 + 1)^2 = 0$
$C = \partial^2 f/\partial y^2 = [(x^2 + y^2 + 1)2 - 2y(2y)]/(x^2 + y^2 + 1)^2 = 2$
$D = B^2 - AC = -4 < 0$ and $A > 0$

Relative minimum at $(0, 0)$. Note that this is not only a relative minimum, but is also an absolute minimum.

21. $\partial f/\partial x = -2xe^{1/(x^2+y^2+1)}/(x^2 + y^2 + 1)$ and $\partial f/\partial y = -2ye^{1/(x^2+y^2+1)}/(x^2 + y^2 + 1)$

These are always defined since the denominators are always positive. Since a power of e is never zero, both $\partial f/\partial x$ and $\partial f/\partial y$ are zero only when both $x = 0$ and $y = 0$. The only critical point is $(0, 0)$. $f(0, 0) = e$. For any point (x, y) in a neighborhood of $(0, 0)$, $x^2 + y^2 + 1 \geq 1$, so $1/(x^2 + y^2 + 1) \leq 1$, and thus $f(x, y) \leq e = f(0, 0)$. Therefore, there is a relative maximum at $(0, 0)$. Note that this is not only a relative maximum, but is also an absolute maximum.

23. At $(0, 0)$, $z = 4 - 0 = 4$. For any other (x, y), $\sqrt{x^2 + y^2}$ exists and is positive, so $z = 4 - \sqrt{x^2 + y^2} < 4$. Therefore, there is a relative maximum at $(0, 0)$.

25. There can be no highest point. The trace in the plane $x = 0$ is the graph of $z = 2y^3$, which increases without bound as $y \to \infty$.

27. A nearby point is of the form $(-1 + h, 2 + k)$ for small h and k.

$f(-1, 2) - f(-1 + h, 2 + k) = 0 - [-(-1 + h)^2 - (2 + k)^2 + 2(-1 + h) + 4(2 + k) - 3]$
$= -[1 - 2h + h^2 - 4 - 4k - k^2 - 2 + 2h + 8 + 4k - 3]$
$= -h^2 + k^2$

Since this is nonnegative when $h = 0$ and nonpositive when $k = 0$, there must be a saddle point at $(-1, 2)$.

29. Let l, w and h be the length, width and height of the box respectively. Let C be the total cost of box, and let $a > 0$ and $b > 0$ be constants representing the costs of the side walls and top/bottom walls respectively. We wish to minimize C. We are given that the volume of the box is $lwh = 16$, that $b = 2a$, and that the total cost is $C = 2alh + 2awh + 2blw$. Substituting $b = 2a$ and $h = 16/(lw)$ into the cost formula produces the function $C(l, w) = (32a/w) + (32a/l) + 4alw$, with domain $\{(l, w)|l > 0 \text{ and } w > 0\}$.

The partial derivatives are $\partial C/\partial l = (-32a/l^2) + 4aw$ and $\partial C/\partial w = (-32a/w^2) + 4al$. These are both defined throughout the domain of C. They are both zero at any solution for the simultaneous equations $(-32a/l^2) + 4aw = 0$ and $(-32a/w^2) + 4al = 0$. Solving for w in the first equation gives $w = 8/l^2$, and substituting into the second equation gives $(-al^4/2) + 4al = 0$. Since $a > 0$ and $l > 0$, we may multiply this equation by $-2/(al)$ in order to obtain $l^3 - 8 = 0$, whose only solution is $l = 2$. For this value of l one has $w = 8/2^2 = 2$. Therefore, $(2, 2)$ is a critical point for $C(l, w)$.

The fact that $C \to \infty$ as $l \to \infty$, $w \to \infty$, $l \to 0^+$ or $w \to 0^+$ will imply that C does in fact have an absolute minimum, which must occur at $(2, 2)$ since there are no other critical points. At $l = 2$, $w = 2$, we have $h = 16/4 = 4$, and therefore the dimensions which produce the least expensive box are length and width of 2 meters, and height of 4 meters.

31. At least one of the coefficients a, b or c must be nonzero. Assume in the following that $c \neq 0$; the cases $a \neq 0$ and $b \neq 0$ are analogous and yield the same result. For any point (x, y, z), let $r = \sqrt{x^2 + y^2 + z^2}$ be the distance from the point to the origin. Minimize r for points on the plane $ax + by + cz = d$. Substituting $z = (d - ax - by)/c$, one wishes to find the minimum for the function $r(x, y) = \sqrt{x^2 + y^2 + (d - ax - by)^2/c^2}$, which is minimized at the same point as the function $f(x, y) = r^2 = x^2 + y^2 + (d - ax - by)^2/c^2$. The partial derivatives are

$$\partial f/\partial x = 2x + 2(d - ax - by)(-a)/c^2 = (2(a^2 + c^2)/c^2)x + (2ab/c^2)y - (2ad/c^2) \quad \text{and}$$
$$\partial f/\partial y = 2y + 2(d - ax - by)(-b)/c^2 = (2ab/c^2)x + (2(b^2 + c^2)/c^2)y - (2bd/c^2).$$

These are both zero at the solution to the simultaneous equations

$$2(a^2 + c^2)x + 2aby = 2ad \quad \text{and} \quad 2abx + 2(b^2 + c^2)y = 2bd.$$

The solution to these equations is $x = ad/(a^2 + b^2 + c^2)$ and $y = bd/(a^2 + b^2 + c^2)$, and for these values $z = (d - ax - by)/c = cd/(a^2 + b^2 + c^2)$ Therefore, the point on the plane closest to the origin is $(a\lambda, b\lambda, c\lambda)$, where $\lambda = d/(a^2 + b^2 + c^2)$.

33. Let l, w and h be the length, width and height of the package, g be its girth, and V be its volume. We wish to maximize $V = lwh$ subject to the constraint that $l + g = 84$ (the maximum will not occur for $l + g < 84$). We also know that $g = 2w + 2h$. Thus $l = 84 - 2w - 2h$, and substituting in the volume formula produces the function $V(w, h) = (84 - 2w - 2h)wh = 84wh - 2w^2h - 2wh^2$ for which we wish to find a maximum. Note that $g \leq 84$ implies that the domain for V is $\{(w, h)|w \geq 0, h \geq 0 \text{ and } w + h \leq 42\}$. Since this domain is closed and bounded, V must possess an absolute maximum, and this maximum will be in the interior since $V = 0$ at all boundary points. The partial derivatives for V are $\partial V/\partial w = 84h - 4wh - 2h^2$ and $\partial V/\partial h = 84w - 2w^2 - 4wh$. These are both zero at the solution to the simultaneous equations $4wh + 2h^2 = 84h$ and $2w^2 + 4wh = 84w$, and dividing by $2h$ in the first case and $2w$ in the second (both of which are nonzero) gives $2w + h = 42$ and $w + 2h = 42$. The solution to these is $w = h = 14$. The corresponding length is $l = 84 - 2(14) - 2(14) = 28$. The desired dimensions are a length of 28 inches, and width and height of 14 inches each.

35. Note that $z = 2x + 2y + 4$ is a plane, which has no relative maxima or minima, and therefore the absolute maximum and minimum must occur on the boundary of D. Let $\mathbf{r}(t) = \cos t\,\mathbf{i} + \sin t\,\mathbf{j}$, and let $g(t) = f(\mathbf{r}(t)) = f(\cos t, \sin t) = 2\cos t + 2\sin t + 4$. The values of g are exactly the values of f on the boundary of D. g has maxima or minima when $0 = g'(t) = -2\sin t + 2\cos t$, $2\sin t = 2\cos t$, $\tan t = 1$, $t = (\pi/4) + n\pi$. The maximum value of g, and thus of f on D, is $g((\pi/4) + 2n\pi) = 2\sqrt{2} + 4$, and the minimum value is $g((5\pi/4) + 2n\pi) = -2\sqrt{2} + 4$.

37. Note that $z = x + 2y + 6$ is a plane, which has no relative maxima or minima, and therefore the absolute maximum and minimum must occur on the ellipse itself. Let $\mathbf{r}(t) = \cos t\,\mathbf{i} + \sqrt{2}\sin t\,\mathbf{j}$, which is a parameterization of the ellipse, and let $g(t) = f(\mathbf{r}(t)) = f(\cos t, \sqrt{2}\sin t) = \cos t + 2\sqrt{2}\sin t + 6$. The values of g are exactly the values of f on the ellipse. g has maxima or minima when:
$0 = g'(t) = -\sin t + 2\sqrt{2}\cos t$,
$\sin t = 2\sqrt{2}\cos t$,
$\tan t = 2\sqrt{2}$.
The maximum value of g, and thus of f on D, is $g(\text{Tan}^{-1}(2\sqrt{2})) = (1/3) + 2\sqrt{2}(2\sqrt{2}/3) + 6 = 9$ and the minimum value is $g(\text{Tan}^{-1}(2\sqrt{2}) + \pi) = (-1/3) + 2\sqrt{2}(-2\sqrt{2}/3) + 6 = 3$.

39. Let C be total cost, R be total revenue, and P be total profit.

$C = 65(1200 - 40x + 20y) + 80(2650 + 10x - 30y) = 290000 - 1800x - 1100y$

$R = x(1200 - 40x + 20y) + y(2650 + 10x - 30y) = 1200x + 2650y + 30xy - 40x^2 - 30y^2$

$P = R - C = 3000x + 3750y + 30xy - 40x^2 - 30y^2 - 290000$

Since $P \to -\infty$ as $x \to \pm\infty$ or $y \to \pm\infty$, P will have an absolute maximum.

$\partial P/\partial x = 3000 + 30y - 80x$ and $\partial P/\partial y = 3750 + 30x - 60y$

A critical point must satisfy the simultaneous equations $80x - 30y = 3000$ and $-30x + 60y = 3750$. The only solution to these equations is $x = 75$ and $y = 100$. The standard model should be priced at \$75 and the deluxe model at \$100.

41. The partial derivatives are:

$$\frac{\partial S}{\partial m} = \sum_{j=1}^{n} 2(y_j - mx_j - b)(-x_j)$$

$$= 2\sum_{j=1}^{n}(mx_j^2 - x_j y_j + bx_j)$$

$$= 2m\sum_{j=1}^{n} x_j^2 - 2\sum_{j=1}^{n} x_j y_j + 2b\sum_{j=1}^{n} x_j$$

$$\frac{\partial S}{\partial b} = \sum_{j=1}^{n} 2(y_j - mx_j - b)(-1)$$

$$= 2\sum_{j=1}^{n}(mx_j - y_j + b)$$

$$= 2m\sum_{j=1}^{n} x_j - 2\sum_{j=1}^{n} y_j + 2nb$$

Define the following values for ease of computation:

$$X = \sum_{j=1}^{n} x_j \quad \text{and} \quad Y = \sum_{j=1}^{n} y_j \quad \text{and} \quad U = \sum_{j=1}^{n} x_j y_j \quad \text{and} \quad V = \sum_{j=1}^{n} x_j^2$$

Setting the partial derivatives equal to zero and dividing by 2 gives the simultaneous equations

$$mV - U + bX = 0 \quad \text{and} \quad mX - Y + nb = 0.$$

Solving for b in the second equation gives $b = (Y - mX)/n$, and substituting this into the first equation gives $mV - U + X(Y - mX)/n = 0$. Multiply by n to get $mnV - nU + XY - mX^2 = 0$, and solving for m, one obtains $m = (nU - XY)/(nV - X^2)$, which agrees with the formula given. The corresponding value of b is:

$b = (Y/n) - m(X/n)$

$= (Y/n) - (nXU - X^2Y)/(n^2V - nX^2)$

$= [Y(nV - X^2) + X^2Y - nXU]/(n^2V - nX^2)$

$= (nVY - nXU)/(n^2V - nX^2)$

$= (VY - XU)/(nV - X^2)$

which agrees with the formula given.

43. $y = (0.95)(60) + 29.31 = 86$, where the prediction is rounded to the nearest integer.

45. a. $\sum_{j=1}^{5} x_j = 45$

$\sum_{j=1}^{5} y_j = 28$

$\sum_{j=1}^{5} x_j y_j = 275$

$\sum_{j=1}^{5} x_j^2 = 425$

$m = [5(275) - (45)(28)]/[5(425) - 45^2]$

$= 115/100 = 1.15$

$b = [(425)(28) - (45)(275)]/[5(425) - 45^2]$

$= -475/100 = -4.75$

$y = 1.15x - 4.75$

b. See the figure.

c. $y = (1.15)4 - 4.75 = -0.15$

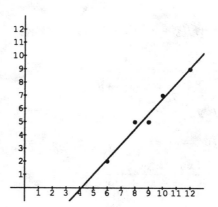

47. Let $X = \sum x_j$, $Y = \sum y_j$, $U = \sum x_j y_j$ and $V = \sum x_j^2$. Then $\bar{x} = X/n$ and $\bar{y} = Y/n$, and the regression line is $y = mx + b$ for $m = (nU - XY)/(nV - X^2)$ and $b = (VY - XU)/(nV - X^2)$. Show that $\bar{y} = m\bar{x} + b$.

$m\bar{x} + b = [(nU - XY)/(nV - X^2)][X/n] + (VY - XU)/(nV - X^2)$

$= [(nU - XY)X + n(VY - XU)]/[n(nV - X^2)]$

$= (nVY - X^2Y)/[n(nV - X^2)]$

$= [(nV - X^2)Y]/[n(nV - X^2)]$

$= Y/n$

$= \bar{y}$

49. $f(x, y) = ax^2 + bxy + cy^2$

$\partial f/\partial x = 2ax + by$ and $\partial f/\partial y = bx + 2cy$

$A = \partial^2 f/\partial x^2 = 2a$, $B = \partial^2 f/(\partial y \partial x) = b$, $C = \partial^2 f/\partial y^2 = 2c$ and $D = B^2 - AC = b^2 - 4ac$

Assume that $D = b^2 - 4ac \neq 0$. A critical point must satisfy $2ax + by = 0$ and $bx + 2cy = 0$. Multiplying the first equation by b and the second equation by $-2a$ and then adding gives $(b^2 - 4ac)y = 0$ for which $y = 0$ since $b^2 - 4ac \neq 0$. At $y = 0$ one has $2ax = 0$ and $bx = 0$, and since $b^2 - 4ac \neq 0$, this implies that $x = 0$. Thus $(0, 0)$ is the only critical point, with $f(0, 0) = 0$.

Consider a point (x, y) in the neighborhood of $(0, 0)$. If $y = 0$, then $f(x, y) = ax^2$ is greater than or equal to $f(0, 0)$ if $a \geq 0$ and less than or equal to $f(0, 0)$ if $a \leq 0$. If $y \neq 0$, then $f(x, y) = y^2[a(x/y)^2 + b(x/y) + c] = y^2[au^2 + bu + c]$, where $u = x/y$. If the discriminant $D = b^2 - 4ac < 0$, then this polynomial has no roots, and the value of $f(x, y)$ is always greater than or equal to $0 = f(0, 0)$ if $a \geq 0$, or less than or equal to $f(0, 0)$ if $a \leq 0$. Therefore, if $D < 0$, then there is a relative minimum at $(0, 0)$ if $A = 2a$ is positive, and a relative maximum there if A is negative. If the discriminant $D = b^2 - 4ac > 0$, then the polynomial has roots, and can thus take on both positive and negative values. Therefore, there are points (x, y) where $f(x, y)$ is positive, and other points where it is negative, and $(0, 0)$ is a saddle point.

18.5   Approximation and Differentiability

1. $df = 2xy^4 \, dx + 4x^2 y^3 \, dy$

3. $\partial f/\partial x = (1/2)(x^2 + y^4)^{-1/2} 2x = x/\sqrt{x^2 + y^4}$

$\partial f/\partial y = (1/2)(x^2 + y^4)^{-1/2} 4y^3 = 2y^3/\sqrt{x^2 + y^4}$

$df = (x/\sqrt{x^2 + y^4})dx + (2y^3/\sqrt{x^2 + y^4})dy$

5. $df = [(e^{\sqrt{x}} \cos y)/(2\sqrt{x})]dx - (e^{\sqrt{x}} \sin y)dy$

7. $df = 2xyz^3 \, dx + x^2 z^3 \, dy + 3x^2 yz^2 \, dz$

9. $df = [1/(y + 3^z)]dx - [x/(y + 3^z)^2]dy - [(x3^z \ln 3)/(y + 3^z)^2]dz$

11. $\partial f/\partial x = \dfrac{(x^2 + y^2 + z^2)1 - (x - y)2x}{(x^2 + y^2 + z^2)^2} = \dfrac{2xy - x^2 + y^2 + z^2}{(x^2 + y^2 + z^2)^2}$

$\partial f/\partial y = \dfrac{(x^2 + y^2 + z^2)(-1) - (x - y)2y}{(x^2 + y^2 + z^2)^2} = \dfrac{y^2 - x^2 - z^2 - 2xy}{(x^2 + y^2 + z^2)^2}$

$\partial f/\partial z = -\dfrac{x - y}{(x^2 + y^2 + z^2)^2}2z = \dfrac{2yz - 2xz}{(x^2 + y^2 + z^2)^2}$

$df = \dfrac{2xy - x^2 + y^2 + z^2}{(x^2 + y^2 + z^2)^2}dx + \dfrac{y^2 - x^2 - z^2 - 2xy}{(x^2 + y^2 + z^2)^2}dy + \dfrac{2yz - 2xz}{(x^2 + y^2 + z^2)^2}dz$

13. a. $f(x_0, y_0) = f(2, 3) = 48$

    $f(x_0 + \Delta x, y_0 + \Delta y) = f(2.1, 3.2) = 57.12$

b. $\Delta f = 57.12 - 48 = 9.12$

c. $\partial f/\partial x = 2y^2 + 2xy$ and $\partial f/\partial x(2, 3) = 30$

    $\partial f/\partial y = 4xy + x^2$ and $\partial f/\partial y(2, 3) = 28$

    $\Delta f \approx 30(.1) + 28(.2) = 3 + 5.6 = 8.6$

d. The approximation would yield a value of

    $f(x_0, y_0) + \Delta f = 48 + 8.6 = 56.6$.

The absolute error is $|57.12 - 56.6| = 0.52$.

The relative error is the ratio of the absolute error to the actual value: $0.52/57.12 \approx 0.0091$.

The percentage error is 0.91%.

15. $f(x, y) = \sqrt{x^2 + y^2}$, $x_0 = 3$, $y_0 = 4$

$\Delta x = .02$, $\Delta y = .08$

$f(3, 4) = 5$

$\partial f/\partial x = x/\sqrt{x^2 + y^2}$

$\partial f/\partial x(3, 4) = 3/5 = .6$

$\partial f/\partial y = y/\sqrt{x^2 + y^2}$

$\partial f/\partial y(3, 4) = 4/5 = .8$

$\Delta f \approx (.6)(.02) + (.8)(.08) = .076$

$f(3.02, 4.08) \approx 5.076$

17. $f(x, y) = \sin x \cos y$

$x_0 = 90° = \pi/2$, $y_0 = 45° = \pi/4$

$\Delta x = -2° \approx -.0349$

$\Delta y = -3° \approx / -.0524$

$f(90°, 45°) = \sqrt{2}/2 \approx 0.7071$

$\partial f/\partial x = \cos x \cos y$

$\partial f/\partial x(90°, 45°) = 0$

$\partial f/\partial y = -\sin x \sin y$

$\partial f/\partial y(90°, 45°) = -\sqrt{2}/2 \approx -0.7071$

$\Delta f \approx (0)(-.0349) + (-0.7071)(-.0524)$

    ≈ 0.037

$f(88°, 42°) \approx 0.7071 + 0.037 = 0.7441$

19. $f(x, y) = x^2y^3$, $x_0 = 5$, $y_0 = 1$

$\Delta x = .03$, $\Delta y = .02$

$f(5, 1) = 25$

$\partial f/\partial x = 2xy^3$ and $\partial f/\partial x(5, 1) = 10$

$\partial f/\partial y = 3x^2y^2$ and $\partial f/\partial y(5, 1) = 75$

$\Delta f \approx 10(.03) + 75(.02) = 1.8$

$f(5.03, 1.02) \approx 25 + 1.8 = 26.8$

21. $f(x, y, z) = xyz$, $x_0 = 6$, $y_0 = 3$, $z_0 = 3$

$\Delta x = .04$, $\Delta y = .1$ and $\Delta z = -.04$

$f(6, 3, 3) = 54$

$\partial f/\partial x = yz$ and $\partial f/\partial x(6, 3, 3) = 9$

$\partial f/\partial y = xz$ and $\partial f/\partial y(6, 3, 3) = 18$

$\partial f/\partial z = xy$ and $\partial f/\partial z(6, 3, 3) = 18$

$\Delta f \approx 9(.04) + 18(.1) + 18(-.04) = 1.44$

$f(6.04, 3.1, 2.96) \approx 54 + 1.44 = 55.44$

23. $V(r, h) = \pi r^2 l$, $r_0 = 2$, $l_0 = 6$

$\Delta r = .04$, $\Delta l = 2(.04) = .08$

$V(2, 6) = 24\pi \approx 75.40 \text{ cm}^3$

$\partial V/\partial r = 2\pi rl$

$\partial V/\partial r(2, 6) = 24\pi \approx 75.40$

$\partial V/\partial l = \pi r^2$

$\partial V/\partial l(2, 6) = 4\pi \approx 12.57$

$\Delta V \approx (75.40)(.04) + (12.57)(.08) \approx 4.02$

$V(2.04, 6.08) \approx 75.40 + 4.02 = 79.42 \text{ cm}^3$

25. $dL = .04L$ and $dK = -.1K$
$$dP = 3\lambda L^2 K^{3/2}\, dL + (3/2)\lambda L^3 K^{1/2}\, dK$$
$$= .12\lambda L^3 K^{3/2} - .15\lambda L^3 K^{3/2}$$
$$= -.03\lambda L^3 K^{3/2}$$
Percentage change is approximately:
$100\, dP/P = -3\%$

27. $ds = .2$ and $dg = -.05$
$$\partial T/\partial s = 2\pi(1/2)(s/g)^{-1/2}(1/g) = \pi\sqrt{1/sg}$$
$$\partial T/\partial g = 2\pi(1/2)(s/g)^{-1/2}(-s/g^2) - \pi\sqrt{s/g^3}$$
$$dT = \pi\sqrt{1/sg}\, ds - \pi\sqrt{s/g^3}\, dg$$
$$= \pi\sqrt{1/98}(.2) - \pi\sqrt{10/(9.8)^3}(-.05)$$
$$\approx 0.08$$
The amount of error is 0.08 seconds.

29. Note that R_1, R_2, R_3 and R are nonnegative.
$$|dR_1| \le .1R_1, \ |dR_2| \le .1R_2 \text{ and } |dR_3| \le .1R_3$$
$$|-R^{-2}dR| = |-R_1^{-2}\, dR_1 - R_2^{-2}\, dR_2 - R_3^{-2}\, dR_3|$$
$$R^{-2}|dR| = |R_1^{-2}\, dR_1 + R_2^{-2}\, dR_2 + R_3^{-2}\, dR_3|$$
$$\le R_1^{-2}|dR_1| + R_2^{-2}|dR_2| + R_3^{-2}|dR_3|$$
$$\le .1R_1^{-1} + .1R_2^{-1} + .1R_3^{-1}$$
$$= .1\left[(1/R_1) + (1/R_2) + (1/R_3)\right]$$
$$= .1(1/R)$$
$$|dR|/R \le R(.1(1/R)) = .1$$
The maximum percentage change in R is 10%.

31. Let $P = (x_0, y_0, z_0)$ and $Q = (x_0 + \Delta x, y_0 + \Delta y, z_0 + \Delta z)$. Since B is a box, the points $S = (x_0 + \Delta x, y_0)$ and $T = (x_0 + \Delta x, y_0 + \Delta y)$ also lies in B, as do the line segments PS, ST and TQ.

Along the line segment PS the variables $y = y_0$ and $z = z_0$ are fixed, so $g_1(x) = f(x, y_0, z_0)$ is a function of x alone, and $g_1'(x) = \partial f/\partial x$ exists at (x, y_0, z_0) for all $x \in (x_0, x_0 + \Delta x)$. Thus, by the Mean Value Theorem, there exists a number $c_1 \in (x_0, x_0 + \Delta x)$ such that

$$\frac{f(x_0 + \Delta x, y_0, z_0) - f(x_0, y_0, z_0)}{\Delta x} = \frac{\partial f}{\partial x}(c_1, y_0, z_0).$$

Similarly, along the line segment ST, the function $g_2(y) = f(x_0 + \Delta x, y, z_0)$ is a function of y alone, and $g_2'(y) = \partial f/\partial y$ exists at $(x_0 + \Delta x, y, z_0)$ for all $y \in (y_0, y_0 + \Delta y)$. By the Mean Value Theorem, there exists a number $c_2 \in (y_0, y_0 + \Delta y)$ such that

$$\frac{f(x_0 + \Delta x, y_0 + \Delta y, z_0) - f(x_0 + \Delta x, y_0, z_0)}{\Delta y} = \frac{\partial f}{\partial y}(x_0 + \Delta x, c_2, z_0).$$

Likewise, along the line segment TQ, the function $g_3(z) = f(x_0 + \Delta x, y_0 + \Delta y, z)$ is a function of z alone, and $g_3'(z) = \partial f/\partial z$ exists at $(x_0 + \Delta x, y_0 + \Delta y, z)$ for all $z \in (z_0, z_0 + \Delta z)$. By the Mean Value Theorem, there exists a number $c_3 \in (z_0, z_0 + \Delta z)$ such that

$$\frac{f(x_0 + \Delta x, y_0 + \Delta y, z_0 + \Delta z) - f(x_0 + \Delta x, y_0 + \Delta y, z_0)}{\Delta z} = \frac{\partial f}{\partial z}(x_0 + \Delta x, y_0 + \Delta y, c_3).$$

Multiplying the three equations above by Δx, Δy and Δz respectively, and adding the resulting equations, gives

$$f(x_0 + \Delta x, y_0 + \Delta y, z_0 + \Delta z) - f(x_0, y_0, z_0) = \frac{\partial f}{\partial x}(c_1, y_0, z_0)\Delta x + \frac{\partial f}{\partial y}(x_0 + \Delta x, c_2, z_0)\Delta y$$
$$+ \frac{\partial f}{\partial z}(x_0 + \Delta x, y_0 + \Delta y, c_3)\Delta z.$$

Note that $c_1 \to x_0$, $c_2 \to y_0$ and $c_3 \to z_0$ as $\Delta x \to 0$, $\Delta y \to 0$ and $\Delta z \to 0$. Thus, as $\Delta x \to 0$, $\Delta y \to 0$ and $\Delta z \to 0$, and since the partial derivatives of f are continuous, it follows that $\partial f/\partial x(c_1, y_0, z_0) \to \partial f/\partial x(x_0, y_0, z_0)$, $\partial f/\partial y(x_0 + \Delta x, c_2, z_0) \to \partial f/\partial y(x_0, y_0, z_0)$ and $\partial f/\partial z(x_0 + \Delta x, y_0 + \Delta y, c_3) \to \partial f/\partial z(x_0, y_0, z_0)$. This may be rewritten as

$$\frac{\partial f}{\partial x}(c_1, y_0, z_0) = \frac{\partial f}{\partial x}(x_0, y_0, z_0) + \epsilon_1,$$

$$\frac{\partial f}{\partial y}(x_0 + \Delta x, c_2, z_0) = \frac{\partial f}{\partial y}(x_0, y_0, z_0) + \epsilon_2, \text{ and}$$

$$\frac{\partial f}{\partial z}(x_0 + \Delta x, y_0 + \Delta y, c_3) = \frac{\partial f}{\partial z}(x_0, y_0, z_0) + \epsilon_3,$$

where $\epsilon_1 \to 0$, $\epsilon_2 \to 0$ and $\epsilon_3 \to 0$ as $\Delta x \to 0$, $\Delta y \to 0$ and $\Delta z \to 0$.

Substituting these values in the equation above, and adding $f(x_0, y_0, z_0)$ to both sides, gives

$$f(x_0 + \Delta x, y_0 + \Delta y, z_0 + \Delta z) = f(x_0, y_0, z_0) + \frac{\partial f}{\partial x}(x_0, y_0, z_0)\Delta x$$
$$+ \frac{\partial f}{\partial y}(x_0, y_0, z_0)\Delta y + \frac{\partial f}{\partial z}(x_0, y_0, z_0)\Delta z$$
$$+ \epsilon_1\Delta x + \epsilon_2\Delta y + \epsilon_3\Delta z$$

where $\epsilon_1 \to 0$, $\epsilon_2 \to 0$ and $\epsilon_3 \to 0$ as $\Delta x \to 0$, $\Delta y \to 0$ and $\Delta z \to 0$, as desired.

18.6 Chain Rules

1. $\partial f/\partial t = (2x)2 + (2y)(-2t) = 4x - 4yt$
$= 4(2t) - 4(6 - t^2)t$
$= 4t^3 - 16t$

3. $\partial f/\partial t = (y^2)(-\sin t) + (2xy)(\cos t)$
$= -\sin^2 t \sin t + 2\cos t \sin t \cos t$
$= -\sin^3 t + 2\cos^2 t \sin t$

5. $\partial f/\partial t = [1/(2\sqrt{x+y})][1/(2\sqrt{t})] + [1/(2\sqrt{x+y})][2te^{t^2}]$
$= \dfrac{1}{2\sqrt{\sqrt{t}+e^{t^2}}}\left(\dfrac{1}{2\sqrt{t}} + 2te^{t^2}\right)$

7. $\partial f/\partial t = (y-1)(2t) + (x)(-2) + (2z)(\cos t)$
$= (-2t-1)(2t) - 2t^2 + 2\sin t \cos t$
$= -6t^2 - 2t + 2\sin t \cos t$

9. $\partial f/\partial t = (-2xz)(a\sinh t) + (2yz)(b\cosh t) + (y^2 - x^2)(-2e^{-2t})$
$= -2(a\cosh t)(e^{-2t})(a\sinh t) + 2(b\sinh t)(e^{-2t})(b\cosh t) - 2(b^2\sinh^2 t - a^2\cosh^2 t)e^{-2t}$
$= 2e^{-2t}[a^2\cosh^2 t + (b^2 - a^2)\cosh t \sinh t - b^2\sinh^2 t]$

11. $\partial f/\partial t = (2xy^3)(-\sin t) + (3x^2y^2)(\sin t + t\cos t)$
$= -2(\cos t)(t^3\sin^3 t)(\sin t) + 3(\cos^2 t)(t^2\sin^2 t)(\sin t + t\cos t)$
$= -2t^3\cos t \sin^4 t + 3t^2\cos^2 t \sin^3 t + 3t^3\cos^3 t \sin^2 t$

13. a. $\partial f/\partial s = (2x)(te^{st}) + (2y)(t) + (-2z)(1)$
$= 2(e^{st})te^{st} + 2(st)t - 2(s - t)$
$= 2te^{2st} + 2st^2 - 2s + 2t$

b. $\partial f/\partial t = (2x)(se^{st}) + (2y)(s) + (-2z)(-1)$
$= 2(e^{st})se^{st} + 2(st)s + 2(s - t)$
$= 2se^{2st} + 2s^2t + 2s - 2t$

15. a. $\partial f/\partial s = (y^2 \cos(xy^2) - 2xy)(2s - t) + (2xy \cos(xy^2) - x^2)(2st^2)$

$= [(s^2t^2)^2 \cos((s^2 - st)(s^2t^2)^2) - 2(s^2 - st)(s^2t^2)](2s - t) +$
$\quad [2(s^2 - st)(s^2t^2) \cos((s^2 - st)(s^2t^2)^2) - (s^2 - st)^2]2st^2$

$= [s^4t^4 \cos(s^6t^4 - s^5t^5) - 2s^4t^2 + 2s^3t^3](2s - t) +$
$\quad [2s^4t^2 \cos(s^6t^4 - s^5t^5) - 2s^3t^3 \cos(s^6t^4 - s^5t^5) - s^4 + 2s^3t - s^2t^2]2st^2$

$= 2s^5t^4 \cos(s^6t^4 - s^5t^5) - 4s^5t^2 + 4s^4t^3 - s^4t^5 \cos(s^6t^4 - s^5t^5) + 2s^4t^3 - 2s^3t^4 +$
$\quad 4s^5t^4 \cos(s^6t^4 - s^5t^5) - 4s^4t^5 \cos(s^6t^4 - s^5t^5) - 2s^5t^2 + 4s^4t^3 - 2s^3t^4$

$= (6s^5t^4 - 5s^4t^5) \cos(s^6t^4 - s^5t^5) - 6s^5t^2 + 10s^4t^3 - 4s^3t^4$

b. $\partial f/\partial t = (y^2 \cos(xy^2) - 2xy)(-s) + (2xy \cos(xy^2) - x^2)(2s^2t)$

$= [(s^2t^2)^2 \cos((s^2 - st)(s^2t^2)^2) - 2(s^2 - st)(s^2t^2)](-s) +$
$\quad [2(s^2 - st)(s^2t^2) \cos((s^2 - st)(s^2t^2)^2) - (s^2 - st)^2]2s^2t$

$= [s^4t^4 \cos(s^6t^4 - s^5t^5) - 2s^4t^2 + 2s^3t^3](-s) +$
$\quad [2s^4t^2 \cos(s^6t^4 - s^5t^5) - 2s^3t^3 \cos(s^6t^4 - s^5t^5) - s^4 + 2s^3t - s^2t^2]2s^2t$

$= -s^5t^4 \cos(s^6t^4 - s^5t^5) + 2s^5t^2 - 2s^4t^3 + 4s^6t^3 \cos(s^6t^4 - s^5t^5) - 4s^5t^4 \cos(s^6t^4 - s^5t^5) -$
$\quad 2s^6t + 4s^5t^2 - 2s^4t^3$

$= 4s^6t^3 \cos(s^6t^4 - s^5t^5) - 5s^5t^4 \cos(s^6t^4 - s^5t^5) + 6s^5t^2 - 4s^4t^3 - 2s^6t$

$= (4s^6t^3 - 5s^5t^4) \cos(s^6t^4 - s^5t^5) + 6s^5t^2 - 4s^4t^3 - 2s^6t$

17. a. $\partial f/\partial s = (y^3z^2)(\sin t) + (3xy^2z^2)(-t \sin s) + (2xy^3z)(-2s)$

$= (t^3 \cos^3 s)(t^2 - s^2)^2 \sin t - 3t(s \sin t)(t^2 \cos^2 s)(t^2 - s^2)^2 \sin s - 4s(s \sin t)(t^3 \cos^3 s)(t^2 - s^2)$

$= t^3(t^4 - 2s^2t^2 + s^4) \cos^3 s \sin t - 3st^3(t^4 - 2s^2t^2 + s^4) \cos^2 s \sin s \sin t - 4s^2t^3(t^2 - s^2) \cos^3 s \sin t$

$= (t^7 - 2s^2t^5 + s^4t^3) \cos^3 s \sin t + (-3st^7 + 6s^3t^5 - 3s^5t^3) \cos^2 s \sin s \sin t + (-4s^2t^5 + 4s^4t^3) \cos^3 s \sin t$

$= (t^7 - 6s^2t^5 + 5s^4t^3) \cos^3 s \sin t + (-3st^7 + 6s^3t^5 - 3s^5t^3) \cos^2 s \sin s \sin t$

b. $\partial f/\partial t = (y^3z^2)(s \cos t) + (3xy^2z^2)(\cos s) + (2xy^3z)(2t)$

$= s(t^3 \cos^3 s)(t^2 - s^2)^2 \cos t + 3(s \sin t)(t^2 \cos^2 s)(t^2 - s^2)^2 \cos s + 4t(s \sin t)(t^3 \cos^3 s)(t^2 - s^2)$

$= st^3(t^4 - 2s^2t^2 + s^4) \cos^3 s \cos t + 3st^2(t^4 - 2s^2t^2 + s^4) \cos^3 s \sin t + 4st^4(t^2 - s^2) \cos^3 s \sin t$

$= (st^7 - 2s^3t^5 + s^5t^3) \cos^3 s \cos t + (3st^6 - 6s^3t^4 + 3s^5t^2) \cos^3 s \sin t + (4st^6 - 4s^3t^4) \cos^3 s \sin t$

$= (st^7 - 2s^3t^5 + s^5t^3) \cos^3 s \cos t + (7st^6 - 10s^3t^4 + 3s^5t^2) \cos^3 s \sin t$

19. a. $\partial f/\partial x = (2ue^{u^2})(y^2) + (-2e^{2v})(2xy)$

$= 2(xy^2)e^{x^2y^4}y^2 - 4e^{2x^2y}xy$

$= 2xy^4 e^{x^2y^4} - 4xy e^{2x^2y}$

b. $\partial f/\partial y = (2ue^{u^2})(2xy) + (-2e^{2v})(x^2)$

$= 4(xy^2)e^{x^2y^4}xy - 2e^{2x^2y}x^2$

$= 4x^2y^3 e^{x^2y^4} - 2x^2 e^{2x^2y}$

21. $\partial f/\partial t = [2r(1 - \cos \theta)](3t^2) + (r^2 \sin \theta)(t/\sqrt{1 + t^2})$

$= 6(1 + t^3)[1 - \cos(\sqrt{1 + t^2})]t^2 + [(1 + t^3)^2 t(\sin \sqrt{1 + t^2})/\sqrt{1 + t^2}]$

$= 6t^2 + 6t^5 - (6t^2 + 6t^5) \cos(\sqrt{1 + t^2}) + [(t + 2t^4 + t^7)(\sin \sqrt{1 + t^2})/\sqrt{1 + t^2}]$

23. The volume is $V = \pi r^2 h/3$.

$dV/dt = (2\pi rh/3)(dr/dt) + (\pi r^2/3)(dh/dt)$
$\quad = [2\pi(6)(10)/3]2 + [\pi(6)^2/3]3$
$\quad = 116\pi$ cm^3/s

25. The area is $A = lw$. Two hours later, the length is 6 inches, and the width is 5 inches.

$dA/dt = (w)(dl/dt) + (l)(dw/dt)$
$\quad = 5(.5) + 6(1)$
$\quad = 8.5$ in^2/h

27. Given $S = \pi r\sqrt{r^2 + h^2}$.

$dS/dt = [\pi\sqrt{r^2 + h^2} + (\pi r^2/\sqrt{r^2 + h^2})](dr/dt) + (\pi rh/\sqrt{r^2 + h^2})(dh/dt)$

$= [\pi\sqrt{3^2 + 10^2} + (\pi 3^2/\sqrt{3^2 + 10^2})](1) + (\pi(3)(10)/\sqrt{3^2 + 10^2})(3)$

$= \pi\sqrt{109} + (9\pi\sqrt{109}/109) + 90\pi\sqrt{109}/109$

$= 208\pi\sqrt{109}/109 \text{ cm}^2/\text{s}$

29. a. $D(x, y, z) = \sqrt{x^2 + y^2 + z^2}$ where $x = \cos t$, $y = \sin t$ and $z = t$.

b. $dD/dt = [x/\sqrt{x^2 + y^2 + z^2}](-\sin t) + [y/\sqrt{x^2 + y^2 + z^2}](\cos t) + [z/\sqrt{x^2 + y^2 + z^2}](1)$

$= (-x\sin t + y\cos t + z)/\sqrt{x^2 + y^2 + z^2}$

$= (-\cos t\sin t + \sin t\cos t + t)/\sqrt{\cos^2 t + \sin^2 t + t^2}$

$= t/\sqrt{1 + t^2}$

c. For $t > 0$, $dD/dt = \sqrt{t^2/(1 + t^2)} = [(1 + t^2)/t^2]^{-1/2} = (t^{-2} + 1)^{-1/2}$

For $t < 0$, $dD/dt = -\sqrt{t^2/(1 + t^2)} = -(t^{-2} + 1)^{-1/2}$

Both results follow from the fact that:

$\lim_{t \to \pm\infty}(t^{-2} + 1)^{-1/2} = 1$

31. Given $M = C + D$, $dC/dt = cC$ and $dD/dt = dD$.

The rate of growth m of M is:

$$m = \frac{dM/dt}{M} = \frac{1}{C + D}\left(\frac{\partial M}{\partial C}\frac{dC}{dt} + \frac{\partial M}{\partial D}\frac{dD}{dt}\right) = \frac{1}{C + D}[1(cC) + 1(dD)] = \frac{cC}{C + D} + \frac{dD}{C + D}$$

33. a. $\partial f/\partial s = (2x)(2st) + (4y^3)(-2s)$

$= 4(s^2 t)st - 8s(t^2 - s^2)^3$

$= 4s^3 t^2 - 8s(t^6 - 3t^4 s^2 + 3t^2 s^4 - s^6)$

$= 4s^3 t^2 - 8st^6 + 24s^3 t^4 - 24s^5 t^2 + 8s^7$

$\partial^2 f/\partial s^2 = 12s^2 t^2 - 8t^6 + 72s^2 t^4 - 120s^4 t^2 + 56s^6$

b. $\partial f/\partial t = (2x)(s^2) + (4y^3)(2t)$

$= 2(s^2 t)s^2 + 8t(t^2 - s^2)^3$

$= 2s^4 t + 8t(t^6 - 3s^2 t^4 + 3s^4 t^2 - s^6)$

$= 2s^4 t + 8t^7 - 24s^2 t^5 + 24s^4 t^3 - 8s^6 t$

$\partial^2 f/\partial t^2 = 2s^4 + 56t^6 - 120s^2 t^4 + 72s^4 t^2 - 8s^6$

35. $\partial f/\partial r = (y)(st) + (x - 2y)(ste^{rst})$

$= (e^{rst})st + (rst - 2e^{rst})ste^{rst}$

$= ste^{rst} + rs^2 t^2 e^{rst} - 2ste^{2rst}$

$\partial^2 f/\partial r^2 = s^2 t^2 e^{rst} + (s^2 t^2 e^{rst} + rs^3 t^3 e^{rst}) - 4s^2 t^2 e^{2rst}$

$= 2s^2 t^2 e^{rst} + rs^3 t^3 e^{rst} - 4s^2 t^2 e^{2rst}$

566

37. From Exercise 36:

$$\frac{\partial^2 f}{\partial s \partial t} = \frac{\partial^2 f}{\partial x^2}\frac{\partial x}{\partial s}\frac{\partial x}{\partial t} + \frac{\partial^2 f}{\partial x \partial y}\left(\frac{\partial x}{\partial t}\frac{\partial y}{\partial s} + \frac{\partial x}{\partial s}\frac{\partial y}{\partial t}\right) + \frac{\partial^2 f}{\partial y^2}\frac{\partial y}{\partial s}\frac{\partial y}{\partial t} + \frac{\partial f}{\partial x}\frac{\partial^2 x}{\partial s \partial t} + \frac{\partial f}{\partial y}\frac{\partial^2 y}{\partial s \partial t}$$

For $f(x,y) = xe^{y^2}$ one has $\partial f/\partial x = e^{y^2}$, $\partial f/\partial y = 2xye^{y^2}$, $\partial^2 f/\partial x^2 = 0$, $\partial^2 f/(\partial x \partial y) = 2ye^{y^2}$ and $\partial^2 f/\partial y^2 = (2x + 4xy^2)e^{y^2}$. For $x(r, \theta) = r\cos\theta$, one has $\partial x/\partial r = \cos\theta$, $\partial x/\partial \theta = -r\sin\theta$ and $\partial^2 x/\partial r\partial\theta = -\sin\theta$. For $y(r, \theta) = r\sin\theta$, one has $\partial y/\partial r = \sin\theta$, $\partial y/\partial\theta = r\cos\theta$ and $\partial^2 y/\partial r\partial\theta = \cos\theta$. Using the formula gives:

$$\partial^2 f/\partial r\partial\theta = (0)(\cos\theta)(-r\sin\theta) + (2ye^{y^2})[(-r\sin\theta)(\sin\theta) + (\cos\theta)(r\cos\theta)] +$$
$$(2x + 4xy^2)e^{y^2}(\sin\theta)(r\cos\theta) + (e^{y^2})(-\sin\theta) + (2xye^{y^2})(\cos\theta)$$

$$= \left[2ry(\cos^2\theta - \sin^2\theta) + r(2x + 4xy^2)\cos\theta\sin\theta - \sin\theta + 2xy\cos\theta\right]e^{y^2}$$

$$= \left[2r^2(\sin\theta)(\cos^2\theta - \sin^2\theta) + r(2r\cos\theta + 4r^3\cos\theta\sin^2\theta)\cos\theta\sin\theta - \right.$$
$$\left.\sin\theta + 2r^2(\cos\theta\sin\theta)\cos\theta\right]e^{r^2\sin^2\theta}$$

$$= \left(6r^2\cos^2\theta\sin\theta - 2r^2\sin^3\theta + 4r^4\cos^2\theta\sin^3\theta - \sin\theta\right)e^{r^2\sin^2\theta}$$

39. Use $x = r\cos\theta$ and $y = r\sin\theta$.

$$\frac{\partial v}{\partial r} = \frac{\partial v}{\partial x}\frac{\partial x}{\partial r} + \frac{\partial v}{\partial y}\frac{\partial x}{\partial r} + \frac{\partial v}{\partial z}\frac{\partial z}{\partial r} = \frac{\partial v}{\partial x}\cos\theta + \frac{\partial v}{\partial y}\sin\theta \quad (\frac{\partial z}{\partial r} = 0)$$

$$\frac{\partial^2 v}{\partial r^2} = \frac{\partial}{\partial r}\frac{\partial v}{\partial r}$$

$$= \left(\frac{\partial}{\partial r}\frac{\partial v}{\partial x}\right)\cos\theta + \left(\frac{\partial}{\partial r}\frac{\partial v}{\partial y}\right)\sin\theta$$

$$= \left(\frac{\partial^2 v}{\partial x^2}\cos\theta + \frac{\partial^2 v}{\partial y\partial x}\sin\theta\right)\cos\theta + \left(\frac{\partial^2 v}{\partial x\partial y}\cos\theta + \frac{\partial^2 v}{\partial y^2}\sin\theta\right)\sin\theta$$

$$= \frac{\partial^2 v}{\partial x^2}\cos^2\theta + 2\frac{\partial^2 v}{\partial x\partial y}\cos\theta\sin\theta + \frac{\partial^2 v}{\partial y^2}\sin^2\theta$$

$$\frac{\partial v}{\partial\theta} = \frac{\partial v}{\partial x}\frac{\partial x}{\partial\theta} + \frac{\partial v}{\partial y}\frac{\partial y}{\partial\theta} + \frac{\partial v}{\partial z}\frac{\partial z}{\partial\theta}$$

$$= \frac{\partial v}{\partial x}(-r\sin\theta) + \frac{\partial v}{\partial y}(r\cos\theta) + 0$$

$$= -\frac{\partial v}{\partial x}r\sin\theta + \frac{\partial v}{\partial y}r\cos\theta$$

$$\frac{\partial^2 v}{\partial\theta^2} = \frac{\partial}{\partial\theta}\frac{\partial v}{\partial\theta}$$

$$= -\left(\frac{\partial}{\partial\theta}\frac{\partial v}{\partial x}\right)r\sin\theta - \frac{\partial v}{\partial x}r\left(\frac{\partial}{\partial\theta}\sin\theta\right) + \left(\frac{\partial}{\partial\theta}\frac{\partial v}{\partial y}\right)r\cos\theta + \frac{\partial v}{\partial y}r\left(\frac{\partial}{\partial\theta}\cos\theta\right)$$

$$= -\left(\frac{\partial^2 v}{\partial x^2}(-r\sin\theta) + \frac{\partial^2 v}{\partial y\partial x}r\cos\theta\right)r\sin\theta - \frac{\partial v}{\partial x}r\cos\theta +$$
$$\left(\frac{\partial^2 v}{\partial x\partial y}(-r\sin\theta) + \frac{\partial^2 v}{\partial y^2}r\cos\theta\right)r\cos\theta + \frac{\partial v}{\partial y}r(-\sin\theta)$$

$$= \frac{\partial^2 v}{\partial x^2}r^2\sin^2\theta - 2\frac{\partial^2 v}{\partial x\partial y}r^2\cos\theta\sin\theta + \frac{\partial^2 v}{\partial y^2}r^2\cos^2\theta - \frac{\partial v}{\partial x}r\cos\theta - \frac{\partial v}{\partial y}r\sin\theta$$

Substituting gives:

$$\frac{\partial^2 v}{\partial r^2} + \frac{1}{r}\frac{\partial v}{\partial r} + \frac{1}{r^2}\frac{\partial^2 v}{\partial \theta^2} + \frac{\partial^2 v}{\partial^2 z}$$

$$= \left(\frac{\partial^2 v}{\partial x^2}\cos^2\theta + 2\frac{\partial^2 v}{\partial x \partial y}\cos\theta\sin\theta + \frac{\partial^2 v}{\partial y^2}\sin^2\theta\right) + \frac{1}{r}\left(\frac{\partial v}{\partial x}\cos\theta + \frac{\partial v}{\partial y}\sin\theta\right) +$$

$$\quad \frac{1}{r^2}\left(\frac{\partial^2 v}{\partial x^2}r^2\sin^2\theta - 2\frac{\partial^2 v}{\partial x \partial y}r^2\cos\theta\sin\theta + \frac{\partial^2 v}{\partial y^2}r^2\cos^2\theta - \frac{\partial v}{\partial x}r\cos\theta - \frac{\partial v}{\partial y}r\sin\theta\right) + \frac{\partial^2 v}{\partial^2 z}$$

$$= \frac{1}{r^2}\left(\frac{\partial^2 v}{\partial x^2}r^2\cos^2\theta + 2\frac{\partial^2 v}{\partial x \partial y}(r\cos\theta)(r\sin\theta) + \frac{\partial^2 v}{\partial y^2}r^2\sin^2\theta\right) + \frac{1}{r^2}\left(\frac{\partial v}{\partial x}r\cos\theta + \frac{\partial v}{\partial y}r\sin\theta\right) +$$

$$\quad \frac{1}{r^2}\left(\frac{\partial^2 v}{\partial x^2}r^2\sin^2\theta - 2\frac{\partial^2 v}{\partial x \partial y}(r\cos\theta)(r\sin\theta) + \frac{\partial^2 v}{\partial y^2}r^2\cos^2\theta - \frac{\partial v}{\partial x}r\cos\theta - \frac{\partial v}{\partial y}r\sin\theta\right) + \frac{\partial^2 v}{\partial^2 z}$$

$$= \frac{1}{x^2+y^2}\left(\frac{\partial^2 v}{\partial x^2}x^2 + 2\frac{\partial^2 v}{\partial x \partial y}xy + \frac{\partial^2 v}{\partial y^2}y^2 + \frac{\partial v}{\partial x}x + \frac{\partial v}{\partial y}y + \frac{\partial^2 v}{\partial x^2}y^2 - 2\frac{\partial^2 v}{\partial x \partial y}xy + \frac{\partial^2 v}{\partial y^2}x^2 - \frac{\partial v}{\partial x}x - \frac{\partial v}{\partial y}y\right)$$

$$\quad + \frac{\partial^2 v}{\partial^2 z}$$

$$= \frac{1}{x^2+y^2}\left(\frac{\partial^2 v}{\partial x^2}(x^2+y^2) + \frac{\partial^2 v}{\partial y^2}(x^2+y^2)\right) + \frac{\partial^2 v}{\partial^2 z}$$

$$= \frac{\partial^2 v}{\partial x^2} + \frac{\partial^2 v}{\partial y^2} + \frac{\partial^2 v}{\partial^2 z}$$

18.7   Directional Derivatives

1. $\nabla f(x,y) = 2xy\mathbf{i} + x^2\mathbf{j}$
$\nabla f(3,1) = 6\mathbf{i} + 9\mathbf{j}$

3. $\nabla f(x,y) = [\cos(y-x) + x\sin(y-x)]\mathbf{i} - x\sin(y-x)\mathbf{j}$
$\nabla f(\pi/2, \pi/4) = [\sqrt{2}/2 + (\pi/2)(-\sqrt{2}/2)]\mathbf{i} - (\pi/2)(-\sqrt{2}/2)\mathbf{j}$
$\quad = [(2-\pi)\sqrt{2}/4]\mathbf{i} + (\pi\sqrt{2}/4)\mathbf{j}$

5. $\nabla f(x,y,z) = (2xy + z^2)\mathbf{i} + x^2\mathbf{j} + 2xz\mathbf{k}$
$\nabla f(1,1,2) = 6\mathbf{i} + \mathbf{j} + 4\mathbf{k}$

7. $\nabla f(x,y,z) = (2xy + z^3)\mathbf{i} + (x^2 - 2yz)\mathbf{j} + (3xz^2 - y^2)\mathbf{k}$
$\nabla f(1,2,1) = 5\mathbf{i} - 3\mathbf{j} - \mathbf{k}$

9. $\nabla f(x,y,z) = \sqrt{y}\cosh z\,\mathbf{i} + [(x\cosh z)/(2\sqrt{y})]\mathbf{j} + x\sqrt{y}\sinh z\,\mathbf{k}$
$\nabla f(2,4,1) = 2[(e + e^{-1})/2]\mathbf{i} + 2[(e + e^{-1})/2](1/4)\mathbf{j} + 4[(e - e^{-1})/2]\mathbf{k}$
$\quad = (e + e^{-1})\mathbf{i} + [(e + e^{-1})/4]\mathbf{j} + 2(e - e^{-1})\mathbf{k}$

11. $\nabla f(x,y) = 2xy^2\mathbf{i} + 2x^2y\mathbf{j}$
$\nabla f(-2,3) = -36\mathbf{i} + 24\mathbf{j}$
$\mathbf{u} = (1/\sqrt{2})(\mathbf{i} - \mathbf{j})$
$D_{\mathbf{u}}f(-2,3) = \nabla f(-2,3) \cdot \mathbf{u}$
$\quad = (1/\sqrt{2})(-36 - 24) = -30\sqrt{2}$

13. $\nabla f(x,y) = ([(x+y)1 - x(1)]/(x+y)^2)\,\mathbf{i} - [x/(x+y)^2]\mathbf{j}$
$\nabla f(1,2) = (2/9)\mathbf{i} - (1/9)\mathbf{j} = (1/9)(2\mathbf{i} - \mathbf{j})$
$\mathbf{u} = (1/2)(\sqrt{3}\,\mathbf{i} + \mathbf{j})$
$D_{\mathbf{u}}f(1,2) = \nabla f(1,2) \cdot \mathbf{u} = (1/18)(2\sqrt{3} - 1) = (2\sqrt{3} - 1)/18$

15. $\nabla f(x, y, z) = (y + z)\mathbf{i} + (x + z)\mathbf{j} + (x + y)\mathbf{k}$
$\nabla f(1, 2, 1) = 3\mathbf{i} + 2\mathbf{j} + 3\mathbf{k}$
$\mathbf{u} = (1/\sqrt{3})(\mathbf{i} + \mathbf{j} - \mathbf{k})$
$D_{\mathbf{u}}f(1, 2, 1) = \nabla f(1, 2, 1) \cdot \mathbf{u}$
$= (1/\sqrt{3})(3 + 2 - 3)$
$= 2\sqrt{3}/3$

17. $\nabla f(x, y, z) = e^{yz}\mathbf{i} + xze^{yz}\mathbf{j} + xye^{yz}\mathbf{k}$
$\nabla f(2, 0, 1) = \mathbf{i} + 2\mathbf{j} \quad \mathbf{u} = (1/\sqrt{11})(\sqrt{3}\mathbf{i} + \sqrt{3}\mathbf{j} - \sqrt{5}\mathbf{k})$
$D_{\mathbf{u}}f(2, 0, 1) = \nabla f(2, 0, 1) \cdot \mathbf{u} = (1/\sqrt{11})(\sqrt{3} + 2\sqrt{3}) = 3\sqrt{33}/11$

19. $\nabla f(x, y, z) = (\cos y)\mathbf{i} + (-x \sin y - \sinh z)\mathbf{j} + (-y \cosh z)\mathbf{k}$
$\nabla f(6, \pi/4, -1) = (\sqrt{2}/2)\mathbf{i} + [(-6\sqrt{2}/2) - (e^{-1} - e)/2]\mathbf{j} - [(\pi/4)(e^{-1} + e)/2]\mathbf{k}$
$= (\sqrt{2}/2)\mathbf{i} + [(-6\sqrt{2} - e^{-1} + e)/2]\mathbf{j} - [\pi(e^{-1} + e)/8]\mathbf{k}$
$\mathbf{u} = (1/3)(\mathbf{i} - 2\mathbf{j} - 2\mathbf{k})$
$D_{\mathbf{u}}f(6, \pi/4, -1) = \nabla f(6, \pi/4, -1) \cdot \mathbf{u} = (1/3)[(\sqrt{2}/2) - 2[(-6\sqrt{2} - e^{-1} + e)/2] + 2[\pi(e^{-1} + e)/8]$
$= (1/3)[(\sqrt{2}/2) + 6\sqrt{2} + e^{-1} - e + \pi(e^{-1} + e)/4]$
$= (26\sqrt{2} + 4e^{-1} + \pi e^{-1} - 4e + \pi e)/12$

21. Let $f(x, y) = 4x^2 - y^2$.
$\nabla f(x, y) = 8x\mathbf{i} - 2y\mathbf{j}$
$\mathbf{n} = \nabla f(2, 3) = 16\mathbf{i} - 6\mathbf{j}$
Normal line: $\mathbf{r}(t) = (2\mathbf{i} + 3\mathbf{j}) + t(16\mathbf{i} - 6\mathbf{j})$
Tangent line: $\mathbf{r}(t) = (2\mathbf{i} + 3\mathbf{j}) + t(6\mathbf{i} + 16\mathbf{j})$

23. Let $f(x, y) = \sqrt{x} + \sqrt{y}$.
$\nabla f(x, y) = (1/2\sqrt{x})\mathbf{i} + (1/2\sqrt{y})\mathbf{j}$
$\mathbf{n} = \nabla f(4, 4) = (1/4)\mathbf{i} + (1/4)\mathbf{j}$
$4\mathbf{n} = \mathbf{i} + \mathbf{j}$
Normal line: $\mathbf{r}(t) = (4\mathbf{i} + 4\mathbf{j}) + t(\mathbf{i} + \mathbf{j})$
Tangent line: $\mathbf{r}(t) = (4\mathbf{i} + 4\mathbf{j}) + t(\mathbf{i} - \mathbf{j})$

25. Let $f(x, y) = \sin x - \cos y$.
$\nabla f(x, y) = \cos x\, \mathbf{i} + \sin y\, \mathbf{j}$
$\mathbf{n} = \nabla f(\pi/2, \pi/2) = \mathbf{j}$
Normal line: $\mathbf{r}(t) = (\pi/2)\mathbf{i} + (\pi/2)\mathbf{j} + t\mathbf{j}$
Tangent line: $\mathbf{r}(t) = (\pi/2)\mathbf{i} + (\pi/2)\mathbf{j} + t\mathbf{i}$

27. Let $f(x, y) = \ln x + 2 \ln y$.
$\nabla f(x, y) = (1/x)\mathbf{i} + (2/y)\mathbf{j}$
$\mathbf{n} = \nabla f(1, \sqrt{e}) = \mathbf{i} + (2/\sqrt{e})\mathbf{j}$
$\sqrt{e}\,\mathbf{n} = \sqrt{e}\,\mathbf{i} + 2\mathbf{j}$
Normal line: $\mathbf{r}(t) = (\mathbf{i} + \sqrt{e}\,\mathbf{j}) + t(\sqrt{e}\,\mathbf{i} + 2\mathbf{j})$
Tangent line: $\mathbf{r}(t) = (\mathbf{i} + \sqrt{e}\,\mathbf{j}) + t(2\mathbf{i} - \sqrt{e}\,\mathbf{j})$

29. $\nabla f(x, y) = (y^2 + ye^x)\mathbf{i} + (2xy + e^x)\mathbf{j}$
$\nabla f(0, 1) = 2\mathbf{i} + \mathbf{j}$
$\mathbf{u}$ is parallel to $\nabla f(0, 1)$
$\mathbf{u} = (1/\sqrt{5})(2\mathbf{i} + \mathbf{j})$
$D_{\mathbf{u}}f(0, 1) = \nabla f(0, 1) \cdot \mathbf{u}$
$= (1/\sqrt{5})(4 + 1)$
$= \sqrt{5}$

31. $\nabla f(x, y) = (y^2 + ye^x)\mathbf{i} + (2xy + e^x)\mathbf{j}$
$\nabla f(0, 1) = 2\mathbf{i} + \mathbf{j}$
$\mathbf{u} = -\mathbf{j}$
$D_{\mathbf{u}}f(0, 1) = \nabla f(0, 1) \cdot \mathbf{u} = -1$

33. $\nabla f(x, y) = e^x \sin y\, \mathbf{i} + e^x \cos y\, \mathbf{j}$
$\nabla f(0, \pi/4) = (\sqrt{2}/2)\mathbf{i} + (\sqrt{2}/2)\mathbf{j} = (\sqrt{2}/2)(\mathbf{i} + \mathbf{j})$
$\mathbf{u}$ is parallel to $(1 - 0)\mathbf{i} + (\pi/2 - \pi/4)\mathbf{j} = \mathbf{i} + (\pi/4)\mathbf{j}$
$\mathbf{u} = (1/\sqrt{16 + \pi^2})(4\mathbf{i} + \pi\mathbf{j})$
$D_{\mathbf{u}}f(0, \pi/4) = \nabla f(0, \pi/4) \cdot \mathbf{u} = (\sqrt{2}/2\sqrt{16 + \pi^2})(4 + \pi) = (4 + \pi)\sqrt{32 + 2\pi^2}/(32 + 2\pi^2)$

35. $\nabla f(x, y) = ([(x + y)1 - x(1)]/(x + y)^2)\mathbf{i} - [x/(x + y)^2]\mathbf{j}$
$\nabla f(1, 1) = (1/4)\mathbf{i} - (1/4)\mathbf{j} = (1/4)(\mathbf{i} - \mathbf{j})$
$\mathbf{u}$ is parallel to $\nabla f(1, 1)$
$\mathbf{u} = (1/\sqrt{2})(\mathbf{i} - \mathbf{j})$
$D_{\mathbf{u}}f(1, 1) = \nabla f(1, 1) \cdot \mathbf{u} = (1/4\sqrt{2})(1 + 1) = \sqrt{2}/4$

37. Let $f(x,y) = x^2 - 6x + 2y^2 - 4y$. The tangent is horizontal when the normal is vertical, which occurs when $\nabla f(x,y)$ is vertical, or parallel to $\mathbf{j}$. The tangent is vertical when the normal is horizontal, which occurs when $\nabla f(x,y)$ is horizontal, or parallel to $\mathbf{i}$.

$\nabla f(x,y) = (2x - 6)\mathbf{i} + (4y - 4)\mathbf{j}$

This is parallel to $\mathbf{j}$ if $2x - 6 = 0$, which occurs when $x = 3$, and is parallel to $\mathbf{i}$ if $4y - 4 = 0$, which occurs when $y = 1$.

a. Substituting $x = 3$ into the equation gives:

$9 - 18 + 2y^2 - 4y = -7$

$2y^2 - 4y - 2 = 0$

$y = (4 \pm \sqrt{16 - 4(2)(-2)})/4 = (4 \pm \sqrt{32})/4 = (4 \pm 4\sqrt{2})/4 = 1 \pm \sqrt{2}$

The two points on the ellipse where the tangent is horizontal are $(3, 1 - \sqrt{2})$ and $(3, 1 + \sqrt{2})$.

b. Substituting $y = 1$ into the equation gives:

$x^2 - 6x + 2 - 4 = -7$

$0 = x^2 - 6x + 5 = (x - 5)(x - 1)$

Thus $x = 1$ or $x = 5$. The two points on the ellipse where the tangent is vertical are $(1, 1)$ and $(5, 1)$.

39. Let $f(x,y) = x \ln y$. The tangent is parallel to the line $y = -x$ when it is parallel to the vector $\mathbf{i} - \mathbf{j}$. When this occurs, the normal is parallel to $\mathbf{i} + \mathbf{j}$, and thus $\nabla f(x,y)$ is also parallel to $\mathbf{i} + \mathbf{j}$.

$\nabla f(x,y) = \ln y \, \mathbf{i} + (x/y)\mathbf{j}$

This is parallel to $\mathbf{i} + \mathbf{j}$ if and only if $\ln y = x/y$. Since x and y must satisfy $x \ln y = 1$, one has $1 = x \ln y = x(x/y) = x^2/y$, or $y = x^2$. Substituting $y = x^2$ into the equation gives $x \ln(x^2) = 1$, or $2x \ln x = 1$. Solving this equation is equivalent to finding a zero for the function $g(x) = 2x \ln x - 1 = 0$. Some relevant information about this function is that:

the domain of $g(x)$ is $x > 0$,

$g'(x) = 2 \ln x + 2$,

the only critical number, which satisfies $g'(x) = 0$, is $x = e^{-1}$,

$g'(x)$ is negative on $(0, e^{-1})$ and positive on (e^{-1}, ∞),

$g(x)$ is decreasing on $(0, e^{-1}]$ with upper bound $\lim_{x \to 0+} g(x) = -1$, and

$g(x)$ is increasing on $[e^{-1}, \infty)$ with $g(e^{-1}) = -2e^{-1} - 1 < 0$ and $\lim_{x \to \infty} g(x) = \infty$.

Consequently, $g(x)$ has a unique zero, which must be in the interval (e^{-1}, ∞). This zero may be approximated by a technique such as Newton's method, yielding $x \approx 1.421529936$, and $y = x^2 \approx 2.020747359$. Therefore, there is a unique point on the graph of $x \ln y = 1$ with tangent parallel to the line $y = -x$, and the approximate coordinates of this point are $(1.421529936, 2.020747359)$.

41. Let $f(x,y,z) = xyz$.

$\nabla f(x,y,z) = yz\mathbf{i} + xz\mathbf{j} + xy\mathbf{k}$

$\mathbf{n} = \nabla f(2,1,3) = 3\mathbf{i} + 6\mathbf{j} + 2\mathbf{k}$

Tangent plane: $3x + 6y + 2z = d$

$d = 3(2) + 6(1) + 2(3) = 18$ since P is on the plane.

Tangent plane: $3x + 6y + 2z = 18$

43. $x^2 - y^3 + xy - z = 0$

Let $f(x,y,z) = x^2 - y^3 + xy - z$.

$\nabla f(x,y,z) = (2x + y)\mathbf{i} + (-3y^2 + x)\mathbf{j} - \mathbf{k}$

$\mathbf{n} = \nabla f(2,1,5) = 5\mathbf{i} - \mathbf{j} - \mathbf{k}$

Tangent plane: $5x - y - z = d$

$d = 5(2) - (1) - (5) = 4$ since P is on the plane.

Tangent plane: $5x - y - z = 4$

45. $y - \sin x = 0$

Let $f(x,y,z) = y - \sin x$.

$\nabla f(x,y,z) = -\cos x \, \mathbf{i} + \mathbf{j}$

$\mathbf{n} = \nabla f(\pi/2, 1, 5) = \mathbf{j}$

Tangent plane: $y = d$

$d = 1$ since P is on the plane.

Tangent plane: $y = 1$

47. Let $f(x,y,z) = xy^2 + x^2 y - xz$.

$\nabla f(x,y,z) = (y^2 + 2xy - z)\mathbf{i} + (2xy + x^2)\mathbf{j} - x\mathbf{k}$

$\mathbf{n} = \nabla f(1,2,3) = 5\mathbf{i} + 5\mathbf{j} - \mathbf{k}$

$|\mathbf{n}| = \sqrt{25 + 25 + 1} = \sqrt{51}$

A unit normal is:

$(5\sqrt{51}/51)\mathbf{i} + (5\sqrt{51}/51)\mathbf{j} - (\sqrt{51}/51)\mathbf{k}$.

49. First note that this point is on both the cylinder and the sphere. Then show that the cylinder and the sphere have parallel normal vectors at this point.

For the cylinder:

Let $f(x, y, z) = x^2 + y^2$

$\nabla f(x, y, z) = 2x\mathbf{i} + 2y\mathbf{j}$

$\mathbf{n}_1 = \nabla f(0, 2, 3) = 4\mathbf{j}$

For the sphere:

Let $g(x, y, z) = x^2 + y^2 + z^2 - 8y - 6z + 21$

$\nabla g(x, y, z) = 2x\mathbf{i} + (2y - 8)\mathbf{j} + (2z - 6)\mathbf{k}$

$\mathbf{n}_2 = \nabla g(0, 2, 3) = -4\mathbf{j}$

$\mathbf{n}_2 = -\mathbf{n}_1$, and therefore they are parallel.

51. First determine the points of intersection. From $z = \sqrt{x^2 + y^2}$, one may replace the $x^2 + y^2$ in the equation by z^2, obtaining $2z^2 - 2\sqrt{2}\, z + 1 = 0$. The left side factors as $(\sqrt{2}\, z - 1)^2 = 0$, and thus $\sqrt{2}\, z - 1 = 0$. Therefore $z = 1/\sqrt{2}$, and the cone and sphere intersect along the circle $x^2 + y^2 = 1/2$ and $z = \sqrt{2}/2$.

Now show that the tangent planes of the cone and sphere along this circle are the same by showing that they have parallel normal vectors along this circle. Let $f(x, y, z) = x^2 + y^2 + z^2 - 2\sqrt{2}\, z + 1$ for the sphere, and $g(x, y, z) = \sqrt{x^2 + y^2} - z$ for the cone.

$\nabla f(x, y, z) = 2x\mathbf{i} + 2y\mathbf{j} + (2z - 2\sqrt{2})\mathbf{k}$

$\nabla g(x, y, z) = (x/\sqrt{x^2 + y^2})\mathbf{i} + (y/\sqrt{x^2 + y^2})\mathbf{j} - \mathbf{k}$

At any point $(x, y, \sqrt{2}/2)$ along the circle, one has $1/\sqrt{x^2 + y^2} = \sqrt{2}$. Therefore:

$\mathbf{n}_1 = \nabla f(x, y, \sqrt{2}/2) = 2x\mathbf{i} + 2y\mathbf{j} - \sqrt{2}\,\mathbf{k}$

$\mathbf{n}_2 = \nabla g(x, y, \sqrt{2}/2) = \sqrt{2}\, x\mathbf{i} + \sqrt{2}\, y\mathbf{j} - \mathbf{k}$

$\mathbf{n}_1 = \sqrt{2}\,\mathbf{n}_2$, and therefore they are parallel.

53. The most rapid decrease of f at $\mathbf{x}$ occurs when $D_{\mathbf{u}}f(\mathbf{x})$ is a minimum.

$$D_{\mathbf{u}}f(\mathbf{x}) = (\nabla f(\mathbf{x}) \cdot \mathbf{u}) = |\nabla f(\mathbf{x})||\mathbf{u}|\cos\theta = |\nabla f(\mathbf{x})|\cos\theta$$

is a minimum when $\cos\theta = -1$, which occurs when $\nabla f(\mathbf{x})$ and $\mathbf{u}$ point in opposite directions, or in other words, when $\mathbf{u}$ points in the direction of $-\nabla f(\mathbf{x})$.

55. $D_{\mathbf{u}}f(\mathbf{x}_0)$ takes on its maximum value in the direction of $\nabla f(\mathbf{x}_0)$. Let $\mathbf{u}$ be the unit vector in this direction, $\mathbf{u} = (1/|\nabla f(\mathbf{x}_0)|)\nabla f(\mathbf{x}_0)$. Then:

$D_{\mathbf{u}}f(\mathbf{x}_0) = \nabla f(\mathbf{x}_0) \cdot \mathbf{u} = \nabla f(\mathbf{x}_0) \cdot (1/|\nabla f(\mathbf{x}_0)|)\nabla f(\mathbf{x}_0)$

$= (1/|\nabla f(\mathbf{x}_0)|)(\nabla f(\mathbf{x}_0) \cdot \nabla f(\mathbf{x}_0))$

$= (1/|\nabla f(\mathbf{x}_0)|)|\nabla f(\mathbf{x}_0)|^2$

$= |\nabla f(\mathbf{x}_0)|$

57. $f(x, y) = e^{x^2 + y^2}$

$\nabla f(x, y) = 2xe^{x^2 + y^2}\mathbf{i} + 2ye^{x^2 + y^2}\mathbf{j}$

$\nabla f(4, 2) = 8e^{20}\mathbf{i} + 4e^{20}\mathbf{j} = 4e^{20}(2\mathbf{i} + \mathbf{j})$

The direction of steepest ascent is $2\mathbf{i} + \mathbf{j}$.

59. $f(x, y) = 4x^2 - 2y^2$

$\nabla f(x, y) = 8x\mathbf{i} - 4y\mathbf{j}$

$\nabla f(1, 2) = 8\mathbf{i} - 8\mathbf{j} = 8(\mathbf{i} - \mathbf{j})$

The direction of steepest ascent is $\mathbf{i} - \mathbf{j}$.

The direction of steepest descent is $-\mathbf{i} + \mathbf{j}$.

61. By definition of the directional derivative, $D_{\mathbf{i}}f(x_0, y_0) = \lim_{t \to 0^+} (f(x_0 + t, y_0) - f(x_0, y_0))/t$, which will be the value of $\partial f/\partial x$ provided that $\partial f/\partial x$ exists. (This matches the definition of $\partial f/\partial x$ with the exception that here one only has $t \to 0^+$ instead of $t \to 0$.)

For the second part, let $f(x, y) = |x|$ and $(x_0, y_0) = (0, 0)$.

$D_{\mathbf{u}} f(0, 0) = \lim_{t \to 0+} \big(f(tu_1, tu_2) - f(0, 0)\big)/t$

$= \lim_{t \to 0+} (|tu_1| - 0)/t$

$= \lim_{t \to 0+} |u_1| \, |t|/t$

$= |u_1|$

Note the importance for this limit that t is approaching 0 strictly *from the right*. Finally, it should be clear that $\partial f/\partial x$ does not exist at $(0, 0)$ (for the same reason that $g'(0)$ does not exist for $g(x) = |x|$).

63. For $\mathbf{u} = u_1 \mathbf{i} + u_2 \mathbf{j}$,

$D_{\mathbf{u}} f(0, 0) = \lim_{t \to 0+} \big(f(tu_1, tu_2) - f(0, 0)\big)/t$

$= \lim_{t \to 0+} \big(t^3 u_1 u_2^2/(t^2 u_1^2 + t^4 u_2^4)\big)/t$

$= \lim_{t \to 0+} u_1 u_2^2/(u_1^2 + t^2 u_2^4)$

If $u_1 = 0$, then this limit is 0. If $u_1 \neq 0$, then this limit is $u_1 u_2^2/u_1^2 = u_2^2/u_1$. In either case, the limit exists, and therefore the directional derivative exists.

Recall that if a function (of more than one variable) is differentiable at a point, then it is also continuous at that point. Show that $f(x, y)$ is not differentiable at $(0, 0)$ by showing that it is not continuous at this point. For any point $(x, y) \neq (0, 0)$ on the parabola $y^2 = x$,

$$f(x, y) = xy^2/(x^2 + y^4) = x^2/(x^2 + x^2) = 1/2.$$

Consequently, $\lim_{(x,y) \to (0,0)} f(x, y) \neq 0 = f(0, 0)$, and therefore $f(x, y)$ is not continuous at $(0, 0)$.

18.8 Constrained Extrema: The Method of Lagrange Multipliers

1. The constraint is $g(x, y) = x^2 + y^2 - 1 = 0$.
The system of equations to solve is:
$4x = 2\lambda x$
$8y = 2\lambda y$
$x^2 + y^2 = 1$
From the first equation, either $x = 0$ or $\lambda = 2$. If $x = 0$, then $y^2 = 1$ and $y = \pm 1$. If $\lambda = 2$, then $8y = 4y$ implies that $y = 0$, so $x^2 = 1$ and $x = \pm 1$.
$f(0, 1) = 4$, $f(0, -1) = 4$, $f(1, 0) = 2$ and $f(-1, 0) = 2$
The maximum value is 4.
The minimum value is 2.

3. The constraint is $g(x, y) = x - y - 2 = 0$.
The system of equations to solve is:
$3x^2 = \lambda$
$-3y^2 = -\lambda$
$x - y = 2$
From the first two equations, one has $3x^2 = 3y^2$, so $x = \pm y$. The possibility $x = y$ is not consistent with $x - y = 2$, thus one must have $x = -y$. Substituting $-y$ for x in $x - y = 2$ gives $y = -1$, and thus $x = 1$. Therefore, the only possible maximum or minimum occurs at the point $(1, -1)$.

For points (x, y) on the line $x - y = 2$, one has $x = y + 2$, and thus $f(x, y) = x^3 - y^3 = (y+2)^3 - y^3 = 6y^2 + 12y + 8$ is a quadratic polynomial with positive leading coefficient. One therefore expects a unique minimum and no maximum. Thus:
The minimum value is $f(1, -1) = 2$.
There is no maximum value.

5. The constraint is $g(x, y) = x^2 + y^2 - 1 = 0$.
The system of equations to solve is:
$$y = 2\lambda x$$
$$x = 2\lambda y$$
$$x^2 + y^2 = 1$$
Squaring the first two equations and adding gives:
$$y^2 + x^2 = 4\lambda^2 x^2 + 4\lambda^2 y^4$$
$$x^2 + y^2 = 4\lambda^2(x^2 + y^2)$$
$$1 = 4\lambda^2$$
$$\lambda = \pm 1/2$$
Thus either $y = x$ or $y = -x$ (from either the first or second equation). Substituting $\pm x$ for y in the third equation gives $x^2 + x^2 = 1$, from which it follows that $x^2 = 1/2$ or $x = \pm\sqrt{2}/2$. Therefore $y = \pm x = \pm\sqrt{2}/2$.
$f(\pm\sqrt{2}/2, \pm\sqrt{2}/2) = \pm 1/2$
The maximum value is $1/2$.
The minimum value is $-1/2$.

7. The constraint is $g(x, y) = x^2 + 2y^2 - 4 = 0$.
The system of equations to solve is:
$$2x + 4 = 2\lambda x$$
$$8y = 4\lambda y$$
$$x^2 + 2y^2 = 4$$
From the second equation, either $y = 0$ or $\lambda = 2$. If $y = 0$, then the third equation gives $x^2 = 4$ and thus $x = \pm 2$. If $\lambda = 2$, then the first equation gives $2x + 4 = 4x$, and thus $4 = 2x$ or $x = 2$. If $x = 2$, then the third equation gives $4 + 2y^2 = 4$, which implies that $y = 0$.
$f(2, 0) = 12$, $f(-2, 0) = -4$
The maximum value is 12.
The minimum value is -4.

9. The constraint is $g(x, y) = x^2 + y^2 - 1 = 0$.
The system of equations to solve is:
$$2x - 4 = 2\lambda x$$
$$8y = 2\lambda y$$
$$x^2 + y^2 = 1$$
From the second equation, either $y = 0$ or $\lambda = 4$. If $y = 0$, then the third equation gives $x^2 = 1$, or $x = \pm 1$. If $\lambda = 4$, then the first equation gives $2x - 4 = 8x$, $-4 = 6x$, $x = -2/3$, and substituting this value for x in the third equation gives $(4/9) + y^2 = 1$, $y^2 = 5/9$, $y = \pm\sqrt{5}/3$.
$f(1, 0) = -3$, $f(-1, 0) = 5$, $f(-2/3, \sqrt{5}/3) = 16/3$, $f(-2/3, -\sqrt{5}/3) = 16/3$
The maximum value is $16/3$.
The minimum value is -3.

11. The constraint is $g(x, y, z) = x^2 + y^2 + z^2 - 1 = 0$.
The system of equations to solve is:
$$1 = 2\lambda x$$
$$2 = 2\lambda y$$
$$-1 = 2\lambda z$$
$$x^2 + y^2 + z^2 = 1$$

Squaring the first three equations and adding gives:
$1^2 + 2^2 + (-1)^2 = 4\lambda^2 x^2 + 4\lambda^2 y^2 + 4\lambda^2 z^2$
$6 = 4\lambda^2(x^2 + y^2 + z^2) = 4\lambda^2(1)$
$\lambda^2 = 3/2$
$\lambda = \pm\sqrt{3/2} = \pm\sqrt{6}/2$
From the first three equations, if $\lambda = \sqrt{6}/2$, then $(x, y, z) = (\sqrt{6}/6, \sqrt{6}/3, -\sqrt{6}/6)$, and if $\lambda = -\sqrt{6}/2$,
then $(x, y, z) = (-\sqrt{6}/6, -\sqrt{6}/3, +\sqrt{6}/6)$.
$f(\sqrt{6}/6, \sqrt{6}/3, -\sqrt{6}/6) = \sqrt{6}$, $f(-\sqrt{6}/6, -\sqrt{6}/3, +\sqrt{6}/6) = -\sqrt{6}$
The maximum value is $\sqrt{6}$.
The minimum value is $-\sqrt{6}$.

13. The constraint is $g(x, y, z) = x^2 + y^2 + z^2 - 12 = 0$.
The system of equations to solve is:
$1 = 2\lambda x$
$1 = 2\lambda y$
$1 = 2\lambda z$
$x^2 + y^2 + z^2 = 12$
Squaring the first three equations and adding gives:
$1^2 + 1^2 + 1^2 = 4\lambda^2 x^2 + 4\lambda^2 y^2 + 4\lambda^2 z^2$
$3 = 4\lambda^2(x^2 + y^2 + z^2)$
$3 = 4\lambda^2(12)$
$\lambda^2 = 1/16$
$\lambda = \pm 1/4$
If $\lambda = 1/4$, then $(x, y, z) = (2, 2, 2)$. If $\lambda = -1/4$, then $(x, y, z) = (-2, -2, -2)$.
$f(2, 2, 2) = 6$, $f(-2, -2, -2) = -6$
The maximum value is 6.
The minimum value is -6.

15. The constraint is $g(x, y) = x - y = 0$.
For points along this line, one has $x = y$, and so $f(x, y) = x \sin y + y \sin x = x \sin x + x \sin x = 2x \sin x$,
where x may be any real number (so long as y is the same real number). Since $2x \sin x$ may be made
an arbitrarily large positive or negative number by choosing x to be a sufficiently large multiple of $\pi/2$,
it follows that $f(x, y)$ has neither a maximum nor a minimum value subject to this constraint.

17. The constraint is $g(x, y, z) = z + \ln x = 0$.
Note that the value of y is unconstrained, and since $y^2 \geq 0$ may therefore be made arbitrarily large for
any fixed x and z satisfying the constraint, the value of $f(x, y, z) = x + y^2 + z$ cannot have a maximum.
One may, however, expect a minimum value.
The system of equations to solve is:
$1 = \lambda/x$
$2y = 0$
$1 = \lambda$
$z + \ln x = 0$
The second equation gives $y = 0$. Substituting 1 for λ in the first equation gives $x = 1$. Substituting 1
for x in the last equation gives $z = 0$.
$f(1, 0, 0) = 1$
There is no maximum value.
The minimum value is 1.

19. One seeks to find a point (x, y, z) such that the distance from this point to the point $(2, -1, 2)$ is a minimum, and subject to the constraint that $g(x, y, z) = (y + 1)^2 - (x - 2)^2 + 1 - z = 0$. Let $d(x, y, z) = \sqrt{(x - 2)^2 + (y + 1)^2 + (z - 2)^2}$ be the distance, for which we want a minimum, and let $f(x, y, z) = d(x, y, z)^2 = (x - 2)^2 + (y + 1)^2 + (z - 2)^2$. Note that for any point (x, y, z) at which d is a minimum, f will also be a minimum. Therefore, look for points (subject to the constraint) where f is a minimum.

The system of equations to solve is:
$2(x - 2) = -2(x - 2)\lambda$
$2(y + 1) = 2(y + 1)\lambda$
$2(z - 2) = -\lambda$
$(y + 1)^2 - (x - 2)^2 + 1 = z$

From the first equation, either $x - 2 = 0$ ($x = 2$) or $\lambda = -1$. From the second equation, either $y + 1 = 0$ ($y = -1$) or $\lambda = 1$. Since λ cannot equal both 1 and -1, it follows that at least one of $x = 2$ or $y = -1$ must be true. If both are true ($x = 2$ and $y = -1$), then the last equation gives $z = 1$. If $x = 2$ and $y \neq -1$, then $\lambda = 1$, and the third equation gives $2z - 4 = -1$ or $z = 3/2$, from which it follows that:
$(y + 1)^2 - 0 + 1 = 3/2$
$(y + 1)^2 = 1/2$
$y + 1 = \pm\sqrt{2}/2$
$y = (-2 \pm \sqrt{2})/2$

If $y = -1$ and $x \neq 2$, then $\lambda = -1$, and the third equation gives $2z - 4 = 1$ or $z = 5/2$, from which it follows that:
$-(x - 2)^2 + 1 = 5/2$
$(x - 2)^2 = -3/2$
This is impossible.

$f(2, -1, 1) = 1$, $f(2, (-2 + \sqrt{2})/2, 3/2) = 3/4$, $f(2, (-2 - \sqrt{2})/2, 3/2) = 3/4$

Since $3/4 < 1$, the points nearest $(2, -1, 2)$ are $(2, (-2 + \sqrt{2})/2, 3/2)$ and $(2, (-2 - \sqrt{2})/2, 3/2)$.

21. Find the points where the minimum of $f(x, y) = x^2 + y^2$ occurs, subject to the constraint $g(x, y) = (x - 3)^2 + (y + 2)^2 - 9 = 0$.

The system of equations to solve is:
$2x = 2(x - 3)\lambda$
$2y = 2(y + 2)\lambda$
$(x - 3)^2 + (y + 2)^2 = 9$

Note that $x - 3 \neq 0$, since if it did, then the first equation would give $x = 0$, which is inconsistent with $x - 3 = 0$. Similarly $y + 2 \neq 0$. From the first two equations, one has $\lambda = x/(x - 3) = y/(y + 2)$, and multiplying by $(x - 3)(y + 2)$ gives $x(y + 2) = y(x - 3)$. Simplifying: $xy + 2x = xy - 3y$, $2x = -3y$, $y = -2x/3$. Substitute $-2x/3$ for y in the third equation:
$(x - 3)^2 + ((-2x/3) + 2)^2 = 9$
$x^2 - 6x + 9 + (4x^2/9) - (8x/3) + 4 = 9$
$(13x^2/9) - (26x/3) + 4 = 0$
$13x^2 - 78x + 36 = 0$
$x = (78 \pm \sqrt{6084 - 1872})/26 = 3 \pm (18\sqrt{13}/26) = 3 \pm (9\sqrt{13}/13)$
$y = -2x/3 = -2 \mp (6\sqrt{13}/13)$
$f(3 + (9\sqrt{13}/13), -2 - (6\sqrt{13}/13)) = 9 + (54\sqrt{13}/13) + (81/13) + 4 + (24\sqrt{13}/13) + (36/13) = 22 + 6\sqrt{13}$
$f(3 - (9\sqrt{13}/13), -2 + (6\sqrt{13}/13)) = 9 - (54\sqrt{13}/13) + (81/13) + 4 - (24\sqrt{13}/13) + (36/13) = 22 - 6\sqrt{13}$
The point closest to the origin is $(3 - (9\sqrt{13}/13), -2 + (6\sqrt{13}/13))$.

23. Find the point where $f(x, y, z) = (x-1)^2 + (y+1)^2 + (z-1)^2$ (the square of the distance) is a minimum, subject to the constraint $g(x, y, z) = 36(x-1)^2 + 9(y-2)^2 + 4(z-1)^2 - 36 = 0$.
The system of equations to solve is:
$2(x - 1) = 72\lambda(x - 1)$
$2(y + 1) = 18\lambda(y - 2)$
$2(z - 1) = 8\lambda(z - 1)$
$36(x-1)^2 + 9(y-2)^2 + 4(z-1)^2 = 36$.
From the first equation, either $x - 1 = 0$ ($x = 1$) or $\lambda = 1/36$. From the third equation, either $z - 1 = 0$ ($z = 1$) or $\lambda = 1/4$. Since λ can have only one value, at least one of $x = 1$ or $z = 1$ must be true. If $x = 1$ and $z = 1$ are both true, then the third equation becomes $9(y-2)^2 = 36$, $(y-2)^2 = 4$, $y - 2 = \pm 2$, $y = 2 \pm 2$, which implies that $y = 4$ or $y = 0$. If $x = 1$ and $z \neq 1$, then $\lambda = 1/4$ and the second equation gives $2y + 2 = (9y/2) - 9$, $y = 22/5$. Substituting $x = 1$ and $y = 22/5$ into the third equation gives:
$9(12/5)^2 + 4(z-1)^2 = 36$
$(z-1)^2 = -99/25$
This cannot be satisfied by any real z. The last possibility is that $z = 1$ and $x \neq 1$, and thus $\lambda = 1/36$. The second equation now gives $2y + 2 = (y/2) - 1$, $y = -6/5$. Substituting $z = 1$ and $y = -6/5$ into the third equation gives:
$36(x-1)^2 + 9(-16/5)^2 = 36$
$(x-1)^2 = -39/25$
This cannot be satisfied by any real x.
Therefore, the only points to be considered are $(1, 4, 1)$ and $(1, 0, 1)$.
$f(1, 4, 1) = 25$, $f(1, 0, 1) = 1$
The nearest point is $(1, 0, 1)$.

25. Note that the point $(2, -1, 1)$ is actually on the surface, and therefore it is the point which is closest to itself.

27. Assume without loss of generality that the sides of the box are parallel to the planes $x = 0$, $y = 0$ and $z = 0$ and that the sphere has its center at the origin. Then one corner of the box is a point (x, y, z) all of whose coordinates are nonnegative, these coordinates must satisfy the equation $x^2 + y^2 + z^2 = r^2$ for the sphere, and the lengths of the edges of the box are $2x$, $2y$ and $2z$. Therefore, find a maximum value for $f(x, y, z) = 8xyz$ subject to the constraint $g(x, y, z) = x^2 + y^2 + z^2 - r^2 = 0$.
The system of equations to solve is:
$8yz = 2\lambda x$
$8xz = 2\lambda y$
$8xy = 2\lambda z$
$x^2 + y^2 + z^2 = r^2$
Multiplying the first three equations by x, y and z respectively gives the following three equations:
$8xyz = 2\lambda x^2$
$8xyz = 2\lambda y^2$
$8xyz = 2\lambda z^2$
Adding these three equations gives:
$24xyz = 2\lambda(x^2 + y^2 + z^2) = 2\lambda r^2$
$\lambda = 12xyz/r^2$
Substituting this expression for λ in the second set of three equations gives the following three equations:
$8xyz = xyz(24x^2/r^2)$
$8xyz = xyz(24y^2/r^2)$
$8xyz = xyz(24z^2/r^2)$

This implies that either $xyz = 0$ (which gives $f(x, y, z) = 0$) or $(x, y, z) = (\pm r\sqrt{3}/3, \pm r\sqrt{3}/3, \pm r\sqrt{3}/3)$. Consider the point $(x, y, z) = (r\sqrt{3}/3, r\sqrt{3}/3, r\sqrt{3}/3)$, since we want a point with all coordinates non-negative. The maximum volume will be $f(r\sqrt{3}/3, r\sqrt{3}/3, r\sqrt{3}/3) = 8r^3\sqrt{3}/9$. For this box, all edges will have length $2r\sqrt{3}/3$.

29. For a barrel of radius r and height h, the volume is $\pi r^2 h = 2$, and the surface area is given by $f(r, h) = 2\pi r^2 + 2\pi r h$. Find a minimum value for f subject to the constraint $g(r, h) = \pi r^2 h - 2 = 0$. The system of equations to solve is:
$4\pi r + 2\pi h = 2\lambda \pi r h \ (\partial/\partial r)$
$2\pi r = \lambda \pi r^2 \ (\partial/\partial h)$
$\pi r^2 h = 2$
From the second equation, either $r = 0$ or $\lambda = 2/r$. If $r = 0$, then the first equation gives $h = 0$, but this is inconsistent with the third equation. Thus $r \neq 0$ and $\lambda = 2/r$. Substituting this expression for λ in the first equation gives:
$4\pi r + 2\pi h = 4\pi h$
$4\pi r = 2\pi h$
$2r = h$
Substituting $2r$ for h in the third equation gives $2\pi r^3 = 2$ or $r = \sqrt[3]{1/\pi} = \pi^{-1/3}$. This gives $h = 2r = 2\pi^{-1/3}$. The barrel should have a radius of $\pi^{-1/3}$ and a height of $2\pi^{-1/3}$.

31. Find the point where $f(x, y, z) = x^2 + y^2 + z^2$ (the square of the distance) is a minimum, subject to the constraint $g(x, y, z) = 2x + 3y + z - 14 = 0$. The system of equations to solve is:
$2x = 2\lambda$
$2y = 3\lambda$
$2z = \lambda$
$2x + 3y + z = 14$
From the first equation, $\lambda = x$. Substituting x for λ in the second and third equations gives $y = 3x/2$ and $z = x/2$. Substituting these expressions for y and z in the fourth equation gives:
$2x + (9x/2) + (x/2) = 14$
$4x + 9x + x = 28$
$x = 2$
Thus $y = 3x/2 = 3$ and $z = x/2 = 1$. The point nearest the origin is $(2, 3, 1)$.

33. Let x, y and z be the length, width and height of the box. Minimize the surface area $f(x, y, z) = 2xy + 2xz + 2yz$ subject to the constraint $g(x, y, z) = xyz - 8000 = 0$. The system of equations to solve is:
$2y + 2z = \lambda yz$
$2x + 2z = \lambda xz$
$2x + 2y = \lambda xy$
$xyz = 8000$
Note that the fourth equation implies that none of x, y or z can equal 0. Multiplying the first, second, and third equations by x, y and z respectively, gives the following three equations:
$2xy + 2xz = \lambda xyz = 8000\lambda$
$2xy + 2yz = \lambda xyz = 8000\lambda$
$2xz + 2yz = \lambda xyz = 8000\lambda$
From the first two of these equations, one has:
$2xy + 2xz = 2xy + 2yz$
$2xz = 2yz$
$x = y$ (recall $z \neq 0$)

577

From the second and third of these equations, one has:

$2xy + 2yz = 2xz + 2yz$

$2xy = 2xz$

$y = z$ (recall $x \neq 0$)

Thus $x = y = z$. Substituting x for y and z in the equation $xyz = 8000$ gives $x^3 = 8000$ or $x = 20$. Thus $y = z = 20$ also. The length, width and height of the box should all be 20 cm.

18.9   Reconstructing a Function from its Gradient

1. $f(x, y) = \int 1 \, dx = x + h(y)$
$-1 = \partial f / \partial y = h'(y)$
$h(y) = -y + C$
$f(x, y) = x - y + C$

3. $\partial M / \partial y = \pi$ and $\partial N / \partial x = -\pi$
No potential.

5. $\partial M / \partial y = -1$ and $\partial N / \partial x = 1$
No potential.

7. $f(x, y) = \int (3x^2 \cos y - 1) dx$
$\quad = x^3 \cos y - x + h(y)$
$2y - x^3 \sin y = \partial f / \partial y = -x^3 \sin y + h'(y)$
$h'(y) = 2y$
$h(y) = y^2 + C$
$f(x, y) = x^3 \cos y - x + y^2 + C$

9. $f(x, y) = \int (2xe^{xy} + x^2 y e^{xy}) \, dx$
$\quad = \int 2xe^{xy} \, dx + \int x^2 y e^{xy} \, dx$
Let $u = x^2$ and $dv = y e^{xy} \, dx$.
Then $du = 2x \, dx$ and $v = e^{xy}$.
$f(x, y) = \int 2xe^{xy} \, dx + x^2 e^{xy} - \int 2xe^{xy} \, dx$
$\quad = x^2 e^{xy} + h(y)$
$x^3 e^{xy} - (1/(2\sqrt{y})) = \partial f / \partial y = x^3 e^{xy} + h'(y)$
$h'(y) = -1/(2\sqrt{y})$
$h(y) = -\sqrt{y} + C$
$f(x, y) = x^2 e^{xy} - \sqrt{y} + C$

11. $f(x, y) = \int (2xy^2 - y \sin x) dx$
$\quad = x^2 y^2 + y \cos x + h(y)$
$2x^2 y + \cos x = \partial f / \partial y = 2x^2 y + \cos x + h'(y)$
$h'(y) = 0$
$h(y) = C$
$f(x, y) = x^2 y^2 + y \cos x + C$

Review Exercises — Chapter 18

1. Does not exist. The limit along the x-axis ($y = 0$) is 1 and the limit along the y-axis ($x = 0$) is -1.

3.

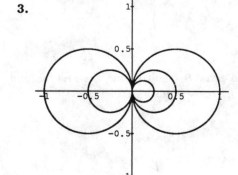

$$f(x, y) = x/(x^2 + y^2)$$

5. Let $f(x, y) = x^3 - 2xy^2 + y + 5$.
$\nabla f(x, y) = (3x^2 - 2y^2)\mathbf{i} + (-4xy + 1)\mathbf{j}$
$\mathbf{n} = \nabla f(1, 2) = -5\mathbf{i} - 7\mathbf{j}$

7. Let $f(x, y, z) = e^x \cos y - z$.
$\nabla f(x, y, z) = e^x \cos y\, \mathbf{i} - e^x \sin y\, \mathbf{j} - \mathbf{k}$
$\mathbf{n} = \nabla f(0, \pi, -5) = -\mathbf{i} - \mathbf{k}$

9. $\partial f/\partial x = \sin(\sqrt{x^2 + y^2}) + x\cos(\sqrt{x^2 + y^2})(1/2)(1/\sqrt{x^2 + y^2})2x$
$\quad = \sin(\sqrt{x^2 + y^2}) + (x^2 \cos(\sqrt{x^2 + y^2})/\sqrt{x^2 + y^2})$
$\partial f/\partial y = x\cos(\sqrt{x^2 + y^2})(1/2)(1/\sqrt{x^2 + y^2})2y$
$\quad = xy\cos(\sqrt{x^2 + y^2})/\sqrt{x^2 + y^2}$

11. For $\mathbf{v} = 2\mathbf{i} + \mathbf{j} + 2\mathbf{k}$, $|\mathbf{v}| = \sqrt{4 + 1 + 4} = 3$. Thus a unit vector in this direction is $\mathbf{u} = (1/3)(2\mathbf{i} + \mathbf{j} + 2\mathbf{k})$.
$\nabla f(x, y, z) = (y^3 + 2x\sqrt{y^2 + z^2})\mathbf{i} + (3xy^2 + (x^2 y/\sqrt{y^2 + z^2}))\mathbf{j} + (x^2 z/\sqrt{y^2 + z^2})\mathbf{k}$
$\nabla f(2, 3, 4) = 47\mathbf{i} + (282/5)\mathbf{j} + (16/5)\mathbf{k}$
$D_{\mathbf{u}}f(2, 3, 4) = \nabla f(2, 3, 4) \cdot \mathbf{u} = (1/3)(94 + (282/5) + (32/5)) = 784/15$

13. $\partial f/\partial r = (\partial f/\partial x)(\partial x/\partial r) + (\partial f/\partial y)(\partial y/\partial r)$
$\quad = (9x^2 + 2y^2)(2) + (4xy - 2y)(1)$
$\quad = 18x^2 + 4y^2 + 4xy - 2y$
$\quad = 18(2r + 5s)^2 + 4(r - 2s^2)^2 + 4(2r + 5s)(r - 2s^2) - 2(r - 2s^2)$
$\quad = 18(4r^2 + 20rs + 25s^2) + 4(r^2 - 4rs^2 + 4s^4) + 4(2r^2 - 4rs^2 + 5rs - 10s^3) - 2r + 4s^2$
$\quad = 72r^2 + 360rs + 450s^2 + 4r^2 - 16rs^2 + 16s^4 + 8r^2 - 16rs^2 + 20rs - 40s^3 - 2r + 4s^2$
$\quad = 84r^2 - 2r + 380rs - 32rs^2 + 454s^2 - 40s^3 + 16s^4$
$\partial f/\partial s = (\partial f/\partial x)(\partial x/\partial s) + (\partial f/\partial y)(\partial y/\partial s)$
$\quad = (9x^2 + 2y^2)(5) + (4xy - 2y)(-4s)$
$\quad = 45x^2 + 10y^2 - 16sxy + 8sy$
$\quad = 45(2r + 5s)^2 + 10(r - 2s^2)^2 - 16s(2r + 5s)(r - 2s^2) + 8s(r - 2s^2)$
$\quad = 45(4r^2 + 20rs + 25s^2) + 10(r^2 - 4rs^2 + 4s^4) - 16s(2r^2 - 4rs^2 + 5rs - 10s^3) + 8rs - 16s^3$
$\quad = 180r^2 + 900rs + 1125s^2 + 10r^2 - 40rs^2 + 40s^4 - 32r^2 s + 64rs^3 - 80rs^2 + 160s^4 + 8rs - 16s^3$
$\quad = 190r^2 - 32r^2 s + 908rs - 120rs^2 + 64rs^3 + 1125s^2 - 16s^3 + 200s^4$

15. Let $f(x, y) = x\cos \pi y + x^2 e^y$.
$\nabla f(x, y) = (\cos \pi y + 2xe^y)\mathbf{i} + (-\pi x \sin \pi y + x^2 e^y)\mathbf{j}$
$\mathbf{n} = \nabla f(2, 0) = 5\mathbf{i} + 4\mathbf{j}$
The equation of the tangent line is of the form $5x + 4y = c$. Since the point $(2, 0)$ is on the line, $c = 5(2) + 4(0) = 10$. Therefore, an equation for the tangent line is $5x + 4y = 10$.

17. Let $h = \alpha f + \beta g$.
$\nabla h = (\partial h/\partial x)\mathbf{i} + (\partial h/\partial y)\mathbf{j}$
$\quad = (\alpha(\partial f/\partial x) + \beta(\partial g/\partial x))\mathbf{i} + (\alpha(\partial f/\partial y) + \beta(\partial g/\partial y))\mathbf{j}$
$\quad = \alpha(\partial f/\partial x)\mathbf{i} + \alpha(\partial f/\partial y)\mathbf{j} + \beta(\partial g/\partial x)\mathbf{i} + \beta(\partial g/\partial y)\mathbf{j}$
$\quad = \alpha((\partial f/\partial x)\mathbf{i} + (\partial f/\partial y)\mathbf{j}) + \beta((\partial g/\partial x)\mathbf{i} + (\partial g/\partial y)\mathbf{j})$
$\quad = \alpha \nabla f + \beta \nabla g$

19. The desired vector is $-\nabla f(1, 2, 3)$.
$\nabla f(x, y, z) = \left(yz - z\dfrac{1}{(y/x)^2 + 1}\left(-\dfrac{y}{x^2}\right)\right)\mathbf{i} + \left(xz - z\dfrac{1}{(y/x)^2 + 1}\dfrac{1}{x}\right)\mathbf{j} + (xy - \text{Tan}^{-1}(y/x))\mathbf{k}$
$\nabla f(1, 2, 3) = (36/5)\mathbf{i} + (12/5)\mathbf{j} + (2 - \text{Tan}^{-1} 2)\mathbf{k}$
$-\nabla f(1, 2, 3) = (-36/5)\mathbf{i} + (-12/5)\mathbf{j} + (-2 + \text{Tan}^{-1} 2)\mathbf{k}$

21. $\partial^2 f/\partial x^2 = \partial f/\partial x = e^{x + ct}$
$\partial f/\partial t = ce^{x + ct}$
$\partial^2 f/\partial t^2 = c^2 e^{x + ct} = c^2(\partial^2 f/\partial x^2)$

23. Assume without loss of generality that the sides of the box are parallel to the planes $x = 0$, $y = 0$ and $z = 0$ and that the ellipsoid has its center at the origin. Then one corner of the box is a point (x, y, z) all of whose coordinates are nonnegative, these coordinates must satisfy the equation $x^2 + (y^2/9) + (z^2/4) = 1$, or $36x^2 + 4y^2 + 9z^2 = 36$, for the ellipsoid, and the lengths of the edges of the box are $2x$, $2y$ and $2z$. Therefore, find a maximum value for $f(x, y, z) = 8xyz$ subject to the constraint $g(x, y, z) = 36x^2 + 4y^2 + 9z^2 - 36 = 0$.
The system of equations to solve is:
$8yz = 72\lambda x$
$8xz = 8\lambda y$
$8xy = 18\lambda z$
$36x^2 + 4y^2 + 9z^2 = 36$
Multiplying the first, second and third equations by x, y and z respectively gives the following three equations:
$8xyz = 72\lambda x^2$
$8xyz = 8\lambda y^2$
$8xyz = 18\lambda z^2$
Adding these three equations gives:
$24xyz = 2\lambda(36x^2 + 4y^2 + 9z^2) = 72\lambda$
$\lambda = xyz/3$
Substituting this expression for λ in the three equations above gives the following three equations:
$8xyz = 24xyz(x^2)$
$8xyz = (8/3)xyz(y^2)$
$8xyz = 6xyz(z^2)$
This implies that either $xyz = 0$ (which gives $f(x, y, z) = 0$) or $(x, y, z) = (\pm\sqrt{3}/3, \pm\sqrt{3}, \pm2\sqrt{3}/3)$, and since we want a point with all coordinates nonnegative, consider the point $(x, y, z) = (\sqrt{3}/3, \sqrt{3}, 2\sqrt{3}/3)$. The maximum volume will be $f(\sqrt{3}/3, \sqrt{3}, 2\sqrt{3}/3) = 16\sqrt{3}/3$. For this box, the dimensions will be $2\sqrt{3}/3$, $2\sqrt{3}$ and $4\sqrt{3}/3$.

25. $\partial f/\partial x = 6x + y$ and $\partial f/\partial y = x - 12y$
Critical points must satisfy $6x + y = 0$ and $x - 12y = 0$, which has the unique solution $x = 0$ and $y = 0$. The only critical point is $(0, 0)$. At this point:
$A = \partial^2 f/\partial x^2 = 6$, $B = \partial^2 f/(\partial y \partial x) = 1$ and $C = \partial^2 f/\partial y^2 = -12$
$D = B^2 - AC = 73 > 0$
Saddle point at $(0, 0)$.

27. $\partial f/\partial x = (2x - 4y)e^{x^2 - 4xy}$ and $\partial f/\partial y = -4xe^{x^2 - 4xy}$
Critical points must satisfy $2x - 4y = 0$ and $4x = 0$. These are only satisfied by $x = 0$ and $y = 0$. The only critical point is $(0, 0)$. At this point:
$A = \partial^2 f/\partial x^2 = 2e^{x^2 - 4xy} + (2x - 4y)^2 e^{x^2 - 4xy} = 2$
$B = \partial^2 f/(\partial y \partial x) = -4e^{x^2 - 4xy} + (2x - 4y)(-4x)e^{x^2 - 4xy} = -4$
$C = \partial^2 f/\partial y^2 = 16x^2 e^{x^2 - 4xy} = 0$
$D = B^2 - AC = 16 > 0$
Saddle point at $(0, 0)$.

29. $\partial f/\partial x = 2xe^{1+x^2-y^2}$ and $\partial f/\partial y = -2ye^{1+x^2-y^2}$
Critical points must satisfy $x = 0$ and $y = 0$. The only critical point is $(0,0)$. At this point:
$A = \partial^2 f/\partial x^2 = (2 + 4x^2)e^{1+x^2-y^2} = 2e$
$B = \partial^2 f/(\partial y \partial x) = -4xye^{1+x^2-y^2} = 0$
$C = \partial^2 f/\partial y^2 = (-2 + 4y^2)e^{1+x^2-y^2} = -2e$
$D = B^2 - AC = 4e^2 > 0$
Saddle point at $(0,0)$.

31. $3 = D_{\mathbf{u_1}} f(P) = (\partial f/\partial y)(P)$
$3 = D_{\mathbf{u_2}} f(P) = (3/5)(\partial f/\partial x)(P) + (4/5)(\partial f/\partial y)(P)$
From the first equation, substitute 3 for $(\partial f/\partial y)(P)$ in the second equation to get $3 = (3/5)(\partial f/\partial x)(P) + (12/5)$, so $(\partial f/\partial x)(P) = 1$.

   a. 1

   b. 3

   c. $\mathbf{i} + 3\mathbf{j}$

33. $f(x,y) = \int 2xe^y \, dx = x^2e^y + h(y)$
$x^2e^y - \sin y = \partial f/\partial y = x^2e^y + h'(y)$
$h'(y) = -\sin y$
$h(y) = \cos y + C$
$f(x,y) = x^2e^y + \cos y + C$

35. At $r_1 = 10$, $r_2 = 20$ and $r_3 = 25$ one has a value of $R = 100/19$. We are given $dr_1 = .1(10) = 1$, $dr_2 = .1(20) = 2$ and $dr_3 = .1(25) = 2.5$.
$-R^{-2}dR = -r_1^{-2}dr_1 - r_2^{-2}dr_2 - r_3^{-2}dr_3$
$(-0.0361)dR = -(0.01)(1) - (0.0025)(2) - (0.0016)(2.5) = -0.019$
$dR \approx 0.5263$
R increases by approximately 0.53 ohms, or 10%.

37. $\partial w/\partial x = e^{x-y} - 1/(2\sqrt{z-x})$
$\partial w/\partial y = -e^{x-y} - \sin(y-z)$
$\partial w/\partial z = \sin(y-z) + 1/(2\sqrt{z-x})$
The sum of these three is clearly 0.

39. The maximum and minimum will occur on the ellipse since the graph of $f(x,y)$ is a plane. At any point in the interior of the ellipse, there would be another point higher on the plane which is also in the interior of the ellipse. Rewrite the equation of the ellipse as $9x^2 + 4y^2 = 36$. So find the maximum and minimum of $f(x,y) = 3x - 2y + 5$ subject to the constraint $g(x,y) = 9x^2 + 4y^2 - 36 = 0$.
The system of equations to solve is:
$3 = 18\lambda x$
$-2 = 8\lambda y$
$9x^2 + 4y^2 = 36$

Multiplying the first equation by 2 and the second equation by -3 gives $36\lambda x = 6 = -24\lambda y$, which in turn implies that $y = -3x/2$ ($\lambda \neq 0$ follows from either the first or second equation). Substitute $-3x/2$ for y in the third equation to get:

$9x^2 + 4(9x^2/4) = 36$

$18x^2 = 36$

$x = \pm\sqrt{2}$

$y = -3x/2 = \mp 3\sqrt{2}/2$

$f(\sqrt{2}, -3\sqrt{2}/2) = 5 + 6\sqrt{2}$, $f(-\sqrt{2}, 3\sqrt{2}/2) = 5 - 6\sqrt{2}$

The maximum value is $5 + 6\sqrt{2}$.

The minimum value is $5 - 6\sqrt{2}$.

41. No, essentially due to the fact that $\mathbf{u} + \mathbf{v}$ is not a unit vector.

$D_{\mathbf{u}+\mathbf{v}}f(x_0, y_0) = \nabla f(x_0, y_0) \cdot (\mathbf{u} + \mathbf{v})(1/|\mathbf{u} + \mathbf{v}|)$

$= \big(\nabla f(x_0, y_0) \cdot \mathbf{u} + \nabla f(x_0, y_0) \cdot \mathbf{v}\big)(1/|\mathbf{u} + \mathbf{v}|)$

$= \big(D_{\mathbf{u}}f(x_0, y_0) + D_{\mathbf{v}}f(x_0, y_0)\big)/|\mathbf{u} + \mathbf{v}|$

43. The constraint is $g(x, y, z) = x^2 + y^2 + z^2 - 16 = 0$.

The system of equations to solve is:

$1 = 2\lambda x$

$2 = 2\lambda y$

$-3 = 2\lambda z$

$x^2 + y^2 + z^2 = 16$

From the first equation, $x = 1/(2\lambda)$. From the second equation, $y = 1/\lambda = 2x$. From the third equation, $z = -3/(2\lambda) = -3x$. Substitute $2x$ for y and $-3x$ for z in the fourth equation to obtain:

$x^2 + 4x^2 + 9x^2 = 16$

$x^2 = 16/14 = 8/7$

$x = \pm 2\sqrt{14}/7$

Thus $y = 2x = \pm 4\sqrt{14}/7$ and $z = -3x = \mp 6\sqrt{14}/7$.

$f(2\sqrt{14}/7, 4\sqrt{14}/7, -6\sqrt{14}/7) = 1 + 4\sqrt{14}$, $f(-2\sqrt{14}/7, -4\sqrt{14}/7, 6\sqrt{14}/7) = 1 - 4\sqrt{14}$

The maximum occurs at the point $(2\sqrt{14}/7, 4\sqrt{14}/7, -6\sqrt{14}/7)$.

45. In Section 4 we were given as a fact that a continuous function always has both an absolute maximum and an absolute minimum on any closed and bounded set. A disk D is a closed and bounded set. The fact that f is differentiable implies that it is continuous. Let M be the larger of the absolute values of the maximum and minimum for f on D. Then for all (x, y) in D, $|f(x, y)| \leq M$.

47. Let $\mathbf{u} = u_1\mathbf{i} + u_2\mathbf{j}$ be a unit vector.

$D_{\mathbf{u}}f(0, 0) = \lim_{t \to 0+} \big(f(0 + tu_1, 0 + tu_2) - f(0, 0)\big)/t$

$= \lim_{t \to 0+} \sqrt{t^2 u_1^2 + t^2 u_2^2}/t$

$= \lim_{t \to 0+} t\sqrt{u_1^2 + u_2^2}/t$ (note $t > 0$)

$= \lim_{t \to 0+} \sqrt{u_1^2 + u_2^2}$

$= \sqrt{u_1^2 + u_2^2}$

$= 1$ since $\mathbf{u}$ is a unit vector.

However, the partial derivatives correspond to the derivatives of $f(0, y) = \sqrt{y^2} = |y|$ with respect to y and of $f(x, 0) = \sqrt{x^2} = |x|$ with respect to x, neither of which exists at 0.

49. Let $f(x, y, z) = x^2 + y^2 + z^2$ (the square of the distance from (x, y, z) to the origin), and find the minimum value of f subject to the constraint $g(x, y, z) = x - 3y + 2z - 6 = 0$.
The system of equations to solve is:
$2x = \lambda$
$2y = -3\lambda$
$2z = 2\lambda$
$x - 3y + 2z = 6$
Substituting $\lambda/2$ for x, $-3\lambda/2$ for y and λ for z in the fourth equation gives $(\lambda/2) + (9\lambda/2) + 2\lambda = 6$ which has solution $\lambda = 6/7$. Thus the minimum occurs at $x = \lambda/2 = 3/7$, $y = -3\lambda/2 = -9/7$ and $z = \lambda = 6/7$, which is the point $(3/7, -9/7, 6/7)$.

51. Let $f(x, y, z) = x^2 + 4y^2 + z^2$.
$\nabla f(x, y, z) = 2x\mathbf{i} + 8y\mathbf{j} + 2z\mathbf{k}$
$\nabla f(2, 1, 2) = 4\mathbf{i} + 8\mathbf{j} + 4\mathbf{k} = 4(\mathbf{i} + 2\mathbf{j} + \mathbf{k})$
Use $\mathbf{n} = \mathbf{i} + 2\mathbf{j} + \mathbf{k}$ as the normal vector for the plane. The equation of the tangent plane is of the form $x + 2y + z = d$. Since $(2, 1, 2)$ is a point on the plane, $d = 2 + 2(1) + 2 = 6$. Therefore, the equation of the tangent plane is $x + 2y + z = 6$.

53. The constraint is $g(x, y, z) = x^2 + 3y^2 - 3 = 0$.
The system of equations to solve is:
$-y + 2x = 2\lambda x$
$6y - x = 6\lambda y$
$x^2 + 3y^2 = 3$
If $x = 0$, then the first equation becomes $-y = 0$, so $y = 0$, but $x = 0$ and $y = 0$ is inconsistent with the third equation. Thus $x \neq 0$. Solving for λ in the first equation gives $\lambda = (2x - y)/2x$, and substituting this into the second equation gives:
$6y - x = 6y(2x - y)/2x$
$12xy - 2x^2 = 12xy - 6y^2$
$x^2 = 3y^2$
$x = \pm\sqrt{3}\,y$
Substituting $\pm\sqrt{3}\,y$ for x in the third equation gives $3y^2 + 3y^2 = 3$, which implies that $y = \pm\sqrt{2}/2$, and thus $x = \pm\sqrt{6}/2$.
$f(\sqrt{6}/2, \sqrt{2}/2) = (6 - \sqrt{3})/2$, $f(\sqrt{6}/2, -\sqrt{2}/2) = (6 + \sqrt{3})/2$,
$f(-\sqrt{6}/2, \sqrt{2}/2) = (6 + \sqrt{3})/2$, $f(-\sqrt{6}/2, -\sqrt{2}/2) = (6 - \sqrt{3})/2$
The minimum value is $(6 - \sqrt{3})/2$.

55. Let $v = xy$, so that $u = f(v)$. Then:
$\partial u/\partial x = (df/dv)(\partial v/\partial x) = (df/dv)y$ and
$\partial u/\partial y = (df/dv)(\partial v/\partial y) = (df/dv)x$.
Hence, $x(\partial u/\partial x) = (df/dv)xy = y(\partial u/\partial y)$.

57. Let $v = t - x$, so that $u = f(v)$. Then:
$\partial u/\partial x = (df/dv)(\partial v/\partial x) = -(df/dv)$
$\partial u/\partial t = (df/dv)(\partial v/\partial t) = df/dv$
$\partial^2 u/\partial x^2 = -(d^2 f/dv^2)(\partial v/\partial x) = d^2 f/dv^2$
$\partial^2 u/(\partial x \partial t) = (d^2 f/dv^2)(\partial v/\partial x) = -d^2 f/dv^2$
$\partial^2 u/\partial t^2 = (d^2 f/dv^2)(\partial v/\partial t) = d^2 f/dv^2$
Therefore $(\partial^2 u/\partial x^2) + 2(\partial^2 u/(\partial x \partial t)) + (\partial^2 u/\partial t^2) = (d^2 f/dv^2) + 2(-d^2 f/dv^2) + (d^2 f/dv^2) = 0$.

59. Let $v = bx - t$, so that $u = f(v)$. Then:
$\partial u/\partial x = (df/dv)(\partial v/\partial x) = (df/dv)b$ and $\partial u/\partial t = (df/dv)(\partial v/\partial t) = -(df/dv)$
Hence, $-b(\partial u/\partial t) = (df/dv)b = \partial u/\partial x$.

Chapter 19

Double and Triple Integrals

19.1   The Double Integral over a Rectangle

1. $\int_0^1 \int_0^2 xy\, dx\, dy$
$= \int_0^1 \left[x^2 y/2 \right]_{x=0}^{x=2} dy$
$= \int_0^1 2y\, dy$
$= y^2 \big|_0^1$
$= 1$

3. $\int_1^3 \int_1^2 (4 + x - y)\, dx\, dy$
$= \int_1^3 \left[4x + (x^2/2) - xy \right]_{x=1}^{x=2} dy$
$= \int_1^3 ((11/2) - y)\, dy$
$= (11y - y^2)/2 \big|_1^3$
$= 7$

5. $\int_0^1 \int_0^{\pi/2} x \sin y\, dy\, dx$
$= \int_0^1 \left[-x \cos y \right]_{y=0}^{y=\pi/2} dx$
$= \int_0^1 x\, dx$
$= x^2/2 \big|_0^1$
$= 1/2$

7. $\int_0^2 \int_1^e y^2 \ln x\, dx\, dy$
$= \int_0^2 \left[y^2 (x \ln x - x) \right]_{x=1}^{x=e} dy$
$= \int_0^2 y^2\, dy$
$= y^3/3 \big|_0^2$
$= 8/3$

9. $\int_0^{\pi/2} \int_0^3 y \sin(xy)\, dx\, dy$
$= \int_0^{\pi/2} \left[-\cos(xy) \right]_{x=0}^{x=3} dy$
$= \int_0^{\pi/2} (1 - \cos 3y)\, dy$
$= y - (1/3) \sin 3y \big|_0^{\pi/2}$
$= (\pi/2) + (1/3) = (3\pi + 2)/6$

11. $\int_0^2 \int_0^2 xy e^{xy^2} dy\, dx$
Let $u = xy^2$, $du = 2xy\, dy$.
$\int_0^2 \int_{y=0}^{y=2} (1/2) e^u\, du\, dx$
$= \int_0^2 \left[(1/2) e^{xy^2} \right]_{y=0}^{y=2} dx$
$= (1/2) \int_0^2 (e^{4x} - 1)\, dx$
$= (1/2) \left[(1/4) e^{4x} - x \right]_0^2$
$= (e^8 - 9)/8$

13. $\int_0^1 \int_0^{\pi/2} xy \sin x\, dx\, dy$
$= \int_0^1 \left[-xy \cos x + y \sin x \right]_{x=0}^{x=\pi/2} dy$
$= \int_0^1 y\, dy$
$= y^2/2 \big|_0^1$
$= 1/2$

15. $\int_0^1 \int_0^4 \sqrt{y}/(1 + x^2)\, dy\, dx$
$= \int_0^1 \left[(2y^{3/2}/3)/(1 + x^2) \right]_{y=0}^{y=4} dx$
$= (16/3) \int_0^1 1/(1 + x^2)\, dx$
$= (16/3) \arctan x \big|_0^1$
$= 4\pi/3$

17. $\int_0^1 \int_0^1 (x + y^2)\, dx\, dy$
$= \int_0^1 \left[(x^2/2) + xy^2 \right]_{x=0}^{x=1} dy$
$= \int_0^1 [(1/2) + y^2]\, dy$
$= (y/2) + (y^3/3) \big|_0^1$
$= 5/6$

19. $\int_0^{\pi/2} \int_0^4 x \cos y\, dx\, dy$
$= \int_0^{\pi/2} \left[(x^2/2) \cos y \right]_{x=0}^{x=4} dy$
$= \int_0^{\pi/2} 8 \cos y\, dy$
$= 8 \sin y \big|_0^{\pi/2}$
$= 8$

21. $\int_0^{\pi/4} \int_0^1 xy \sec^2(xy^2)\, dy\, dx$
Let $u = xy^2$, $du = 2xy\, dy$.
$(1/2) \int_0^{\pi/4} \int_{y=0}^{y=1} \sec^2 u\, du\, dx$
$= (1/2) \int_0^{\pi/4} \left[\tan(xy^2) \right]_{y=0}^{y=1} dx$
$= (1/2) \int_0^{\pi/4} \tan x\, dx$
$= (1/2) \ln \sec x \big|_0^{\pi/4}$
$= (1/2) \ln \sqrt{2}$
$= (1/4) \ln 2$

23. $\int_0^1 \int_0^2 ye^{xy}\, dx\, dy$
$= \int_0^1 \left[e^{xy} \right]_{x=0}^{x=2} dy$
$= \int_0^1 (e^{2y} - 1)\, dy$
$= (1/2)e^{2y} - y \big|_0^1$
$= (e^2 - 3)/2$

25. $\int_0^4 \int_4^8 (x + y)^{-1/2}\, dx\, dy$
$= \int_0^4 \left[2(x + y)^{1/2} \right]_{x=4}^{x=8} dy$
$= 2 \int_0^4 [(8 + y)^{1/2} - (4 + y)^{1/2}]\, dy$
$= 4[(8 + y)^{3/2} - (4 + y)^{3/2}]/3 \big|_0^4$
$= (96\sqrt{3} - 128\sqrt{2} + 32)/3$

27. $\int_0^3 \int_0^2 x\, dx\, dy$
$= \int_0^3 \left[x^2/2 \right]_{x=0}^{x=2} dy$
$= \int_0^3 2\, dy$
$= 2y \big|_0^3$
$= 6$

29. $\int_0^1 \int_0^{\pi/2} x \sin y\, dy\, dx$
$= \int_0^1 \left[-x \cos y \right]_{y=0}^{y=\pi/2} dx$
$= \int_0^1 x\, dx$
$= x^2/2 \big|_0^1$
$= 1/2$

31. $\int_1^4 \int_1^9 e^{\sqrt{x}} x^{-1/2} y^{-1/2}\, dy\, dx$
$= \int_1^4 \left[2e^{\sqrt{x}} x^{-1/2} y^{1/2} \right]_{y=1}^{y=9} dx$
$= \int_1^4 4e^{\sqrt{x}} x^{-1/2}\, dx$
Let $u = \sqrt{x}$, $du = (1/2)x^{-1/2}\, dx$.
$8 \int_1^2 e^u\, du$
$= 8e^u \big|_1^2$
$= 8e^2 - 8e$

33. $z = 1$ and $z = 2 - y^2$ intersect along the lines $y = -1$ and $y = 1$.
$\int_{-1}^1 \int_{-1}^1 [(2 - y^2) - 1]\, dy\, dx$
$= \int_{-1}^1 \int_{-1}^1 (1 - y^2)\, dy\, dx$
$= \int_{-1}^1 \left[y - (y^3/3) \right]_{y=-1}^{y=1} dx$
$= \int_{-1}^1 (4/3)\, dx$
$= 4x/3 \big|_{-1}^1$
$= 8/3$

35. $z = x^2 - 1$ and $z = 1$ intersect along the lines $x = -\sqrt{2}$ and $x = \sqrt{2}$.
$\int_0^2 \int_{-\sqrt{2}}^{\sqrt{2}} [1 - (x^2 - 1)]\, dx\, dy$
$= \int_0^2 \int_{-\sqrt{2}}^{\sqrt{2}} (2 - x^2)\, dx\, dy$
$= \int_0^2 \left[2x - (x^3/3) \right]_{x=-\sqrt{2}}^{x=\sqrt{2}} dy$
$= \int_0^2 (8\sqrt{2}/3)\, dy$
$= (8\sqrt{2}/3)y \big|_0^2$
$= 16\sqrt{2}/3$

37. If f and g are continuous on the rectangle $R = \{(x, y) \mid a \leq x \leq b, c \leq y \leq d\}$, then:

$$\iint\limits_{R} [f(x, y) + g(x, y)] \, dA = \int_c^d \int_a^b [f(x, y) + g(x, y)] \, dx \, dy$$

$$= \int_c^d \left[\int_a^b f(x, y) \, dx + \int_a^b g(x, y) \, dx \right] dy$$

$$= \left[\int_c^d \int_a^b f(x, y) \, dx \, dy \right] + \left[\int_c^d \int_a^b g(x, y) \, dx \, dy \right]$$

$$= \iint\limits_{R} f(x, y) \, dA + \iint\limits_{R} g(x, y) \, dA$$

39. $\int_a^b \left[\int_c^d f(x, y) \, dy \right] dx = - \int_b^a \left[\int_c^d f(x, y) \, dy \right] dx$ follows immediately from Definition 4 in Chapter 6.

$$\int_a^b \int_c^d f(x, y) \, dy \, dx = \int_a^b \left[- \int_d^c f(x, y) \, dy \right] dx = - \int_a^b \int_d^c f(x, y) \, dy \, dx$$

$$\int_a^b \int_c^d f(x, y) \, dy \, dx = - \int_b^a \int_c^d f(x, y) \, dy \, dx = - \left(- \int_b^a \int_d^c f(x, y) \, dy \, dx \right) = \int_b^a \int_d^c f(x, y) \, dy \, dx$$

41. Change line 110 of BASIC Program 8 to "`LET S = S + EXP((A+J*D1)^2+(C+K*D2)^2)`". For the first line of input, type "0,1,0,2" as the values for a, b, c and d. For $n = 100$ subintervals, one obtains the approximation 25.0058. For larger values of n, one should obtain values closer to the approximation 24.0645.

43. Change line 110 of BASIC Program 8 to "`LET S = S + SQR((A+J*D1)^3+(C+K*D2)^3)`". For the first line of input, type "2,4,1,3" as the values for a, b, c and d. For $n = 100$ subintervals, one obtains the approximation 24.7670. For larger values of n, one should obtain values closer to the approximation 24.6361.

45. $\iint\limits_{R} f(x) g(y) \, dA = \int_c^d \int_a^b f(x) g(y) \, dx \, dy$

$$= \int_c^d \left[g(y) \int_a^b f(x) \, dx \right] dy \text{ since } g(y) \text{ is constant with respect to } x$$

$$= \left[\int_a^b f(x) \, dx \right] \left[\int_c^d g(y) \, dy \right] \text{ since the integral of } f(x) \text{ is constant with respect to } y.$$

19.2 Double Integrals over More General Regions

1.

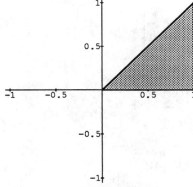

$$0 \leq x \leq 1,\, 0 \leq y \leq x$$
$$\int_0^1 \int_0^x (y^2 - x^2)\, dy\, dx$$
$$= \int_0^1 \left[(y^3/3) - x^2 y \right]_{y=0}^{y=x} dx$$
$$= \int_0^1 (-2x^3/3)\, dx$$
$$= -x^4/6 \Big|_0^1$$
$$= -1/6$$

3.

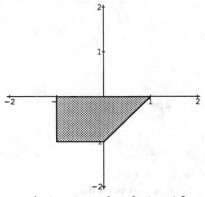

$$-1 \leq x \leq y+1,\, -1 \leq y \leq 0$$
$$\int_{-1}^0 \int_{-1}^{y+1} (xy - x)\, dx\, dy$$
$$= \int_{-1}^0 \left[(x^2/2)y - (x^2/2) \right]_{x=-1}^{x=y+1} dy$$
$$= (1/2) \int_{-1}^0 (y^3 + y^2 - 2y)\, dy$$
$$= (1/2) \left[(y^4/4) + (y^3/3) - y^2 \right]_{-1}^0$$
$$= 13/24$$

5.

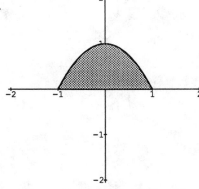

$$-1 \leq x \leq 1,\, 0 \leq y \leq 1 - x^2$$
$$\int_{-1}^1 \int_0^{1-x^2} xy\, dy\, dx$$
$$= \int_{-1}^1 \left[xy^2/2 \right]_{y=0}^{y=1-x^2} dx$$
$$= (1/2) \int_{-1}^1 (x - 2x^3 + x^5)\, dx$$
$$= (1/2) \left[(x^2/2) - (x^4/2) + (x^6/6) \right]_{-1}^1$$
$$= 0$$

7.

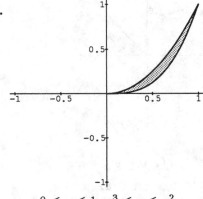

$$0 \leq x \leq 1,\, x^3 \leq y \leq x^2$$
$$\int_0^1 \int_{x^3}^{x^2} x\, dy\, dx$$
$$= \int_0^1 \left[xy \right]_{y=x^3}^{y=x^2} dx$$
$$= \int_0^1 (x^3 - x^4)\, dx$$
$$= (x^4/4) - (x^5/5) \Big|_0^1$$
$$= 1/20$$

9.

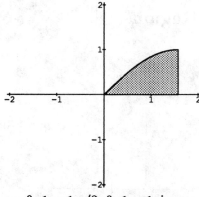

$$0 \le x \le \pi/2, \; 0 \le y \le \sin x$$
$$\int_0^{\pi/2} \int_0^{\sin x} 2y \cos x \, dy \, dx$$
$$= \int_0^{\pi/2} \left[y^2 \cos x\right]_{y=0}^{y=\sin x} dx$$
$$= \int_0^{\pi/2} \sin^2 x \cos x \, dx$$
Let $u = \sin x$, $du = \cos x \, dx$.
$$\int_0^1 u^2 \, du$$
$$= u^3/3 \Big|_0^1$$
$$= 1/3$$

11.

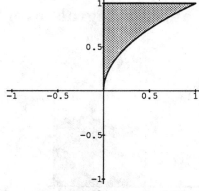

$$0 \le x \le y^2, \; 0 \le y \le 1$$
$$\int_0^1 \int_0^{y^2} e^{x/y} \, dx \, dy$$
$$= \int_0^1 \left[y e^{x/y}\right]_{x=0}^{x=y^2} dy$$
$$= \int_0^1 (y e^y - y) \, dy$$
$$= y e^y - e^y - (y^2/2) \Big|_0^1$$
$$= 1/2$$

13.

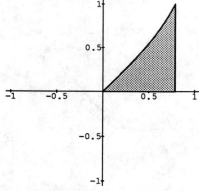

$$0 \le x \le \pi/4, \; 0 \le y \le \tan x$$
$$\int_0^{\pi/4} \int_0^{\tan x} 3y \sec^2 x \, dy \, dx$$
$$= \int_0^{\pi/4} \left[(3y^2/2) \sec^2 x\right]_{y=0}^{y=\tan x} dx$$
$$= (3/2) \int_0^{\pi/4} \tan^2 x \sec^2 x \, dx$$
Let $u = \tan x$, $du = \sec^2 x \, dx$.
$$(3/2) \int_0^1 u^2 \, du$$
$$= (3/2)(u^3/3) \Big|_0^1$$
$$= 1/2$$

15. $\int_0^1 \int_0^{y^2} y \sqrt{x} \, dx \, dy$
$$= \int_0^1 \left[(2/3)y x^{3/2}\right]_{x=0}^{x=y^2} dy$$
$$= \int_0^1 (2/3) y^4 \, dy$$
$$= (2/3)(y^5/5) \Big|_0^1$$
$$= 2/15$$

17. $\int_0^1 \int_0^{e^{y/2}} (x+2)\sqrt{1+e^y}\, dx\, dy$

$= \int_0^1 \left[((x^2/2) + 2x)\sqrt{1+e^y} \right]_{x=0}^{x=e^{y/2}} dy$

$= (1/2) \int_0^1 (e^y + 4e^{y/2})\sqrt{1+e^y}\, dy$

$= (1/2) \int_0^1 e^y \sqrt{1+e^y}\, dy + 2 \int_0^1 e^{y/2}\sqrt{1+e^y}\, dy$

Let $u = 1 + e^y$, $du = e^y\, dy$ and let $v = e^{y/2}$, $dv = (1/2)e^{y/2}\, dy$.

$(1/2) \int_2^{1+e} \sqrt{u}\, du + 4 \int_1^{\sqrt{e}} \sqrt{1+v^2}\, dv$

$= (1/2)\left[(2/3)u^{3/2} \right]_{u=2}^{u=1+e} + 4(1/2)\left[v\sqrt{1+v^2} + \ln(v + \sqrt{1+v^2}) \right]_{v=1}^{v=\sqrt{e}}$

$= \left[(1+e)^{3/2}/3 \right] - (8\sqrt{2}/3) + 2\sqrt{e+e^2} + 2\ln(\sqrt{e} + \sqrt{1+e}) - 2\ln(1+\sqrt{2})$

19. $\int_0^1 \int_y^{\sqrt{y}} xy\, dx\, dy$

$= \int_0^1 \left[x^2 y/2 \right]_{x=y}^{x=\sqrt{y}} dy$

$= (1/2) \int_0^1 (y^2 - y^3)\, dy$

$= (1/2)\left[(y^3/3) - (y^4/4) \right]_0^1$

$= 1/24$

21. $\int_{-1}^1 \int_{e^x}^e y\, dy\, dx$

$= \int_{-1}^1 \left[y^2/2 \right]_{y=e^x}^{y=e} dx$

$= (1/2) \int_{-1}^1 (e^2 - e^{2x})\, dx$

$= (1/2)\left[e^2 x - (1/2)e^{2x} \right]_{-1}^1$

$= (3e^2 + e^{-2})/4$

23. $\int_0^1 \int_0^{\sqrt{1-y^2}} x\, dx\, dy$

$= \int_0^1 \left[x^2/2 \right]_{x=0}^{x=\sqrt{1-y^2}} dy$

$= (1/2) \int_0^1 (1 - y^2)\, dy$

$= (1/2)\left[y - (y^3/3) \right]_0^1$

$= 1/3$

25. $\int_0^{(\pi/3)^{1/3}} \int_0^{x^2} 3\sec x^3\, dy\, dx$

$= \int_0^{(\pi/3)^{1/3}} \left[3y\sec x^3 \right]_{y=0}^{y=x^2} dx$

$= \int_0^{(\pi/3)^{1/3}} 3x^2 \sec x^3\, dx$

Let $u = x^3$, $du = 3x^2\, dx$.

$\int_0^{\pi/3} \sec u\, du$

$= \ln|\tan u + \sec u|\,\Big|_0^{\pi/3}$

$= \ln(2 + \sqrt{3})$

27. $\int_1^4 \int_0^{\ln y} e^x e^{e^x}\, dx\, dy$

Let $u = e^x$, $du = e^x\, dx$.

$\int_1^4 \int_1^y e^u\, du\, dy$

$= \int_1^4 \left[e^u \right]_{u=1}^{u=y} dy$

$= \int_1^4 (e^y - e)\, dy$

$= e^y - ey\,\Big|_1^4$

$= e^4 - 4e$

29.

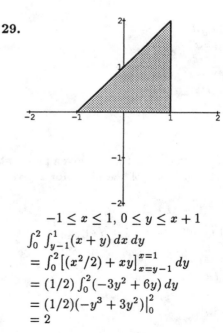

$-1 \le x \le 1$, $0 \le y \le x + 1$

$\int_0^2 \int_{y-1}^1 (x+y)\, dx\, dy$

$= \int_0^2 \left[(x^2/2) + xy \right]_{x=y-1}^{x=1} dy$

$= (1/2) \int_0^2 (-3y^2 + 6y)\, dy$

$= (1/2)(-y^3 + 3y^2)\,\Big|_0^2$

$= 2$

31.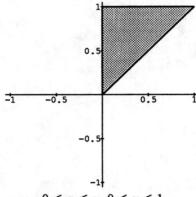

$$0 \le x \le y, \ 0 \le y \le 1$$
$$\int_0^1 \int_x^1 xy^2 \, dy \, dx$$
$$= \int_0^1 \left[xy^3/3 \right]_{y=x}^{y=1} dx$$
$$= (1/3) \int_0^1 (x - x^4) \, dx$$
$$= (1/3)\left[(x^2/2) - (x^5/5) \right]_0^1$$
$$= 1/10$$

33.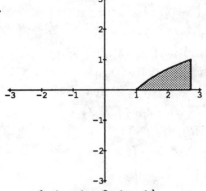

$$1 \le x \le e, \ 0 \le y \le \ln x$$
$$\int_0^1 \int_{e^y}^e f(x,y) \, dx \, dy$$

35. $\int_{-2}^1 \int_{x+2}^{4-x^2} 1 \, dy \, dx$
$$= \int_{-2}^1 [y]_{y=x+2}^{y=4-x^2} dx$$
$$= \int_{-2}^1 (2 - x^2 - x) \, dx$$
$$= 2x - (x^3/3) - (x^2/2)\big|_{-2}^1$$
$$= 9/2$$

37. $\int_0^{\pi/4} \int_{\sin x}^{\cos x} 1 \, dy \, dx$
$$= \int_0^{\pi/4} [y]_{y=\sin x}^{y=\cos x} dx$$
$$= \int_0^{\pi/4} (\cos x - \sin x) \, dx$$
$$= \sin x + \cos x \big|_0^{\pi/4}$$
$$= \sqrt{2} - 1$$

39. $\int_0^1 \int_{x^2}^{\sqrt{x}} 1 \, dy \, dx$
$$= \int_0^1 [y]_{y=x^2}^{y=\sqrt{x}} dx$$
$$= \int_0^1 (\sqrt{x} - x^2) \, dx$$
$$= (2/3)x^{3/2} - (x^3/3)\big|_0^1$$
$$= 1/3$$

41. $\int_0^1 \int_{1+x^2}^{1+x} (1/y^2) \, dy \, dx$
$$= \int_0^1 [-1/y]_{y=1+x^2}^{y=1+x} dx$$
$$= \int_0^1 [-(1/(1+x)) + (1/(1+x^2))] \, dx$$
$$= -\ln|1+x| + \text{Tan}^{-1} x \big|_0^1$$
$$= (\pi/4) - \ln 2$$

43. The x-intercept is $(1, 0, 0)$. The given plane intersects the xy-plane in the line with equation $y = -2x+2$. Solving the equation of the plane for z gives $z = (6 - 6x - 3y)/2$. The volume is given by the double integral:
$$\int_0^1 \int_0^{-2x+2} (6 - 6x - 3y)/2 \, dy \, dx$$
$$= (1/2) \int_0^1 \left[6y - 6xy - (3y^2/2) \right]_{y=0}^{y=-2x+2} dx$$
$$= 3 \int_0^1 (x^2 - 2x + 1) \, dx$$
$$= 3\left[(x^3/3) - x^2 + x \right]_0^1$$
$$= 1$$

45. Cut the solid along the plane $x = y$, compute the volume for half of the solid, then multiply by 2.

$2 \int_0^1 \int_0^x \sqrt{1 - x^2} \, dy \, dx$

$= 2 \int_0^1 \left[y\sqrt{1 - x^2} \right]_{y=0}^{y=x} dx$

$= 2 \int_0^1 x\sqrt{1 - x^2} \, dx$

Let $u = 1 - x^2$, $du = -2x \, dx$

$-\int_1^0 \sqrt{u} \, du$

$= \int_0^1 \sqrt{u} \, du$

$= (2/3)u^{3/2} \big|_0^1$

$= 2/3$

$z^2 = 1 - x^2$ and $z^2 = 1 - y^2$

19.3   Double Integrals in Polar Coordinates

1.

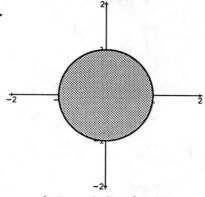

$0 \le r \le 1,\ 0 \le \theta \le 2\pi$

$\int_0^{2\pi} \int_0^1 r^2 \, dr \, d\theta$

$= \int_0^{2\pi} \left[r^3/3 \right]_{r=0}^{r=1} d\theta$

$= \int_0^{2\pi} (1/3) \, d\theta$

$= \theta/3 \big|_0^{2\pi}$

$= 2\pi/3$

3.

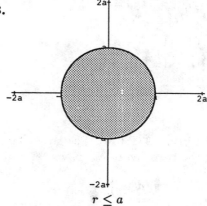

$r \le a$

$\int_0^{2\pi} \int_0^a re^{r^2} \, dr \, d\theta$

$= \int_0^{2\pi} \left[(1/2)e^{r^2} \right]_{r=0}^{r=a} d\theta$

$= (1/2) \int_0^{2\pi} (e^{a^2} - 1) \, d\theta$

$= (1/2)(e^{a^2} - 1)\theta \big|_0^{2\pi}$

$= \pi(e^{a^2} - 1)$

5.

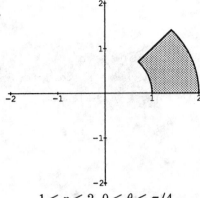

$$1 \le r \le 2, \; 0 \le \theta \le \pi/4$$

$\int_0^{\pi/4} \int_1^2 3r^3 \, dr \, d\theta$

$= \int_0^{\pi/4} \left[3r^4/4 \right]_{r=1}^{r=2} d\theta$

$= \int_0^{\pi/4} (45/4) \, d\theta$

$= 45\theta/4 \Big|_0^{\pi/4}$

$= 45\pi/16$

7. $\int_0^{\pi/3} \int_0^{\sin 3\theta} r^2 \, dr \, d\theta$

$= \int_0^{\pi/3} \left[r^3/3 \right]_{r=0}^{r=\sin 3\theta} d\theta$

$= (1/3) \int_0^{\pi/3} \sin^3 3\theta \, d\theta$

Let $u = \cos 3\theta$, $du = -3 \sin 3\theta \, d\theta$. Note that
$\sin^2 3\theta = 1 - \cos^2 3\theta = 1 - u^2$.

$-(1/9) \int_1^{-1} (1 - u^2) \, du$

$= (1/9) \int_{-1}^1 (1 - u^2) \, du$

$= (1/9) \left[u - (u^3/3) \right]_{-1}^1$

$= 4/27$

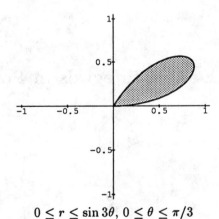

$$0 \le r \le \sin 3\theta, \; 0 \le \theta \le \pi/3$$

9. $3 \int_0^{\pi/3} \int_0^{\sin 3\theta} r \, dr \, d\theta$

$= 3 \int_0^{\pi/3} \left[r^2/2 \right]_{r=0}^{r=\sin 3\theta} d\theta$

$= (3/2) \int_0^{\pi/3} \sin^2 3\theta \, d\theta$

$= (3/4) \int_0^{\pi/3} (1 - \cos 6\theta) \, d\theta$

$= (3/4) \left[\theta - (1/6) \sin 6\theta \right]_0^{\pi/3}$

$= \pi/4$

11. $\int_{-\pi/2}^{\pi/2} \int_1^{1+\cos\theta} r \, dr \, d\theta$

$= \int_{-\pi/2}^{\pi/2} \left[r^2/2 \right]_{r=1}^{r=1+\cos\theta} d\theta$

$= (1/2) \int_{-\pi/2}^{\pi/2} (2\cos\theta + \cos^2\theta) \, d\theta$

$= \left[\sin\theta \right]_{-\pi/2}^{\pi/2} + (1/4) \int_{-\pi/2}^{\pi/2} (1 + \cos 2\theta) \, d\theta$

$= 2 + (1/4) \left[\theta + (1/2) \sin 2\theta \right]_{-\pi/2}^{\pi/2}$

$= 2 + (\pi/4)$

13. The points of intersection occur at $\theta = \pi/6$ and $\theta = 5\pi/6$. Compute the area between $\theta = -\pi/2$ and $\theta = \pi/6$, then double. The area between $\theta = -\pi/2$ and $\theta = 0$ is the area of a quarter circle of radius 5, or $25\pi/4$.

$2\left((25\pi/4) + \int_0^{\pi/6} \int_{10\sin\theta}^5 r\,dr\,d\theta\right)$
$= (25\pi/2) + 2\int_0^{\pi/6} \left[r^2/2\right]_{r=10\sin\theta}^{r=5} d\theta$
$= (25\pi/2) + 25\int_0^{\pi/6}(1 - 4\sin^2\theta)\,d\theta$
$= (25\pi/2) + 25\int_0^{\pi/6}(-1 + 2\cos 2\theta)\,d\theta$
$= (25\pi/2) + 25\left[-\theta + \sin 2\theta\right]_0^{\pi/6}$
$= (50\pi + 75\sqrt{3})/6$

15. $\int_0^{2\pi} \int_0^{2-\sin\theta} r\,dr\,d\theta$
$= \int_0^{2\pi} \left[r^2/2\right]_{r=0}^{r=2-\sin\theta} d\theta$
$= (1/2)\int_0^{2\pi}(4 - 4\sin\theta + \sin^2\theta)\,d\theta$
$= (1/4)\int_0^{2\pi}(8 - 8\sin\theta + 1 - \cos 2\theta)\,d\theta$
$= (1/4)\left[9\theta + 8\cos\theta - (1/2)\sin 2\theta\right]_0^{2\pi}$
$= 9\pi/2$

17. $\int_0^{\pi/2} \int_0^1 2r\,dr\,d\theta$
$= \int_0^{\pi/2} \left[r^2\right]_{r=0}^{r=1} d\theta$
$= \int_0^{\pi/2} 1\,d\theta$
$= \theta\Big|_0^{\pi/2}$
$= \pi/2$

19. $\int_0^{\pi/2} \int_0^1 re^{r^2}\,dr\,d\theta$
$= \int_0^{\pi/2} \left[(1/2)e^{r^2}\right]_{r=0}^{r=1} d\theta$
$= (1/2)\int_0^{\pi/2}(e - 1)\,d\theta$
$= (1/2)(e - 1)\theta\Big|_0^{\pi/2}$
$= \pi(e - 1)/4$

21. $\int_{-\pi/3}^{\pi/3} \int_{\sec\theta}^2 r\cos^2\theta\,dr\,d\theta$
$= \int_{-\pi/3}^{\pi/3} \left[(r^2/2)\cos^2\theta\right]_{r=\sec\theta}^{r=2} d\theta$
$= (1/2)\int_{-\pi/3}^{\pi/3}(4\cos^2\theta - 1)\,d\theta$
$= (1/2)\int_{-\pi/3}^{\pi/3}(1 + 2\cos 2\theta)\,d\theta$
$= (1/2)\left[\theta + \sin 2\theta\right]_{-\pi/3}^{\pi/3}$
$= (2\pi + 3\sqrt{3})/6$

23. $\int_0^{\pi/2} \int_0^{2\cos\theta} 1\,dr\,d\theta$
$= \int_0^{\pi/2} \left[r\right]_{r=0}^{r=2\cos\theta} d\theta$
$= \int_0^{\pi/2} 2\cos\theta\,d\theta$
$= 2\sin\theta\Big|_0^{\pi/2}$
$= 2$

25. $\int_0^\pi \int_0^2 r\cos\theta\,dr\,d\theta$
$= \int_0^\pi \left[(r^2/2)\cos\theta\right]_{r=0}^{r=2} d\theta$
$= \int_0^\pi 2\cos\theta\,d\theta$
$= 2\sin\theta\Big|_0^\pi$
$= 0$

27. If we move the cylinder in the y direction by -1, then the equation of the cylinder becomes $x^2 + y^2 = 4$, or $r = 2$, and the bottom and top are still bounded by the xy-plane and the plane $z = x + 4$ (since the latter does not depend upon y). This latter equation may be rewritten as $z = r\cos\theta + 4$. The volume is now given by:

$\int_0^{2\pi} \int_0^2 r(r\cos\theta + 4)\,dr\,d\theta$
$= \int_0^{2\pi} \int_0^2 (r^2\cos\theta + 4r)\,dr\,d\theta$
$= \int_0^{2\pi} \left[(r^3/3)\cos\theta + 2r^2\right]_{r=0}^{r=2} d\theta$
$= \int_0^{2\pi} \left[(8/3)\cos\theta + 8\right] d\theta$
$= (8/3)\sin\theta + 8\theta\Big|_0^{2\pi}$
$= 16\pi$

29. Compute the volume for the upper half of this region, and then multiply by 2. This region lies above the annulus between $r = 1$ and $r = 2$ in the xy-plane. It is bounded above by the sphere, whose equation may be rewritten as $r^2 + z^2 = 4$, or $z = \sqrt{4 - r^2}$.

$2 \int_0^{2\pi} \int_1^2 r\sqrt{4 - r^2} \, dr \, d\theta$ Let $u = 4 - r^2$, $du = -2r \, dr$.

$= 2 \int_0^{2\pi} \int_3^0 -(1/2)\sqrt{u} \, du \, d\theta$

$= \int_0^{2\pi} \int_0^3 \sqrt{u} \, du \, d\theta$

$= \int_0^{2\pi} \left[(2/3)u^{3/2} \right]_{u=0}^{u=3} d\theta$

$= \int_0^{2\pi} 2\sqrt{3} \, d\theta$

$= 2\sqrt{3}\,\theta \big|_0^{2\pi}$

$= 4\pi\sqrt{3}$

31. The equation for the plane which bounds the solid above may be rewritten as $z = r\sin\theta + 2$.

$\int_0^{2\pi} \int_0^{1+\cos\theta} (r\sin\theta + 2)r \, dr \, d\theta$

$= \int_0^{2\pi} \int_0^{1+\cos\theta} (r^2\sin\theta + 2r) \, dr \, d\theta$

$= \int_0^{2\pi} \left[(r^3/3)\sin\theta + r^2 \right]_{r=0}^{r=1+\cos\theta} d\theta$

$= (1/3)\int_0^{2\pi} [(1 + \cos\theta)^3 \sin\theta + 3(1 + 2\cos\theta + \cos^2\theta)] \, d\theta$

$= (1/3)\int_0^{2\pi} (1 + \cos\theta)^3 \sin\theta \, d\theta + \int_0^{2\pi} [1 + 2\cos\theta + (1/2)(1 + \cos 2\theta)] \, d\theta$

For the first integral, let $u = 1 + \cos\theta$ and $du = -\sin\theta \, d\theta$.

$-(1/3)\int_2^2 u^3 \, du + \left[\theta + 2\sin\theta + (\theta/2) + (1/4)\sin 2\theta \right]_0^{2\pi}$

$= 0 + 3\pi$

$= 3\pi$

33. Picture the cone with its base in the xy-plane and its axis on the positive z-axis. We will refer to the radius of the cone as R for the time being. The surface of the cone satisfies a linear function $z = mr + b$ where $z = h$ at $r = 0$ and $z = 0$ at $r = R$, since the vertex of the cone is at height h and the base of the cone is at height 0. The z-intercept is therefore $b = h$, and the slope is $m = -h/R$. Thus, the equation describing the surface of the cone is $z = (-h/R)r + h = (h/R)(-r + R)$. In order to determine the volume of the cone, we integrate this expression over the base of the cone, or in other words, over the region in the xy-plane bounded by the circle of radius R with center at the origin.

$\int_0^{2\pi} \int_0^R (h/R)(R - r)r \, dr \, d\theta$

$= (h/R) \int_0^{2\pi} \int_0^R (Rr - r^2) \, dr \, d\theta$

$= (h/R) \int_0^{2\pi} \left[(Rr^2/2) - (r^3/3) \right]_{r=0}^{r=R} d\theta$

$= (h/R) \int_0^{2\pi} (R^3/6) \, d\theta$

$= (h/R)(R^3/6)\theta \big|_0^{2\pi}$

$= hR^2\pi/3$

The volume is $\pi r^2 h/3$.

35. The two paraboloids intersect in the circle $x^2 + y^2 = 4$ (or $r = 2$) at height $z = 5$. The equations of the paraboloids may be rewritten as $z = 9 - r^2$ and $z = 1 + r^2$.

$\int_0^{2\pi} \int_0^2 [(9 - r^2) - (1 + r^2)]r \, dr \, d\theta$

$= \int_0^{2\pi} \int_0^2 (8r - 2r^3) \, dr \, d\theta$

$= \int_0^{2\pi} \left[4r^2 - (r^4/2) \right]_{r=0}^{r=2} d\theta$

$= \int_0^{2\pi} 8 \, d\theta$

$= 8\theta \big|_0^{2\pi}$

$= 16\pi$

37. $\int_0^1 \int_{\text{Sin}^{-1}r}^{\pi/2} \sin\theta \, d\theta \, dr$

$= \int_0^1 \left[-\cos\theta \right]_{\theta=\text{Sin}^{-1}r}^{\theta=\pi/2} dr$

$= \int_0^1 \cos\text{Sin}^{-1} r \, dr$

$= \int_0^1 \sqrt{1-r^2} \, dr$

$= (1/2)\left[r\sqrt{1-r^2} + \text{Sin}^{-1} r \right]_0^1$

$= \pi/4$

39. This is the solid above the region inside $r = 2$ in the xy-plane, and bounded above by $z = r$.

$\int_0^{2\pi} \int_0^2 r^2 \, dr \, d\theta$

$= \int_0^{2\pi} \left[r^3/3 \right]_{r=0}^{r=2} d\theta$

$= \int_0^{2\pi} (8/3) \, d\theta$

$= 8\theta/3 \Big|_0^{2\pi}$

$= 16\pi/3$

19.4   Calculating Mass and Centers of Mass

1. $\int_0^2 \int_0^1 (x + y) \, dy \, dx$

$= \int_0^2 \left[xy + (y^2/2) \right]_{y=0}^{y=1} dx$

$= \int_0^2 \left[x + (1/2) \right] dx$

$= (x^2/2) + (x/2) \Big|_0^2$

$= 3$

3. $\int_{-1}^1 \int_0^{1-x^2} (6 + x) \, dy \, dx$

$= \int_{-1}^1 \left[6y + xy \right]_{y=0}^{y=1-x^2} dx$

$= \int_{-1}^1 (6 + x - 6x^2 - x^3) \, dx$

$= 6x + (x^2/2) - 2x^3 - (x^4/4) \Big|_{-1}^1$

$= 8$

5. $\int_0^{\pi/4} \int_0^{\pi/4} \sin(x + y) \, dy \, dx$

$= \int_0^{\pi/4} \left[-\cos(x + y) \right]_{y=0}^{y=\pi/4} dx$

$= \int_0^{\pi/4} \left[\cos x - \cos(x + (\pi/4)) \right] dx$

$= \sin x - \sin(x + (\pi/4)) \Big|_0^{\pi/4}$

$= \sqrt{2} - 1$

7. $\int_{-1}^1 \int_0^{\sqrt{1-x^2}} (x^2 + y^2) \, dy \, dx$

$= \int_0^{\pi} \int_0^1 r^3 \, dr \, d\theta$

$= \int_0^{\pi} \left[r^4/4 \right]_{r=0}^{r=1} d\theta$

$= \int_0^{\pi} (1/4) \, d\theta$

$= \theta/4 \Big|_0^{\pi}$

$= \pi/4$

9. $\int_0^{2\pi} \int_1^2 1 \, dr \, d\theta$

$= \int_0^{2\pi} \left[r \right]_{r=1}^{r=2} d\theta$

$= \int_0^{2\pi} 1 \, d\theta$

$= \theta \Big|_0^{2\pi}$

$= 2\pi$

11. $M = 3$

$M_y = \int_0^2 \int_0^1 (x^2 + xy) \, dy \, dx$

$= \int_0^2 \left[x^2 y + (xy^2/2) \right]_{y=0}^{y=1} dx$

$= \int_0^2 \left[x^2 + (x/2) \right] dx$

$= (x^3/3) + (x^2/4) \Big|_0^2$

$= 11/3$

$M_x = \int_0^2 \int_0^1 (xy + y^2) \, dy \, dx$

$= \int_0^2 \left[(xy^2/2) + (y^3/3) \right]_{y=0}^{y=1} dx$

$= \int_0^2 \left[(x/2) + (1/3) \right] dx$

$= (x^2/4) + (x/3) \Big|_0^2$

$= 5/3$

$\bar{x} = M_y/M = (11/3)/3 = 11/9$

$\bar{y} = M_x/M = (5/3)/3 = 5/9$

The center of mass is $(11/9, 5/9)$.

13. $M = 4\pi/3$

$M_y = \int_0^{\pi/2} \int_0^2 x r^2 \, dr \, d\theta$

$= \int_0^{\pi/2} \int_0^2 r^3 \cos\theta \, dr \, d\theta$

$= \int_0^{\pi/2} \left[(r^4/4) \cos\theta \right]_{r=0}^{r=2} d\theta$

$= \int_0^{\pi/2} 4 \cos\theta \, d\theta$

$= 4 \sin\theta \big|_0^{\pi/2}$

$= 4$

$M_x = \int_0^{\pi/2} \int_0^2 y r^2 \, dr \, d\theta$

$= \int_0^{\pi/2} \int_0^2 r^3 \sin\theta \, dr \, d\theta$

$M_x = \int_0^{\pi/2} \left[(r^4/4) \sin\theta \right]_{r=0}^{r=2} d\theta$

$= \int_0^{\pi/2} 4 \sin\theta \, d\theta$

$= -4 \cos\theta \big|_0^{\pi/2}$

$= 4$

$\bar{x} = M_y/M = 4/(4\pi/3) = 3/\pi$

$\bar{y} = M_x/M = 4/(4\pi/3) = 3/\pi$

The center of mass is $(3/\pi, 3/\pi)$ in rectangular coordinates, or $(3\sqrt{2}/\pi, \pi/4)$ in polar coordinates.

15. $M = 18$

$M_y = \int_0^1 \int_0^{-4x+4} (xy - x^2 + 8x) \, dy \, dx$

$= \int_0^1 \left[(xy^2/2) - x^2 y + 8xy \right]_{y=0}^{y=-4x+4} dx$

$= \int_0^1 (12x^3 - 52x^2 + 40x) \, dx$

$= 3x^4 - (52x^3/3) + 20x^2 \big|_0^1$

$= 17/3$

$M_x = \int_0^1 \int_0^{-4x+4} (y^2 - xy + 8y) \, dy \, dx$

$= \int_0^1 \left[(y^3/3) - (xy^2/2) + 4y^2 \right]_{y=0}^{y=-4x+4} dx$

$= (1/3) \int_0^1 (-88x^3 + 432x^2 - 600x + 256) \, dx$

$= (1/3) \left[-22x^4 + 144x^3 - 300x^2 + 256x \right]_0^1$

$= 26$

$\bar{x} = M_y/M = (17/3)/18 = 17/54$

$\bar{y} = M_x/M = 26/18 = 13/9$

The center of mass is $(17/54, 13/9)$.

17. $M = \int_{-2}^2 \int_0^{4-x^2} 1 \, dy \, dx$

$= \int_{-2}^2 [y]_{y=0}^{y=4-x^2} dx$

$= \int_{-2}^2 (4 - x^2) \, dx$

$= 4x - (x^3/3) \big|_{-2}^2$

$= 32/3$

$M_y = \int_{-2}^2 \int_0^{4-x^2} x \, dy \, dx$

$= \int_{-2}^2 [xy]_{y=0}^{y=4-x^2} dx$

$= \int_{-2}^2 (4x - x^3) \, dx$

$= 2x^2 - (x^4/4) \big|_{-2}^2$

$= 0$

$M_x = \int_{-2}^2 \int_0^{4-x^2} y \, dy \, dx$

$= \int_{-2}^2 [y^2/2]_{y=0}^{y=4-x^2} dx$

$= (1/2) \int_{-2}^2 (16 - 8x^2 + x^4) \, dx$

$= (1/2) \left[16x - (8x^3/3) + (x^5/5) \right]_{-2}^2$

$= 256/15$

$\bar{x} = M_y/M = 0$

$\bar{y} = M_x/M = (256/15)/(32/3) = 8/5$

The centroid is $(0, 8/5)$.

19. $0 \le x \le 4$ and $0 \le y \le f(x)$ where $f(x) = \begin{cases} 1 & \text{if } 1 \le x < 2 \\ 4 & \text{if } 2 \le x \le 4 \end{cases}$

$M = \int_0^4 \int_0^{f(x)} 1 \, dy \, dx$

$= \int_0^2 \int_0^1 1 \, dy \, dx + \int_2^4 \int_0^4 1 \, dy \, dx$

$= \int_0^2 [y]_{y=0}^{y=1} \, dx + \int_2^4 [y]_{y=0}^{y=4} \, dx$

$= \int_0^2 1 \, dx + \int_2^4 4 \, dx$

$= [x]_0^2 + [4x]_2^4$

$= 10$

$M_y = \int_0^4 \int_0^{f(x)} x \, dy \, dx$

$= \int_0^2 \int_0^1 x \, dy \, dx + \int_2^4 \int_0^4 x \, dy \, dx$

$= \int_0^2 [xy]_{y=0}^{y=1} \, dx + \int_2^4 [xy]_{y=0}^{y=4} \, dx$

$= \int_0^2 x \, dx + \int_2^4 4x \, dx$

$= [x^2/2]_0^2 + [2x^2]_2^4$

$= 26$

$M_x = \int_0^4 \int_0^{f(x)} y \, dy \, dx$

$= \int_0^2 \int_0^1 y \, dy \, dx + \int_2^4 \int_0^4 y \, dy \, dx$

$= \int_0^2 [y^2/2]_{y=0}^{y=1} \, dx + \int_2^4 [y^2/2]_{y=0}^{y=4} \, dx$

$= \int_0^2 (1/2) \, dx + \int_2^4 8 \, dx$

$= [x/2]_0^2 + [8x]_2^4$

$= 17$

$\bar{x} = M_y/M = 26/10 = 13/5$

$\bar{y} = M_x/M = 17/10$

The centroid is $(13/5, 17/10)$.

21. $M = \int_0^a \int_0^{\sqrt{a^2-x^2}} 1 \, dy \, dx$

$= \int_0^{\pi/2} \int_0^a r \, dr \, d\theta$

$= \int_0^{\pi/2} [r^2/2]_{r=0}^{r=a} \, d\theta$

$= \int_0^{\pi/2} (a^2/2) \, d\theta$

$= (a^2/2)\theta \big|_0^{\pi/2}$

$= \pi a^2/4$

$M_y = \int_0^a \int_0^{\sqrt{a^2-x^2}} x \, dy \, dx$

$= \int_0^{\pi/2} \int_0^a xr \, dr \, d\theta$

$= \int_0^{\pi/2} \int_0^a r^2 \cos\theta \, dr \, d\theta$

$= \int_0^{\pi/2} [(r^3/3)\cos\theta]_{r=0}^{r=a} \, d\theta$

$= (a^3/3) \int_0^{\pi/2} \cos\theta \, d\theta$

$M_y = (a^3/3) \sin\theta \big|_0^{\pi/2}$

$= a^3/3$

$M_x = \int_0^a \int_0^{\sqrt{a^2-x^2}} y \, dy \, dx$

$= \int_0^{\pi/2} \int_0^a yr \, dr \, d\theta$

$= \int_0^{\pi/2} \int_0^a r^2 \sin\theta \, dr \, d\theta$

$= \int_0^{\pi/2} [(r^3/3)\sin\theta]_{r=0}^{r=a} \, d\theta$

$= (a^3/3) \int_0^{\pi/2} \sin\theta \, d\theta$

$= -(a^3/3)\cos\theta \big|_0^{\pi/2}$

$= a^3/3$

$\bar{x} = M_y/M = (a^3/3)/(\pi a^2/4) = (4a)/(3\pi)$

$\bar{y} = M_x/M = (a^3/3)/(\pi a^2/4) = (4a)/(3\pi)$

23. $M = \int_0^4 \int_0^{y(4-y)} 1 \, dx \, dy$

$= \int_0^4 [x]_{x=0}^{x=y(4-y)} \, dy$

$= \int_0^4 (4y - y^2) \, dy$

$= 2y^2 - (y^3/3) \big|_0^4$

$= 32/3$

$M_y = \int_0^4 \int_0^{y(4-y)} x \, dx \, dy$

$= \int_0^4 [x^2/2]_{x=0}^{x=y(4-y)} \, dy$

$= (1/2) \int_0^4 (16y^2 - 8y^3 + y^4) \, dy$

$M_y = (1/2) [(16y^3/3) - 2y^4 + (y^5/5)]_0^4$

$= 256/15$

$M_x = \int_0^4 \int_0^{y(4-y)} y \, dx \, dy$

$= \int_0^4 [xy]_{x=0}^{x=y(4-y)} \, dy$

$= \int_0^4 (4y^2 - y^3) \, dy$

$= (4y^3/3) - (y^4/4) \big|_0^4$

$= 64/3$

$\bar{x} = M_y/M = (256/15)/(32/3) = 8/5$

$\bar{y} = M_x/M = (64/3)/(32/3) = 2$

The centroid is $(8/5, 2)$.

25. Let the right triangle be the region given by $R = \{(x, y) \mid 0 \le x \le 6, 0 \le y \le x/3\}$. The density is then given by $\rho(x, y) = \lambda y$.

$$\int_0^6 \int_0^{x/3} \lambda y \, dy \, dx$$
$$= \int_0^6 \left[\lambda y^2/2 \right]_{y=0}^{y=x/3} dx$$
$$= \int_0^6 (\lambda x^2/18) \, dx$$
$$= \lambda x^3/54 \Big|_0^6$$
$$= 4\lambda$$

27. The equation of the hypotenuse can be written as $x = -2y + 10$, and the mass is given by

$$\int_0^5 \int_0^{-2y+10} \lambda(x^2 + y^2) \, dx \, dy.$$

This is the same integral that appears in Example 1 with the variables x and y exchanged, and thus produces the same value.

29. The points of intersection occur at $\theta = \pm\pi/3$.

$$M = \int_{-\pi/3}^{\pi/3} \int_{1+\cos\theta}^{3\cos\theta} (1 + r) r \, dr \, d\theta$$
$$= \int_{-\pi/3}^{\pi/3} \int_{1+\cos\theta}^{3\cos\theta} (r + r^2) \, dr \, d\theta$$
$$= \int_{-\pi/3}^{\pi/3} \left[(r^2/2) + (r^3/3) \right]_{r=1+\cos\theta}^{r=3\cos\theta} d\theta$$
$$= (1/6) \int_{-\pi/3}^{\pi/3} (52\cos^3\theta + 18\cos^2\theta - 12\cos\theta - 5) \, d\theta$$
$$= (26/3) \int_{-\pi/3}^{\pi/3} \cos^3\theta \, d\theta + 3 \int_{-\pi/3}^{\pi/3} \cos^2\theta \, d\theta - (1/6) \int_{-\pi/3}^{\pi/3} (12\cos\theta + 5) \, d\theta$$
For the first integral, use the substitution $u = \sin\theta$, $du = \cos\theta \, d\theta$.
Note that $\cos^2\theta = 1 - \sin^2\theta = 1 - u^2$.
$$M = (26/3) \int_{-\sqrt{3}/2}^{\sqrt{3}/2} (1 - u^2) \, du + (3/2) \int_{-\pi/3}^{\pi/3} (1 + \cos 2\theta) \, d\theta - (1/6) \left[12\sin\theta + 5\theta \right]_{-\pi/3}^{\pi/3}$$
$$= (26/3) \left[u - (u^3/3) \right]_{-\sqrt{3}/2}^{\sqrt{3}/2} + (3/2) \left[\theta + (1/2)\sin 2\theta \right]_{-\pi/3}^{\pi/3} - 2\sqrt{3} - (5\pi/9)$$
$$= (189\sqrt{3} + 16\pi)/36$$
$$M_y = \int_{-\pi/3}^{\pi/3} \int_{1+\cos\theta}^{3\cos\theta} x(1 + r) r \, dr \, d\theta$$
$$= \int_{-\pi/3}^{\pi/3} \int_{1+\cos\theta}^{3\cos\theta} (r^2 + r^3) \cos\theta \, dr \, d\theta$$
$$= \int_{-\pi/3}^{\pi/3} \left[((r^3/3) + (r^4/4)) \cos\theta \right]_{r=1+\cos\theta}^{r=3\cos\theta} d\theta$$
$$= (1/12) \int_{-\pi/3}^{\pi/3} (240\cos^5\theta + 92\cos^4\theta - 30\cos^3\theta - 24\cos^2\theta - 7\cos\theta) \, d\theta$$
$$= (1/12) \int_{-\pi/3}^{\pi/3} (240\cos^5\theta - 30\cos^3\theta - 7\cos\theta) \, d\theta + (23/3) \int_{-\pi/3}^{\pi/3} \cos^4\theta \, d\theta - 2 \int_{-\pi/3}^{\pi/3} \cos^2\theta \, d\theta$$
For the first integral, let $u = \sin\theta$, $du = \cos\theta \, d\theta$.
$$M_y = (1/12) \int_{-\sqrt{3}/2}^{\sqrt{3}/2} [240(1 - u^2)^2 - 30(1 - u^2) - 7] \, du + (23/12) \int_{-\pi/3}^{\pi/3} (1 + \cos 2\theta)^2 \, d\theta - \int_{-\pi/3}^{\pi/3} (1 + \cos 2\theta) \, d\theta$$
$$= (1/12) \int_{-\sqrt{3}/2}^{\sqrt{3}/2} (240u^4 - 450u^2 + 203) \, du + (23/12) \int_{-\pi/3}^{\pi/3} (1 + 2\cos 2\theta + \cos^2 2\theta) \, d\theta - \left[\theta + (1/2)\sin 2\theta \right]_{-\pi/3}^{\pi/3}$$
$$= (1/12) \left[48u^5 - 150u^3 + 203u \right]_{-\sqrt{3}/2}^{\sqrt{3}/2} + (23/12) \int_{-\pi/3}^{\pi/3} [1 + 2\cos 2\theta + (1/2) + (1/2)\cos 4\theta] \, d\theta - (2\pi/3) - (\sqrt{3}/2)$$
$$= (235\sqrt{3}/24) + (23/12) \left[(3\theta/2) + \sin 2\theta + (1/8)\sin 4\theta \right]_{-\pi/3}^{\pi/3} - (2\pi/3) - (\sqrt{3}/2)$$
$$= (351\sqrt{3} + 40\pi)/32$$

$M_x = \int_{-\pi/3}^{\pi/3} \int_{1+\cos\theta}^{3\cos\theta} y(1+r)r\,dr\,d\theta$

$= \int_{-\pi/3}^{\pi/3} \int_{1+\cos\theta}^{3\cos\theta} (r^2 + r^3)\sin\theta\,dr\,d\theta$

$= \int_{-\pi/3}^{\pi/3} \left[((r^3/3) + (r^4/4))\sin\theta \right]_{r=1+\cos\theta}^{r=3\cos\theta} d\theta$

$= (1/12) \int_{-\pi/3}^{\pi/3} (240\cos^4\theta + 92\cos^3\theta - 30\cos^2\theta - 24\cos\theta - 7)\sin\theta\,d\theta$

Let $u = \cos\theta$, $du = -\sin\theta\,d\theta$.

$M_x = -(1/12)\int_{1/2}^{1/2}(240u^4 + 92u^3 - 30u^2 - 24u - 7)\,du$

$= 0$

$\bar{x} = M_y/M = [(351\sqrt{3} + 40\pi)/32][36/(189\sqrt{3} + 16\pi)] = \dfrac{3159\sqrt{3} + 360\pi}{1512\sqrt{3} + 128\pi}$

$\bar{y} = M_x/M = 0$

19.5 Surface Area

1. $\partial f/\partial x = 1$ and $\partial f/\partial y = 1$

$\sqrt{(\partial f/\partial x)^2 + (\partial f/\partial y)^2 + 1} = \sqrt{3}$

$A_S = \int_0^1 \int_0^1 \sqrt{3}\,dy\,dx$

$= \int_0^1 \left[y\sqrt{3} \right]_{y=0}^{y=1} dx$

$= \int_0^1 \sqrt{3}\,dx$

$= x\sqrt{3}\Big|_0^1$

$= \sqrt{3}$

3. $\partial f/\partial x = -2x$ and $\partial f/\partial y = -2y$

$\sqrt{(\partial f/\partial x)^2 + (\partial f/\partial y)^2 + 1} = \sqrt{4x^2 + 4y^2 + 1} = \sqrt{4r^2 + 1}$

$A_S = \int_0^{2\pi} \int_0^{\sqrt{3}} r\sqrt{4r^2 + 1}\,dr\,d\theta$ Let $u = 4r^2 + 1$, $du = 8r\,dr$.

$= (1/8)\int_0^{2\pi}\int_1^{13} \sqrt{u}\,du\,d\theta$

$= (1/8)\int_0^{2\pi} \left[(2/3)u^{3/2} \right]_{u=1}^{u=13} d\theta$

$= (1/12)\int_0^{2\pi}(13\sqrt{13} - 1)\,d\theta$

$= (1/12)\left[(13\sqrt{13} - 1)\theta \right]_0^{2\pi}$

$= (13\sqrt{13} - 1)\pi/6$

5. $\partial f/\partial x = -1$ and $\partial f/\partial y = -1$

$\sqrt{(\partial f/\partial x)^2 + (\partial f/\partial y)^2 + 1} = \sqrt{3}$

$A_S = \int_0^2 \int_0^{2-x} \sqrt{3}\,dy\,dx$

$= \int_0^2 \left[y\sqrt{3} \right]_{y=0}^{y=2-x} dx$

$= \sqrt{3}\int_0^2 (2 - x)\,dx$

$= \sqrt{3}\left[2x - (x^2/2) \right]_0^2$

$= 2\sqrt{3}$

7. $\partial f/\partial x = x/\sqrt{x^2 + y^2} = x/r$ and $\partial f/\partial y = y/\sqrt{x^2 + y^2} = y/r$
$\sqrt{(\partial f/\partial x)^2 + (\partial f/\partial y)^2 + 1} = \sqrt{((x^2 + y^2)/r^2) + 1} = \sqrt{2}$

$A_S = \int_0^{2\pi} \int_1^2 r\sqrt{2} \, dr \, d\theta$

$ = \sqrt{2} \int_0^{2\pi} [r^2/2]_1^2 \, d\theta$

$ = (\sqrt{2}/2) \int_0^{2\pi} 3 \, d\theta$

$ = (3\sqrt{2}/2)\theta \big|_0^{2\pi}$

$ = 3\pi\sqrt{2}$

9. $\partial f/\partial x = -2x$ and $\partial f/\partial y = \sqrt{3}$
$\sqrt{(\partial f/\partial x)^2 + (\partial f/\partial y)^2 + 1} = \sqrt{4x^2 + 4} = 2\sqrt{x^2 + 1}$

$A_S = \int_0^1 \int_0^1 2\sqrt{x^2 + 1} \, dy \, dx$

$ = \int_0^1 [2y\sqrt{x^2 + 1}]_{y=0}^{y=1} \, dx$

$ = 2\int_0^1 \sqrt{x^2 + 1} \, dx$

$ = x\sqrt{x^2 + 1} + \ln(x + \sqrt{x^2 + 1}) \big|_0^1$

$ = \sqrt{2} + \ln(1 + \sqrt{2})$

11. $\partial f/\partial x = \tan x$ and $\partial f/\partial y = 0$
$\sqrt{(\partial f/\partial x)^2 + (\partial f/\partial y)^2 + 1} = \sqrt{\tan^2 x + 1} = \sqrt{\sec^2 x} = \sec x$ since $\sec x > 0$ for $0 \le x \le \pi/4$.

$A_S = \int_0^{\pi/4} \int_0^{\sec^2 x} \sec x \, dy \, dx$

$ = \int_0^{\pi/4} [y \sec x]_{y=0}^{y=\sec^2 x} \, dx$

$ = \int_0^{\pi/4} \sec^3 x \, dx \quad$ Use Integration by parts, eventually obtaining:

$ = (1/2) \sec x \tan x + (1/2) \ln|\sec x + \tan x| \, \big|_0^{\pi/4}$

$ = (\sqrt{2}/2) + (1/2)\ln(1 + \sqrt{2})$

13. $\partial z/\partial x = 4x$ and $\partial z/\partial y = 1$
$\sqrt{(\partial z/\partial x)^2 + (\partial z/\partial y)^2 + 1} = \sqrt{16x^2 + 2}$

$A_S = \int_0^1 \int_x^1 \sqrt{16x^2 + 2} \, dy \, dx$

$ = \int_0^1 [y\sqrt{16x^2 + 2}]_{y=x}^{y=1} \, dx$

$ = \int_0^1 (1 - x)\sqrt{16x^2 + 2} \, dx$

$ = 4\int_0^1 \sqrt{x^2 + (1/8)} \, dx - \int_0^1 x\sqrt{16x^2 + 2} \, dx$

Use the Table of Integrals for the first integral, and the substitution $u = 16x^2 + 2$, $du = 32x \, dx$ for the second integral.

$A_S = 2[x\sqrt{x^2 + (1/8)} + (1/8)\ln(x + \sqrt{x^2 + (1/8)})]_0^1 - (1/32)\int_2^{18} \sqrt{u} \, du$

$ = (3\sqrt{2}/2) + (1/4)\ln((4 + 3\sqrt{2})/4) - (1/4)\ln(\sqrt{2}/4) - (1/32)[(2/3)u^{3/2}]_2^{18}$

$ = (3\sqrt{2}/2) + (1/4)\ln(4 + 3\sqrt{2}) - (1/4)\ln 4 - (1/4)\ln(\sqrt{2}) + (1/4)\ln 4 - (9\sqrt{2}/8) + (\sqrt{2}/24)$

$ = (5\sqrt{2}/12) + (1/4)\ln(4 + 3\sqrt{2}) - (1/8)\ln 2$

15. This part of the paraboloid lies above the region in the xy-plane between the circles $x^2+y^2 = 7$ ($r = \sqrt{7}$) and $x^2 + y^2 = 12$ ($r = 2\sqrt{3}$).

$\partial z/\partial x = -2x$ and $\partial z/\partial y = -2y$

$\sqrt{(\partial z/\partial x)^2 + (\partial z/\partial y)^2 + 1} = \sqrt{4x^2 + 4y^2 + 1} = \sqrt{4r^2 + 1}$

$A_S = \int_0^{2\pi} \int_{\sqrt{7}}^{2\sqrt{3}} r\sqrt{4r^2 + 1} \, dr \, d\theta$ Let $u = 4r^2 + 1$, $du = 8r \, dr$.

$= (1/8) \int_0^{2\pi} \int_{29}^{49} \sqrt{u} \, du \, d\theta$

$= (1/8) \int_0^{2\pi} \left[(2/3)u^{3/2} \right]_{u=29}^{u=49} d\theta$

$= (1/12) \int_0^{2\pi} (343 - 29\sqrt{29}) \, d\theta$

$= (1/12)(343 - 29\sqrt{29})\theta \Big|_0^{2\pi}$

$= (343 - 29\sqrt{29})\pi/6$

17. $\partial z/\partial x = -x/\sqrt{4 - x^2 - y^2}$ and $\partial z/\partial y = -y/\sqrt{4 - x^2 - y^2}$

$\sqrt{(\partial f/\partial z)^2 + (\partial f/\partial z)^2 + 1} = \sqrt{(x^2 + y^2)/(4 - x^2 - y^2) + 1} = \sqrt{4/(4 - x^2 - y^2)} = 2/\sqrt{4 - r^2}$

$A_S = \int_0^{2\pi} \int_0^1 (2r/\sqrt{4 - r^2}) \, dr \, d\theta$ Let $u = 4 - r^2$, $du = -2r \, dr$.

$= -\int_0^{2\pi} \int_4^3 (1/\sqrt{u}) \, du \, d\theta$

$= \int_0^{2\pi} \int_3^4 u^{-1/2} \, du \, d\theta$

$= \int_0^{2\pi} \left[2u^{1/2} \right]_{u=3}^{u=4} d\theta$

$= \int_0^{2\pi} (4 - 2\sqrt{3}) \, d\theta$

$= (4 - 2\sqrt{3})\theta \Big|_0^{2\pi}$

$= 8\pi - 4\pi\sqrt{3}$

19. The surface intersects the xy-plane in the two lines $y = \pm x$, and the region in the xy-plane above which the graph lies is the region where $|y| > |x|$, or in other words, the two regions bounded by the lines $y = \pm x$ which contain the y-axis. In polar coordinates, these are the regions between $\theta = \pi/4$ and $\theta = 3\pi/4$, and between $\theta = 5\pi/4$ and $\theta = 7\pi/4$.

$\partial z/\partial x = -2x$ and $\partial z/\partial y = 2y$

$\sqrt{(\partial z/\partial x)^2 + (\partial z/\partial y)^2 + 1} = \sqrt{4x^2 + 4y^2 + 1} = \sqrt{4r^2 + 1}$

$A_S = \int_{\pi/4}^{3\pi/4} \int_0^5 r\sqrt{4r^2 + 1} \, dr \, d\theta + \int_{5\pi/4}^{7\pi/4} \int_0^5 r\sqrt{4r^2 + 1} \, dr \, d\theta$

$= \int_0^5 \int_{\pi/4}^{3\pi/4} r\sqrt{4r^2 + 1} \, d\theta \, dr + \int_0^5 \int_{5\pi/4}^{7\pi/4} r\sqrt{4r^2 + 1} \, d\theta \, dr$

$= \int_0^5 \left[\theta r\sqrt{4r^2 + 1} \right]_{\theta=\pi/4}^{\theta=3\pi/4} dr + \int_0^5 \left[\theta r\sqrt{4r^2 + 1} \right]_{\theta=5\pi/4}^{\theta=7\pi/4} dr$

$= \int_0^5 (\pi/2)r\sqrt{4r^2 + 1} \, dr + \int_0^5 (\pi/2)r\sqrt{4r^2 + 1} \, dr$

$= \pi \int_0^5 r\sqrt{4r^2 + 1} \, dr$ Let $u = 4r^2 + 1$, $du = 8r \, dr$.

$= (\pi/8) \int_1^{101} \sqrt{u} \, du$

$= (\pi/8) \left[(2/3)u^{3/2} \right]_1^{101}$

$= (101\sqrt{101} - 1)\pi/12$

21. The cylinder and sphere intersect in the circles $y^2 + z^2 = 1$ and $x = \pm 7$. Find the area of that part of the cylinder which lies in the first octant and then multiply by 8. The equation of the upper half of the cylinder is $z = \sqrt{1 - y^2}$.

$\partial z/\partial x = 0$ and $\partial z/\partial y = -y/\sqrt{1 - y^2}$

$\sqrt{(\partial z/\partial x)^2 + (\partial z/\partial y)^2 + 1} = \sqrt{y^2/(1 - y^2) + 1} = 1/\sqrt{1 - y^2}$

$A_S = 8 \int_0^7 \int_0^1 (1/\sqrt{1 - y^2}) \, dy \, dx$

Note that this is an improper integral, since the integrand is undefined at $y = 1$.

601

$A_S = 8 \int_0^7 \left[\lim_{b \to 1^-} \int_0^b (1/\sqrt{1-y^2})\, dy \right] dx$

$= 8 \int_0^7 \left[\lim_{b \to 1^-} [\operatorname{Sin}^{-1}(y)]_{y=0}^{y=b} \right] dx$

$= 8 \int_0^7 (\pi/2)\, dx$

$= 28\pi$

23. This portion of the paraboloid lies above the region inside of the circle $x^2 + y^2 = 1$.

$\partial z/\partial x = -2x$ and $\partial z/\partial y = -2y + 2 = -2(y-1)$, $\sqrt{(\partial z/\partial x)^2 + (\partial z/\partial y)^2 + 1} = \sqrt{4x^2 + 4(y-1)^2 + 1}$

$A_S = \int_{-1}^1 \int_{-\sqrt{1-x^2}}^{\sqrt{1-x^2}} \sqrt{4x^2 + 4(y-1)^2 + 1}\, dy\, dx$ Let $u = y - 1$, $du = dy$.

$= \int_{-1}^1 \int_{-1-\sqrt{1-x^2}}^{-1+\sqrt{1-x^2}} \sqrt{4x^2 + 4u^2 + 1}\, du\, dx$ Replace the variable u by the variable y.

$= \int_{-1}^1 \int_{-1-\sqrt{1-x^2}}^{-1+\sqrt{1-x^2}} \sqrt{4x^2 + 4y^2 + 1}\, dy\, dx$ Change to polar coordinates.

$= \int_{\pi}^{2\pi} \int_0^{-2\sin\theta} r\sqrt{4r^2 + 1}\, dr\, d\theta$ Let $u = 4r^2 + 1$, $du = 8r\, dr$.

$= (1/8) \int_{\pi}^{2\pi} \int_1^{16\sin^2\theta + 1} \sqrt{u}\, du\, d\theta$

$= (1/8) \int_{\pi}^{2\pi} \left[(2/3)u^{3/2} \right]_{u=1}^{u=16\sin^2\theta + 1} d\theta$

$= (1/12) \int_{\pi}^{2\pi} [(16\sin^2\theta + 1)^{3/2} - 1]\, d\theta$

Use Simpson's Rule on this integral. The value of the integral is approximately 7.90418.

25. Find the area of the upper hemisphere and multiply by 2. The equation for the upper hemisphere of a sphere of radius R may be written $z = \sqrt{R^2 - x^2 - y^2}$.

$\partial z/\partial x = -x/\sqrt{R^2 - x^2 - y^2}$ and $\partial z/\partial y = -y/\sqrt{R^2 - x^2 - y^2}$

$\sqrt{(\partial z/\partial x)^2 + (\partial z/\partial y)^2 + 1} = \sqrt{(x^2 + y^2)/(R^2 - x^2 - y^2) + 1} = R/\sqrt{R^2 - x^2 - y^2} = R/\sqrt{R^2 - r^2}$

$A_S = 2 \int_0^{2\pi} \int_0^R (rR/\sqrt{R^2 - r^2})\, dr\, d\theta$

The inner integral is an improper integral, since the integrand is undefined at $r = R$.

$A_S = 2 \int_0^{2\pi} \left[\lim_{b \to R^-} \int_0^b (rR/\sqrt{R^2 - r^2})\, dr \right] d\theta$ Let $u = R^2 - r^2$, $du = -2r\, dr$.

$= 2 \int_0^{2\pi} \left[\lim_{b \to R^-} (-1/2) \int_{R^2}^{R^2 - b^2} (R/\sqrt{u})\, du \right] d\theta$

$= \int_0^{2\pi} \left[\lim_{b \to R^-} \int_{R^2 - b^2}^{R^2} Ru^{-1/2}\, du \right] d\theta$

$= \int_0^{2\pi} \left[\lim_{b \to R^-} [2Ru^{1/2}]_{u=R^2-b^2}^{u=R^2} \right] d\theta$

$= \int_0^{2\pi} \left[\lim_{b \to R^-} 2R^2 - 2R\sqrt{R^2 - b^2} \right] d\theta$

$= 2R^2 \int_0^{2\pi} 1\, d\theta = 2R^2 \theta \big|_0^{2\pi}$

$= 4\pi R^2$

27. $\partial x/\partial y = -2y$ and $\partial x/\partial z = 0$

$\sqrt{(\partial x/\partial y)^2 + (\partial x/\partial z)^2 + 1} = \sqrt{4y^2 + 1}$

$A_S = \int_0^1 \int_0^{1-y} \sqrt{4y^2 + 1}\, dz\, dy$

$= \int_0^1 [z\sqrt{4y^2 + 1}]_{z=0}^{z=1-y}\, dy$

$= \int_0^1 (1-y)\sqrt{4y^2 + 1}\, dy$

$= 2 \int_0^1 \sqrt{y^2 + (1/4)}\, dy - \int_0^1 y\sqrt{4y^2 + 1}\, dy$

Use the Table of Integrals for the first integral, and for the second, use the substitution $u = 4y^2 + 1$, $du = 8y\, dy$.

$A_S = \left[y\sqrt{y^2 + (1/4)} + (1/4)\ln(y + \sqrt{y^2 + (1/4)}) \right]_0^1 - (1/8)\int_1^5 \sqrt{u}\, du$

$= (\sqrt{5}/2) + (1/4)\ln((2 + \sqrt{5})/2) - (1/4)\ln(1/2) - \left[(1/12)u^{3/2} \right]_1^5$

$= (\sqrt{5}/2) + (1/4)\ln(2 + \sqrt{5}) - (1/4)\ln 2 + (1/4)\ln 2 - (5\sqrt{5}/12) + (1/12)$

$= [(1 + \sqrt{5})/12] + (1/4)\ln(2 + \sqrt{5})$

29. The surface and cylinder intersect in the circle $x^2 + z^2 = 4$ at $y = 4$.

$\partial y/\partial x = 2x$ and $\partial y/\partial z = 2z$

$\sqrt{(\partial y/\partial x)^2 + (\partial y/ddz)^2 + 1} = \sqrt{4x^2 + 4z^2 + 1}$

$A_S = \int_0^1 \int_0^{\sqrt{4-x^2}} \sqrt{4x^2 + 4z^2 + 1}\, dz\, dx$ Replace the variable z by the variable y.

$= \int_0^1 \int_0^{\sqrt{4-x^2}} \sqrt{4x^2 + 4y^2 + 1}\, dy\, dx$ Convert to polar coordinates.

$= \int_0^{\pi/2} \int_0^2 r\sqrt{4r^2 + 1}\, dr\, d\theta$ Let $u = 4r^2 + 1$, $du = 8r\, dr$.

$= (1/8) \int_0^{\pi/2} \int_1^{17} \sqrt{u}\, du\, d\theta$

$= (1/8) \int_0^{\pi/2} \left[(2/3)u^{3/2}\right]_{u=1}^{u=17} d\theta$

$= (1/12) \int_0^{\pi/2} (17\sqrt{17} - 1)\, d\theta$

$= (1/12)(17\sqrt{17} - 1)\theta \big|_0^{\pi/2}$

$= (17\sqrt{17} - 1)\pi/24$

31. a. From Chapter 18, we know that one normal vector to the surface at P_m is $\mathbf{N} = \mathbf{u}_x \times \mathbf{u}_y$, where
$\mathbf{u}_x = \mathbf{i} + (\partial f/\partial x)(x_m, y_m)\mathbf{k}$ and $\mathbf{u}_y = \mathbf{j} + (\partial f/\partial y)(x_m, y_m)\mathbf{k}$. But $\mathbf{u}_m = (\Delta x)\mathbf{u}_x$ and $\mathbf{v}_m = (\Delta y)\mathbf{u}_y$.
Therefore, $\mathbf{N}_m = \mathbf{u}_m \times \mathbf{v}_m = (\Delta x \Delta y)(\mathbf{u}_x \times \mathbf{u}_y) = (\Delta x \Delta y)\mathbf{N}$ is normal to the surface at P_m.

b. $\mathbf{N} = \mathbf{u}_x \times \mathbf{u}_y = -(\partial f/\partial x)(x_m, y_m)\mathbf{i} - (\partial f/\partial y)(x_m, y_m)\mathbf{j} + \mathbf{k}$
$\mathbf{N} \cdot \mathbf{k} = 1$
$\mathbf{N}_m \cdot \mathbf{k} = (\Delta x \Delta y)\mathbf{N} \cdot \mathbf{k} = \Delta x \Delta y = \Delta A$

c. $\mathbf{N}_m \cdot \mathbf{k} = |\mathbf{N}_m|\,|\mathbf{k}| \cos\theta = |\mathbf{u}_m \times \mathbf{v}_m| \cos\theta$ since $|\mathbf{k}| = 1$.

d. This is obvious.

e. $A_S \approx \sum_{m=1}^n \Delta T_m = \sum_{m=1}^n \sec\theta\, \Delta A$
This Riemann sum converges to $\iint\limits_Q \sec\theta\, dA$.

19.6   Triple Integrals

1. $\int_0^1 \int_0^1 \int_0^1 xyz\, dx\, dy\, dz$

$= \int_0^1 \int_0^1 \left[(x^2/2)yz\right]_{x=0}^{x=1} dy\, dz$

$= (1/2) \int_0^1 \int_0^1 yz\, dy\, dz$

$= (1/2) \int_0^1 \left[(y^2/2)z\right]_{y=0}^{y=1} dz$

$= (1/4) \int_0^1 z\, dz$

$= z^2/8 \big|_0^1$

$= 1/8$

3. $\int_0^1 \int_0^y \int_0^x 3\, dz\, dx\, dy$

$= \int_0^1 \int_0^y \left[3z\right]_{z=0}^{z=x} dx\, dy$

$= \int_0^1 \int_0^y 3x\, dx\, dy$

$= \int_0^1 \left[3x^2/2\right]_{x=0}^{x=y} dy$

$= \int_0^1 (3y^2/2)\, dy$

$= y^3/2 \big|_0^1$

$= 1/2$

5. $\int_0^2 \int_0^x \int_0^{x+y} z \, dz \, dy \, dx$

$= \int_0^2 \int_0^x [z^2/2]_{z=0}^{z=x+y} \, dy \, dx$

$= (1/2) \int_0^2 \int_0^x (x^2 + 2xy + y^2) \, dy \, dx$

$= (1/2) \int_0^2 [x^2 y + xy^2 + (y^3/3)]_{y=0}^{y=x} \, dx$

$= (7/6) \int_0^2 x^3 \, dx$

$= (7/6)(x^4/4)\big|_0^2$

$= 14/3$

7. $\int_{-1}^1 \int_0^y \int_0^x ye^{x^2+y^2} \, dz \, dx \, dy$

$= \int_{-1}^1 \int_0^y [yze^{x^2+y^2}]_{z=0}^{z=x} \, dx \, dy$

$= \int_{-1}^1 \int_0^y xye^{x^2+y^2} \, dx \, dy$

Let $u = x^2 + y^2$, $du = 2x \, dx$.

$(1/2) \int_{-1}^1 \int_{y^2}^{2y^2} ye^u \, du \, dy$

$= (1/2) \int_{-1}^1 [ye^u]_{u=y^2}^{u=2y^2} \, dy$

$= (1/2) \int_{-1}^1 (ye^{2y^2} - ye^{y^2}) \, dy$

$= (1/2)[(1/4)e^{2y^2} - (1/2)e^{y^2}]_{-1}^1$

$= 0$

9. $\int_0^2 \int_{y/2}^1 \int_0^{x+y} f(x,y,z) \, dz \, dx \, dy$

11. $\int_{-2}^2 \int_{-\sqrt{4-x^2}}^{\sqrt{4-x^2}} \int_{x^2+y^2}^4 f(x,y,z) \, dz \, dy \, dx$

13.

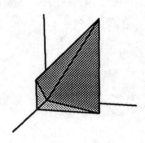

$\int_0^2 \int_0^{2x} \int_0^{x+y} dz \, dy \, dx$

For the sake of clarity, the solid has been drawn with its front face transparent.

15. $\int_{-1}^1 \int_{-\sqrt{1-x^2}}^{\sqrt{1-x^2}} \int_{\sqrt{x^2+y^2}}^{2-\sqrt{x^2+y^2}} dz \, dy \, dx$

$= \int_{-1}^1 \int_{-\sqrt{1-x^2}}^{\sqrt{1-x^2}} [z]_{z=\sqrt{x^2+y^2}}^{z=2-\sqrt{x^2+y^2}} \, dy \, dx$

$= \int_{-1}^1 \int_{-\sqrt{1-x^2}}^{\sqrt{1-x^2}} (2 - 2\sqrt{x^2+y^2}) \, dy \, dx$

$= \int_0^{2\pi} \int_0^1 (2r - 2r^2) \, dr \, d\theta$

$= \int_0^{2\pi} [r^2 - (2r^3/3)]_{r=0}^{r=1} \, d\theta$

$= \int_0^{2\pi} (1/3) \, d\theta$

$= \theta/3 \big|_0^{2\pi}$

$= 2\pi/3$

17. $\int_{-3}^3 \int_{-\sqrt{9-x^2}}^{\sqrt{9-x^2}} \int_0^{y+3} 1 \, dz \, dy \, dx$

$= \int_{-3}^3 \int_{-\sqrt{9-x^2}}^{\sqrt{9-x^2}} [z]_{z=0}^{z=y+3} \, dy \, dx$

$= \int_{-3}^3 \int_{-\sqrt{9-x^2}}^{\sqrt{9-x^2}} (y+3) \, dy \, dx$

$= \int_0^{2\pi} \int_0^3 (r \sin\theta + 3) r \, dr \, d\theta$

$= \int_0^{2\pi} \int_0^3 (r^2 \sin\theta + 3r) \, dr \, d\theta$

$= \int_0^{2\pi} [(r^3/3)\sin\theta + (3r^2/2)]_{r=0}^{r=3} \, d\theta$

$= (1/2) \int_0^{2\pi} (18\sin\theta + 27) \, d\theta$

$= (1/2)[-18\cos\theta + 27\theta]_0^{2\pi}$

$= 27\pi$

19. $\int_{-3}^3 \int_{-\sqrt{9-x^2}}^{\sqrt{9-x^2}} \int_0^{3-z} (3x+xz) \, dy \, dz \, dx$

$= \int_{-3}^3 \int_{-\sqrt{9-x^2}}^{\sqrt{9-x^2}} [(3x+xz)y]_{y=0}^{y=3-z} \, dz \, dx$

$= \int_{-3}^3 \int_{-\sqrt{9-x^2}}^{\sqrt{9-x^2}} (9x - xz^2) \, dz \, dx$

Replace the variable z by the variable y.

$\int_{-3}^3 \int_{-\sqrt{9-x^2}}^{\sqrt{9-x^2}} (9x - xy^2) \, dy \, dx$

Change to polar coordinates.

$\int_0^{2\pi} \int_0^3 (9r\cos\theta - r^3\cos\theta\sin^2\theta) r \, dr \, d\theta$

$= \int_0^{2\pi} \int_0^3 (9r^2\cos\theta - r^4\cos\theta\sin^2\theta) \, dr \, d\theta$

$= \int_0^{2\pi} [3r^3\cos\theta - (r^5/5)\cos\theta\sin^2\theta]_{r=0}^{r=3} \, d\theta$

$= \int_0^{2\pi} [81\cos\theta - (243/5)\cos\theta\sin^2\theta] \, d\theta$

$= [81\sin\theta]_0^{2\pi} - (243/5)\int_0^{2\pi} \cos\theta\sin^2\theta \, d\theta$

Let $u = \sin\theta$, $du = \cos\theta \, d\theta$.

$-(243/5)\int_0^0 u^2 \, du$

$= 0$

21. $\int_{-4\sqrt{2}}^{4\sqrt{2}} \int_{-\sqrt{32-z^2}}^{\sqrt{32-z^2}} \int_{1}^{9-(y^2+z^2)/4} 1 \, dx \, dy \, dz$

$= \int_{-4\sqrt{2}}^{4\sqrt{2}} \int_{-\sqrt{32-z^2}}^{\sqrt{32-z^2}} [x]_{x=1}^{x=9-(y^2+z^2)/4} \, dy \, dz$

$= (1/4) \int_{-4\sqrt{2}}^{4\sqrt{2}} \int_{-\sqrt{32-z^2}}^{\sqrt{32-z^2}} (32 - y^2 - z^2) \, dy \, dz$

Replace the variable z by the variable x.

$(1/4) \int_{-4\sqrt{2}}^{4\sqrt{2}} \int_{-\sqrt{32-x^2}}^{\sqrt{32-x^2}} (32 - y^2 - x^2) \, dy \, dx$

Change to polar coordinates.

$(1/4) \int_{0}^{2\pi} \int_{0}^{4\sqrt{2}} (32 - r^2) r \, dr \, d\theta$

$= (1/4) \int_{0}^{2\pi} \int_{0}^{4\sqrt{2}} (32r - r^3) \, dr \, d\theta$

$= (1/4) \int_{0}^{2\pi} [16r^2 - (r^4/4)]_{r=0}^{r=4\sqrt{2}} \, d\theta$

$= (1/4) \int_{0}^{2\pi} 256 \, d\theta$

$= 64\theta \big|_{0}^{2\pi}$

$= 128\pi$

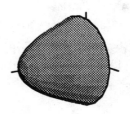

23. The two paraboloids intersect in the circle $y^2 + z^2 = 1$ at $x = 1$.

$\int_{-1}^{1} \int_{-\sqrt{1-y^2}}^{\sqrt{1-y^2}} \int_{y^2+z^2}^{2-y^2-z^2} 1 \, dx \, dz \, dy$

$= \int_{-1}^{1} \int_{-\sqrt{1-y^2}}^{\sqrt{1-y^2}} [x]_{x=y^2+z^2}^{x=2-y^2-z^2} \, dz \, dy$

$= \int_{-1}^{1} \int_{-\sqrt{1-y^2}}^{\sqrt{1-y^2}} (2 - 2y^2 - 2z^2) \, dz \, dy$

Replace the variable z by the variable x, then change to polar coordinates.

$\int_{0}^{2\pi} \int_{0}^{1} (2 - 2r^2) r \, dr \, d\theta$

$= \int_{0}^{2\pi} \int_{0}^{1} (2r - 2r^3) \, dr \, d\theta$

$= \int_{0}^{2\pi} [r^2 - (r^4/2)]_{r=0}^{r=1} \, d\theta$

$= \int_{0}^{2\pi} (1/2) \, d\theta$

$= \theta/2 \big|_{0}^{2\pi}$

$= \pi$

25. $\int_{-2}^{2} \int_{x^2}^{4} \int_{z}^{4} 1 \, dy \, dz \, dx$

$= \int_{-2}^{2} \int_{x^2}^{4} [y]_{y=z}^{y=4} \, dz \, dx$

$= \int_{-2}^{2} \int_{x^2}^{4} (4 - z) \, dz \, dx$

$= \int_{-2}^{2} [4z - (z^2/2)]_{z=x^2}^{z=4} \, dx$

$= (1/2) \int_{-2}^{2} (16 - 8x^2 + x^4) \, dx$

$= (1/2) [16x - (8x^3/3) + (x^5/5)]_{-2}^{2}$

$= 256/15$

27. $\int_{-3}^{3} \int_{-\sqrt{9-x^2}}^{\sqrt{9-x^2}} \int_{0}^{3-x} z \, dz \, dy \, dx$

$= \int_{-3}^{3} \int_{-\sqrt{9-x^2}}^{\sqrt{9-x^2}} [z^2/2]_{z=0}^{z=3-x} \, dy \, dx$

$= (1/2) \int_{-3}^{3} \int_{-\sqrt{9-x^2}}^{\sqrt{9-x^2}} (x^2 - 6x + 9) \, dy \, dx$

$= (1/2) \int_{0}^{2\pi} \int_{0}^{3} (r^2 \cos^2 \theta - 6r \cos \theta + 9) r \, dr \, d\theta$

$= (1/2) \int_{0}^{2\pi} \int_{0}^{3} (r^3 \cos^2 \theta - 6r^2 \cos \theta + 9r) \, dr \, d\theta$

$= (1/2) \int_{0}^{2\pi} [(r^4/4) \cos^2 \theta - 2r^3 \cos \theta + (9r^2/2)]_{r=0}^{r=3} \, d\theta$

$= (27/8) \int_{0}^{2\pi} (3 \cos^2 \theta - 8 \cos \theta + 6) \, d\theta$

$= (27/16) \int_{0}^{2\pi} (3 + 3 \cos 2\theta - 16 \cos \theta + 12) \, d\theta$

$= (27/16) [15\theta + (3/2) \sin 2\theta - 16 \sin \theta]_{0}^{2\pi}$

$= 405\pi/8$

29. $\int_0^1 \int_0^2 \int_0^2 \lambda(x^2 + y^2 + z^2)\, dz\, dy\, dx$

$= \lambda \int_0^1 \int_0^2 \left[x^2 z + y^2 z + (z^3/3)\right]_{z=0}^{z=2} dy\, dx$

$= (\lambda/3) \int_0^1 \int_0^2 (6x^2 + 6y^2 + 8)\, dy\, dx$

$= (\lambda/3) \int_0^1 \left[6x^2 y + 2y^3 + 8y\right]_0^2 dx$

$= (\lambda/3) \int_0^1 (12x^2 + 32)\, dx$

$= (\lambda/3) \left[4x^3 + 32x\right]_0^1$

$= 12\lambda$

31. Let ρ represent the constant density. From the answer to Exercise 17, the mass of the region is $27\rho\pi$.

$M_{yz} = \int_{-3}^3 \int_{-\sqrt{9-x^2}}^{\sqrt{9-x^2}} \int_0^{y+3} x\rho\, dz\, dy\, dx$

$= \rho \int_{-3}^3 \int_{-\sqrt{9-x^2}}^{\sqrt{9-x^2}} \left[xz\right]_{z=0}^{z=y+3} dy\, dx$

$= \rho \int_{-3}^3 \int_{-\sqrt{9-x^2}}^{\sqrt{9-x^2}} (xy + 3x)\, dy\, dx$

$= \rho \int_0^{2\pi} \int_0^3 (r^2 \cos\theta \sin\theta + 3r\cos\theta)r\, dr\, d\theta$

$= \rho \int_0^{2\pi} \int_0^3 (r^3 \cos\theta \sin\theta + 3r^2 \cos\theta)\, dr\, d\theta$

$= \rho \int_0^{2\pi} \left[(r^4/4)\cos\theta\sin\theta + r^3\cos\theta)\right]_{r=0}^{r=3} d\theta$

$= (\rho/4) \int_0^{2\pi} (81\cos\theta\sin\theta + 108\cos\theta)\, d\theta$

$= (\rho/8) \int_0^{2\pi} (81\sin 2\theta + 216\cos\theta)\, d\theta$

$= (\rho/8)\left[-(81/2)\cos 2\theta + 216\sin\theta\right]_0^{2\pi}$

$= 0$

$M_{xz} = \int_{-3}^3 \int_{-\sqrt{9-x^2}}^{\sqrt{9-x^2}} \int_0^{y+3} y\rho\, dz\, dy\, dx$

$= \rho \int_{-3}^3 \int_{-\sqrt{9-x^2}}^{\sqrt{9-x^2}} \left[yz\right]_{z=0}^{z=y+3} dy\, dx$

$= \rho \int_{-3}^3 \int_{-\sqrt{9-x^2}}^{\sqrt{9-x^2}} (y^2 + 3y)\, dy\, dx$

$= \rho \int_0^{2\pi} \int_0^3 (r^2 \sin^2\theta + 3r\sin\theta)r\, dr\, d\theta$

$= \rho \int_0^{2\pi} \int_0^3 (r^3 \sin^2\theta + 3r^2\sin\theta)\, dr\, d\theta$

$= \rho \int_0^{2\pi} \left[(r^4/4)\sin^2\theta + r^3\sin\theta\right]_{r=0}^{r=3} d\theta$

$= (\rho/4) \int_0^{2\pi} (81\sin^2\theta + 108\sin\theta)\, d\theta$

$= (\rho/8) \int_0^{2\pi} (81 - 81\cos 2\theta + 216\sin\theta)\, d\theta$

$= (\rho/8)\left[81\theta - (81/2)\sin 2\theta - 216\cos\theta\right]_0^{2\pi}$

$= 81\rho\pi/4$

$M_{xy} = \int_{-3}^3 \int_{-\sqrt{9-x^2}}^{\sqrt{9-x^2}} \int_0^{y+3} z\rho\, dz\, dy\, dx$

$= \rho \int_{-3}^3 \int_{-\sqrt{9-x^2}}^{\sqrt{9-x^2}} \left[z^2/2\right]_{z=0}^{z=y+3} dy\, dx$

$= (\rho/2) \int_{-3}^3 \int_{-\sqrt{9-x^2}}^{\sqrt{9-x^2}} (y^2 + 6y + 9)\, dy\, dx$

$= (\rho/2) \int_0^{2\pi} \int_0^3 (r^2 \sin^2\theta + 6r\sin\theta + 9)r\, dr\, d\theta$

$= (\rho/2) \int_0^{2\pi} \int_0^3 (r^3 \sin^2\theta + 6r^2\sin\theta + 9r)\, dr\, d\theta$

$= (\rho/2) \int_0^{2\pi} \left[(r^4/4)\sin^2\theta + 2r^3\sin\theta + (9r^2/2)\right]_{r=0}^{r=3} d\theta$

$= (\rho/8) \int_0^{2\pi} (81\sin^2\theta + 216\sin\theta + 162)\, d\theta$

$= (\rho/16) \int_0^{2\pi} (81 - 81\cos 2\theta + 432\sin\theta + 324)\, d\theta$

$= (\rho/16)\left[405\theta - (81/2)\sin 2\theta - 432\cos\theta\right]_0^{2\pi}$

$= 405\rho\pi/8$

$\bar{x} = M_{yz}/M = 0$

$\bar{y} = M_{xz}/M = (81\rho\pi/4)/(27\rho\pi) = 3/4$

$\bar{z} = M_{xy}/M = (405\rho\pi/8)/(27\rho\pi) = 15/8$

The center of mass is $(0, 3/4, 15/8)$.

33. Let ρ be the constant density.

$M = \int_0^2 \int_0^{\sqrt{4-y^2}} \int_0^{\sqrt{4-x^2-y^2}} \rho\,dz\,dx\,dy$

$\quad = \rho \int_0^2 \int_0^{\sqrt{4-y^2}} [z]_{z=0}^{z=\sqrt{4-x^2-y^2}}\,dx\,dy$

$\quad = \rho \int_0^2 \int_0^{\sqrt{4-y^2}} \sqrt{4-x^2-y^2}\,dx\,dy$

$\quad = \rho \int_0^{\pi/2} \int_0^2 r\sqrt{4-r^2}\,dr\,d\theta$

Let $u = 4 - r^2$, $du = -2r\,dr$.

$M = -(\rho/2) \int_0^{\pi/2} \int_4^0 \sqrt{u}\,du\,d\theta$

$\quad = (\rho/2) \int_0^{\pi/2} \int_0^4 \sqrt{u}\,du\,d\theta$

$\quad = (\rho/2) \int_0^{\pi/2} [(2/3)u^{3/2}]_{u=0}^{u=4}\,d\theta$

$\quad = (\rho/3) \int_0^{\pi/2} 8\,d\theta$

$\quad = (8\rho/3)\theta\big|_0^{\pi/2}$

$\quad = 4\rho\pi/3$

$M_{yz} = \int_0^2 \int_0^{\sqrt{4-y^2}} \int_0^{\sqrt{4-x^2-y^2}} x\rho\,dz\,dx\,dy$

$\quad = \rho \int_0^2 \int_0^{\sqrt{4-y^2}} [xz]_{z=0}^{z=\sqrt{4-x^2-y^2}}\,dx\,dy$

$\quad = \rho \int_0^2 \int_0^{\sqrt{4-y^2}} x\sqrt{4-x^2-y^2}\,dx\,dy$

Let $u = 4 - x^2 - y^2$, $du = -2x\,dx$

35. $\int_0^2 \int_0^{(2-y)/2} \int_0^{4-4x-2y} (x-y+z)\,dz\,dx\,dy$

$\int_0^1 \int_0^{4-4x} \int_0^{(4-4x-z)/2} (x-y+z)\,dy\,dz\,dx$

$\int_0^4 \int_0^{(4-z)/4} \int_0^{(4-4x-z)/2} (x-y+z)\,dy\,dx\,dz$

$\int_0^2 \int_0^{4-2y} \int_0^{(4-2y-z)/4} (x-y+z)\,dx\,dz\,dy$

$\int_0^4 \int_0^{(4-z)/2} \int_0^{(4-2y-z)/4} (x-y+z)\,dx\,dy\,dz$

$M_{yz} = -(\rho/2) \int_0^2 \int_{4-y^2}^0 \sqrt{u}\,dy$

$\quad = (\rho/2) \int_0^2 \int_0^{4-y^2} \sqrt{u}\,dy$

$\quad = (\rho/2) \int_0^2 [(2/3)u^{3/2}]_{u=0}^{u=4-y^2}\,dy$

$\quad = (\rho/3) \int_0^2 (4-y^2)^{3/2}\,dy$

Let $y = 2\sin\theta$, $dy = 2\cos\theta\,d\theta$.

$M_{yz} = (\rho/3) \int_0^{\pi/2} (4 - 4\sin^2\theta)^{3/2} 2\cos\theta\,d\theta$

$\quad = (16\rho/3) \int_0^{\pi/2} (\cos^2\theta)^{3/2} \cos\theta\,d\theta$

$\quad = (16\rho/3) \int_0^{\pi/2} \cos^4\theta\,d\theta$

$\quad = (4\rho/3) \int_0^{\pi/2} (1 + \cos 2\theta)^2\,d\theta$

$\quad = (4\rho/3) \int_0^{\pi/2} (1 + 2\cos 2\theta + \cos^2 2\theta)\,d\theta$

$\quad = (2\rho/3) \int_0^{\pi/2} (2 + 4\cos 2\theta + 1 + \cos 4\theta)\,d\theta$

$\quad = (2\rho/3)[3\theta + 2\sin 2\theta + (1/4)\sin 4\theta]_0^{\pi/2}$

$\quad = \rho\pi$

By the symmetry of the situation, one also has $M_{xz} = M_{xy} = M_{yz} = \rho\pi$.

$\bar{x} = \bar{y} = \bar{z} = (\rho\pi)/(4\rho\pi/3) = 3/4$

The center of mass is $(3/4, 3/4, 3/4)$.

19.7 Triple Integrals in Cylindrical and Spherical Coordinates

1. $\int_0^{2\pi} \int_0^3 \int_{r^2}^9 r\,dz\,dr\,d\theta$

$\quad = \int_0^{2\pi} \int_0^3 [rz]_{z=r^2}^{z=9}\,dr\,d\theta$

$\quad = \int_0^{2\pi} \int_0^3 (9r - r^3)\,dr\,d\theta$

$\quad = \int_0^{2\pi} [(9r^2/2) - (r^4/4)]_{r=0}^{r=3}\,d\theta$

$\quad = \int_0^{2\pi} (81/4)\,d\theta$

$\quad = (81/4)\theta\big|_0^{2\pi}$

$\quad = 81\pi/2$

3. Let ρ be the constant density. The mass is $81\rho\pi/2$.

$M_{yz} = \int_0^{2\pi}\int_0^3\int_{r^2}^9 rx\rho\,dz\,dr\,d\theta$
$= \rho\int_0^{2\pi}\int_0^3\int_{r^2}^9 r^2\cos\theta\,dz\,dr\,d\theta$
$= \rho\int_0^{2\pi}\int_0^3\left[r^2 z\cos\theta\right]_{z=r^2}^{z=9} dr\,d\theta$
$= \rho\int_0^{2\pi}\int_0^3 (9r^2 - r^4)\cos\theta\,dr\,d\theta$
$= \rho\int_0^{2\pi}\left[(3r^3 - (r^5/5))\cos\theta\right]_{r=0}^{r=3} d\theta$
$= \rho\int_0^{2\pi}(162/5)\cos\theta\,d\theta$
$= (162\rho/5)\sin\theta\Big|_0^{2\pi}$
$= 0$

Due to symmetry of the situation, one may conclude that $M_{xz} = M_{yz} = 0$.

$M_{xy} = \int_0^{2\pi}\int_0^3\int_{r^2}^9 rz\rho\,dz\,dr\,d\theta$
$= \rho\int_0^{2\pi}\int_0^3\left[rz^2/2\right]_{z=r^2}^{z=9} dr\,d\theta$
$= (\rho/2)\int_0^{2\pi}\int_0^3 (81r - r^5)\,dr\,d\theta$
$= (\rho/2)\int_0^{2\pi}\left[(81r^2/2) - (r^6/6)\right]_{r=0}^{r=3} d\theta$
$= (\rho/2)\int_0^{2\pi} 243\,d\theta$
$= (243\rho/2)\theta\Big|_0^{2\pi}$
$= 243\rho\pi$
$\bar{x} = \bar{y} = 0$
$\bar{z} = (243\rho\pi)/(81\rho\pi/2) = 6$

The center of mass is $(0, 0, 6)$.

5. $\rho(x, y, z) = \lambda\sqrt{y^2 + z^2}$
$M = \int_0^4\int_{-2}^2\int_{-\sqrt{4-y^2}}^{\sqrt{4-y^2}} \lambda\sqrt{y^2+z^2}\,dz\,dy\,dx$

Exchange the variables x and z.

$M = \int_0^4\int_{-2}^2\int_{-\sqrt{4-y^2}}^{\sqrt{4-y^2}} \lambda\sqrt{y^2+x^2}\,dx\,dy\,dz$

Change to cylindrical coordinates.

$M = \int_0^{2\pi}\int_0^2\int_0^4 \lambda r^2\,dz\,dr\,d\theta$
$= \lambda\int_0^{2\pi}\int_0^2\left[r^2 z\right]_{z=0}^{z=4} dr\,d\theta$
$= \lambda\int_0^{2\pi}\int_0^2 4r^2\,dr\,d\theta$
$= \lambda\int_0^{2\pi}\left[4r^3/3\right]_{r=0}^{r=2} d\theta$
$= \lambda\int_0^{2\pi}(32/3)\,d\theta$
$= \lambda(32/3)\theta\Big|_0^{2\pi}$
$= 64\lambda\pi/3$

7. $(4/3)\pi 3^3 - 2\int_0^{\pi/2}\int_0^{3\sin\theta}\int_{-\sqrt{9-r^2}}^{\sqrt{9-r^2}} r\,dz\,dr\,d\theta$
$= 36\pi - 2\int_0^{\pi/2}\int_0^{3\sin\theta}\left[rz\right]_{z=-\sqrt{9-r^2}}^{z=\sqrt{9-r^2}} dr\,d\theta$
$= 36\pi - 2\int_0^{\pi/2}\int_0^{3\sin\theta} 2r\sqrt{9-r^2}\,dr\,d\theta$

Let $u = 9 - r^2$, $du = -2r\,dr$.

$36\pi + 2\int_0^{\pi/2}\int_9^{9-9\sin^2\theta}\sqrt{u}\,du\,d\theta$
$= 36\pi + 2\int_0^{\pi/2}\int_9^{9\cos^2\theta}\sqrt{u}\,du\,d\theta$
$= 36\pi + 2\int_0^{\pi/2}\left[(2/3)u^{3/2}\right]_{u=9}^{u=9\cos^2\theta} d\theta$
$= 36\pi + 36\int_0^{\pi/2}(\cos^3\theta - 1)\,d\theta$
$= 36\pi + 36\int_0^{\pi/2}\cos^3\theta\,d\theta - 36\int_0^{\pi/2} 1\,d\theta$

For the first integral, let $u = \sin\theta$, $du = \cos\theta\,d\theta$. Note that $\cos^2\theta = 1 - \sin^2\theta = 1 - u^2$.

$36\pi + 36\int_0^1(1-u^2)\,du - \left[36\theta\right]_0^{\pi/2}$
$= 36\pi + 36[u - (u^3/3)]_{u=0}^{u=1} - 18\pi$
$= 18\pi + 24$

9. The graph of $x^2 + y^2 - 4y = 0$ in the xy-plane is a circle with radius 2 and center $(0, 2)$, and it has polar equation $r = 4\sin\theta$.

$\int_0^{\pi}\int_0^{4\sin\theta}\int_{-r\sqrt{2}}^{r\sqrt{2}} r\,dz\,dr\,d\theta$
$= \int_0^{\pi}\int_0^{4\sin\theta}\left[rz\right]_{z=-r\sqrt{2}}^{z=r\sqrt{2}} dr\,d\theta$
$= \int_0^{\pi}\int_0^{4\sin\theta} 2r^2\sqrt{2}\,dr\,d\theta$
$= 2\sqrt{2}\int_0^{\pi}\left[r^3/3\right]_{r=0}^{r=4\sin\theta} d\theta$
$= (128\sqrt{2}/3)\int_0^{\pi}\sin^3\theta\,d\theta$

Let $u = \cos\theta$, $du = -\sin\theta\,d\theta$. Note that $\sin^2\theta = 1 - \cos^2\theta = 1 - u^2$.

$-(128\sqrt{2}/3)\int_1^{-1}(1-u^2)\,du$
$= (128\sqrt{2}/3)\int_{-1}^1(1-u^2)\,du$
$= (128\sqrt{2}/3)\left[u - (u^3/3)\right]_{-1}^1$
$= 512\sqrt{2}/9$

11. $\int_0^{2\pi}\int_0^1\int_r^1 (r^2\cos^2\theta)r\,dz\,dr\,d\theta$
$= \int_0^{2\pi}\int_0^1\left[r^3 z\cos^2\theta\right]_{z=r}^{z=1} dr\,d\theta$
$= \int_0^{2\pi}\int_0^1 (r^3 - r^4)\cos^2\theta\,dr\,d\theta$
$= \int_0^{2\pi}\left[((r^4/4) - (r^5/5))\cos^2\theta\right]_{r=0}^{r=1} d\theta$
$= (1/20)\int_0^{2\pi}\cos^2\theta\,d\theta$
$= (1/40)\int_0^{2\pi}(1 + \cos 2\theta)\,d\theta$
$= (1/40)\left[\theta + (1/2)\sin 2\theta\right]_0^{2\pi}$
$= \pi/20$

13. $\int_0^\pi \int_0^1 \int_{r^2}^1 zr \, dz \, dr \, d\theta$

$= \int_0^\pi \int_0^1 [z^2 r/2]_{z=r^2}^{z=1} \, dr \, d\theta$

$= (1/2) \int_0^\pi \int_0^1 (r - r^5) \, dr \, d\theta$

$= (1/2) \int_0^\pi [(r^2/2) - (r^6/6)]_{r=0}^{r=1} \, d\theta$

$= (1/2) \int_0^\pi (1/3) \, d\theta$

$= (1/6)\theta \big|_0^\pi$

$= \pi/6$

15. $x = \sqrt{3y - y^2}$ is the semicircle in the first quadrant with equation $x^2 + (y - (3/2))^2 = 9/4$, with radius $3/2$ and center $(0, 3/2)$. It has polar equation $r = 3\sin\theta$ for $0 \le \theta \le \pi/2$.

$\int_0^{\pi/2} \int_0^{3\sin\theta} \int_r^4 2r \, dz \, dr \, d\theta$

$= \int_0^{\pi/2} \int_0^{3\sin\theta} [2rz]_{z=r}^{z=4} \, dr \, d\theta$

$= \int_0^{\pi/2} \int_0^{3\sin\theta} (8r - 2r^2) \, dr \, d\theta$

$= \int_0^{\pi/2} [4r^2 - (2r^3/3)]_{r=0}^{r=3\sin\theta} \, d\theta$

$= 18 \int_0^{\pi/2} (2\sin^2\theta - \sin^3\theta) \, d\theta$

$= 18 \int_0^{\pi/2} (1 - \cos 2\theta) \, d\theta - 18 \int_0^{\pi/2} \sin^3\theta \, d\theta$

Let $u = \cos\theta$, $du = -\sin\theta \, d\theta$. Note that $\sin^2\theta = 1 - \cos^2\theta = 1 - u^2$.

$18[\theta - (1/2)\sin 2\theta]_0^{\pi/2} + 18 \int_1^0 (1 - u^2) \, du$

$= 9\pi + 18[u - (u^3/3)]_1^0$

$= 9\pi - 12$

17. $\int_0^{2\pi} \int_0^\pi \int_r^R \rho^2 \sin\phi \, d\rho \, d\phi \, \theta$

$= \int_0^{2\pi} \int_0^\pi [(\rho^3/3) \sin\phi]_{\rho=r}^{\rho=R} \, d\phi \, d\theta$

$= [(R^3 - r^3)/3] \int_0^{2\pi} \int_0^\pi \sin\phi \, d\phi \, d\theta$

$= [(R^3 - r^3)/3] \int_0^{2\pi} [-\cos\phi]_{\phi=0}^{\phi=\pi} \, d\theta$

$= [(R^3 - r^3)/3] \int_0^{2\pi} 2 \, d\theta$

$= [(R^3 - r^3)/3] [2\theta]_0^{2\pi}$

$= 4\pi(R^3 - r^3)/3$

19. Converting the equation of the sphere to spherical coordinates gives $\rho = 2\cos\phi$.

$\int_0^{2\pi} \int_0^{\pi/4} \int_0^{2\cos\phi} \rho^2 \sin\phi \, d\rho \, d\phi \, d\theta$

$= \int_0^{2\pi} \int_0^{\pi/4} [(\rho^3/3) \sin\phi]_{\rho=0}^{\rho=2\cos\phi} \, d\phi \, d\theta$

$= (8/3) \int_0^{2\pi} \int_0^{\pi/4} \cos^3\phi \sin\phi \, d\phi \, d\theta$

Let $u = \cos\phi$, $du = -\sin\phi \, d\phi$.

$-(8/3) \int_0^{2\pi} \int_1^{\sqrt{2}/2} u^3 \, du \, d\theta$

$= -(8/3) \int_0^{2\pi} [u^4/4]_{u=1}^{u=\sqrt{2}/2} \, d\theta$

$= (1/2) \int_0^{2\pi} 1 \, d\theta$

$= (1/2)\theta \big|_0^{2\pi}$

$= \pi$

21. $M = \int_0^{2\pi} \int_0^\pi \int_1^2 (\lambda/\rho) \rho^2 \sin\phi \, d\rho \, d\phi \, d\theta$

$= \lambda \int_0^{2\pi} \int_0^\pi \int_1^2 \rho \sin\phi \, d\rho \, d\phi \, d\theta$

$= \lambda \int_0^{2\pi} \int_0^\pi [(\rho^2/2) \sin\phi]_{\rho=1}^{\rho=2} \, d\phi \, d\theta$

$= (3\lambda/2) \int_0^{2\pi} \int_0^\pi \sin\phi \, d\phi \, d\theta$

$= (3\lambda/2) \int_0^{2\pi} [-\cos\phi]_{\phi=0}^{\phi=\pi} \, d\theta$

$= (3\lambda/2) \int_0^{2\pi} 2 \, d\theta$

$= 3\lambda\theta \big|_0^{2\pi}$

$= 6\lambda\pi$

23. Let the constant density be λ.

$M = \int_0^\pi \int_0^{\pi/2} \int_0^1 \lambda \rho^2 \sin \phi \, d\rho \, d\phi \, d\theta$

$= \lambda \int_0^\pi \int_0^{\pi/2} [(\rho^3/3) \sin \phi]_{\rho=0}^{\rho=1} d\phi \, d\theta$

$= (\lambda/3) \int_0^\pi \int_0^{\pi/2} \sin \phi \, d\phi \, d\theta$

$= (\lambda/3) \int_0^\pi [-\cos \phi]_{\phi=0}^{\phi=\pi/2} d\theta$

$= (\lambda/3) \int_0^\pi 1 \, d\theta$

$= (\lambda/3)\theta \big|_0^\pi$

$= \lambda \pi/3$

$M_{yz} = \int_0^\pi \int_0^{\pi/2} \int_0^1 x \lambda \rho^2 \sin \phi \, d\rho \, d\phi \, d\theta$

$= \lambda \int_0^\pi \int_0^{\pi/2} \int_0^1 \rho^3 \sin^2 \phi \cos \theta \, d\rho \, d\phi \, d\theta$

$= \lambda \int_0^\pi \int_0^{\pi/2} [(\rho^4/4) \sin^2 \phi \cos \theta]_{\rho=0}^{\rho=1} d\phi \, d\theta$

$= (\lambda/4) \int_0^\pi \int_0^{\pi/2} \sin^2 \phi \cos \theta \, d\phi \, d\theta$

$= (\lambda/8) \int_0^\pi \int_0^{\pi/2} (1 - \cos 2\phi) \cos \theta \, d\phi \, d\theta$

$= (\lambda/8) \int_0^\pi [(\phi - (1/2) \sin 2\phi) \cos \theta]_{\phi=0}^{\phi=\pi/2} d\theta$

$= (\lambda \pi/16) \int_0^\pi \cos \theta \, d\theta$

$= (\lambda \pi/16) \sin \theta \big|_0^\pi$

$= 0$

$M_{xz} = \int_0^\pi \int_0^{\pi/2} \int_0^1 y \lambda \rho^2 \sin \phi \, d\rho \, d\phi \, d\theta$

$= \lambda \int_0^\pi \int_0^{\pi/2} \int_0^1 \rho^3 \sin^2 \phi \sin \theta \, d\rho \, d\phi \, d\theta$

$M_{xz} = \lambda \int_0^\pi \int_0^{\pi/2} [(\rho^4/4) \sin^2 \phi \sin \theta]_{\rho=0}^{\rho=1} d\phi \, d\theta$

$= (\lambda/4) \int_0^\pi \int_0^{\pi/2} \sin^2 \phi \sin \theta \, d\phi \, d\theta$

$= (\lambda/8) \int_0^\pi \int_0^{\pi/2} (1 - \cos 2\phi) \sin \theta \, d\phi \, d\theta$

$= (\lambda/8) \int_0^\pi [(\phi - (1/2) \sin 2\phi) \sin \theta]_{\phi=0}^{\phi=\pi/2} d\theta$

$= (\lambda \pi/16) \int_0^\pi \sin \theta \, d\theta$

$= -(\lambda \pi/16) \cos \theta \big|_0^\pi$

$= \lambda \pi/8$

$M_{xy} = \int_0^\pi \int_0^{\pi/2} \int_0^1 z \lambda \rho^2 \sin \phi \, d\rho \, d\phi \, d\theta$

$= \lambda \int_0^\pi \int_0^{\pi/2} \int_0^1 \rho^3 \cos \phi \sin \phi \, d\rho \, d\phi \, d\theta$

$= \lambda \int_0^\pi \int_0^{\pi/2} [(\rho^4/4) \cos \phi \sin \phi]_{\rho=0}^{\rho=1} d\phi \, d\theta$

$= (\lambda/4) \int_0^\pi \int_0^{\pi/2} \cos \phi \sin \phi \, d\phi \, d\theta$

$= (\lambda/8) \int_0^\pi \int_0^{\pi/2} \sin 2\phi \, d\phi \, d\theta$

$= (\lambda/8) \int_0^\pi [-(1/2) \cos 2\phi]_{\phi=0}^{\phi=\pi/2} d\theta$

$= (\lambda/8) \int_0^\pi 1 \, d\theta$

$= (\lambda/8)\theta \big|_0^\pi$

$= \lambda \pi/8$

$\bar{x} = M_{yz}/M = 0$

$\bar{y} = M_{xz}/M = (\lambda \pi/8)/(\lambda \pi/3) = 3/8$

$\bar{z} = M_{xy}/M = (\lambda \pi/8)/(\lambda \pi/3) = 3/8$

The center of mass is $(0, 3/8, 3/8)$.

25. $\int_0^{2\pi} \int_0^{\pi/2} \int_0^1 3\rho^2 \sin \phi \, d\rho \, d\phi \, d\theta$

$= \int_0^{2\pi} \int_0^{\pi/2} [\rho^3 \sin \phi]_{\rho=0}^{\rho=1} d\phi \, d\theta$

$= \int_0^{2\pi} \int_0^{\pi/2} \sin \phi \, d\phi \, d\theta$

$= \int_0^{2\pi} [-\cos \phi]_{\phi=0}^{\phi=\pi/2} d\theta$

$= \int_0^{2\pi} 1 \, d\theta$

$= \theta \big|_0^{2\pi}$

$= 2\pi$

27. $\sqrt{x^2 + y^2 + z^2} = \rho$

$\int_0^{\pi/2} \int_0^{\pi/4} \int_0^{\sqrt{2}} \rho^3 \sin \phi \, d\rho \, d\phi \, d\theta$

$= \int_0^{\pi/2} \int_0^{\pi/4} [(\rho^4/4) \sin \phi]_{\rho=0}^{\rho=\sqrt{2}} d\phi \, d\theta$

$= \int_0^{\pi/2} \int_0^{\pi/4} \sin \phi \, d\phi \, d\theta$

$= \int_0^{\pi/2} [-\cos \phi]_{\phi=0}^{\phi=\pi/4} d\theta$

$= \int_0^{\pi/2} [(2 - \sqrt{2})/2] \, d\theta$

$= [(2 - \sqrt{2})\theta/2]_0^{\pi/2}$

$= (2 - \sqrt{2})\pi/4$

29. Assume that the cylinders are $x^2 + y^2 = R^2$ and $x^2 + z^2 = R^2$. Find the volume of that portion of the solid which is in the first octant and below the plane $z = y$, and then multiply by 16. In cylindrical coordinates, the equation of the plane $z = y$ becomes $z = r \sin \theta$.

$16 \int_0^{\pi/2} \int_0^R \int_0^{r \sin \theta} r \, dz \, dr \, d\theta$

$= 16 \int_0^{\pi/2} \int_0^R [rz]_{z=0}^{z=r \sin \theta} \, dr \, d\theta$

$= 16 \int_0^{\pi/2} \int_0^R r^2 \sin \theta \, dr \, d\theta$

$= 16 \int_0^{\pi/2} [(r^3/3) \sin \theta]_{r=0}^{r=R} \, d\theta$

$= (16R^3/3) \int_0^{\pi/2} \sin \theta \, d\theta$

$= -(16R^3/3) \cos \theta \big|_0^{\pi/2}$

$= 16R^3/3$

31. $z/(x^2 + y^2)^{3/2} = z/r^3$

$\int_0^{2\pi} \int_1^{\sqrt 3} \int_0^3 (z/r^3) r \, dz \, dr \, d\theta$

$= \int_0^{2\pi} \int_1^{\sqrt 3} \int_0^3 (z/r^2) \, dz \, dr \, d\theta$

$= \int_0^{2\pi} \int_1^{\sqrt 3} [z^2/(2r^2)]_{z=0}^{z=3} \, dr \, d\theta$

$= (9/2) \int_0^{2\pi} \int_1^{\sqrt 3} (1/r^2) \, dr \, d\theta$

$= (9/2) \int_0^{2\pi} [-1/r]_{r=1}^{r=\sqrt 3} \, d\theta$

$= (3/2) \int_0^{2\pi} (3 - \sqrt 3) \, d\theta$

$= (3/2)(3 - \sqrt 3)\theta \big|_0^{2\pi}$

$= 9\pi - 3\pi\sqrt 3$

33. Exchanging the variables x and z gives the triple integral

$$\iiint\limits_Q \sqrt{z/(x^2 + y^2)} \, dV,$$

where Q is the region bounded by the cone $x^2 + y^2 = z^2$, the cylinder $x^2 + y^2 = 4$, and the planes $z = 0$ and $z = 2$.

$\int_0^{2\pi} \int_0^2 \int_0^r (\sqrt{z/r^2}) r \, dz \, dr \, d\theta$

$= \int_0^{2\pi} \int_0^2 \int_0^r \sqrt z \, dz \, dr \, d\theta$

$= \int_0^{2\pi} \int_0^2 [(2/3)z^{3/2}]_{z=0}^{z=r} \, dr \, d\theta$

$= (2/3) \int_0^{2\pi} \int_0^2 r^{3/2} \, dr \, d\theta$

$= (2/3) \int_0^{2\pi} [(2/5)r^{5/2}]_{r=0}^{r=2} \, d\theta$

$= (16\sqrt 2/15) \int_0^{2\pi} 1 \, d\theta$

$= (16\sqrt 2/15)\theta \big|_0^{2\pi}$

$= 32\pi\sqrt 2/15$

35. This is the same as the volume of the smaller of the two parts of the sphere determined by the plane $z = 2$. At the intersection of the sphere and the plane, $16 = \rho^2 = z^2 + r^2 = 4 + r^2$, $r = 2\sqrt 3$, $\tan \phi = r/z = \sqrt 3$, and $\phi = \pi/3$. The spherical equation of the plane is $\rho = 2 \sec \phi$.

$\int_0^{2\pi} \int_0^{\pi/3} \int_{2 \sec \phi}^4 \rho^2 \sin \phi \, d\rho \, d\phi \, d\theta$

$= \int_0^{2\pi} \int_0^{\pi/3} [(\rho^3/3) \sin \phi]_{\rho=2 \sec \phi}^{\rho=4} \, d\phi \, d\theta$

$= (8/3) \int_0^{2\pi} \int_0^{\pi/3} (8 - \sec^3 \phi) \sin \phi \, d\phi \, d\theta$

Let $u = \cos \phi$, $du = -\sin \phi \, d\phi$.

$-(8/3) \int_0^{2\pi} \int_1^{1/2} (8 - u^{-3}) \, du \, d\theta$

$= -(8/3) \int_0^{2\pi} [8u + (1/2)u^{-2}]_{u=1}^{u=1/2} \, d\theta$

$= (8/3) \int_0^{2\pi} (5/2) \, d\theta$

$= (20/3)\theta \big|_0^{2\pi}$

$= 40\pi/3$

37. The volume is the same using the cone $z^2 = x^2 + y^2$. Find the volume for the upper half, and then multiply by 2.

$2 \int_0^{2\pi} \int_0^{\pi/4} \int_1^3 \rho^2 \sin \phi \, d\rho \, d\phi \, d\theta$

$= 2 \int_0^{2\pi} \int_0^{\pi/4} [(\rho^3/3) \sin \phi]_{\rho=1}^{\rho=3} \, d\phi \, d\theta$

$= (52/3) \int_0^{2\pi} \int_0^{\pi/4} \sin \phi \, d\phi \, d\theta$

$= (52/3) \int_0^{2\pi} [-\cos \phi]_{\phi=0}^{\phi=\pi/4} \, d\theta$

$= (26/3)(2 - \sqrt 2) \int_0^{2\pi} 1 \, d\theta$

$= (52 - 26\sqrt 2)\theta/3 \big|_0^{2\pi}$

$= (104 - 52\sqrt 2)\pi/3$

39. a. An arc which subtends an angle of t radians on a circle of radius R has length Rt. Since the circle in this case has radius ρ_i and the angle in question is $\Delta\phi_k$, the length of the arc is $\rho_i \Delta\phi_k$.

611

b. The arc lies in a horizontal circle on the sphere. If r is the radius of this circle, then $\sin \phi_k = r/\rho_i$, and thus $r = \rho_i \sin \phi_k$. The angle subtended by the arc is $\Delta \theta_j$. Thus the length of the arc is $\rho_i \sin \phi_k \, \Delta \theta_j$.

c. The length of PQ is $\Delta \rho_i$. Thus the approximation to the volume is

$$(\Delta \rho_i)(\rho_i \Delta \phi_k)(\rho_i \sin \phi_k \, \Delta \theta_j) = \rho_i^2 \sin \phi_k \, \Delta \rho_i \, \Delta \theta_j \, \Delta \phi_k.$$

19.8 Change of Variables and Jacobians

1.

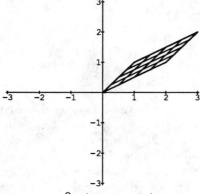

$x = 2u + v, \; y = u + v$

3.

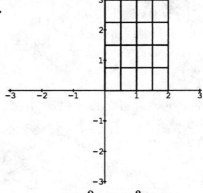

$x = 2u, \; y = 3v$

5.

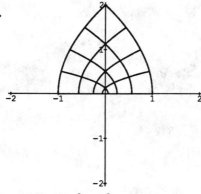

$x = u^2 - v^2, \; y = 2uv$

7. P' is the square with vertices $(0,0)$, $(1,0)$, $(1,1)$, $(0,1)$.

$\dfrac{\partial(x,y)}{\partial(u,v)} = \det \begin{bmatrix} 3 & 0 \\ 1 & 1 \end{bmatrix} = 3$

$\int_0^1 \int_0^1 3u(u+v)3 \, du \, dv$

$= \int_0^1 \int_0^1 (9u^2 + 9uv) \, du \, dv$

$= \int_0^1 \left[3u^3 + (9u^2 v/2) \right]_{u=0}^{u=1} dv$

$= (1/2) \int_0^1 (6 + 9v) \, dv$

$= (1/2) \left[6v + (9v^2/2) \right]_0^1$

$= 21/4$

9. E' is the circle $u^2 + v^2 \le 4$.

$\dfrac{\partial(x,y)}{\partial(u,v)} = \det \begin{bmatrix} 3 & 0 \\ 0 & 2 \end{bmatrix} = 6$

$\int_{-2}^2 \int_{-\sqrt{4-v^2}}^{\sqrt{4-v^2}} (u^2 + v^2) 6 \, du \, dv$ Convert to polar coordinates.

$= \int_0^{2\pi} \int_0^2 6r^3 \, dr \, d\theta$

$= \int_0^{2\pi} \left[3r^4/2 \right]_{r=0}^{r=2} d\theta$

$= \int_0^{2\pi} 24 \, d\theta$

$= 24\theta \Big|_0^{2\pi}$

$= 48\pi$

11. Let $x = 2u \cos v, \; y = 3u \sin v, \; z = w$. Then $(x^2/4) + (y^2/9) = u^2$.

$C' = \{(u,v,w) \mid 0 \le u \le 1, 0 \le v \le 2\pi, 0 \le w \le 2\}$

$$\frac{\partial(x,y,z)}{\partial(u,v,w)} = \det \begin{bmatrix} 2\cos v & -2u\sin v & 0 \\ 3\sin v & 3u\cos v & 0 \\ 0 & 0 & 1 \end{bmatrix} = 6u\cos^2 v + 6u\sin^2 v = 6u$$

$\int_0^2 \int_0^{2\pi} \int_0^1 |6u|\, du\, dv\, dw$

$= \int_0^2 \int_0^{2\pi} \int_0^1 6u\, du\, dv\, dw$

$= \int_0^2 \int_0^{2\pi} [3u^2]_{u=0}^{u=1}\, dv\, dw$

$= \int_0^2 \int_0^{2\pi} 3\, dv\, dw$

$= \int_0^2 [3v]_{v=0}^{v=2\pi}\, dw$

$= \int_0^2 6\pi\, dw$

$= 6\pi w\big|_0^2$

$= 12\pi$

13. Let $x = 2u\cos v$, $y = 3u\sin v$, $z = w$. Then $1 - (x^2/4) - (y^2/9) = 1 - u^2$.
$Q' = \{(u,v,w) \mid 0 \le u \le 1, 0 \le v \le \pi/2, 0 \le w \le 1 - u^2\}$

$$\frac{\partial(x,y,z)}{\partial(u,v,w)} = \det \begin{bmatrix} 2\cos v & -2u\sin v & 0 \\ 3\sin v & 3u\cos v & 0 \\ 0 & 0 & 1 \end{bmatrix} = 6u\cos^2 v + 6u\sin^2 v = 6u$$

$\int_0^{\pi/2} \int_0^1 \int_0^{1-u^2} |6u|\, dw\, du\, dv$

$= \int_0^{\pi/2} \int_0^1 \int_0^{1-u^2} 6u\, dw\, du\, dv$

$= \int_0^{\pi/2} \int_0^1 [6uw]_{w=0}^{w=1-u^2}\, du\, dv$

$= \int_0^{\pi/2} \int_0^1 (6u - 6u^3)\, du\, dv$

$= \int_0^{\pi/2} [3u^2 - (3u^4/2)]_{u=0}^{u=1}\, dv$

$= \int_0^{\pi/2} (3/2)\, dv$

$= (3/2)v\big|_0^{\pi/2}$

$= 3\pi/4$

15. Let $x = r\cos\theta$, $y = r\sin\theta$. The region Q in the xy-plane for which one wishes to find the area corresponds to the following region in the $r\theta$-plane: $Q' = \{(r,\theta) \mid 0 \le r \le f(\theta), a \le \theta \le b\}$.

$$\frac{\partial(x,y)}{\partial(r,\theta)} = \det \begin{bmatrix} \cos\theta & -r\sin\theta \\ \sin\theta & r\cos\theta \end{bmatrix} = r\cos^2\theta + r\sin^2\theta = r$$

$\iint_Q 1\, dA = \int_a^b \int_0^{f(\theta)} |r|\, dr\, d\theta$

$= \int_a^b \int_0^{f(\theta)} r\, dr\, d\theta$

$= \int_a^b [r^2/2]_{r=0}^{r=f(\theta)}\, d\theta$

$= \int_a^b (1/2)[f(\theta)]^2\, d\theta$

17. Let $x = \rho\sin\phi\cos\theta$, $y = \rho\sin\phi\sin\theta$, $z = \rho\cos\phi$.

$$\frac{\partial(x,y,z)}{\partial(\rho,\theta,\phi)} = \det \begin{bmatrix} \sin\phi\cos\theta & -\rho\sin\phi\sin\theta & \rho\cos\phi\cos\theta \\ \sin\phi\sin\theta & \rho\sin\phi\cos\theta & \rho\cos\phi\sin\theta \\ \cos\phi & 0 & -\rho\sin\phi \end{bmatrix}$$

$= -\rho^2\sin^3\phi\cos^2\theta - \rho^2\sin\phi\cos^2\phi\sin^2\theta + 0 - \rho^2\sin\phi\cos^2\phi\cos^2\theta - 0 - \rho^2\sin^3\phi\sin^2\theta$

$= -\rho^2\sin^3\phi(\cos^2\theta + \sin^2\theta) - \rho^2\sin\phi\cos^2\phi(\sin^2\theta + \cos^2\theta)$

$= -\rho^2\sin^3\phi - \rho^2\sin\phi\cos^2\phi$

$= -\rho^2\sin\phi(\sin^2\phi + \cos^2\phi)$

$= -\rho^2\sin\phi$

$$\iiint\limits_{Q} f(\rho, \theta, \phi)\, dV = \int_a^b \int_{h_1(\theta)}^{h_2(\theta)} \int_{g_1(\theta,\phi)}^{g_2(\theta,\phi)} f(\rho, \theta, \phi)| - \rho^2 \sin\phi|\, d\rho\, d\phi\, d\theta$$

$$= \int_a^b \int_{h_1(\theta)}^{h_2(\theta)} \int_{g_1(\theta,\phi)}^{g_2(\theta,\phi)} f(\rho, \theta, \phi)\rho^2 \sin\phi\, d\rho\, d\phi\, d\theta$$

since ρ and $\sin\phi$ are always nonnegative.

Review Exercises — Chapter 19

1. $\int_0^2 \int_0^{(2-x)/2} \int_0^{xy^2} 1\, dz\, dy\, dx$

$\quad = \int_0^2 \int_0^{(2-x)/2} [z]_{z=0}^{z=xy^2}\, dy\, dx$

$\quad = \int_0^2 \int_0^{(2-x)/2} xy^2\, dy\, dx$

$\quad = \int_0^2 [xy^3/3]_{y=0}^{y=(2-x)/2}\, dx$

$\quad = (1/24) \int_0^2 (8x - 12x^2 + 6x^3 - x^4)\, dx$

$\quad = (1/24)[4x^2 - 4x^3 + (3x^4/2) - (x^5/5)]_0^2$

$\quad = 1/15$

3. $\int_0^1 \int_0^{2x} e^{-x^2/2}\, dy\, dx$

$\quad = \int_0^1 [ye^{-x^2/2}]_{y=0}^{y=2x}\, dx$

$\quad = \int_0^1 2xe^{-x^2/2}\, dx$

Let $u = -x^2/2$, $du = -x\, dx$.

$\quad -\int_0^{-1/2} 2e^u\, du$

$\quad = \int_{-1/2}^0 2e^u\, du$

$\quad = 2e^u\big|_{-1/2}^0$

$\quad = 2 - 2e^{-1/2}$

5. $\int_0^4 \int_x^4 \int_0^{y^2} 1\, dz\, dy\, dx$

$\quad = \int_0^4 \int_x^4 [z]_{z=0}^{z=y^2}\, dy\, dx$

$\quad = \int_0^4 \int_x^4 y^2\, dy\, dx$

$\quad = \int_0^4 [y^3/3]_{y=x}^{y=4}\, dx$

$\quad = (1/3) \int_0^4 (64 - x^3)\, dx$

$\quad = (1/3)[64x - (x^4/4)]_0^4$

$\quad = 64$

7. Find the area for $z \geq 0$ and $y \geq 0$, and then multiply by 4. Rewrite the equation of the sphere as $z = \sqrt{4 - x^2 - y^2}$. In cylindrical coordinates, the equation of the cylinder becomes $r = 2\cos\theta$.

$\partial z/\partial x = -x/\sqrt{4 - x^2 - y^2}$ and $\partial z/\partial y = -y/\sqrt{4 - x^2 - y^2}$

$\sqrt{(\partial z/\partial x)^2 + (\partial z/\partial y)^2 + 1} = \sqrt{(x^2 + y^2)/(4 - x^2 - y^2) + 1} = 2/\sqrt{4 - x^2 - y^2} = 2/\sqrt{4 - r^2}$

$A_S = 4 \int_0^{\pi/2} \int_0^{2\cos\theta} (2r/\sqrt{4 - r^2})\, dr\, d\theta$

Note that this is an improper integral, since the integrand is undefined at $(r, \theta) = (2, 0)$. However, one may obtain the area in question as the limit, as $a \to 0^+$, of the area within that portion of the cylinder for which $a \leq \theta \leq \pi$.

$A_S = 4 \lim_{a \to 0^+} \int_a^{\pi/2} \int_0^{2\cos\theta} (2r/\sqrt{4 - r^2})\, dr\, d\theta$

Let $u = 4 - r^2$, $du = -2r\, dr$. At $r = 2\cos\theta$, $u = 4 - 4\cos^2\theta = 4\sin^2\theta$.

$A_S = -4 \lim_{a \to 0^+} \int_a^{\pi/2} \int_4^{4\sin^2\theta} u^{-1/2}\, du\, d\theta$

$\quad = -4 \lim_{a \to 0^+} \int_a^{\pi/2} [2u^{1/2}]_{u=4}^{u=4\sin^2\theta}\, d\theta$

$\quad = -4 \lim_{a \to 0^+} \int_a^{\pi/2} (4\sin\theta - 4)\, d\theta$

$\quad = -4 \int_0^{\pi/2} (4\sin\theta - 4)\, d\theta$

$\quad = 4(4\cos\theta + 4\theta)\big|_0^{\pi/2}$

$\quad = 8\pi - 16$

9. $\int_{-\pi/2}^{\pi/2} \int_0^{2\cos\theta} \int_0^{2r^2} r\,dz\,dr\,d\theta$

$= \int_{-\pi/2}^{\pi/2} \int_0^{2\cos\theta} [rz]_{z=0}^{z=2r^2} \, dr\,d\theta$

$= \int_{-\pi/2}^{\pi/2} \int_0^{2\cos\theta} 2r^3 \, dr\,d\theta$

$= \int_{-\pi/2}^{\pi/2} [r^4/2]_{r=0}^{r=2\cos\theta} \, d\theta$

$= \int_{-\pi/2}^{\pi/2} 8\cos^4\theta \, d\theta$

$= \int_{-\pi/2}^{\pi/2} 2(1+\cos 2\theta)^2 \, d\theta$

$= \int_{-\pi/2}^{\pi/2} (2 + 4\cos 2\theta + 1 + \cos 4\theta) \, d\theta$

$= 3\theta + 2\sin 2\theta + (1/4)\sin 4\theta \big|_{-\pi/2}^{\pi/2}$

$= 3\pi$

11. The x-, y- and z-intercepts are a, b and c respectively.

$\int_0^a \int_0^{b[1-(x/a)]} \int_0^{c[1-(x/a)-(y/b)]} 1 \, dz\,dy\,dx$

$= \int_0^a \int_0^{b[1-(x/a)]} [z]_{z=0}^{z=c[1-(x/a)-(y/b)]} \, dy\,dx$

$= [1/(ab)] \int_0^a \int_0^{b[1-(x/a)]} (abc - bcx - acy) \, dy\,dx$

$= [1/(ab)] \int_0^a [abcy - bcxy - (acy^2/2)]_{y=0}^{y=b[1-(x/a)]} \, dx$

$= [bc/(2a^2)] \int_0^a (a^2 - 2ax + x^2) \, dx$

$= [bc/(2a^2)] [a^2 x - ax^2 + (x^3/3)]_0^a$

$= abc/6$

13. The equation of the plane $z = 9 - y$ may be rewritten as $9 - r\sin\theta$.

$\int_0^{2\pi} \int_0^2 \int_0^{9-r\sin\theta} r\,dz\,dr\,d\theta$

$= \int_0^{2\pi} \int_0^2 [rz]_{z=0}^{z=9-r\sin\theta} \, dr\,d\theta$

$= \int_0^{2\pi} \int_0^2 (9r - r^2\sin\theta) \, dr\,d\theta$

$= \int_0^{2\pi} [(9r^2/2) - (r^3/3)\sin\theta]_{r=0}^{r=2} \, d\theta$

$= (1/3) \int_0^{2\pi} (54 - 8\sin\theta) \, d\theta$

$= (1/3) [54\theta + 8\cos\theta]_0^{2\pi}$

$= 36\pi$

15. Let ρ represent the uniform density.

$M = \int_0^{2\pi} \int_0^2 \int_0^r \rho r \, dz \, dr \, d\theta$

$= \rho \int_0^{2\pi} \int_0^2 [rz]_{z=0}^{z=r} \, dr \, d\theta$

$= \rho \int_0^{2\pi} \int_0^2 r^2 \, dr \, d\theta$

$= \rho \int_0^{2\pi} [r^3/3]_{r=0}^{r=2} \, d\theta$

$= \rho \int_0^{2\pi} (8/3) \, d\theta$

$= \rho(8/3)\theta \big|_0^{2\pi}$

$= 16\rho\pi/3$

$M_{yz} = \int_0^{2\pi} \int_0^2 \int_0^r x\rho r \, dz \, dr \, d\theta$

$= \rho \int_0^{2\pi} \int_0^2 \int_0^r r^2 \cos\theta \, dz \, dr \, d\theta$

$= \rho \int_0^{2\pi} \int_0^2 [r^2 z \cos\theta]_{z=0}^{z=r} \, dr \, d\theta$

$= \rho \int_0^{2\pi} \int_0^2 r^3 \cos\theta \, dr \, d\theta$

$= \rho \int_0^{2\pi} [(r^4/4)\cos\theta]_{r=0}^{r=2} \, d\theta$

$= \rho \int_0^{2\pi} 4\cos\theta \, d\theta$

$= 4\rho \sin\theta \big|_0^{2\pi}$

$= 0$

Due to symmetry of the situation, one may conclude that $M_{xz} = M_{yz} = 0$.

$M_{xy} = \int_0^{2\pi} \int_0^2 \int_0^r z\rho r \, dz \, dr \, d\theta$

$= \rho \int_0^{2\pi} \int_0^2 [rz^2/2]_{z=0}^{z=r} \, dr \, d\theta$

$= (\rho/2) \int_0^{2\pi} \int_0^2 r^3 \, dr \, d\theta$

$= (\rho/2) \int_0^{2\pi} [r^4/4]_{r=0}^{r=2} \, d\theta$

$= (\rho/2) \int_0^{2\pi} 4 \, d\theta$

$= 2\rho\theta \big|_0^{2\pi}$

$= 4\rho\pi$

$\bar{x} = \bar{y} = 0$

$\bar{z} = M_{xy}/M = (4\rho\pi)/(16\rho\pi/3) = 3/4$

The center of mass is $(0, 0, 3/4)$.

17. Determine the volume for $z \geq 0$ and $\theta \geq 0$, and then multiply by 4.

$4 \int_0^{\pi/2} \int_0^{a\cos\theta} \int_0^{\sqrt{a^2-r^2}} r \, dz \, dr \, d\theta$

$= 4 \int_0^{\pi/2} \int_0^{a\cos\theta} [rz]_{z=0}^{z=\sqrt{a^2-r^2}} \, dr \, d\theta$

$= 4 \int_0^{\pi/2} \int_0^{a\cos\theta} r\sqrt{a^2-r^2} \, dr \, d\theta$

Let $u = a^2 - r^2$, $du = -2r \, dr$. At $r = a\cos\theta$, $u = a^2 - a^2\cos^2\theta = a^2\sin^2\theta$.

$-2 \int_0^{\pi/2} \int_{a^2}^{a^2\sin^2\theta} \sqrt{u} \, du \, d\theta$

$= -2 \int_0^{\pi/2} [(2/3)u^{3/2}]_{u=a^2}^{u=a^2\sin^2\theta} \, d\theta$

$= -(4a^3/3) \int_0^{\pi/2} (\sin^3\theta - 1) \, d\theta$

$= -(4a^3/3) \int_0^{\pi/2} \sin^3\theta \, d\theta + (4a^3/3) \int_0^{\pi/2} 1 \, d\theta$

For the first integral, let $u = \cos\theta$, $du = -\sin\theta \, d\theta$. Note that $\sin^2\theta = 1 - \cos^2\theta = 1 - u^2$.

$(4a^3/3) \int_1^0 (1 - u^2) \, du + [(4a^3/3)\theta]_0^{\pi/2}$

$= (4a^3/3)[u - (u^3/3)]_1^0 + (2a^3\pi/3)$

$= (6\pi - 8)a^3/9$

19. $\partial z/\partial x = 0$ and $\partial z/\partial y = 4$

$\sqrt{(\partial z/\partial x)^2 + (\partial z/\partial y)^2 + 1} = \sqrt{17}$

$A_S = \int_0^{2\pi} \int_0^2 r\sqrt{17} \, dr \, d\theta$

$= \sqrt{17} \int_0^{2\pi} [r^2/2]_{r=0}^{r=2} \, d\theta$

$= \sqrt{17} \int_0^{2\pi} 2 \, d\theta$

$= (2\sqrt{17})\theta \big|_0^{2\pi}$

$= 4\pi\sqrt{17}$

21. One obtains the same volume by considering the cylinder $x^2 + y^2 = 4$.

$\int_0^{2\pi} \int_0^2 \int_{-\sqrt{16-r^2}}^{\sqrt{16-r^2}} r \, dz \, dr \, d\theta$

$= \int_0^{2\pi} \int_0^2 [rz]_{z=-\sqrt{16-r^2}}^{r=\sqrt{16-r^2}} dr \, d\theta$

$= \int_0^{2\pi} \int_0^2 2r\sqrt{16 - r^2} \, dr \, d\theta$

Let $u = 16 - r^2$, $du = -2r \, dr$.

$-\int_0^{2\pi} \int_{16}^{12} \sqrt{u} \, du \, d\theta$

$= \int_0^{2\pi} \int_{12}^{16} \sqrt{u} \, du \, d\theta$

$= \int_0^{2\pi} [(2/3)u^{3/2}]_{u=12}^{u=16} d\theta$

$= (2/3) \int_0^{2\pi} (64 - 24\sqrt{3}) \, d\theta$

$= (2/3)(64 - 24\sqrt{3})\theta \big|_0^{2\pi}$

$= (256 - 96\sqrt{3})\pi/3$

25. Assume a constant density of 1.

$M = \int_0^{\pi} \int_0^4 r \, dr \, d\theta$

$= \int_0^{\pi} [r^2/2]_{r=0}^{r=4} d\theta$

$= \int_0^{\pi} 8 \, d\theta$

$= 8\theta \big|_0^{\pi}$

$= 8\pi$

$M_y = \int_0^{\pi} \int_0^4 xr \, dr \, d\theta$

$= \int_0^{\pi} \int_0^4 r^2 \cos\theta \, dr \, d\theta$

$= \int_0^{\pi} [(r^3/3)\cos\theta]_{r=0}^{r=4} d\theta$

$= (64/3) \int_0^{\pi} \cos\theta \, d\theta$

$= (64/3)\sin\theta \big|_0^{\pi}$

$= 0$

27. $\int_0^{2\pi} \int_0^{\pi/2} \int_0^1 e^{(\rho^2)^{3/2}} \rho^2 \sin\phi \, d\rho \, d\phi \, d\theta$

$= \int_0^{2\pi} \int_0^{\pi/2} \int_0^1 \rho^2 e^{\rho^3} \sin\phi \, d\rho \, d\phi \, d\theta$

Let $u = \rho^3$, $du = 3\rho^2 \, d\rho$.

$(1/3) \int_0^{2\pi} \int_0^{\pi/2} \int_0^1 e^u \sin\phi \, du \, d\phi \, d\theta$

$= (1/3) \int_0^{2\pi} \int_0^{\pi/2} [e^u \sin\phi]_{u=0}^{u=1} d\phi \, d\theta$

$= (1/3) \int_0^{2\pi} \int_0^{\pi/2} (e - 1) \sin\phi \, d\phi \, d\theta$

$= [(e-1)/3] \int_0^{2\pi} [-\cos\phi]_{\phi=0}^{\phi=\pi/2} d\theta$

$= [(e-1)/3] \int_0^{2\pi} 1 \, d\theta$

$= [(e-1)/3]\theta \big|_0^{2\pi}$

$= (2e - 2)\pi/3$

29. Area $= \int_0^{2\pi} \int_0^{1+\cos\theta} r \, dr \, d\theta$

$= \int_0^{2\pi} [r^2/2]_{r=0}^{r=1+\cos\theta} d\theta$

$= (1/2) \int_0^{2\pi} (1 + 2\cos\theta + \cos^2\theta) \, d\theta$

$= (1/4) \int_0^{2\pi} (2 + 4\cos\theta + 1 + \cos 2\theta) \, d\theta$

$= (1/4)[3\theta + 4\sin\theta + (1/2)\sin 2\theta]_0^{2\pi}$

$= 3\pi/2$

23. $\int_{-1}^2 \int_{x^2-2}^x \int_0^{10+x-2y} 1 \, dz \, dy \, dx$

$= \int_{-1}^2 \int_{x^2-2}^x [z]_{z=0}^{z=10+x-2y} dy \, dx$

$= \int_{-1}^2 \int_{x^2-2}^x (10 + x - 2y) \, dy \, dx$

$= \int_{-1}^2 [10y + xy - y^2]_{y=x^2-2}^{y=x} dx$

$= \int_{-1}^2 (x^4 - x^3 - 14x^2 + 12x + 24) \, dx$

$= (x^5/5) - (x^4/4) - (14x^3/3) + 6x^2 + 24x \big|_{-1}^2$

$= 1017/20$

$M_x = \int_0^{\pi} \int_0^4 yr \, dr \, d\theta$

$= \int_0^{\pi} \int_0^4 r^2 \sin\theta \, dr \, d\theta$

$= \int_0^{\pi} [(r^3/3)\sin\theta]_{r=0}^{r=4} d\theta$

$= (64/3) \int_0^{\pi} \sin\theta \, d\theta$

$= -(64/3)\cos\theta \big|_0^{\pi}$

$= 128/3$

$\bar{x} = M_y/M = 0$

$\bar{y} = M_x/M = (128/3)/(8\pi) = 16/(3\pi)$

The centroid is $(0, 16/(3\pi))$.

$\bar{x} = [2/(3\pi)] \int_0^{2\pi} \int_0^{1+\cos\theta} r^2 \cos\theta\, dr\, d\theta$

$= [2/(3\pi)] \int_0^{2\pi} [(r^3/3) \cos\theta]_{r=0}^{r=1+\cos\theta} d\theta$

$= [2/(9\pi)] \int_0^{2\pi} (1 + \cos\theta)^3 \cos\theta\, d\theta$

$= [2/(9\pi)] \int_0^{2\pi} (\cos\theta + 3\cos^2\theta + 3\cos^3\theta + \cos^4\theta)\, d\theta$

$= [2/(9\pi)] \int_0^{2\pi} (\cos\theta + 3\cos^3\theta)\, d\theta + [1/(18\pi)] \int_0^{2\pi} (7 + 8\cos 2\theta + \cos^2 2\theta)\, d\theta$

For the first integral, let $u = \sin\theta$, $du = \cos\theta\, d\theta$.

$\bar{x} = [2/(9\pi)] \int_0^0 [1 + 3(1 - u^2)]\, du + [1/(36\pi)] \int_0^{2\pi} (14 + 16\cos 2\theta + 1 + \cos 4\theta)\, d\theta$

$= [1/(36\pi)] [15\theta + 8\sin 2\theta + (1/4)\sin 4\theta]_0^{2\pi}$

$= 5/6$

$\bar{y} = [2/(3\pi)] \int_0^{2\pi} \int_0^{1+\cos\theta} r^2 \sin\theta\, dr\, d\theta$

$= [2/(3\pi)] \int_0^{2\pi} [(r^3/3) \sin\theta]_{r=0}^{r=1+\cos\theta} d\theta$

$= [2/(9\pi)] \int_0^{2\pi} (1 + \cos\theta)^3 \sin\theta\, d\theta$ Let $u = \cos\theta$, $du = -\sin\theta\, d\theta$.

$= -[2/(9\pi)] \int_1^1 (1 + u^3)\, du = 0$

The center of mass is $(5/6, 0)$.

31. The equations of the paraboloids may be rewritten as $z = 9 - (x-1)^2 - (y+2)^2$ and $z = (x-1)^2 + (y+2)^2$. These intersect when $(x-1)^2 + (y+2)^2 = 9/2$. Use a transformation which moves the point $(1, -2)$ to the origin, and switches to cylindrical coordinates. Let $x = 1 + u\cos v$, $y = -2 + u\sin v$, $z = w$. The region in question is the image of a region which is bounded by $0 \le u \le 3/\sqrt{2}$, $0 \le v \le 2\pi$ and $u^2 \le w \le 9 - u^2$.

$\dfrac{\partial(x,y,z)}{\partial(u,v,w)} = \det \begin{bmatrix} \cos v & -u\sin v & 0 \\ \sin v & u\cos v & 0 \\ 0 & 0 & 1 \end{bmatrix} = u\cos^2 v + u\sin^2 v = u$

$V = \int_0^{2\pi} \int_0^{3/\sqrt{2}} \int_{u^2}^{9-u^2} |u|\, dw\, du\, dv$

$= \int_0^{2\pi} \int_0^{3/\sqrt{2}} [uw]_{w=u^2}^{w=9-u^2} du\, dv$

$= \int_0^{2\pi} \int_0^{3/\sqrt{2}} (9u - 2u^3)\, du\, dv$

$= \int_0^{2\pi} [(9u^2/2) - (u^4/2)]_{u=0}^{u=3/\sqrt{2}} dv$

$= \int_0^{2\pi} (81/8)\, dv$

$= (81/8)v \big|_0^{2\pi}$

$= 81\pi/4$

33. The density is λz.

$\int_0^{2\pi} \int_0^2 \int_0^4 \lambda zr\, dz\, dr\, d\theta$

$= \lambda \int_0^{2\pi} \int_0^2 [z^2 r/2]_{z=0}^{z=4} dr\, d\theta$

$= \lambda \int_0^{2\pi} \int_0^2 8r\, dr\, d\theta$

$= \lambda \int_0^{2\pi} [4r^2]_{r=0}^{r=2} d\theta$

$= \lambda \int_0^{2\pi} 16\, d\theta$

$= 16\lambda\theta \big|_0^{2\pi}$

$= 32\lambda\pi$

35. $\int_0^{\pi/2} \int_0^1 (r/\sqrt{1+r^2})\, dr\, d\theta$

Let $u = 1 + r^2$, $du = 2r\, dr$.

$(1/2) \int_0^{\pi/2} \int_1^2 (1/\sqrt{u})\, du\, d\theta$

$= (1/2) \int_0^{\pi/2} [2\sqrt{u}]_{u=1}^{u=2} d\theta$

$= \int_0^{\pi/2} (\sqrt{2} - 1)\, d\theta$

$= (\sqrt{2} - 1)\theta \big|_0^{\pi/2}$

$= (\sqrt{2} - 1)\pi/2$

37. In cylindrical coordinates, the equation of the upper half of the cone is $z = r$, and the equation of the cylinder is $r = 4 \sin \theta$.

$\int_0^\pi \int_0^{4\sin\theta} \int_0^r r \, dz \, dr \, d\theta$

$= \int_0^\pi \int_0^{4\sin\theta} [rz]_{z=0}^{z=r} \, dr \, d\theta$

$= \int_0^\pi \int_0^{4\sin\theta} r^2 \, dr \, d\theta$

$= \int_0^\pi [r^3/3]_{r=0}^{r=4\sin\theta} \, d\theta$

$= (64/3) \int_0^\pi \sin^3 \theta \, d\theta$ Let $u = -\cos\theta$, $du = \sin\theta \, d\theta$.

$= (64/3) \int_{-1}^1 (1 - u^2) \, du$

$= (64/3) [u - (u^3/3)]_{-1}^1$

$= 256/9$

39. $\int_0^2 \int_1^3 \sin^2 \sqrt{x - y^2} \, dy \, dx$

This integral is undefined since a portion of the rectangle $\{(x, y) \mid 0 \le x \le 2, 1 \le y \le 3\}$ is not in the domain of the integrand.

In the second printing, this problem is to be changed to $\int_0^2 \int_1^3 \sin^2 \sqrt{x + y^2} \, dy \, dx$. Change line 110 of BASIC Program 8 to

```
LET S = S + SIN(SQR(A+J*D1+(C+K*D2)^2))^2.
```

Use the values 0, 2, 1, 3 for a, b, c and d respectively (since 0 and 2 are the x values, and 1 and 3 are the y values). Using $n = 10$ gives a value of 1.991516, and using $n = 100$ gives a value of 2.200004.

41. In cylindrical coordinates, the equations of the lower and upper halves of the cone are $z = -r$ and $z = r$, and the equation of the cylinder is $r = 3$.

$\int_0^{2\pi} \int_0^3 \int_{-r}^r r \, dz \, dr \, d\theta$

$= \int_0^{2\pi} \int_0^3 [rz]_{z=-r}^{z=r} \, dr \, d\theta$

$= \int_0^{2\pi} \int_0^3 2r^2 \, dr \, d\theta$

$= \int_0^{2\pi} [2r^3/3]_{r=0}^{r=3} \, d\theta$

$= \int_0^{2\pi} 18 \, d\theta$

$= 18\theta \big|_0^{2\pi}$

$= 36\pi$

43. If one directly computes the area of the surface outside of the cylinder, then one will obtain an improper integral. In order to avoid this situation, compute the area of the surface inside the cylinder, and then subtract from the total surface area of the sphere ($4\pi 2^2 = 16\pi$). Proceed by computing the area of that part of the surface above the xy-plane, and then multiplying by 2. The equation of the upper hemisphere is $z = \sqrt{4 - x^2 - y^2}$.

$\partial z/\partial x = -x/\sqrt{4 - x^2 - y^2}$ and $\partial z/\partial y = -y/\sqrt{4 - x^2 - y^2}$

$\sqrt{(\partial z/\partial x)^2 + (\partial z/\partial y)^2 + 1} = \sqrt{(x^2 + y^2)/(4 - x^2 - y^2) + 1} = 2/\sqrt{4 - x^2 - y^2} = 2/\sqrt{4 - r^2}$

$A_S = 16\pi - 2 \int_0^{2\pi} \int_0^1 (2r/\sqrt{4 - r^2})\, dr\, d\theta$ \qquad Let $u = 4 - r^2$, $du = -2r\, dr$.

$= 16\pi + 2 \int_0^{2\pi} \int_4^3 (1/\sqrt{u})\, du\, d\theta$

$= 16\pi - 2 \int_0^{2\pi} \int_3^4 u^{-1/2}\, du\, d\theta$

$= 16\pi - 2 \int_0^{2\pi} \left[2\sqrt{u} \right]_{u=3}^{u=4} d\theta$

$= 16\pi - 4 \int_0^{2\pi} (2 - \sqrt{3})\, d\theta$

$= 16\pi - \left[(8 - 4\sqrt{3})\theta \right]_0^{2\pi}$

$= 16\pi - (16\pi - 8\pi\sqrt{3})$

$= 8\pi\sqrt{3}$

Chapter 20

Topics in Vector Analysis

20.1 Vector Fields

1.

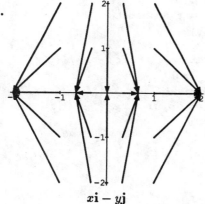

$x\mathbf{i} - y\mathbf{j}$

3.

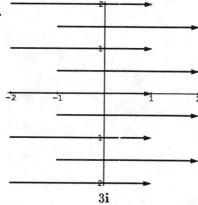

$3\mathbf{i}$

5.

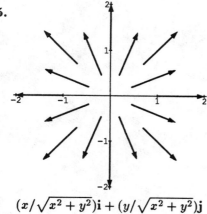

$(x/\sqrt{x^2 + y^2})\mathbf{i} + (y/\sqrt{x^2 + y^2})\mathbf{j}$

7.

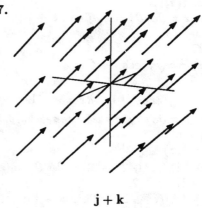

$\mathbf{j} + \mathbf{k}$

9.

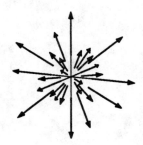

$$2x\mathbf{i} + 2y\mathbf{j} + 2z\mathbf{k}$$

11. 4: $\mathbf{F}(x,y) = (-1)(x\mathbf{i} + y\mathbf{j})$
  5: $\mathbf{F}(x,y) = (1/\sqrt{x^2 + y^2})(x\mathbf{i} + y\mathbf{j})$
  9: $\mathbf{F}(x,y,z) = 2(x\mathbf{i} + y\mathbf{j} + z\mathbf{k})$

13. $\mathbf{F}(x,y) = 2x\mathbf{i} - 2y\mathbf{j}$

15. $\mathbf{F}(x,y,z) = (1/\sqrt{x^2 + y^2 + z^2})(x\mathbf{i} + y\mathbf{j} + z\mathbf{k})$

17. $\mathbf{F}(x,y,z) = ze^{x-y}\mathbf{i} - ze^{x-y}\mathbf{j} + e^{x-y}\mathbf{k}$

19. $\phi(x,y) = \int(\tan xy + xy\sec^2 xy)\,dx$
  $= x\tan xy + h(y)$
  $x^2\sec^2 xy = \partial\phi/\partial y = x^2\sec^2 xy + h'(y)$
  $h'(y) = 0$
  $h(y) = C$
  $\phi(x,y) = x\tan xy + C$

21. No.
  $\partial M/\partial y = (y - 2x)/(y - x)^2$ and
  $\partial N/\partial x = y/(y - x)^2$.

23. $\phi(x,y,z) = \int yze^{xyz}\,dx = e^{xyz} + h(y,z)$
  $xze^{xyz} = \partial\phi/\partial y = xze^{xyz} + h_y(y,z)$
  $h_y(y,z) = 0$
  $h(y,z) = k(z)$
  $xye^{xyz} = \partial\phi/\partial z = xye^{xyz} + k'(z)$
  $k'(z) = 0$
  $k(z) = C$
  $\phi(x,y,z) = e^{xyz} + C$

25. $\phi(x,y,z) = \int \sqrt{y}\cosh z\,dx = x\sqrt{y}\cosh z + h(y,z)$
  $(x/(2\sqrt{y}))\cosh z = \partial\phi/\partial y = (x/(2\sqrt{y}))\cosh z + h_y(y,z)$
  $h_y(y,z) = 0$
  $h(y,z) = k(z)$
  $x\sqrt{y}\sinh z = \partial\phi/\partial z = x\sqrt{y}\sinh z + k'(z)$
  $k'(z) = 0$
  $k(z) = C$
  $\phi(x,y,z) = x\sqrt{y}\cosh z + C$

27. $\phi(x,y,z) = \int 2xye^{x^2 y}\ln z\,dx = e^{x^2 y}\ln z + h(y,z)$
  $x^2 e^{x^2 y}\ln z = \partial\phi/\partial y = x^2 e^{x^2 y}\ln z + h_y(y,z)$
  $h_y(y,z) = 0$
  $h(y,z) = k(z)$
  $(e^{x^2 y}/z) - [1/(z^2 + 2z + 2)] = \partial\phi/\partial z = (e^{x^2 y}/z) + k'(z)$
  $k'(z) = -1/(z^2 + 2z + 2)$
  $k(z) = -\int 1/(z^2 + 2z + 2)\,dz = -\operatorname{Tan}^{-1}(z + 1) + C$
  $\phi(x,y,z) = e^{x^2 y}\ln z - \operatorname{Tan}^{-1}(z + 1) + C$

29. Let $A(x)$ be the area of a cross-section of the pipe at position x. Let r be the rate of flow of the water in, e.g., cm^3/sec; r is assumed to be constant. Then $|\mathbf{F}(x, y, z)| = r/A(x)$, so when $A(x)$ is smaller, the magnitude of $\mathbf{F}(x, y, z)$ is larger.

31. For parts (a) and (b), let $\mathbf{r}_1$ and $\mathbf{r}_2$ be vectors whose terminal points are the given point, with $(-1, 0, 0)$ as the initial point of $\mathbf{r}_1$ and $(1, 0, 0)$ as the initial point of $\mathbf{r}_2$.

    a. $\mathbf{r}_1 = \mathbf{i}$ and $\mathbf{r}_2 = -\mathbf{i}$.
       $\mathbf{F}(0, 0, 0) = (kq_1q/|\mathbf{r}_1|^3)\mathbf{r}_1 + (kq_2q/|\mathbf{r}_2|^3)\mathbf{r}_2 = 3k\mathbf{i} - k(-\mathbf{i}) = 4k\mathbf{i}$

    b. $\mathbf{r}_1 = -\mathbf{i}$ and $\mathbf{r}_2 = -3\mathbf{i}$.
       $\mathbf{F}(-2, 0, 0) = (kq_1q/|\mathbf{r}_1|^3)\mathbf{r}_1 + (kq_2q/|\mathbf{r}_2|^3)\mathbf{r}_2 = 3k(-\mathbf{i}) - (k/27)(-3\mathbf{i}) = (-26k/9)\mathbf{i}$

    c. The left figure below.

    d. The middle and right figures below. The middle figure shows the flow lines only in the xy-plane.

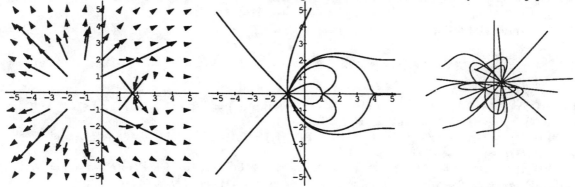

20.2   Work and Line Integrals

1. $\mathbf{F}(\mathbf{r}(t)) = t^2\mathbf{i} + (3 + t)\mathbf{j}$
$\mathbf{r}'(t) = 2t\mathbf{i} + \mathbf{j}$
$\mathbf{F}(\mathbf{r}(t)) \cdot \mathbf{r}'(t) = 2t^3 + t + 3$
$W = \int_0^2 (2t^3 + t + 3)\, dt$
$\quad = (t^4/2) + (t^2/2) + 3t\big|_0^2$
$\quad = 16$

3. $\mathbf{F}(\mathbf{r}(t)) = t\mathbf{i} + 9t^2\mathbf{j}$
$\mathbf{r}'(t) = (1/2)t^{-1/2}\mathbf{i} + 3\mathbf{j}$
$\mathbf{F}(\mathbf{r}(t)) \cdot \mathbf{r}'(t) = (\sqrt{t}/2) + 27t^2$
$W = \int_1^4 [(\sqrt{t}/2) + 27t^2]\, dt$
$\quad = (t^{3/2}/3) + 9t^3\big|_1^4$
$\quad = 1708/3$

5. $\mathbf{F}(\mathbf{r}(t)) = (t^2 + t^4)\mathbf{i} + t^3\mathbf{j}$
$\mathbf{r}'(t) = \mathbf{i} + 2t\mathbf{j}$
$\mathbf{F}(\mathbf{r}(t)) \cdot \mathbf{r}'(t) = t^2 + 3t^4$
$W = \int_0^2 (t^2 + 3t^4)\, dt$
$\quad = (t^3/3) + (3t^5/5)\big|_0^2$
$\quad = 328/15$

7. $\mathbf{F}(\mathbf{r}(t)) = \cos t \sin t\, \mathbf{i} + t \cos t\, \mathbf{j} + t \sin t\, \mathbf{k}$
$\mathbf{r}'(t) = -\sin t\, \mathbf{i} + \cos t\, \mathbf{j} + \mathbf{k}$
$\mathbf{F}(\mathbf{r}(t)) \cdot \mathbf{r}'(t) = -\cos t \sin^2 t + t \cos^2 t + t \sin t$
$W = \int_0^\pi [-\cos t \sin^2 t + (t/2) + (t/2)\cos 2t + t \sin t]\, dt$
$\quad = -(1/3)\sin^3 t + (t^2/4) + (t/4)\sin 2t + (1/8)\cos 2t - t \cos t + \sin t\big|_0^\pi$
$\quad = \pi + (\pi^2/4)$

9. $\mathbf{F}(\mathbf{r}(t)) = (36 - 6t^3)\mathbf{i} + (9t^4 - 216t)\mathbf{j} + 72t^2\mathbf{k}$
$\mathbf{r}'(t) = 6t\mathbf{i} - \mathbf{j}$
$\mathbf{F}(\mathbf{r}(t)) \cdot \mathbf{r}'(t) = 432t - 45t^4$
$W = \int_{-1}^{1}(432t - 45t^4)\, dt$
$\quad = 216t^2 - 9t^5\big|_{-1}^{1}$
$\quad = -18$

11. $\mathbf{r}(t) = 4t\mathbf{i} + 2t\mathbf{j},\ 0 \le t \le 1$
$\mathbf{r}'(t) = 4\mathbf{i} + 2\mathbf{j}$
$\mathbf{F}(\mathbf{r}(t)) = 2t\mathbf{i} + 8t\mathbf{j}$
$\mathbf{F}(\mathbf{r}(t)) \cdot \mathbf{r}'(t) = 24t$
$\int_C \mathbf{F} \cdot d\mathbf{r} = \int_0^1 24t\, dt = 12$

13. $\mathbf{r}(t) = \cos t\, \mathbf{i} + \sin t\, \mathbf{j},\ 0 \le t \le 2\pi,$ and $\mathbf{r}'(t) = -\sin t\, \mathbf{i} + \cos t\, \mathbf{j}$
$\mathbf{F}(\mathbf{r}(t)) = (\sin t - \cos t)\mathbf{i} + \cos t \sin t\, \mathbf{j}$
$\mathbf{F}(\mathbf{r}(t)) \cdot \mathbf{r}'(t) = -\sin^2 t + \cos t \sin t + \cos^2 t \sin t$
$\int_C \mathbf{F} \cdot d\mathbf{r} = \int_0^{2\pi}[-(1/2) + (1/2)\cos 2t + \cos t \sin t + \cos^2 t \sin t]\, dt$
$\quad = -(t/2) + (1/4)\sin 2t + (1/2)\sin^2 t - (1/3)\cos^3 t\,\big|_0^{2\pi}$
$\quad = -\pi$

15. $\mathbf{r}_1(t) = (2t - 1)\mathbf{i} - 4t\mathbf{j},\ \mathbf{r}_2(t) = (1 - t)\mathbf{i} + (6t - 4)\mathbf{j},\ \mathbf{r}_3(t) = -t\mathbf{i} + (2 - 2t)\mathbf{j}$ for $0 \le t \le 1$
$\mathbf{r}_1'(t) = 2\mathbf{i} - 4\mathbf{j},\ \mathbf{r}_2'(t) = -\mathbf{i} + 6\mathbf{j},\ \mathbf{r}_3'(t) = -\mathbf{i} - 2\mathbf{j}$
$\mathbf{F}(\mathbf{r}_1(t)) = (-2t - 1)\mathbf{i} + (12t^2 + 4t - 1)\mathbf{j},\ \mathbf{F}(\mathbf{r}_2(t)) = (5t - 3)\mathbf{i} + (35t^2 - 46t + 15)\mathbf{j}$, and
$\mathbf{F}(\mathbf{r}_3(t)) = (2 - 3t)\mathbf{i} + (3t^2 - 8t + 4)\mathbf{j}$
$\mathbf{F}(\mathbf{r}_1(t)) \cdot \mathbf{r}_1'(t) = 2 - 20t - 48t^2,\ \mathbf{F}(\mathbf{r}_2(t)) \cdot \mathbf{r}_2'(t) = 210t^2 - 281t + 93,\ \mathbf{F}(\mathbf{r}_3(t)) \cdot \mathbf{r}_3'(t) = -6t^2 + 19t - 10$
$\int_C \mathbf{F} \cdot d\mathbf{r} = \int_0^1(2 - 20t - 48t^2)\, dt + \int_0^1(210t^2 - 281t + 93)\, dt + \int_0^1(-6t^2 + 19t - 10)\, dt$
$\quad = \int_0^1(156t^2 - 282t + 85)\, dt$
$\quad = 52t^3 - 141t^2 + 85t\,\big|_0^1$
$\quad = -4$

17. $\mathbf{F}(\mathbf{r}(t)) = \sin t\, \mathbf{i} - \cos t\, \mathbf{j} + 2t\mathbf{k}$
$\mathbf{r}'(t) = -\sin t\, \mathbf{i} + \cos t\, \mathbf{j} + \mathbf{k}$
$\mathbf{F}(\mathbf{r}(t)) \cdot \mathbf{r}'(t) = -\sin^2 t - \cos^2 t + 2t = 2t - 1$
$\int_C \mathbf{F} \cdot d\mathbf{r} = \int_0^{\pi/2}(2t - 1)\, dt$
$\quad = t^2 - t\,\big|_0^{\pi/2}$
$\quad = (\pi^2 - 2\pi)/4$

19. $\mathbf{r}(t) = 2t\mathbf{i} + 12t\mathbf{j} - 4t\mathbf{k},\ 0 \le t \le 1$
$\mathbf{r}'(t) = 2\mathbf{i} + 12\mathbf{j} - 4\mathbf{k}$
$\mathbf{F}(\mathbf{r}(t)) = 4t^2\mathbf{i} + 144t^2\mathbf{j} + 16t^2\mathbf{k}$
$\mathbf{F}(\mathbf{r}(t)) \cdot \mathbf{r}'(t) = 1672t^2$
$\int_C \mathbf{F} \cdot d\mathbf{r} = \int_0^1 1672t^2\, dt$
$\quad = 1672t^3/3\,\big|_0^1$
$\quad = 1672/3$

21. $\mathbf{r}(t) = \cos t\, \mathbf{i} + \sin t\, \mathbf{j},\ \pi/2 \le t \le \pi$
$\mathbf{r}'(t) = -\sin t\, \mathbf{i} + \cos t\, \mathbf{j}$
$\mathbf{F}(x, y) = (x^2 + y^2)\mathbf{j}$
$\mathbf{F}(\mathbf{r}(t)) = \mathbf{j}$
$\mathbf{F}(\mathbf{r}(t)) \cdot \mathbf{r}'(t) = \cos t$
$\int_C(x^2 + y^2)\, dy = \int_{\pi/2}^{\pi}\cos t\, dt$
$\quad = \sin t\,\big|_{\pi/2}^{\pi}$
$\quad = -1$

23. $\mathbf{r}(t) = -t\mathbf{i} + (t + 1)\mathbf{j} + t\mathbf{k},\ 0 \le t \le 1$
$\mathbf{r}'(t) = -\mathbf{i} + \mathbf{j} + \mathbf{k}$
$\mathbf{F}(x, y, z) = (x^2 + y^2 + z^2)\mathbf{j}$
$\mathbf{F}(\mathbf{r}(t)) = (3t^2 + 2t + 1)\mathbf{j}$
$\mathbf{F}(\mathbf{r}(t)) \cdot \mathbf{r}'(t) = 3t^2 + 2t + 1$
$\int_C(x^2 + y^2 + z^2)\, dy = \int_0^1(3t^2 + 2t + 1)\, dt$
$\quad = t^3 + t^2 + t\,\big|_0^1$
$\quad = 3$

25. $\mathbf{r}_1(t) = \cos t\,\mathbf{i} + \sin t\,\mathbf{j}$ for $0 \le t \le \pi/2$, $\mathbf{r}_2(t) = -t\mathbf{i} + \mathbf{j}$ for $0 \le t \le 1$
$\mathbf{r}_1'(t) = -\sin t\,\mathbf{i} + \cos t\,\mathbf{j}$, $\mathbf{r}_2'(t) = -\mathbf{i}$
$\mathbf{F}(x, y) = xy\mathbf{i} + xy^2\mathbf{j}$
$\mathbf{F}(\mathbf{r}_1(t)) = \cos t \sin t\,\mathbf{i} + \cos t \sin^2 t\,\mathbf{j}$, $\mathbf{F}(\mathbf{r}_2(t)) = -t\mathbf{i} - t\mathbf{j}$
$\mathbf{F}(\mathbf{r}_1(t)) \cdot \mathbf{r}_1'(t) = -\cos t \sin^2 t + \cos^2 t \sin^2 t$
$= -\cos t \sin^2 t + (1/4) \sin^2 2t$
$= -\cos t \sin^2 t + (1/8) - (1/8) \cos 4t$
$\mathbf{F}(\mathbf{r}_2(t)) \cdot \mathbf{r}_2'(t) = t$
$\int_C xy\,dx + xy^2\,dy = \int_0^{\pi/2} (-\cos t \sin^2 t + (1/8) - (1/8) \cos 4t)\,dt + \int_0^1 t\,dt$
$= [-(1/3) \sin^3 t + (t/8) - (1/32) \sin 4t]_0^{\pi/2} + [t^2/2]_0^1$
$= (8 + 3\pi)/48$

27. $\mathbf{r}(t) = t\mathbf{i} + 2t^2\mathbf{j}$, $0 \le t \le 2$
$\mathbf{r}'(t) = \mathbf{i} + 4t\mathbf{j}$
$\mathbf{F}(x, y, z) = (x + y)\mathbf{i} + (x^2 - y^2)\mathbf{j}$
$\mathbf{F}(\mathbf{r}(t)) = (2t^2 + t)\mathbf{i} + (t^2 - 4t^4)\mathbf{j}$
$\mathbf{F}(\mathbf{r}(t)) \cdot \mathbf{r}'(t) = t + 2t^2 + 4t^3 - 16t^5$
$\int_C (x + y)\,dx + (x^2 - y^2)\,dy = \int_0^2 (t + 2t^2 + 4t^3 - 16t^5)\,dt$
$= (t^2/2) + (2t^3/3) + t^4 - (8t^6/3)\big|_0^2$
$= -442/3$

29. $\mathbf{r}_1(t) = -2t\mathbf{j}$, $\mathbf{r}_2(t) = 4\mathbf{i}$, $\mathbf{r}_3(t) = 4\mathbf{i} + t\mathbf{j}$ for $0 \le t \le 1$
$\mathbf{r}_1'(t) = -2\mathbf{j}$, $\mathbf{r}_2'(t) = 0$, $\mathbf{r}_3'(t) = \mathbf{j}$
$\mathbf{F}(x, y, z) = x^2 y\mathbf{i} + xy^2\mathbf{j}$
$\mathbf{F}(\mathbf{r}_1(t)) = 0$, $\mathbf{F}(\mathbf{r}_3(t)) = 16t\mathbf{i} + 4t^2\mathbf{j}$
$\mathbf{F}(\mathbf{r}_1(t)) \cdot \mathbf{r}_1'(t) = 0$, $\mathbf{F}(\mathbf{r}_2(t)) \cdot \mathbf{r}_2'(t) = 0$, $\mathbf{F}(\mathbf{r}_3(t)) \cdot \mathbf{r}_3'(t) = 4t^2$
$\int_C x^2 y\,dx + xy^2\,dy = \int_0^1 0\,dt + \int_0^1 0\,dt + \int_0^1 4t^2\,dt$
$= 4t^3/3\big|_0^1$
$= 4/3$

31. If $f(x)$ is the force applied to a particle moving along the x-axis, then from Section 7.6 we have the result that $W = \int_a^b f(x)\,dx$. Next, compute the work using Definition 2:
$\mathbf{r}(t) = t\mathbf{i}$ for $a \le t \le b$ and $\mathbf{r}'(t) = \mathbf{i}$
$\mathbf{F}(x) = f(x)\mathbf{i}$ and $\mathbf{F}(\mathbf{r}(t)) = f(t)\mathbf{i}$
$\mathbf{F}(\mathbf{r}(t)) \cdot \mathbf{r}'(t) = f(t)$
$W = \int_a^b \mathbf{F}(\mathbf{r}(t)) \cdot \mathbf{r}'(t)\,dt = \int_a^b f(t)\,dt$, which is the same integral as above.

33. $x(t) = 2t$ and $y(t) = -4t$
$x'(t) = 2$ and $y'(t) = -4$
$[x'(t)]^2 + [y'(t)]^2 = 20$
$\int_C (x + 2y^2)\,ds = \int_0^1 (2t + 32t^2)\sqrt{20}\,dt$
$= 2\sqrt{5}[t^2 + (32t^3/3)]_0^1$
$= 70\sqrt{5}/3$

35. $x(t) = t$, $y(t) = 3$ and $z(t) = t^2$
$x'(t) = 1$, $y'(t) = 0$ and $z'(t) = 2t$
$[x'(t)]^2 + [y'(t)]^2 + [z'(t)]^2 = 1 + 4t^2$
$\int_C (8x + y - 3)\,ds = \int_0^1 8t\sqrt{4t^2 + 1}\,dt$
Let $u = 4t^2 + 1$, $du = 8t\,dt$.
$\int_1^5 \sqrt{u}\,du$
$= (2/3)u^{3/2}\big|_1^5$
$= (10\sqrt{5} - 2)/3$

37. $x(t) = 3\sin t$, $y(t) = 3\cos t$ and $z(t) = -3t$
$x'(t) = 3\cos t$, $y'(t) = -3\sin t$ and $z'(t) = -3$
$[x'(t)]^2 + [y'(t)]^2 + [z'(t)]^2 = 18$
$\int_C (x^2 + y^2 - 9)\, ds = \int_0^{100} (9\sin^2 t + 9\cos^2 t - 9)\sqrt{18}\, dt = \int_0^{100} 0\, dt = 0$

39. $x(t) = \cos t$, $y(t) = -\sin t$ and $z(t) = 2$
$x'(t) = -\sin t$, $y'(t) = -\cos t$ and $z'(t) = 0$
$M = \int_0^\pi \left([x(t)]^2 + [y(t)]^2 + [z(t)]^2\right) \sqrt{[x'(t)]^2 + [y'(t)]^2 + [z'(t)]^2}\, dt$
$= \int_0^\pi (\cos^2 t + \sin^2 t + 4)\sqrt{\sin^2 t + \cos^2 t}\, dt$
$= \int_0^\pi 5\sqrt{1}\, dt$
$= 5\pi$

20.3 Line Integrals: Independence of Path

1. $\phi(x,y) = \int y\, dx = xy + h(y)$
$x = \partial\phi/\partial y = x + h'(y)$
$h'(y) = 0$, so use $h(y) = 0$
$\phi(x,y) = xy$
$\int_C \mathbf{F} \cdot d\mathbf{r} = \phi(3,1) - \phi(0,0) = 3$

3. $\phi(x,y) = \int ye^{xy}\, dx = e^{xy} + h(y)$
$xe^{xy} = \partial\phi/\partial y = xe^{xy} + h'(y)$
$h'(y) = 0$, so use $h(y) = 0$
$\phi(x,y) = e^{xy}$
$\int_C \mathbf{F} \cdot d\mathbf{r} = \phi(1,2) - \phi(0,0) = e^2 - 1$

5. $\phi(x,y) = \int 2x\, dx = x^2 + h(y)$
$2y = \partial\phi/\partial y = h'(y)$
Use $h(y) = y^2$
$\phi(x,y) = x^2 + y^2$
$\int_C \mathbf{F} \cdot d\mathbf{r} = \phi(-1,0) - \phi(1,0) = 1 - 1 = 0$

7. $\phi(x,y) = \int e^{y^2}\, dx = xe^{y^2} + h(y)$
$2xye^{y^2} = \partial\phi/\partial y = 2xye^{y^2} + h'(y)$
$h'(y) = 0$, so use $h(y) = 0$
$\phi(x,y) = xe^{y^2}$
$\int_{(0,0)}^{(2,1)} e^{y^2}\, dx + 2xye^{y^2}\, dy$
$= \phi(2,1) - \phi(0,0) = 2e$

9. $\phi(x,y) = \int e^x \sin y\, dx = e^x \sin y + h(y)$
$e^x \cos y = \partial\phi/\partial y = e^x \cos y + h'(y)$
$h'(y) = 0$, so use $h(y) = 0$
$\phi(x,y) = e^x \sin y$
$\int_{(0,0)}^{(1,\pi/4)} e^x \sin y\, dx + e^x \cos y\, dy = \phi(1, \pi/4) - \phi(0,0) = e\sqrt{2}/2$

11. $\phi(x,y,z) = \int (2xy + z^2)\, dx = x^2 y + xz^2 + h(y,z)$
$x^2 - 2 = \partial\phi/\partial y = x^2 + h_y(y,z)$
$h_y(y,z) = -2$
$h(y,z) = -2y + k(z)$
$2xz + 1 = \partial\phi/\partial z = 2xz + k'(z)$
$k'(z) = 1$
Use $k(z) = z$
$\phi(x,y,z) = x^2 y + xz^2 - 2y + z$
$\int_{(1,3,-4)}^{(7,14,0)} (2xy + z^2)\, dx + (x^2 - 2)\, dy + (2xz + 1)\, dz = \phi(7,14,0) - \phi(1,3,-4) = 658 - 9 = 649$

13. Let $u = x + y + z$. Then the integrand becomes $(dx + dy + dz)/(u\sqrt{u^2 - 1})$.

$\phi(x, y, z) = \int [1/(u\sqrt{u^2 - 1})]\, dx = \int [1/(u\sqrt{u^2 - 1})]\, du = \text{Sec}^{-1} u + h(y, z)$

Note that $(\partial/\partial y)(\text{Sec}^{-1} u) = [1/(u\sqrt{u^2 - 1})](\partial u/\partial y) = 1/(u\sqrt{u^2 - 1})$. Similarly for $\partial/\partial z$.

Therefore, use $h(y, z) = 0$.

$\phi(x, y, z) = \text{Sec}^{-1} u = \text{Sec}^{-1}(x + y + z)$

$\displaystyle\int_{(0,0,2\sqrt{3}/3)}^{(1,1,0)} \frac{dx + dy + dz}{(x + y + z)\sqrt{x^2 + y^2 + z^2 + 2xy + 2xz + 2yz - 1}} = \phi(1, 1, 0) - \phi(0, 0, 2\sqrt{3}/3) = \frac{\pi}{3} - \frac{\pi}{6} = \frac{\pi}{6}$

15. $-a$. Since traversing C_1 and then traversing C_2 is a closed path, one must have:

$\int_{C_1} \mathbf{F} \cdot d\mathbf{r} + \int_{C_2} \mathbf{F} \cdot d\mathbf{r} = 0$.

17. First suppose that the vector field $\mathbf{F}$ is conservative in D. By Theorem 2, line integrals in D are independent of path. For any closed path C in D, partition C into two paths C_1 and C_2 such that traversing C is equivalent to first traversing C_1 and then traversing C_2. Since the first endpoint of C_1 is the second endpoint of C_2, and vice versa, and since line integrals in D are independent of path, one has $\int_{C_1} \mathbf{F} \cdot d\mathbf{r} = \int_{-C_2} \mathbf{F} \cdot d\mathbf{r} = -\int_{C_2} \mathbf{F} \cdot d\mathbf{r}$. Therefore, $\int_C \mathbf{F} \cdot d\mathbf{r} = \int_{C_1} \mathbf{F} \cdot d\mathbf{r} + \int_{C_2} \mathbf{F} \cdot d\mathbf{r} = 0$.

Next suppose that $\int_C \mathbf{F} \cdot d\mathbf{r} = 0$ for every closed path C in D. Let C_1 and C_2 be two paths in D which start and end at the same points. Let C be the path obtained by first traversing C_1, and then traversing $-C_2$ (traversing C_2 in the opposite direction). C is a closed path, and therefore

$$0 = \int_C \mathbf{F} \cdot d\mathbf{r} = \int_{C_1} \mathbf{F} \cdot d\mathbf{r} + \int_{-C_2} \mathbf{F} \cdot d\mathbf{r} = \int_{C_1} \mathbf{F} \cdot d\mathbf{r} - \int_{C_2} \mathbf{F} \cdot d\mathbf{r}.$$

Consequently, $\int_{C_1} \mathbf{F} \cdot d\mathbf{r} = \int_{C_2} \mathbf{F} \cdot d\mathbf{r}$. By the arbitrary nature of C_1 and C_2, line integrals in D are independent of path. By Theorem 2, this implies that the vector field $\mathbf{F}$ is conservative in D.

19. Let $\mathbf{r}_1(t) = -\cos t\,\mathbf{i} + \sin t\,\mathbf{j}$ for $0 \le t \le \pi$, and let $\mathbf{r}_2(t) = (2t - 1)\mathbf{i}$ for $0 \le t \le 1$. $\mathbf{r}_1(t)$ traces out the curve C_1 and $\mathbf{r}_2(t)$ traces out the curve C_2.

$\mathbf{r}_1'(t) = \sin t\,\mathbf{i} + \cos t\,\mathbf{j}$

$\mathbf{F}(\mathbf{r}_1(t)) = -2\cos t \sin t e^{\cos^2 t \sin t}\mathbf{i} + \cos^2 t e^{\cos^2 t \sin t}\mathbf{j}$

$\mathbf{F}(\mathbf{r}_1(t)) \cdot \mathbf{r}_1'(t) = -2\cos t \sin^2 t e^{\cos^2 t \sin t} + \cos^3 t e^{\cos^2 t \sin t}$

$\int_{C_1} \mathbf{F} \cdot d\mathbf{r}_1 = \int_0^\pi (-2\cos t \sin^2 t + \cos^3 t)e^{\cos^2 t \sin t}\, dt$ Let $u = \cos^2 t \sin t$, $du = (-2\cos t \sin^2 t + \cos^3 t)\, dt$.

$= \int_0^0 e^u\, du = 0$

$\mathbf{r}_2'(t) = 2\mathbf{i}$

$\mathbf{F}(\mathbf{r}_2(t)) = (2t - 1)^2\mathbf{j}$

$\mathbf{F}(\mathbf{r}_2(t)) \cdot \mathbf{r}_2'(t) = 0$

$\int_{C_2} \mathbf{F} \cdot d\mathbf{r}_2 = \int_0^1 0\, dt = 0$

The line integrals are equal. This is to be expected since $\mathbf{F}$ is a conservative vector field with potential $\phi(x, y) = e^{x^2 y}$.

21. Let $\mathbf{F}(x, y, z)$ be the conservative force field, $\phi(x, y, z)$ be its potential, and $U(x, y, z)$ be the associated potential energy function. By definition, the difference between the potential energies at points A and B is $U(B) - U(A) = -\int_A^B \mathbf{F} \cdot d\mathbf{r} = -(\phi(B) - \phi(A))$. Therefore, choosing some fixed point A, one has $U(x, y, z) = U(A) + \phi(A) - \phi(x, y, z) = C - \phi(x, y, z)$ for some constant C. Therefore $\nabla U(x, y, z) = -\nabla\phi(x, y, z) = -\mathbf{F}(x, y, z)$, or $\mathbf{F}(x, y, z) = -\nabla U(x, y, z)$.

23. No, since $\mathbf{F}$ is not conservative: $\partial M/\partial y = -1$ and $\partial N/\partial x = 1$ are not equal.

25. If C is piecewise smooth, then there exists a partition $a = t_0 < t_1 < \ldots < t_n = b$ such that each piece $C_i = \{\mathbf{r}(t) \mid t_{i-1} \le t \le t_i\}$ of the curve is smooth. By the partial proof given in the section, $\int_{C_i} \mathbf{F} \cdot d\mathbf{r} = \phi(\mathbf{r}(t_i)) - \phi(\mathbf{r}(t_{i-1}))$. Therefore:

$$\int_C \mathbf{F} \cdot d\mathbf{r} = \sum_{i=1}^n \int_{C_i} \mathbf{F} \cdot d\mathbf{r}$$
$$= \sum_{i=1}^n \phi(\mathbf{r}(t_i)) - \phi(\mathbf{r}(t_{i-1}))$$
$$= \phi(\mathbf{r}(t_1)) - \phi(\mathbf{r}(t_0)) + \phi(\mathbf{r}(t_2)) - \phi(\mathbf{r}(t_1)) + \ldots + \phi(\mathbf{r}(t_n)) - \phi(\mathbf{r}(t_{n-1}))$$
$$= \phi(\mathbf{r}(t_n)) - \phi(\mathbf{r}(t_0))$$
$$= \phi(\mathbf{r}(b)) - \phi(\mathbf{r}(a))$$

20.4 Green's Theorem

1. $\partial N/\partial x = 1$ and $\partial M/\partial y = x$
$$\int_0^1 \int_0^1 (1 - x)\, dy\, dx$$
$$= \int_0^1 \left[(1-x)y\right]_{y=0}^{y=1} dx$$
$$= \int_0^1 (1-x)\, dx$$
$$= x - (x^2/2)\big|_0^1$$
$$= 1/2$$

3. $\partial N/\partial x = 2x$ and $\partial M/\partial y = 3x^2 y^2$
$$\int_0^1 \int_0^1 (2x - 3x^2 y^2)\, dy\, dx$$
$$= \int_0^1 \left[2xy - x^2 y^3\right]_{y=0}^{y=1} dx$$
$$= \int_0^1 (2x - x^2)\, dx$$
$$= x^2 - (x^3/3)\big|_0^1$$
$$= 2/3$$

5. Note $M(x, y) = -x^2 y$ and $N(x, y) = xy^2$.
$\partial N/\partial x = y^2$ and $\partial M/\partial y = -x^2$
$$\int_0^1 \int_0^{\sqrt{1-x^2}} (y^2 + x^2)\, dy\, dx$$
$$= \int_0^{\pi/2} \int_0^1 r^3\, dr\, d\theta$$
$$= \int_0^{\pi/2} \left[(r^4/4)\right]_{r=0}^{r=1} d\theta$$
$$= \int_0^{\pi/2} (1/4)\, d\theta$$
$$= (1/4)\theta\big|_0^{\pi/2}$$
$$= \pi/8$$

7. $\partial N/\partial x = -2x$ and $\partial M/\partial y = 1$
$$\int_0^1 \int_{x^3}^{x^2} (-2x - 1)\, dy\, dx$$
$$= -\int_0^1 \left[(2x+1)y\right]_{y=x^3}^{y=x^2} dx$$
$$= -\int_0^1 (x^2 + x^3 - 2x^4)\, dx$$
$$= -(x^3/3) - (x^4/4) + (2x^5/5)\big|_0^1$$
$$= -11/60$$

9. $\int_C (\cos^5 x + \sqrt{x})\, dx + \mathrm{Tan}^{-1} y\, dy$ and C is the ellipse $4x^2 + y^2 = 1$.

The line integral is undefined since $M(x, y)$ is undefined on that part of the ellipse for which $x < 0$.

In the second printing, the integral is to be changed to:
$\int_C (\cos^5 x + \sqrt{x+1})\, dx + \mathrm{Tan}^{-1} y\, dy$.
$\partial N/\partial x = 0$ and $\partial M/\partial y = 0$
$$\iint_Q 0\, dA = 0$$

11. $\partial N/\partial x = -3$ and $\partial M/\partial y = 2$
$$\iint_Q (-5)\, dA = -5\pi,$$
since the area is π.

13. $\partial N/\partial x = 2x^2 y$ and $\partial M/\partial y = 2x^2 y$
$$\iint_Q 0\, dA = 0$$

15. $\partial N/\partial x = -1$ and $\partial M/\partial y = 2$
$$\int_C \mathbf{F} \cdot d\mathbf{r} = \iint_Q (-3)\, dA$$
$$= \int_0^\pi \int_1^2 (-3r)\, dr\, d\theta$$
$$= \int_0^\pi \left[-3r^2/2\right]_{r=1}^{r=2} d\theta$$
$$= \int_0^\pi (-9/2)\, d\theta$$
$$= (-9/2)\theta\big|_0^\pi$$
$$= -9\pi/2$$

17. $\mathbf{F}(x, y) = -y\mathbf{i} + x\mathbf{j}$
$\mathbf{r}_1(t) = t\mathbf{i} + t\mathbf{j}$, $\mathbf{r}_2(t) = (1-t)^2\mathbf{i} + (1-t)\mathbf{j}$ for $0 \leq t \leq 1$
$\mathbf{r}_1'(t) = \mathbf{i} + \mathbf{j}$, $\mathbf{r}_2'(t) = (-2 + 2t)\mathbf{i} - \mathbf{j}$
$\mathbf{F}(\mathbf{r}_1(t)) = -t\mathbf{i} + t\mathbf{j}$, $\mathbf{F}(\mathbf{r}_2(t)) = (t-1)\mathbf{i} + (1-t)^2\mathbf{j}$
$\mathbf{F}(\mathbf{r}_1(t)) \cdot \mathbf{r}_1'(t) = 0$, $\mathbf{F}(\mathbf{r}_2(t)) \cdot \mathbf{r}_2'(t) = t^2 - 2t + 1$
$A = (1/2) \int_C x\, dy - y\, dx$
$= (1/2)[\int_0^1 0\, dt + \int_0^1 (t^2 - 2t + 1)\, dt]$
$= (1/2)[(t^3/3) - t^2 + t]_0^1$
$= 1/6$

19. $dx = -3a\cos^2 t \sin t$ and $dy = 3a\cos t \sin^2 t$
$A = (1/2) \int_C x\, dy - y\, dx$
$= (1/2) \int_0^{2\pi} 3a^2 \cos^4 t \sin^2 t\, dt + (1/2) \int_0^{2\pi} 3a^2 \cos^2 t \sin^4 t\, dt$
$= (3a^2/2) \int_0^{2\pi} \cos^2 t \sin^2 t (\cos^2 t + \sin^2 t)\, dt$
$= (3a^2/2) \int_0^{2\pi} \cos^2 t \sin^2 t\, dt$
$= (3a^2/8) \int_0^{2\pi} \sin^2 2t\, dt$
$= (3a^2/16) \int_0^{2\pi} (1 - \cos 4t)\, dt$
$= (3a^2/16)[t - (1/4)\sin 4t]_0^{2\pi}$
$= 3a^2\pi/8$

21. If $\mathbf{F}$ is a conservative vector field, then $\partial M/\partial y = \partial N/\partial x$, and therefore $\partial N/\partial x - \partial M/\partial y = 0$. If two curves C_1 and C_2 have the same initial and final endpoints, are not self-intersecting and only intersect each other at their endpoints, then let C be the curve which results from first traversing C_1 and then traversing $-C_2$ (traversing C_2 from the final to the initial endpoint). By Green's Theorem,

$$0 = \iint\limits_Q [(\partial N/\partial x) - (\partial M/\partial y)]\, dA = \int_C \mathbf{F} \cdot d\mathbf{r} = \int_{C_1} \mathbf{F} \cdot d\mathbf{r} + \int_{-C_2} \mathbf{F} \cdot d\mathbf{r} = \int_{C_1} \mathbf{F} \cdot d\mathbf{r} - \int_{C_2} \mathbf{F} \cdot d\mathbf{r},$$

and therefore $\int_{C_1} \mathbf{F} \cdot d\mathbf{r} = \int_{C_2} \mathbf{F} \cdot d\mathbf{r}$.

Cases in which C_1 and C_2 are self-intersecting and intersect each other, possibly infinitely often, would be much more difficult to handle by this approach.

23. $\mathbf{F}(x, y) = M(x, y)\mathbf{i} + N(x, y)\mathbf{j}$ is conservative in D if and only if $\partial M/\partial y = \partial N/\partial x$ throughout D, if and only if curl $\mathbf{F} = (\partial N/\partial x) - (\partial M/\partial y) = 0$ throughout D.

20.5 Surface Integrals

1. $z = (4 - x + y)/2$
$\partial z/\partial x = -1/2$ and $\partial z/\partial y = 1/2$
$\iint\limits_S (x + 3y + z)\, dS = \iint\limits_Q [x + 3y + (4 - x + y)/2]\sqrt{(-1/2)^2 + (1/2)^2 + 1}\, dA$
$= \sqrt{6}/4 \int_0^1 \int_0^2 (4 + x + 7y)\, dy\, dx$
$= \sqrt{6}/4 \int_0^1 [4y + xy + (7y^2/2)]_{y=0}^{y=2}\, dx$
$= \sqrt{6}/4 \int_0^1 (22 + 2x)\, dx$
$= \sqrt{6}/4 [22x + x^2]_0^1$
$= 23\sqrt{6}/4$

3. $z = 6 - 2x - 3y$, $\partial z/\partial x = -2$ and $\partial z/\partial y = -3$

$\iint\limits_{S} x^2 z \, dS = \iint\limits_{Q} x^2(6 - 2x - 3y)\sqrt{4 + 9 + 1} \, dA$

$= \sqrt{14} \int_0^3 \int_0^{(6-2x)/3} (6x^2 - 2x^3 - 3x^2 y) \, dy \, dx$

$= \sqrt{14} \int_0^3 \left[6x^2 y - 2x^3 y - (3x^2 y^2/2) \right]_{y=0}^{y=(6-2x)/3} dx$

$= (2\sqrt{14}/3) \int_0^3 (x^4 - 6x^3 + 9x^2) \, dx$

$= (2\sqrt{14}/3) \left[(x^5/5) - (3x^4/2) + 3x^3 \right]_0^3$

$= 27\sqrt{14}/5$

5. $\partial z/\partial x = 2x$ and $\partial z/\partial y = 0$

$\iint\limits_{S} (x^2 + 5y - z) \, dS = \iint\limits_{Q} (x^2 + 5y - x^2)\sqrt{4x^2 + 0 + 1} \, dA$

$= \int_{-1}^1 \int_{-1}^1 5y\sqrt{4x^2 + 1} \, dy \, dx$

$= \int_{-1}^1 \left[(5y^2/2)\sqrt{4x^2 + 1} \right]_{y=-1}^{y=1} dx$

$= \int_{-1}^1 0 \, dx$

$= 0$

7. Find the mass for the upper part of the sphere and then multiply by 2.

$z = \sqrt{9 - x^2 - y^2}$, $\partial z/\partial x = -x/\sqrt{9 - x^2 - y^2}$ and $\partial z/\partial y = -y/\sqrt{9 - x^2 - y^2}$

$M = 2 \iint\limits_{S} \delta \, dS = 2 \iint\limits_{Q} \delta \sqrt{(x^2/(9 - x^2 - y^2)) + (y^2/(9 - x^2 - y^2)) + 1} \, dA$

$= 2\delta \iint\limits_{Q} \sqrt{9/(9 - x^2 - y^2)} \, dA$

$= 2\delta \int_0^{2\pi} \int_0^2 3r\sqrt{1/(9 - r^2)} \, dr \, d\theta$ Let $u = 9 - r^2$, $du = -2r \, dr$.

$= -3\delta \int_0^{2\pi} \int_9^5 (1/\sqrt{u}) \, du \, d\theta$

$= -3\delta \int_0^{2\pi} \left[2\sqrt{u} \right]_{u=9}^{u=5} d\theta$

$= -3\delta \int_0^{2\pi} (2\sqrt{5} - 6) \, d\theta$

$= -3\delta(2\sqrt{5} - 6)\theta \big|_0^{2\pi}$

$= \delta\pi(36 - 12\sqrt{5})$

9. $\partial z/\partial x = -x/\sqrt{4 - x^2 - y^2}$ and $\partial z/\partial y = -y/\sqrt{4 - x^2 - y^2}$

$\iint\limits_{S} (x + y + z^2) \, dS = \iint\limits_{Q} (x + y + 4 - x^2 - y^2)\sqrt{(x^2/(4 - x^2 - y^2)) + (y^2/(4 - x^2 - y^2)) + 1} \, dA$

$= \iint\limits_{Q} (x + y + 4 - x^2 - y^2)\sqrt{4/(4 - x^2 - y^2)} \, dA$

$= \int_0^2 \int_0^{2\pi} (r\cos\theta + r\sin\theta + 4 - r^2) 2r\sqrt{1/(4 - r^2)} \, d\theta \, dr$ This is an improper integral.

$= \lim_{b \to 2-} \int_0^b \int_0^{2\pi} (r\cos\theta + r\sin\theta + 4 - r^2)(2r/\sqrt{4 - r^2}) \, d\theta \, dr$

$= \lim_{b \to 2-} \int_0^b \left[(r\sin\theta - r\cos\theta + (4 - r^2)\theta)(2r/\sqrt{4 - r^2}) \right]_{\theta=0}^{\theta=2\pi} dr$

$= 2\pi \lim_{b \to 2-} \int_0^b (4 - r^2) 2r/\sqrt{4 - r^2} \, dr$ Let $u = 4 - r^2$, $du = -2r \, dr$.

$= -2\pi \lim_{c \to 0+} \int_4^c (u/\sqrt{u}) \, du$

$= 2\pi \lim_{c \to 0+} \int_c^4 \sqrt{u} \, du$

$= 2\pi \lim_{c \to 0+} \left[(2/3)u^{3/2} \right]_c^4$

$= 32\pi/3$

11. $\partial z/\partial x = -x/\sqrt{1-x^2-y^2}$ and $\partial z/\partial y = -y/\sqrt{1-x^2-y^2}$

$-M(x,y,z)(\partial z/\partial x) - N(x,y,z)(\partial z/\partial y) + P(x,y,z)$

$= (x^2/\sqrt{1-x^2-y^2}) + (y^2/\sqrt{1-x^2-y^2}) + \sqrt{1-x^2-y^2}$

$= 1/\sqrt{1-x^2-y^2}$

$\iint\limits_S \mathbf{F}\cdot\mathbf{N}\,dS = \int_0^{2\pi}\int_0^1 (r/\sqrt{1-r^2})\,dr\,d\theta$ This is an improper integral.

$= \lim_{b\to 1^-}\int_0^{2\pi}\int_0^b (r/\sqrt{1-r^2})\,dr\,d\theta$ Let $u = 1-r^2$, $du = -2r\,dr$.

$= \lim_{c\to 0^+}(-1/2)\int_0^{2\pi}\int_1^c (1/\sqrt{u})\,du\,d\theta$

$= \lim_{c\to 0^+}(-1/2)\int_0^{2\pi}\left[2\sqrt{u}\right]_{u=1}^{u=c}\,d\theta$

$= \int_0^{2\pi} 1\,d\theta$

$= 2\pi$

13. In this case, the outward unit normal is the downward unit normal.

$\partial z/\partial x = 2x$ and $\partial z/\partial y = 2y$

$M(x,y,z)(\partial z/\partial x) + N(x,y,z)(\partial z/\partial y) - P(x,y,z) = 4x^2 + 2y^2 - (x^2+y^2) = 3x^2 + y^2$

$\iint\limits_S \mathbf{F}\cdot\mathbf{N}\,dS = \int_0^{2\pi}\int_0^2 (3r^2\cos^2\theta + r^2\sin^2\theta)r\,dr\,d\theta$

$= \int_0^{2\pi}\int_0^2 r^3(3\cos^2\theta + \sin^2\theta)\,dr\,d\theta$

$= \int_0^{2\pi}\left[(r^4/4)(3\cos^2\theta + \sin^2\theta)\right]_{r=0}^{r=2}\,d\theta$

$= 4\int_0^{2\pi}(3\cos^2\theta + \sin^2\theta)\,d\theta$

$= 2\int_0^{2\pi}(3 + 3\cos 2\theta + 1 - \cos 2\theta)\,d\theta$

$= \int_0^{2\pi}(8 + 4\cos 2\theta)\,d\theta$

$= 8\theta + 2\sin 2\theta\big|_0^{2\pi}$

$= 16\pi$

15. First consider $\mathbf{F}(x,y,z) = z\mathbf{k}$. Let $\delta = 1$ when considering the upper hemisphere, and let $\delta = -1$ for the lower hemisphere.

$z = \delta\sqrt{a^2-x^2-y^2}$

$\partial z/\partial x = -\delta x/\sqrt{a^2-x^2-y^2}$ and $\partial z/\partial y = -\delta y/\sqrt{a^2-x^2-y^2}$

Since the outward unit normal is the upward unit normal for the upper hemisphere and the downward unit normal for the lower hemisphere, in either case one has:

$-\delta M(x,y,z)(\partial z/\partial x) - \delta N(x,y,z)(\partial z/\partial y) + \delta P(x,y,z) = 0 - 0 + \delta(\delta\sqrt{a^2-x^2-y^2})$

$= \sqrt{a^2-x^2-y^2}$ since $\delta^2 = 1$.

For either the upper or lower hemisphere:

$\iint\limits_S \mathbf{F}\cdot\mathbf{N}\,dS = \int_0^{2\pi}\int_0^a r\sqrt{a^2-r^2}\,dr\,d\theta$ Let $u = a^2 - r^2$, $du = -2r\,dr$.

$= (-1/2)\int_0^{2\pi}\int_{a^2}^0 \sqrt{u}\,du\,d\theta$

$= (-1/2)\int_0^{2\pi}\left[(2/3)u^{3/2}\right]_{u=a^2}^{u=0}\,d\theta$

$= (a^3/3)\int_0^{2\pi} 1\,d\theta$

$= 2a^3\pi/3$

Therefore, the total flux outward through the sphere is twice this amount, or $4a^3\pi/3$. Note that this is the volume of the sphere. Since the flow at any point is perpendicular to the xy-plane outward through the sphere, and is equal in magnitude to the distance from that point to the xy-plane, it should be expected that a quantity equal to the volume of the sphere flows outward per unit time.

The situation for $\mathbf{F}(x,y,z) = y\mathbf{j}$ is symmetric to the above situation, with the variables y and z exchanged.

631

17. Due to the symmetry of the situation, one may compute the flux for the two faces S_1 and S_2 in the planes $z = 0$ and $z = 1$ respectively, and then triple this result. For S_1, one has $z = 0$, $\partial z/\partial x = 0$, $\partial z/\partial y = 0$, $M(x, y, z)(\partial z/\partial x) + N(x, y, z)(\partial z/\partial y) - P(x, y, z) = 0$ and therefore $\iint\limits_{S_1} \mathbf{F} \cdot \mathbf{N}\, dS_1 = 0$. (This is unsurprising since the flow is parallel to this face.) For S_2, one has $z = 1$, $\partial z/\partial x = 0$, $\partial z/\partial y = 0$, $-M(x, y, z)(\partial z/\partial x) - N(x, y, z)(\partial z/\partial y) + P(x, y, z) = 1$ and therefore $\iint\limits_{S_2} \mathbf{F} \cdot \mathbf{N}\, dS_2 = \int_0^1 \int_0^1 1\, dy\, dx = 1$.

The total flux is therefore 3.

19. As in Example 8, one may calculate the flux for the upper hemisphere, and then double the result.
$\mathbf{F}(x, y, z) = (1/(x^2 + y^2 + z^2)^2)(x\mathbf{i} + y\mathbf{j} + z\mathbf{k}) = (1/R^4)(x\mathbf{i} + y\mathbf{j} + z\mathbf{k})$
$z = \sqrt{R^2 - x^2 - y^2}$, $\partial z/\partial x = -x/\sqrt{R^2 - x^2 - y^2}$ and $\partial z/\partial y = -y/\sqrt{R^2 - x^2 - y^2}$
$-M(x, y, z)(\partial z/\partial x) - N(x, y, z)(\partial z/\partial y) + P(x, y, z)$
$= [x^2/(R^4\sqrt{R^2 - x^2 - y^2})] + [y^2/(R^4\sqrt{R^2 - x^2 - y^2})] + (\sqrt{R^2 - x^2 - y^2}/R^4)$
$= 1/(R^2\sqrt{R^2 - x^2 - y^2})$
$2\iint\limits_{S} \mathbf{F} \cdot \mathbf{N}\, dS = 2\int_0^{2\pi} \int_0^R r/(R^2\sqrt{R^2 - r^2})\, dr\, d\theta$ This is an improper integral.

$= 2\lim_{b \to R^-} \int_0^{2\pi} \int_0^b r/(R^2\sqrt{R^2 - r^2})\, dr\, d\theta$ Let $u = R^2 - r^2$, $du = -2r\, dr$.
$= (-1/R^2)\lim_{c \to 0^+} \int_0^{2\pi} \int_{R^2}^c (1/\sqrt{u})\, du\, d\theta$
$= (-1/R^2)\lim_{c \to 0^+} \int_0^{2\pi} [2\sqrt{u}]_{u=R^2}^{u=c}\, d\theta$
$= (-1/R^2)\int_0^{2\pi} (-2R)\, d\theta$
$= (2/R)\theta\big|_0^{2\pi}$
$= 4\pi/R$
The result is not the same; in this case the flux *does* depend on the radius of the sphere.

21. From the symmetry of the situation, is it clear that $\bar{x} = \bar{y} = 0$.
$\partial z/\partial x = -x/\sqrt{a^2 - x^2 - y^2} = -x/z$
$\partial z/\partial y = -y/\sqrt{a^2 - x^2 - y^2} = -y/z$
The area of the hemisphere is $A = 2\pi a^2$.
$\bar{z} = (1/A)\iint\limits_{S} z\, dS$
$= [1/(2\pi a^2)]\iint\limits_{Q} z\sqrt{(x^2/z^2) + (y^2/z^2) + 1}\, dA$
$= [1/(2\pi a^2)]\iint\limits_{Q} z(\sqrt{x^2 + y^2 + z^2}/z)\, dA$
$= [1/(2\pi a^2)]\iint\limits_{Q} a\, dA$
$= [1/(2\pi a)]\iint\limits_{Q} dA$
$= [1/(2\pi a)](\pi a^2)$ since the area of Q is πa^2
$= a/2$
The centroid is $(0, 0, a/2)$.

632

23. Due to the symmetry of the situation, one will have $\bar{x} = \bar{y} = \bar{z}$.

The area of the surface is $A = (1/8)4\pi r^2 = \pi/2$.

$z = \sqrt{1 - x^2 - y^2}$, $\partial z/\partial x = -x/\sqrt{1 - x^2 - y^2} = -x/z$ and $\partial z/\partial y = -y/\sqrt{1 - x^2 - y^2} = -y/z$

$\bar{z} = (1/A) \iint\limits_{S} z \, dS = (2/\pi) \iint\limits_{Q} z\sqrt{(x^2/z^2) + (y^2/z^2) + 1} \, dA$

$= (2/\pi) \iint\limits_{Q} z(\sqrt{x^2 + y^2 + z^2}/z) \, dA$

$= (2/\pi) \iint\limits_{Q} dA$

$= (2/\pi)(\pi/4)$ since the area of Q is $\pi/4$

$= 1/2$

The centroid is $(1/2, 1/2, 1/2)$.

25. Let $f_1(x, y, z)$ be the function in rectangular coordinates, and $z = g_1(x, y)$ be the equation for the surface S, such that $f(r, \theta, z) = f_1(r\cos\theta, r\sin\theta, z)$ and $g(r, \theta) = g_1(r\cos\theta, r\sin\theta)$. Start with Definition 5:

$\iint\limits_{S} f_1(x, y, z) \, dS = \iint\limits_{Q} f_1(x, y, g_1(x, y))\sqrt{(\partial g_1/\partial x)^2 + (\partial g_1/\partial y)^2 + 1} \, dA$

Now use the change of variables $x = r\cos\theta$ and $y = r\sin\theta$. The Jacobian is r. Note the following:

$\partial g/\partial r = (\partial g_1/\partial x)(\partial x/\partial r) + (\partial g_1/\partial y)(\partial y/\partial r) = (\partial g_1/\partial x)\cos\theta + (\partial g_1/\partial y)\sin\theta$

$\partial g/\partial \theta = (\partial g_1/\partial x)(\partial x/\partial \theta) + (\partial g_1/\partial y)(\partial y/\partial \theta) = -(\partial g_1/\partial x)(r\sin\theta) + (\partial g_1/\partial y)(r\cos\theta)$

$r^2(\partial g/\partial r)^2 + (\partial g/\partial \theta)^2$

$= r^2(\partial g_1/\partial x)^2(\cos^2\theta + \sin^2\theta) + 2r^2(\partial g_1/\partial x)(\partial g_1/\partial y)\cos\theta\sin\theta$

$\quad - 2r^2(\partial g_1/\partial x)(\partial g_1/\partial y)\cos\theta\sin\theta + r^2(\partial g_1/\partial y)^2(\cos^2\theta + \sin^2\theta)$

$= r^2[(\partial g_1/\partial x)^2 + (\partial g_1/\partial y)^2]$

Therefore, the integral becomes:

$\iint\limits_{S} f(r, \theta, z) \, dS = \iint\limits_{Q} f(r, \theta, g(r, \theta))\sqrt{\dfrac{r^2(\partial g/\partial r)^2 + (\partial g/\partial \theta)^2}{r^2} + 1}\, r \, dr \, d\theta$

$= \iint\limits_{Q} f(r, \theta, g(r, \theta))\sqrt{r^2\left(\dfrac{\partial g}{\partial r}\right)^2 + \left(\dfrac{\partial g}{\partial \theta}\right)^2 + r^2} \, dr \, d\theta$

20.6 Stokes' Theorem

1. $\operatorname{curl}\mathbf{F} = [(\partial P/\partial y) - (\partial N/\partial z)]\mathbf{i} + [(\partial M/\partial z) - (\partial P/\partial x)]\mathbf{j} + [(\partial N/\partial x) - (\partial M/\partial y)]\mathbf{k}$
$= 0$

3. $\operatorname{curl}\mathbf{F} = [(\partial P/\partial y) - (\partial N/\partial z)]\mathbf{i} + [(\partial M/\partial z) - (\partial P/\partial x)]\mathbf{j} + [(\partial N/\partial x) - (\partial M/\partial y)]\mathbf{k}$
$= z\cos(yz)\mathbf{i} + xy\mathbf{j} + [y\sin(xy) - xz]\mathbf{k}$

5. $\operatorname{curl}\mathbf{F} = (x - x)\mathbf{i} + (y - y)\mathbf{j} + (z - z)\mathbf{k} = 0$
Therefore $\mathbf{F}$ is conservative.

7. $\operatorname{curl}\mathbf{F} = [x/(2\sqrt{xy})]\mathbf{i} - [y/(2\sqrt{xy})]\mathbf{j} + (-\sin x - \cos y)\mathbf{k} \neq 0$
Therefore $\mathbf{F}$ is not conservative.

9. For the line integral around the boundary C, represent C by $\mathbf{r}(t) = \cos t\,\mathbf{i} + \sin t\,\mathbf{j}$, for $0 \le t \le 2\pi$.

$\mathbf{r}'(t) = -\sin t\,\mathbf{i} + \cos t\,\mathbf{j}$

$\mathbf{F}(\mathbf{r}(t)) = \cos t\,\mathbf{j} + \sin t\,\mathbf{k}$

$\mathbf{F}(\mathbf{r}(t)) \cdot \mathbf{r}'(t) = \cos^2 t$

$\int_C \mathbf{F} \cdot d\mathbf{r} = \int_0^{2\pi} \cos^2 t\, dt$

$\quad = (1/2)\int_0^{2\pi}(1 + \cos 2t)\, dt$

$\quad = (1/2)\big[t + (1/2)\sin 2t\big]_0^{2\pi}$

$\quad = \pi$

For the surface integral, note that $\operatorname{curl}\mathbf{F} = \mathbf{i} + \mathbf{j} + \mathbf{k}$ and $\mathbf{N} = x\mathbf{i} + y\mathbf{j} + z\mathbf{k}$

$(\operatorname{curl}\mathbf{F}) \cdot \mathbf{N} = x + y + z$

Also, $\partial z/\partial x = -x/\sqrt{1 - x^2 - y^2} = -x/z$ and $\partial z/\partial y = -y/\sqrt{1 - x^2 - y^2} = -y/z$.

$\iint\limits_S (\operatorname{curl}\mathbf{F}) \cdot \mathbf{N}\, dS = \iint\limits_S (x + y + z)\, dS$

$\quad = \iint\limits_Q (x + y + z)\sqrt{(x^2/z^2) + (y^2/z^2) + 1}\, dA$

$\quad = \iint\limits_Q (x + y + z)(\sqrt{x^2 + y^2 + z^2}/z)\, dA$

$\quad = \iint\limits_Q [(x + y + z)/z]\, dA$

$\quad = \int_0^1 \int_0^{2\pi} [(r\cos\theta + r\sin\theta + \sqrt{1 - r^2})/\sqrt{1 - r^2}]r\, d\theta\, dr$

$\quad = \int_0^1 \big[(r^2\sin\theta - r^2\cos\theta + r\theta\sqrt{1 - r^2})/\sqrt{1 - r^2}\big]_{\theta=0}^{\theta=2\pi}\, dr$

$\quad = \int_0^1 2\pi r\, dr$

$\quad = \pi r^2\big|_0^1$

$\quad = \pi$

Thus, $\int_C \mathbf{F} \cdot d\mathbf{r} = \iint\limits_S (\operatorname{curl}\mathbf{F}) \cdot \mathbf{N}\, dS$

11. Note that $z = 0$ for both C and Q, and consequently $\mathbf{F}(x, y, z) = x^2 y\mathbf{i}$ for all points on C.

For the line integral around the boundary C, represent C by $\mathbf{r}_1(t) = t\mathbf{i}$, $\mathbf{r}_2(t) = \mathbf{i} + 2t\mathbf{j}$, $\mathbf{r}_3(t) = (1-t)\mathbf{i} + 2\mathbf{j}$, and $\mathbf{r}_4(t) = (2 - 2t)\mathbf{j}$ for $0 \le t \le 1$.

$\mathbf{r}_1'(t) = \mathbf{i}$, $\mathbf{r}_2'(t) = 2\mathbf{j}$, $\mathbf{r}_3'(t) = -\mathbf{i}$, $\mathbf{r}_4'(t) = -2\mathbf{j}$

$\mathbf{F}(\mathbf{r}_1(t)) = \mathbf{0}$, $\mathbf{F}(\mathbf{r}_2(t)) = 2t\mathbf{i}$, $\mathbf{F}(\mathbf{r}_3(t)) = (2t^2 - 4t + 2)\mathbf{i}$, $\mathbf{F}(\mathbf{r}_4(t)) = \mathbf{0}$

$\mathbf{F}(\mathbf{r}_1(t)) \cdot \mathbf{r}_1'(t) = 0$, $\mathbf{F}(\mathbf{r}_2(t)) \cdot \mathbf{r}_2'(t) = 0$, $\mathbf{F}(\mathbf{r}_3(t)) \cdot \mathbf{r}_3'(t) = -2t^2 + 4t - 2$, $\mathbf{F}(\mathbf{r}_4(t)) \cdot \mathbf{r}_4'(t) = 0$

$\int_C \mathbf{F} \cdot d\mathbf{r} = \int_0^1 (-2t^2 + 4t - 2)\, dt = (-2t^3/3) + 2t^2 - 2t\big|_0^1 = -2/3$

For the surface integral, note that $\operatorname{curl}\mathbf{F} = -y^2\mathbf{i} - z\mathbf{j} - x^2\mathbf{k}$ and $\mathbf{N} = \mathbf{k}$

$(\operatorname{curl}\mathbf{F}) \cdot \mathbf{N} = -x^2$

$\iint\limits_Q (\operatorname{curl}\mathbf{F}) \cdot \mathbf{N}\, dA = \int_0^1 \int_0^2 (-x^2)\, dy\, dx = \int_0^2 (-2x^2)\, dx = -2x^3/3\big|_0^1 = -2/3$

Thus, $\int_C \mathbf{F} \cdot d\mathbf{r} = \iint\limits_Q (\operatorname{curl}\mathbf{F}) \cdot \mathbf{N}\, dA$

13. The x-, y- and z-intercepts are 2, 1 and 2, respectively. For the line integral around the boundary C, represent C by $\mathbf{r}_1(t) = 2t\mathbf{i} + (2 - 2t)\mathbf{k}$, $\mathbf{r}_2(t) = (2 - 2t)\mathbf{i} + t\mathbf{j}$ and $\mathbf{r}_3(t) = (1 - t)\mathbf{j} + 2t\mathbf{k}$, for $0 \le t \le 1$.

$\mathbf{r}_1'(t) = 2\mathbf{i} - 2\mathbf{k}$, $\mathbf{r}_2'(t) = -2\mathbf{i} + \mathbf{j}$, $\mathbf{r}_3'(t) = -\mathbf{j} + 2\mathbf{k}$

$\mathbf{F}(\mathbf{r}_1(t)) = (2 - 2t)\mathbf{i} + 2t\mathbf{j}$, $\mathbf{F}(\mathbf{r}_2(t)) = (2 - 2t)\mathbf{j} + t\mathbf{k}$, $\mathbf{F}(\mathbf{r}_3(t)) = 2t\mathbf{i} + (1 - t)\mathbf{k}$

$\mathbf{F}(\mathbf{r}_1(t)) \cdot \mathbf{r}_1'(t) = 4 - 4t$, $\mathbf{F}(\mathbf{r}_2(t)) \cdot \mathbf{r}_2'(t) = 2 - 2t$, $\mathbf{F}(\mathbf{r}_3(t)) \cdot \mathbf{r}_3'(t) = 2 - 2t$

$\int_C \mathbf{F} \cdot d\mathbf{r} = \int_0^1 (4 - 4t)\, dt + \int_0^1 (2 - 2t)\, dt + \int_0^1 (2 - 2t)\, dt = \int_0^1 (8 - 8t)\, dt = 8t - 4t^2\big|_0^1 = 4$

For the surface integral, note that a normal vector for the plane is $\mathbf{i} + 2\mathbf{j} + \mathbf{k}$, which is upward, and thus the upward unit normal vector is $\mathbf{N} = (1/\sqrt{6})(\mathbf{i} + 2\mathbf{j} + \mathbf{k})$.

$\operatorname{curl} \mathbf{F} = \mathbf{i} + \mathbf{j} + \mathbf{k}$

$(\operatorname{curl} \mathbf{F}) \cdot \mathbf{N} = 4/\sqrt{6}$

From the equation of the plane, $z = 2 - x - 2y$, $\partial z/\partial x = -1$ and $\partial z/\partial y = -2$.

$\iint\limits_{S}(\operatorname{curl} \mathbf{F}) \cdot \mathbf{N}\, dS = \iint\limits_{S}(4/\sqrt{6})\, dS = (4/\sqrt{6}) \iint\limits_{Q} \sqrt{1+4+1}\, dA = 4 \iint\limits_{Q} dA = 4,$

since the area of the triangle Q is 1.

Thus, $\int_{C} \mathbf{F} \cdot d\mathbf{r} = \iint\limits_{S}(\operatorname{curl} \mathbf{F}) \cdot \mathbf{N}\, dS$

15. $\mathbf{F}(x, y, z) = (\sqrt{x} + y)\mathbf{i} + (e^y - x)\mathbf{j} + (\sin z + y)\mathbf{k}$

The line integral is undefined since $\mathbf{F}(x, y, z)$ is undefined on that part of the unit circle for which $x < 0$.

In the second printing, the vector field is to be changed to $\mathbf{F}(x, y, z) = (x+y)\mathbf{i} + (e^y - x)\mathbf{j} + (\sin z + y)\mathbf{k}$.

The upward unit normal vector is $\mathbf{N} = \mathbf{k}$.

$\operatorname{curl} \mathbf{F} = \mathbf{i} - 2\mathbf{k}$

$(\operatorname{curl} \mathbf{F}) \cdot \mathbf{N} = -2$

$\int_{C} \mathbf{F} \cdot d\mathbf{r} = \iint\limits_{Q}(-2)\, dA = -2 \iint\limits_{Q} dA = -2\pi$

Note that the area of Q is $\pi r^2 = \pi$.

17. For the line integral around the boundary C, represent C by $\mathbf{r}(t) = 3\cos t\, \mathbf{i} + 3\sin t\, \mathbf{j}$, for $0 \le t \le 2\pi$.

$\mathbf{r}'(t) = -3\sin t\, \mathbf{i} + 3\cos t\, \mathbf{j}$

$\mathbf{F}(\mathbf{r}(t)) = 9\cos^2 t\, \mathbf{i} + 9\sin^2 t\, \mathbf{j}$

$\mathbf{F}(\mathbf{r}(t)) \cdot \mathbf{r}'(t) = -27\cos^2 t \sin t + 27\cos t \sin^2 t$

$\int_{C} \mathbf{F} \cdot d\mathbf{r} = \int_{0}^{2\pi}(-27\cos^2 t \sin t + 27\cos t \sin^2 t)\, dt$

Let $u = \cos t$, $du = -\sin t\, dt$ and let $v = \sin t$, $dv = \cos t\, dt$.

$\int_{C} \mathbf{F} \cdot d\mathbf{r} = \int_{1}^{1} 27u^2\, du + \int_{0}^{0} 27v^2\, dv = 0$

For the surface integral, note that $\operatorname{curl} \mathbf{F} = \mathbf{0}$ and thus $(\operatorname{curl} \mathbf{F}) \cdot \mathbf{N} = 0$.

$\iint\limits_{S}(\operatorname{curl} \mathbf{F}) \cdot \mathbf{N}\, dS = \iint\limits_{S} 0\, dS = 0$

Thus, $\int_{C} \mathbf{F} \cdot d\mathbf{r} = \iint\limits_{S}(\operatorname{curl} \mathbf{F}) \cdot \mathbf{N}\, dS$

19. Given $\mathbf{F}(x, y) = M(x, y)\mathbf{i} + N(x, y)\mathbf{j}$ where $\partial M/\partial y = \partial N/\partial x$,

and $\mathbf{G}(x, y) = \mathbf{F}(x, y) + \lambda z\mathbf{k} = M(x, y)\mathbf{i} + N(x, y)\mathbf{j} + \lambda z\mathbf{k}$.

$\operatorname{curl} \mathbf{G} = [(\partial P/\partial y) - (\partial N/\partial z)]\mathbf{i} + [(\partial M/\partial z) - (\partial P/\partial x)]\mathbf{j} + [(\partial N/\partial x) - (\partial M/\partial y)]\mathbf{k}$

$= 0\mathbf{i} + 0\mathbf{j} + [(\partial N/\partial x) - (\partial M/\partial y)]\mathbf{k}$

$= \mathbf{0}$ since $\partial N/\partial x = \partial M/\partial y$.

Therefore, $\mathbf{G}$ is conservative. The result is exactly the same if λz is replaced by any other function of z.

21. First deal with the case where one has a curve C with no self-intersections. Let $\mathbf{r}_1(t)$ for $a_1 \le t \le b_1$ and $\mathbf{r}_2(t)$ for $a_2 \le t \le b_2$ be parameterizations for C (with the same orientation). Let S be a small strip, like a curved "rectangle", with C as one of its "edges". Let C' be the remaining boundary of S, traversed from the terminal point of C to its initial point, and let $\mathbf{r}_3(t)$ for $a_3 \le t \le b_3$ be a parameterization for C'. Let $\mathbf{N}$ be a normal vector for S which agrees with the orientation given by C and C'. Then for any vector field $\mathbf{F}$, one has:

$$\int_{C} \mathbf{F} \cdot d\mathbf{r}_1 + \int_{C'} \mathbf{F} \cdot d\mathbf{r}_3 = \iint\limits_{S}(\operatorname{curl} \mathbf{F}) \cdot \mathbf{N}\, dS = \int_{C} \mathbf{F} \cdot d\mathbf{r}_2 + \int_{C'} \mathbf{F} \cdot d\mathbf{r}_3.$$

Subtracting $\int_{C'} \mathbf{F} \cdot d\mathbf{r}_3$ from both sides gives $\int_{C} \mathbf{F} \cdot d\mathbf{r}_1 = \int_{C} \mathbf{F} \cdot d\mathbf{r}_2$.

If the curve C has a finite number of self intersections, then one may divide it into parts $C_1, C_2, \ldots,$ C_n such that each C_i has no self-intersections, and partition the intervals $[a_1, b_1]$ and $[a_2, b_2]$ such that $\mathbf{r}_1$ and $\mathbf{r}_2$ give parameterizations for the C_i. By the above result, for each i one has $\int_{C_i} \mathbf{F} \cdot d\mathbf{r}_1 = \int_{C_i} \mathbf{F} \cdot d\mathbf{r}_2$, and therefore

$$\int_C \mathbf{F} \cdot d\mathbf{r}_1 = \sum_{i=1}^n \int_{C_i} \mathbf{F} \cdot d\mathbf{r}_1 = \sum_{i=1}^n \int_{C_i} \mathbf{F} \cdot d\mathbf{r}_2 = \int_C \mathbf{F} \cdot d\mathbf{r}_2.$$

20.7 The Divergence Theorem

1. $\operatorname{div} \mathbf{F} = (\partial M / \partial x) + (\partial N / \partial y) + (\partial P / \partial z) = 2x + 2y + 2z$

3. $\operatorname{div} \mathbf{F} = (\partial M / \partial x) + (\partial N / \partial y) + (\partial P / \partial z) = 1 + x + xy$

5. $\operatorname{div} \mathbf{F} = (\partial M / \partial x) + (\partial N / \partial y) + (\partial P / \partial z) = -y \sin(xy) + xze^{xyz} + xy \cos(xz)$

7. Let S be the ellipsoid and R be the region which it bounds.
$\operatorname{div} \mathbf{F} = 0$
$\iint\limits_S \mathbf{F} \cdot \mathbf{N} \, dS = \iiint\limits_R \operatorname{div} \mathbf{F} \, dV = 0$

9. $\operatorname{div} \mathbf{F} = 3 - z^2 + 2 = 5 - z^2$
$\iint\limits_S \mathbf{F} \cdot \mathbf{N} \, dS = \iiint\limits_R \operatorname{div} \mathbf{F} \, dV$
$= \int_{-2}^1 \int_0^3 \int_0^5 (5 - z^2) \, dz \, dy \, dx$
$= \int_{-2}^1 \int_0^3 \left[5z - (z^3/3)\right]_{z=0}^{z=5} dy \, dx$
$= -(50/3) \int_{-2}^1 \int_0^3 dy \, dx$
$= -(50/3)9$ (area of rectangle is 9)
$= -150$

11. $\operatorname{div} \mathbf{F} = 2y + z^2 + x$
$\iint\limits_S \mathbf{F} \cdot \mathbf{N} \, dS = \iiint\limits_R \operatorname{div} \mathbf{F} \, dV$
$= \int_0^1 \int_0^1 \int_0^1 (2y + z^2 + x) \, dz \, dy \, dz$
$= \int_0^1 \int_0^1 \left[2yz + (z^3/3) + xz\right]_{z=0}^{z=1} dy \, dz$
$= (1/3) \int_0^1 \int_0^1 (6y + 1 + 3x) \, dy \, dx$
$= (1/3) \int_0^1 \left[3y^2 + y + 3xy\right]_{y=0}^{y=1} dx$
$= (1/3) \int_0^1 (4 + 3x) \, dx$
$= (1/3) \left[4x + (3x^2/2)\right]_0^1$
$= 11/6$

13. $\operatorname{div} \mathbf{F} = 2xy^2 + 3xy^2 + 0 = 5xy^2$
$\iint\limits_S \mathbf{F} \cdot \mathbf{N} \, dS = \iiint\limits_R \operatorname{div} \mathbf{F} \, dV$
$= \int_0^2 \int_0^2 \int_0^{2\pi} (5r^3 \cos\theta \sin^2\theta) r \, d\theta \, dz \, dr$
$= \int_0^2 \int_0^2 \left[(5/3)r^4 \sin^3\theta\right]_0^{2\pi} dz \, dr$
$= 0$

15. $\operatorname{div} \mathbf{F} = 3x^2 + 0 + x^2 = 4x^2$
$\iint\limits_S \mathbf{F} \cdot \mathbf{N} \, dS = \iiint\limits_R \operatorname{div} \mathbf{F} \, dV$
$= \int_0^{2\pi} \int_0^\pi \int_0^2 (4\rho^2 \sin^2\phi \cos^2\theta)\rho^2 \sin\phi \, d\rho \, d\phi \, d\theta$
$= \int_0^{2\pi} \int_0^\pi \int_0^2 4\rho^4 \sin^3\phi \cos^2\theta \, d\rho \, d\phi \, d\theta$
$= \int_0^{2\pi} \int_0^\pi \left[(4\rho^5/5) \sin^3\phi \cos^2\theta\right]_{\rho=0}^{\rho=2} d\phi \, d\theta$
$= (128/5) \int_0^{2\pi} \int_0^\pi \sin^3\phi \cos^2\theta \, d\phi \, d\theta$ Let $u = -\cos\phi$, $du = \sin\phi \, d\phi$.

$= (128/5) \int_0^{2\pi} \int_{-1}^1 (1 - u^2) \cos^2 \theta \, du \, d\theta$

$= (128/5) \int_0^{2\pi} \left[(u - (u^3/3)) \cos^2 \theta \right]_{u=-1}^{u=1} d\theta$

$= (512/15) \int_0^{2\pi} \cos^2 \theta \, d\theta$

$= (256/15) \int_0^{2\pi} (1 + \cos 2\theta) \, d\theta$

$= (256/15) \left[\theta + (1/2) \sin 2\theta \right]_0^{2\pi}$

$= 512\pi/15$

17. $\operatorname{div} \mathbf{F} = 6 - 1 + 0 = 5$

$\iint\limits_S \mathbf{F} \cdot \mathbf{N} \, dS = \iiint\limits_R \operatorname{div} \mathbf{F} \, dV = 5 \iiint\limits_R dV$

The value of this integral is $5V$, where V is the volume of the solid bounded by the ellipsoid. Note that cross sections of the ellipsoid in planes $y = c$ are circles ($x^2 + z^2 = 4 - 4c^2$), and thus the solid may be viewed as a solid of revolution generated by revolving the graph of $x = 2\sqrt{1 - y^2}$ around the y-axis. One therefore has:

$\iint\limits_S \mathbf{F} \cdot \mathbf{N} \, dS = 5V = 5 \int_{-1}^1 \pi (2\sqrt{1 - y^2})^2 \, dy$

$= 20\pi \int_{-1}^1 (1 - y^2) \, dy$

$= 20\pi \left[y - (y^3/3) \right]_{-1}^1$

$= 80\pi/3$

19. Let $\mathbf{F} = M_1 \mathbf{i} + N_1 \mathbf{j} + P_1 \mathbf{k}$ and $\mathbf{G} = M_2 \mathbf{i} + N_2 \mathbf{j} + P_2 \mathbf{k}$.

$\nabla \times (\mathbf{F} + \mathbf{G}) = \nabla \times [(M_1 + M_2)\mathbf{i} + (N_1 + N_2)\mathbf{j} + (P_1 + P_2)\mathbf{k}]$

$= [(\partial/\partial y)(P_1 + P_2) - (\partial/\partial z)(N_1 + N_2)]\mathbf{i} + [(\partial/\partial z)(M_1 + M_2) - (\partial/\partial x)(P_1 + P_2)]\mathbf{j}$
$\quad + [(\partial/\partial x)(N_1 + N_2) - (\partial/\partial y)(M_1 + M_2)]\mathbf{k}$

$= [(\partial P_1/\partial y) + (\partial P_2/\partial y) - (\partial N_1/\partial z) - (\partial N_2/\partial z)]\mathbf{i} + [(\partial M_1/\partial z) + (\partial M_2/\partial z) - (\partial P_1/\partial x) - (\partial P_2/\partial x)]\mathbf{j}$
$\quad + [(\partial N_1/\partial x) + (\partial N_2/\partial x) - (\partial M_1/\partial y) - (\partial M_2/\partial y)]\mathbf{k}$

$= [(\partial P_1/\partial y) - (\partial N_1/\partial z)]\mathbf{i} + [(\partial P_2/\partial y) - (\partial N_2/\partial z)]\mathbf{i} + [(\partial M_1/\partial z) - (\partial P_1/\partial x)]\mathbf{j} + [(\partial M_2/\partial z) - (\partial P_2/\partial x)]\mathbf{j}$
$\quad + [(\partial N_1/\partial x) - (\partial M_1/\partial y)]\mathbf{k} + [(\partial N_2/\partial x) - (\partial M_2/\partial y)]\mathbf{k}$

$= \Big([(\partial P_1/\partial y) - (\partial N_1/\partial z)]\mathbf{i} + [(\partial M_1/\partial z) - (\partial P_1/\partial x)]\mathbf{j} + [(\partial N_1/\partial x) - (\partial M_1/\partial y)]\mathbf{k} \Big)$

$\quad + \Big([(\partial P_2/\partial y) - (\partial N_2/\partial z)]\mathbf{i} + [(\partial M_2/\partial z) - (\partial P_2/\partial x)]\mathbf{j} + [(\partial N_2/\partial x) - (\partial M_2/\partial y)]\mathbf{k} \Big)$

$= (\nabla \times \mathbf{F}) + (\nabla \times \mathbf{G})$

21. Let $\mathbf{F} = M_1 \mathbf{i} + N_1 \mathbf{j} + P_1 \mathbf{k}$ and $\mathbf{G} = M_2 \mathbf{i} + N_2 \mathbf{j} + P_2 \mathbf{k}$.

$(\nabla \times \mathbf{F}) \cdot \mathbf{G} - (\nabla \times \mathbf{G}) \cdot \mathbf{F}$

$= \big([(\partial P_1/\partial y) - (\partial N_1/\partial z)]\mathbf{i} + [(\partial M_1/\partial z) - (\partial P_1/\partial x)]\mathbf{j} + [(\partial N_1/\partial x) - (\partial M_1/\partial y)]\mathbf{k} \big) \cdot \mathbf{G}$
$\quad - \big([(\partial P_2/\partial y) - (\partial N_2/\partial z)]\mathbf{i} + [(\partial M_2/\partial z) - (\partial P_2/\partial x)]\mathbf{j} + [(\partial N_2/\partial x) - (\partial M_2/\partial y)]\mathbf{k} \big) \cdot \mathbf{F}$

$= (\partial P_1/\partial y)M_2 - (\partial N_1/\partial z)M_2 + (\partial M_1/\partial z)N_2 - (\partial P_1/\partial x)N_2 + (\partial N_1/\partial x)P_2 - (\partial M_1/\partial y)P_2$
$\quad - (\partial P_2/\partial y)M_1 + (\partial N_2/\partial z)M_1 - (\partial M_2/\partial z)N_1 + (\partial P_2/\partial x)N_1 - (\partial N_2/\partial x)P_1 + (\partial M_2/\partial y)P_1$

$= (\partial N_1/\partial x)P_2 + (\partial P_2/\partial x)N_1 - (\partial N_2/\partial x)P_1 - (\partial P_1/\partial x)N_2 + (\partial P_1/\partial y)M_2 + (\partial M_2/\partial y)P_1$
$\quad - (\partial P_2/\partial y)M_1 - (\partial M_1/\partial y)P_2 + (\partial M_1/\partial z)N_2 + (\partial N_2/\partial z)M_1 - (\partial M_2/\partial z)N_1 - (\partial N_1/\partial z)M_2$

$= (\partial/\partial x)(N_1 P_2 - P_1 N_2) + (\partial/\partial y)(P_1 M_2 - M_1 P_2) + (\partial/\partial z)(M_1 N_2 - N_1 M_2)$

$= \nabla \cdot [(N_1 P_2 - P_1 N_2)\mathbf{i} + (P_1 M_2 - M_1 P_2)\mathbf{j} + (M_1 N_2 - N_1 M_2)\mathbf{k}]$

$= \nabla \cdot (\mathbf{F} \times \mathbf{G})$

23. The answer is yes if both the curl and the divergence are zero. (The curl in order to show that it is a gradient, and the divergence to show that it is harmonic.)

a. $\operatorname{curl} \mathbf{F} = \mathbf{0}$ and $\operatorname{div} \mathbf{F} = 0 + 0 + 0 = 0$. Yes.

b. $\operatorname{curl} \mathbf{F} = \mathbf{0}$, but $\operatorname{div} \mathbf{F} = 2ye^z + 0 + x^2 ye^z$. No.

c. curl $\mathbf{F} = xy\cos(yz)\mathbf{j} - xz\cos(yz)\mathbf{k}$. This is not even a gradient.
(Also, div $\mathbf{F} = \sin(yz) - z^2\sin(yz) - y^2\sin(yz) \neq 0$.)

25. Since div $\mathbf{F} = 0$, one has $\iint\limits_{S} \mathbf{F} \cdot \mathbf{N}\, dS = \iiint\limits_{R} \operatorname{div}\mathbf{F}\, dV = \iiint\limits_{R} 0\, dV = 0$

27. One must assume that $\mathbf{F}$ is incompressible throughout the region R bounded by S, not just on the points of S itself. In this case, div $\mathbf{F} = 0$ for all points in R by the definition of incompressibility, so $\iint\limits_{S} \mathbf{F} \cdot \mathbf{N}\, dS = \iiint\limits_{R} \operatorname{div}\mathbf{F}\, dV = \iiint\limits_{R} 0\, dV = 0$.

29. False. It is possible to have $0 = \iint\limits_{S} \mathbf{F} \cdot \mathbf{N}\, dS = \iiint\limits_{R} \operatorname{div}\mathbf{F}\, dV$ if one has both sources and sinks in R, and if the positive contribution of the sources cancels with the negative contribution of the sinks. Examples of such behavior include Exercises 13 and 16.

Review Exercises — Chapter 20

1. a. $\nabla f(x,y,z) = -y(\sin x)\ln z\,\mathbf{i} + (\cos x)\ln z\,\mathbf{j} + [(y\cos x)/z]\mathbf{k}$
   b. $\nabla f(x,y,z) = (3x^2yz^2 + y\cos xy)\mathbf{i} + (x^3z^2 + x\cos xy)\mathbf{j} + 2x^3yz\,\mathbf{k}$

3. $\partial M/\partial y = [(x^2 + y^2)1 - y(2y)]/(x^2 + y^2)^2 = (x^2 - y^2)/(x^2 + y^2)^2$
$\partial N/\partial x = -[(x^2 + y^2)1 - x(2x)]/(x^2 + y^2)^2 = (x^2 - y^2)/(x^2 + y^2)^2$
These are equal, and therefore $\mathbf{F}$ is conservative.

5. Let $\mathbf{r}(t) = \cos t\,\mathbf{i} + \sin t\,\mathbf{j}$ for $0 \leq t \leq \pi$.
$\mathbf{r}'(t) = -\sin t\,\mathbf{i} + \cos t\,\mathbf{j}$
$\mathbf{F}(\mathbf{r}(t)) = \sin t\,\mathbf{i} - \cos t\,\mathbf{j}$
$\mathbf{F}(\mathbf{r}(t)) \cdot \mathbf{r}'(t) = -\sin^2 t - \cos^2 t = -1$
$\int_C \mathbf{F} \cdot d\mathbf{r} = \int_0^{2\pi} (-1)\, dt = -2\pi \neq 0$
While it is true that curl $\mathbf{F} = 0$, this does not contradict Stoke's Theorem since $\mathbf{F}$ is not continuous throughout Q (specifically, at the origin).

7. curl $\mathbf{F} = (2xy - 2xy)\mathbf{i} + (y^2 - y^2)\mathbf{j} + (2yz - 2yz)\mathbf{k} = 0$
$\mathbf{F}$ is conservative since curl $\mathbf{F} = 0$.

9. div $\mathbf{F} = z + x - x^2$
$\iint\limits_{S} \mathbf{F} \cdot \mathbf{N}\, dS = \iiint\limits_{R} \operatorname{div}\mathbf{F}\, dV$
$= \int_0^1 \int_0^{1-x} \int_0^{1-x-y} (z + x - x^2)\, dz\, dy\, dx$
$= \int_0^1 \int_0^{1-x} \left[(z^2/2) + xz - x^2z\right]_{z=0}^{z=1-x-y} dy\, dx$
$= (1/2)\int_0^1 \int_0^{1-x} (1 + 2x^3 - 3x^2 + 2x^2y - 2y + y^2)\, dy\, dx$
$= (1/2)\int_0^1 \left[y + 2x^3y - 3x^2y + x^2y^2 - y^2 + (y^3/3)\right]_{y=0}^{y=1-x} dx$
$= (1/6)\int_0^1 (1 - 6x^2 + 8x^3 - 3x^4)\, dx$
$= (1/6)\left[x - 2x^3 + 2x^4 - (3x^5/5)\right]_0^1$
$= 1/15$

11. ∇f is a conservative vector field, and the curl of a conservative vector field is zero.

13. True. This follows immediately from Theorem 5 of Section 20.6.

15. $\text{div }\mathbf{F} = a + b + c$

$\iint\limits_{S} \mathbf{F} \cdot \mathbf{N}\, dS = \iiint\limits_{R} \text{div }\mathbf{F}\, dV$

$= (a + b + c) \iiint\limits_{R} dV$

$= (a + b + c)V$

17. $\mathbf{r}(t) = \cos t\,\mathbf{i} + \sin t\,\mathbf{j}$ for $0 \le t \le \pi$ is a parameterization for C.

$\mathbf{r}'(t) = -\sin t\,\mathbf{i} + \cos t\,\mathbf{j}$

$\mathbf{F}(\mathbf{r}(t)) = \cos^2 t\,\mathbf{i} - \cos t \sin t\,\mathbf{j}$

$\mathbf{F}(\mathbf{r}(t)) \cdot \mathbf{r}'(t) = -2\cos^2 t \sin t$

$\int_C \mathbf{F} \cdot d\mathbf{r} = \int_0^\pi (-2\cos^2 t \sin t)\, dt$

$= (2/3)\cos^3 t \big|_0^\pi$

$= -4/3$

19. $\mathbf{F}(x, y, z) = xz\mathbf{i} + yz\mathbf{j} + z\mathbf{k}$

$\mathbf{F}(\mathbf{r}(t)) = t\sin t\,\mathbf{i} + \sin t\,\mathbf{j} + \sin t\,\mathbf{k}$

$\mathbf{r}'(t) = \mathbf{i} + \cos t\,\mathbf{k}$

$\mathbf{F}(\mathbf{r}(t)) \cdot \mathbf{r}'(t) = t\sin t + \cos t \sin t$

$\int_C \mathbf{F} \cdot d\mathbf{r} = \int_0^\pi (t\sin t + \cos t \sin t)\, dt$

$= -t\cos t + \sin t + (1/2)\sin^2 t \big|_0^\pi$

$= \pi$

21. $\mathbf{F}(\mathbf{r}(t)) = -6t\mathbf{i} - t\mathbf{j} + 3\mathbf{k}$

$\mathbf{r}'(t) = \mathbf{i} - 3\mathbf{j}$

$\mathbf{F}(\mathbf{r}(t)) \cdot \mathbf{r}'(t) = -6t + 3t = -3t$

$\int_C \mathbf{F} \cdot d\mathbf{r} = \int_0^1 (-3t)\, dt$

$= -3t^2/2 \big|_0^1$

$= -3/2$

23. $\mathbf{r}(t) = \cos t\,\mathbf{i} + \sin t\,\mathbf{j}$ for $0 \le t \le \pi$ is a parameterization for C.

$\mathbf{r}'(t) = -\sin t\,\mathbf{i} + \cos t\,\mathbf{j}$

$\mathbf{F}(\mathbf{r}(t)) = \cos t \sin^2 t\,\mathbf{i} + \cos^3 t \sin t\,\mathbf{j}$

$\mathbf{F}(\mathbf{r}(t)) \cdot \mathbf{r}'(t) = -\cos t \sin^3 t + \cos^4 t \sin t$

$\int_C \mathbf{F} \cdot d\mathbf{r} = \int_0^\pi (-\cos t \sin^3 t + \cos^4 t \sin t)\, dt$

$= -(1/4)\sin^4 t - (1/5)\cos^5 t \big|_0^\pi$

$= 2/5$

25. $z = 8 - x + 2y$, $\partial z/\partial x = -1$ and $\partial z/\partial y = 2$

$\iint\limits_{S}(x + y)\, dS = \iint\limits_{Q}(x + y)\sqrt{1 + 4 + 1}\, dA$

$= \sqrt{6} \int_0^1 \int_0^1 (x + y)\, dy\, dx$

$= \sqrt{6} \int_0^1 [xy + (y^2/2)]_{y=0}^{y=1}\, dx$

$= (\sqrt{6}/2) \int_0^1 (2x + 1)\, dx$

$= (\sqrt{6}/2)[x^2 + x]_0^1$

$= \sqrt{6}$

27. $z = \sqrt{x^2 + y^2}$

$\partial z/\partial x = x/\sqrt{x^2 + y^2} = x/z$

$\partial z/\partial y = y/\sqrt{x^2 + y^2} = y/z$

$\iint\limits_{S} x^2 z\, dS = \iint\limits_{Q} x^2 z \sqrt{(x^2/z^2) + (y^2/z^2) + 1}\, dA$

$= \iint\limits_{Q} x^2 z\sqrt{1 + 1}\, dA$

$= \sqrt{2} \iint\limits_{Q} x^2 \sqrt{x^2 + y^2}\, dA$

$= \sqrt{2} \int_0^{2\pi} \int_1^4 r^2 \cos^2 \theta \sqrt{r^2}\, r\, dr\, d\theta$

$= \sqrt{2} \int_0^{2\pi} \int_1^4 r^4 \cos^2 \theta\, dr\, d\theta$

$= \sqrt{2} \int_0^{2\pi} [(r^5/5)\cos^2 \theta]_{r=1}^{r=4}\, d\theta$

$= (1023\sqrt{2}/5) \int_0^{2\pi} \cos^2 \theta\, d\theta$

$= (1023\sqrt{2}/10) \int_0^{2\pi} (1 + \cos 2\theta)\, d\theta$

$= (1023\sqrt{2}/10)[\theta + (1/2)\sin 2\theta]_0^{2\pi}$

$= 1023\pi\sqrt{2}/5$

29. $\partial z/\partial x = 2x$ and $\partial z/\partial y = 2y$

$M(x,y,z)(\partial z/\partial x) + N(x,y,z)(\partial z/\partial y) - P(x,y,z) = xy(2x) + y(2y) - 2z = 2x^2y + 2y^2 - 2z$

$\iint\limits_{S} \mathbf{F} \cdot \mathbf{N}\, dS = \iint\limits_{Q}[2x^2y + 2y^2 - 2(x^2 + y^2)]\, dA$

$= \iint\limits_{Q}(2x^2y - 2x^2)\, dA$

$= \int_0^{2\pi} \int_1^2 (2r^3 \cos^2\theta \sin\theta - 2r^2 \cos^2\theta) r\, dr\, d\theta$

$= \int_0^{2\pi} [(2r^5/5)\cos^2\theta \sin\theta - (r^4/2)\cos^2\theta]_{r=1}^{r=2}\, d\theta$

$= \int_0^{2\pi} [(62/5)\cos^2\theta \sin\theta - (15/2)\cos^2\theta]\, d\theta$

$= \int_0^{2\pi} [(62/5)\cos^2\theta \sin\theta - (15/4) - (15/4)\cos 2\theta]\, d\theta$

$= -(62/15)\cos^3\theta - (15/4)\theta - (15/8)\sin 2\theta \big|_0^{2\pi}$

$= -15\pi/2$

31. $\operatorname{div} \mathbf{F} = 0$

$\iint\limits_{S} \mathbf{F} \cdot \mathbf{N}\, dS = \iiint\limits_{R} \operatorname{div} \mathbf{F}\, dV = \iiint\limits_{R} 0\, dV = 0$

33. $\partial z/\partial x = -2x$ and $\partial z/\partial y = -2y$

$-M(x,y,z)(\partial z/\partial x) - N(x,y,z)(\partial z/\partial y) + P(x,y,z) = -ax(-2x) - by(-2y) + cz$

$= 2ax^2 + 2by^2 + cz$

$\iint\limits_{S} \mathbf{F} \cdot \mathbf{N}\, dS = \iint\limits_{Q}[2ax^2 + 2by^2 + c(4 - x^2 - y^2)]\, dA$

$= \iint\limits_{Q}[4c + 2ax^2 + 2by^2 - c(x^2 + y^2)]\, dA$

$= \int_0^{2\pi} \int_0^2 [4c + 2ar^2 \cos^2\theta + 2br^2 \sin^2\theta - cr^2]r\, dr\, d\theta$

$= \int_0^{2\pi} [2cr^2 + a(r^4/2)\cos^2\theta + b(r^4/2)\sin^2\theta - c(r^4/4)]_{r=0}^{r=2}\, d\theta$

$= \int_0^{2\pi} (4c + 8a \cos^2\theta + 8b \sin^2\theta)\, d\theta$

$= \int_0^{2\pi} (4c + 4a + 4a \cos 2\theta + 4b - 4b \cos 2\theta)\, d\theta$

$= (4c + 4a + 4b)\theta + 2a \sin 2\theta - 2b \sin 2\theta \big|_0^{2\pi}$

$= 8\pi(a + b + c)$

35. $\operatorname{curl} \mathbf{F} = 3\mathbf{i} - 2\mathbf{j} - 4y\mathbf{k}$

The triangle lies in the plane $x - z = 0$, for which one normal vector is $\mathbf{i} - \mathbf{k}$. The desired unit normal vector is $\mathbf{N} = (1/\sqrt{2})(-\mathbf{i} + \mathbf{k})$.

$(\operatorname{curl} \mathbf{F}) \cdot \mathbf{N} = (1/\sqrt{2})(-3 - 4y)$

From the equation for the plane, one has for the triangle that $z = x$.

$\partial z/\partial x = 1$ and $\partial z/\partial y = 0$.

$\int_C \mathbf{F} \cdot d\mathbf{r} = \iint\limits_{S}(\operatorname{curl} \mathbf{F}) \cdot \mathbf{N}\, dS$

$= -(1/\sqrt{2}) \iint\limits_{S}(3 + 4y)\, dS$

$= -(1/\sqrt{2}) \iint\limits_{Q}(3 + 4y)\sqrt{1 + 0 + 1}\, dA$

$= -\int_0^4 \int_0^{y/2}(3 + 4y)\, dx\, dy$

$= -\int_0^4 [3x + 4xy]_{x=0}^{x=y/2}\, dy$

$= -(1/2)\int_0^4 (3y + 4y^2)\, dy$

$= -(1/2)[(3y^2/2) + (4y^3/3)]_0^4$

$= -164/3$

37. $\operatorname{curl} \mathbf{F} = 0\mathbf{i} + 0\mathbf{j} + [(-1) - (-1)]\mathbf{k} = \mathbf{0}$

$\int_C \mathbf{F} \cdot d\mathbf{r} = \iint\limits_S (\operatorname{curl} \mathbf{F}) \cdot \mathbf{N} \, dS = \iint\limits_S 0 \, dS = 0$

39. $\mathbf{F} = \nabla \phi$ where $\phi(x, y, z) = (x^4/4) + yz$

Since $\mathbf{F}$ is conservative, the result is independent of path.

$\int_C \mathbf{F} \cdot d\mathbf{r} = \phi(-1, 0, \pi) - \phi(1, 0, 0) = 0$

Chapter 21

Differential Equations

21.1 First Order Linear Differential Equations

1. $\int 1/(y+1)\,dy = \int dt$
$\ln|y+1| = t + C$
$|y+1| = e^{t+C}$
$y + 1 = Ae^t$
$y = Ae^t - 1$
$1 = y(0) = A - 1$, so $A = 2$
$y = 2e^t - 1$
steady state: -1
transient: $2e^t$

3. $dy/dt = -1(y - \pi)$
$\int 1/(y-\pi)\,dy = \int(-1)\,dt$
$\ln|y-\pi| = -t + C$
$|y-\pi| = e^{-t+C}$
$y - \pi = Ae^{-t}$
$y = Ae^{-t} + \pi$
$\pi/2 = y(0) = A + \pi$, so $A = -\pi/2$
$y = -(\pi/2)e^{-t} + \pi$
steady state: π
transient: $-(\pi/2)e^{-t}$

5. $dy/dt = -2(y - 2)$
$\int 1/(y-2)\,dy = \int(-2)\,dt$
$\ln|y-2| = -2t + C$
$|y-2| = e^{-2t+C}$
$y - 2 = Ae^{-2t}$
$y = Ae^{-2t} + 2$

7. $u(t) = \int(-2/t)\,dt = -2\ln t$
Integrating factor: $e^{-2\ln t} = t^{-2}$
$t^{-2}y' - 2t^{-3}y = t^2$
$(d/dt)(t^{-2}y) = t^2$
$t^{-2}y = \int t^2\,dt = (t^3/3) + C$
$y = (t^5/3) + Ct^2$

9. $u(t) = \int(-2)\,dt = -2t$
Integrating factor: e^{-2t}
$e^{-2t}y' - 2e^{-2t}y = \sin t$
$(d/dt)(e^{-2t}y) = \sin t$
$e^{-2t}y = \int \sin t\,dt = -\cos t + C$
$y = -e^{2t}\cos t + Ce^{2t}$

11. $u(t) = \int 1\,dt = t$
Integrating factor: e^t
$e^t y' + e^t y = e^{2t}$
$(d/dt)(e^t y) = e^{2t}$
$e^t y = \int e^{2t}\,dt = (1/2)e^{2t} + C$
$y = (1/2)e^t + Ce^{-t}$

13. $(d/dt)(ty) = t$
$ty = \int t\,dt = (t^2/2) + C$
$y = (t/2) + (C/t)$

15. $\int(1/y)\,dy = \int ax\,dx$
$\ln|y| = (ax^2/2) + C$
$|y| = e^{(ax^2/2)+C}$
$y = Ae^{ax^2/2}$

17. a. The rate at which salt enters the solution is 0 kg/min, and the rate at which salt is removed is $[(p/200)\text{ kg/L}][3\text{ L/min}] = (3p/200)\text{ kg/min}$. Therefore, the rate of change of salt is $0 - (3p/200)$. The differential equation is $p'(t) = -3p(t)/200$.

b. Solve the differential equation, which is of the form $p' = kp$ where $k = -3/200 = -.015$. The solution is of the form $p(t) = Ae^{-.015t}$. Since $p(0) = [.5 \text{ kg/L}][200 \text{ L}] = 100 \text{ kg}$, one has $100 = p(0) = A$. The solution is therefore $p(t) = 100e^{-.015t}$. The answer to the question is therefore $p(20) = 100e^{-.3} \approx 74.08 \text{ kg}$.

c. 0 kg/L

19. a. The rate of increase of P due to interest is $.1P$ dollars/yr (10%), and the rate of increase due to additional deposits is 100 dollars/yr. The differential equation is therefore $P'(t) = .1P(t) + 100$.

b. First, solve the differential equation.
$P' = .1(P + 1000)$
$\int 1/(P + 1000)\, dP = \int .1\, dt$
$\ln|P + 1000| = .1t + C$
$|P + 1000| = e^{.1t + C}$
$P + 1000 = Ae^{.1t}$
$P(t) = Ae^{.1t} - 1000$
From the initial deposit, one determines that $100 = P(0) = A - 1000$, so $A = 1100$.
$P(7) = 1100e^{.7} - 1000 \approx \$1215.13$

21. Solve $R(dI/dt) + (I/C) = 0$
$dI/dt = -I/(RC)$
$\int (1/I)\, dI = -\int 1/(RC)\, dt$
$\ln|I| = -(t/(RC)) + k$
$|I| = e^{-(t/(RC)) + k}$
$I(t) = Ae^{-t/(RC)}$
Note that $A = I(0)$.
$I(t) = I(0)e^{-t/(RC)}$

23. Let $y(t)$ be the temperature of the pan after t minutes.
$y' = k(y - 30)$
$\int 1/(y - 30)\, dy = \int k\, dt$
$\ln|y - 30| = kt + C$
$|y - 30| = e^{kt + C}$
$y - 30 = Ae^{kt}$
$y(t) = 30 + Ae^{kt}$
$100 = y(0) = 30 + A$, so $A = 70$
$80 = y(3) = 30 + 70e^{3k}$, so $k = (1/3)\ln(5/7)$
$y(10) = 30 + 70e^{10k} \approx 52.80°\text{C}$

25. Let $y(t)$ be the concentration of salt in the mixture after t minutes. Note that $y(0) = 2$. It will be easier to do the initial work with a function $f(t)$ which represents the number of pounds of salt in the mixture after t minutes. Note that $y(t) = f(t)/500$ and $y'(t) = f'(t)/500$. The rate at which salt enters the tank is $[6 \text{ lb/gal}][10 \text{ gal/min}] = 60 \text{ lb/min}$. The rate at which salt leave the tank is $[y(t) \text{ lb/gal}][10 \text{ gal/min}] = 10y(t) \text{ lb/min}$. The rate of change of salt in the tank is $f'(t) = 60 - 10y(t)$, so $y'(t) = (60 - 10y)/500 = .12 - .02y$. Solve this differential equation.
$y' = .12 - .02y$
$y' = -.02(y - 6)$
$\int 1/(y - 6)\, dy = -\int .02\, dt$
$\ln|y - 6| = -.02t + C$
$|y - 6| = e^{-.02t + C}$
$y - 6 = Ae^{-.02t}$
$y(t) = 6 + Ae^{-.02t}$
$2 = y(0) = 6 + A$, so $A = -4$
$y(t) = 6 - 4e^{-.02t}$

27. Let $y(t)$ be the amount present in the annuity after t years. Note that $y(0) = 10000$. The rate at which the annuity changes due to interest is $.1y$ dollars/yr, and the rate at which it changes due to withdrawal from the annuity is -2000 dollars/yr. Therefore, $y' = .1y - 2000$. Solve this differential equation.

$y' = .1(y - 20000)$
$\int 1/(y - 20000)\,dy = \int .1\,dt$
$\ln|y - 20000| = .1t + C$
$|y - 20000| = e^{.1t+C}$
$y - 20000 = Ae^{.1t}$
$y(t) = 20000 + Ae^{.1t}$
$10000 = y(0) = 20000 + A$, so $A = -10000$
$y(t) = 20000 - 10000e^{.1t}$
Find the value of t when $y(t) = 0$.
$0 = 20000 - 10000e^{.1t}$
$e^{.1t} = 2$
$t = 10\ln(2) \approx 6.93$ years

29. $y' = ky/s$
$\int(1/y)\,dy = \int(k/s)\,ds$
$\ln|y| = k\ln|s| + C$
$|y| = e^{k\ln|s|+C}$
$y = Ae^{\ln|s|^k} = A|s|^k$
$y = As^k$ if one may assume $s \geq 0$

21.2 Exact Equations

1. $f(x,y) = \int 2xy\,dx = x^2y + h(y)$
$x^2 = \partial f/\partial y = x^2 + h'(y)$
$h'(y) = 0$, so use $h(y) = 0$
$x^2y = C$

3. $f(x,y) = \int \dfrac{x}{\sqrt{x^2+y^2}}\,dx = \sqrt{x^2+y^2} + h(y)$
$\dfrac{y}{\sqrt{x^2+y^2}} = \dfrac{\partial f}{\partial y} = \dfrac{y}{\sqrt{x^2+y^2}} + h'(y)$
$h'(y) = 0$, so use $h(y) = 0$
$\sqrt{x^2+y^2} = C$

5. $\partial M/\partial y = -2x/y^3$
$\partial N/\partial x = -2y/x^3$
Not exact.

7. $\partial M/\partial y = -4xy$
$\partial N/\partial x = 2xy$
Not exact.

9. $\partial M/\partial y = -xy/(x + y^2)^{3/2}$
$\partial N/\partial x = -y/[2(x + y^2)^{3/2}]$
Not exact.

11. The solution was $x^2y = C$.
At $x = 1$, $y = 3$ one has $C = 3$.
$x^2y = 3$, or $y = 3/x^2$

13. The solution was $xy^2 - x^2y = C$.
At $x = 1$, $y = 2$ one has $C = 2$.
$xy^2 - x^2y = 2$

15. $\partial M/\partial y = 2x^2y + 1$ and $\partial N/\partial x = 1$. Therefore, the original differential equation is not exact. After multiplying by the integrating factor, one has:

$$\frac{1 + x^2y^2 + y}{1 + x^2y^2}\,dx + \frac{x}{1 + x^2y^2}\,dy = 0$$

$$\left(1 + \frac{y}{1 + x^2y^2}\right)dx + \frac{x}{1 + x^2y^2}\,dy = 0$$

$$\frac{\partial M}{\partial y} = \frac{(1 + x^2y^2)1 - y(2x^2y)}{(1 + x^2y^2)^2} = \frac{1 - x^2y^2}{(1 + x^2y^2)^2}$$

$$\frac{\partial N}{\partial x} = \frac{(1 + x^2y^2)1 - x(2xy^2)}{(1 + x^2y^2)^2} = \frac{1 - x^2y^2}{(1 + x^2y^2)^2}$$

Therefore, the resulting differential equation is exact. Any solution of this exact equation must also be a solution of the original differential equation.

21.3   Second Order Linear Equations

1. $r^2 + 5r + 6 = 0$
$(r + 2)(r + 3) = 0$
$r = -2$ or $r = -3$
$y = Ae^{-2t} + Be^{-3t}$

3. $r^2 - 1 = 0$
$r = \pm 1$
$y = Ae^t + Be^{-t}$

5. $r^2 + 3r - 10 = 0$
$(r - 2)(r + 5) = 0$
$r = 2$ or $r = -5$
$y = Ae^{2t} + Be^{-5t}$

7. $r^2 + 2r + 1 = 0$
$(r + 1)^2 = 0$
$r = -1$
$y = Ae^{-t} + Bte^{-t}$

9. $r^2 + 6r + 9 = 0$
$(r + 3)^2 = 0$
$r = -3$
$y = Ae^{-3t} + Bte^{-3t}$

11. $\sqrt{b} = \sqrt{4} = 2$
$y = A\sin 2t + B\cos 2t$

13. $y'' + 5y = 0$
$\sqrt{b} = \sqrt{5}$
$y = A\sin\sqrt{5}\,t + B\cos\sqrt{5}\,t$

15. $r^2 - 3r - 10 = 0$
$(r - 5)(r + 2) = 0$
$r = 5$ or $r = -2$
$y = Ae^{5t} + Be^{-2t}$

17. $r^2 - 4r + 1 = 0$
$r = (4 \pm \sqrt{16 - 4})/2 = 2 \pm \sqrt{3}$
$y = Ae^{(2+\sqrt{3})t} + Be^{(2-\sqrt{3})t}$

19. $r^2 - 2r + 6 = 0$
$r = (2 \pm \sqrt{4 - 24})/2 = 1 \pm i\sqrt{5}$
$y = Ae^t\sin\sqrt{5}\,t + Be^t\cos\sqrt{5}\,t$

21. $r^2 - r - 12 = 0$
$(r - 4)(r + 3) = 0$
$r = 4$ or $r = -3$
$y = Ae^{4t} + Be^{-3t}$

23. $r^2 + 2r + 4 = 0$
$r = (-2 \pm \sqrt{4 - 16})/2 = -1 \pm i\sqrt{3}$
$y = Ae^{-t}\sin\sqrt{3}\,t + Be^{-t}\cos\sqrt{3}\,t$

25. $r^2 - r - 6 = 0$
$(r - 3)(r + 2) = 0$
$r = 3$ or $r = -2$
$y = Ae^{3t} + Be^{-2t}$
$y' = 3Ae^{3t} - 2Be^{-2t}$
$0 = y(0) = A + B$
$5 = y'(0) = 3A - 2B$
$A = 1$ and $B = -1$
$y = e^{3t} - e^{-2t}$

27. $r^2 + 2r + 1 = 0$
$(r + 1)^2 = 0$
$r = -1$
$y = Ae^{-t} + Bte^{-t}$
$y' = -Ae^{-t} + Be^{-t} - Bte^{-t}$
$3 = y(0) = A$
$1 = y'(0) = -A + B = -3 + B$, so $B = 4$
$y = 3e^{-t} + 4te^{-t} = (4t + 3)e^{-t}$

29. $r^2 + 9 = 0$
$r = \pm 3i$
$y = A \sin 3t + B \cos 3t$
$y' = 3A \cos 3t - 3B \sin 3t$
$3 = y(0) = B$
$-3 = y'(0) = 3A$, so $A = -1$
$y = -\sin 3t + 3 \cos 3t$

31. $x'' = -y' = -4x$
$x'' + 4x = 0$
$x = A \sin 2t + B \cos 2t$
$-y = dx/dt = 2A \cos 2t - 2B \sin 2t$
The general solution is:
$x = A \sin 2t + B \cos 2t$
$y = -2A \cos 2t + 2B \sin 2t$.

33. $x'' + (k/m)x = 0$
$k/m = 2/.5 = 4$
$x'' + 4x = 0$
$x = A \sin 2t + B \cos 2t$
$x' = 2A \cos 2t - 2B \sin 2t$
$.1 = x(0) = B$
$0 = x'(0) = 2A$, so $A = 0$
$x = .1 \cos 2t$

35. See the figure below.
$k = -9$: $x(t) = (1/2)e^{3t} + (1/2)e^{-3t} = \cosh(3t)$ $k = 1$: $x(t) = \cos t$
$k = -4$: $x(t) = (1/2)e^{2t} + (1/2)e^{-2t} = \cosh(2t)$ $k = 4$: $x(t) = \cos 2t$
$k = -1$: $x(t) = (1/2)e^{t} + (1/2)e^{-t} = \cosh(t)$ $k = 9$: $x(t) = \cos 3t$
$k = 0$: $x(t) = 1$

a. As $k \to \infty$, one obtains wave functions of amplitude 1 which oscillate with ever greater frequency.

b. As $k \to -\infty$, one obtains graphs which always intersect the y-axis at 1, but which rise ever more steeply on either side of this axis. At any particular value of t, the corresponding values $x(t)$ approach ∞ as a limit.

c. As $k \to 0^+$, one obtains wave functions of amplitude 1 and with ever greater wavelengths. At any particular value of t, the corresponding values $x(t)$ eventually begin to approach 1 as a limit.

d. As $k \to 0^-$, one obtains graphs which always intersect the y-axis at 1, and which rise ever more slowly on either side of this axis. At any particular value of t, the corresponding values $x(t)$ approach 1 as a limit.

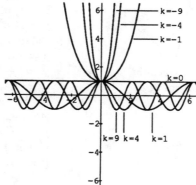

21.4 Nonhomogeneous Linear Equations

1. Homogeneous solution:
$r^2 + 2r - 1 = 0$
$r = (-2 \pm \sqrt{4+4})/2 = -1 \pm \sqrt{2}$
$y = C_1 e^{(-1+\sqrt{2})t} + C_2 e^{(-1-\sqrt{2})t}$
Particular solution:
$y = A \sin 2t + B \cos 2t$
$y' = 2A \cos 2t - 2B \sin 2t$
$y'' = -4A \sin 2t - 4B \cos 2t$
$\sin 2t = y'' + 2y' - y$
$\quad = (-5A - 4B) \sin 2t + (4A - 5B) \cos 2t$
$4A - 5B = 0$ and $-5A - 4B = 1$
$A = -5/41$ and $B = -4/41$
General solution:
$y = C_1 e^{(-1+\sqrt{2})t} + C_2 e^{(-1-\sqrt{2})t}$
$\quad - (5/41) \sin 2t - (4/41) \cos 2t$

3. Homogeneous solution:
$r^2 + 1 = 0$
$r = \pm i$
$y = C_1 \sin t + C_2 \cos t$
Particular solution:
$y = A e^{-t}, \, y' = -A e^{-t}, \, y'' = A e^{-t}$
$e^{-t} = y'' + y = 2A e^{-t}$
$2A = 1$, so $A = 1/2$
$y = (1/2) e^{-t}$
General solution:
$y = C_1 \sin t + C_2 \cos t + (1/2) e^{-t}$

5. Homogeneous solution:
$r^2 - 4 = 0$
$r = \pm 2$
$y = C_1 e^{2t} + C_2 e^{-2t}$
Particular solution:
$y = At + B, \, y' = A, \, y'' = 0$
$4t = y'' - 4y = -4At - 4B$
$A = -1$ and $B = 0$
$y = -t$
General solution:
$y = C_1 e^{2t} + C_2 e^{-2t} - t$

7. Homogeneous solution:
$r^2 - 3r - 10 = 0$
$(r - 5)(r + 2) = 0$
$r = 5$ or $r = -2$
$y = C_1 e^{5t} + C_2 e^{-2t}$
Particular solution:
$y = A e^{2t}, \, y' = 2A e^{2t}, \, y'' = 4A e^{2t}$
$5e^{2t} = y'' - 3y' - 10y = -12A e^{2t}$
$A = -5/12$
$y = -(5/12) e^{2t}$
General solution:
$y = C_1 e^{5t} + C_2 e^{-2t} - (5/12) e^{2t}$

9. Homogeneous solution:
$r^2 + 9 = 0$
$r = \pm 3i$
$y = C_1 \sin 3t + C_2 \cos 3t$
Particular solution:
$y = At \sin 3t + Bt \cos 3t$
$y' = (A - 3Bt) \sin 3t + (3At + B) \cos 3t$
$y'' = (-9At - 6B) \sin 3t + (6A - 9Bt) \cos 3t$
$\cos 3t = y'' + 9y = -6B \sin 3t + 6A \cos 3t$
$A = 1/6$ and $B = 0$
$y = (t/6) \sin 3t$
General solution:
$y = C_1 \sin 3t + C_2 \cos 3t + (t/6) \sin 3t$

11. Homogeneous solution:
$r^2 - 3r - 4 = 0$
$(r - 4)(r + 1) = 0$
$r = 4$ or $r = -1$
$y = C_1 e^{4t} + C_2 e^{-t}$
Particular solution:
$y = Ate^{-t}$
$y' = (-At + A)e^{-t}$
$y'' = (At - 2A)e^{-t}$
$2e^{-t} = y'' - 3y' - 4y = -5Ae^{-t}$
$A = -2/5$
$y = -(2t/5)e^{-t}$
General solution:
$y = C_1 e^{4t} + C_2 e^{-t} - (2t/5)e^{-t}$

13. Homogeneous solution:
$r^2 + 2r + 1 = 0$
$(r + 1)^2 = 0$
$r = -1$
$y = C_1 e^{-t} + C_2 te^{-t}$
Particular solution:
$y = 3Ae^t + Bt^2 + Ct + D$
$y' = 3Ae^t + 2Bt + C$
$y'' = 3Ae^t + 2B$
$3e^t + t^2 = y'' + 2y' + y = 12Ae^t + Bt^2 + (4B + C)t + (2B + 2C + D)$
$A = 1/4$, $B = 1$, $4B + C = 0$, $2B + 2C + D = 0$
$C = -4$, $D = 6$
$y = (3/4)e^t + t^2 - 4t + 6$
General solution:
$y = C_1 e^{-t} + C_2 te^{-t} + (3/4)e^t + t^2 - 4t + 6$

15. Homogeneous solution:
$r^2 - 2r + 1 = 0$
$(r - 1)^2 = 0$
$r = 1$
$y = C_1 e^t + C_2 te^t$
Particular solution:
$y = Ae^{-t} + B \sin 2t + C \cos 2t$
$y' = -Ae^{-t} + 2B \cos 2t - 2C \sin 2t$
$y'' = Ae^{-t} - 4B \sin 2t - 4C \cos 2t$
$2e^{-t} + \sin 2t + 2 \cos 2t = y'' - 2y' + y = 4Ae^{-t} + (-3B + 4C) \sin 2t + (-3C - 4B) \cos 2t$
$A = 1/2$, $-3B + 4C = 1$, $-4B - 3C = 2$
$B = -11/25$ and $C = -2/25$
$y = (1/2)e^{-t} - (11/25) \sin 2t - (2/25) \cos 2t$
General solution:
$y = C_1 e^t + C_2 te^t + (1/2)e^{-t} - (11/25) \sin 2t - (2/25) \cos 2t$

17. Homogeneous solution:
$r^2 + r + 1 = 0$
$r = (-1 \pm \sqrt{1 - 4})/2 = -(1/2) \pm i(\sqrt{3}/2)$
$y = C_1 e^{-t/2} \sin(t\sqrt{3}/2) + C_2 e^{-t/2} \cos(t\sqrt{3}/2)$

Particular solution:
$$y = At^3 + Bt^2 + Ct + D$$
$$y' = 3At^2 + 2Bt + C$$
$$y'' = 6At + 2B$$
$$t^3 + 2t^2 = y'' + y' + y = At^3 + (3A + B)t^2 + (6A + 2B + C)t + (2B + C + D)$$
$$A = 1,\ 3A + B = 2,\ 6A + 2B + C = 0,\ 2B + C + D = 0$$
$$B = -1,\ C = -4,\ D = 6$$
$$y = t^3 - t^2 - 4t + 6$$
General solution:
$$y = C_1 e^{-t/2} \sin(t\sqrt{3}/2) + C_2 e^{-t/2} \cos(t\sqrt{3}/2) + t^3 - t^2 - 4t + 6$$

19. a. Given that y_p and z_p are particular solutions of $y'' + ay' + by = f(t)$. Let $y = y_p - z_p$. Then:
$$y'' + ay' + by = (y_p'' - z_p'') + a(y_p' - z_p') + b(y_p - z_p)$$
$$= (y_p'' + ay_p' + by_p) - (z_p'' + az_p' + bz_p)$$
$$= f(t) - f(t)$$
$$= 0$$
Therefore, $y = y_p - z_p$ is a solution of $y'' + ay' + by = 0$.

b. Since $y_p - z_p$ is a solution of the homogeneous equation, there must be constants C_1' and C_2' such that $y_p - z_p = C_1' y_1 + C_2' y_2$. Letting $C_1 = -C_1'$ and $C_2 = -C_2'$, one has:
$$z_p = y_p - C_1' y_1 - C_2' y_2 = C_1 y_1 + C_2 y_2 + y_p.$$

c. By the arbitrary nature of y_p and z_p, for any two particular solutions y_p and z_p, there must exist constants C_1 and C_2 such that $z_p = C_1 y_1 + C_2 y_2 + y_p$. Hence, given any one particular solution y_p, every other particular solution is always of the form $z_p = y_g + y_p$.

21. Given that $y_1'' + ay_1' + by_1 = f$ and $y_2'' + ay_2' + by_2 = f$, and that $f(t) \neq 0$. Let $y = y_1 + y_2$. Then:
$$y'' + ay' + by = (y_1'' + y_2'') + a(y_1' + y_2') + b(y_1 + y_2)$$
$$= (y_1'' + ay_1' + by_1) + (y_2'' + ay_2' + by_2)$$
$$= f + f$$
$$= 2f$$
and $2f(t) \neq f(t)$ since $f(t) \neq 0$. Note that $y = y_1 + y_2$ is a solution of the differential equation $y'' + ay' + by = 2f(t)$.

23. Particular solution:
$$y = At + B,\ y' = A$$
$$6t - 14 = y' - 3y = -3At + (A - 3B)$$
$$A = -2 \text{ and } A - 3B = -14$$
$$B = 4$$
$$y = -2t + 4$$
General solution:
$$y = Ce^{3t} - 2t + 4$$

25. Particular solution:
$$y = At^2 + Bt + C,\ y' = 2At + B$$
$$1 - 4t^2 = y' - 4y$$
$$= -4At^2 + (2A - 4B)t + (B - 4C)$$
$$A = 1,\ 2A - 4B = 0,\ B - 4C = 1$$
$$B = 1/2 \text{ and } C = -1/8$$
$$y = t^2 + (1/2)t - (1/8)$$
General solution:
$$y = Ce^{4t} + t^2 + (1/2)t - (1/8)$$

27. Particular solution:
$$y = Ae^{3t},\ y' = 3Ae^{3t}$$
$$2e^{3t} = y' - 4y = -Ae^{3t}$$
$$A = -2$$
$$y = -2e^{3t}$$
General solution:
$$y = Ce^{4t} - 2e^{3t}$$

29. Particular solution:

$y = Ae^t + Bt + C$, $y' = Ae^t + B$

$4e^t + 6t - 14 = y' - 3y = -2Ae^t - 3Bt + (B - 3C)$

$A = -2$, $B = -2$, $B - 3C = -14$, so $C = 4$

$y = -2e^t - 2t + 4$

General solution:

$y = Ce^{3t} - 2e^t - 2t + 4$

31. General solution:

$y = Ce^{-4t}$

The solution:

$3 = y(0) = C$

$y = 3e^{-4t}$

33. Particular solution:

$y = Ae^{2t}$, $y' = 2Ae^{2t}$

$5e^{2t} = y' + 3y = 5Ae^{2t}$

$A = 1$

$y = e^{2t}$

General solution:

$y = Ce^{-3t} + e^{2t}$

The solution:

$7 = y(0) = C + 1$, so $C = 6$

$y = 6e^{-3t} + e^{2t}$

21.5   Power Series Solutions of Differential Equations

1. $y' = -y$

$\sum_{k=0}^{\infty} ka_k x^{k-1} = -\sum_{k=0}^{\infty} a_k x^k$

$a_1 + 2a_2 x + 3a_3 x^2 + \ldots + ka_k x^{k-1} + \ldots = -a_0 - a_1 x - a_2 x^2 - \ldots - a_{k-1} x^{k-1} - \ldots$

$ka_k = -a_{k-1}$

$a_k = -\dfrac{1}{k} a_{k-1} = \dfrac{1}{k} \dfrac{1}{k-1} a_{k-2} = \ldots = \dfrac{(-1)^k}{k!} a_0$

$y = a_0 \sum_{k=0}^{\infty} \dfrac{(-1)^k}{k!} x^k$

Check: $y' + ay = 0$ has general solution $y = Ce^{-ax}$, so the general solution for $y' + y = 0$ is $y = Ce^{-x}$.
The power series representation for this function is the power series above.

3. $y' = 6y$

$\sum_{k=0}^{\infty} ka_k x^{k-1} = 6 \sum_{k=0}^{\infty} a_k x^k$

$a_1 + 2a_2 x + 3a_3 x^2 + \ldots + ka_k x^{k-1} + \ldots = 6a_0 + 6a_1 x + 6a_2 x^2 + \ldots + 6a_{k-1} x^{k-1} + \ldots$

$ka_k = 6a_{k-1}$

$a_k = \dfrac{6}{k} a_{k-1} = \dfrac{6}{k} \dfrac{6}{k-1} a_{k-2} = \ldots = \dfrac{6^k}{k!} a_0$

$y = a_0 \sum_{k=0}^{\infty} \dfrac{6^k}{k!} x^k$

Check: $y' + ay = 0$ has general solution $y = Ce^{-ax}$, so the general solution for $y' - 6y = 0$ is $y = Ce^{6x}$.
The power series representation for this function is the power series above.

5. $y' = 2xy$

$\sum_{k=0}^{\infty} k a_k x^{k-1} = 2x \sum_{k=0}^{\infty} a_k x^k$

$a_1 + 2a_2 x + 3a_3 x^2 + \ldots + k a_k x^{k-1} + \ldots = 2a_0 x + 2a_1 x^2 + 2a_2 x^3 + \ldots + 2a_{k-2} x^{k-1} + \ldots$

$k a_k = 2a_{k-2}$ for $k \geq 2$, and $a_1 = 0$

$a_k = \dfrac{2}{k} a_{k-2} = \dfrac{2}{k} \dfrac{2}{k-2} a_{k-4} = \ldots$

This result is 0 if k is odd, since one eventually reaches $a_1 = 0$. If $k = 2n$ is even, then one eventually obtains:

$a_k = \dfrac{2^n}{k(k-2)(k-4)\cdots 2} a_0 = \dfrac{2^n}{(2n)(2n-2)\cdots 2} a_0 = \dfrac{2^n}{2^n n!} a_0 = \dfrac{1}{n!} a_0$

$y = a_0 \sum_{n=0}^{\infty} \dfrac{1}{n!} x^{2n}$

Check by separation of variables:

$\int (1/y)\, dy = \int 2x\, dx$

$\ln|y| = x^2 + C$

$|y| = e^{x^2 + C}$

$y = Ae^{x^2}$

The power series representation for this function is the power series above.

7. $y' = -4y$

$\sum_{k=0}^{\infty} k a_k x^{k-1} = -4 \sum_{k=0}^{\infty} a_k x^k$

$a_1 + 2a_2 x + 3a_3 x^2 + \ldots + k a_k x^{k-1} + \ldots = -4a_0 - 4a_1 x - 4a_2 x^2 - \ldots - 4a_{k-1} x^{k-1} - \ldots$

$k a_k = -4a_{k-1}$

$a_k = -\dfrac{4}{k} a_{k-1} = \dfrac{4}{k} \dfrac{4}{k-1} a_{k-2} = \ldots = \dfrac{(-1)^k 4^k}{k!} a_0$

$y = a_0 \sum_{k=0}^{\infty} \dfrac{(-1)^k 4^k}{k!} x^k$

$1 = y(0) = a_0$

$y = \sum_{k=0}^{\infty} \dfrac{(-1)^k 4^k}{k!} x^k$

Check: $y' + ay = 0$ has general solution $y = Ce^{-ax}$, so the general solution for $y' + 4y = 0$ is $y = Ce^{-4x}$. From the initial condition, $1 = y(0) = C$, so $y = e^{-4x}$. The power series representation for this function is the power series above.

9. $xy'' = -y$

$x \sum_{k=0}^{\infty} k(k-1) a_k x^{k-2} = - \sum_{k=0}^{\infty} a_k x^k$

$2 \cdot 1 a_2 x + 3 \cdot 2a_3 x^2 + 4 \cdot 3a_4 x^3 + \ldots + k(k-1) a_k x^{k-1} + \ldots = -a_0 - a_1 x - a_2 x^2 - \ldots - a_{k-1} x^{k-1} - \ldots$

$k(k-1) a_k = -a_{k-1}$ for $k \geq 2$, and $a_0 = 0$

$a_k = -\dfrac{1}{k(k-1)} a_{k-1} = \dfrac{1}{k(k-1)} \dfrac{1}{(k-1)(k-2)} a_{k-2} = \ldots$

$= \dfrac{(-1)^{k-1}}{k(k-1)^2 (k-2)^2 \cdots (2^2) 1} a_1 = \dfrac{(-1)^{k-1}}{[k(k-1)(k-2)\cdots 2][(k-1)(k-2)\cdots 1]} a_1 = \dfrac{(-1)^{k-1}}{k!(k-1)!} a_1$

$y = a_1 \sum_{k=1}^{\infty} \dfrac{(-1)^{k-1}}{k!(k-1)!} x^k \text{ or } a_1 \sum_{k=1}^{\infty} \dfrac{(-1)^{k-1}}{k[(k-1)!]^2} x^k$

11. For $y = \sum_{k=0}^{\infty} a_k x^k$ and $y' = \sum_{k=0}^{\infty} k a_k x^{k-1}$ one has $y(0) = a_0$ and $y'(0) = a_1$. Therefore, $a_0 = 1$ and $a_1 = 0$.

$y'' = -y$

$\sum_{k=0}^{\infty} k(k-1) a_k x^{k-2} = -\sum_{k=0}^{\infty} a_k x^k$

$2 \cdot 1 a_2 + 3 \cdot 2 a_3 x + 4 \cdot 3 a_4 x^2 + \ldots + k(k-1) a_k x^{k-2} + \ldots = -a_0 - a_1 x - a_2 x^2 - \ldots - a_{k-2} x^{k-2} - \ldots$

$k(k-1) a_k = -a_{k-2}$ for $k \geq 2$

$a_k = -\dfrac{1}{k(k-1)} a_{k-2} = \dfrac{1}{k(k-1)} \dfrac{1}{(k-2)(k-3)} a_{k-4} = \ldots$

This result is 0 if k is odd, since one eventually reaches $a_1 = 0$. If $k = 2n$ is even, and remembering that $a_0 = 1$, one eventually obtains:

$a_k = \dfrac{(-1)^n}{k!} = \dfrac{(-1)^n}{(2n)!}$

$y = \sum_{n=0}^{\infty} \dfrac{(-1)^n}{(2n)!} x^{2n} = \cos x$

21.6 Approximating Solutions of Differential Equations

1. $f(t, y) = 2y$

t_j	0	0.25	0.5	0.75	1
y_j	1	1.5	2.25	3.375	5.0625

3. $f(t, y) = 4 - 2y$

t_j	0	0.25	0.5	0.75	1
y_j	1	1.5	1.75	1.875	1.9375

5. $f(x, y) = x(y + 1)$

x_j	0	0.25	0.5	0.75	1
y_j	1	1	1.125	1.3906	1.8389

7. $f(t, y) = t - 2ty$

t_j	0	0.25	0.5	0.75	1
y_j	1.5	1.5	1.375	1.1563	0.9102

9. $f(t, y) = 2y - 2t$

t_j	0	0.25	0.5	0.75	1
y_j	1.5	2.25	3.25	4.625	6.5625

11. The general solution to the homogeneous equation is $y = Ce^{-2t}$, and a particular solution is $y = 2$. The general solution is therefore $y = 2 + Ce^{-2t}$. From the initial condition, one has $1 = y(0) = 2 + C$, and thus $C = -1$ and $y = 2 - e^{-2t}$.

t_j	0	0.25	0.5	0.75	1
y_j	1	1.5	1.75	1.875	1.9375
$y(t_j)$	1	1.3935	1.6321	1.7769	1.8647

Review Exercises — Chapter 21

1. $y = \int (x - 3)\, dx = (1/2)x^2 - 3x + C$

3. $y = \int (\sec \sqrt{t} \tan \sqrt{t})/\sqrt{t}\, dt = 2 \sec \sqrt{t} + C$

5. $\int (1/y^2)\, dy = \int 3t\, dt$
$-1/y = (3/2)t^2 + C$ or $(3t^2 + C)/2$
$y = -2/(3t^2 + C)$

7. $\int 1/(2 + y)\, dy = \int (1 + x)\, dx$
$\ln|2 + y| = x + (1/2)x^2 + C$
$|2 + y| = e^{x + (1/2)x^2 + C}$
$2 + y = Ae^{x + (1/2)x^2}$
$y = Ae^{x + (1/2)x^2} - 2$

9. $y' = 4 - 2y = -2(y - 2)$
$\int 1/(y - 2)\, dy = \int (-2)\, dt$
$\ln|y - 2| = -2t + C$
$|y - 2| = e^{-2t + C}$
$y - 2 = Ae^{-2t}$
$y = 2 + Ae^{-2t}$

11. $\int 1/[y(4 + y)]\, dy = \int 1\, dt$
$(1/4)\int 1/y\, dy - (1/4)\int 1/(4 + y)\, dy = t + C$
$(1/4)\ln|y| - (1/4)\ln|4 + y| = t + C$
$\ln|y/(4 + y)| = 4t + C$
$|y/(4 + y)| = e^{4t + C}$
$y/(4 + y) = Ae^{4t}$
$y = 4Ae^{4t} + yAe^{4t}$
$y(1 - Ae^{4t}) = 4Ae^{4t}$
$y = 4Ae^{4t}/(1 - Ae^{4t}) = 4e^{4t}/(C - e^{4t})$

13. $y' = -y\cos t$
$\int 1/y\, dy = \int (-\cos t)\, dt$
$\ln|y| = -\sin t + C$
$|y| = e^{-\sin t + C}$
$y = Ae^{-\sin t}$

15. $y' = 4(y + 2)$
$\int 1/(y + 2)\, dy = \int 4\, dt$
$\ln|y + 2| = 4t + C$
$|y + 2| = e^{4t + C}$
$y + 2 = Ae^{4t}$
$y = Ae^{4t} - 2$

17. $r^2 + 4r + 4 = 0$
$(r + 2)^2 = 0$
$r = -2$
$y = Ae^{-2t} + Bte^{-2t}$

19. Homogeneous solution:
$r^2 - 2r - 15 = 0$
$(r - 5)(r + 3) = 0$
$r = 5$ or $r = -3$
$y = C_1 e^{5t} + C_2 e^{-3t}$
Particular solution:
$y = A\sin 2t + B\cos 2t$
$y' = 2A\cos 2t - 2B\sin 2t$
$y'' = -4A\sin 2t - 4B\cos 2t$
$\cos 2t = y'' - 2y' - 15y$
$= (-19A + 4B)\sin 2t + (-4A - 19B)\cos 2t$
$-19A + 4B = 0$ and $-4A - 19B = 1$
$A = -4/377$ and $B = -19/377$
$y = -(4/377)\sin 2t - (19/377)\cos 2t$
General solution:
$y = C_1 e^{5t} + C_2 e^{-3t}$
$\qquad - (4/377)\sin 2t - (19/377)\cos 2t$

21. Homogeneous solution:
$r^2 - 9 = 0$
$r = \pm 3$
$y = C_1 e^{3t} + C_2 e^{-3t}$
Particular solution:
$y = A + Be^{2t},\ y' = 2Be^{2t},\ y'' = 4Be^{2t}$
$5 + e^{2t} = y'' - 9y = -9A - 5Be^{2t}$
$A = -5/9$ and $B = -1/5$
$y = -(5/9) - (1/5)e^{2t}$
General solution:
$y = C_1 e^{3t} + C_2 e^{-3t} - (5/9) - (1/5)e^{2t}$

23. $y' = -y/t^2$
$\int 1/y\, dy = - \int 1/t^2\, dt$
$\ln|y| = (1/t) + C$
$|y| = e^{(1/t) + C}$
$y = Ae^{1/t}$

25. Homogeneous solution:
$r^2 - 5r - 14 = 0$
$(r - 7)(r + 2) = 0$
$r = 7$ or $r = -2$
$y = C_1 e^{7t} + C_2 e^{-2t}$
Particular solution:
$y = At + B$, $y' = A$, $y'' = 0$
$2 + t = y'' - 5y' - 14y = -14At + (-5A - 14B)$
$A = -1/14$ and $-5A - 14B = 2$
$B = -23/196$
$y = -(1/14)t - (23/196)$
General solution:
$y = C_1 e^{7t} + C_2 e^{-2t} - (1/14)t - (23/196)$

27. This equation is exact.
$f(x, y) = \int 2xy^3 \, dx = x^2 y^3 + h(y)$
$3x^2 y^2 = \partial f / \partial y = 3x^2 y^2 + h'(y)$
$h'(y) = 0$, so use $h(y) = 0$
$x^2 y^3 = C$, or $y = Cx^{-2/3}$

29. This equation is not exact. Solve by separation of variables.
$2x \, dy = (2y - y^2) \, dx$
$\int 1/(2y - y^2) \, dy = \int 1/(2x) \, dx$
$(1/2) \int 1/y \, dy + (1/2) \int 1/(2 - y) \, dy = (1/2) \ln |x| + C$
$(1/2) \ln |y| - (1/2) \ln |2 - y| = (1/2) \ln |x| + C$
$\ln |y/(2 - y)| = \ln |x| + C$
$|y/(2 - y)| = e^{\ln |x| + C}$
$y/(2 - y) = Ae^{\ln |x|} = A|x|$
$y = 2A|x| - yA|x|$
$y(1 + A|x|) = 2A|x|$
$y = 2A|x|/(1 + A|x|) = 2|x|/(C + |x|)$
Note that for $x > 0$ one has $y = 2x/(C + x)$, and for $x < 0$ one has $y = 2(-x)/[C + (-x)] = 2x/(B + x)$
where $B = -C$. Since the constant C (and B) has arbitrary value, these answers are of the same form.
One may therefore use $y = 2x/(C + x)$ for the general solution, since on any interval not containing zero
this solution will match the previous solution for an appropriate choice of the arbitrary constants, and
this solution has the added advantage of being differentiable at $x = 0$.

An alternative, but less obvious, approach is to make the original differential equation exact by
multiplying the equation by the integrating factor $1/y^2$.

31. $y = \int (x/\sqrt{1 + x^2}) \, dx = \sqrt{1 + x^2} + C$
$3 = y(0) = 1 + C$, so $C = 2$
$y = \sqrt{1 + x^2} + 2$

33. $\int 1/y \, dy = \int t \, dt$
$\ln |y| = (t^2/2) + C$
$|y| = e^{(t^2/2) + C}$
$y = Ae^{t^2/2}$
$3 = y(0) = A$
$y = 3e^{t^2/2}$

35. $y'' + 9y = 0$
$r^2 + 9 = 0$
$r = \pm 3i$
$y = A \sin 3t + B \cos 3t$
$y' = 3A \cos 3t - 3B \sin 3t$
$3 = y(0) = B$
$9 = y'(0) = 3A$, so $A = 3$
$y = 3 \sin 3t + 3 \cos 3t$

37. $r^2 - 6r + 9 = 0$
$(r - 3)^2 = 0$
$r = 3$
$y = Ae^{3t} + Bte^{3t}$
$y' = (3A + B)e^{3t} + 3Bte^{3t}$
$0 = y(0) = A$
$6 = y'(0) = 3A + B = B$
$y = 6te^{3t}$

39. a. $y' = 2Ce^{2t} = 2y$
$\qquad y' - 2y = 0$

b. $y' = C_1 e^t - C_2 e^{-t}$
$\qquad y'' = C_1 e^t + C_2 e^{-t} = y$
$\qquad y'' - y = 0$

41. Let $y(t)$ be the concentration of antifreeze after t minutes, and for convenience, let $f(t)$ be the amount of antifreeze after t minutes. Note that $y(0) = 5/5 = 1$, $y(t) = f(t)/5$ and $y'(t) = f'(t)/5$. The rate at which antifreeze is added to the radiator is 0, and the rate at which it is removed is $y(t)[1 \text{ gal/min}] = y(t)$ gal/min. Consequently, the rate of change in the amount of antifreeze present in the radiator is given by $f'(t) = 0 - y(t) = -y(t)$. Thus, we need to solve the differential equation $y' = -y/5 = -.2y$.
$\int 1/y \, dy = \int (-.2) \, dt$
$\ln|y| = -.2t + C$
$|y| = e^{-.2t+C}$
$y(t) = Ae^{-.2t}$
$1 = y(0) = A$
$y(t) = e^{-.2t}$

43. $y' = k(y - 90)$
$\int 1/(y - 90) \, dy = \int k \, dt$
$\ln|y - 90| = kt + C$
$|y - 90| = e^{kt+C}$
$y - 90 = Ae^{kt}$
$y(t) = 90 + Ae^{kt}$
$40 = y(0) = 90 + A$, so $A = -50$
$50 = y(5) = 90 - 50e^{5k}$, so $k = (1/5)\ln(4/5)$
$y(10) = 90 - 50e^{10k} = 58°\text{F}$

45. $xy'' = y$
$x \sum_{k=0}^{\infty} k(k-1)a_k x^{k-2} = \sum_{k=0}^{\infty} a_k x^k$
$2 \cdot 1a_2 x + 3 \cdot 2a_3 x^2 + \ldots + k(k-1)a_k x^{k-1} + \ldots = a_0 + a_1 x + a_2 x^2 + \ldots + a_{k-1} x^{k-1} + \ldots$
$k(k-1)a_k = a_{k-1}$ for $k \geq 2$, and $a_0 = 0$
$$a_k = \frac{1}{k(k-1)} a_{k-1} = \frac{1}{k(k-1)} \frac{1}{(k-1)(k-2)} a_{k-2} = \ldots$$
$$= \frac{1}{k(k-1)^2(k-2)^2 \cdots (2^2)1} a_1$$
$$= \frac{1}{[k(k-1)(k-2)\cdots 2][(k-1)(k-2)\cdots 1]} a_1$$
$$= \frac{1}{k!(k-1)!} a_1$$
$$y = a_1 \sum_{k=1}^{\infty} \frac{1}{k!(k-1)!} x^k \text{ or } a_1 \sum_{k=1}^{\infty} \frac{1}{k[(k-1)!]^2} x^k$$

47. $r^2 - 4r + 6 = 0$
$r = (4 \pm \sqrt{16 - 24})/2 = 2 \pm i\sqrt{2}$
$y = Ae^{2t} \sin \sqrt{2}\, t + Be^{2t} \cos \sqrt{2}\, t$